ECOLOGY

The Experimental Analysis
of Distribution and Abundance

FIFTH EDITION

Charles J. Krebs

The University of British Columbia

An imprint of Addison Wesley Longman, Inc.

San Francisco   Boston   New York
Capetown   Hong Kong   London   Madrid   Mexico City
Montreal   Munich   Paris   Singapore   Sydney   Tokyo   Toronto

Acquisitions Editor: Elizabeth Fogarty
Project Editor: Heather Dutton
Managing Editor: Wendy Earl
Design Manager: Bradley Birch
Production Editor: Todd Tedesco
Text Designers: Electronic Publishing Services Inc., N.Y.C. and Yvo Riczebos
Cover Designer: Yvo Riczebos
Art Studio: Electronic Publishing Services Inc., N.Y.C.
Photo Researcher: Electronic Publishing Services Inc., N.Y.C.
Manufacturing Coordinator: Vivian McDougal
Project Coordination and Electronic Page Makeup: Electronic Publishing Services Inc., N.Y.C.

Library of Congress Cataloging-in-Publication Data
Krebs, Charles J.
    Ecology: the experimental analysis of distribution and abundance/Charles J. Krebs.—
5th Ed.
      p. cm.
    Includes bibliographical references (p.) and index.
    ISBN 0-321-04289-1
    1. Ecology.   2. Population biology.   I. Title.
QH541.K67   2001   00-064356
577.8—dc21

ISBN 0-321-04289-1

1 2 3 4 5 6 7 8 9 10 - QWT - 04 03 02 01

The text paper of this book is recycled paper containing 10% post-consumer fiber by fiber
weight. The paper is produced from harvested forests managed under the tenets
of the Sustainable Forestry Initiative under third party verification.

Benjamin
Cummings

1301 Sansome Street
San Francisco, California 94111

To the Unknown Ecologist
who works without recognition to preserve the
ecological integrity of the Earth for our grandchildren
and who measures riches not in stocks and bonds
but in biodiversity

CONTENTS IN BRIEF

PART 1   What Is Ecology?   1

CHAPTER 1   Introduction to the Science of Ecology   2

CHAPTER 2   Evolution and Ecology   17

PART 2   The Problem of Distribution: Populations   31

CHAPTER 3   Methods for Analyzing Distributions   32

CHAPTER 4   Factors That Limit Distributions: Dispersal   41

CHAPTER 5   Factors That Limit Distributions: Habitat Selection   57

CHAPTER 6   Factors That Limit Distributions: Interrelations with Other Species   71

CHAPTER 7   Factors That Limit Distributions: Temperature, Moisture, and Other Physical-Chemical Factors   86

CHAPTER 8   The Relationship Between Distribution and Abundance   106

PART 3   The Problem of Abundance: Populations   115

CHAPTER 9   Population Parameters   116

CHAPTER 10   Demographic Techniques: Vital Statistics   133

CHAPTER 11   Population Growth   157

CHAPTER 12   Species Interactions: Competition   179

CHAPTER 13   Species Interactions: Predation   206

CHAPTER 14   Species Interactions: Herbivory and Mutualism   235

CHAPTER 15   Species Interactions: Disease and Parasitism   258

CHAPTER 16   Population Regulation   280

CHAPTER 17   Applied Problems I: Harvesting Populations   305

CHAPTER 18   Applied Problems II: Pest Control   331

CHAPTER 19   Applied Problems III: Conservation Biology   355

PART 4   **Distribution and Abundance at the Community Level   385**

CHAPTER 20   The Nature of the Community   386

CHAPTER 21   Community Change   403

CHAPTER 22   Community Organization I: Biodiversity   434

CHAPTER 23   Community Organization II:
Predation and Competition in Equilibrial Communities   459

CHAPTER 24   Community Organization III: Disturbance and Nonequilibrium
Communities   485

CHAPTER 25   Ecosystem Metabolism I: Primary Production   513

CHAPTER 26   Ecosystem Metabolism II: Secondary Production   537

CHAPTER 27   Ecosystem Metabolism III: Nutrient Cycles   560

CHAPTER 28   Ecosystem Health: Human Impacts   583

EPILOGUE   *609*

APPENDIX I   A Primer on Population Genetics   610

APPENDIX II   Estimation of the Size of the Marked Popluation
in Capture-Recapture Studies   612

APPENDIX III   Instantaneous and Finite Rates   614

APPENDIX IV   Species Diversity Measures of Heterogeneity   617

GLOSSARY   *619*

BIBLIOGRAPHY   *623*

CREDITS   *664*

SPECIES INDEX   *673*

SUBJECT INDEX   *680*

CONTENTS

Preface   xviii

PART 1   WHAT IS ECOLOGY?   1

CHAPTER 1   INTRODUCTION TO THE SCIENCE OF ECOLOGY   2

Definition of Ecology   2
History of Ecology   3
Basic Problems and Approaches to Ecology   7
Levels of Integration   10
Methods of Approach to Ecology   11
Application of the Scientific Method to Ecology   11
Essay 1.1 Science and Values in Ecology   12
Box 1.1 Scientific Method: Definitions of Terms   13
Essay 1.2 On Ecological Truth   14
Key Concepts   15
Selected References   15
Questions and Problems   15

CHAPTER 2   EVOLUTION AND ECOLOGY   17

What Is Evolution?   17
Adaptation   18
Box 2.1 What Is Fitness?   19
 Clutch Size in Birds   22
Coevolution   25
Essay 2.1 Evolution and "Arms Races"   26
Units of Selection   26
 Gametic Selection   27
 Kin Selection   27
 Group Selection   27
Summary   27
Key Concepts   28
Selected References   28
Questions and Problems   28

PART 2 THE PROBLEM OF DISTRIBUTION: POPULATIONS 31

CHAPTER 3 METHODS FOR ANALYZING DISTRIBUTIONS 32

Transplant Experiments 32
Essay 3.1 Liebig's Law of the Minimum 34
Physiological Ecology 35
Adaptation 36
Summary 38
Key Concepts 39
Selected References 39
Questions and Problems 39

CHAPTER 4 FACTORS THAT LIMIT DISTRIBUTIONS: DISPERSAL 41

Examples of Dispersal 41
Zebra Mussel (Dreissena polymorphis) 41
Gypsy Moth (Lymantria dispar) 42
Chestnut Blight (Cryphonectria parasitica) 44
California Sea Otter (Enhydra lutris) 45
The Three Modes of Dispersal 46
Essay 4.1 Ships, Ballast Water, and Marine Dispersal 47
Box 4.1 Definition of Terms for Introduced Nonnative Species 50
Colonization and Extinction 50
Evolutionary Advantages of Dispersal 53
Summary 55
Key Concepts 55
Selected References 56
Questions and Problems 56

CHAPTER 5 FACTORS THAT LIMIT DISTRIBUTIONS: HABITAT SELECTION 57

Behavioral Mechanisms of Habitat Selection 57
Evolution of Habitat Preferences 65
A Theory of Habitat Selection 67
Summary 69
Key Concepts 69
Selected References 69
Questions and Problems 70

CHAPTER 6 FACTORS THAT LIMIT DISTRIBUTIONS: INTERRELATIONS WITH OTHER SPECIES 71

Predation 71
Restriction of Prey by Predators 71
Restriction of Predators by Prey 75

(continued)

(Chapter 6, continued)

Disease and Parasitism   77
Allelopathy   78
Competition   80
Essay 6.1 What is Competition?   *82*
Summary   83
Key Concepts   83
Selected References   84
Questions and Problems   84

CHAPTER 7   FACTORS THAT LIMIT DISTRIBUTIONS: TEMPERATURE, MOISTURE, AND OTHER PHYSICAL-CHEMICAL FACTORS   86

Climatology   86
Temperature and Moisture as Limiting Factors   88
Interaction Between Temperature and Moisture   *91*
Adaptations to Temperature and Moisture   *94*
Light as a Limiting Factor   97
Climate Change and Species Distributions   102
Summary   103
Key Concepts   104
Selected References   104
Questions and Problems   104

CHAPTER 8   THE RELATIONSHIP BETWEEN DISTRIBUTION AND ABUNDANCE   106

The Spatial Scale of Geographic Ranges   106
Variations in Geographic Range Size   108
Range Size and Abundance   111
Summary   113
Key Concepts   114
Selected References   114
Questions and Problems   114

PART 3   THE PROBLEM OF ABUNDANCE   115

CHAPTER 9   POPULATION PARAMETERS   116

The Population as a Unit of Study   116
Box 9.1 Definitions of Population Parameters   *00*
Unitary and Modular Organisms   117
Estimation of Population Parameters   119
Box 9.2 Calculation of Expected Population Density from the Regression Data Given in Table 9.1   *120*
Measurements of Absolute Density   *120*
Indices of Relative Density   *125*

Essay 9.1 A Historical Essay in Ecology: The Cormack-Jolly-Seber
   Mark-Recapture Model   *126*
     Natality   *128*
     Mortality   *128*
     Immigration and Emigration   *129*
Limitations of the Population Approach   130
Composition of Populations   130
Summary   131
Key Concepts   131
Selected References   131
Questions and Problems   132

CHAPTER 10   **DEMOGRAPHIC TECHNIQUES: VITAL STATISTICS**   **133**
Life Tables   133
Box 10.1 Calculation of Per Capita Rates   *135*
Intrinsic Capacity for Increase in Numbers   138
Essay 10.1 Demographic Projections and Predictions   *143*
Box 10.2 Calculation of the Intrinsic Capacity for Increase
   from Lotka's Characteristic Equation   145
Reproductive Value   145
Age Distributions   146
Evolution of Demographic Traits   149
Summary   154
Key Concepts   154
Selected References   154
Questions and Problems   155

CHAPTER 11   **POPULATION GROWTH**   **157**
Mathematical Theory   157
   Growth in Populations with Discrete Generations   *157*
   Growth in Populations with Overlapping Generations   *160*
*Box 11.1 What is Little-*r *and Why Is It So Confusing?*   *162*
Laboratory Tests of the Logistics Theory   163
Field Data on Population Growth   164
Essay 11.1 What Is a "Good" Population Growth Model?   *166*
Box 11.2 A Simple Time-Lag Model of Population Growth   *168*
Time-Lag Models of Population Growth   169
Stochastic Models of Population Growth   169
Population Projection Matrices   173
Summary   176
Key Concepts   176
Selected References   177
Questions and Problems   177

CHAPTER 12   **SPECIES INTERACTIONS: COMPETITION   179**

Classification of Species Interactions   179
Theories on Competition for Resources   180
    Mathematical Model of Lotka and Volterra   180
Tilman's Model   182
Essay 12.1 What Is a Phase Plane, and What Is an Isocline?   184
Competition in Experimental Laboratory Populations   185
Competition in Natural Populations   190
Evolution of Competitive Ability   199
    Theory of r-Selection and K-Selection   199
    Grime's Theory of Plant Strategies   201
Character Displacement   201
Diffuse Competition and Indirect Effects   202
Summary   203
Key Concepts   204
Selected References   204
Questions and Problems   205

CHAPTER 13   **SPECIES INTERACTIONS: PREDATION   206**

Mathematical Models of Predation   207
    Discrete Generations   207
    Continuous Generations   209
Laboratory Studies of Predation   212
Field Studies of Predation   216
Essay 13.1 Laboratory Studies and Field Studies   217
Optimal Foraging Theory   225
Evolution of Predator-Prey Systems   228
    Warning Coloration   229
    Group Living   231
Summary   232
Key Concepts   232
Selected References   232
Questions and Problems   233

CHAPTER 14   **SPECIES INTERACTIONS: HERBIVORY AND MUTUALISM   235**

Defense Mechanisms in Plants   235
    Tannins in Oak Trees   238
    Ants and Acacias   239
    Spines in a Marine Bryozoan   241
    Spines and Thorns in Terrestrial Plants   242
Herbivores on the Serengeti Plains   242
Can Grazing Benefit Plants?   246
Essay 14.1 Herbivory, Economics, and Land Use   247
Dynamics of Herbivore Populations   248
    Interactive Grazing: Ungulate Irruptions   248
    Noninteractive Grazing: Finch Populations   251

Seed Dispersal: An Example of Mutualism   253
Complex Species Interactions   254
Summary   256
Key Concepts   256
Selected References   256
Questions and Problems   257

CHAPTER 15   SPECIES INTERACTIONS: DISEASE AND PARASITISM   258

Mathematical Models of Host-Disease Interaction   258
    Compartment Models with Constant Population Size   *259*
Essay 15.1 What Is the Transmission Coefficient (ß),
    and How Can We Measure It?   *261*
    Compartment Models with Variable Population Size   *262*
Box 15.1 How Can Determine R_0? A Mathematical Excursion   *263*
Effects of Disease on Individuals   264
    Effects on Reproductions   *264*
Effects on Mortality   265
Effects of Disease on Populations   267
    Brucellosis in Ungulates   *267*
    Rabies in Wildlife   *268*
    Myxomatosis in the European Rabbit   *272*
Box 15.2 A Simple Rabies Model   *273*
Evolution of Host-Parasite Systems   275
Essay 15.2 What is the Red Queen Hypothesis?   *276*
Summary   277
Key Concepts   278
Selected References   278
Questions and Problems   279

CHAPTER 16   POPULATION REGULATION   280

A Simple Model of Population Regulation   281
Historical Views of Population Regulation   282
Essay 16.1 Definitions in Population Regulation   *283*
A Modern Synthesis of Population Regulation   288
Essay 16.2 Why Is Population Regulation So Controversial?   *290*
Two Approaches to Studying Population Dynamics   293
    Key Factor Analysis   *293*
    Experimental Analysis   *296*
Plant Population Regulation   297
Source and Sink Populations   299
Evolutionary Implications of Population Regulation   300
Summary   302
Key Concepts   302
Selected References   303
Questions and Problems   303

CHAPTER 17   APPLIED PROBLEMS I: HARVESTING POPULATIONS   305

Logistic Models   307
Dynamic Pool Models   309
Laboratory Studies on Harvesting Theory   316
The Concept of Optimum Yield   318
Case Study: The King Crab Fishery   319
Case Study: The Northern Cod Fishery   321
Case Study: Antarctic Whaling   324
Risk-Aversive Management Strategies   325
Essay 17.1 Principles of Effective Resource Management   326
Box 17.1 What Are the Harvest Strategies for a Fishery?   327
Summary   328
Key Concepts   328
Selected References   328
Questions and Problems   329

CHAPTER 18   APPLIED PROBLEMS II: PEST CONTROL   331

Examples of Biological Control   333
   Cottony-Cushion Scale (Icerya purchasi)   *333*
   Prickly Pear (Opuntia spp.)   *334*
   Floating Fern (Salvinia molesta)   *336*
Theory of Biological Control   337
Genetic Controls of Pests   342
Immunocontraception   344
Integrated Control   346
Generalizations About Biological Control   349
Risks of Biological Control   351
Summary   352
Key Concepts   353
Selected References   353
Questions and Problems   353

CHAPTER 19   APPLIED PROBLEMS III: CONSERVATION BIOLOGY   355

Small-Population Paradigm   355
   Minimum Viable Populations   356
Box 19.1 What Is Effective Population Size?   359
Essay 19.1 Diagnosing a Declining Population   360
   The Declining-Population Paradigm   360
   Overkill   361
   Habitat Destruction and Fragmentation   362
Essay 19.2 Fragmentation of Habitats and Area-Sensitive Species   367
Impacts of Introduced Species   371
   Chains of Extinctions   373
Reserve Design and Reserve Selection   373
*Box 19.2 Recovery of Petrels After Eradication of Feral Cats on Marion
   Island, Indian Ocean   374*
Box 19.3 An Algorithm for Choosing Reserves for a Taxonomic Group   375

Examples of Conservation Problems 377
    Furbish's Lousewart 377
    The Northern Spotted Owl 379
Summary 381
Conclusion 382
Key Concepts 382
Selected References 382
Question and Problems 383

PART 4   DISTRIBUTION AND ABUNDANCE AT THE COMMUNITY LEVEL 385

CHAPTER 20   THE NATURE OF THE COMMUNITY 386

Dynamic Relations Between Populations 386
Essay 20.1 What Is the Gaia Hypothesis? 388
Community Characteristics 392
    Community Boundaries? 392
    Distributional Relations of Species in Communities 395
    Indicator Species in Communities 398
Box 20.1 Criteria for Indicator Species 399
Summary 400
Key Concepts 401
Selected References 401
Questions and Problems 401

CHAPTER 21   COMMUNITY CHANGE 403

Primary Succession on Mount St. Helens 403
Concepts of Succession 406
A Simple Mathematical Model of Succession 409
Case Studies of Succession 412
    Glacial Moraine Succession in Southeastern Alaska 413
    Lake Michigan Sand-Dune Succession 416
Essay 21.1 Why Is Sphagnum Moss So Common? 417
    Abandoned Farmland in North Carolina 419
The Climax State 423
Patch Dynamics 424
Summary 431
Key Concepts 431
Selected References 431
Questions and Problems 432

CHAPTER 22   COMMUNITY ORGANIZATION I: BIODIVERSITY 434

Measurement of Biodiversity 434
Essay 22.1 Biodiversity: A Brief History 436
Some Examples of Diversity Gradients 438

(continued)

(Chapter 22, continued)

    Box 22.1 A Simple Model of Latitudinal Gradients in Biodiversity 443
    Factors That Might Cause Diversity Gradients 443
        History Factory 444
        Spatial Heterogeneity 445
        Competition 447
        Predation 448
        Climate and Climatic Variability 450
        Productivity 452
        Disturbance 452
    Local and Regional Diversity 454
    Summary 456
    Key Concepts 456
    Selected References 457
    Questions and Problems 457

CHAPTER 23 COMMUNITY ORGANIZATION II: PREDATION AND COMPETITION IN EQUILIBRIAL COMMUNITIES 459
    Box 23.1 Measuring Community Importance 461
    Food Chains and Trophic Levels 463
    Essay 23.1 Use of Stable Isotopes to Analyze Food Chains 466
    Functional Roles and Guilds 469
    Keystone Species 471
    Dominant Species 474
    Essay 23.2 Fishing Down Food Webs 477
    Community Stability 477
    Restoration Ecology 480
    Summary 482
    Key Concepts 483
    Selected References 483
    Questions and Problems 484

CHAPTER 24 COMMUNITY ORGANIZATION III: DISTURBANCE AND NONEQUILIBRIUM COMMUNITIES 485
    Patches and Disturbance 485
    The Role of Disturbance in Communities 486
        Coral Reef Communities 486
    Essay 24.1 Why Are Corals Bleaching? 489
        Rocky Intertidal Communities 491
    Theoretical Nonequilibrium Models 492
    Conceptual Models of Community Organization 495
    Essay 24.2 Biomanipulation of Lakes 499
    The Special Case of Island Species 501
    Box 24.1 Measuring Immigration and Extinction Rates 507
    Multiple Stable States in Communities 508
    Summary 509
    Key Concepts 510
    Selected References 510
    Questions and Problems 511

CHAPTER 25   ECOSYSTEM METABOLISM I: PRIMARY PRODUCTION   513
Primary Production   515
Factors That Limit Primary Productivity   517
Aquatic Communities   518
Marine Communities   518
Freshwater Communities   523
Essay 25.1 Nutrient Ratios and Phytoplankton   526
Terrestrial Communities   527
Box 25.1 Estimating Primary Production from Satellite Data   528
Essay 25.2 Why Does Primary Production Decline with Age in Trees?   532
Plant Diversity and Productivity   533
Summary   534
Key Concepts   534
Selected References   535
Questions and Problems   535

CHAPTER 26   ECOSYSTEM METABOLISM II: SECONDARY PRODUCTION   537
Measure of Secondary Production   537
Box 26.1 Estimating Energy Expenditure with Doubly Labeled Water   540
Essay 26.1 Thermodynamics and Ecology   543
Problems in Estimating Secondary Production   544
Ecological Efficiencies   544
What Limits Secondary Production?   549
Grassland Ecosystems   549
Essay 26.2 Why Is the World Green?   551
Game Ranching in Africa   554
Sustainable Energy Budgets   556
Summary   557
Key Concepts   557
Selected References   558
Questions and Problems   558

CHAPTER 27   ECOSYSTEM METABOLISM III: NUTRIENT CYCLES   560
Nutrient Pools and Exchanges   560
Nutrient Cycles in Forests   562
Box 27.1 Estimating Turnover Time for Nutrients   564
Efficiency of Nutrient Use   568
Acid Rain: The Sulfur Cycle   572
Essay 27.1 Acid Rain and the Sudbury Experience   575
The Nitrogen Cycle   576
Summary   580
Key Concepts   581
Selected References   581
Questions and Problems   582

CHAPTER 28   ECOSYSTEM HEALTH: HUMAN IMPACTS   583
Human Population Growth   583
Current Patterns of Population Growth   584

(continued)

(Chapter 28, continued)

Box 28.1 How Large Is a Billion Anyway? 585
    Carrying Capacity of the Earth 586
Essay 28.1 The Demographic Transition: An Evolutionary Dilemma 587
Box 28.2 How to Calculate an Ecological Footprint 589
The Carbon Cycle 590
Individual Plant Responses to CO_2 593
    Plant Community Responses to CO_2 594
Climate Change 596
Essay 28.2 El Niño and the Southern Oscillation 598
Changes in Land Use 600
Biotic Invasions and Species Ranges 601
Ecosystem Services 603
Box 28.3 How Is Biodiversity Related to Ecosystem Function? 604
Essay 28.3 On Corals and Climate Change 605
Essay 28.4 Economics of Ecosystem Services 606
Summary 606
Key Concepts 607
Selected References 607
Questions and Problems 608

EPILOGUE 609

APPENDIX I   A Primer on Population Genetics 610

APPENDIX II   Estimation of the Size of the Marked Popluation
in Capture-Recapture Studies 612

APPENDIX III   Instantaneous and Finite Rates 614

APPENDIX IV   Species Diversity Measures of Heterogeneity 617

GLOSSARY 619

BIBLIOGRAPHY 623

CREDITS 664

SPECIES INDEX 673

SUBJECT INDEX 680

ABOUT THE AUTHOR

WHILE GROWING UP IN ST. LOUIS, I became interested in polar exploration and read every book in the library about the arctic. I was fortunate to get a summer job working on the Pribilof Islands in the Bering Sea, and have been a northern ecologist ever since. I began my studies in wildlife management at the University of Minnesota, and moved to Canada to do graduate work at the University of British Columbia in 1957. After a Ph.D. on lemming populations in the Canadian arctic, I moved to Berkeley to do a Miller Postdoctoral Fellowship with Frank Pitelka. I finally got a proper job at Indiana University in Bloomington in 1964 and began teaching ecology to undergraduates. It was clear to me that there was a shortage of teaching material in ecology, and in particular there was no text that captured the Eltonian approach to ecology through population and community dynamics. I began writing this book in 1967 and the first edition appeared in 1972. One of the great joys of writing a textbook is meeting people all over the world who used my text during their education. Through five editions I have tried to track the progress of ecological science, and it is a sign of progress that ecologists are now recognized all over the world for their contributions to wise management of our natural heritage.

I am currently Professor of Zoology at the University of British Columbia in Vancouver. In addition to teaching ecology, I have worked extensively on the population of rodents in northern Canada, the United States, and Australia, trying to understand the mechanisms behind population fluctuations. I have written three ecology textbooks including *Ecology: The Experimental Analysis of Distribution* and *Abundance, Fifth Edition* and *Ecological Methodology, Second Edition* both published by Benjamin Cummings

PREFACE

YOU ARE LIVING IN THE AGE OF ECOLOGY, and as a citizen you ought to learn something about this subject. There has been a revolution of human thinking in the last 30 years that has centered on the relationship between humans and their environment. The broader policy problems this revolution has brought forward are the focus of the environment movement, the applied scientific problems, and the focus of environmental science. The basic science behind it all is the science of ecology. Sustainability is the mantra of all our politicians, and environmental problems are now a common subject in the daily newspapers.

Just as it is useful to know something about physics if you wish to be an engineer, it is useful to learn something about ecology if you wish to understand the problems humans face with sustaining their environment. This text is dedicated to presenting to you the outlines of the science of ecology. If you understand how the natural world works, you will be better poised to understand the Age of Ecology as it unfolds, to think with an ecological conscience.

Two dilemmas face the textbook writer. First, the writer must plot a course that will place the book serenely between the pitfalls of the past and the bandwagons of the present. We all recognize the pitfalls of the past, and any text has an obligation to point out some of these lest history repeat itself. We do not do as well at recognizing the pitfalls of the present-at recognizing which of the current bandwagons in ecology are enduring and which are ephemeral. Science, like most subjects of human endeavor, is subject to bandwagons, only some of which are useful for our long-term understanding. Second, the writer must create a textbook that pleases not only the students but also their instructors. I have strived to make this book readable, and the greatest compliment any text can get is that students think it readable and interesting.

Each chapter in this book attempts to raise a question about how populations and communities operate in nature, and to give you enough information that you can think about it intelligently. If you need more information, a list of suggested readings is a good starting point. Each chapter ends with a series of questions and problems that are devised to stimulate thought. I have not provided answers to these questions. For many of them the answer is not yet known. An overview question at the end of each chapter is a still more general question that may be the focus of a class discussion. Many overview questions are action-oriented. A key focus of much ecological thought ought to be "What are the practical consequences of this idea?"

In this edition I have added chapters on the population dynamics of disease and parasitism, and on ecosystem health and human impacts. I have tried to emphasize the historical development of ecology by adding photos of famous ecologists in each chapter. Science is a human activity and the scientists who have built ecology and are building it today are themselves interesting characters worthy of more recognition. I have extensively revised many of the chapters, particularly those on conservation biology, community organization, and primary production, and strengthened the integration of evolutionary and functional ecology. Conservation biology is a focus for practical problems that cry out for ecological understanding and is one of the strong growth points of ecological science. Many of the attempts to converse biodiversity hinge on concepts of community organization that need careful thought and analysis. Many chapters deal with ecological attributes and their evolutionary background. Ecologists can benefit by stepping back and looking at ecological systems in a revolutionary perspective, and students of evolution can benefit from knowing how ecological systems function, for they cannot otherwise understand natural selection. I have added essays in many chapters to illustrate some of the kinds of problems and questions ecologists deal with in their attempt to understand nature.

This book is my own attempt to present modern ecology as an interesting and dynamic subject. Beneath the variety of approaches that characterize modern ecology lie a few basic problems that I have attempted to sketch. I have placed special emphasis on problems and have illustrated them by examples chosen as diversely as possible from the plant and animal kingdoms. This book is not an encyclopedia of ecology but an introduction to its problems. It is not descriptive ecology and will not tell students about the ecology of the seashore or the ecology of the alpine tundra. It approaches ecology as a series of problems, problems that confined neither to the seashore nor to the alpine tundra but are sufficiently general to be studied in either area.

To understand the problems of ecology, students must have some background in biology and mathematics. Students will find that they can understand ecology without knowing any mathematics but that mathematics is necessary for those who wish to proceed beyond the simplest level of analysis. Ecology is not a haven for people who cannot do mathematics, and in this respect it is no different from chemistry and physics. Statistics and calculus are useful but not essential for an understanding of this book. I present mathematical analyses step by step and illustrate them with graphs. Students who cannot follow the mathematics should be able to get the essence of the arguments from the graphs.

The problems of ecology are "biological" problems and will be solved not by mathematicians but by biologists. Students will find that, contrary to the impression they get from other sources, the problems of ecology have not all been solved. A start has been made in solving many ecological problems, and I cannot give the "answer" to many of the problems I discuss. Controversies are common in ecology, and an important part of ecological training is appreciating the controversies and trying to understand why people may look at the same data and yet reach opposite conclusions.

Students can learn much about the science of ecology by analyzing one of its controversies. If you think that ecologists have the answer to most of our environmental questions, you will be surprised when you look into the variety of ecological controversies. Controversy is not a sign of weak science, and to appreciate controversies you should try to find out the scope of the controversy and what kinds of observations are needed to solve it. Many of the environmental controversies of our day, like climate change, involve a mixture of scientific facts and policy decisions. Scientific facts alone do not determine policy, but policy without a solid scientific grounding is doomed. The relevant scientific facts are never completely known for many environmental problems, and we must decide what to do in the face of uncertainty. Interim policy decisions always point out the need for more scientific analysis, and there must be a continuous feedback loop between policy and all the environmental sciences, including ecology.

Good ecology is quantitative. At the end of many of the quantitative chapters, problems are included because no one can appreciate the quantitative aspects of ecology without going through some of the calculations. Most of the calculations are simple, but I have tried to leave some of them open-ended so that interested students can carry on under their own steam.

If there is a message in this book, it is a simple one: Progress is answering ecological questions comes when experimental techniques are used. The habit of asking, "What experiment could answer this question?" is the most basic aspect of scientific method that students should learn to cultivate. When there is controversy, asking this question can cut to the heart of the matter.

Technical terms in this book are kept to a minimum; labeling with words should not be confused with understanding. The glossary of technical words, together with the indexes, should be adequate to cover technical definitions.

I thank my many friends and colleagues who have contributed to formulating and clarifying the material presented here. In particular I thank my colleagues Dennis Chitty, Judy Myers, Jamie Smith, Carl Walters, and Tony Sinclair for their assistance, and Brian Walker and the many ecologists at CSIRO Wildlife and Ecology in Canberra who answered endless queries during this version. For a detailed critique of the revision I am indebted to John C. Horn, St. Ambrose University, Davenport, IA; Alan Stam, Capital University, Columbus, OH; Merrill Sweet, Texas A&M University, College Station, TX; John Baccus, Southwest Texas State University, San Marcos, TX; Ralph J. Larson, San Francisco State University, San Francisco, CA; Mary Wicksten, Texas A&M University, College Statuibm TX; Stephen G. Tilley, Smith College, Northampton, MA; Robert Bailey, Central

Michigan University, Mt Pleasant, MI; Patricia L. Kennedy, Colorado State University, Ft. Collins, CO; Henry Merchant, George Washington University, Washington DC; Kathy Williams, San Diego State University, San Diego, CA; S. Sweet, Texas A&M University, College Station, TX. To all of these I am most grateful.

Thanks to Elizabeth Fogerty, Heather Dutton, and Chriscelle Merquillo at Addison Wesley who did more than their share of editorial work to help improve this edition. Finally I want to thank the real authors of this book, the hundreds of ecologists who have toiled in the field and laboratory to extract from the study of organisms the concepts discussed here. A person's life work may be boiled down to a few sentences in a book, and we ecologists owe a debt that we cannot pay to our intellectual ancestors.

Charley Krebs

P A R T   O N E

What Is Ecology?

CHAPTER 1   INTRODUCTION TO THE SCIENCE OF ECOLOGY
CHAPTER 2   EVOLUTION AND ECOLOGY

CHAPTER 1

Introduction to the Science of Ecology

YOU ARE EMBARKING ON A STUDY OF ECOLOGY, the most integrative discipline in the biological sciences. The purpose of this introductory chapter is to get you started by defining the subject, providing a small amount of background history, and introducing the broad concepts that will serve as a road map for the details explained in the next 27 chapters.

Definition of Ecology

The word *ecology* came into use in the second half of the nineteenth century. Ernst Haeckel in 1869 defined *ecology* as the total relations of the animal to both its organic and its inorganic environment. This very broad definition has provoked some authors to point out that if this is ecology, there is very little that is *not* ecology. Since four biological disciplines are closely related to ecology—genetics, evolution, physiology, and behavior—the problem of defining ecology may be viewed schematically in the following way:

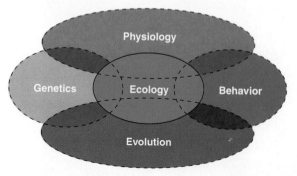

Broadly interpreted, ecology overlaps each of these subjects; hence we need a more restrictive definition.

Charles Elton (1927) in his pioneering book *Animal Ecology* defined ecology as scientific natural history. Although this definition points out the origin of many of our ecological problems, it is again uncomfortably vague. Eugene Odum (1963) defined ecology as the study of the structure and function of nature. This statement has the merit of emphasizing the form-and-function idea that permeates biology, but it is still not a completely clear definition. A clear and restrictive definition of ecology is this: Ecology is the scientific study of the distribution and abundance of organisms (Andrewartha 1961). This definition is static and leaves out the important idea of *relationships*. Because ecology is about relationships, we can modify Andrewartha's definition to make a precise definition of ecology as follows: *Ecology is the scientific study of the interactions that determine the distribution and abundance of organisms.*

This definition of ecology restricts the scope of our quest to a manageable level and is the way we use this word in this book. To understand what ecology is, we need to know what is special about *scientific* studies, and what we mean by distribution and abundance. We have an intuitive feeling about *distribution* (where organisms are found) and about *abundance* (how many occur there), and once we know these facts we must answer the most difficult question: *why* this particular distribution, *why* this abundance? We seek the cause and effect relationships that govern distribution and abundance.

History of Ecology

The historical roots of ecology are varied, and in this section we explore briefly some of the origins of ecological ideas. We are not the first humans to think about ecological problems. The roots of ecology lie in natural history, which is as old as humans. Primitive tribes, which depended on hunting, fishing, and food gathering, needed detailed knowledge of where and when their quarry might be found. The establishment of agriculture increased the need to learn about the practical ecology of plants and domestic animals. Agriculture today is a special form of applied ecology.

We read about locust outbreaks in our newspapers and plagues of rats in rice crops in Asia. These are not new. Spectacular plagues of animals attracted the attention of the earliest writers. The Egyptians and Babylonians feared locust plagues, and supernatural powers were often believed to cause these outbreaks. Exodus (7:14–12:30) describes the plagues that God called down upon the Egyptians. In the fourth century B.C., Aristotle tried to explain plagues of field mice and locusts in his *Historia Animalium*. He pointed out that the high reproductive rate of field mice could produce more mice than could be reduced by their natural predators, such as foxes and ferrets, or by the control efforts of humans. Nothing succeeded in reducing these mouse plagues, Aristotle stated, except the rain, and after heavy rains the mice disappeared rapidly. Australian wheat farmers currently face plagues of house mice and ask the same questions: How can we get rid of these pests?

Pests have always been a problem for people because they violate our feeling of harmony or balance in the universe. Ecological harmony was a guiding principle basic to the Greeks' understanding of nature, and Egerton (1968a) has traced this concept from ancient times to the modern term *balance of nature*. This concept of "providential ecology," in which nature is designed to benefit and preserve each species, was implicit in the writings of Herodotus and Plato. The assumptions of this worldview were that the numbers of every species remain essentially constant. Outbreaks of some populations might occur, but these could usually be traced to divine intervention for the punishment of evil-doers. Each species had a special place in nature, and extinction did not occur because it would disrupt the balance and harmony in nature. We are in the midst of a major extinction crisis in the world today, and these early ideas about harmony in nature are difficult for us to grasp.

How did we get from these early Greek and Roman ideas about harmony to our modern understanding? A combination of mathematics and natural history paved the way. By the seventeenth century students of natural history and human ecology began to focus on population ecology and to construct a quantitative framework. Graunt (1662), who described human population change in quantitative terms, can be called the "father of demography"[1] (Cole 1958). He recognized the importance of measuring birth rates, death rates, and age structure of human populations, and he complained about the inadequate census data available in England in the seventeenth century. Graunt estimated the potential rate of population growth for London and concluded that even without immigration, London's population would double in 64 years.

Human population growth occupies our newspapers every week, but population growth was not always considered quantitatively for animals and plants. Leeuwenhoek studied the reproductive rate of grain beetles, carrion flies, and human lice. In 1687 he counted the number of eggs laid by female carrion flies and calculated that one pair of flies could produce 746,496 flies in three months. Leeuwenhoek made one of the first attempts to calculate theoretical rates of increase for an animal species (Egerton 1968b).

By the eighteenth century, natural history had become an important cultural occupation. Buffon in his *Natural History* (1756) touched on many of our modern ecological problems and recognized that populations of humans, other animals, and plants are subjected to the same processes. Buffon discussed, for example, how the great fertility of every species was counterbalanced by innumerable agents of destruction. He believed that plague populations of field mice were checked partly by diseases and scarcity of food. Buffon did not accept Aristotle's idea that heavy rains caused the decline of dense mouse populations but thought instead that control was achieved by biological agents. Rabbits, he stated, would reduce the countryside to a desert if it were not for their predators. If the Australians had listened to Buffon before they introduced rabbits in 1859, they could have saved their rangelands from destruction by this agricultural pest. Buffon in 1756 was dealing with problems of population regulation that are still unsolved today.

---

[1] Demography originated as the study of human population growth and decline. It is now used as a more general term that includes plant and animal population changes.

Malthus, the most famous of the early demographers, published one of the earliest controversial books on demography. In his *Essay on Population* (1798), he calculated that although the numbers of organisms can increase geometrically (1, 2, 4, 8, 16,…), their food supply may never increase faster than arithmetically (1, 2, 3, 4,…). The arithmetic rate of increase in food production seems to be somewhat arbitrary. The great disproportion between these two powers of increase led Malthus to infer that reproduction must eventually be checked by food production. The thrust of Malthus's ideas was negative: What prevents populations from reaching the point at which they deplete their food supply? What checks operate against the tendency toward a geometric rate of increase? Two centuries later we still ask these questions. These ideas were not new; Machiavelli had said much the same thing around 1525, as did Buffon in 1751, and several others had anticipated Malthus. It was Malthus, however, who brought these ideas to general attention. Darwin used the reasoning of Malthus as one of the bases for his theory of natural selection.

Other workers questioned the ideas of Malthus and made different predictions for human populations. For example, in 1841 Doubleday brought out his *True Law of Population*. He believed that whenever a species was threatened, nature made a corresponding effort to preserve it by increasing the fertility of its members. Human populations that were undernourished had the highest fertility; those that were well fed had the lowest fertility. You can make the same observations by looking around the world today (Table 1.1). Doubleday explained these effects by the oversupply of mineral nutrients in well-fed populations. Doubleday observed a basic fact that we recognize today, low birth rates occur in wealthy countries—although his explanations were completely wrong.

Interest in the mathematical aspects of demography increased after Malthus. Can we describe a mathematical law of population growth? Quetelet, a Belgian statistician, suggested in 1835 that the growth of a population was checked by factors opposing population growth. In 1838 his student Pierre-François Verhulst derived an equation describing the initial rapid growth and eventual leveling off of a population over time. This S-shaped curve he called the logistic curve. This work was overlooked until modern times, but it is of fundamental importance, and we shall return to it later in detail.

Until the nineteenth century, philosophical thinking had not changed from the idea of Plato's day that there was harmony in nature. Providential design was

TABLE 1.1 Total fertility rate of human populations and gross national product (GNP) per person in selected countries of the globe in 1997.

The total fertility rate is the average number of children a woman would have, assuming no change in birth rates. The GNP is in U.S. dollars per person. (Data from 1999 World Population Data Sheet, Population Reference Bureau.)

Country	Total fertility rate	Gross national product per person
Sudan	4.6	290
Gambia	5.6	340
Niger	7.5	200
Tanzania	5.7	210
Botswana	4.1	3,310
South Africa	3.3	3,210
Canada	1.5	19,640
United States	2.0	29,080
Costa Rica	2.7	2,680
Mexico	3.0	3,700
Haiti	4.8	380
Brazil	2.3	4,790
Peru	3.5	2,610
Turkey	2.6	3,130
India	3.4	370
Pakistan	5.6	500
Indonesia	2.8	1,110
China	1.8	860
Japan	1.4	38,160
Sweden	1.5	26,210
Switzerland	1.5	43,060
Russia	1.2	2,680
Italy	1.2	20,170
Solomon Islands	5.4	870

still the guiding light. In the late eighteenth and early nineteenth centuries, two ideas that undermined the idea of the balance of nature gradually gained support: (1) that many species had become extinct and (2) that resources are limited, and that competition caused by population pressure is important in nature. The consequences of these two ideas became clear with the work of Malthus, Lyell, Spencer, and Darwin in the nineteenth century. Providential ecology and the balance of nature were replaced by natural selection and the struggle for existence (Egerton 1968c).

The balance of nature idea, redefined after Darwin, has continued to persist in modern ecology (Pimm 1991). The idea that natural systems are stable and in equilibrium with their environments unless humans disturb them is still accepted by many ecologists and theoreticians. We discuss the equilibrium view in Chapters 24 and 25.

Humans must eat, and many of the early developments in ecology came from the applied fields of agriculture and fisheries. Insect pests of crops have been one focus of work. Before the advent of modern chemistry, biological control was the only feasible approach. In 1762 the mynah bird was introduced from India to the island of Mauritius to control the red locust; by 1770 the locust threat was a negligible problem (Moutia and Mamet 1946). Forskål wrote in 1775 about the introduction of predatory ants from nearby mountains into date-palm orchards to control other species of ants feeding on the palms in southwestern Arabia. In subsequent years an increasing knowledge of insect parasitism and predation led to many such introductions all over the world in the hope of controlling nonnative and native agricultural pests (De Bach 1974). We discuss this problem of *biological* control in Chapter 18.

Medical work on infectious diseases such as malaria in the late 1800s gave rise to the study of epidemiology and interest in the spread of disease through a population. Malaria is still one of the great scourges of humans. In 1900 no one even knew the cause of the disease. Once mosquitoes were pinpointed as the vectors, medical workers realized that it was necessary to know in detail the ecology of mosquitoes. The pioneering work of Robert Ross (1911) attempted to describe in mathematical terms the propagation of malaria, which is transmitted by mosquitoes. In an infected area, the propagation of malaria is determined by two continuous and simultaneous processes: (1) The number of new infections among people depends on the number and infectivity of mosquitoes, and (2) the infectivity of mosquitoes depends on the number of people in the locality and the frequency of malaria among them. Ross could write these two processes as two simultaneous differential equations:

$$\begin{pmatrix} \text{Rate of increase} \\ \text{of infected humans} \end{pmatrix} = \begin{pmatrix} \text{New infections} \\ \text{per unit time} \end{pmatrix} - \begin{pmatrix} \text{Recoveries per} \\ \text{unit time} \end{pmatrix}$$

(Depends on number of infected mosquitoes)

$$\begin{pmatrix} \text{Rate of increase} \\ \text{of infected mosquitoes} \end{pmatrix} = \begin{pmatrix} \text{New infections} \\ \text{per unit time} \end{pmatrix} - \begin{pmatrix} \text{Death of infected per} \\ \text{unit time} \end{pmatrix}$$

(Depends on number of infected humans)

Ross had described an ecological process with a mathematical model, and his work represents a pioneering parasite-host model of species interactions (see Chapter 15). Such models can help us to clarify the problem—we can analyze the components of the model—and predict the spread of malaria or other diseases.

Production ecology, the study of the harvestable yields of plants and animals, had its beginnings in agriculture, and Egerton (1969) traced this back to the eighteenth-century botanist Richard Bradley. Bradley recognized the fundamental similarities of animal and plant production, and he proposed methods of maximizing agricultural yields (and hence profits) for wine grapes, trees, poultry, rabbits, and fish. The conceptual framework that Bradley used—monetary investment versus profit—is now called the *optimum-yield* problem and is a central issue in applied ecology (see Chapter 17).

Individual species do not exist in a vacuum, but instead in a matrix of other species with which they interact. Recognition of communities of living organisms in nature is very old, but specific recognition of the interrelations of the organisms in a community is relatively recent. Edward Forbes in 1844 described the distribution of animals in British coastal waters and part of the Mediterranean Sea, and he wrote of zones of differing depths that were distinguished by the associations of species they contained. Forbes noted that some species are found only in one zone, and that other species have a maximum of development in one zone but occur sparsely in other adjacent zones. Mingled in are stragglers that do not fit the zonation pattern. Forbes recognized the dynamic aspect of the interrelations between these organisms and their environment. As the environment changed, one species might die out, and another might increase in abundance. Karl Möbius expressed similar ideas in 1877 in a classic essay on the oyster-bed community as a unified collection of species.

Studies of communities were greatly influenced by the Danish botanist J. E. B. Warming (1895, 1909), one of the fathers of plant ecology. Warming was the first plant ecologist to ask questions about the composition of plant communities and the associations of species that made up these communities. The dynamics of vegetation change was emphasized first by North American plant ecologists. In 1899 H. C. Cowles described *plant succession* on the sand dunes at the southern end of Lake Michigan. The development of vegetation was analyzed by the American ecologist Frederick Clements (1916) in a classic book that began a long controversy about the nature of the community (see Chapter 20).

With the recognition of the broad problems of populations and communities, ecology was by 1900 on the road to becoming a science. Its roots lay in natural history, human demography, biometry (statistical approach), and applied problems of agriculture and medicine.

The development of ecology during the twentieth century has followed the lines developed by naturalists during the nineteenth century. The struggle to understand how nature works has been carried on by a collection of colorful characters quite unlike the mythical stereotypes of scientists. From Alfred Lotka, who worked for the Metropolitan Life Insurance Company in New York while laying the groundwork of mathematical ecology (Kingsland 1995), to Charles Elton, the British ecologist who wrote the first animal ecology textbook in 1927 and founded the Bureau of Animal Population at Oxford (Crowcroft 1991), ecology has blossomed with an increasing understanding of our

and its social ramifications. It is important to distinguish ecology from *environmental studies*.

Ecology is focused on the natural world of animals and plants, and includes humans as a very significant species by virtue of its impacts. Environmental studies is the analysis of these human impacts on the environment of the Earth—physical, chemical, and biological. Environmental studies as a discipline is much broader than ecology because it deals with many sciences, including ecology, geology, and climatology, as well as with the social sciences of sociology, economics, anthropology, political science, and philosophy. The science of ecology is not solely concerned with human impact on the environment but with the interrelations

Alfred J. Lotka *(1880–1949) Founder of Theoretical Ecology*

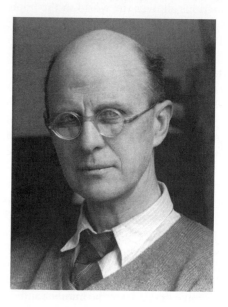

Charles Elton *(1900–1991) Founder of Animal Ecology, Oxford University*

world and how we humans affect its ecological systems (McIntosh 1985).

Until the 1960s ecology was not considered by society to be an important science. The continuing increase of the human population and the associated destruction of natural environments with pesticides and pollutants awakened the public to the world of ecology. Much of this recent interest centers on the human environment and human ecology and is called *environmentalism*. Unfortunately, the word *ecology* became identified in the public mind with the much narrower problems of the human environment, and "ecology" came to mean everything and anything about the environment, especially human impact on the environment

of all plants and animals. As such, ecology has much to contribute to some of the broad questions about humans and their environment that are an important scientific component of environmental studies.

Environmental studies have led to "environmentalism" and "deep ecology," social movements with an important agenda for political and social change intended to minimize human impacts on the Earth. These social and political movements are indeed important and are supported by many ecologists, but they are not the science of ecology. Ecology should be to environmental science as physics is to engineering. Just as we humans are constrained by the laws of physics when we build airplanes and bridges, so also are we constrained by the principles of ecology when altering the environment.

Ecological research can help to tell us what will happen when we increase global temperatures by increasing CO_2 emissions, but it will not tell us what we *ought* to do about these emissions, or whether increased global temperature is a good or a bad thing. Ecological scientists are not policy makers or moral authorities, and should not, as scientists, make ethical or political recommendations. As human beings, of course, ecologists do make these judgements, and virtually every ecologist is concerned about the extinction of species and would like to prevent extinctions. Many ecologists work hard in the political arena to achieve the social goals of environmentalism.

Basic Problems and Approaches to Ecology

We can approach the study of ecology from three points of view: *descriptive*, *functional*, or *evolutionary*. The descriptive point of view is mainly natural history and proceeds by describing the vegetation groups of the world, such as the temperate deciduous forests, tropical rain forests, grasslands, and tundra, and by describing the animals and plants and their interactions within each of these ecosystems. The descriptive approach is the foundation of all of ecological science. The functional point of view, on the other hand, is oriented more toward dynamics and relationships and seeks to identify and analyze general problems common to most or all of the different ecosystems. Functional studies deal with populations and communities as they exist and can be measured now. Functional ecology studies *proximate* causes—the dynamic responses of populations and communities to immediate factors of the environment. Evolutionary ecology studies *ultimate* causes—the historical reasons why natural selection has favored the particular adaptations we now see. The evolutionary point of view considers organisms and the relationships between organisms as historical products of evolution. Functional ecologists ask *how:* How does the system operate? Evolutionary ecologists ask *why:* Why does natural selection favor this particular ecological solution? Since evolution has occurred not only in the past but is also going on in the present, the evolutionary ecologist must work closely with the functional ecologist to understand ecological systems (Pianka 1994). Because the environment of an organism contains all

the selective forces that shape its evolution, ecology and evolution are two viewpoints of the same reality.

All three approaches to ecology have their strengths, but the important point is that we need all three to produce good science. The descriptive approach is absolutely fundamental, because, unless we have a good description of nature, we cannot construct good theories or good explanations. The descriptive approach provides us maps of geographical distributions and estimates of relative abundances of different species. With the functional approach, we need the detailed biological knowledge that natural history brings if we are to discover how ecological systems operate. The evolutionary approach needs good natural history and good functional ecology to speculate about past events and to suggest hypotheses that can be tested in the real world. No single approach can encompass all ecological questions.

This book uses a mixture of all three approaches and emphasizes the general problems ecologists try to understand.

The basic problem of ecology is to determine the causes of the distribution and abundance of organisms. Every organism lives in a matrix of space and time. Consequently, the concepts of distribution and abundance are closely related, although at first glance they may seem quite distinct. What we observe for many species is that the numbers of individuals in an area vary in space, so that if we make a contour map of a species's geographical distribution we might get something like this schematic illustration:

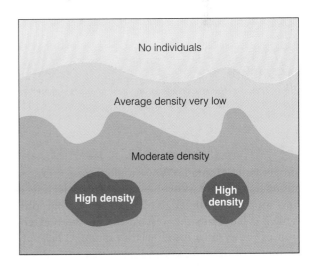

Figure 1.1 illustrates this idea for the eastern meadowlark of North America. Eastern meadowlarks are most common in the central United States, rare in western Kansas and North Dakota, and absent

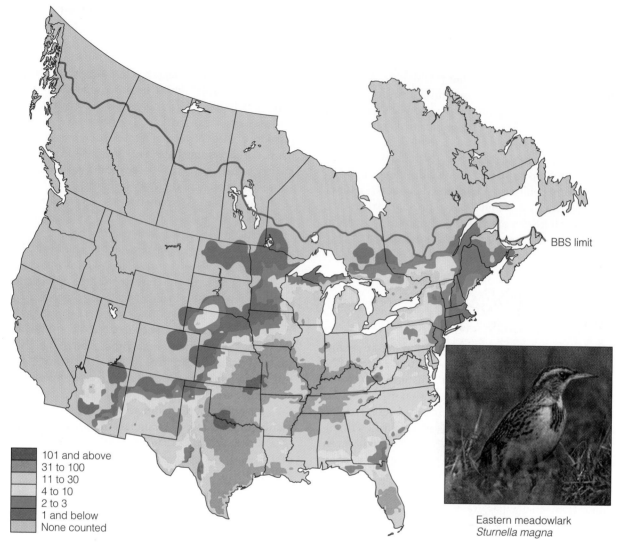

101 and above
31 to 100
11 to 30
4 to 10
2 to 3
1 and below
None counted

BBS limit

Eastern meadowlark
Sturnella magna

FIGURE 1.1
Abundance of the eastern meadowlark in North America, as measured by the Breeding Bird Survey (BBS), 1966–1996. (From Sauer et al. 1997. Photo courtesy of A. Wilson.)

altogether in Montana. Why should these patterns of abundance occur? Why does abundance decline as one approaches the edge of a species's geographic range? What limits the western and northern extension of the eastern meadowlark's range? These are examples of the fundamental questions an ecologist brings to nature.

Similarly, the red kangaroo occurs throughout the arid zone of Australia (Figure 1.2). It is absent from the tropical areas of northern Australia and most

common in western New South Wales and central Queensland. Why are there no red kangaroos in tropical Australia? Why is this species absent from Victoria in southern Australia and from Tasmania?

We can view the average density of any species as a contour map, with the provision that the contour map may change with time. Throughout the area of distribution, the abundance of an organism must be greater than zero, and the limit of distribution equals the contour of zero abundance. Distribution may be

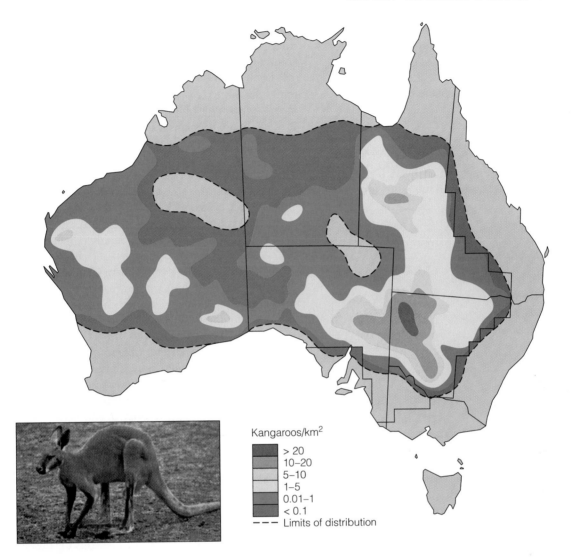

Kangaroos/km^2

- \> 20
- 10–20
- 5–10
- 1–5
- 0.01–1
- \< 0.1
- – – – Limits of distribution

FIGURE 1.2
Distribution and abundance of the red kangaroo in Australia, from aerial surveys, 1980–1982. (From Caughley et al. 1987a.)

considered a facet of abundance, and distribution and abundance may be said to be reverse sides of the same coin (Andrewartha and Birch 1954). The factors that affect the distribution of a species may also affect its abundance.

The problems of distribution and abundance can be analyzed at the level of the population of a single species or at the level of the community, which contains many species. The complexity of the analysis may increase as more and more species are considered in a community;

consequently, in this book we first consider the simpler problems involving single-species populations.

Considerable overlap exists between ecology and its related disciplines, which we cannot cover thoroughly in this book. *Environmental physiology* has developed a wealth of information that is needed to analyze problems of distribution and abundance. *Population genetics* and *ecological genetics* are two additional foci of interest that we touch only peripherally. *Behavioral ecology* is another interdisciplinary area that has implications for the study of distribution and abundance. *Evolutionary ecology* is an important focus for problems of adaptation and studies of natural selection in populations. Each of these disciplines can become an area of study entirely on its own.

Levels of Integration

In ecology we are dealing primarily with the five starred (*) levels of integration in the following:

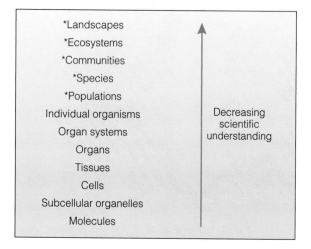

*Landscapes
*Ecosystems
*Communities
*Species
*Populations
Individual organisms
Organ systems
Organs
Tissues
Cells
Subcellular organelles
Molecules

Decreasing scientific understanding

At one end of the spectrum, ecology overlaps with environmental physiology and behavioral studies of individual organisms, and at the other end, ecology merges into meteorology, geology, and geochemistry when we consider landscapes. Landscapes can be aggregated to include the whole-earth ecosystem, which is sometimes called the *ecosphere* or the *biosphere*. The important message is that the boundaries of the sciences are not sharp but diffuse, and nature does not come in discrete packages.

Each level of integration involves a separate and distinct series of attributes and problems. For example, a population has a *density* (e.g., number of deer per square kilometer), a property that cannot be attributed to an individual organism. A community has *biodiversity* (or species richness), an attribute without meaning at the population level. In general, a scientist dealing with a particular level of integration seeks explanatory mechanisms from lower levels of integration and biological significance from higher levels. For example, to understand mechanisms of changes in a population, an ecologist might study mechanisms that operate on the behavior and physi-

ology of individual organisms and might try to view the significance of these population events within a community and ecosystem framework.

Some ecologists have suggested that the ecosystem—the biotic community and its abiotic environment—is the basic unit of ecology (Evans 1956, Rowe 1961). A particular significance may be attached to the ecosystem level from the viewpoint of human ecology, but this is only one of the levels of organization at which ecologists operate. There are meaningful and important questions to be asked at each level of integration, and none of them should be neglected or regarded as more fundamental to the study of ecology.

Much of modern biology is highly reductionistic, as it attempts to work out the physical-chemical basis of life. A good example is the Human Genome Project, an expensive and highly targeted research program to sequence all the genes on human chromosomes. The Human Genome Project is nearly completed, yet we do not know how many species of beetles live on the Earth, or how many species of trees there are in the Amazon basin. It should not surprise you that the amount of scientific understanding varies with the level of integration. We know an enormous amount about the molecular and cellular levels of organisms, we know very much about organs and organ systems and about whole organisms, but we know relatively little about populations and even less about communities and ecosystems. This point is illustrated very nicely by looking at the levels of integration: Ecology constitutes about one-third of the levels of biology, but no biology curriculum can be one-third ecology and do justice to *current* biological knowledge. The reasons for this are not hard to find—they include the increasing complexity of these higher levels and the difficulties involved in dealing with them in the laboratory.

Whatever the reasons for this decrease in understanding at the higher levels, it has serious implications for the study that we are about to undertake. You will not find in ecology the strong theoretical framework that you find in physics, chemistry, molecular biology, or genetics. It is not always easy to see where the pieces fit in ecology, and we shall encounter many isolated parts of ecology that are well developed theoretically but are not clearly connected to anything else. This is typical of a young science. Many students unfortunately think of science as a monumental pile of facts that must be memorized. But science is more than a pile of precise facts; it is a search for systematic relations, for explanations to problems in the physical world, and for unifying concepts. This is the growing end of science, so evident in a young science like ecology. It involves many unanswered questions and much more controversy.

The theoretical framework of ecology may be weaker than we would like at the present time, but this must not be interpreted as a terminal condition. Chemistry in the eighteenth century was perhaps in a comparable state of theoretical development as ecology at the present time. Sciences are not static, and ecology is in a strong growth phase.

Methods of Approach to Ecology

Ecology has been approached on three broad fronts: the *theoretical*, the *laboratory*, and the *field*. These three approaches are interrelated, but some problems have arisen when the results of one approach fail to verify those of another. For example, theoretical predictions may not be borne out by field data. We are primarily interested in understanding the distribution and abundance of organisms in *nature*—that is, in the *field*. Consequently, the descriptive ecology of populations, communities, and ecosystems will always be our basis for comparison, our basic standard.

Plant and animal ecology have tended to develop along separate paths. Historically, plant ecology got off to a faster start than animal ecology, despite the early interest in human demography. Because animals are highly dependent on plants, many of the concepts of animal ecology are patterned on those of plant ecology. *Succession* is one example. Also, since plants are the source of energy for many animals, to understand animal ecology we must also know a good deal of plant ecology. This is illustrated particularly well in the study of community relationships.

Some important differences, however, separate plant and animal ecology. First, because animals tend to be highly mobile whereas plants are stationary, a whole series of new techniques and ideas must be applied to animals—for example, to determine population density. Second, animals fulfill a greater variety of functional roles in nature—some are herbivores, some are carnivores, some are parasites. This distinction is not complete because there are carnivorous plants and parasitic plants, but the possible interactions are on the average more numerous for animals than for plants.

During the 1960s population ecology was stimulated by the experimental field approach in which natural populations were manipulated to test specific predictions arising from controversial ecological theory. During these years ecology was transformed from a static, descriptive science to a dynamic, experimental one in which theoretical predictions and field experiments were linked. At the same time, ecologists realized that populations were only parts of larger ecosystems, and that we needed to study communities and ecosystems in the same experimental way as populations. To study a complex ecosystem, teams of ecologists had to be organized and integrated, and work of this scale was first attempted during the late 1960s and the 1970s.

Modern ecology is advancing particularly strongly in three major areas. First, communities and ecosystems are being studied with experimental techniques and analyzed as systems of interacting species that process nutrients and energy. Insights into ecosystems have been provided by the comparative studies of communities on different continents. Second, modern evolutionary thinking is being combined with ecological studies to provide an explanation of how evolution by natural selection has molded the ecological patterns we observe today. Behavioral ecology is a particularly strong and expanding area combining evolutionary insights with the ecology of individual animals. Third, conservation biology is becoming a dominant theme in scientific and political arenas, and this has increased the need for ecological input in habitat management. All of these developments are providing excitement for students of ecology in the 2000s.

Application of the Scientific Method to Ecology

The essential features of the scientific method are the same in ecology as in other sciences (Figure 1.3 and Box 1.1). An ecologist begins with a problem, often based on natural history observations. For example, pine tree seedlings do not occur in mature hardwood forests on the Piedmont of North Carolina. If the problem is not based on correct observations, all subsequent stages will be useless, and thus accurate natural history is a prerequisite for all

E S S A Y   1 . 1

SCIENCE AND VALUES IN ECOLOGY

Science is thought by many people to be value free, but this is certainly not the case. Values are woven all through the tapestry of science. All applied science is done because of value judgments. Medical research is a good example of basic research applied to human health that virtually everyone supports. Weapons research is carried out because countries wish to be able to defend themselves against military aggression.

In ecology the strongest discussions about values have involved conservation biology. Should conservation biologists be objective scientists studying biodiversity, or should they be public advocates for preserving biodiversity? The preservation of biodiversity is a value that often conflicts with other values—for example, clear-cut logging that produces jobs and wood products. The pages of the journal *Conservation Biology* are peppered with this discussion about advocacy (see, for example, the June 1996 issue, page 904).

Scientists in fact have a dual role. First, they carry out objective science that both obtains data and tests hypotheses about ecological systems. They can also be advocates for particular policies that attempt to change society, like the use of electric cars to reduce air pollution. But it is crucial to separate these two kinds of activities.

Science is a way of knowing, a method for determining the principles by which systems like ecological systems operate. The key scientific virtues are honesty and objectivity in the search for truth. Scientists assume that once we know these scientific principles we can devise effective policies to achieve social goals. All members of society collectively decide on what social goals we will pursue, and civic responsibility is part of the job of everyone, scientists included. There will always be a healthy tension between scientific knowledge and public policy in environmental matters because there are always several ways of reaching a particular policy goal. The debates over public policy in research funding and environmental matters will continue, so please join in.

ecological studies. Given a problem, an ecologist suggests a possible answer. This answer is called a *hypothesis* and is basically a statement of cause and effect. In many cases, several answers might be possible, and several different hypotheses can be proposed to explain the observations. Hypotheses arise from previous research, intuition, or inspiration. The origin of a hypothesis tells us nothing about its likelihood of being correct.

A hypothesis makes predictions, and the more precise predictions it makes the better. Predictions follow logically from the hypothesis, and mathematical reasoning is most useful here in checking on the logic of the predictions. An example of a hypothesis is that pines do not grow under hardwoods because of a shortage of light. Alternative hypotheses might be that the cause is a shortage of pine seeds, or a shortage of soil water. Predictions from simple hypotheses like these are often straightforward: If you provide more light, pine seedlings will grow (under the light hypothesis). A hypothesis is tested by making observations to check the predictions—an *experiment*. An experiment is defined as any set of observations that test a hypothesis. Experiments can be manipulative or natural. We could provide light artificially under the mature forest canopy, or we could look for natural gaps in the forest canopy. The protocol for the experiments and the data to be obtained are called the experimental design. From the data that result from the experiments we either accept or reject the hypothesis. And so the cycle begins again (see Figure 1.3).

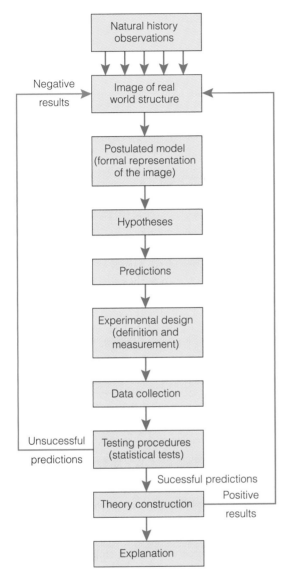

FIGURE 1.3
Schematic illustration of the scientific method as applied to ecological questions.

Many qualifications need to be attached to this simple scheme. Popper (1963) pointed out that we should always look for evidence that falsifies a hypothesis, and that progress in science consists of getting rid of incorrect ideas. In the real world we cannot achieve this ideal. We should also prefer simple hypotheses over complex ones, according to Popper, because we can reject simple hypotheses more quickly. This does not mean that we must be simple minded. On the contrary, in ecology we must deal with complex hypotheses because the natural world is not simple. Every hypothesis must predict something and forbid other things from happening. The predictions of a hypothesis must say exactly what it allows and what it forbids. If a hypothesis predicts everything and forbids nothing, it is quite useless in science. The light hypothesis for pine seedlings both predicts more seedlings if you add more light and forbids more seedlings if you add more water.

E S S A Y   1 . 2

ON ECOLOGICAL TRUTH

We wish our scientists to speak the truth, and when politicians bend the truth they lose credibility. What is truth, and what in particular is the hallmark of ecological truth? The notion of truth is a profound one that philosophers discuss in detail and scientists just assume is simple.

Truth consists of correspondence with the facts. If we say that there are 23 elephants in a particular herd in the Serengeti, we are stating an ecological truth because we assume that if another person counted the elephants, he or she would get the same number. These kinds of facts are relatively simple, and scientists rarely get into arguments about them. Where arguments start is in the inferences that are drawn from whole sets of facts. For example, if we had counts of the same elephant herd over 20 years, and numbers were continually falling, we could say that this elephant population is declining in size. This statement is also an ecological truth if we have done our counting well and recorded all the data correctly.

But now suppose we wish to state that the elephant population is declining and that a disease is the cause of this decline. Is this statement an ecological truth? It is better to consider it an ecological hypothesis and to outline the predictions it makes about what we will find if we search for a disease organism in elephants dying in this particular area. We now enter a gray zone in which ecological truth is approximately equivalent to a supported hypothesis, one in which we checked the predictions and found them to be correct. But if a scientist wished to extend this argument to state that elephant populations all over east Africa are collapsing because of this disease, this is a more general hypothesis, and before we can consider it an ecological truth we would need to test its predictions by studying many more populations of elephants and their diseases. Many of our ecological ideas are in this incomplete stage because we have lacked the time, money, or personnel to gather the data to decide whether the general hypothesis is correct or not. So ecologists, like other scientists, must then face the key question of how to deal with uncertainty when we do not know if we have an ecological truth or not.

The key resolution to this dilemma for environmental management has been the *precautionary principle*: "Look before you leap," or "An ounce of prevention is worth a pound of cure." The precautionary principle is the ecological equivalent of part of the Hippocratic Oath in medicine: "Physician, do no harm." The central idea of this principle is to do no harm to the environment, to take no action that is not reversible, and to avoid risk. Ecological truth is never obvious in complex environmental issues and emerges more slowly than we might like, so we cannot wait for truth or certainty before deciding what to do about emerging problems in the environment, whether they concern declining elephant populations or introduced pest species.

Ecological systems are complex, and this causes difficulty in applying the simple method outlined in Figure 1.3. In some cases factors operate together, so it may not be an "either light or water" situation for pine seedlings but "both light and water" together. Systems in which many factors operate together are most difficult to analyze, and ecologists must be alert for their presence (Quinn and Dunham 1983). The principle, however, remains—no matter how complex the hypothesis, it must make some predictions that we can check in the real world.

All ecological systems have an evolutionary history, and this provides another fertile source of possible explanations. There is controversy in ecology about whether one needs to invoke evolutionary history to explain present-day population and community dynamics. Evolutionary hypotheses can be tested as Darwin did, by comparative methods but not by manipulative experiments (Diamond 1986).

Ecological hypotheses may be statistical in nature, but they do not fall into the "either A or B" category of hypotheses. Statistical hypotheses postulate quantitative relationships. For example, in North Carolina forests pine seedling abundance (per m^2) is linearly related to incident light in summer. Tests of statistical hypotheses are well understood and are discussed in all statistics textbooks. They are tested in the same way as indicated in Figure 1.3.

Some ecological hypotheses have been very fruitful in stimulating work, even though they are known to be incorrect. The progress of ecology, and of sci-ence in general, occurs in many ways using mathe-matical models, laboratory experiments, and field studies. The following chapters illustrate the variety of problems and approaches that characterize the young science of ecology.

Key Concepts

1. Ecology is the scientific study of the interactions that determine the distribution and abundance of organisms.

2. Descriptive ecology forms the essential foundation for functional ecology, which asks how systems work, and for evolutionary ecology, which asks why natural selection has favored this particular solution.

3. Ecological problems can be analyzed using a theoretical approach, a laboratory approach, or a field approach.

4. Like other scientists, ecologists observe problems, make hypotheses, and test the predictions of each hypothesis by field or laboratory observations.

5. Ecological systems are complex, and simple cause/effect relationships are rare.

Selected References

Aarssen, L. W. 1997. On the progress of ecology. *Oikos* 80: 177–178.

Berry, R. J. 1991. Charles Elton FRS: The book that most... *Biologist* 38(4):127–129.

Dann, K. and G. Mitman. 1997. Exploring the borders of environmental history and the history of ecology. *Journal of the History of Biology* 30:291–302.

Egerton, F. N., III. 1973. Changing concepts of the balance of nature. *Quarterly Review of Biology* 48:322–350.

Grime, J. P. 1993. Ecology sans frontières. *Oikos* 68:385–392.

Hage, J. B. 1992. *An Entangled Bank: The Origins of Ecosystem Ecology*. Rutgers University Press, New Brunswick, N. J. 245 pp.

Kingsland, S. E. 1995. *Modeling Nature: Episodes in the History of Population Ecology*, 2nd ed. University of Chicago Press, Chicago. 306 pp.

Loehle, C. 1987. Hypothesis testing in ecology: psychological aspects and the importance of theory maturation. *Quarterly Review of Biology* 62:397–409.

McIntosh, R. P. 1985. *The Background of Ecology: Concept and Theory*. Cambridge University Press, Cambridge. 383 pp.

Platt, J. R. 1964. Strong inference. *Science* 146:347–353.

Popper, K. R. 1963. *Conjectures and Refutations*, Chapter 1. Routledge & Kegan Paul, London.

Questions and Problems

1.1 Discuss the connotation of the words *ecologist* and *environmentalist*. Would you like to be labeled either of these names? Where in a public ranking of preferred professions would these two fall?

1.2 "The definition 'ecology is the branch of biological science that deals with relations of organisms and environments' would provide the title for an encyclopedia but does not delimit a scientific discipline" (Richards 1939, p. 388). Discuss.

1.3 Is it necessary to define a scientific subject before one can begin to discuss it? Contrast the introduction to several ecology textbooks with those of some areas of physics and chemistry, as well as other biological areas such as genetics and physiology.

1.4 A plant ecologist proposed the following hypothesis to explain the absence of trees from a grassland area: Periodic fires may prevent tree seedlings from becoming established in grassland. Is this a suitable hypothesis? How could you improve it?

1.5 Is it necessary to study the scientific method and the philosophy of science in order to understand how science works? Consider this question before and after reading the essays by Popper (1963) and Platt (1964).

1.6 Discuss the application of the distribution and abundance model on page 7 to the human population.

1.7 Quinn and Dunham (1983) argue that the conventional methods of science cannot be applied to ecological questions because there is not just one cause; one effect and many factors act together to produce ecological changes. Discuss the problem of "multiple causes" and how scientists can deal with complex systems that have multiple causes.

1.8 Debate the following resolution: Resolved that our government should distribute more research funding to ecological and environmental research, and less funding to space research.

1.9 Plot the data in Table 1.1 graphically, with gross national product (*x*-axis) versus total fertility rate (*y*-axis). How tight is the relationship between these two variables? Discuss the reasons for the overall form of this relationship, and the reasons why there might be variation or spread in the data?

Overview Question

Does ecology progress as rapidly as physics? How can we measure progress in the sciences, and what might limit the rate of progress in different sciences? Will there be an "end to science"?

CHAPTER 2

Evolution and Ecology

CHARLES DARWIN WAS AN ECOLOGIST before the term had even been coined, and he is an appropriate patron for the science of ecology because he recognized the intricate connection between ecology and evolution. As we discuss ecological ideas throughout this book, we will use evolutionary concepts. This chapter provides a brief survey of the basic principles of evolution that are important in evolutionary ecology.

What Is Evolution?

Evolution is change, and biological evolution might be defined as changes in any attribute of a population over time. But we must be more specific than this. Evolutionary changes lead to adaptation and must involve a change in the frequency of individual genes in a population from generation to generation. What produces evolutionary changes?

Natural selection, said Charles Darwin and Alfred Wallace independently in 1858, is the mechanism that drives adaptive evolution. Natural selection operates through the following steps:

- Variation occurs in every group of plants and animals. Individuals of the same species are not identical in any population, as was observed in the breeding of domestic animals.

- Every population of organisms produces an excess of offspring. (The high reproductive capacity of plants and animals was well known to Malthus and Buffon long before Darwin.)

- Life is difficult, and not all individuals will survive and reproduce.

- Among all the offspring competing for limited resources, only those individuals best able to obtain and use these resources will survive and reproduce.

- If the characteristics of these organisms are inherited, the favored traits will be more frequent in the next generation.

Natural selection will favor traits that allow individuals possessing those traits to leave more descendants. These individuals are said to be fitter, and evolution in general maximizes fitness (see Box 2.1). The process of natural selection is the end result of the processes of ecology in action. The environments that organisms inhabit shape the evolution that occurs. The present distribution, abundance and diversity of animals and plants are set by the evolutionary processes of the past impinging on the environment of the present.

A simple example of natural selection is shown in Figure 2.1. The moth *Biston betularia* shows variation in the amount of black color on the wings. The typical moth is white with black speckling on the wings. The black form, *carbonaria*, was first described near Manchester in central England in 1848, and it spread over most of England during the next 50 years. When industrial pollution in central England caused lichens on tree bark to die, black-colored moths survived better because bird predators could not see them against this dark background (see Figure 2.1). Black wing color is inherited in these moths, and the result was an

FIGURE 2.1
Evolution in the peppered moth
Biston betularia *in England and*
North America. The photo shows
both phenotypes of the American
peppered moth posed on birch bark
and lichens. The black form,
carbonaria *has been declining in*
abundance since 1950 with the
decline in industrial pollution in
central England, and the same
change has occurred in eastern
North America. Differential bird
predation is believed to be the
major mechanism of selection.
(Data from Cambridge, England,
from Majerus 1998, p. 152, and
photo courtesy of Professor Bruce
Grant.)

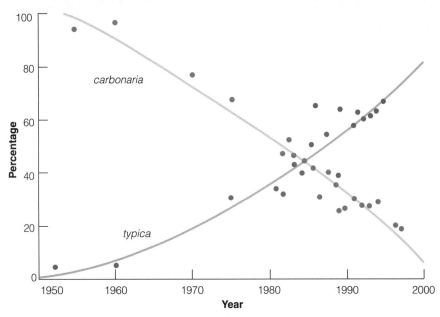

increase in the frequency of black moths during indus-trialization (Majerus 1998). Because industrial pollu-tion has decreased in England during the past 50 years, this process of natural selection is reversing (see Figure 2.1). The same changes have occurred in the American form of the peppered moth as air quality has improved in the eastern United States (Grant et al. 1996).

Evolution through natural selection results in *adaptation*, and under appropriate conditions produces new species (*speciation*). Adaptation has important eco-logical implications because it sets limits to the life cycle traits that determine distribution and abundance.

Adaptation

Natural selection acts on *phenotypes*, the observable attributes of individuals. Different genotypes give rise

B O X 2 . 1

WHAT IS FITNESS?

Evolutionary ecologists discuss fitness in many forms, and we need to have a clear idea of what fitness means. *Fitness* is a measure of the contribution of an individual to future generations and can also be called *adaptive value*. Individuals have higher fitness if they leave more descendents. Individuals can be fitter for three reasons: They may reproduce at a high rate, they may survive longer, or both. A fish that reproduces rapidly and dies young may be fitter than another fish of the same species that lives a long time but reproduces slowly. From this definition it should be clear that fitness is a *relative* term and applies to individual organisms within the same species. One individual may be fitter than another of the same species, or less fit. Ecologists tend to assume that there are traits that allow greater fitness and that these traits have a genetic basis. Evolution will act to maximize fitness.

We should also be clear what fitness is *not*:

- **Fitness is not absolute.** Measures of fitness are specific for a given environment. Individuals with genes that make them fit for cold environments may not be fit if the climate changes and they must live in warm environments.

- **Fitness cannot be compared across species.** We cannot compare the fitness of an elephant with that of an oak tree. Fitness is a measure that is defined only within a single species.

- **Fitness is not only about reproduction.** High reproductive rates may not by themself confer high fitness if survival rates of these young are poor.

- **Fitness is not a short-term measure.** Fitness should be measured across several generations, although this is difficult for studies of long-lived plants and animals. Ecologists often study short-term measures that they hope will correlate with fitness in the long term.

- **Fitness is not about individual traits.** Evolution is a whole-organism affair. Individual traits like large body size or fast growth rates may be components of fitness, but the test of fitness is the test of whole organism survival and reproduction.

Peter and Rosemary Grant *Department of Ecology and Evolutionary Biology, Princeton University*

to different phenotypes, but not in a simple way because embryological and subsequent development is affected in many ways by environmental factors such as temperature. Consequently, it is simpler to observe the effect of natural selection directly on the phenotype and to ignore the underlying genotype. Ecologists, like plant and animal breeders, are primarily interested in phenotypic characters like seed numbers or body size.

Three types of selection can operate on phenotypic characters (Figure 2.2). The simplest form is *directional selection*, in which phenotypes at one extreme are selected against. Directional selection produces genotypic changes more rapidly than any other form, so most artificial selection is of this type. Darwin's finches on the Galápagos Islands have been the best studied example of directional selection. Peter and Rosemary Grant from Princeton University have spent more than 20 years studying these finches on the Galápagos. Figure 2.3 illustrates directional selection in one of Darwin's finches, the Galápagos ground finch *Geospiza fortis*. During a prolonged drought the birds that survived were predominantly those with large beaks that could crack large seeds (Boag and Grant 1981). Birds with large beaks can eat both large and small seeds, while birds with small beaks can eat only small seeds. Directional selection probably accounts for most of the phenotypic changes that occur during evolution. In wild populations resistance of pests to insecticides or herbicides is produced by directional selection.

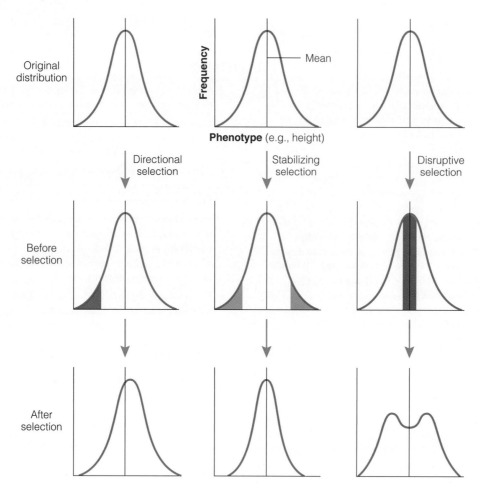

FIGURE 2.2
Three types of selection on phenotypic characters. Individuals in the colored areas are selected against. (After Tamarin 1999.)

FIGURE 2.3
Directional selection for beak size in the Galápagos ground finch Geospiza fortis. *From 1976 to 1978 a severe drought in the Galápagos Islands caused an 85% drop in the population, and birds with larger beaks survived better because they could crack larger, harder seeds. (From Grant 1986, p. 213.)*

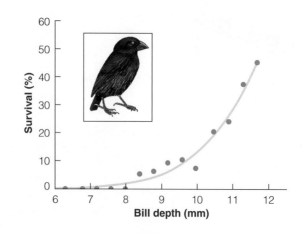

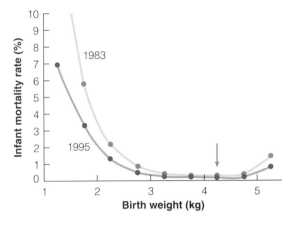

FIGURE 2.4

Stabilizing selection for birth weight in humans. Data from United States for infants between 1983 and 1995. The optimal birth weight (red arrow) is 4.25 kg, with a broad range of minimal mortality between 3.2 kg and 4.8 kg. Because of medical advances infant mortality has been falling steadily, so the 1983 curve is higher than the 1995 curve. (Data from National Center for Health Statistics 1998.)

Stabilizing selection (see Figure 2.2) is very common in present-day populations. In stabilizing selection, phenotypes near the mean of the population are fitter than those at either extreme, and thus the population mean value does not change. Figure 2.4 illustrates stabilizing selection for birth weight in humans in the United States. Early mortality is lowest for babies around 4.2 kg, slightly above the observed mean birth weight of 3.4 kg for the population. Very small babies die more frequently, and very large babies also suffer higher losses even with modern medical care.

Figure 2.5 shows another example of stabilizing selection in lesser snow geese *Anser caerulescens caerulescens*. Snow geese nest in colonies in northern Canada, and clutches hatch over a two-week period in early summer. Because predation is concentrated on whole colonies, eggs hatching synchronously confer a "safety-in-numbers" advantage against predators like foxes. Females whose eggs hatch synchronously on or near the mean date for the colony are more likely to raise their young successfully. Nests that hatch early suffer greater predation loss, as do nests that hatch later. The result is natural selection favoring an optimum hatching time (Cooke and Findlay 1982).

In the third type of selection, *disruptive selection* (Figure 2.2), the extremes are favored over the mean.

But because the extreme forms breed with one another, every generation will produce many intermediate forms doomed to be eliminated. In any environment favoring the extremes, any mechanism that would prevent the opposite extremes from breeding with one another would be advantageous. Isolating mechanisms are thus an important adjunct of disruptive selection. Few examples of disruptive selection have been found in natural populations. Smith (1990) showed that bill width in juvenile African finches was subject to disruptive selection because they could feed on two main seed types, one very hard and one very soft. Bill width determines how well a bird can feed on hard seeds, and in the African finch two bill-size morphs have specialized on hard and soft seeds.

The net result of all this selection is that organisms are adapted to their environment, and the great diversity of biological forms is a graphic essay on the power of adaptation by natural selection. But we must be careful to note that adaptation does not produce the "best" phenotypes or "optimal" phenotypes (defined as phenotypes that are theoretically the most efficient in surviving and reproducing). The "better" survive, not the "best," and the biological world can never be described as "the best of all possible worlds."

Adaptation is constrained in real populations by four major forces. First, genetic forces prevent perfect

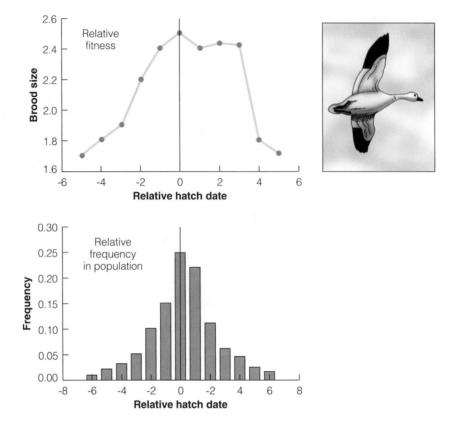

FIGURE 2.5
Stabilizing selection for hatching synchrony in lesser snow geese (Anser caerulescens caerulescens) *at La Perouse Bay in northern Manitoba, Canada. Relative hatch date is the number of days a female's eggs hatched before or after the mean date for the colony. (From Cooke and Findlay 1982.)*

adaptation because of mutation and gene flow. Mutation is always occurring, generating variation in populations, and most mutations are detrimental to organisms rather than adaptive. The immigration of individuals into an area where local environments differ will add other alleles to the gene pool and acts to smooth out local adaptations. Second, environments are continually changing, and this is the most significant short-term constraint on adaptation. Third, adaptation is always a compromise because organisms have at their disposal only a limited amount of time and energy. There are trade-offs between adaptations like wing-shape in birds. A loon's wings are efficient for diving but not so efficient for flying. Fourth, historical constraints are always present because organisms have a history and change in small increments. Let us look in detail at one example of adaptation to illustrate some of these principles.

Clutch Size in Birds

Each year Emperor penguins lay one egg, pigeons lay one or two eggs, gulls typically lay three eggs, the Canada goose four to six eggs, and the American merganser 10 or 11 eggs. What determines clutch size in birds? We must distinguish two different aspects of this question: proximate and ultimate.

Proximate factors explain *how* a trait is regulated by an individual. Proximate factors that determine clutch size are the physiological factors that control ovulation and egg laying. Ultimate factors are always selective factors, and ultimate explanations for clutch size differences always involve evolutionary arguments about adaptations. Proximate factors affecting clutch size have to do with how an individual bird decodes its genetic information on egg laying; ultimate factors have to do with changes in this genetic program through time and with the reason for these changes (Mayr 1982). Clutch size may be modified by the age of the female, spring weather, population density, and habitat suitability. The *ultimate factors* that determine clutch size are the requirements for long-term (evolutionary) survival. Clutch size is viewed as an adaptation under the control of natural selection, and we seek the selective forces that have shaped the reproductive rates of birds. We shall not be concerned here with the proximate factors determining clutch size, which are reviewed by Murphy and Haukioja (1986) and by Carey (1996).

Natural selection will favor those birds that leave the most descendants to future generations. At first thought we might hypothesize that natural selection favors a clutch size that is the physiological maximum the bird can lay. We can test this hypothesis by taking eggs from nests as they are laid. When we do this, we find that some birds, such as the common pigeon, are *determinate* layers; they lay a given number of eggs, no matter what. The pigeon lays two eggs; if you take away the first, it will incubate the second egg only. If you add a third egg, it will incubate all three. But many other birds are *indeterminate* layers; they will continue to lay eggs until the nest is "full." If eggs are removed once they are laid, these birds will continue laying. When this subterfuge was used on a mallard female, she continued to lay one egg per day until she had laid 100 of them. In other experiments, herring gull females laid up to 16 eggs (normal clutch: 2–3), a yellow-shafted flicker female 71 eggs (normal clutch: 6–8), and a house sparrow 50 eggs (normal clutch: 3–5) (Klomp 1970, Carey 1996). This evidence suggests that most birds under normal circumstances do not lay their physiological limit of eggs but that ovulation is stopped long before this limit is reached.

The British ornithologist David Lack was one of the first ecologists to recognize the importance of evolutionary thinking in understanding adaptations in life history traits. In 1947 Lack put forward the idea that clutch size in birds was determined ultimately by the number of young that parents can provide with food. This hypothesis stimulated much research on birds because it immediately suggested experimental manipulations. If this hypothesis is correct, the total production of young ought to be highest at the normal clutch size, and if one experimentally increased clutch size by adding eggs to nests, increased clutches should suffer greater losses because the parents could not feed the extra young in the nest.

One way to think about this problem of optimum clutch size is to use a simple economic approach. Everything an organism does has some costs and some *benefits*. Organisms integrate these *costs* and benefits in evolutionary time. The benefits of laying more eggs are very clear—more descendants in the next generation. The costs are less clear. There is an energy cost to make each additional egg, and there is a further cost to feeding each additional nestling. If the adult birds must work harder to feed their young, there is also a potential cost in adult survival—the adults may not live until the next breeding season. If adults are unable to work harder, there is a potential reduction in offspring quality. A cost-benefit model of this general type is

David Lack *(1910–1973) Edward Grey Institute, Oxford University*

shown in Figure 2.6. Models of this type are called *optimality models*. They are useful because they help us think about what the costs and benefits are for a particular ecological strategy.

No organism has an infinite amount of energy to spend on its activities. The reproductive rate of birds can be viewed as one sector of a bird's energy balance, and the needs of reproduction must be maximized within the constraints of other energy requirements. The total requirements involve metabolic maintenance, growth, and energy used for predator avoidance, competitive interactions, and reproduction. Lack's (1947) hypothesis—that the clutch size of birds

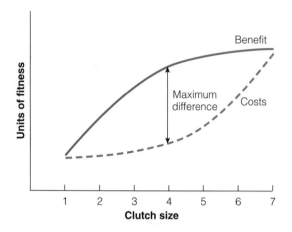

FIGURE 2.6
A cost-benefit model for the evolution of clutch size in birds. An individual benefits from laying more eggs because it will have more descendents, but it incurs costs because of increasing parental care required for larger clutches. The clutch size with the maximum difference between benefits and costs is the optimal clutch size for that individual. (The Lack clutch size, named after David Lack.)

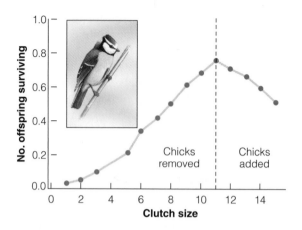

FIGURE 2.7
Production of young blue tits (Parus caeruleus) *in relation to clutch size in Wytham Wood, Oxford, England. Only females that had laid 11 eggs in previous years are shown here, since we expect these individuals to have their highest fitness at a clutch size of 11. These results fit Lack's hypothesis because adding more chicks just after hatching does not increase fitness. (Data from Pettifor 1993, p. 136.)*

that feed their young in the nest was adapted by natural selection to correspond to the largest number of young for which the parents can provide enough food—has been a very fertile hypothesis in evolutionary ecology because it has stimulated a variety of experiments. According to this idea, if enough additional eggs are placed in a bird's nest, the whole brood will suffer from starvation so that, in fact, fewer young birds will fledge from nests containing larger numbers of eggs. In other words, clutch size is postulated to be under stabilizing selection (see Figure 2.2). Let us look at a few examples to test this idea.

In England, the blue tit normally lays a clutch of 9–11 eggs. What would happen if blue tits had a brood of 12 or 13? Pettifor (1993) artificially manipulated broods at hatching by adding or subtracting chicks and found that the survival of the young blue tits in manipulated broods was poor (Figure 2.7). Blue tits feed on insects and apparently cannot feed additional young adequately, so more of the young starve. Consequently, it would not benefit a blue tit in the evolutionary sense to lay more eggs, and the results are consistent with Lack's hypothesis. Individual birds appear to produce the clutch size that maximizes their reproductive potential.

Tropical birds usually lay small clutches, and Skutch (1967) argued that this was an adaptation against nest predators. If the intensity of nest predation increases with the number of parental feeding trips away from the nest, natural selection would favor a reduced clutch size. Low clutch size and low predation rates are associated. Parents would leave more descendents if they had smaller broods and did not need to feed them as often. Exactly the same argument was used by Martin (1995) to explain the pattern of clutch size in hole-nesting birds. Hole-nesting passerine birds lay fewer eggs than comparable species that nest in the open, and predation rates are much lower for hole-nesting species (Martin 1995). So again, low clutch size and low predation rates are correlated. This suggests that a high risk of predation on the whole brood in the nest is a strong selective factor that increased clutch size in open-nesting birds, and also favors a shortened nesting period, independent of the ability of the parents to provide food to the nestlings. Open nesting is a gamble because of high predation rates, and passerine birds gamble on large clutches and short nesting periods.

Natural selection would seem to operate to maximize reproductive rate, subject to the constraints

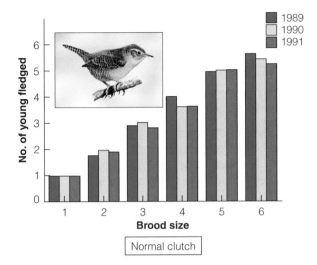

1989
1990
1991

Normal clutch

FIGURE 2.8
Number of house wren chicks fledged from broods manipulated to have larger and smaller than average brood size. These results do not agree with the predictions from Lack's hypothesis that larger broods should fledge fewer young than average-size broods. (After Young 1996.)

imposed by feeding and predator avoidance. This is called the theory of *maximum reproduction*, and Lack's hypothesis is part of this theory. It is a good example of how stabilizing selection can operate on a phenotypic trait such as reproductive rate. The maximum clutch size is called the *Lack clutch size* (see Figure 2.6), after David Lack.

Not all manipulation experiments confirm Lack's hypothesis. Young (1996) manipulated clutch size in tropical house wrens (*Troglodytes aedon*) in Costa Rica to produce clutches ranging from one to six. House wrens in the tropics typically lay three or four eggs. Figure 2.8 shows the resulting offspring produced. The number of surviving offspring per brood was maximized for broods of six eggs. Since mean brood size was 3.5, the most common clutch size was smaller than the most productive clutch size. Vanderwerf (1992) surveyed 77 experiments in which clutches had been manipulated and found that 69% of these were like the house wren—the most productive clutch size was larger than the most common clutch size. Why should this be?

The presence of trade-offs is one explanation of why clutches are smaller than the Lack clutch size. Clutch size may affect the chances of the adult birds

surviving to breed again. Birds may become exhausted by rearing large clutches; such exhaustion is a delayed cost of reproduction. Alternatively, laying a large clutch may postpone the next breeding attempt, leading to reduced lifetime reproduction. Laying a large clutch is energetically costly for birds, and this cost is not usually measured in brood manipulation experiments (Monaghan and Nager 1997).

The Lack clutch size may not be a constant for a species, and different individuals may vary in their parenting abilities and have a personal Lack clutch size. This is called the *individual optimization hypothesis*, and there is clear evidence for individual optimization in birds (Pettifor 1993) and in mammals (Risch et al. 1995).

An alternative explanation of why the average clutch may be smaller than the Lack clutch size is that observed clutch sizes are a nonadaptive compromise. If gene flow occurs between two habitats, one good and one poor, clutches may be larger than optimal in poor habitats and smaller than optimal in good habitats. Blue tits and great tits in Belgium rarely breed in woodlands where they were born and show this nonadaptive compromise (Dhondt et al. 1990).

Recent work on bird reproduction investigates how individual parents adjust their reproductive costs in relation to environmental conditions to maximize the output of young (Godfray et al. 1991). The proximate controls of reproduction operate through the energy available to reproducing birds, and the role of female condition is critical in determining reproductive effort. Reproductive effort this year may affect the chances of surviving until next year, and parents must balance the short-term and long-term costs of breeding.

Coevolution

The term *coevolution* was popularized by Ehrlich and Raven (1964) to describe the reciprocal evolutionary influences that plants and plant-eating insects have had on each other. Coevolution occurs when a trait of species *A* has evolved in response to a trait of species *B*, which has in turn evolved in response to the trait in species *A*. Coevolution is specific and reciprocal. In the more general case, several species may be involved instead of just two, and this is called diffuse coevolution (Thompson 1994).

Coevolution is simply a part of evolution, and it provides important linkages to ecology. The

E S S A Y   2 . 1

EVOLUTION AND "ARMS RACES"

If you look up "arms race" on the Web you will find much discussion of military strategies and little of biological evolution. An arms race in evolution is a reciprocal interaction between species, in which as species *A* evolves better adaptations to exploiting species *B*, the latter fights back by evolving adaptations to thwart the improvements in species *A*. The best examples of arms races occur between hosts and parasites. The brown-headed cowbird in North America and the European cuckoo in Britain are good examples of parasitic birds that lay their eggs in the nests of other species. The host species then raise the cowbird or cuckoo chick, often to the detriment of its own young.

The brown-headed cowbird has greatly expanded its geographic range in North America because of agriculture and is invading new areas and utilizing new host species, so it has become a major conservation problem. Parasitic birds like the cowbird often lay eggs that have the same color and pattern as the host species in order to avoid detection and the possibility that the host species will remove the parasite's eggs from their nests. The host species on the other hand should evolve the ability to discriminate cowbird eggs from its own eggs. Host individuals that discriminate more will leave more offspring (and raise fewer cowbirds). Consequently, an evolutionary arms race can develop in which both the parasite and the host are continually evolving counterstrategies in a tit-for-tat manner (Abrams 1986, Takasu 1998).

Coevolution shapes the characteristics of coevolving pairs of species, while diffuse coevolution might also occur in communities of many species. There is

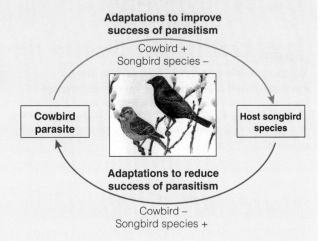

considerable controversy about whether community-level patterns could be caused by coevolution (Orians and Paine 1983), and most ecologists believe that coevolution is restricted to interactions between only a few species that interact tightly (Benkman 1999).

interactions between herbivores and their food plants have been emphasized as a critical coevolutionary interaction (Chapter 14). Predator-prey interactions (Chapter 13) can also be coevolutionary, and in some cases can lead to "arms races" between species.

Units of Selection

Darwin conceived natural selection as operating through the reproduction and survival of *individuals* who differ in their genetic constitution. Most discussion of natural selection operates at this level of Darwinian selection, or individual selection.

But natural selection is not restricted to individuals. It can act on any biological units so long as these units meet the following criteria: (1) They have the ability to replicate; (2) they produce an excess number of units above replacement needs; (3) survival depends on some attribute (size, color, behavior); and (4) a mechanism exists for the transmission of these attributes. Three units of selection other than the individual can fulfill these criteria: *gametic*, *kin*, and *group selection*.

Gametic Selection

Gametes (eggs and sperm) have a genetic composition that differs from the diploid organisms that produce them. Gametes are produced in vast excess and may have characteristics that they transmit through the zygote and adult organism to the next generation of gametes. Consequently, natural selection can act on a population of gametes independently from the natural selection that operates on the parent organisms. Many different characteristics of gametes could be under natural selection. Sperm mobility, for example, may be under strong selection. In plants, pollen grains that produce a faster-growing pollen tube have a better chance of releasing their sperm nuclei and fertilizing an egg. Gametic selection is an interesting and important aspect of natural selection, but it does not directly impinge on ecological relationships.

Kin Selection

If an individual is able to increase the survival or reproduction of its relatives with whom it shares some of the same genes, natural selection can operate through kin selection. Kin selection and individual selection may act together, and this action is described by the concept of inclusive fitness. Natural selection not only favors alleles that benefit an individual but also alleles that benefit close relatives of that individual because close relatives share many alleles. All relatives can help pass copies of an individual's genes to future generations.

Kin selection was recognized as one way of explaining the existence of altruistic traits such as the sounding of alarm calls. When ground squirrels sight a predator, they give an alarm call. As a result, the individual calling (1) draws attention to itself and thus may be attacked by the predator (detrimental to the individual) and (2) warns nearby squirrels to run for cover (beneficial to relatives nearby).

Kin selection has important consequences for ecological relationships because of its effects on social organization and population dynamics. Competition between individual organisms will be affected by the proximity of close relatives; thus it can be important for an ecologist to know the degree of kinship among members of a population.

Group Selection

Group selection can occur when populations of a species are broken up into discrete groups more or less isolated from other such groups. Groups that contain less adaptive genes can become extinct, and the conditions for natural selection could occur at the level of the group as well as at the level of the individual organism.

Group selection is highly controversial, and most biologists consider it to be rare in nature. Most of the characteristics of organisms that are favorable to groups can also be explained by individual or kin selection. Controversy erupts over traits that appear to be good for the group but bad for the individual. A classic example is the evolution of reproductive rates in birds. Group selectionists argue that many birds reproduce at less than maximal rates because populations with low reproductive rates will not overpopulate their habitats. Any populations with higher reproductive rates will overpopulate their habitats and become extinct, so restraint is selected for at the group level. But low reproductive rates are bad for the individual, and individual selection will act to favor higher reproductive rates, so group selection and individual selection are operating at cross purposes. In a group in which all members restrain themselves, a cheater will always be favored.

The alternative argument is that all reproductive rates are in fact maximal and have responded only to individual selection favoring individuals that leave the most offspring for future generations. Restraint does not exist, according to this view.

Group selection may occur, but at present it is not believed to be an important force shaping the adaptations that ecologists observe while trying to understand the distribution and abundance of organisms.

Summary

Organisms survive and reproduce and, because not all individuals are equally successful at these activities, natural selection occurs. The fitter individuals leave more descendents to future generations either because of higher survival or higher reproductive rates. Natural selection is ecology in action, and the ecologist asks which traits of individuals improve their chances of survival or reproduction.

The clutch size of birds is a classic problem in evolutionary ecology—why don't birds lay more eggs? David Lack suggested in 1947 that clutch size was limited by the number of chicks the adults could feed successfully. Experimental additions of eggs and chicks to nests have often shown that bird parents can in fact rear more young than they usually do. This anomaly is probably due to the higher costs of reproduction for birds rearing large broods, and adults may die or lay fewer eggs in subsequent years as a cost of breeding performance in the current year. Clutch size is thus under stabilizing selection.

Individual, or *Darwinian*, selection is the classic form of selection on individual phenotypes, and it is the level of selection responsible for most of the adaptations we see in nature. Some adaptations may evolve by *kin* selection for actions that favor the survival or reproduction of close relatives carrying the same genes. *Group* selection might also occur if whole groups or populations become extinct because of genetic characteristics present in the group. Group selection is probably uncommon in nature.

Key Concepts

1. Evolution is the genetic adaptation of organisms to the environment.

2. Ecology and evolution are intricately connected because evolution operates through natural selection, which is ecology in action.

3. Natural selection may act by *directional selection, stabilizing selection*, or *disruptive selection*.

4. Evolution results from directional selection, but for most ecological situations stabilizing selection is most common.

5. Natural selection may operate on four different levels: gametic, individual, kin, or group. Individual or Darwinian selection is probably most important in nature.

Selected References

Cooke, F., P. Taylor, C. Francis, and R. Rockwell. 1990. Directional selection and clutch size in birds. *American Naturalist* 136:261–267.

Futuyma, D. J. 1998. Adaptation. In *Evolutionary Biology*, 3rd ed. Sinauer Associates, Sunderland, Mass.

Godfray, H. C. J., L. Partridge, and P. H. Harvey. 1991. Clutch size. *Annual Review of Ecology and Systematics* 22:409–429.

Grant, B. S., D. F. Owen, and C. A. Clarke. 1996. Parallel rise and fall of melanic peppered moths in America and Britain. *Journal of Heredity* 87:351–357.

Grant, P. R. 1999. *Ecology and Evolution of Darwin's Finches*. Princeton University Press, Princeton, N.J. 492 pp.

Monaghan, P. and R. G. Nager. 1997. Why don't birds lay more eggs? *Trends in Ecology and Evolution* 12:271–274.

Price, T., M. Kirkpatrick, and S. J. Arnold. 1988. Directional selection and the evolution of breeding date in birds. *Science* 240:798–799.

Primack, R. B. and H. Kang. 1989. Measuring fitness and natural selection in wild plant populations. *Annual Review of Ecology and Systematics* 20:367–396.

Terrenato, L., M. F. Gravina, A. San Martini, and L. Ulizzi. 1981. Natural selection associated with birth weight. III. Changes over the last 20 years. *Annals of Human Genetics* 45:267–278.

Wills, C. 1996. *Yellow Fever, Black Goddess: The Coevolution of People and Plagues*. Addison-Wesley, Reading, Mass. 324 pp.

Questions and Problems

2.1 Birds living on oceanic islands tend to have a smaller clutch size than the same species (or close relatives) breeding on the mainland (Klomp 1970, p. 85). Explain this on the basis of Lack's hypothesis.

2.2 Will medical advances in the future eliminate stabilizing natural selection for birth weight in humans (see Figure 2.4)? Are humans still subject to natural selection?

2.3 Ladybird beetles are distasteful to predators because of toxic chemicals they secrete, yet they also have dark melanic forms (Majerus 1998, p. 221). Melanic ladybirds have declined in frequency in central England along with the peppered moth during the past 50 years as air quality has improved. If ladybirds are not eaten by predators, how might you explain these changes in melanic frequency?

2.4 In many ungulate species, males have horns or antlers that they use in combat with other males during the breeding season. How can natural selection limit the "arms race" in species where combat is part of the reproductive cycle? See Geist (1978) for a discussion.

2.5 Royama (1970, pp. 641–642) states:

> Natural selection favors those individuals in a population with the most efficient reproductive capacity (in terms of the number of offspring contributed to the next generation), which means that the present-day generations consist of those individuals with the highest level of reproduction possible in their environment.

Is this correct? Discuss.

2.6 In many temperate zone birds, those individuals that breed earlier in the season have higher reproductive success than those that breed later in the season. Explain why this might or might not lead to directional selection for earlier breeding dates. Read Price et al. (1988) for a discussion of this problem.

2.7 Some birds such as grouse and geese have young that are mobile and able to feed themselves at hatching (precocial chicks). Discuss which factors might limit clutch size in these bird species. Winkler and Walters (1983) have reviewed studies on clutch size in precocial birds.

2.8 In arctic ground squirrels. adult females are more likely to give alarm calls than adult males. If alarm calls are favored by kin selection, why might this difference occur? Could alarm calls be explained by group selection? Why or why not?

2.9 Apply the cost-benefit model in Figure 2.6 to seed production in a herbaceous plant. Discuss biological reasons for the general shape of these curves. Can you apply this model to both annual and perennial plants in the same way?

2.10 A research scientist obtained the following data on the fitness of seven females in a small population of house sparrows:

Female Band Number

Letter		A	B	C	D	E	F	G
1995	No. eggs laid	8	5	4	5	7	6	7
	No. young fledged	5	1	3	4	2	4	3
1996	No. eggs laid	5	6	4	5	7	7	9
	No. young fledged	4	2	3	5	3	3	0
1997	No. eggs laid	6	10	5	5	8	10	10
	No. young fledged	4	8	5	4	6	8	9
1998	No. eggs laid	6	6	3	5	7	8	10
	No. young fledged	2	5	3	5	3	4	4

Rank the fitness of these seven female sparrows. What data might you collect to improve on this measure of fitness for these birds?

2.11 Discuss how the concept of time applies to evolutionary changes and to ecological situations. Do ecological time and evolutionary time ever correspond?

2.12 A hypothetical population of frogs consists of 50 individuals in each of two ponds. In one pond, all of the individuals are green; in the other pond, half are green and half are brown. During a drought, the first pond dries up, and all the frogs in it die. In the population as a whole, the frequency of the brown phenotype has gone from 25% to 50%. Has evolution occurred? Has there been natural selection for the brown color morph?

Overview Question

Humans in industrialized countries increased in average body size during the twentieth century. List several possible explanations for this change, and discuss how you could decide if an evolutionary explanation is needed to interpret it. How does a physiological explanation for this change differ from an evolutionary explanation?

PART TWO

The Problem of Distribution: Populations

CHAPTER 3   METHODS FOR ANALYZING DISTRIBUTIONS

CHAPTER 4   FACTORS THAT LIMIT DISTRIBUTIONS: DISPERSAL

CHAPTER 5   FACTORS THAT LIMIT DISTRIBUTIONS: HABITAT SELECTION

CHAPTER 6   FACTORS THAT LIMIT DISTRIBUTIONS: INTERRELATIONS
WITH OTHER SPECIES

CHAPTER 7   FACTORS THAT LIMIT DISTRIBUTIONS: TEMPERATURE, MOISTURE,
AND OTHER PHYSICAL-CHEMICAL FACTORS

CHAPTER 8   THE RELATIONSHIP BETWEEN DISTRIBUTION AND ABUNDANCE

CHAPTER 3

Methods for Analyzing Distributions

WHY ARE ORGANISMS OF A PARTICULAR species present in some places and absent from others? This is the simplest ecological question one can ask, and hence it forms a good starting point for introducing you to ecology. This simple question about the distributions of species can be of enormous practical importance. Two examples illustrate why. Five species of Pacific salmon live in the North Pacific Ocean and spawn in the river systems of western North America, Asia, and Japan. Salmon are valuable fish for commercial fishermen and sport fishermen alike. Why not transplant such valuable fish to other areas—for example, to the North Atlantic or to the Southern Hemisphere? Why are five species of salmon present in the Pacific but only one species in the Atlantic? Sockeye salmon have been transplanted to France, Denmark, Mexico, and Argentina but did not survive there (Lever 1996, Baltz 1991). Why?

The African honey bee is a second example that illustrates the practical consequences of species distributions. The African honey bee (*Apis mellifera scutellata*) is a very aggressive subspecies of honey bee that was brought to Brazil in 1956 in order to develop a tropical strain with improved honey productivity. They escaped by accident, and the spread of the African bee since 1956 is shown in Figure 3.1. Because African bees are aggressive, they may drive out established colonies of the Italian honey bee (*Apis mellifera ligustica*). In other situations, hybrids may be formed between the African and Italian subspecies.

In 1982 the African honey bee crossed the Panama Canal. It reached Mexico in 1985, and southern Texas in 1990. Moving roughly 110 km per year, it crossed the border into California in 1994 and is currently spreading into the southern United States. At present the African bee has reached northern Texas, southern California, southern Nevada, and all of Arizona. Unfortunately, African bees are aggressive toward humans and domestic animals, and accounts of severe stinging and even deaths have served to highlight the spread of the African bee. By 1998 seven people had been killed by these bees in the United States, and beekeepers are understandably worried that the African bee will damage the established honey bee industry. What factors limit the distribution of the African honey bee? Will this species be able to live as far north as northern California and North Carolina?

Transplant Experiments

To answer questions concerning distribution, we must first determine whether the limitation on distribution results from the inaccessibility of the particular area to the species. One way to determine the source of limitation is a transplant experiment. In a transplant experiment, we move individuals of a species to an unoccupied area and determine whether they can survive and reproduce successfully in the

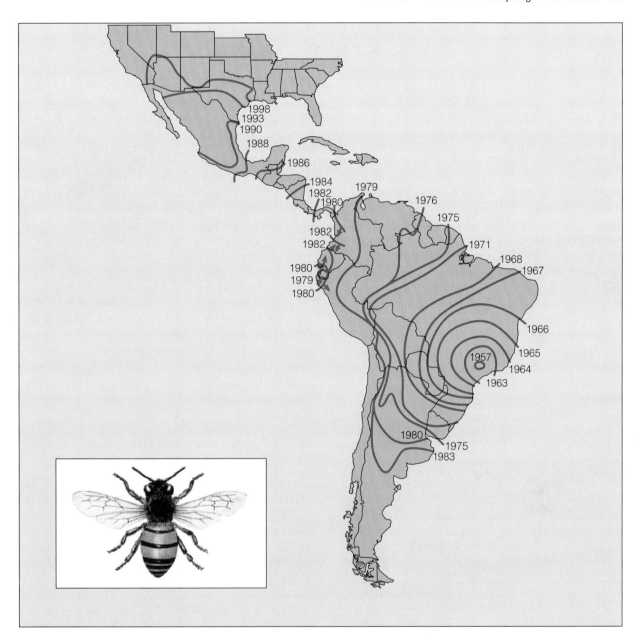

FIGURE 3.1
Spread of the African honey bee in the Americas since 1956. Southward and westward expansion in South America has been slight since 1983. The African bee reached southern Texas in October 1990 and southern California in October 1994. (Data provided by O. R. Taylor, personal communication.)

E S S A Y   3 . 1

LIEBIG'S LAW OF THE MINIMUM

Justus von Liebig (1803–1873) was a nineteenth-century German chemist. After working in organic chemistry in the first part of his life, he became interested in biochemistry and in particular how plants transform inorganic matter in the soil and atmosphere into organic matter. He postulated in 1840 what has been called Liebig's Law of the Minimum, which states that the rate of any biological process is limited by that factor in least amount relative to requirements. Crop yields were limited, according to Liebig, by a single nutrient, and if one added the limiting nutrient in fertilizer, production would increase. Liebig's work with artificial fertilizers was revolutionary in its time because of its linkage of chemistry and biology and its achievements in producing higher crop yields.

Liebig's Law has been attacked as too simplistic because it postulates that at any point in time *there is only one limiting factor* for any process. The modern view is that a number of nutrients may be limiting simultaneously, and that there may be combined enhancements from mixtures of nutrients. Although Liebig's Law is not applicable in all situations, it forms a very useful starting point for understanding what limits ecological processes. Consider one relatively simple question: What limits the distribution of African bees? The key to answering this question is experimental work in physiological ecology: Vary temperature and describe how survival and reproduction change with changing temperature. If we find that temperature is limiting in one area, however, this does not mean that moisture is not limiting in another area, or that in yet another area a combination of temperature and moisture are not the key factors. In this manner, we built complexity experimentally, but on a foundation established by a German chemist 150 years ago.

new environment. Some organisms can survive in areas but cannot reproduce there, so we should follow transplant experiments through at least one complete generation. The two possible outcomes of a transplant experiment are the following:

Outcome	Interpretation
Transplant successful	Distribution limited either because the area is inaccessible, time has been too short to reach the area, or because the species fails to recognize the area as suitable living space.
Transplant unsuccessful	Distribution limited either by other species or by physical and chemical factors.

A proper transplant experiment should have a *control*, transplants done within the distribution to provide data on the effects of handling and transplanting the individual plants or animals.

If a transplant is successful, it indicates that the potential range of the species is larger than its actual range. Figure 3.2 shows this schematically for a hypothetical plant or animal. The results of successful transplant experiments thus direct our further investigations in one of two ways. If a species does not seem to occupy all of its potential range, we must determine if it can move into its potential range or if it lacks suitable means of reaching new areas. We discuss the problem of movement, or *dispersal*, in Chapter 4. Some animal species could move into new areas but do not do so. For these species, we must study their mechanisms of *habitat selection* (Chapter 5).

If the species cannot survive and reproduce in the transplant areas, we ask whether interactions with other species or abiotic factors exclude it from these areas. Limits imposed by other species may involve either the negative effects of predators, parasites, dis-

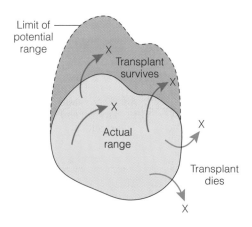

FIGURE 3.2
Hypothetical sets of transplant experiments applied to the same species. The yellow area represents the actual distribution of the population. The limit of the potential range is shown in light blue. Many separate transplant experiments are needed to define the limits of the potential range; only two are shown in this illustration.

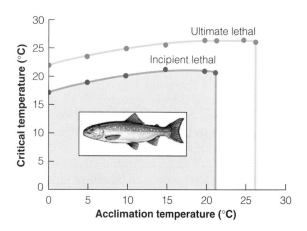

FIGURE 3.3
Temperature tolerance polygon for newly-hatched arctic charr. Incipient lethal temperature measures long-term survival over seven days, and ultimate lethal temperature measures short-term survival over only 10 minutes. Acclimation temperature is the temperature at which the fish had been living before the experimental tests. Arctic charr eggs can hatch successfully in the blue zone of the graph. (From Baroudy and Elliott 1994.)

ease organisms, or competitors, or the positive effects of interdependent species within the actual range. We can often determine if other species are restricting distribution by conducting transplant experiments with protective devices such as cages designed to exclude the suspected predators or competitors. For example, we can transplant barnacles in mesh cages to deeper waters along the coast to see if they can survive in deep water if starfish or gastropod predators are kept away. Some examples of such experiments are described in Chapter 6.

If other species do not set limits on the actual range, we are left with the possibility that some physical or chemical factors set the range limits (Chapter 7). For example, many tropical plant species cannot withstand freezing temperatures, and the frost line effectively limits their distributions. Limitations imposed by physical or chemical factors have been studied extensively and are the subject of a whole discipline called *physiological ecology*.

Physiological Ecology

Physiological ecologists study the reactions of organisms to physical and chemical factors. To live in a

given environment, an organism must be able to survive, grow, and reproduce, and consequently physiological ecologists must try to measure the effect of environmental factors on survival, growth, and reproduction. Victor Shelford, one of the earliest North American animal ecologists, was the first to formalize the ideas of physiological ecology with the view to understanding the distribution of species in natural communities. Working at the University of Illinois, Shelford developed the major conceptual tool of physiological ecologists, *Shelford's law of tolerance*, which can be stated as follows: *The distribution of a species will be controlled by that environmental factor for which the organism has the narrowest range of tolerance.*

The job of physiological ecologists is to determine the tolerances of organisms to a range of environmental factors. This is not a simple chore. We could, for example, determine the range of temperatures over which a species can survive. Figure 3.3 shows some data of this type for newly hatched arctic charr (*Salvelinus alpinus*). There is an upper lethal temperature beyond which these juvenile charr cannot survive. This upper lethal temperature is slightly sensitive to the acclimation temperature at which the juvenile fish have been living. Clearly, if the water temperature rises above about 20°C because of climate change or human impacts, this fish species will disappear. Many species also have a lower lethal temperature, but for arctic charr, which are adapted to life in cold waters, cold is not a problem until the water freezes.

We can repeat these tolerance studies for oxygen, pH, and salinity and build up a detailed picture of the tolerance limits of any particular species of plant or animal. Figure 3.4 illustrates hypothetical limitation by two factors. To these simple Shelford models we can then add complications. The stages of the life cycle may differ in their tolerance limits, and consequently we should measure the most sensitive stage when constructing these models. The young stages of both plants and animals are often most sensitive to environmental factors. These tolerance limits define the fundamental niche of the species (see page 191).

Victor E. Shelford *(1877–1968) Professor of Zoology, University of Illinois*

Two other factors complicate the determination of tolerance limits. First, species can acclimate physiologically to some environmental factors. Figure 3.3 illustrates this concept: The lethal temperatures depend on the acclimation temperature, the temperature at which the fish have been living. Second, tolerance limits for one environmental factor will depend on the levels of other environmental factors. Thus in many fish, for example, pH differences will affect temperature tolerances.

Another problem arises when we try to apply these tolerance limits to situations in the real world. Animals are particularly difficult because they are

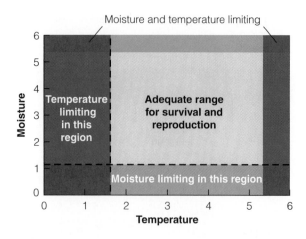

FIGURE 3.4
Idealized plot of Shelford's Law of Tolerance for two factors that might limit geographical distributions. In this hypothetical example, if temperature is below 1.7 units, or moisture level is below 1.15 units, the organism cannot survive. The geographical range must occur in the yellow zone for these two factors. Too much moisture and excessively high temperatures set the upper limits of tolerance in this simple example.

mobile and can resort to a variety of tactics that help them avoid lethal environmental conditions. Both plants and animals have evolved many types of escape mechanisms. Many birds and some insects migrate from polar to temperate or equatorial regions to avoid the polar winters. Some mammals, such as the arctic ground squirrel, hibernate during the winter and thereby avoid the necessity of feeding during the cold months of the year. Plants become dormant and resistant to cold temperatures in winter, while many insects enter a cold-tolerant diapause.

Adaptation

The organisms whose distributions we study today are products of a long history of evolution, and the physiological ecologist studies their adaptations in much the same way that we might study a single frame of a motion picture. The tolerances of species can change via the process of natural selection. One good example is the adaptation of plants to heavy-metal toxicity.

Heavy metals such as lead are extremely toxic to plants: 0.001% of lead and 0.00005% of copper will

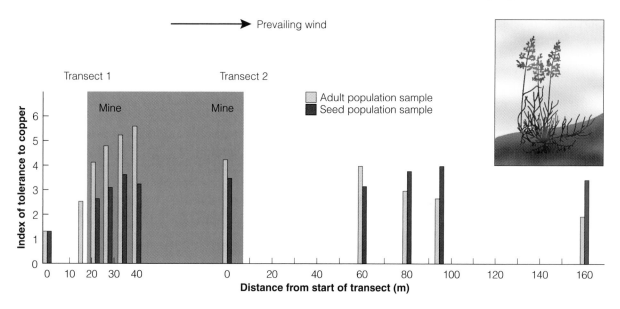

FIGURE 3.5
Copper tolerance of the grass Agrostis tenuis *along transects at the edge of the Drws y Coed mine in North Wales. The tolerance of both adult plants and seeds produced in situ is shown. The contaminated area is shaded. (After Macnair 1981.)*

kill most plants within a week. In mining wastes contaminated soils often contain 1% of lead, copper, and zinc and thus should kill all plant life. But in less than 50 years the grass *Agrostis tenuis* has evolved populations that live on mine wastes in Great Britain (Antonovics et al. 1971). A few species adapt to these high concentrations of lead, zinc, and copper, but most plants from pastures will not survive on mine soil. Normal *Agrostis tenuis* populations, however, contain a few tolerant individuals. If one sows 2000 seeds on mine soil, only four or five grass plants will grow (Wu et al. 1975). On toxic mine soils only these tolerant genotypes survive.

Figure 3.5 illustrates the differences in copper tolerance that occur along transects through a copper mine in North Wales. Genes for copper tolerance have spread in a downwind direction in this valley. Copper-tolerant populations of *Agrostis* have evolved by rapid natural selection acting on very rare individual grass plants that are partly tolerant to high copper levels (Wu et al. 1975). Current populations are maintained by strong disruptive selection dictated by contaminated soils. Not all plant species are able to evolve metal tolerances; many species do not have the appropriate genetic varia-

tion in their normal populations (Bradshaw and Hardwick 1989).

There is a cost to being tolerant to heavy-metal pollution. The tolerant genotypes of *Agrostis tenuis* grow poorly compared with normal genotypes when they are grown in normal soil under crowded conditions (Macnair 1981). One possible reason is that tolerant plants require more than trace amounts of heavy metals to be able to grow properly. Tolerant plants are at a selective disadvantage away from contaminated soils.

Such evolutionary changes further complicate the task of the ecologist who is trying to understand the distribution of a species. We must ask the question, What factor sets the current limitation on the geographic distribution of this species? But then we must ask further: Why for many species has natural selection not been able to increase the tolerance limits of a species and thereby expand its geographic range? If *Agrostis tenuis* has been able to increase its limits of tolerance to heavy metals, why has this not occurred in many other plant species? Is there simply a lack of genetic variability in nontolerant species so that natural selection has no impact on adaptation to tolerating heavy metals?

The grass *Agrostis tenuis* has thus increased its geographic range on a microscale by adapting to contaminated soils. When we observe geographic range changes we typically think that such shifts are due to changes in the environment, but the *Agrostis tenuis* example shows that some range shifts are caused by evolutionary changes in the physiological attributes of the individuals in a population.

Transplant experiments, such as that illustrated in Figure 3.2, can be disastrous when pests are introduced to new areas, as we will see in Chapter 4. It is critical that all transplant experiments be done safely, with due regard for the ecosystem. Indiscriminant transplanting of organisms contains all the seeds of ecological disaster (Ruesink et al. 1995). Most governments have stringent rules prohibiting the importation of plants and animals from other regions.

The next four chapters contain many examples to illustrate the ideas presented in this chapter. One cautionary note: We will begin by assuming that the factors affecting geographic distributions operate in isolation from one another, as Liebig first suggested in 1840. But we know that this is not true from our personal experience—a spring day at 15°C will be pleasant if there is no wind, but it will seem cold if a strong wind is blowing. The effects of temperature and wind, of temperature and moisture, and of moisture and soil nutrients are not independent but often interact. We will begin simply and see how much we can understand by treating factors as separate effects, and then we will add factors together when necessary.

Summary

Why are organisms of a particular species present in some places and absent from others? This simple ecological question has significant practical consequences and thus deserves careful analysis. A transplant experiment is the major technique used to analyze the factors that limit geographic ranges. This technique leads sequentially through the following steps:

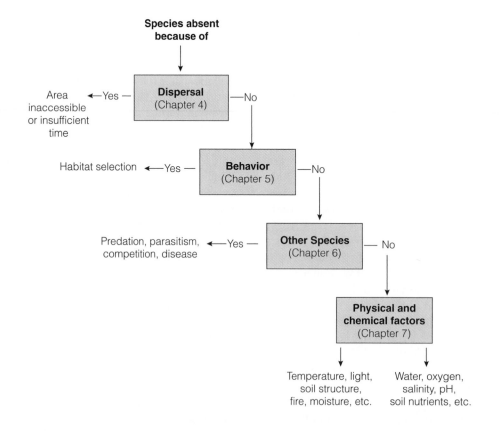

To examine any particular problem of distribution, ecologists proceed down this chain, eliminating things one by one. In the next four chapters we will see many examples in which part of this chain has been experimentally analyzed, but in no case has this chain been studied completely for a species.

The analytical question—What limits distribution now?—is complementary to the evolutionary question:

Why has there not been more adaptation? Thus we are led to investigate the genetic variation within populations and to look for range extensions or contractions that are associated with evolutionary shifts in the adaptations of organisms to their environment.

Key Concepts

1. All species have a limited geographical range, and the problem is to discover what causes these limits.

2. Transplant experiments can help to identify the potential range of a species.

3. Shelford's Law of Tolerance can be used to define the critical environmental limits to survival and reproduction and thus the potential geographical range for a species.

4. The tolerance ranges of species can change via natural selection.

Selected References

Bennett, W. A., R. J. Currie, P. F. Wagner, and T. L. Beitinger. 1997. Cold tolerance and potential overwintering of the red-bellied piranha *Pygocentrus nattereri* in the United States. *Transactions of the American Fisheries Society* 126:841–849.

Brown, J. H. and M. V. Lomolino. 1998. Distributions of single species. In *Biogeography*. Sinauer, Sunderland, Mass.

Gaston, K. J. 1991. How large is a species' geographic range? *Oikos* 61:434–438.

Hengeveld, R. 1989. *Dynamics of Biological Invasions*. Chapman and Hall, London. 160 pp.

Macnair, M. R. 1987. Heavy metal tolerance in plants: A model evolutionary system. *Trends in Ecology and Evolution* 2:254–259.

Ruesink, J. L., I. M. Parker, M. J. Groom, and P. M. Kareiva. 1995. Reducing the risks of nonindigenous species introductions. *BioScience* 45:465–477.

Taylor, O. R., Jr. 1985. African bees: Potential impact in the United States. *Bulletin of the Entomological Society of America* 31(4):15–28.

Vekemans, X. and C. Lefebvre. 1997. On the evolution of heavy-metal tolerant populations of *Armeria maritima*: Evidence from allozyme variation and reproductive barriers. *Journal of Evolutionary Biology* 10:175–191.

Questions and Problems

3.1 What assumptions must be made in mapping the potential spread of the African honey bee in North America? Discuss how to test these assumptions. Taylor (1985) evaluates this problem.

3.2 In discussing the role of physiology in studying distributions of animals, Bartholomew (1958, p. 84) states:

It usually develops that after much laborious and frustrating effort the investigator of environmental physiology succeeds in proving that the animal in question can actually exist where it lives.

Is this a problem for analyzing the factors limiting geographic distributions?

3.3 Discuss the problem of defining exactly the "geographic distribution" of a plant or animal. Gaston (1991) reviews this problem.

3.4 Discuss the application of general methods for studying distributions to the problems of what limits the geographic range of humans both today and early in our evolutionary history.

3.5 In discussing Liebig's Law of the Minimum, Colinvaux (1973, p. 278) states:

> The idea of critically limiting physical factors may serve only to obstruct a theoretical ecologist in his quest for a true understanding of nature....To say that animals live where their tolerances let them live has an uninteresting sound to it. It implies that animals have been designed by some arbitrary engineer according to some preconceived sets of tolerances, and that they then have to make do with whatever places on the face of the earth will provide enough of the required factors.

Evaluate this critique.

3.6 Why are the genes for copper tolerance more common in *Agrostis tenuis* growing on non-mine soils that are downwind of the mine site in Figure 3.5?

Overview Questions

Find a field guide to local flowers, birds, mammals, or amphibians, and discuss what the maps showing geographic ranges mean. On what scale would you map these ranges? Consult the Patuxent Wildlife Research Center of the U.S. Department of the Interior at *www.pwrc.usgs.gov* and look at the bird distribution maps for the "Breeding Bird Survey Results and Analysis" section. At what scale are these ranges mapped?

CHAPTER 4

Factors That Limit Distributions: Dispersal

I̲N THIS CHAPTER WE BEGIN TO UNRAVEL the first possible explanation of what limits geographic distributions of plants and animals. Some organisms do not occupy all of their potential range, and if transplanted outside their normal range they survive, reproduce, and spread. The simplest explanation for the absence of an organism from a particular area may be the species's failure to reach the area being studied, and thus this failure should be examined before more involved possibilities.

The transport, or *dispersal*, of organisms is a vast subject that has been of primary interest not only to ecologists but also to *biogeographers*, who seek to understand the historical changes in distributions of animals and plants. Some very difficult problems are associated with the study of dispersal. For one thing, the detailed distribution is known for so few species that most dispersals are probably not noticed. Dispersal of individuals between different parts of a species's range may occur often. Second, an organism may disperse to a new area but not colonize it because of biotic or physical factors.

If colonization is successful, dispersal will result in *gene flow* and thus affect the genetic structure of a population. If the dispersing individuals are not a random sample of the population, dispersal will result in a founder effect, and the new population may be genetically quite distinct from the source population. Not all dispersing individuals survive to breed, so gene flow may be quite restricted in many species (Slatkin 1987). Dispersal is thus simultaneously an ecological process affecting distributions, and a genetic process affecting geographic differentiation.

Examples of Dispersal

The most spectacular examples of dispersal affecting distribution involve species that are introduced by humans and proliferate to occupy a new area. Other examples are exploited species recolonizing their original range. Next we look into two examples of each of these situations.

Zebra Mussel (*Dreissena polymorpha*)

In 1988 a fingernail-sized mussel native to the Caspian Sea of Asia was discovered in Lake St. Claire near Detroit. No one knows how these mussels got transplanted there, but the best guess is that around 1985 a ship from a freshwater port in Europe arrived at the Great Lakes, where its ballast water containing mussel larvae was dumped with no concern about what organisms it might contain. The zebra mussel quickly became a pest because it forms dense clusters on hard surfaces and grows rapidly. The mussels were noticed when they reached densities of 750,000 per m^2 in water pipes in Lake Erie, clogging the water intakes of city water systems, electrical power stations, and other industrial facilities in the Great Lakes (MacIsaac 1996).

Since 1988 zebra mussels have spread rapidly in the river systems of the central United States (Figure 4.1). Because they are such efficient filter-feeders, zebra mussels have a positive impact on water quality, making Lake Erie, for example, much clearer than it had been previously. By feeding on phytoplankton

41

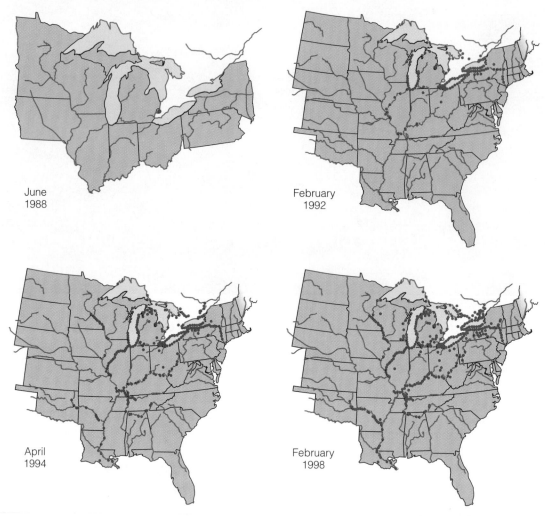

FIGURE 4.1
Expansion of the geographic range of the zebra mussel (Dreissena polymorpha) *from its discovery near Detroit in 1988 to February 1998. (Source: National Zebra Mussel Information Clearinghouse, NOAA, U.S. Department of Commerce.)*

they depress populations of zooplankton, and by making the water clearer they increase the growth of rooted aquatic plants in shallow waters. In the Hudson River in New York, phytoplankton biomass was reduced 80–90% after zebra mussels invaded, and zooplankton that feed on phytoplankton declined by more than 70% after the invasion (Pace et al. 1998). Zebra mussels also physically smother other native clam species as they colonize all available hard surfaces, including the shells of other clams, possibly causing local extinction of some native clam species.

Gypsy Moth (*Lymantria dispar*)

In 1850 a French astronomer employed at Harvard University near Boston brought some eggs of the European gypsy moth (*Lymantria dispar*) to his house

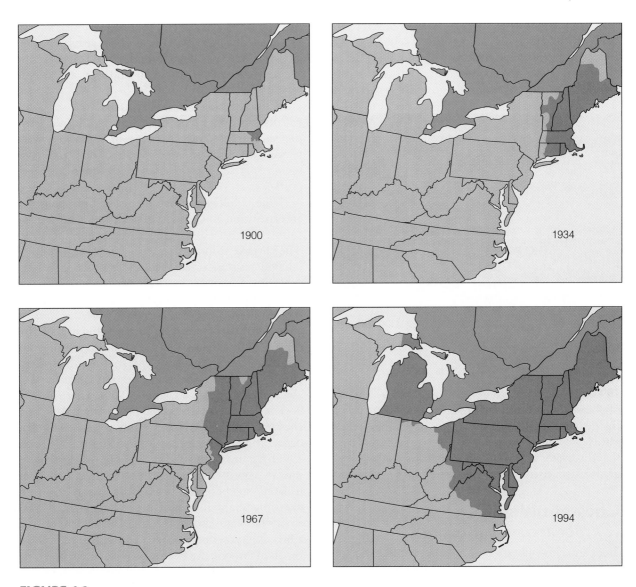

FIGURE 4.2
Spread of the gypsy moth (Lymantria dispar) *in the United States since its introduction in Boston in 1868. Most of the forests in the eastern United States are at risk from attack as this pest spreads south and west. (Maps courtesy of S. Liebhold, U.S. Forest Service Northeastern Research Station.)*

in Massachusetts. A few of the emerging caterpillars escaped in 1868, beginning one of the most devastating caterpillar plagues of the New England states.

Gypsy moths defoliate a great variety of deciduous and coniferous trees, including apple, alder, basswood, oaks, poplars, willows, and birches. Because of severe defoliation of deciduous trees, Massachusetts in 1889 initiated a control program, and by 1900 the

severity of the outbreaks was reduced, so the program was terminated by the state. After 1900 gypsy moths began to spread in a wave across New England (Figure 4.2). The spread slowed around 1950, but then gypsy moths were accidentally carried to Michigan in the early 1960s, starting a spread from that locus. About 1980 they began to spread west and south. The rate of spread from 1900 to 1915 was about 10 km per year, from 1916 to 1965 about 3 km per year, and from 1966 to 1989 about 21 km per year (Liebhold et al. 1992). The gypsy moth defoliated 13.8 million acres of forest in 1981, and another major outbreak began in 1989. The U.S. Forest Service began a program in

Gyspy moth fifth instar larva (©Marizio Lanini/Corbis)

1990 to slow the spread of the gypsy moth by the use of pheromone traps along the expanding boundary of the distribution (Sharov et al. 1998). Timber losses associated with these outbreaks are sufficiently large that this major gypsy moth research and control program has a benefit/cost ratio of more than 4:1.

Chestnut Blight (*Cryphonectria parasitica*)

The American chestnut (*Castanea dentata*) was an important component of many deciduous forests of eastern North America, from Maine and southern Ontario south to Georgia and west to Illinois (Figure 4.3). Chestnut made up more than 40% of the overstory trees in the climax forests of this area. In 40 years this species has been eliminated as a canopy tree from its entire range by chestnut blight.

Chestnut blight is a fungal disease that attacks chestnut trees. The disease was first noticed about 1900 in the area around New York City, where it killed all its hosts. The fungus was apparently introduced on nursery stock from Asia. Although found on other species of trees, *Cryphonectria parasitica* is lethal only to the American chestnut. The fungus enters the host tree through a wound in the bark, grows chiefly in the cambium, penetrates only short distances into

the wood, and kills the tree by girdling. Once a chestnut tree is attacked, it is killed in 2–10 years. The fungus kills only the aboveground part of the tree, and the root systems of trunks killed 60 years ago are still sending up shoots.

Closely related native species of *Cryphonectria* are usually saprophytic on chestnuts and do not harm their host tree. *Cryphonectria parasitica* is a very weak parasite on closely related species of oaks and never seems to attack the closely related American beech (Parker et al. 1993).

The U.S. Department of Agriculture sponsored an expedition to China, Japan, Korea, and Taiwan from 1927 to 1930 to collect Asiatic chestnut seed. The purpose of importing seeds was (1) to determine if blight-resistant oriental chestnuts could replace the vanishing American chestnut and (2) to establish oriental chestnuts for crossbreeding with the American species. Both Chinese chestnuts (*Castanea mollissima*) and hybrid trees have been successful in areas of the central Appalachians and the Ohio River valley but not in northern New York and southern New England, where the American chestnut once lived in abundance (Diller and Clapper 1969). There has also been a search for a native blight-resistant American chestnut. Large surviving chestnut trees are found on occasion, and some are partly blight-resistant (Brewer 1995). Strains of the chestnut blight fungus that are infective but not lethal to chestnut trees, have been isolated, so there has been some evolution in the chestnut blight fungus. Unfortunately these weak strains of fungus do not spread in the field. At present a breeding program has been established to produce resistant American chestnuts by backcrossing with resistant Chinese chestnuts. Two genes seem to control blight resistance, and if these can be incorporated into the American chestnut genome we may once again see the chestnut become a common canopy tree in eastern United States (Burnham 1988, Anagnostakis and Hillman 1992, Brewer 1995).

Most of the chestnuts killed by the blight have been replaced by codominant trees, especially oak species but also beech, hickories, and red maple (Good 1968, Keever 1953). The oak-chestnut forests have now become oak forests or oak-hickory forests. The large crops of chestnuts were presumably important food items for squirrels and other animals, but there is no information on how their populations may have suffered with the disappearance of the American chestnut from eastern North America.

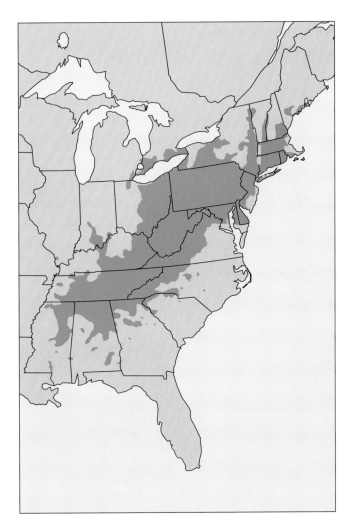

FIGURE 4.3
Original distribution of the American chestnut (Castanea dentata). *A few trees still occur throughout this range, and root shoots from old trees are killed by chestnut blight once they get aboveground. In the southern Appalachians, chestnut trees up to 6 m diameter were important timber trees until the early 1900s. Chestnut wood is highly resistant to decay, and the bark was an important source of tannins for the leather industry. (Photos courtesy of the Forest History Society, Inc.)*

California Sea Otter
(*Enhydra lutris*)

Sea otters were hunted by fur traders around the North Pacific to very low numbers by 1900. The few remaining small populations were protected by international treaty in 1911, and the California subpopulation of the sea otter was believed to be extinct at that time. In 1914 a small population was discovered at Point Sur in central California. Since then, otters on the central California coast have increased in numbers and expanded their geographic range to reoccupy areas from which they had been exterminated in the nineteenth century (Figure 4.4). The rate of spread of the sea otter is easy to estimate because it lives along the coastline in a linear habitat. The southern range

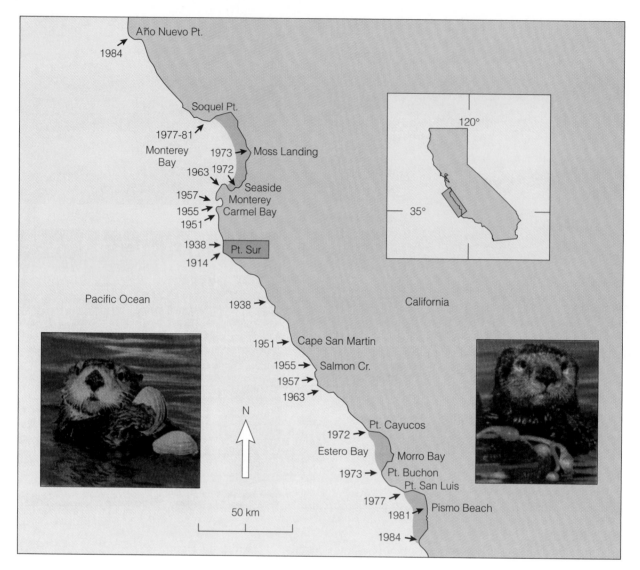

FIGURE 4.4
Expansion of the range of the California sea otter (Enhydra lutris) *along the California coast. The current range expansion began from Point Sur (red), where 50 sea otters were rediscovered in 1914. (After Lubina and Levin 1988.)*

expanded 3.1 km/year between 1938 and 1972, and the northern range expanded 1.4 km/year. These differences could result from the southern otters moving more as individuals, or from the northern otters suffering greater mortality (Lubina and Levin 1988).

The Three Modes of Dispersal

These spectacular cases undoubtedly constitute a biased sample, but the important point they illustrate is how rapidly some organisms can spread to new

E S S A Y 4 . 1

SHIPS, BALLAST WATER, AND MARINE DISPERSAL

Species invasions in terrestrial habitats have long been recognized as a source of environmental problems, but much less attention has been paid to marine invasions. Many marine invasions have been due to human-assisted dispersal either as organisms attached to the bottoms of ships or by the release of ballast water (Ruiz et al. 1997).

During the nineteenth century many organisms reached new ports attached to the bottoms of wooden ships, and this was the main means of human-assisted marine introductions. But the advent of metal ships, antifouling paints, and faster ships has eliminated this transport mechanism. At the same time, ballast water discharge has increased dramatically as ships have become larger. Chesapeake Bay received 10 million metric tons of ballast water discharge in 1991, mostly from ships originating in Europe and the Mediterranean. The zebra mussel is one of the best known examples in North America of a species brought in ballast water from elsewhere. A single ship can now carry 150,000 tons of ballast water to maintain trim and stability. The ballast water of five container ships entering Hong Kong contained 81 species from eight animal phyla and five protist phyla (Chu et al. 1997).

The biological results of these dispersal movements are significant. Chesapeake Bay now has 116 introduced marine species. San Francisco Bay has had 212 species added to its marine ecosystem. Some of these introduced species, such as the zebra mussel and the Asian clam in San Francisco Bay, have become dominant members of the community. Other introductions have not been studied in detail, so their impact is not known. Two health risks of introduced species have been detected. Toxic red-tide dinoflagellates are transferred worldwide in ballast water and may serve to trigger these algal outbreaks. The cholera bacterium *Vibrio cholerae* occurs in the ballast tanks of some ships and can survive for up to 240 days in seawater at 18°C (McCarthy 1996). When released into an estuary cholera bacteria can attach to a variety of marine organisms and thus enter the human food chain.

This story contains two ecological messages: Many marine species were originally limited in their global distribution, and action to reduce the global transport of potentially harmful organisms in ballast water is urgently needed.

areas if conditions are favorable. Before we discuss the ecological consequences of dispersal, let us define more carefully what we mean by dispersal.

The three ways in which species spread geographically, all loosely labeled as *dispersal*, are the following (Pielou 1979):

1. **Diffusion:** Diffusion is the gradual movement of a population across hospitable terrain for a period of several generations. This common form of dispersal is illustrated by the gypsy moth and the chestnut blight after their introduction to North America.

2. **Jump dispersal:** Jump dispersal is the movement of individual organisms across large distances followed by the successful establishment of a population in the new area. This form of dispersal occurs in a short time during the life of an individual, and the movement usually occurs across unsuitable terrain. Island colonization is achieved by jump dispersal, and human introductions such as the African honey bee (see page 33) can be viewed as an assisted form of jump dispersal.

3. **Secular dispersal:** If diffusion occurs in evolutionary time, the species that is spreading undergoes extensive evolutionary change in the process. The geographic range of a secularly dispersing species expands over geologic time, but at the same time natural selection is causing the migrants to diverge from the ancestral population. Secular dispersal is an important process in

biogeography, but, since it occurs in evolutionary time, it is rarely of immediate interest for ecologists working in ecological time.

One of the most spectacular colonizations occurred at the end of the Ice Age when glaciers retreated from Europe and North America. In 1899 a British botanist, Charles Reid, raised the question of how trees recolonized the British Isles after the Ice Age. Reid (1899) identified a great discrepancy between the life history characteristics of trees before and after the Ice Age, and pointed out they spread quickly after the ice melted. From the melting of the ice about 10,000 years ago until the Romans occupied Britain about 50 A.D., trees such as oaks expanded their range 1000 km northwards. Reid calculated that this would take a million years. Oaks, like most deciduous temperate zone trees, mature at 10–50 years of age and drop seeds that on average fall 30 m from the parent tree. If trees migrate by simple diffusion, the migration rate is set by the following simple equation (Skellam 1951):

$$\text{Distance moved} = D\,n\,\sqrt{\log_e R_0} \qquad (4.1)$$

where D = average dispersal distance

   n = number of generations

   R_0 = reproductive rate per generation

Thus if a tree produces 10^7 seeds per generation over 300 generations, and seeds disperse 30 m with each generation, this simple diffusion model predicts a range extension of 36 km, far short of the observed recolonization of 1000 km since the end of the Ice Age. This discrepancy is now called *Reid's paradox*.

Paleoecologists have calculated that to repopulate Britain or northern parts of North America since the glaciers melted, trees had to migrate 100–1000 m per year (Clark 1998). How can we resolve Reid's paradox between the expected slow rates of tree diffusion and the observed rates of range expansion?

Tree seed dispersal can be mapped by putting out seed traps at different distances from the parent tree or by mapping the locations of seedlings that have been produced by isolated trees. Figure 4.5 illustrates the typical pattern of the seed shadow from a deciduous tree. Most seeds fall near the parent tree, and a few are carried farther by wind or by animals. Clark (1998) measured the average seed dispersal distances

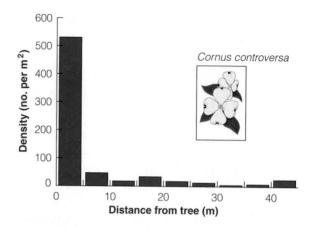

FIGURE 4.5
Seed dispersal distances for the dogwood Cornus controversa. *This tree has fruits that are dispersed by birds, but the vast bulk of the seeds fall near the parent tree. Mean seed dispersal distance for these trees was 6.7 m. (Data from Masaki et al. 1994, Table 2.)*

of 12 species of temperate zone trees in the southern Appalachians and found a range of 4–34 m, distances far too small to account for recolonization by simple diffusion after the ice melted.

The answer to Reid's paradox seems to lie in haphazard, long-range dispersal of seeds. Even though the mean dispersal distance is small, colonization rates are driven not by the mean dispersal distance but by extreme dispersal events. A few seeds are blown by wind or moved by animals a long distance from the parent tree. In the frequency diagram in Figure 4.5, these long-range dispersers would be off the scale to the right. Extreme long-distance events are difficult to record and measure, since less than one seed in 10,000 might be blown a long distance by wind or moved a great distance by animals.

Dispersal can be affected by *barriers*. Many kinds of barriers exist for different kinds of animals and plants. Let us consider a few cases in which barriers on a local scale are important in restricting jump dispersal.

Freshwater organisms to which both land and sea are barriers might be expected to show local distributions strongly affected by jump dispersal. Water can also be a barrier to some terrestrial animals. Ruffed grouse (*Bonasa umbellus*) were originally found only on three Michigan islands in the Great Lakes, all within 800 meters of the mainland. All islands more than 800 meters from shore were uninhabited by this

TABLE 4.1 Historical success rates for introducing terrestrial and freshwater birds to some selected islands and mainland locations.

This list includes only exotic species that increased and spread beyond the point of introduction. The success rate is generally higher on islands than on continents.

Location	No. of successfully introduced spp.	No. of spp. introduced	Success rate (%)	No. of native spp. present
Australia (Victoria)	16	48	33	271
Bermuda	7	17	41	9
Continental USA	13	98	13	553
Great Britain	9	30	30	146
Hawaii (Kauai)	27	52	52	18
Mauritius	19	44	43	13
New Zealand	41	149	28	52
Tahiti	11	54	20	12
Tasmania	13	16	81	104

Source: Data from Case (1996, Table 2).

bird. Palmer (1962) suggested that lack of dispersal explained this island pattern of distribution, and he tested the flight capacity of several grouse over water. None could fly as much as 800 meters over water, and Palmer concluded that ruffed grouse are not capable of flying across more than 800 meters of open water to colonize offshore islands by jump dispersal. Some of the offshore islands were artificially stocked with ruffed grouse, and populations have been very successful (Moran and Palmer 1963). There is a certain ecological irony in the fact that a bird capable of flight cannot colonize many offshore islands yet immobile trees did so with great speed after the ice retreated in North America.

On a local or global scale, dispersal may not limit distribution, because introduced species may be unable to survive. Humans have moved many species around the globe during the past 200 years, often with disastrous consequences. Long (1981) and Ebenhard (1988) list some of the early attempts to introduce nonnative birds to North America. Unfortunately, failures to establish a species are rarely studied to obtain an explanation, and accidental introductions are often recorded only when they are successful. The wool trade has been responsible for the introduction of 348 plant species into England resulting from seeds adhering to wool fleeces shipped in from overseas, but of these only four species have become established (Salisbury 1961). Few plant species introduced into continental areas can become established except in disturbed areas.

Bird introductions into continental areas are usually failures (Case 1996). Table 4.1 gives some data for terrestrial and freshwater birds. In the continental United States, only 13 species of introduced birds are common, although 98 species have been introduced. In Great Britain, only 9 successful establishments of birds are recorded from 30 species introduced. About 204 species of breeding birds live around Sydney, Australia, and 50 or more bird species were introduced to this area. Only 15 species got established, and only eight species are common. Thus continental bird introductions are successful about 10–30% of the time.

Can we make any statistical generalizations about the success of introduced species? Williamson and Fitter (1996) have proposed the *tens rule*, which makes the statistical prediction that 1 species in 10 imported into a country become introduced, 1 in 10 of the introduced species becomes established, and 1 in 10 of the established species becomes a pest. To interpret the tens rule we need some precise definitions of these terms (see Box 4.1). Exceptions to the tens rule occur—for example, among Hawaiian birds and island mammals. Islands are particularly vulnerable to introduced species, and areas like Hawaii that have been largely cleared for agriculture contain habitats that often have become unsuitable for native birds but are suitable for introduced species.

Humans have increased dispersal on a continental scale, but on a local scale many species have good to excellent dispersal mechanisms. Plants disperse primarily by means of seeds and spores, and transport is

BOX 4.1

DEFINITION OF TERMS FOR INTRODUCED NONNATIVE SPECIES

Introduced species (also called nonindigenous species or NIS) occur in four "states" and undergo three "transitions," as follows (Williamson and Fitter 1996):

States	Transition	Definition
Imported		Brought into the country
↓	Escaping	Transition from imported to introduced
Introduced		Found in the wild; feral
↓	Establishing	Transition from introduced to established
Established		Has a self-sustaining population
↓	Becoming a pest	Transition from established to pest
Pest		Has a negative economic impact

In the terminology now applied to genetically engineered organisms, introduced species are *released*, whereas imported species are *contained*.

The *tens rule* states that each transition in the table has a probability of about 10% (between 5% and 20%). The tens rule is only approximate and seems to describe a wide variety of plant and animal groups.

mosquitoes disperse determine the limits to which a given breeding location may allow contact with people and the area where control work must be done if a given human habitationa are to be protected from diseases like malaria. Morris et al. (1991) marked 451,000 mosquitoes with fluorescent dust in central Florida in 1987 to see how far individuals would disperse into human habitations. They found that 10% of the marked individuals moved over 2.2 km within 2 days, and the maximum distance moved was 4.2 km. Malaria control zones in tropical countries typically use a 2-km barrier zone surrounding human habitations as a rule-of-thumb for control since mosquitoes rarely move that far. Wind can move mosquitoes much farther than 2 km, and there are many records of mosquitoes being carried long distances by wind (Service 1997). Salt marsh mosquitoes in Louisiana have been captured on oil rigs 74–106 km offshore, and in Australia salt marsh species have been collected 96 km inland. One marked mosquito in California was collected 61 km from the release point. Mosquitoes are serious pests in northern Canada, Alaska, and Eurasia, and local control efforts are of limited success because of dispersal. Hocking (1953) calculated the maximum flight range of four species of northern mosquitoes to be between 22 and 53 km. Dispersal in mosquitoes is clearly very effective in colonizing vacant areas.

Colonization and Extinction

If dispersal occurs rapidly on a local scale, one would expect areas that are cleared of organisms to be recolonized rapidly. Some large-scale colonization experiments have occurred naturally. On August 26, 1883, the small volcanic island of Krakatau in the East Indies was completely destroyed by a volcanic eruption. Six cubic miles (25 km³) of rock was blown away, and all that remained of the original island was a smaller peak covered with ashes. Two islands within a few kilometers of the volcano were buried in ashes. These sterilized islands in effect constituted a large natural experiment on dispersal. The nearest island not destroyed by the explosion was 40 kilometers away. Nine months after the eruption, only one species—a spider—could be found on the island. After only three years, the ground was thickly covered with blue-green algae, and 11 species of ferns and 15 species of flowering plants were found. Ten years after

rarely an important factor limiting distributions of plants on a local scale. Few experimental data are available to substantiate this general conclusion. *Rumex crispus* var. *littoreus* is a British plant confined to a narrow zone of the seashore at the level of the highest tides. Cavers and Harper (1967) sowed seed of this species in a variety of nonmaritime habitats and found that whereas seedlings became established in large numbers, most did not survive the first year. Thus the local distribution of this plant is not limited by dispersal or by seed-germination requirements. Salisbury (1961, p. 100) discusses the invasion of weeds into bombed areas of London during World War II. Seeds and spores that were wind dispersed colonized these areas very quickly.

Small animals often have a life cycle stage that can be transported by wind, and these species resemble plants in that their distributions are rarely limited by lack of dispersal. Many insect species are transported by wind for long distances. Mosquitoes are a good example. The flight patterns of disease-carrying mosquitoes have been studied to enable the implementation of adequate control measures. The distances

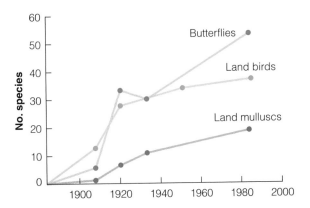

FIGURE 4.6
Colonization of the Krakatau Islands by butterflies, birds, and land molluscs after the volcanic eruption of 1883. Colonization continues even after 100 years. (Data from Thornton et al. 1990.)

the explosion, coconut trees began growing on the island. After 25 years, 263 species of animals lived on the island, which was covered by a dense forest. Figure 4.6 illustrates the colonization curves for three groups of organisms. Bird colonization of the islands has depended on vegetation colonization, and the flora of the islands has continued to increase (Whittaker et al. 1989). There is some controversy about the methods of transport, but the majority of the plants and animals were probably transported by wind. Larger vertebrates probably arrived on driftwood rafts or in a few cases by swimming. The suggestion that emerges from these observations is that when there is vacant space, animals and plants are not long in finding it.

In 1980 the eruption of Mount St. Helens in Washington provided a more recent opportunity to study recolonization of an area sterilized by volcanic eruption. During the first six years following the eruption, vascular plant invasion of the barren pumice plains landscape was limited, despite the proximity of seed sources. Was colonization limited by the dispersal of seed, or by a lack of tolerance to the volcanic ash covering the soil? Wood and del Moral (1988) studied 22 species of grasses and herbs on Mt. St. Helens. They recognized two groups: "dispersers," whose seeds were lighter and moved by the wind, and "nondispersers," whose seeds rarely got as far as 3 meters from the parent plant. From the 16,000 seeds they planted in barren sites at the volcano, only 1745 seedlings emerged. Species with the largest seeds had

the highest chances of growing. The slow colonization of the barren soils resulted from the lack of dispersal of heavy seeds onto the devastated area. Plants like Aster that have light seeds can reach the barren soil, but because their seeds are small they die from drought stress. Recovery on Mt. St. Helens will be patchy and slow because of limited seed dispersal from plants that can effectively colonize the bare soil.

The subalpine and alpine flora of Mount St. Helens in 1999 consisted of 95 species of vascular plants. Three nearby volcanoes contain two or three times as many plants. More than half of the plant species missing from Mount St. Helens seem to be absent because of a lack of dispersal (del Moral and Wood 1988). The subalpine zone of Mount St. Helens is isolated by 50–80 km of lowland forest, and as such it resembles an oceanic island. Furthermore, volcanic eruptions have also eliminated species. About 20 species of subalpine and alpine plants became locally extinct with the 1980 eruption.

These examples suggest that dispersal may limit *local* distributions of a few plants and some animals, but in most cases empty places get filled rapidly. Let us now look at the other extreme and consider *global* distribution patterns before humans began to move organisms on a large scale.

Terrestrial mammals other than bats do not easily cross saltwater barriers (Darlington 1965), so whole faunas can diverge if they are isolated by ocean. Marsupials, for example, became isolated in South America and in Australia early in the Tertiary period (60 million years ago). Of the placental mammals, only rodents and bats were able to colonize Australia before the arrival of humans. South America was also isolated by a water gap across Central America for most of the Tertiary and became connected to North America only during the last 2 million years. Once a land connection was established, a flood of dispersing mammals moved in both directions. The results for North America were relatively minor—the arrival of the opossum, the porcupine, and the armadillo as additions to the mammal fauna. But in South America the results of colonization were dramatic. Many South American mammals became extinct and were replaced by North American species. Carnivores from North America have completely replaced the carnivorous marsupials that previously occupied South America. Ungulates from North America have entirely replaced the unique set of South American ungulates (Darlington 1965).

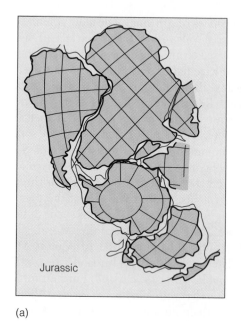
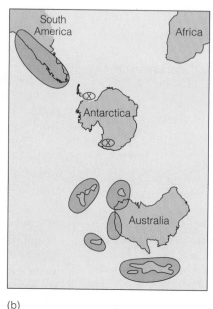

(a)   (b)

FIGURE 4.7

Effects of continental drift on geographic distributions. (a) Fit of the Gondwana continents during the Jurassic period, about 135 million years ago, before breakup. (b) Modern distribution of the genus Nothofagus *(Antarctic beech).* Nothofagus *pollen (inset) of the Oligocene age (30 million years ago) has been found in Antarctica at the two sites indicated by x in part (b). (After Pielou 1979).*

The faunas and floras of oceanic islands also show in graphic detail the limitations of distribution on a global scale. New Zealand had no native marsupials or other land mammals except for two species of bats at the time Europeans first arrived. All of the plants and animals that colonize New Zealand or any oceanic island must do so across water. The unique combination of difficult access, limited dispersal powers of different species, and adaptive radiation has produced island floras and faunas of an unusual nature, such as the plants and animals of Hawaii and the species Charles Darwin found on the Galápagos Islands off Ecuador.

The antarctic beech (*Nothofagus* spp.) is a good example of how present geographic distributions are set by geologic events. Until 135 million years ago the southern continents were connected in a large land mass called Gondwana (Figure 4.7a). Groups present on Gondwana now have a very disjunct distribution—*Nothofagus* is a good example (Figure 4.7b). *Nothofagus* seeds are heavy and poorly adapted for jump dispersal. Species of *Nothofagus* have probably spread slowly overland by diffusion and have been stopped

by the sea, so their present distribution is a byproduct of continental drift.

Continental drift not only takes certain continents farther apart, it also brings some continents closer together. As the Australian tectonic plate, for example, drifted northward after becoming detached from Antarctica, it made contact with the Asian plate about 20 million years ago. As distances over water decreased, jump dispersal of plants between Australia and Asia has become steadily easier.

The Quaternary Ice Age is a more recent example of how geographic distributions are affected by geologic events. Chris Pielou (1991), working in Canada, has integrated much of the data on how the Ice Age affected the flora and fauna of North America. The Ice Age began about 2 million years ago. During the past 500,000 years, ice sheets in North America and Eurasia have undergone great oscillations, waxing and waning at least four times. We are now in the fourth interglacial period. At the height of the last glaciation—about 20,000 years ago—the ice volume was 77 million km^3, three times the current amount. Sea level at the height of the last glaciation was 130 meters below its present level. If all the present ice melted, sea level would rise 70 meters (Pielou 1991). The biological effects of glaciations are spectacular but slow. Dropping sea levels open up migration routes for terrestrial organisms and may restrict dispersal of marine organisms.

E. Chris Pielou (1924–) *Professor of Zoology, Dalhousie University*

Disjunct distributions, such as that of *Nothofagus* (see Figure 4.7b), can be explained historically in two quite different ways. *Dispersal explanations* assume that the organism dispersed across preexisting barriers such as mountains or rivers and then underwent speciation. *Vicariance explanations*, on the other hand, assume that a species was present on the entire area and subsequently was fragmented by the formation of barriers (Brown and Lomolino 1998). If the timing of barrier formation can be determined and the phylogeny of the disjunct species is well known, one can test vicariance explanations explicitly. *Nothofagus* is a good example of a geographical pattern of distribution that is better explained by vicariance than by dispersal.

The flora and fauna of the world today have been strongly affected both by the dispersal of species and the geological formation of barriers that prevent organisms from colonizing all of their potential range. The great sweep of evolutionary history is a prolonged essay on the role of dispersal and barrier formation in limiting species distributions.

Evolutionary Advantages of Dispersal

Why disperse? The answer seems obvious: to find and colonize a new area. Natural selection will clearly favor an individual that leaves a relatively crowded habitat and colonizes an empty one in which it can leave many descendants. But the evolutionary problem is this: Most dispersing organisms die, and only a few are successful. In an evolutionary sense, an individual can do one of two things: stay at home and live in a suitable place but have only a few descendants (if any), or disperse and take a chance on surviving, colonizing a new habitat, and having many descendants.

Very few species have abandoned dispersal altogether. The best-known examples are flightless birds and insects on remote islands. Darwin noted the high frequency of flightless animals on oceanic islands during his voyage on the Beagle, and dispersal ability is reduced in many plants on islands. Birds often evolve flightlessness on islands that have no mammalian predators. In New Zealand 30–35 species of birds, about one-third of the total bird fauna, were

Dandelion seed head (©Buddy Mays/Corbis)

flightless when humans first arrived on these islands around 900 A.D. (McNab 1994). On subantarctic islands an average of 76% of the insect species are flightless (Carlquist, 1974, p. 494). Flightless insects are also found on ecological islands such as the alpine zone of tropical mountains. Figure 4.8a shows a striking example of a flightless crane fly from Mt. Kilimanjaro.

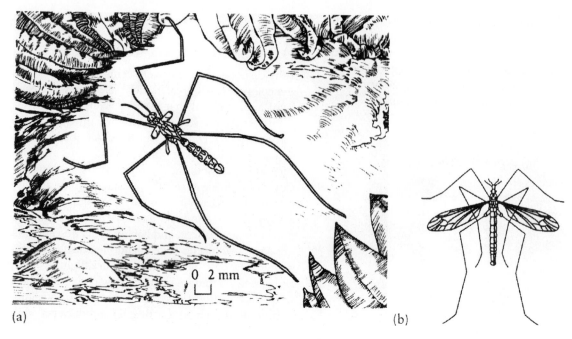

FIGURE 4.8
(a) Tipula subaptera, *a flightless crane fly from the alpine zone of Mount Kilimanjaro, Tanzania. (From Carlquist 1974.)*
(b) Tipula trivittata, *a typical crane fly from eastern North America. (From Swann and Papp 1972).*

Other organisms devote most of their effort to dispersal. Fugitive species are one extreme. These are the "weeds" of the plant and animal kingdoms that colonize temporary habitats, reproduce, and leave quickly before the temporary habitat disappears or competition with other organisms overwhelms them. Fugitive plant species grow entirely or predominantly in disturbed areas, and they produce large numbers of seeds adapted to long-distance dispersal by wind or by animals. The common dandelion (*Taraxacum officinale*) is an example of a fugitive or weedy plant species.

Strategies of dispersal in insects have been studied extensively because many pest species have high powers of dispersal (Service 1997, Dingle 1978). Dispersal in insects is largely a prereproductive phenomenon of the adult stage of the life cycle. Energy is first put into flight muscles and migration, and only after migration has occurred does egg formation start. The more highly migratory insect species are those associated with temporary habitats. Brown (1951) was one of the first to recognize this effect in the water boatman (*Corixidae*). A temporary pond received many immigrant insects of species that normally occurred in ponds but few immigrants from species that normally lived in large lakes and streams:

Species of water boatman	No. of immigrants to temporary pond	Percentage of total numbers collected in temporary habitats
Corixa nigrolineata (pond species)	209	81
Corixa falleni (large lake species)	11	23

Southwood (1962) summarized an abundance of evidence showing that dispersal occurs most often in insects that occupy temporary habitats. These insects are good examples of fugitive animal species.

The life cycle stages of some insects alternate between winged and wingless forms. Aphids are good examples. When conditions are favorable for reproduction, wingless forms are produced. When the environment deteriorates, winged forms are produced that disperse to a new habitat. In aphids, the tactile stimuli associated with high density stimulate the production of winged forms and subsequent dispersal. Water striders (gerrids) are another insect group in which both winged and wingless forms occur within the same species (Roff 1986, Fairbairn 1986).

Dispersal is clearly advantageous when habitats are patchy or unstable. But not all organisms live in unstable environments, and it is important to ask whether dispersal can also be adaptive in uniform and predictable environments. Hamilton and May (1977) have shown with a simple model that dispersal is highly adaptive in uniform environments, as long as potentially empty sites are available away from the parent organism, even if the likelihood of successful dispersal and colonization is low. This result agrees with natural history observations that nearly all organisms devote considerable resources to the dispersal of propagules.

We reach this general conclusion: Natural selection has molded the anatomy, physiology, and behavior of organisms to provide the dispersal powers needed to complete the life cycle. The many adaptations for dispersal in plants and animals illustrate in a graphic way the importance of dispersal in the lives of most species.

Summary

Dispersal operates in three general ways. *Diffusion* is the slow spread of a species across hospitable terrain over several generations. *Jump dispersal* operates quickly across great expanses of inhospitable terrain. *Secular dispersal* is spread in geological time, with extensive evolutionary change in the process; it may be associated with continental drift.

A species may not occur in an area because it has not been able to disperse there. This hypothesis can be tested by artificial introductions of the organism into unoccupied habitats. Some species introduced by humans from one continent to another, such as the zebra mussel, the African honey bee, and the gypsy moth, have spread very rapidly. The *tens rule* describes the statistical regularity that about 10% of species introduced become established, and about 10% of these become pests. Most introduced species (approximately 90%) die out and are not successful in new areas. On a local scale, few introduced species, once they have become established, seem to be restricted in distribution by poor powers of dispersal.

Thus dispersal is rarely an important factor limiting the *local* distribution of plants and animals. Organisms have many special adaptations for dispersal, and consequently they rapidly colonize new areas. On a global scale, however, dispersal is a critical factor in biogeography, and barriers to dispersal help to determine distribution patterns among continents and islands.

Dispersal is adaptive if it permits individuals to colonize new areas successfully. Some species—*fugitive* species—inhabit temporary habitats and exist only because of their great powers of dispersal. Other species live in more permanent habitats and disperse less. Dispersal ability has been molded by natural selection to maximize the chances of colonizing vacant areas, so it is rare that dispersal limits geographic distributions on a local scale.

Key Concepts

1. Some species do not inhabit an area because they have not yet been able to disperse there.

2. Dispersal can be analyzed on several time scales from the geological to the immediate, and on many spatial scales from continental to local.

3. Dispersal limitation can be tested by transplant experiments. Humans have conducted many of these experiments both deliberately and inadvertently.

4. The tens rule summarizes an approximate regularity—that 10% of introduced species become established and 10% of these become pests.

5. Adaptations for dispersal are common such that few species are limited in distribution on a local scale by a failure to disperse.

Selected References

Case, T. J. 1996. Global patterns in the establishment and distribution of exotic birds. *Biological Conservation* 78:69–96.

Clark, J. S. et al. 1998. Reid's paradox of rapid plant migration. *Bioscience* 48 (1):13–24.

Elton, C. S. 1958. *The Ecology of Invasions by Animals and Plants.* Methuen, London. 181 pp.

Lodge, D. M. 1993. Biological invasions: Lessons for ecology. *Trends in Ecology and Evolution* 8:133–136.

Lubina, J. A. and S. A. Levin. 1988. The spread of a reinvading species: Range expansion in the California sea otter. *American Naturalist* 131:526–543.

Pielou, E. C. 1991. *After the Ice Age: The Return of Life to Glaciated North America.* University of Chicago Press, Chicago. 366 pp.

Richardson, D. M. and W. J. Bond. 1991. Determinants of plant distribution: Evidence from pine invasions. *American Naturalist* 137:639–668.

Roff, D. A. 1986. The evolution of wing dimorphism in insects. *Evolution* 40:1009–1020.

Williamson, M. and A. Fitter. 1996. The varying success of invaders. *Ecology* 77:1661–1666.

Wood, D. M. and R. del Moral. 1987. Mechanisms of early primary succession in subalpine habitats on Mount St. Helens. *Ecology* 68:780–790.

Questions and Problems

4.1 Assume dispersal by simple diffusion (Eq. 4.1, page 48). How far would a plant be expected to move in 50 generations if the average dispersal distance was 100 m and the plant produced 10^3 seeds per generation? Is the distance moved more sensitive to the number of seeds produced or to the average dispersal distance? Double or triple each of these parameters and discuss the impact on the distance colonized by these life cycle changes.

4.2 Species introduced to marine environments often have planktonic larvae that are spread by ocean currents, and one might expect these introduced species to spread more rapidly than terrestrial species. In fact, marine introductions such as barnacles and mussels seem to spread slightly more slowly than terrestrial introductions such as gypsy moths and starlings (Grosholz 1996). Why might this be?

4.3 One of the recurrent themes in studying introduced species is that introductions are more successful when more individuals are released (Green 1997). Are there cases of successful introductions by humans when only a few individuals were released? What might account for this pattern?

4.4 Dutch elm disease is a viral disease spread by bark beetles that is lethal to the American elm (*Ulmus americana*). Compare the American chestnut–chestnut blight interaction to that of the American elm–Dutch elm disease interaction. Discuss in particular the factors that promote the spread of the two diseases. Swinton and Gilligan (1996), Anderbrant and Schlyter (1987), and Brasier (1988) provide some literature references on Dutch elm disease, and more recent developments may be found on the Web.

4.5 In discussing dispersal and colonization after human introductions, many authors beginning with Elton (1958) have concluded that island faunas offer far less resistance to immigrants than do mainland faunas. Consider the data in Table 4.1 on page 49 and obtain additional data on island and mainland faunas to evaluate this conclusion (see Case 1996). List some possible reasons for these facts.

4.6 Review the global distributions of flightless birds and comment on the ecological situations that lead to flightlessness in birds. Carlquist (1965, pp. 224–241) provides a general background, and McNab (1994) discusses this problem.

4.7 Review the introduction and spread of purple loosestrife (*Lythrum salicaria*), which was introduced from Eurasia to North America in the early 1800s (Thompson 1991, Ruesink et al. 1995). Is this invading species a pest? What can be done now to alleviate the problem?

4.8 The spread of human diseases is often more carefully documented than the spread of diseases of plants and animals. Review the spread of cholera in the United States from 1832 to 1866, and the spread of measles in Iceland from 1945 to 1970 (see references in Hengeveld 1989, pp. 18–24).

Overview Question

Would you expect the same dispersal abilities in plants from tropical rain forests and from boreal conifer forests? In animals? Why or why not?

CHAPTER 5

Factors That Limit Distributions:
Habitat Selection

SOME ORGANISMS DO NOT OCCUPY all their potential range, and in Chapter 4 we discussed cases in which limited dispersal was the reason for the absence of a species. In this chapter we discuss cases in which organisms do not occupy all their potential range even though they are physically able to disperse into the unoccupied areas. Thus individuals "choose" not to live in certain habitats, and the distribution of a species may be limited by the behavior of individuals in selecting their habitat. We define a *habitat* as any part of the biosphere where a particular species can live, either temporarily or permanently. Habitat selection is typically thought of only with respect to animals that can in some sense choose where to live by moving among habitats.

Habitat selection is one of the most poorly understood ecological processes. If we assume that an animal cannot live everywhere, natural selection will favor the development of sensory systems that can recognize suitable habitats. What elements of the habitat do animals recognize as relevant? We must be careful here to define the perceptual world of the animal in question before we begin to postulate the mechanism of habitat selection. Areas that appear "similar" to a human observer may appear very different to a mosquito or a fish. Conversely, habitats we think are very different may be treated as the same by a bird.

There are two broad approaches to the study of habitat selection. The proximal approach views habitat choice as a result of behavioral mechanisms and asks how, in a physiological sense, an animal chooses its habitat. The ultimate or evolutionary approach looks at the adaptive reasons for habitat choice and at the evolutionary significance of the behaviors involved. In this chapter we first consider the behavioral mechanisms of habitat selection and then discuss the evolution of habitat preference.

Plants show habitat preferences in quite different ways than animals because they cannot actively move from one habitat to another. Seeds or spores arrive in different habitats through dispersal, and then either survive and grow or die because of biological or physical factors discussed in the next three chapters. Animals use behavioral mechanisms to choose their habitats, and individual movements are an essential component of the resulting habitat selection.

Behavioral Mechanisms of Habitat Selection

In many invertebrates, habitat selection is accomplished in a simple manner. For example, when the isopod *Porcellio scaber* is placed in a humidity gradient, it moves at random with respect to the gradient, but it moves much more rapidly in dry air than in moist air. An individual that in the course of its random movements happens to find moist air will slow down and become motionless. The result is that most of the animals eventually come to rest at the moist end of the gradient, a very simple form of habitat selection (Fraenkel and Gunn 1940).

A more complex form of stereotyped behavior is involved in the choice of oviposition sites by many insects. Although we assume that in most insects the choice of oviposition sites is the female's decision, this is not always the case in territorial species. In some dragonflies, the males occupy territories that determine where females will oviposit (Macan 1974). But in most other insects, the female's behavior determines the placement of eggs. European corn borer larvae will feed on a wide variety of plants, but they occur mainly on corn because the ovipositing females are attracted by volitile odors produced by the corn plant (Schoonhoven 1968). Leaf beetles of the tribe Luperini (Coleoptera: Chrysomelidae: Galerucinae) use the secondary chemicals cucurbitacins as cues for host selection (Metcalf 1986). This association is present worldwide in about 1500 species of beetles feeding on plants of the family Cucurbitaceae (squash, cucumbers, melons). The southern corn rootworm (*Diabrotica undecimpunctata*), one example from this tribe, it can feed on many species of plants but is most frequently found on cucurbits because of its attraction to cucurbitacins. These chemical cues for oviposition sites can thus result in a species being more locally restricted than its potential range might allow.

Anopheline mosquitoes are often important disease vectors, and their ecology has been studied a great deal because of the practical problems of malaria eradication. Each mosquito species is usually associated with a particular type of breeding site, and one of the striking observations that a student of malaria first makes is that large areas of water seem to be completely free of dangerous mosquitoes. Large areas of rice fields on the Malay Peninsula are free of *Anopheles maculatus*, as are the majority of shallow pools in some breeding grounds of *Anopheles gambiae* (Muirhead-Thomson 1951). Why are some habitats occupied by larvae and others not? Early workers assumed that something in the water prevented the larvae from surviving, and they neglected to study the behavior of females in selecting sites in which to lay eggs. More recent work has emphasized the role of habitat selection for oviposition sites in female mosquitoes and shown that larvae can develop successfully over a much wider range of conditions than those in which eggs are laid (Bentley and Day 1989). Thus, although we presume that the female selects a type of habitat most suitable for the larvae, many of the places she avoids are suitable for larval growth and development.

In Belize, the malaria-transmitting mosquito *Anopheles albimanus* oviposits only on floating mats of blue-green algae (Rejmánková et al. 1996). In marshes with dense cattail growth, blue-green algae are shaded and do not produce mats. In cattail marshes no larval mosquitoes were found, and in oviposition tests no larvae were produced unless algal mats were present. Larval mosquitoes are not found in open waters, and in Belize *A. albimanus* is limited to marshes with algal mats.

In southern India, the mosquito *Anopheles culicifacies* (a malaria vector) does not occur in rice fields after the plants grow to a height of 12 inches (30 cm) or more, even though these older rice fields support two other *Anopheles* species. Russell and Rao (1942) could find no eggs of *A. culicifacies* in old rice fields, yet when they transplanted this mosquito's eggs into old rice fields, the larvae survived and produced normal numbers of adults. The absence of *A. culicifacies* from this particular habitat is apparently due to the selection of oviposition sites by females. In a series of simple experiments, Russell and Rao were able to show that the main limiting factor was the physical barrier posed by rice plants of a certain height. Glass rods placed vertically in small ponds also deterred female *A. culicifacies* from laying eggs, as did barriers of vertical bamboo strips. Shade did not influence egg laying. This mosquito oviposits while flying and performing a hovering dance, never touching the water but remaining 2–4 inches (5–10 cm) above it. Physical obstructions seem to prevent the female mosquitoes from the free performance of this ovipositing dance and thereby restrict the species to a smaller habitat range than it could otherwise occupy.

Habitat selection is a process that operates at the level of the individual animal. Decision making or choices by mobile individuals such as migratory birds occur in a hierarchical manner (Figure 5.1) from a larger spatial scale to the local microhabitat. We should recognize that the "explanation" of habitat selection is really a series of explanations appropriate to the spatial scale of interest. John Wiens at Colorado State University is a leading ecologist who has considered habitat selection and its consequences for populations and communities. Most studies of habitat selection are done on the small spatial scale of microhabitats. Figure 5.2 presents his conceptual model of the hierarchy of variables that can influence the observed habitat choice of a species for studies of individual habitat selection (Wiens 1985).

Habitat selection in birds has been studied in greater detail than in most other groups, and most of the examples in this chapter involve birds. Two kinds

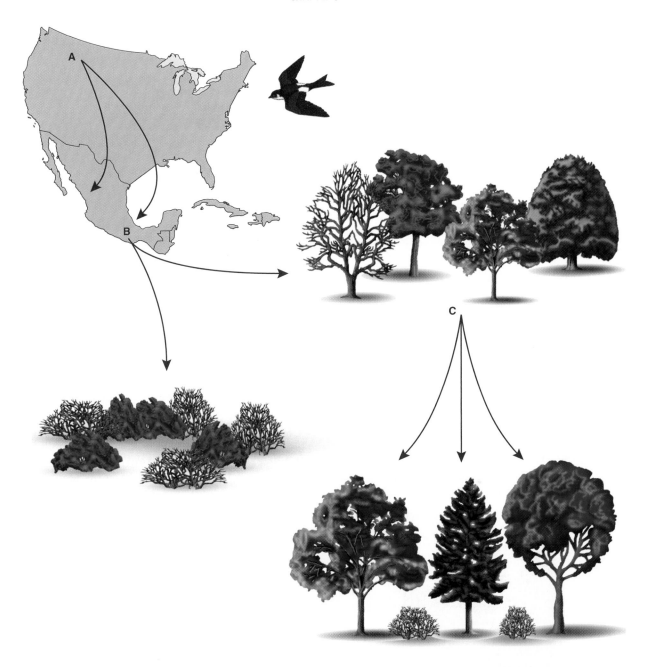

FIGURE 5.1
An illustration of the hierarchical decision making process involved in the choice of nonbreeding habitats by a migratory bird in Mexico. At A the bird must decide whether to go to southern Mexico or western Mexico to overwinter. At B it must decide whether to move into woodland or shrub habitats. At C it must decide which of the tree types it will occupy. At these different spatial scales the reasons for the bird's choice between shrub habitats and forest habitats, or among tree habitats, may differ. (Modified from Hutto 1985.)

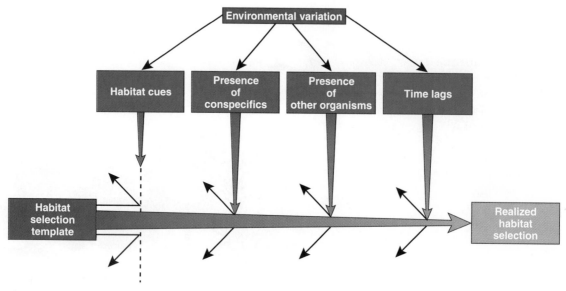

FIGURE 5.2

A conceptual model of the various factors that may affect habitat selection at the local level. These constraints operate to limit the range of habitats occupied by a species. (From Wiens 1985.)

of factors must be kept separate in discussing habitat selection: (1) evolutionary factors, conferring survival value on habitat selection, and (2) behavioral factors, giving the mechanism by which birds select areas. Habitat selection is the result of stimuli from landscape and terrain; nest, song, watch, feeding, and drinking sites; food; and other animals (Cody 1985).

John A. Wiens *(1939–) Professor of Ecology, Colorado State University*

Habitat cues for birds of prey may involve perch sites. Three species of buteos (broad-winged hawks) breed in grassland and shrub-steppe areas of western North America. The red-tailed hawk selects areas

with many perch trees or bluffs, while the Swainson's hawk and ferruginous hawk select more open areas with few trees (Janes 1985). These three hawks eat much the same prey (ground squirrels, jackrabbits), and their habitat choice corresponds with their foraging behavior. Red-tailed hawks sit on perches and look for prey; their wings are less suited to soaring. Swainson's hawks are best at soaring and hunt from the air much more than from perches. Ferruginous hawks are intermediate in soaring abilities. Flapping flight is uncommon in all these hawks, and habitat selection is closely tied to their hunting methods.

The distribution of many species can be affected by human changes to habitats. In Britain the tree pipit (*Anthus trivialis*) and the meadow pipit (*Anthus pratensis*) have similar requirements except that the tree pipit breeds only in areas having one or more tall trees. For this reason, the tree pipit is absent from many treeless areas in Britain that the meadow pipit inhabits. Lack (1933) found tree pipits breeding in one treeless area close to a telegraph pole. The only use to which the pole was put by the tree pipit was as a perch on which to land at the end of its aerial song. The meadow pipit has a similar song but ends it on the ground. Thus the tree pipit is excluded from colonizing heathlands only because it needs a perch on which to land after singing. Lack (1933) concluded that the distribution of bird species in the Breckland heaths of England and their associated pine plantations was largely a result of idiosyncratic behaviors and specific habitat selection restricting each bird to a habitat range smaller than it could potentially occupy. Adding telegraph poles to heathlands can change this habitat from unsuitable to suitable for the tree pipet.

Habitat selection in birds is partly a genetic trait, although it can be modified somewhat by learning and experience. The genetic basis of habitat selection is probably responsible for a slow response by some birds to human changes in the environment. The lack of genetic variability for habitat selection could be responsible for the lack of response to environmental changes. The original habitat selected by a bird is often reinforced by tenacity of individuals to their sites. Many old birds return year after year to the same nesting site, even if the habitat at that site is deteriorating. Unfortunately, there has been little experimental analysis of habitat selection in birds. Klopfer (1963) has shown that chipping sparrows (*Spizella passerina*) raised in the laboratory preferred to spend their time in pine branches rather than in oak branches, just as wild birds do. However, laboratory birds reared with oak branches showed a decreased preference for pine as adults. Thus the innate preference for staying in pine could be modified somewhat by early experience:

Chipping sparrows	*Time spent (%) in* Pine	Oak
Wild caught adults	71	29
Laboratory-reared, no foliage exposure	67	33
Laboratory-reared, oak foliage exposure only	46	54

Songbirds may have dialects that distinguish subpopulations, and these song dialects can influence habitat selection. White-crowned sparrows in the coastal scrub along Point Reyes National Seashore in California are subdivided into four dialects (Figure 5.3). The habitat is homogeneous, and these sparrows can clearly fly over the whole area in a few minutes, yet the dialects persist with stable boundaries (Baker et al. 1982). Dialects in bird songs are learned and are culturally transmitted from parents to offspring. Individual birds prefer to mate with birds of the same song dialect, so the socially determined dialect boundaries also become a genetic boundary that restricts gene flow. Habitat selection can thus have a cultural component independent of other biological, physical, or chemical factors.

Tree pipit

Desert rodents have been well studied in both North America and Asia. Habitat selection for these animals is related to body size—larger species prefer open habitats, and smaller species prefer bush microhabitats (Figure 5.4). These preferences have been verified experimentally. When Price (1978) removed half of the bushes on half of her study grid, the kangaroo rat *Dipodomys merriami* increased in density at the more open sites within six weeks. Thompson (1982) did the opposite experiment: he added cardboard shelters to desert habitat to increase the cover in open areas (Figure 5.5). The results were as predicted—a decline in the numbers of the large kangaroo rat (*Dipodomys*) that prefers open areas, and an increase in the numbers of the smaller pocket mice (*Perognathus*[1]). Since cardboard shelters clearly do not add food to the system, these responses of nocturnal desert rodents must be to the risk of predation in open versus sheltered areas. The large kangaroo rats have adapted to living in open patches. They hop on their enlarged hind feet, and this speedy locomotion is coupled with well-developed hearing to detect the approach of predators (Kotler and Brown 1988).

Other good examples of habitat selection can be found in small mammals. The deer mouse (*Peromyscus maniculatus*) has received the most attention in this respect. The deer mouse is very widespread in North America and can be divided into two ecologically adapted types: (1) the long-tailed, long-eared forest form and (2) the short-tailed, short-eared prairie form. The prairie form that has been most intensively studied is *P. m. bairdi*. This subspecies avoids forested areas, even those with a grassy substratum, as can be readily seen from the results of trapping lines (Harris 1952).

It was first necessary to determine whether *P. m. bairdi* could live in woody areas, and this was done

---

[1] *Perognathus*, the old genus name for pocket mice, has been changed to *Chaetodipus*.

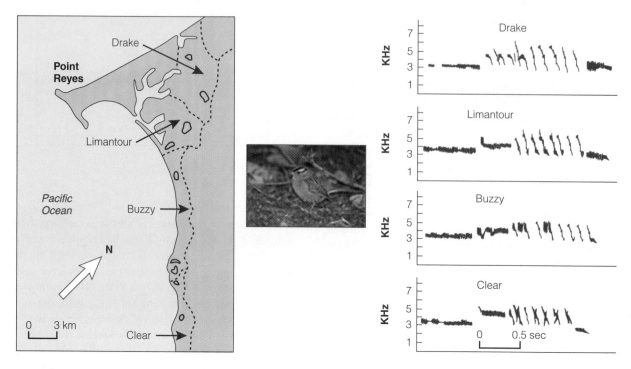

FIGURE 5.3
Song dialect populations of white-crowned sparrows (Zonotrichia leucophrys nuttali) *at Point Reyes National Seashore, California. Suitable habitat is bounded on the west by the ocean and on the east by Douglas fir forests (dashed line). Dotted lines indicate boundaries between dialects. Representative sound spectrograms of each of the four dialects are on the right. The sound spectrogram displays frequency in kilohertz on the vertical axis and time on the horizontal axis. (From Baker et al. 1982.)*

using field enclosures. Other studies with *Peromyscus* have shown little difference in the food preferences or temperature requirements of grassland subspecies compared with forest subspecies. Harris (1952) therefore concluded that *P. m. bairdi*'s absence from forested areas is due to its behavior, not to the unsuitability of the habitat.

Experiments in which the mice were given a choice between "woods" and "grassland" showed that *P. m. bairdi* usually chose the grassland. This behavior might be either genetically determined or learned from early experiences, and an attempt was made to distinguish between these causes. Early experience can play an important part in the development of adult behavioral characteristics, and perhaps the early experience of young mice in a particular habitat could determine their reactions to different habitats as adults.

Wecker (1963) studied laboratory stock and wild mice in three types of situations: (1) direct testing of adults, (2) testing of offspring reared in grassland habitat, and (3) testing of offspring reared in woodland habitat. The results are given in Table 5.1. The

wild-caught animals showed an overwhelming preference for the grassland habitat. The laboratory stock, held in laboratory conditions for 12–20 generations, had apparently lost some of its preference for grassland habitat. Exposure of this stock to early experience in a grassland greatly increased the preference for the grassland habitat, but exposure to woods caused only an insignificant increase in the preference for woodland. The conclusion was that habitat selection in the prairie deer mouse is normally predetermined by heredity and results in the animal's occupying a more restricted range of habitats than is theoretically possible.

Evolution of Habitat Preferences

Why do organisms prefer some habitats and avoid others? Natural selection will favor individuals that use the habitats in which the most progeny can be

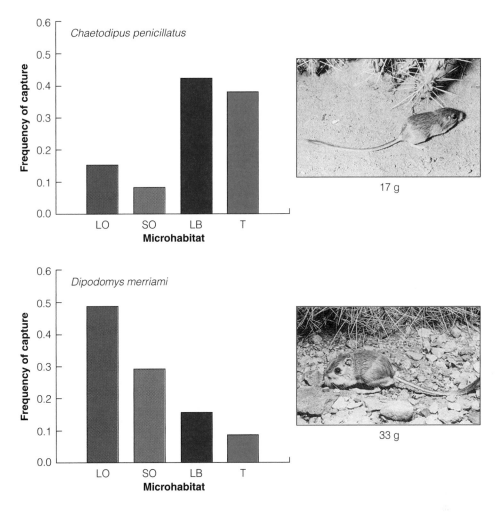

FIGURE 5.4
Summer microhabitat distribution of two heteromyid rodent species near Tucson, Arizona. Relative capture frequencies in live traps are given four microhabitats: LO = large open, SO = small open, LB = large bush, T = tree. The desert pocket mouse (Chaetodipus penicillatus) *prefers closed habitats, while the larger kangaroo rat* (Dipodomys merriami) *prefers open areas. (Data from Price 1978; photos courtesy R. D. Porter, Texas Parks and Wildlife.)*

raised successfully. Individuals that choose the poorer, marginal habitats will not raise as many progeny and consequently will be selected against. Populations in marginal habitats may thus be sustained only by a net outflow of individuals from the preferred habitats. A variety of physical clues can be adopted by organisms as the proximate stimuli in choosing a particular type of habitat. Natural selection may act directly upon the behaviors that result in habitat choice, or it may select for individuals that have the capacity to learn which habitat is appropriate.

Habitat recognition may be very imprecise or very exacting. Aphids, such as the crop pest *Myzus persicae*, for example, land on host and nonhost plants with equal frequency (Emden et al. 1969). Aphids test the chemical suitability of the plant after landing by probing the leaf in several places with their mouthparts. If the plant is of the wrong species or in the wrong condition, the aphid flies to another plant. Many aphids land on apparently suitable plants, test them, and leave anyway. Larvae of some butterflies will die of starvation before they will eat food plants

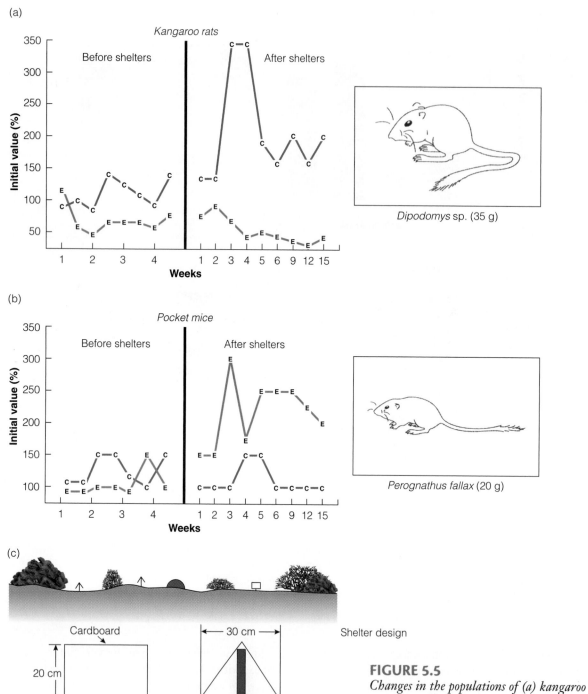

FIGURE 5.5
Changes in the populations of (a) kangaroo rats (Dipodomys sp.) *and (b) pocket mice* (Perognathus fallax) *before (left) and after (right) the addition of cardboard shelters (c) to open areas in the Mojave Desert, California. E = experimental areas; C = control areas. (After Thompson 1982.)*

TABLE 5.1 Habitat selection in prairie deer mice (*Peromyscus maniculatus bairdi*)[a]

	Time spent in grassy part of pen	
Treatment	Active time (%)	Inactive time (%)
Laboratory stock		
Tested directly	**48**	**40**
Offspring reared in grassland	85	97
Offspring reared in woods	59	44
Wild-caught animals		
Tested directly	**84**	**86**
Offspring reared in grassland	72	61
Offspring reared in woods	77	53

[a]Mice in each group were placed in a 100 ft × 16 ft (30 m × 5 m) pen that was half in a grassy field and half in an oak-hickory woodlot. The time spent in each half of the pen was recorded for both the part of the day in which the mice were active and the part in which they were inactive. Test results for adult controls are shown in boldfaced type.
Source: After Wecker (1963).

other than those to which they have become accustomed, even if this food is nutritionally adequate (Jermy 1987). Pacific salmon by contrast go to sea as juveniles and return one to five years later to exactly the same stream or lake in which they hatched. Olfactory cues in the water seem to be the critical stimuli, and if eggs are transferred from a natural spawning bed to a distant hatchery on a different river system, the salmon will return to the hatchery when they are adults (Courtenay et al. 1997).

Thus habitat recognition can be extremely specific if it benefits individuals to act accordingly. Salmon stocks adapt to the peculiar seasonality, temperature, and water-flow conditions of their own spawning grounds, and this precise adaptation can be preserved only if an excellent homing behavior has developed (Dittman and Quinn 1996). At the other extreme, animals like aphids may face habitats that are favorable one month but not the next. Organisms faced with habitat unpredictability must adopt more flexible habitat selection behavior.

How can we explain the idiosyncratic behaviors of animals like the tree pipit that reject perfectly good habitats? We have seen several examples of this behavior earlier in this chapter. The important point illustrated by these examples is that evolution does not produce perfect organisms. Not all behaviors are adaptive, and environmental conditions may change such that behaviors that were formerly adaptive are now maladaptive. Ground-nesting birds on islands, for example, are threatened when new predators such as rats colonize the island (Reed 1999). The genetic blueprint carried by a species cannot change overnight, and unless suitable genetic variation is present in a population, natural selection cannot operate quickly to adjust habitat selection behaviors.

For birds, survival and reproductive success can depend on nest-site choices, and can be the bases for the evolution of nest-site preferences. When there are habitat differences between successful and unsuccessful nests, the process of natural selection can operate to ultimately change nest-site distribution (Clark and Shutler 1999). Figure 5.6 illustrates how natural selection might operate to affect nest-site selection in birds. In this case there is directional selection for specific habitat features that increase the probability of successful nesting. An example of selection for nest-site habitat is given in Figure 5.7. Blue-winged teal (*Anas discors*) are more successful if they nest in areas that have more vegetation and are farther away from habitat edges (such as shrub/grassland borders). This differential nesting success gives rise to directional selection that should alter female nesting-habitat choice in these ducks, if this selection continues for many generations (Clark and Shutler 1999).

A Theory of Habitat Selection

A simple theory of habitat selection can be used to illustrate how habitat selection may operate in a natural population (Fretwell 1972). Recall that for any particular species, we define a *habitat* as any part of the Earth where that species can live, either temporarily or permanently. Each habitat is assumed to have a *suitability* for that species, and in this example we assume that three habitats of different suitabilities are available to a species. Suitability is equivalent to *fitness* in evolutionary time, and we will assume that females

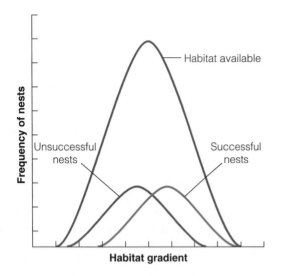

FIGURE 5.6

A hypothetical gradient (e.g., cover) showing the nest habitat available to a bird species (blue), the frequency distribution of unsuccessful nests (red), and the frequency distribution of successful nests (green). In this hypothetical example there is directional selection along the habitat gradient to the right. If this gradient is related to cover, birds would be selected to prefer areas of higher cover in the long term. (Modified after Clark and Shutler 1999.)

produce more young in more suitable habitats than they do in less suitable habitats. Suitability is not constant but will be affected by many factors in the habitat, such as the food supply, shelter, and predators. But in addition, suitability in any habitat is usually a function of the density of other individuals of the species, so that overcrowding reduces suitability (Figure 5.8). We assume in this simple model that all individuals are free to move into any habitat without any constraints, what Fretwell called the *ideal free distribution* (Fretwell 1972). As a population fills up the best habitat (A in Figure 5.8), it reaches a point where the suitability of the intermediate habitat (B) is equal to that in habitat A, so individuals will now enter both habitats A and B. As these two habitats fill even more, the poor habitat (C) finally has a suitability equal to that of habitats A and B. The prediction that arises from this simple model of habitat selection is interesting because it is counterintuitive. We would predict from the model that when density is high, good and poor habitats would have equal suitabilities (but different densities), that individuals would be crowded in the best habitats and at low density in the poor habitats (see Figure 5.8). Is there any evidence that this might in fact be the case in natural populations?

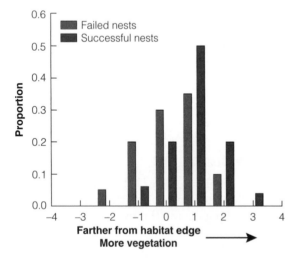

FIGURE 5.7

Breeding success of blue-winged teal in a habitat gradient in Saskatchewan from 1983 to 1997. Successful nests (n = 52) were farther from habitat edges and contained more vegetation than failed nests (n = 81). There is potential directional selection for nest-sites in this duck species, similar to that illustrated in Figure 5.6. (Data from Clark and Shutler 1999.)

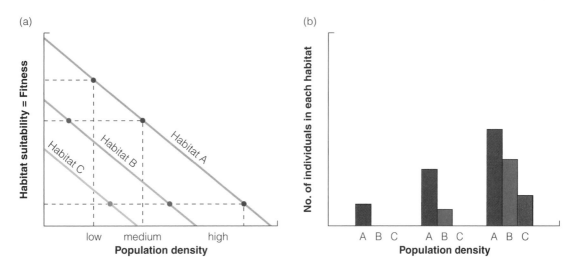

FIGURE 5.8
The ideal free distribution model of habitat selection. Three habitats are used for illustration (A = good habitat, B = intermediate, C = poor habitat). Habitat suitability is measured by the fitness of individuals living in that habitat. For illustrative purposes, three levels of population density are indicated. (a) The model assumes that in all habitats fitness declines as population density increases and crowding occurs. At low density an individual can achieve the highest fitness by living in habitat A, and habitats B and C will be empty; see (b). At high density an individual can choose to live in habitat A under crowded conditions, in habitat B under less crowded conditions, or in the poorest habitat C with the least crowding. If individuals choose their habitat as in this simple model, fitnesses of individuals will be equal in all three habitats at high density. (Modified from Fretwell 1972.)

Yellow-legged gulls nest in two habitats on the Medes Islands of northeastern Spain: shrubby areas and grassy meadows (Bosch and Sol 1998). Gulls prefer to nest in shrubby areas and occupy these areas first. Reproductive parameters did not differ between the two habitats. Gulls in the preferred shrubby habitats laid 2.8 eggs on average, while gulls in the grassy areas laid 2.9 eggs. By all the available criteria, a shrubby habitat is a good habitat for nesting gulls, and shrubs provide not only some protection from predators but also shade to reduce heat stress. Many chicks are killed by adult gulls and other bird predators in gull colonies. Shrubby areas should provide more hiding places for gull chicks.

Figure 5.9 shows the breeding success of yellow-legged gulls from the two habitats. Fitness (measured by breeding success) is nearly equal in both habitats, exactly as predicted by Fretwell's model for the ideal free distribution (Figure 5.8). Gulls crowd into the preferred shrubby habitat until the expected fitness of

a breeding pair falls to the level that can be reached on the poorer grassy areas.

Fretwell suggested a second model of habitat selection that could be applied to organisms that show territorial behavior: the *ideal despotic distribution*. If individuals are not free to move among all the available habitats but are constrained by the aggressive behavior of other individuals, then subordinate animals can be forced into the more marginal habitats. The ideal despotic distribution predicts that the density will not be lower in the marginal habitats and may in fact be higher if individuals are forced into these habitats. Most importantly, the ideal despotic distribution predicts that fitness will be lower in the poorer habitats.

The key to understanding habitat selection is to determine the rules by which individual animals decide which habitat to utilize. For birds, two variations of the ideal free distribution have been postulated. The first model is similar to the ideal despotic

FIGURE 5.9

Breeding success of yellow-legged gulls (Larus cachinnans) *in two habitats on the Medes Islands, Spain. The shrub habitat is preferred and settled first, but in spite of this habitat preference the breeding success was equal in these two habitats in each of two years. This result is one prediction of the ideal free distribution model of habitat selection shown in Figure 5.8. (Data from Bosch and Sol 1998, Table 4; photo courtesy of J. A. Spendelow, U.S. Fish and Wildlife Service)*

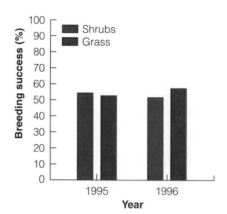

distribution in postulating that individuals preempt breeding places and displace the next arriving individual into lower-quality sites (Pulliam and Danielson 1991). Habitat selection is thus like a game of musical chairs in which the high-quality chairs are chosen first. Like the ideal despotic model, breeding success in marginal habitats will be lower than in the best habitats. The second model postulates that in fact individuals are attracted to conspecifics, and that individuals come to occupied sites rather than avoid them. Pöysä et al. (1998) tested these models in mallards by introducing tame flightless mallards into lakes before wild birds returned in the spring. They found that mallards did not avoid occupied lakes but in fact seemed to be attracted to conspecifics. At least for mallards, habitat selection involves an attraction to places already occupied. Individuals may benefit by selecting occupied sites because they know the site is suitable and because of potential benefits in protection from predators (Stamps 1991).

The proximate mechanisms by which habitats are selected are thus underlain by evolutionary expectations in fitness. We do not know how rapidly organisms can change the genetic and behavioral machinery that results in habitat selection.

Problems can arise whenever habitats change, and this has been a source of difficulty for many organisms since humans have modified the face of the Earth. People provide many new habitats and destroy others. Some species, but not all, have responded by colonizing *Homo sapiens*'s habitats. Other natural events, such as ice ages, cause slower habitat changes. Organisms with carefully fixed, genetically programmed habitat selection may require considerable time to evolve the necessary machinery to select a new habitat that is suitable for them. Adaptation can never be exact and instantaneous, and we must be careful not to expect perfection in organisms. We may judge the mosquito *Anopheles culicifacies* deficient for not laying eggs in all suitable habitats, but this may only reflect the fact that rice fields are recent habitats in evolutionary time. Species evolve under a given set of environmental conditions, and populations adapt to their own particular environment. No species can truly be a "jack of all trades." Adapting to one type of habitat may make it impossible to live in another.

Summary

If individuals are introduced into areas not occupied by their species and are able to survive, grow, and reproduce in their new habitat, the distribution of the species must be restricted either by lack of dispersal (discussed in Chapter 4) or by behavioral reactions (discussed in this chapter). Habitat selection of ovipositing insects provides some good examples in which a species can survive in a wider range of habitats than it usually occupies. Birds have been studied extensively from this point of view, and the rules by which individuals decide to select habitats are critical for many issues in conservation and management. Habitat selection operates on many spatial scales, and most research has been done on the local scale. Experimental manipulations of habitats by adding shelters have highlighted the constraints on habitat choice dictated by risk of predation.

Behavioral limitations on distribution are usually subtle and may be the most difficult to study. At present, few animal distributions are restricted on the landscape scale by behavioral reactions, but at the microhabitat scale habitat selection may be a critical limitation to distributions. Because humans have altered so many of the Earth's communities, the evolutionary history of a species may not adapt it to selecting the best habitats in human landscapes, or alternatively it may permit it to select modified habitats and become a pest.

Habitat selection has been studied most in vertebrates and economically important insects. Plants do not have this mechanism at their disposal, and plant distributions must be limited in other ways.

In a predictable environment, habitat selection may be very exact. When habitats change, some species are not able to adapt quickly and therefore inhabit only a portion of their potential habitat range.

Key Concepts

1. Animal species may be limited in their geographic distribution by selecting a range of habitats that is more restricted than the range they could occupy successfully.

2. Habitat selection operates through a series of behavioral decisions at several spatial scales and is therefore difficult to study.

3. Few species of animals are believed to be limited in their geographic range by habitat selection.

4. Habitat selection can evolve because organisms using some habitats leave more descendents than organisms using other habitats.

5. Human changes to landscapes and environments may make some habitat selection maladaptive.

Selected References

Clark, R. G., and D. Shutler. 1999. Avian habitat selection: Pattern from process in nest-site use by ducks? *Ecology* 80:272–287.

Cody, M .L. 1985. An introduction to habitat selection in birds. In *Habitat Selection in Birds*, M .L. Cody, ed. pp. 3–56. Academic Press, New York.

Danchin, E. and R. H. Wagner. 1997. The evolution of coloniality: The emergence of new perspectives. *Trends in Ecology and Evolution* 12:342–347.

Downes, S. and R. Shine. 1998. Heat, safety or solitude? Using habitat selection experiments to identify a lizard's priorities. *Animal Behaviour* 55:1387–1396.

Jaenike, J. and R. D. Holt. 1991. Genetic variation for habitat preference: Evidence and explanations. *American Naturalist* 137(suppl.): S67–S90.

Stamps, J. A. 1991. The effect of conspecifics on habitat selection in territorial species. *Behavioral Ecology and Sociobiology* 28:29–36.

Sutherland, W. J. 1996. *From Individual Behaviour to Population Ecology*. Oxford University Press, New York. 213 pp.

Tregenza, T. 1995. Building on the ideal free distribution. *Advances in Ecological Research* 26:253–307.

Ward, J. P. J., R. J. Gutierrez, and B. R. Noon. 1998. Habitat selection by northern spotted owls: The consequences of prey selection and distribution. *Condor* 100:79–92.

Wiens, J. A., R .L. Schooley, and R. D. Weeks, Jr. 1997. Patchy landscapes and animal movements: Do beetles percolate? *Oikos* 78:257–264.

Questions and Problems

5.1 Morse (1980, p. 111) states:

> This is a very good time to study the mechanisms of habitat selection, in that many natural habitats are undergoing changes as a result of man's activities. These changes provide animals with potential opportunities and challenges, ones that in many instances they may never have previously experienced.

Discuss.

5.2 After a theoretical analysis of the evolution of habitat selection, Holt (1985, p. 200) concluded:

> Natural selection, by favoring habitat selection by individuals, may thus have the long term evolutionary consequence of restricting the geographical range eventually occupied by a species.

How might this occur?

5.3 Discuss the evolution of habitat selection and try to relate your conclusions to some of the examples presented in this chapter. How can natural selection maintain the particular ovipositing dance of *Anopheles culicifacies*, for example, if it results in suitable habitats being left unoccupied? Does natural selection always favor the broadest possible habitat range for a species?

5.4 Review the factors affecting the settlement of species with open populations, such as many marine organisms with a planktonic larval stage (see references in Gutierrez 1998, Sale 1991). Discuss the ecological events needed for successful settlement of coral reef fishes. If you were studying a coral reef fish, what observations and predictions would you make for a coral reef fish if you hypothesized that its distribution is limited by habitat selection? What alternative hypotheses would you consider?

5.5 The aphid *Brevicoryne brassicae* avoids red cabbage varieties when selecting a host plant (Emden et al. 1969, p. 205). How would you determine whether this habitat-selection behavior was adaptive for the aphid?

5.6 Is it possible for a transplant experiment to be successful and yet lead to the conclusion that neither dispersal nor habitat selection is responsible for range limitation? Discuss the transplant experiment of Dayton et al. (1982) on the antarctic acorn barnacle, and comment on the author's conclusions.

5.7 How are the predictions of the model given in Figure 5.8 affected if the habitat relationships are not straight lines but instead are curves? Describe a situation in which these lines in Figure 5.8 might cross. See Rosenzweig (1985, p. 523) for a discussion.

5.8 Habitat selection has rarely been considered in plants. Why might this be the case? Read Bazzaz's (1991) review of habitat selection in plants and discuss his use of these concepts for plants.

5.9 Discuss three reasons why directional selection for nest sites in blue-winged teal (shown in Figure 5.7) might not result in long-term evolutionary changes in habitat selection by females in this species.

Overview Question

Suppose you discovered a new species of lizard living in a desert area. Discuss how you would go about describing the habitat of this species. What would you measure, and how could you decide when you had a complete understanding of its habitat selection?

CHAPTER 6

Factors That Limit Distributions:
Interrelations with Other Species

UP TO THIS POINT WE HAVE DISCUSSED cases in which an organism could actually live in places that it did not occupy. From now on we will be considering cases in which the organism cannot complete its full life cycle if transplanted to areas it did not originally occupy. The reason for this inability to survive and reproduce could be negative interactions with other organisms, including predation, disease, and competition, or positive interactions such as mutualism or symbiosis. First we examine predation, one of the clearest interactions between species because predators eat their prey.

Predation

We begin our discussion of predation by considering the role of predators in affecting the geographical distributions of their prey. Note that we define predation very broadly. Typical predators like lions kill their prey. Herbivores prey on their food plants and usually do not kill the plants. Parasites live on or inside other organisms and again do not usually kill them.

Restriction of Prey by Predators

The local distribution of some species seems to be limited by predation. Work on intertidal invertebrates has provided some graphic examples of the influence of predation on distribution. Kitching and Ebling

(1967) have summarized a series of studies at Lough Ine, an arm of the sea on the south coast of Ireland, and their studies are an excellent example of ecological work on distribution. Lough Ine is connected to the Atlantic Ocean by a narrow channel called the Rapids, through which the tide ebbs and flows approximately twice a day (Figure 6.1).

The common mussel (*Mytilus edulis*) is a widespread species on exposed rocky coasts in southern Ireland and throughout the world. Small mussels (less than 25 mm long) are abundant on the exposed rocky Atlantic coast but within Lough Ine and the more protected parts of the coast, this mussel is rare or absent (see Figure 6. 1). The only abundant populations are in the northern end of the lough, but these animals are typically very large (30–70 mm long).

Kitching and his coworkers transferred pieces of rock with *Mytilus* attached from various parts of the lough to others. Figure 6.2 presents some typical results. Small *Mytilus* disappeared quickly from all stations to which they had been transferred within the lough, the Rapids, and the protected bays; they survived only on the open coast. The rapid loss, shown in Figure 6.2a, suggested that predators were responsible. Large mussels that were transplanted around the lough also disappeared rapidly from most stations (see Figure 6.2b), except places where they occurred naturally (C in Figure 6.1). Continuous observations on the transplanted mussels showed that three species of crabs and one starfish were the principal agents of mortality. By placing mussels of various sizes and crabs

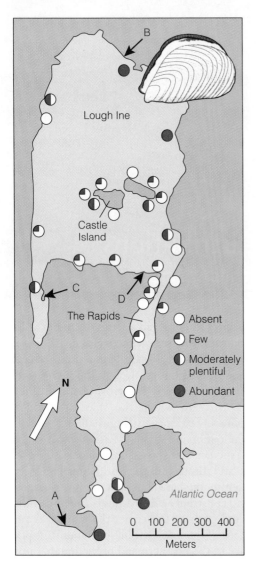

FIGURE 6.1
Map of Lough Ine, the Rapids, and the adjacent Atlantic coast of Ireland, showing the distribution of the common mussel Mytilus edulis. *For explanations of points A–D, see Figure 6.2. (After Kitching and Ebling 1967.)*

restricted in their distribution by wave action, strong currents, and low salinity. Crabs also require an escape habitat in which they spend the day.

The distribution of this mussel in the intertidal zone at Lough Ine is thus controlled as follows: on the open coast, heavy wave action restricts the size of mussels and prevents predators from eliminating small mussels. In sheltered waters, predators eliminate most of the small mussels, and *Mytilus* survive only in areas safe from predators (such as steep rock faces), where they may grow to large sizes.

A common observation in coastal subtidal areas is that the presence of large algae such as kelp is inversely related to the presence of sea urchins. Where sea urchins are common, kelp cannot get established because of urchin grazing, so the local distribution of kelps can be limited by sea urchins. This kind of plant-herbivore interaction can be tested by removal and addition experiments. If sea urchins are removed from an area, kelp should invade (if this hypothesis is correct). Conversely, if urchins are added to an area of kelp, the kelp should be eliminated. Several experiments of this type have been conducted (Hawkins and Hartnoll 1983) .

One complication is that often several other herbivores in addition to sea urchins are present, and thus careful manipulative experiments are needed to explain the absence of algae.

On the eastern coast of Australia the subtidal area has large areas where leafy algae are absent and invertebrate grazers are common. Fletcher (1987) removed sea urchins, limpets, and gastropods in controlled experiments over three years near Sydney, Australia. In one set of experiments, when he removed both the sea urchin *Centrostephanus rodgersii* and four species of limpets, algae immediately began to colonize the cleared areas as follows (data are percent of leafy algal cover in plots):

Time	Unmanipulated sites	Only limpets removed	Only sea urchins removed	Both sea urchins and limpets removed
August 1982	1	2	6	5
February 1983	2	3	25	60
August 1983	1	1.5	60	93
February 1984	0.5	1	85	95

of the three species together in wire cages, Kitching and Ebling were able to show that one of the smaller species of crabs could not kill large *Mytilus* but that the other crabs could open all sizes of mussels. The areas of the lough where large *Mytilus* survive have few large crabs, and where the large crabs are common, *Mytilus* are scarce or absent. Predatory crabs are probably

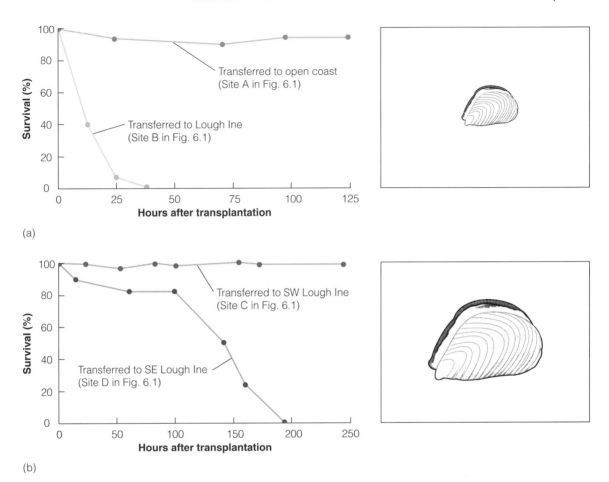

(a)

(b)

FIGURE 6.2

Percentage survival of mussels in transplant experiments in and near Lough Ine. Small mussels (a) disappear rapidly when transplanted anywhere in Lough Ine but do not disappear if transplanted to the open coast (A in Figure 6.1). Large mussels (b) disappear if transplanted to some parts of Lough Ine such as the southeastern part (D in Figure 6.1) but do not disappear if transplanted to other parts of the Lough, such as the southwestern part (C in Figure 6.1), where they occur naturally. (After Kitching and Ebling 1967.)

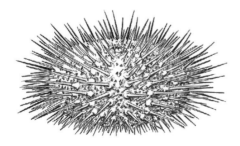

Sea urchin

Adjacent control areas containing both sea urchins and limpets continued to have almost no visible algal growth. Sea urchins were clearly the dominant influence on algal abundance, since limpet removal by itself had almost no effect on algal cover. Fletcher (1987) concluded that leafy algae are absent from these subtidal areas because of grazing by sea urchins.

These kinds of experiments illustrate four criteria that must be fulfilled before one can conclude that

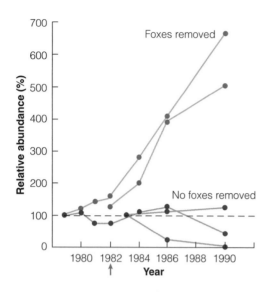

FIGURE 6.3
Changes in rock wallaby (Petrogale lateralis) *abundance at two sites with red fox control and three sites without red fox control. Fox control on the experimental sites began in 1982 (red arrow). The rock wallaby population on one of the three unmanipulated sites went extinct in 1990. (After Kinnear et al. 1998.)*

a predator restricts the distribution of its prey (Kitching and Ebling 1967):

- Prey individuals will survive when transplanted to a site where they do not normally occur if they are protected from predators.
- The distributions of prey organisms and suspected predator(s) are inversely correlated.
- The suspected predator is able to kill the prey, both in the field and in the laboratory.
- The suspected predator can be shown to be responsible for the destruction of the prey in transplantation experiments.

In Australia, several species of small kangaroos have been driven to near extinction by predation from the introduced red fox. Rock wallabies are small kangaroos that live in rocky hill habitats throughout Australia. Their numbers have been declining for nearly a century, and numerous colonies have become extinct. Kinnear et al. (1998) tested the hypothesis that red fox predation was sufficient to limit the population size and distribution of rock wallabies in Western Australia. By poisoning red foxes around two colonies, they showed

that populations of wallabies could recover dramatically in the absence of foxes (Figure 6.3). Red foxes not only kill wallabies directly but also reduce the area available for safe feeding to sites near rocky escape habitat. Without foxes in the area, wallabies ranged farther from the rocky areas to feed. This is a good example in which native species can suffer range reduction and even extinction because of introduced predators.

The burrowing rat kangaroo (*Bettongia lesueur*) provides another example from Australia of a distribution limited by predation (Figure 6.4). This species was driven extinct over its entire range on the Australian mainland, where it was originally so common as to be a pest, and until recently it lived on only three islands off the Western Australia coast where there are no red foxes (Short and Smith 1994). In 1992 it was reintroduced into a peninsula in Shark Bay, Western Australia, which was isolated with an electric fence to prevent fox immigration and then poisoned to eliminate foxes. The reintroduction was successful, in contrast to many reintroduction programs that have failed because of heavy predation (Sinclair et al. 1998).

In the cases just discussed, the predator is believed to restrict the distribution of its prey; con-

FIGURE 6.4
Geographic distribution of the burrowing bettong (rat-kangaroo) Bettongia lesueur *in Australia. This small bettong was extinct on the mainland until 1992, when it was reintroduced successfully to Western Australia. It had previously survived only on three islands (arrows) off the coast that lacked red foxes. (Modified from Ovington 1978.)*

sequently, the reasons for the predator's distributional limits must be sought elsewhere. In these situations, the predator may feed on a variety of prey species, and each prey species may in turn be fed upon by many predatory species. The relationship may also operate in the other direction, and the prey may restrict the distribution of its predator.

Restriction of Predators by Prey

The "prey" may be a food plant and the "predator" a herbivore; alternatively, the prey may be a herbivore and the predator a carnivore. But if the prey is to restrict the predator's range, the predator must be

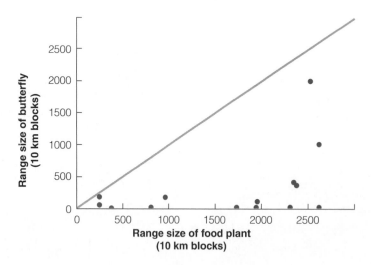

FIGURE 6.5
Relationship between the range sizes of 14 species of monophagous butterflies in Britain and the range sizes of their host plants. If butterflies were limited in their distribution by host plant distribution, the points should fall along the blue line on the diagonal. In most cases the range of these butterflies is not limited by the geographical distribution of their food plants. (Data from Quinn et al. 1998.)

very specialized and feed on only one or two species of prey. Such a predator is called a specialist or a monophagous predator. Many insect predators are specialists, but most vertebrate predators are not.

One example of a species whose distribution is limited by its food source is *Drosophila pachea*, a rare fruit fly, that breeds only in the stems of senita cactus (*Lophocereus schottii*) throughout the Sonoran Desert of the southwestern United States and northern Mexico. This fly will not breed on the standard laboratory medium for *Drosophila* unless the medium is supplemented by a cube of fresh or autoclaved senita cactus. Conversely, medium supplemented by the cactus is toxic in varying degrees to the adults and larvae of other local species of *Drosophila*. The initial hypothesis suggested by these observations is that the senita cactus contains a factor that is necessary for the development of *D. pachea* and a factor that is toxic to other species.

Heed and Kircher (1965) demonstrated that a unique sterol, schottenol (Δ^7-stigmasten-3β-ol), is the factor required by *D. pachea* for growth and reproduction. Every insect species that has been investigated requires a sterol in its diet. *Drosophila pachea* requires Δ^7-sterols, which are intermediates in the phytosterol biosynthetic pathway (Fogleman et al. 1982) and accumulate in senita cactus. The function of sterols in the diet is probably twofold: They are precursors for the

molting hormone ecdysone, and they affect female fertility in some way.

Senita cacti are rich in alkaloids, and 3%–15% of the dry weight of the cactus consists of alkaloids. These alkaloids, particularly pilocereine, are toxic to eight species of *Drosophila* that live in the Sonoran Desert. The process of adaptation of *D. pachea* to this habitat has been possible only because of its tolerance of these potentially toxic alkaloids (Heed 1978).

The leaf-feeding beetle *Chrysolina quadrigemina* was introduced into the United States to control the Klamath weed (*Hypericum perforatum*). In any introduction of this type to control weeds, it is important that the insect should not eat crop plants. Holloway (1964) discusses this problem for the *Chrysolina* beetle. The feeding habits of the beetle are very specific. Adult and larval beetles will not feed and will die when confined with plants other than those of the genus *Hypericum*. Adult beetles display an obligatory feeding response to the chemical hypericin in the foliage (Schoeps et al. 1996). Beetles often refuse to stand on the leaf surfaces of plants that have leaves of different surface texture than that of *Hypericum*. The adults explore the leaf edges with their antennae, and if they encounter a serrated (toothed) leaf edge rather than a smooth one, they drop off the plant. The life history of this beetle is synchronized with that of its host plant and includes a summer (dry season) aestivation of the adults and subsequent response to fall rains. Thus the feeding habits, behavior, and life history of this leaf-eating beetle restrict it to a single host plant, and the range of distribution of *Chrysolina quadrigemina* is thereby restricted.

Insects that feed on only one host plant (monophagous insects) could be limited in their distribution by the host plant. But for the groups studied to date there is no indication that the ranges of food plants and their monophagous insect herbivores coincide (Quinn et al. 1998). Figure 6.5 shows that for butterflies in Britain, no correspondence exists between food plant distributions and butterfly distributions. Even for widespread species of butterflies, the host plant occurs in many areas in which the butterfly does not. Something else must limit butterfly distributions.

Predation is a major process affecting the distribution and the abundance of many organisms. It will be discussed in greater detail in Chapter 13 and in Chapters 24 and 25.

Disease and Parasitism

In addition to predators, enemies include parasites and organisms that cause diseases. Pathogens may eliminate species from areas and thereby restrict geographical distributions. We have already discussed one example of this—chestnut blight, which eliminated the American chestnut tree from eastern North America (see pp. 44–45). Another example involves the native bird fauna of Hawaii.

A large fraction of the endemic bird species of the Hawaiian Islands has become extinct in historical times, and one possible reason for these losses is introduced diseases. Warner (1968) postulated that both avian malaria and avian pox were instrumental in causing extinctions in the Hawaiian Islands. The idea that diseases might be involved arose from the observation that native birds in Hawaii are relatively common only at elevations above 1500 m (Figure 6.6a) while introduced birds occupy the lowland areas. The main malarial vector, the mosquito *Culex*

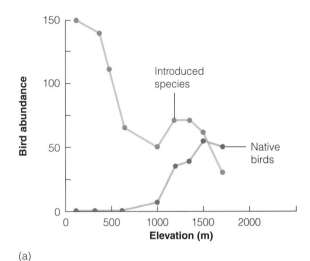

(a)

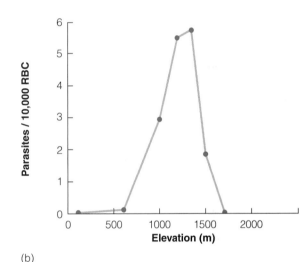

(b)

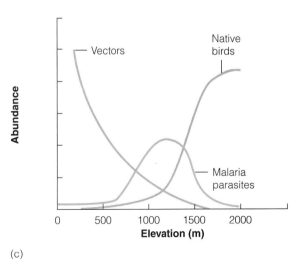

(c)

FIGURE 6.6
(a) Abundance of introduced and native birds in 1978–79 at 16 sampling stations on Mauna Loa, Hawaii. (b) Avian malaria parasite loads (parasites per 10,000 red blood cells) along this same altitudinal gradient on Mauna Loa. (c) A generalized model of native bird abundance, malaria parasite prevalence, and mosquito vector levels on Mauna Loa. (From Van Riper et al. 1986.)

quinquefasciatus, is conversely most common in the lowland areas (Fig. 6.6c). Because the native birds are much more susceptible to malaria than the introduced species, the malaria parasite is most common at intermediate elevations (Fig 6.6b), where the geographical distributions of vectors and hosts overlap (Van Riper et al. 1986).

The extinction of the native Hawaiian bird fauna occurred in two pulses. Before 1900 many of the low-elevation bird species disappeared coincident with extensive habitat clearing for agriculture and the introduction of rats, cats, and pigs. It is possible that other introduced diseases such as avian pox played a role in the early extinctions, but avian malaria did not because it was uncommon before 1900 (Moulton and Pimm 1986). The second period of extinction in Hawaiian birds began in the early 1900s and was most likely the result of avian malaria. Birds that went extinct at this time lived in the mid-elevation forests where malaria parasites are most prevalent (see Figure 6.6b and c). At the same time the geographical distribution of many native birds was also reduced as they retreated to forests at the highest elevations forests where mosquitoes are rare.

Diseases and parasites have always been a major factor in the ecology of humans (Desowitz 1991, Diamond 1997). Their role in the geographical ecology of plants and animals has been studied far less than their potential importance would warrant. We will discuss the impact of disease and parasitism on populations in Chapter 15.

Allelopathy

Some organisms, plants in particular, may be limited in local distribution by poisons or antibiotics, also called *allelopathic* agents. The action of penicillin among microorganisms is a classical case (Brock and Madigan 1988). Interest in toxic secretions of plants arose from a consideration of *soil sickness*. It was observed in the nineteenth century that, as one piece of ground was continuously planted in one crop, the yields decreased and could not be improved by additional fertilizer. As early as 1832, DeCandolle suggested that the deleterious effects of continuous one-crop agriculture might be due to toxic secretions from roots. Several cases were also observed of detrimental effects of plants growing with one another—for example, grass and apple trees (Pickering 1917).

Experiments of the general type shown in Figure 6.7 were performed. Apple seedlings were grown with three different sources of water: tap water, water that had passed through grass growing in soil, and water that had passed through soil only. The growth of the young apple trees was apparently inhibited by something produced by the grass and carried by the water.

In the early 1900s several agronomists commented on the effect of black walnut trees (*Juglans nigra*) on nearby grass and alfalfa plants. Massey (1925) observed that the zone of dead alfalfa around a walnut tree extended over an area two to three times greater than that covered by the tree's canopy and suggested that this zone was determined by the outer limits of the tree's roots. The walnut roots were suspected of secreting a toxin to which some crop plants—for example, alfalfa (lucerne) and tomatoes—were susceptible; other plants—for example, corn (maize) and beets—showed no ill effects. Schneiderhan (1927) showed that black walnut trees injured and killed apple trees up to 25 meters away. The average maximum radius of the toxic zone was about 15 meters from the walnut trunk. In every case, the toxic zone was greater than the area covered by the tree's canopy, but larger walnut trees did not necessarily have much larger toxic zones.

Davis (1928) extracted a crystalline substance called *juglone* (5-hydroxy-α-naphthaquinone) from the roots and hulls of the black walnut and showed that this chemical would kill tomato and alfalfa plants. Ponder (1987) found evidence of antagonism between walnut and some other timber tree species such as European alder, and he emphasized the selective effects of the walnut toxin: Some species, such as alfalfa, are killed; others, such as Kentucky bluegrass, become more abundant than usual near walnut trees. Not all walnut species secrete toxic chemicals. The closely related English walnut (*Juglans regia*) and the California walnuts *Juglans hindsii* and *Juglans californica* apparently do not secrete growth inhibitors.

Agriculturalists have recognized the action of *smother crops* as weed suppressors. These smother crops include barley, rye, sorghum, millet, sweet clover, alfalfa, soybeans, and sunflowers. Their inhibition of weed growth was assumed to be due to competition for water, light, or nutrients. Barley, for example, is rated as a good smother crop and has extensive root growth. Overland (1966) showed that barley (*Hordeum vulgare*) inhibited the germination and growth of several weeds, even in the absence of competition for nutrients or

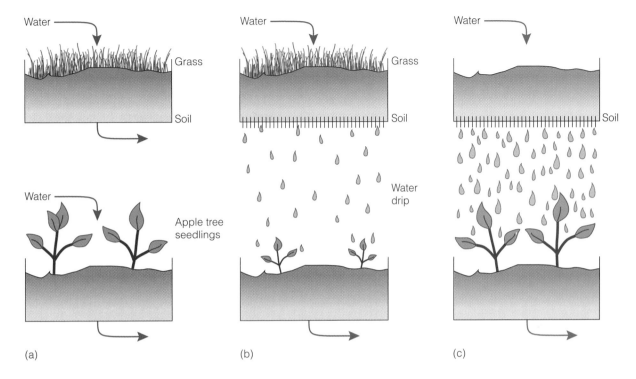

FIGURE 6.7
Experiments that demonstrated the detrimental effects of grass on apple tree seedlings. Grass and apple seedlings are grown in separate flats in a greenhouse. Water is provided either (a) independently from the tap to both grass and seedlings, (b) to grass growing in soil so that the water drips through onto the seedlings, or (c) to soil only so that the water drips onto the seedlings. Apple tree seedlings do not grow properly and often die when the water has passed through grass first (b).

water. Growth experiments with barley and chickweed (*Stellaria media*) gave the following results:

Average Dry Weight per Plant (g) After 2 Months of Growth

Barley	Chickweed	No. of Chickweed Flowers
4.15	3.20	100+
Controls (each grown alone, competition absent)		
4.85	1.43	10
1:1 barley/chickweed mixture (competition present)		

Extracts of living barley roots were more inhibitory than extracts of dead roots. The active inhibitory agent was found to be an alkaloid, but its specific chemical nature is not known. Thus the adverse effect of barley is partly due to the secretion by its roots of chemicals that reduce growth of weeds and germination of weed seeds.

There is great interest among agricultural scientists in the potential uses of allelopathy for weed control in crops (Seigler 1996, Weston 1996). Since allelopathic chemicals often are highly specific, they could be used in agricultural systems in much the same manner as synthetic herbicides. This possibility is premised on our understanding the physiological mechanisms by which allelopathic chemicals operate to suppress weeds.

Whether or not allelopathy is a significant factor affecting the local distribution of plants in natural vegetation is controversial. Many plant ecologists accept the laboratory data on allelopathy but question whether or not it is effective in natural plant communities (Weidenhamer 1996).

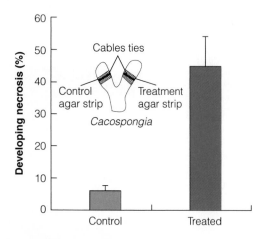

FIGURE 6.8
Percentage of areas on Cacospongia *branches that developed necrosis after one week of field exposure to agar strips containing 7-deacetoxyolepupuane. Agar strips were attached to the sponge branches as illustrated. Error bars represent 1 standard error. (From Thacker et al. 1998)*

Tropical reefs are rich with plant and animal species, and space is a limiting resource for all the attached corals, algae, and sessile invertebrates such as sponges. Competition for space may be mediated by chemicals that produce allelopathic effects (Hay 1996). On tropical reefs at Guam the sponge *Dysidea* sp. overgrows the sponge *Cacospongia* sp. and causes necrosis (Thacker et al. 1998). *Dysidea* produces a sesquiterpene 7-deacetoxyolepupuane that when isolated and placed on agar strips causes necrosis in *Cacospongia* (Figure 6.8). In this way *Dysidea* can displace *Cacospongia* in competition for space on the reef. *Dysidea* does not produce this chemical only in the presence of *Cacospongia*, but the sesquiterpene seems to be a metabolite with multiple functions, including defense against predators. *Cacospongia* itself produces terpenoids that could be used as allelopathic chemicals, but none of these chemicals seems to stop overgrowth by *Dysidea*, so the effects of chemical warfare are one-sided in this particular system.

The chemistry and the ecology of allelopathic interactions have been difficult to unravel in marine systems and in terrestrial systems alike (Gopal and Goel 1993, Inderjit and Dakshini 1994). The complexity of the chemistry of allelopathic interactions has limited precise experiments to the simplified conditions of the laboratory. The challenge of taking chemical ecology into the field is an important one for the next decade.

Competition

The presence of other organisms may limit the distribution of some species through competition. Allelopathy is one specific type of competition for living space. But competition can occur between any two species that use the same types of resources and live in the same sorts of places. Note that two species do not need to be closely related to be involved in competition. For example, birds, rodents, and ants may compete for seeds in desert environments, and herbs and shrubs may compete for water in dry chaparral stands. Competition among animals is often over food. Plants can compete for light, water, nutrients, or even pollinators.

Competition is an important process affecting the distribution and abundance of plants and animals, and it will be discussed in more detail in Chapter 12 and in Chapters 23 and 24.

How can we determine whether competition could be restricting geographical distributions? One indication of competition may be the observation that when species *A* is absent, species *B* lives in a wider range of habitats. In extreme cases a habitat will contain only species *A* or species *B* and never both together. The principal difficulty in understanding these situations is that competition is only one of several hypotheses that can account for the observed distributions.

The geographical distribution of closely related species has provided many examples of distributional patterns that could result from competition. Diamond (1975) described several examples of *checkerboard distributions*. Two ecologically similar species may have mutually exclusive but interdigitating distributions in an island archipelago, each island supporting only one species. Figure 6.9 gives an example of a checkerboard distribution for two fruit pigeons from the western Pacific. The identity of the successful colonist on an island may have been determined on a first-come-first-served basis or on the basis of slight competitive advantages between the two species.

A particularly well-studied case of competition occurs between two terrestrial salamanders of the eastern United States, *Plethodon jordani* and *P. glutinosus* (Hairston 1980). These species have altitudinal distributions that overlap very little (only 70–120 m on any one transect up the Black Mountains in North Car-

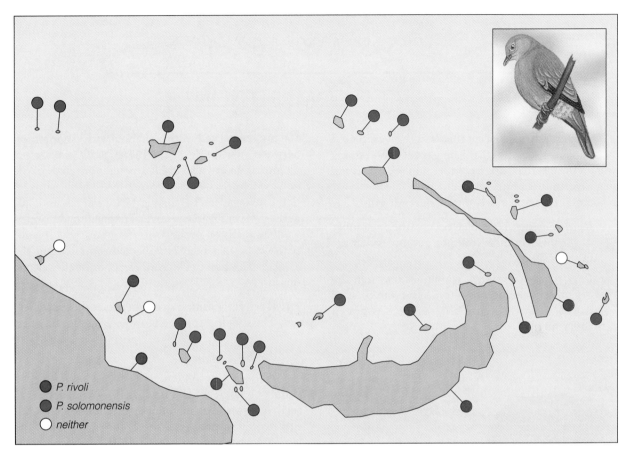

FIGURE 6.9
Checkerboard distribution of two closely related species of fruit pigeons in the Bismarck Archipelago of the western Pacific Ocean off New Guinea. Note that either Ptilinopus pivoli *(R) or* P. solomonensis *(S) occurs on most of these islands, and only one or two islands have both. (From Diamond 1975.)*

olina). Intensive competition between these two species was investigated by removing one salamander species from plots in the overlap zone for five years, and observing an increase in the abundance of the other species. Using proper long-term experiments like this, Hairston (1980) could demonstrate the role of competition in limiting altitudinal ranges of these salamanders. These salamanders compete both for habitats with moist soil and for food, and the superior competitor lives in the more favorable low-altitude areas.

When two species compete for resources, one species will always be better than the other in gathering or utilizing the resource that is scarce. In the long run, one species must lose out and disappear, unless it evolves some adaptation to escape from competition.

A species can adopt one of two general evolutionary strategies: (1) avoid the superior competitor by selecting a different part of the habitat or (2) avoid the superior competitor by making a change in diet. Let us look at an example in which possible competition is avoided by a diet shift.

Crossbills are finches that have curved crossed tips on the mandibles (see Figure 6.10c). Crossbills extract seeds from closed conifer cones by lateral movements of the lower jaw, and the jaw muscles are asymmetrically developed to provide the necessary leverage. Three species of crossbills live in Eurasia, and they are adapted for eating different foods (Newton 1972). The smallest crossbill is the white-winged crossbill, which has a small bill and feeds

ESSAY 6.1

WHAT IS COMPETITION?

Competition is a concept that is so familiar to us in capitalist societies that it might seem odd to ask what it means. In ecology, competition is defined as a negative interaction between two species over resources. It can take two quite different forms:

- *Resource competition*, which occurs when a number of organisms utilize common resources that are in short supply

- *Interference competition*, which occurs when the organisms seeking a resource harm one another in the process, even if the resource is not in short supply

We are concerned in this chapter with competition between two species, called *interspecific competition*. Competition can also occur among individuals of the same species, as we see every day in the business pages of the newspapers. In Chapter 12 we will discuss competition from the perspective of population dynamics.

Competition occurs over resources, and, if competition is suspected as a mechanism affecting the local distribution of two species, we must answer two questions:

1. Does competition occur between these species? The simplest approach to answering this question

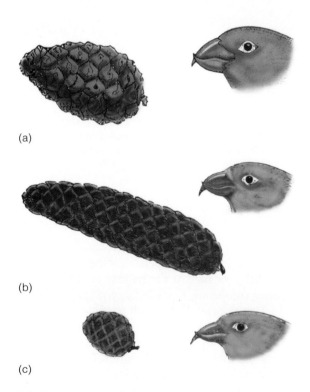

(a)

(b)

(c)

FIGURE 6.10
Heads of the three European crossbill species and the main conifer cones each species feeds on: (a) parrot crossbill and Scotch pine cone, (b) common crossbill and spruce cone, (c) white-winged crossbill and larch cone. (After Newton 1972.)

mainly on larch seeds (Figure 6.10). Larch cones are relatively soft. The medium-sized common crossbill eats mainly spruce seeds (see Figure 6.10b), and the larger parrot crossbill feeds on the hard cones of Scotch pine (see Figure 6.10a). These dietary differences are not necessarily preserved when the species live in isolation. Thus the common crossbill has evolved a Scottish subspecies, that has a large bill and feeds on pine cones, and an Asiatic subspecies, that has a small bill and feeds on larch seeds. The white-winged crossbill has an isolated subspecies on Hispaniola in the West Indies that feeds on pine seeds and has a large beak. The bill adaptations of crossbills can thus be interpreted as devices for minimizing dietary overlap in regions where all three possible competitors live.

Four types of red crossbills (*Loxia curvirostra*) live in the Pacific Northwest of North America, and like their Eurasian counterparts they all represent adaptive peaks concentrating on four conifers with different seed and cone sizes (western hemlock, Douglas fir, ponderosa pine, and lodgepole pine). Benkman (1993) showed in a series of laboratory studies of foraging efficiency that the best bill size for feeding on one of these conifers was only one-half as efficient for feeding on the other conifer seeds. He postulated that disruptive selection maintained these four types of red crossbills in western North America.

is to do a removal experiment. If we remove the dominant competitor, the other species should expand its local geographic range.

2. What are the resources for which competition occurs? We tend to assume that water or nutrients are the limiting resources for plants, and food supplies the limiting resources for animals, but as always in ecology we should assume nothing without an experimental test.

Experimental studies of competition are always asking about the immediate interactions between species in ecological time, and that is our concern in this chapter. In evolutionary time we may see traces of the "ghost of competition" in adaptations that exist now because of intense competition between two species in the past and subsequent evolutionary divergence.

Summary

Many animals and plants are limited in their local distribution by the presence of other organisms—their food plants, predators, diseases, and competitors. Experimental transfers of organisms can test for these factors, and cages or other protective devices can be used to identify the critical interactions. Much of the evidence currently available is only observational and not experimental.

Predators can affect the local distribution of their prey, and studies on intertidal organisms have illustrated this influence. The converse can also occur, in which the prey's distribution determines the distribution of its predators, but this interaction does not seem to be common. In some cases an animal depends on a single food source and may have its distribution limited by the distribution of the food. Few such cases have been described, but this limitation might be common for insect parasites.

Diseases and parasites may restrict geographical distributions, but few such cases have been studied in natural systems. They may play a larger role than we currently suspect in species-rich tropical communities. Much more work needs to be done on the role of disease in limiting geographic distributions.

Some organisms poison the environment for other species, and these chemical poisons, or allelopathic agents, may affect local distributions. The action of penicillin is a classic example. Chemical interactions have been described in a variety of crop plants and in shrubs, marine algae, and sponges. Chemical ecology studies are difficult to do, and in most cases chemical interactions are clear in the laboratory but hard to evaluate in the field, where several factors may be affecting local distributions. Allelopathic interactions may have great practical importance in weed control in agriculture.

Competition among organisms for resources may also restrict local distributions. Some species drive others out by aggressive interactions. In other examples the distributions of closely related species do not overlap, suggesting possible competitive interactions. Species may evolve differences in diet or habitat preferences as a result of competitive pressures.

Key Concepts

1. The presence of other organisms—predators, parasites, pathogens, or competitors—may limit the geographic distributions of many species.

2. Predator limitations on prey distributions often operate on a local scale. Prey rarely limit the distributions of their predators.

3. Diseases and parasites may affect geographic distributions, but this impact has been little studied.

4. Some organisms poison the environment for other species, and this form of interaction has been exploited by humans in developing antibiotics.

5. Competition between species for limiting resources may affect local distributions.

Selected References

Bull, C. M. 1991. Ecology of parapatric distributions. *Annual Review of Ecology and Systematics* 22:19–36.

Diamond, J. M. 1997. *Guns, Germs, and Steel: The Fates of Human Societies*. Norton, New York. 480 pp.

Hairston, N. G. 1980. The experimental test of an analysis of field distributions: Competition in terrestrial salamanders. *Ecology* 61:817–826.

Inderjit and K. M. M. Dakshini. 1994. Algal allelopathy. *Botanical Review* 60:182–196.

Jackson, J. B. 1981. Interspecific competition and species distributions: The ghosts of theories and data past. *American Zoologist* 21:889–902.

Kitching, J. A., and F. J. Ebling. 1967. Ecological studies at Lough Ine. *Advances in Ecological Research* 4:197–291.

McClanahan, T. R. 1998. Predation and the distribution and abundance of tropical sea urchin populations. *Journal of Experimental Marine Biology and Ecology* 221:231–255.

Skelly, D. K. 1996. Pond drying, predators, and the distribution of *Pseudacris* tadpoles. *Copeia* 1996 (3) :599–605.

Van Riper III, C., S. G. Van Riper, M. L. Goff, and M. Laird. 1986. The epizootiology and ecological significance of malaria in Hawaiian land birds. *Ecological Monographs* 56:327–344.

Questions and Problems

6.1 In mutualism, species interactions are beneficial for all involved. Do mutualistic interactions ever limit the distribution of a species? See Jordano et al. (1992) and Noble (1991) for references.

6.2 Norwegian lemmings do not live in lowland forests in Scandinavia even though they are regularly seen in these areas when their alpine populations are at high density. Suggest three hypotheses to explain the failure of lemmings to establish permanent populations in lowland forest, and discuss experiments to test these ideas. Oksanen and Oksanen (1992) discuss this question.

6.3 Macan (1974, p. 124), in discussing aquatic organisms, states:

> All species are probably limited to places that offer refuges from predators, unless they live in waters which, because they are temporary or offer extremes of some factor such as salinity, harbour no predators.

Can you suggest any exceptions to this generalization? Does it apply to terrestrial animals? To plants?

6.4 English yew (*Taxus baccata*) is an evergreen tree with an average life span of 500 years (Hulme 1996). The regeneration potential of local sites will determine the future distribution of this tree, and seed predators, seedling herbivores, or suitable microsites for germination and growth are the three factors that may limit yew distribution on a local scale. Discuss what observations could distinguish between biotic limitation and abiotic microsite limitation of yew distributions.

6.5 *Ixodes ricinus*-like ticks (deer ticks) transmit Lyme disease to humans in the northeastern United States. Since about 1950 these ticks have been spreading south and carrying Lyme disease into new regions (Rich et al. 1995). Discuss how you could determine what limits the geographic distribution of these ticks.

6.6 Pulliainen (1972) studied the summer diet of the three European crossbills (see Figure 6.10) in Lapland and found no differences in the food items being eaten by the three species. Reconcile these observations with the interpretation given on pages 81–82.

6.7 One criterion of interspecific competition is that closely related species having mutually exclusive ranges are in competition at their zones of contact. Discuss this criterion in relation to the factors that limit distribution. Bull (1991) reviews this question.

6.8 Discuss the conditions under which you might expect chemical interactions between species to affect the local distributions of marine organisms. Hay (1996) reviews the chemical ecology of marine organisms and provides references.

6.9 At Point Pelee National Park in Ontario, frog surveys showed that the bullfrog (*Rana catesbeiana*) disappeared in 1990, and from 1990 to 1994 the green frog (*Rana clamitans*) has increased in numbers four-fold (Hecnar and M'Closkey 1997). Suggest three possible interpretations for these natural history observations, and indicate how you would test these hypotheses experimentally.

6.10 Cheetahs in the Serengeti National Park in east Africa compete with hyenas and lions for prey such as Thomson's gazelle and wildebeest. Since both hyenas and lions live in social groups while cheetahs are solitary, the cheetah is thought to be subordinate to these two predators and in danger of being killed by both lions and hyenas (Durant 1998). Discuss the possible strategies cheetahs might adopt in these circumstances and how these strategies might affect their local distribution.

6.11 In crossbills the lower mandible may cross from the left or from the right. In all populations studied there is an equal frequency of bills that cross from the two directions (Benkman 1996). Assume that bill crossing direction is controlled genetically. Why might natural selection favor an equal frequency of left- and right-cross types of individuals, and how might this frequency be maintained at 50:50?

Overview Question

Are the principles by which humans limit the distributions of other species the same as those by which predators limit prey distributions? Classify and describe the various ways humans can affect other species' geographic distributions.

CHAPTER 7

Factors That Limit Distributions:
Temperature, Moisture, and Other Physical-Chemical Factors

Temperature and moisture are the two master limiting factors to the distribution of life on Earth, so it is not surprising that an enormous body of literature addresses the effects of temperature and moisture on organisms. Before we analyze the ecological effects of these two factors, as well as other physical-chemical limiting factors, let us look at the global temperature and moisture conditions to which organisms must adapt.

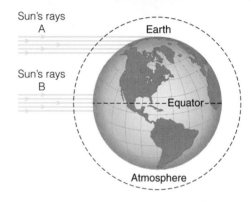

Climatology

The large temperature differentials over the Earth are a reflection of two basic variables: incoming solar radiation and the distributions of land and water. Solar radiation lands obliquely in the higher latitudes (Figure 7.1) and thus delivers less heat energy per unit of surface area. Increased day length in summer partially compensates for the reduced heat input at high latitudes, but total annual insolation is still lower in the polar regions. The amount of heat delivered to the poles is only about 40% of that delivered to the equator.

Land and sea absorb heat differently, and this effect produces more contrasts, even within the same latitude. Land heats quickly but cools rapidly as well, so land-controlled, or *continental*, climates have large daily and seasonal temperature fluctuations. Water heats and cools more slowly because of vertical mixing and a high specific heat. The net result, shown in

FIGURE 7.1
The suns rays strike the Earth at an oblique angle in the polar regions (A) and vertically at the equator (B). Sunlight delivers less energy to the Earth's surface at the poles because its energy is spread over a larger surface area and because it passes through a thicker layer of absorbing, scattering, and reflecting atmosphere.

Figure 7.2, is that annual temperature ranges are greatest over the large continental landmasses.

Water, alone or in conjunction with temperature, is probably the most important physical factor affecting the ecology of terrestrial organisms. Land animals and plants are affected by moisture in a variety of ways. Humidity of the air is important in controlling water loss through the skin and lungs of animals. All animals require some form of water intake (in food or as drink) in order to operate their excretory systems. Plants are affected by the soil water levels as well as

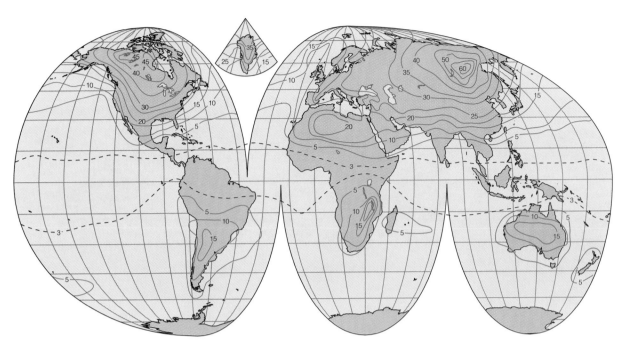

FIGURE 7.2
*Average annual temperature ranges (°C) for the Earth. The annual temperature is
defined as the difference between the average temperatures of the warmest and the coldest
months. Temperature ranges are smallest in the low latitudes and over oceans, and largest
over continents. (After Hidore and Oliver 1993.)*

the humidity of the air around leaf surfaces. Cells are
85–90% water, and without adequate moisture there
can be no life.

Moisture circulates from the ocean and the land
back into clouds only to fall again as rain in a continuous cycle. The global distribution of rainfall resulting from these processes is shown in Figure 7.3. A belt
of high precipitation in equatorial regions is apparent
in the Amazon, West Africa, and Indonesia. Low precipitation around latitude 30°N and S is associated
with the distribution of deserts around the world. The
distribution of continents and oceans also has a strong
effect on the pattern shown in Figure 7.3. More rain
falls over oceans than over land. The average ocean
weather station for the globe records 110 cm of precipitation, compared with 66 cm for the average land
weather station. Finally, mountains and highland
areas intercept more rainfall and also leave a "rain
shadow," or area of reduced precipitation, on their
leeward side.

Water that falls on the land circulates back to the
ocean as runoff or back to the air directly by evapora-

tion or transpiration from plants. Only about 30% of
precipitation is returned via runoff, and hence the
remaining 70% must move directly back into the air
by evaporation and transpiration. The rates of evaporation and transpiration depend primarily on temperature; consequently, a strong interaction between
temperature and moisture affects the water relations
of animals and plants. The absolute amounts of rainfall and evaporation are less important than the relationship between the two variables. Polar areas, for
example, have low precipitation but are not arid
because the amount of evaporation is also low. About
one-third of global land area has a rain deficit (evaporation exceeds precipitation), and about 12% of the
land surface is extremely arid (evaporation at least
twice as great as precipitation).

The vegetation of any site is usually considered a
product of the area's climate. This implies that climatic factors, temperature and moisture primarily, are
the main factors controlling the distribution of vegetation. Geographers have often adopted this viewpoint
and then turned it around to set up a classification of

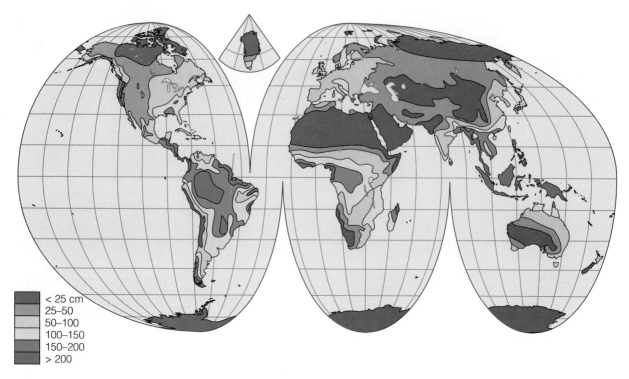

FIGURE 7.3
World distribution of mean annual precipitation. (From Hidore and Oliver 1993.)

climate on the basis of vegetation. Native vegetation is assumed to be a meteorological instrument capable of measuring all the integrated climatic elements.

Some geographers have tried to set climatic boundaries independent of vegetation. Thornthwaite (1948) developed one classification. The basis of his climatic classification is *precipitation*, which is balanced against *potential evapotranspiration*, the amount of water that would be lost from the ground by evaporation and from the vegetation by transpiration if an unlimited supply of water were available. There is no way of measuring potential evapotranspiration directly, and it is normally computed as a function of temperature. Climatic diagrams can then be constructed; two diverse examples are shown in Figure 7.4 to illustrate the climatic regime at a desert station and a temperate deciduous forest station. Vegetation patterns can be described more accurately if we use *actual evapotranspiration*—the evaporative water loss from a site covered by a standard crop, given the precipitation. Major vegetational types such as grassland, temperate deciduous forest, and tundra are closely

associated with certain climatic types defined by the water balance (Stephenson 1990).

Temperature and Moisture as Limiting Factors

Organisms have two options in dealing with the climatic conditions of their habitat: They can simply tolerate the temperature and moisture as they are, or they can escape via some evolutionary adaptation. We begin our consideration of the effects of temperature and moisture by first examining how well organisms tolerate these two factors. Every organism has an upper and a lower lethal temperature, but these parameters are not constants for each species. Organisms can acclimate physiologically to different conditions (Hoar 1983, Chapters 9–10). We illustrated this idea for a fish species in Figure 3.3 on page 35. The resistance of

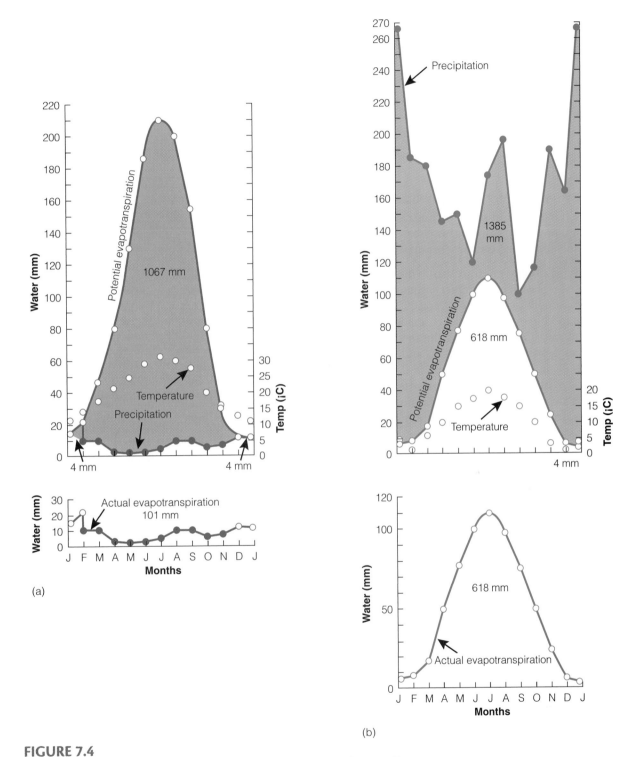

FIGURE 7.4

(a) Climatic diagram for Blythe, California, a station in the hot Sonoran Desert. Available moisture limits plant activity. (b) Climatic diagram for a station in a temperate deciduous forest in Great Smoky Mountain National Park, Tennessee (1160 meters elevation), where available heat limits plant activity. (After Major 1963.)

woody plants to freezing temperatures is another example. Willow twigs (*Salix* spp.) collected in winter can survive freezing at temperatures below –150°C, while the same twigs in summer are killed by –5°C temperatures (Hietala et al. 1998).

Temperature and moisture may act on any stage of the life cycle and can limit the distribution of a species through their effects on one or more of the following:

- Survival
- Reproduction
- Development of young organisms
- Interactions with other organisms (competition, predation, parasitism, diseases) near the limits of temperature or moisture tolerance

If temperature or moisture acts to limit a distribution, what aspect of temperature or moisture is relevant—maximums, minimums, averages, or the level of variability? No overall rule can be applied here; the important measure depends on the mechanism by which temperature or moisture acts and the species involved. Plants (and animals) respond differently to a given environmental variable during different phases of their life cycle. For this reason, mean temperatures or average precipitation will not always be correlated with the limits of distributions, even if temperature or moisture is the critical variable.

To show that temperature or moisture limits the distribution of an organism, we should proceed as follows:

- Determine which phase of the life cycle is most sensitive to temperature or moisture.
- Identify the physiological tolerance range of the organism for this life cycle phase.
- Show that the temperature or moisture range in the microclimate where the organism lives is permissible for sites within the geographic range, and lethal for sites outside the normal geographic range (Figure 7.5).

We will now consider a set of examples that illustrate this approach and show some of the biological complications that may occur.

The range limits of warm-blooded animals may correlate with climatic variables. Winter distributions of passerine birds in North America often correlate with minimum January temperature (Root 1988). One example is the Eastern phoebe, which has a northern winter range limit at –4°C (Figure 7.6).

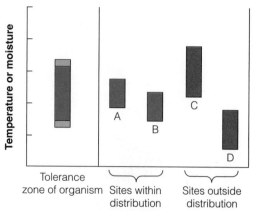

FIGURE 7.5
Hypothetical comparison of the tolerance zone of an organism and the temperature or moisture ranges of the microclimates where it lives. The tolerance zone is measured for the stage of the life cycle that is most sensitive to temperature or moisture and is subdivided into two zones, the optimal zone (dark blue) and the marginal zone (light blue). In this example the organism can live at A and B, but cannot tolerate C or D. The same principle can be applied to other physical-chemical factors such as pH.

Such climatic limitations in temperate zone birds are directly linked with the energetic demands associated with cold temperatures. Higher metabolic rates are needed to maintain body temperature in cold weather, and presumably there is a limit on the amount of food available in winter for passerines that sets these range limits (Root 1988).

It is less common for geographic range limits to coincide with rainfall contours. Few geographic distributions of animals are likely set directly by precipitation. But for plants, moisture is of direct importance, and the water relations of plants is an important area of research for plants of economic value. The water balance of plants is difficult to measure directly, and botanists usually measure the water content of plant tissues as an index of water balance. The leaves are particularly sensitive because most evaporation occurs there. Different plants vary greatly in their ability to withstand water shortages.

Drought resistance is achieved by (1) improvement of water uptake by roots; (2) reduction of water loss by stomatal closure, prevention of cuticular respiration, and reduction of leaf surface; and (3) storage

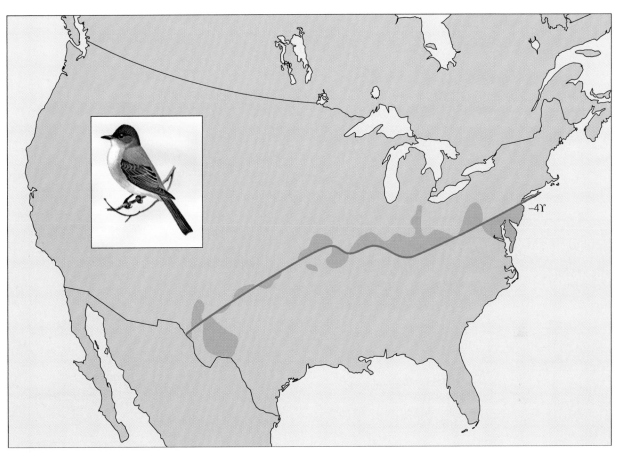

FIGURE 7.6
Contour map of the winter distribution of the Eastern phoebe (Sayornis phoebe) *from* Christmas Bird Counts of 1962–1972. *The bold blue line marks the edge of the winter range, which is closely associated with the –4°C isotherm of average minimum January temperatures. The ligh blue area indicates the area of deviation between the range boundary and the isotherm. (From Root 1988).*

of water. Rapid root growth into deeper areas of the soil is often effective in increasing drought resistance; young plants with little energy reserve will consequently suffer the worst from drought. Leaves of plants subject to poor water supply often have smaller surface areas and thicker cuticles, both of which reduce evaporation losses. By shedding their leaves in the drought season, plants have another very effective means of reducing water loss. *Xerophytes* (plants that live in dry areas) show many of these special adaptations for decreasing water loss. Additionally, leaves may be oriented vertically, which reduces the amount of absorbed radiation and resultant evaporation.

Other xerophytes, such as cacti, store water in their stems and thereby overcome drought.

Interaction Between Temperature and Moisture

In some cases, the moisture requirements of plants can restrict their geographic distributions. In other cases, moisture and temperature interact to limit geographic distributions, and the ecologist must consider explanations like "both-temperature-and-moisture" rather than "either-temperature-or-moisture." Parker (1969) has reviewed drought resistance in woody

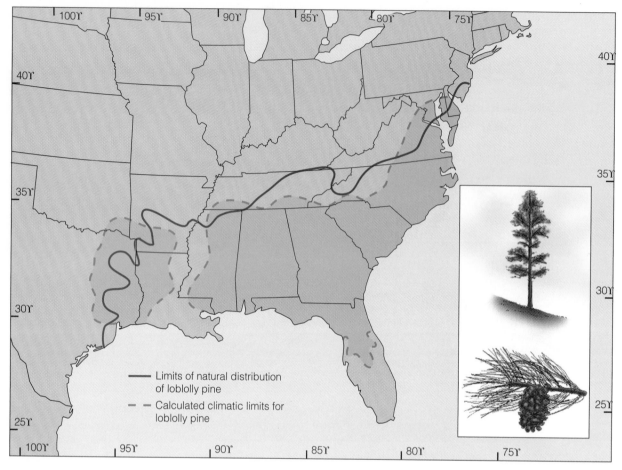

FIGURE 7.7
Natural distribution limits (dashed line) and calculated climatic limits (solid line) of loblolly pine (Pinus taeda) *in the southeastern United States. Winter temperature and rainfall set the northern and western limits of this pine. (After Hocker 1956.)*

plants and concluded that *frost drought* and *soil drought* are both critical in determining ranges of species. Soil drought is the common notion of drought in which soil moisture is deficient (as in the desert); it can usually be described as an *absolute* shortage of water in the soil. Frost drought or winter drought in plants occurs when water is present but unavailable because at low soil temperature (such as occur in the tundra in winter), and the roots are unable to take up water while the leaves continue to lose water by transpiration; it can be described as a *relative* shortage of water for plants. In both situations, water loss from the plant's leaves and stems is greater than water intake through the roots. Thus low temperatures can produce symptoms of drought. This fact emphasizes that water

availability is the critical variable and has led to considerable research on how to measure "available" water in the soil. Many of the distributional effects attributed to temperature may in fact operate through the water balance of plants.

Hocker (1956) sought to describe the distribution range of the loblolly pine (*Pinus taeda*) from the meteorological data available from 207 weather stations in the southeastern United States. He included values for (1) average monthly temperature, (2) average monthly range of temperature, (3) number of days per month of measurable rainfall, (4) number of days per month with rainfall over 13 mm, (5) average monthly precipitation, and (6) average length of frost-free period. Weather stations were divided into two groups, one within the natural range of the pine and the other outside the range; from the difference between these two groups Hocker mapped the climatic limits for loblolly pine (Figure 7.7). There is good agreement between observed limits of range and the limits mapped from these meteorological data. Winter temperature and rainfall probably set the northern limit of this pine. The rate of water uptake in loblolly pine roots decreases rapidly at lower temperatures, and this would accentuate winter drought in more northerly areas. Hocker predicted that a northern extension of the limits of loblolly pine was not feasible because of these basic climatic limitations.

As one moves up large mountains like those in the Rocky Mountains, or north or south toward the poles, one reaches the limit of trees as a vegetation type. This is called the *tree line*, or *timberline*, and is a particularly graphic illustration of the limitation on plant distribution imposed by the physical environment. Not all tree lines are controlled by the same factors; Stevens and Fox (1991) listed nine factors that have been suggested to affect timberlines:

- Lack of soil
- Desiccation of leaves in cold weather
- Short growing season
- Lack of snow, exposing plants to winter drying
- Excessive snow lasting through the summer
- Mechanical effects of high winds
- Rapid heat loss at night
- Excessive soil temperatures during the day
- Drought

These factors can be boiled down into three primary variables: temperature, moisture, and wind. Proceeding up a mountain, temperature decreases, precipitation increases, and wind velocity increases. Because of freezing temperatures during much of the year, available soil moisture decreases. How can we ascertain the effects of temperature, moisture, and wind?

Daubenmire (1954) analyzed alpine timberlines in North America and reviewed the various factors that might affect them. Upper timberlines in North America decrease about 100 meters in altitude for every degree of latitude one moves north, except between the equator and 30°N, where timberlines are approximately constant at 3500 to 4000 meters. In North America, timberlines for any given latitude are lowest in the Appalachians and highest in the Rocky

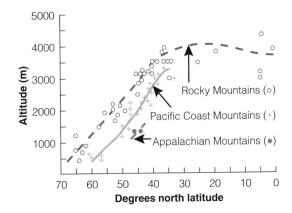

FIGURE 7.8
Altitudes of timberlines in North America. The tree line is highest in the Rocky Mountains and decreases uniformly as one moves north of 30°N latitude. (After Daubenmire 1954.)

Mountains (Figure 7.8). The uniform change of timberlines with latitude is surprising because many different tree species are involved.

Snow depth can affect the local distribution of trees near the timberline but cannot explain the existence of the timberline. In depressions where snow accumulates early and stays late, tree seedlings cannot become established. Only ridges will support trees in these circumstances, but these ridges also have a timberline; consequently, snow depth cannot be a primary factor.

Trees at the upper timberline in the Northern Hemisphere are often windblown and dwarfed, suggesting that wind is a major factor limiting trees on mountains. Within the tropics and in the Southern Hemisphere, wind effects seem to be absent. One difficulty with the wind hypothesis is that all the evidence is relevant to old trees, whereas it is the establishment of very young seedlings that is crucial to timberline formation. Daubenmire (1954) suggests that wind has secondary effects in altering timberlines in local situations, but, like snow depth, wind does not seem to be the primary cause of timberlines.

In spite of the broad correlation of tree lines with temperature, alpine timberlines are determined not by *summer* temperatures but by *winter* desiccation, or frost drought (Tranquillini 1979). Winter desiccation is severe above the timberline because new leaves cannot grow enough in the brief summers to enable a tree to mature and thus become drought-resistant by laying down a thick cuticle.

Little experimental work has been done to determine the causes for timberlines. In New Zealand, the beech (*Nothofagus*) forms an evergreen forest that stops abruptly at the timberline, which is between 900 and 1500 meters above sea level. Wardle (1965) showed that this timberline in New Zealand was produced by factors that reduce seedling establishment. Good seed years near the timberline are uncommon for *Nothofagus*, and at the timberline seed has poor germination rates (0–3%). Seedling survival is poor, and the death of seedlings is associated with drying out of the stems and leaves. When Wardle transplanted small seedlings above the timberline, seedlings planted in the open all died in their first year, but shaded seedlings survived well and became established 180 meters above the timberline.

Since the Earth's climate is warming, we should expect timberlines to move upward in elevation. Wardle and Coleman (1992) found that this was occurring in New Zealand with Antarctic beech, but that the rate of advance was very slow due to limited seed dispersal.

The intertidal zone of rocky coastlines is a zone of tension between sea and land, and, as is the case for the tree line, the distributional boundaries are very clear. The upper and lower limits of dominant invertebrates and algae are often very sharply defined in the intertidal zone, and on rocky shores in the British Isles this zonation is a particularly graphic example of distribution limitations on a local scale (Figure 7.9). Two barnacles dominate the British coasts. *Chthamalus stellatus* is a "southern" species that is absent from the colder waters of the British east coast and is the common barnacle of the upper intertidal zone of western Britain and Ireland. Going farther north in the British Isles, one finds it restricted to a zone higher and higher on the intertidal rocks. *Chthamalus* is relatively tolerant of long periods of exposure to air, and the upper limit of its distribution on the shore is set by desiccation. This basic limitation does not seem to change over its range. Its lower limit on the shore is often determined by competition for space with *Balanus balanoides*, a northern species. Connell (1961b) showed that *Balanus* grew faster than *Chthamalus* in the middle part of the intertidal zone and simply squeezed *Chthamalus* out. He also showed that *Chthamalus* could survive in the *Balanus* zone if *Balanus* were removed.

The upper limit of *B. balanoides* is also set by weather factors, but since this barnacle is less tolerant of desic-

cation and high temperatures than *Chthamalus*, there is a zone high on the shore where *Chthamalus* can survive but *Balanus* cannot (Connell 1961a). The sensitivity of young barnacles sets this upper limit. The lower limit of *Balanus* is set by competition for space with algae and by predation, particularly by a gastropod, *Thais lapillus*.

The distribution of these barnacles is a striking example of limitations imposed by both physical factors (temperature, desiccation) at the upper intertidal limits and biotic factors (competition, predation) at the lower limits.

Adaptations to Temperature and Moisture

We have begun by assuming that certain physiological tolerances are built into all the individuals of a particular species. But we know that local adaptation can occur and that genetic and physiological uniformity cannot be assumed throughout the range of a species. Darwin recognized that species could extend their distribution by local adaptation to limiting environmental factors such as temperature, but the full implications of Darwin's ideas were not appreciated until the early 1900s, when a Swedish botanist, Göte Turesson, began looking at adaptations to local environmental conditions in plants. Turesson (1922) coined the word *ecotype* to describe genetic varieties within a single species. He recognized that much of ecology had been pursued as if genetic diversity within species did not exist. In a series of publications he described some variation associated with climate and soil in a variety of plant species (Turesson 1925). The basic technique was to collect plants from a variety of areas and grow them together in field or laboratory plots at one site, a *common garden*. The type of result he obtained in this early work can be illustrated with an example. *Plantago maritima* grows both as a tall, robust plant (30-40 cm) in marshes along the coast of Sweden and as a dwarf plant (5-10 cm) on exposed sea cliffs in the Faeroe Islands. When plants from marshes and from sea cliffs are grown side by side in a common

FIGURE 7.9
A very common barnacle-dominated slope on moderately exposed rocky shores of northwestern Scotland and northwestern Ireland. MHWS = mean high water, spring; MHWN = mean high water, neap (i.e., minimum tide); MLWN = mean low water, neap; MLWS = mean low water, spring. (After Lewis 1972.)

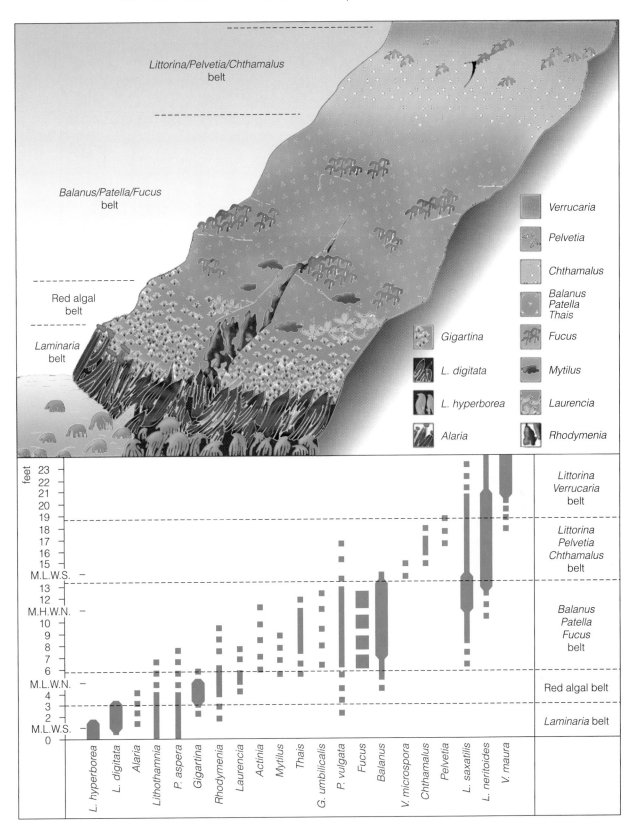

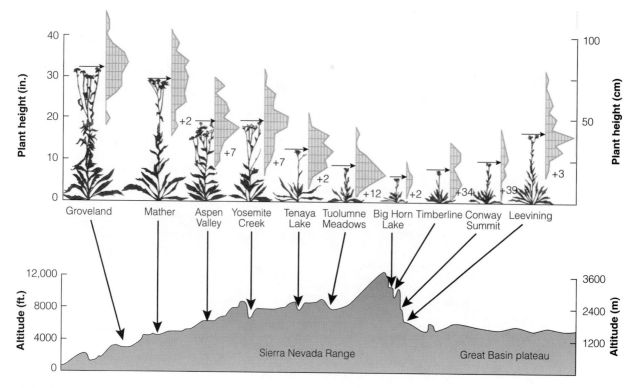

FIGURE 7.10

Representatives of populations of Achillea lanulosa *as grown in a common garden at Stanford. These originated in the localities shown in the profile of a transect across central California at approximately 38°N latitude. Altitudes are to scale, but horizontal distances are not. The plants are herbarium specimens, each representing a population of approximately 60 individuals. The frequency diagrams show variation in height within each population; the horizontal lines separate class intervals of 5 cm according to the marginal scale, and the distance between the vertical lines represents two individuals. The numbers to the right of some frequency diagrams indicate the number of nonflowering plants. The arrows point to the mean heights.*

garden, this height difference is not as extreme but remains significant (Turesson 1930):

Plantago maritima *Source*	*Mean height (cm) in garden*
Marsh population	31.5
Cliff population	20.7

Turesson's early studies on ecotypes such as these helped to create a new research field of ecological genetics.

This common garden technique is an attempt to separate the *phenotypic* (environmental) and *genotypic* (genetic) components of variation. Plants of the same species growing in such diverse environments as sea

cliffs and marshes can differ in morphology and physiology in three ways: (1) all differences are phenotypic, and seeds transplanted from one situation to the other will respond exactly as the resident individuals; (2) all differences are genotypic, and if seeds are transplanted between areas, the mature plants will retain the form and physiology typical for their original habitat; or (3) some combination of phenotypic and genotypic determination produces an intermediate result. In natural situations, the third case is most common. Many examples are now described in the literature, particularly in plants (Sultan 1995).

A classic set of ecotypic races occurs in the perennial herb *Achillea* (yarrow), analyzed by Clausen, Keck, and Hiesey (1948) in a pioneering paper.

Clausen and his colleagues studied two North American species in detail. A maritime form of *Achillea borealis* lives in coastal areas of California as a low succulent evergreen plant that grows throughout the winter. Slightly farther inland grows an evergreen race that is similar but taller. A third race lives in the Pacific Coast Range; it grows during the mild winter and flowers quickly by April, becoming dormant during the hot, dry summer. In the Central Valley of California a giant race of *A. borealis* occurs that survives under high summer temperatures, a long growing season, and ample moisture.

In the Sierra Nevada, races of *Achillea lanulosa* occur. As one proceeds up these mountains, the average winter temperature decreases below freezing, so winter dormancy is necessary and plants are smaller. On the eastern slope of the Sierra Nevada, plants of *A. lanulosa* are late flowering and adapted to cold, dry conditions. Clausen, Keck, and Hiesey collected seeds from a series of populations of *A. lanulosa* across California and raised plants in a greenhouse at Stanford, with the results shown in Figure 7.10. The major attributes of these races are maintained when plants are grown under uniform conditions in the same place.

Many species expanded their geographical range during the twentieth century (Hengeveld 1989), but in nearly all cases we do not know if genotypic changes accompanied these range changes. If we could study a species in the midst of a range extension, we might obtain some insight as to how organisms can extend their tolerance limits. This opportunity may become more frequent in the future, as climatic warming occurs.

Light as a Limiting Factor

Light may be another factor limiting the local distribution of plants. Light is important to organisms for two quite different reasons: it is used as a cue for the timing of daily and seasonal rhythms in both animals and plants, and it is essential for photosynthesis in plants.

Timing, the first reason light is important, is a central issue in the life cycles of organisms. Nocturnal desert animals, for example, use light as a cue for their activity cycles. The breeding seasons of many animals and plants are set by the organisms' responses to day-length changes. The seasonal impact of day length on physiological responses, called *photoperiodism*, has been an important focus of work in environmental physiology (Eckert et al. 1988).

The second reason light is important to organisms is that it is essential for *photosynthesis*, the process by which plants convert radiant energy from the sun into energy in chemical bonds. Photosynthesis is remarkably inefficient. During the growing season, about 0.5–1% of the incoming radiation is captured and stored by photosynthesis. In this process, carbon in the form of CO_2 is taken up from the air (or the water in the case of aquatic plants) and converted into organic compounds. We can measure the rate of photosynthesis by measuring the rate of uptake of CO_2.

Plants show a great diversity of photosynthetic responses to variations in light intensity. Some plants reach maximal photosynthesis at one-quarter full sunlight, and other species such as sugarcane never reach a maximum but continue to increase photosynthetic rate as light intensity rises. We recognize this ecologically by noting that plants in general can be divided into two groups: *shade-tolerant* species and *shade-intolerant* species. This classification is commonly used in forestry and horticulture. Plant physiologists have discovered that shade tolerance is a complex of traits, and that it is not fixed for each species but varies with plant age, microclimate, and geographical area (Kozlowski et al. 1991, p. 139). Shade-tolerant plants have lower photosynthetic rates and hence would be expected to have lower growth rates than shade-intolerant species. The metabolic rate of shade-tolerant seedlings is apparently lower than that of shade-intolerant seedlings.

Plant species become adapted to live in a certain kind of habitat and in the process evolve a series of characteristics (an "adaptive syndrome") that prevent them from occupying other habitats. Grime (1979) suggests that light may be one of the major components directing these adaptations. For example, eastern hemlock seedlings are shade-tolerant and can survive in the forest understory under very low light levels. Hemlock seedlings grow slowly and have a low metabolic rate that allows them to survive low light conditions. One consequence of these adaptations is that hemlock seedlings die easily in droughts because their roots do not grow quickly enough to penetrate deep into the soil. Failure of seedlings in shaded situations is often associated with fungal attack, and part of adaptation to shade involves becoming resistant to fungal infections.

An exceedingly important principle in evolutionary ecology *is that individuals of a species cannot do everything in the best possible way.* Adaptations to live in one ecological habitat make it difficult or impossible to live in a different habitat. Thus life cycles have evolved as tradeoffs between contrasting habitat requirements. Adaptations are always compromises, and there can be no superanimals or superplants.

Seaweeds grow in a variety of forms and display an interesting array of adaptations to light (Hay 1986). Seaweeds differ in the maximum water depth they can tolerate. Since light is attenuated very quickly with increasing depth in the oceans, some of the morphological adaptations of seaweeds are to light levels. Two general types of seaweed occur: Some species have flat, wide, monolayered thalli, while others have highly dissected, narrow thalli that are multilayered (Figure 7.11a). The same distinction has been applied to the leaves of trees (Horn 1971). If light intensity is important in controlling thallus shape and layering, we would predict that multilayered seaweeds would be at an advantage at high light intensity and that monolayered seaweeds would be at an advantage in low light levels in deeper water (Figure 7.11b). Figure 7.12 shows the relationship of thallus width to depth beneath the surface for seven species of seaweeds from the genus *Sargassum* in the Caribbean. Species that have wider blades occur in deeper waters where light is limiting, and thus the local distribution of these seaweeds is determined by light. For these seaweeds the upper (shallow) limits of distribution could be determined by biotic interactions involving competition for light, or by other biotic or abiotic factors that have yet

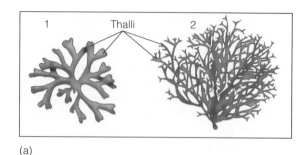

(a)

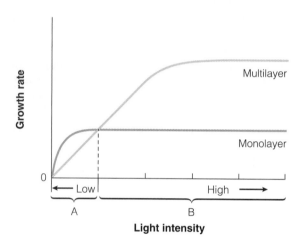

(b)

FIGURE 7.11
(a) Morphology of seaweeds: (1) flat, monolayered forms with opaque wide thalli; (2) upright, multilayered forms with narrow thalli. (b) Proposed model for the effects of light intensity on net photosynthesis for whole plants of monolayered and multilayered seaweeds. In range A, monolayers will be favored over multilayers. In range B with high light intensity, multilayers will be favored. (Modified from Hay 1986.)

FIGURE 7.12
Relationship of thallus morphology to depth beneath the surface for seven species of Sargassum *seaweeds from the Caribbean Sea. Species with wide blades (monolayered) occur in deeper waters, while those with narrow blades (multilayered) occur in shallow waters where light intensity is highest. (From Hay 1986.)*

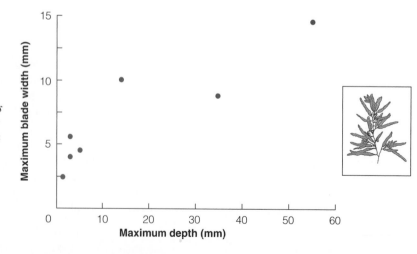

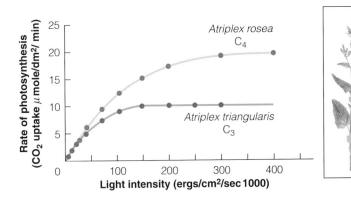

FIGURE 7.13
Comparative photosynthetic production of the C_3 species Atriplex triangularis *and the related C_4 species* Atriplex rosea. *The plants were grown under identical controlled conditions of 25°C during the day and 20°C at night, 16-hour days, and ample water and nutrients. (After Björkman 1975.)*

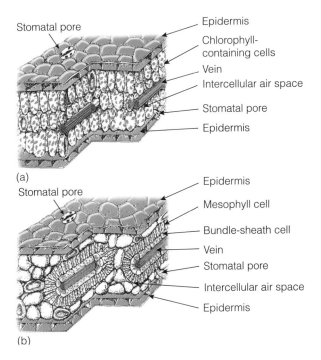

FIGURE 7.14
Leaf anatomy of C_3 and C_4 plants. (a) Leaf structure of a typical C_3 plant, Atriplex triangularis, *in which the cells containing chlorophyll, chloroplasts (red), are of a single type and are found throughout the interior of the leaf. (b)* Atriplex rosea, *a C_4 plant, illustrating the modified leaf structure of C_4 species. The specialized leaf of* A. rosea *has nearly all its chlorophyll in two types of cells that form concentric cylinders around the fine veins of the leaf. The cells of the outer cylinder are mesophyll cells; those of the inner cylinder are bundle-sheath cells. (From Björkman and Berry 1973.)*

to be studied. With plants that require light for photosynthesis, the rate of photosynthesis under given light conditions can affect whether the plant can survive in different environments.

One reason photosynthetic rate varies among plants is that they have evolved three photosynthetic strategies: the C_3 pathway, the C_4 pathway, and crassulacean acid metabolism. Most plants use the C_3 pathway, first described by Calvin and often called the *Calvin cycle*. In the C_3 pathway, CO_2 from the air is first converted to 3-phosphoglyceric acid, a three-carbon molecule (hence the name C_3). Until the mid-1960s this pathway was believed to be the only important means of fixing carbon in the initial steps of photosynthesis. In 1965 sugar cane was found to fix CO_2 by first producing malic and aspartic acids (four-carbon acids), and a the C_4 pathway of photosynthesis was discovered (Björkman and Berry 1973). C_4 plants have all the biochemical elements of the C_3 pathway, so they can use either method to fix CO_2.

The ecological consequences of the C_4 pathway are profound. Figure 7.13 shows the rates of photosynthesis of a pair of closely related species of C_3 and C_4 plants. C_4 plants do not reach saturation light levels even under the brightest sunlight, and they always produce more photosynthate per unit area of leaf than C_3 plants. C_4 plants are thus more efficient than C_3 plants. Leaf anatomy differs in typical C_3 and C_4 plants (Figure 7.14). Chlorophyll in C_3 leaves is found throughout the leaf, but in C_4 leaves the chloroplasts are concentrated in two-layered bundles around the veins of the leaf (called *Krantz anatomy*). The bundle sheath cells in C_4 plants also have a high concentration of mitochondria. The C_4 leaf anatomy is more efficient for utilizing low CO_2 concentrations, for recycling the CO_2 produced in respiration, for rapidly translocating starches to other parts of the leaf. The biochemical reason for this

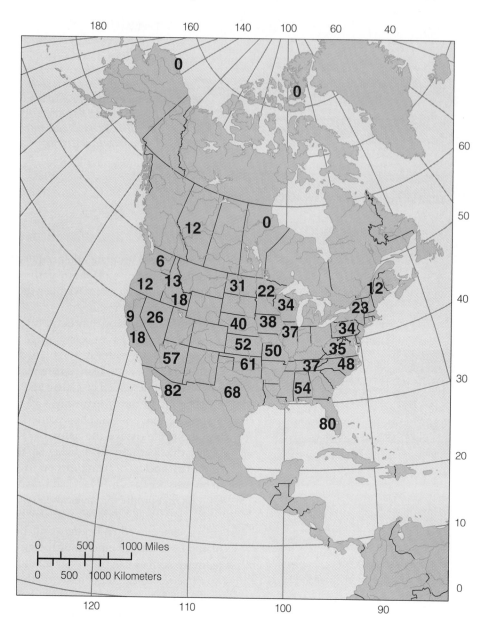

FIGURE 7.15
Percentage of C$_4$ species in the grass floras of 32 regions of North America. (From Teeri and Stowe 1976.)

anatomical difference is simple—the first step in fixing CO$_2$ in these two types of plants differs:

$$C_3: \text{Atmospheric } CO_2 + \begin{array}{c}\text{ribulose - diphosphate}\\ \text{(RuDP)}\end{array} \xrightarrow{\text{RuDP carboxylase}} \begin{array}{c}\text{phosphoglyceric}\\ \text{acid}\end{array}$$

$$C_4: \text{Atmospheric } CO_2 + \begin{array}{c}\text{phospho - enolpyruvate}\\ \text{(PEP)}\end{array} \xrightarrow{\text{PEP carboxylase}} \begin{array}{c}\text{malic acid} +\\ \text{aspartic acid}\end{array}$$

TABLE 7.1 Characteristics of photosynthesis in three groups of higher plants

	Type of photosynthesis		
Characteristics of plants	C₃	C₄	CAM
Leaf anatomy (cross section	Diffuse mesophyll	Mesophyll compact around vascular bundles	Spongy appearance, mesophyll variable
Enzymes used in CO_2 fixation in leaf	RuDP	PEP carboxylase and then RuDP carboxylase	Both PEP and RuDP carboxylases
CO_2 compensation point (ppm CO_2)	30–70	0–10	0–5 in dark, 0–200 with daily rhythm
Transpiration rate (water loss)	High	Low	Very low
Maximum rate of photosynthesis (mg CO_2/dm² leaf surface/hr)	15–40	40–80	1–4
Respiration in light	High rate	Apparently none	Difficult to detect
Optimum day temperature for growth	20–25°C	30–35°C	Approx. 35°C
Response of photosynthesis to increasing light intensity at optimum temperature	Saturation about 1/4 to 1/3 full sunlight	Saturation at full sunlight or at even higher light levels	Saturation uncertain but probably well below full sunlight
Dry matter produced (t/ha/yr)	22	39	Extremely variable

The compensation point is the CO_2 concentration at which photosynthesis just balances respiration so that there is no net oxygen generated and no net CO_2 taken up.

Source: From Black (1971).

The enzyme RuDP carboxylase is inhibited by oxygen in the air and has a lower affinity for CO_2. The enzyme PEP carboxylase is not inhibited by oxygen and has a higher affinity for CO_2. From this biochemical information we can predict that C₄ plants would be at an advantage when photosynthesis is limited by CO_2 concentration. This occurs under high light intensities and high temperatures and when water is in short supply (Epstein et al. 1997, Qi and Redmann 1993).

C₄ grasses, sedges, and dicotyledons are all more common in tropical areas than in temperate or polar areas (Hattersley 1983). Figure 7.15 shows the percentage of grass species that are C₄ plants in different parts of North America and confirms the suggestion that C₄ grasses are at a selective advantage in warmer areas with high solar radiation. On Hawaiian mountains, which have small seasonal changes in temperature, C₃ grasses predominate at high elevations and C₄ grasses at low elevations (Rundel 1980). In Japan, C₄ plants are more prominent in areas with higher temperatures (Ueno and Takeda 1992).

Some desert succulents, such as cacti of the genus *Opuntia*, have evolved a third modification of photosynthesis, *crassulacean acid metabolism (CAM)*. These plants are the opposite of typical plants in that they open their stomata to take up CO_2 at night, presumably as an adaptation for minimizing water loss through the stomata. This CO_2 is stored as malic acid, which is then used to complete photosynthesis during the day. CAM plants have a very low rate of photosynthesis and can switch to the C₃ mode during daytime. They are adapted to live in very dry desert areas where little else can grow. Table 7.1 summarizes the main characteristics of C₃, C₄, and CAM plants.

The C₃ pathway is presumably the ancestral method of photosynthesis since no algae, bryophytes, ferns, gymnosperms, or more primitive angiosperms have the C₄ pathway or the capacity for CAM (Pearcy and Ehleringer 1984, Monson 1989). Almost half of the C₄ plant species are grasses (Black 1971), and this pathway has apparently increased their competitive ability.

We do not know how the different photosynthetic pathways may interact with other factors to affect the geographic distribution of plant species. It is clear that the response of a plant species to temperature and moisture is strongly affected by the type of photosynthetic process it uses. Implications for animal distributions have yet to be considered. Plants possessing the C_4 pathway seem to be of lower nutritional value for herbivorous insect (Barbehenn and Bernays 1992). Further work is needed on the ecological consequences of the three different photosynthetic strategies, both for the plants and for the animals that depend on them.

If C_4 plants are more productive than C_3 plants, why do they not displace C_3 plants everywhere? The photosynthetic productivity plotted in Figure 7.13 on page 99 cannot be directly translated into productivity in natural vegetation (Snaydon 1991). Competition in natural stands is not always for light, and mineral nutrients and water are often limiting to plants. Competition for soil resources can result in plants developing large root systems rather than large aboveground structures. Soil texture also affects C_3 and C_4 grasses differently; clay soils favor C_3 plants while sandy soils favor C_4 grasses in the Great Plains (Epstein et al. 1997). In spite of these differences, as climatic warming occurs in the future, C_3 grasses are predicted to shift their geographical ranges to the north in the Northern Hemisphere.

Climate Change and Species Distributions

If temperature and moisture are the master limiting factors for the geographical ranges of plants and animals, the climatic warming that is now occurring will have profound effects on the Earth's biota. One way to get a glimpse of the kind of changes that may occur is to look back at the changes that have occurred in temperate regions since the end of the last Ice Age.

After the last continental glaciers began retreating in North America and Eurasia about 16,000 years ago, the northward expansion of tree distributions lagged behind the retreat of the ice. A detailed record of these migrations is captured in fossilized pollen deposited in lakes and ponds. Margaret Davis and her students have been leaders in deciphering the record left in fossilized pollen deposits (Davis 1986). In North America oaks and maples moved rapidly in a northeasterly direction from the Mississippi Valley, while hickories advanced more slowly (Delcourt and

Delcourt 1987). Hemlocks and white pines moved rapidly northwest from refuges along the Atlantic Coast. The important finding of this paleoecological work is that the range of each species advanced individualistically. If you were sitting in New Hampshire, you would have seen sugar maple arrive 9000 years ago, hemlock 7500 years ago, and beech 6500 years ago (Davis 1986).

If we can determine the climatic limits of current geographical distributions, we can make predictions about how distributions will change with climatic

Margaret B. Davis *(1931–) Regents Professor of Ecology, University of Minnesota*

warming. A major assumption of using this approach for plants is that seed dispersal is adequate to sustain the migrations of each species. Davis (1986) suggested that hemlock was delayed nearly 2500 years in its movement north at the end of the Ice Age, in part because of slow seed dispersal. If we use climate-change models to predict temperature and rainfall changes over the next 100 years, we can begin to estimate the size of the problem animals and plants will face over the next few centuries.

Figure 7.16 shows the current and potential geographical range of American beech (*Fagus grandifolia*) under two climate-change models. These models predict that the potential northern range limit of beech will move 700–900 km north in the next century, and that the southern range limit will move to the north as well. If left to natural processes, beech must move 7–9 km per year to the north. By contrast, since the end of the Ice Age, beech migrated into its present range at a rate of 0.2 km per year. If these predictions are even approximately correct, slowly colonizing

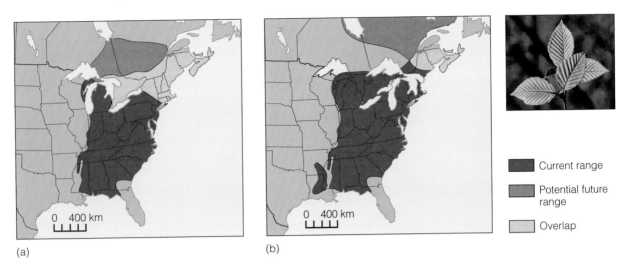

(a)   (b)

FIGURE 7.16
Current geographic range and predicted future range for American beech (Fagus
grandifolia) *under two climatic-change scenarios: (a) the milder scenario of 4.5°C
warming, and (b) the more severe scenario of 6.5°C warming. (From Roberts 1989.)*

species like trees will require human assistance to
move into their new ranges. If this does not occur,
species like beech may become extinct.

These effects of climate change will not appear
immediately. Long-lived plant species such as trees
will survive for many years as adults in inappropriate
places. As the climate changes, their seed production
will decline until finally they are unable to produce
viable seedlings (Roberts 1989). All of these effects are
complicated by ecotypic variation within tree species.
If there are specific ecotypes adapted to northern or
southern climatic conditions, then the range shifts

depicted in Figure 7.16 must be accomplished without the loss of ecotypic variation.

The geographic ranges of species are thus not static but dynamic, and as climate changes in the future,
species will (if time permits) move into new areas that
become climatically suitable. The major concerns of
ecologists are first that the speed of climate change in
the next 100 years may be too great for slowly colonizing forms to move, and second that genetic adaptation to local temperature and rainfall patterns may
be lost for some species.

Summary

Temperature and moisture are the major factors that limit
the distributions of animals and plants. These factors may
act on any stage of the life cycle and affect survival, reproduction, or development. Temperature and moisture may
also indirectly limit distributions through their joint effects
on competitive ability, disease resistance, predation, or parasitism. Other physical and chemical factors, such as light
and pH, can also affect the distributions of plants and animals, but they operate at a local scale.

From a global viewpoint, the distribution of plants can
be associated with climate. Geographers have recognized
this by making climatic classifications based on vegetational
distributions. Tropical rain forest and tundra, for example,
occupy areas with different temperature and moisture
regimes. The effects of climate are less clearly seen at the
level of the distribution of individual species. In only a few
cases has experimental work been done in local populations,
first to pinpoint the life-cycle stage affected by climate and
then to describe the physiological processes involved.

Water availability is the key to moisture effects on
plants, and drought occurs when adequate amounts of water
are not present or are unavailable to the plant. The soil may
be saturated with water, but if all of it is frozen, none may
be taken up by plants, and they may suffer frost drought.
Many of the distributional effects attributed to temperature
may operate through the water balance of plants.

Species may adapt to temperature, moisture, or light
levels phenotypically or genotypically and thereby circumvent some of the restrictions imposed by climate. Turesson
was one of the first to recognize the importance of ecotypes,

genetic varieties within a single species. By transplanting individuals from a variety of habitats into a common garden, Turesson showed that many of the adaptations of plant forms were genotypic. Many ecotypes have now been described, particularly in plants, and these may involve adaptations to any environmental factor, including temperature and moisture. Ecotypic differentiation has often proceeded to the point where one ecotype cannot survive in the habitat of another ecotype of the same species.

Organisms have evolved an array of adaptations to overcome the limitations of high and low temperatures, drought, or other physical factors. Some adaptations might allow a species to extend its geographic range. Many species are known to have extended or reduced their geographic range in historical times, but few cases have been studied in detail. Consequently, we rarely know whether a species changed its distribution because the climate changed or because the genetic makeup of the species altered.

The climatic warming that is currently under way will have strong effects on geographic distributions. The major concerns are that species will not be able to migrate fast enough to keep pace with global warming, and that genetic adaptations to local environments may be lost.

Key Concepts

1. Temperature and moisture are the main limiting factors for both plants and animals on a global scale.

2. Light, fire, pH, and other physical and chemical factors can limit distributions on a local scale.

3. Species may evolve adaptations that overcome the limitations set by physical and chemical factors.

4. Some of these adaptations allow a species to extend its geographical range.

5. Climatic warming in the twenty-first century will have major impacts on the geographical ranges of species that are currently limited by temperature and moisture.

Selected References

Barton, A. M. 1993. Factors controlling plant distributions: Drought, competition, and fire in montane pines in Arizona. *Ecological Monographs* 63:367–398.

Collins, S. L. and L. L. Wallace. 1990. *Fire in North American Tallgrass Prairies*. University of Oklahoma Press, Norman. 175 pp.

Pearcy, R. W. and J. R. Ehleringer. 1984. Comparative ecophysiology of C_3 and C_4 plants. *Plant, Cell, and Environment* 7:1–13.

Richardson, D. M. and W. J. Bond. 1991. Determinants of plant distribution: Evidence from pine invasions. *American Naturalist* 137:639–668.

Root, T. 1988. Energy constraints on avian distributions and abundances. *Ecology* 69:330–339.

Spence, D. H. N. 1982. The zonation of plants in freshwater lakes. *Advances in Ecological Research* 12:37–125.

Stephenson, N. L. 1990. Climatic control of vegetation distribution: The role of the water balance. *American Naturalist* 135:649–670.

Stevens, G. C. and J. F. Fox. 1991. The causes of treeline. *Annual Review of Ecology and Systematics* 22:177–191.

Sutherst, R. W., R. B. Floyd, and G. F. Maywald. 1996. The potential geographical distribution of the cane toad, *Bufo marinus* L. in Australia. *Conservation Biology* 10:294–299.

Turesson, G. 1930. The selective effect of climate upon the plant species. *Hereditas* 14:99–152.

Questions and Problems

7.1 "To become widespread, a species must develop many ecological races" (Clausen, Keck, and Hiesey 1948, p. 121). Is this true in both animals and plants? Discuss with reference to colonizing species such as the starling.

7.2 There is only one known C_4 tree species (Pearcy 1983). Explain why this is the case?

7.3 Cain (1944) states:

Physiological processes are multi-conditioned, and an investigation of the effects of variation of a single factor, when all others are controlled, cannot be applied directly to an interpretation of the role of that factor in nature. It is impossible, then, to speak of a single condition of a factor as being the cause of an observed effect in an organism.

Discuss the implications of this principle—that the factors of the environment act collectively and simultaneously—with regard to methods for studying species distributions.

7.4 Review the effects of acid rain on the local distributions of fishes and aquatic invertebrates in freshwater lakes. What are the biological reasons for the disappearance of these animals as pH falls? Schindler (1988) reviews this topic.

7.5 "The frost line… is probably the most important of all climatic demarcations in plants" (Good 1964, p. 353). Locate the frost line in a climatological atlas, and compare the distributions of some tropical and temperate species of any particular taxonomic group with respect to this boundary.

7.6 Are the grasses that have evolved to live on contaminated mine-waste soils (see Chapter 3, page 37) an example of ecotypes? What types of selection (see Figure 2.2) would you expect to see among ecotypes?

7.7 The British barnacle *Elminius modestus* extends higher on the shore in the intertidal zone than does the barnacle *Balanus balanoides* when the two species occur together. However, these two species have similar tolerances to desiccation, salinity, and temperature. The range of initial settlement of young barnacles is the same for the two species. Given these facts, can you suggest an explanation for the observation that *E. modestus* extends higher on the shore than *B. balanoides*?

7.8 Adult male dark-eye juncos (*Junco hyemalis*) remain farther north in winter than females and juveniles (Ketterson and Nolan 1982) . Review the arguments of Root (1988) on energy balance and winter bird ranges, and suggest an explanation for these observations.

7.9 Elfinwood or krummholz are prostrate, low-growing shrub like forms of trees found near timberline. How would you test the suggestion that elfinwood trees are ecotypes that differ genetically from normal trees?

7.10 The cane toad (*Bufo marinus*) is an introduced pest in parts of Australia. Sutherst et al. (1995) predict the geographic limits of the possible spread of this pest within Australia. Analyze their approach, and list the critical assumptions they use to make these predictions. Compare their predictions with those of Beurden (1981).

Overview Question

An insect is to be brought into your country to control a noxious weed. Describe how you would determine the factors that limit the insects distribution and how you might predict its new geographic range in your country.

CHAPTER 8

The Relationship Between Distribution and Abundance

W E HAVE CONSIDERED in the past five chapters the ways in which ecologists answer the two simple ecological questions *Who lives where?* and *What constrains geographic distributions?* In the next part of the book we will investigate the dynamics of populations within this occupied zone, but before we consider these questions we will discuss the broad question of whether there is any relationship between distribution and abundance. This question was first raised in a general way by the Australian ecologists H. G. Andrewartha and L. C. Birch in their classic 1954 book *The Distribution and Abundance of Animals.* Their ideas, which tied together the two concepts of distribution and abundance, had a strong impact on ecological thinking in the past 50 years. In this chapter we focus on one aspect of this interaction between distribution and abundance—whether species that have large geographic ranges are any more or less abundant than species that have small geographic ranges.

H. G. Andrewartha *(1907–1992) Professor of Zoology, University of Adelaide*

The Spatial Scale of Geographic Ranges

We began our analysis of distribution in Chapter 1 by assuming that we can easily map the geographic range of a species, but this simple assumption breaks down as we map the detailed distribution of a species in a local area. No species occurs everywhere. Figure 8.1 illustrates the range of spatial scales at which one can

describe a species's geographic range. At one extreme the range of a species is defined by the worldwide extent of its occurrence, a line drawn on a map demarcating the outermost points at which the species has been observed. This is the scale of geographic range used in field guides and other regional natural history guides. At the other extreme, we could measure a much smaller area within the larger geographic range and

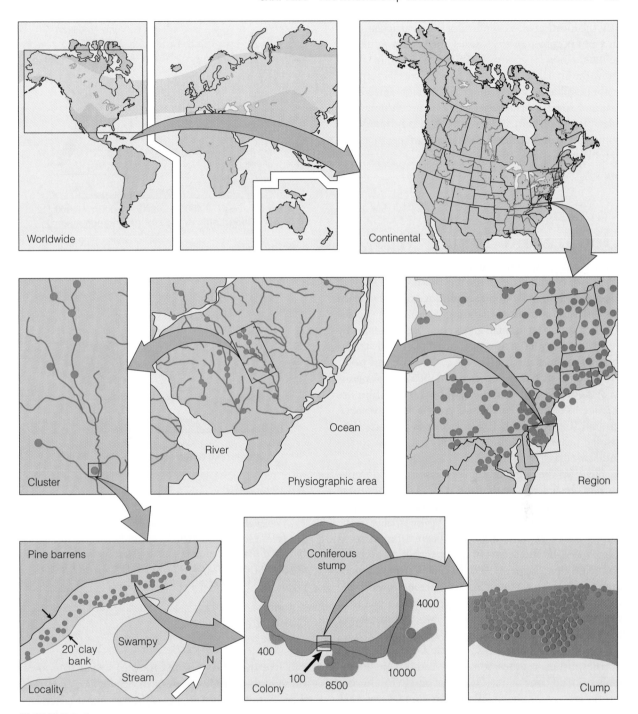

FIGURE 8.1

A hierarchy of scales for analyzing the geographic distribution of the moss Tetraphis. *The answer to the question "What limits geographic distribution?" may have different answers when analyzed at the continental scale versus the local scale of the individual tree stump. (After Forman 1964.)*

map the location of each individual. If a particular habitat is not occupied by the species, this region would not be included in its geographic range. We would like to know the actual area occupied by each species, but this is not possible because ecologists have not collected or mapped species occurrences in this much detail for most plants and animals (Gaston 1991). The important point shown in Figure 8.1 is that one can measure geographic ranges at several spatial scales.

Variations in Geographic Range Size

After we have decided on a measure of geographic range, we can investigate the spread of range sizes of species within a taxonomic group. A general pattern has emerged in many separate groups: Most species within a group have small geographic ranges and only a few have very large ranges. The frequency distributions of range sizes in Figure 8.2 illustrate this point for the birds of North America and the vascular plants of Britain. This pattern, the "hollow curve" of Figure 8.2, seems to be the rule for all groups that have been studied (Gaston et al. 1998). These range-size data display other interesting patterns in addition to the "hollow curve" shape.

In 1975 the Argentinean ecologist Eduardo Rapoport suggested that within the mammals geographic range sizes decreased as one moved from polar to equatorial latitudes, such that range sizes were smaller in the tropics. This generalization has been referred to as *Rapoport's Rule* and seems to apply within many taxonomic groups (Stevens 1989). In North America, geographic ranges of mammals are smaller toward the tropics (Figure 8.3). The average Canadian mammal species inhabit ranges that are an average 25 times larger than those of Mexican mammals (Pagel et al. 1991).

Support for Rapoport's Rule has been widespread in trees, fishes, reptiles, some birds, and many mammals from all continents (Gaston et al. 1998). But not all studies have supported Rapoport, and we must ask why Rapoport's Rule should hold in some species but not in others—what are the ecological mechanisms behind this pattern?

Three ecological explanations for Rapoport's Rule have been put forward. First, climatic variability is greater at high latitudes, and only organisms that have a broad range of tolerance for variable climates

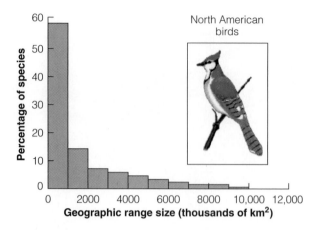

(a)

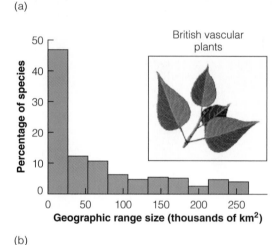

(b)

FIGURE 8.2
Frequency distribution of geographic range sizes for (a) 1370 species of North American birds and (b) 1499 species of British vascular plants. Most species have small geographic ranges. (Data from Anderson 1985 for (a) and from Gaston et al. 1998 for (b).)

can live there. As a side effect of broad tolerance, these high-latitude species can occupy larger ranges. This hypothesis makes two interesting predictions. For terrestrial animals and plants, climatic tolerance should increase from tropical to polar areas. This seems to be true for amphibians (Snyder and Weathers 1975). For marine organisms more interesting patterns can be predicted. For marine fish, temperature variation is greatest in the temperate zone and much smaller in polar waters and in tropical waters.

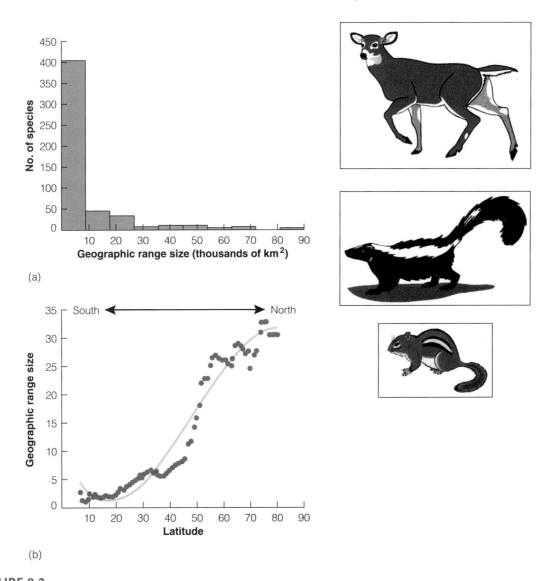

FIGURE 8.3
(a) Geographic range size for 523 species of North American mammals, measured as the percentage of the total land area of North America. (b) Relationship between range size and latitude. Low-latitude species have smaller ranges than high-latitude species, following Rapoport's Rule. (From Pagel et al. 1991.)

Thus the range of temperature tolerances should be minimal in both tropical and polar waters. Figure 8.4 shows that this appears to be the case for shallow-water marine fish. If we go deeper in the oceans, temperature variability becomes minimal, and the climatic variability hypothesis would predict no relationship between latitude and range size for these deep-sea organisms. Unfortunately no data are yet available to test this prediction for the deep sea.

A second explanation of Rapoport's Rule is that it is a product of glaciation, particularly in the Northern Hemisphere (Brown 1995). When the glaciers retreated, only those species with high dispersal capacity were able to repopulate northern areas, and

FIGURE 8.4
Critical temperature limits for marine fish from shallow waters. According to the climatic variability hypothesis, temperate fish should have the widest temperature limits. Upper critical temperatures are in blue, and lower critical temperatures are in red. These data from Brett (1970) fit this explanation well for the polar-temperate comparison and may fit the temperate-tropical comparison as well, but more data are needed from marine fish living in the 0–10° latitude range.

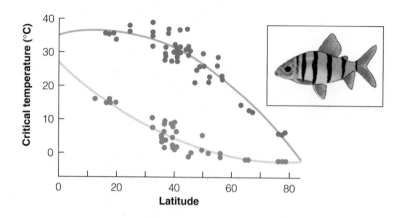

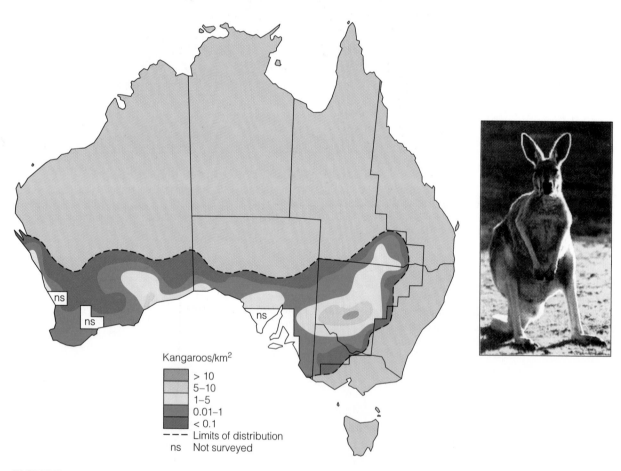

Kangaroos/km²

- > 10
- 5–10
- 1–5
- 0.01–1
- < 0.1
- – – – Limits of distribution
- ns   Not surveyed

FIGURE 8.5
Abundance of the western gray kangaroo (Macropus fuliginosus) *from aerial surveys, 1980–1982. (From Caughley et al. 1987.)*

these species thus have large geographic ranges. The glaciation hypothesis may explain some of the patterns found in the Northern Hemisphere, but it cannot explain Rapoport's Rule in the Southern Hemisphere, where glaciation was much less prominent. Glaciation is probably a contributing factor but not the major cause for these distributional patterns.

A third explanation of Rapoport's Rule is that it arises from a lack of competition in polar communities. Because fewer species live in polar areas, the level of competition may be lower. There is no support at present for this mechanism because we do not have a simple measure of competition that can be applied across global species patterns. It remains a possibility as yet untested.

The boundaries of geographic ranges can be abrupt or gradual (Caughley et al. 1988). In many cases a species's geographic distribution is a contour map with a gradual falloff in density as one approaches the edge of the range, as in the case in Figure 8.5, which shows the distribution of the western gray kangaroo in Australia. Brown (1984) showed that in North America, many species have this type of geographic pattern in which abundance gradually declines from the center of the range to the boundaries. But this is not always the case, and there may be abrupt boundaries in species ranges that are caused by physical or biological factors. The question raised by these observations is how might distribution and abundance be related?

Range Size and Abundance

Is there any relationship between geographic range size and the abundance of a species? If a species is widespread, is it always an abundant species? Or conversely, if a species is rare or threatened, does it have a small geographic range? The data ecologists have collected on a wide variety of plants and animals reveal a correlation between distribution and abundance such that more widespread species are typically more abundant (Brown 1984, Gaston 1990). Figure 8.6 shows data for 263 species of British moths. Moths were collected at light traps in 50 sites throughout Britain, and the geographic distribution was measured as the number of these sites occupied by a given species. Abundance data for each species

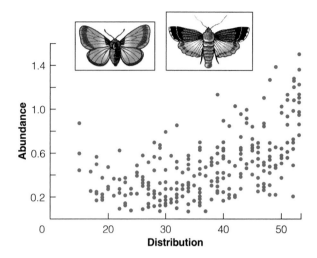

FIGURE 8.6
Relationship between distribution and abundance for 263 species of British moths. Distribution is the number of trap sites across Britain at which the species were caught. Abundance is averaged across all sites for all years. Each dot represents one species. In general, species with wider distributions are more abundant. (From Gaston 1988.)

were averaged over 6–14 years at each light trap site. (Not all light traps could be operated every year.) There is much variability in these moth data, but a clear trend exists: More widespread moth species tend to be more abundant. Similar patterns can be found in birds, plants, and many other groups, such that this positive relationship between distribution and abundance is another ecological generalization that can be called *Hanski's Rule*, after Ilkka Hanski.

Hanski (1982) was one of the first to call attention to the association between distribution and abundance, and there has been extensive discussion of the possible mechanisms behind the simple pattern shown illustrated in Figure 8.6 (Gaston et al. 1997). There are three main explanations of why distribution and abundance may be correlated.

The first explanation is the *sampling model*, which argues that the observed relationship is an artifact of sampling and does not require a biological explanation. Rare species are more difficult to find than common species, and thus if the appropriate biological studies are not done carefully, one will automatically observe the patterns seen in Figures 8.6. This explanation is difficult to evaluate because of the problems

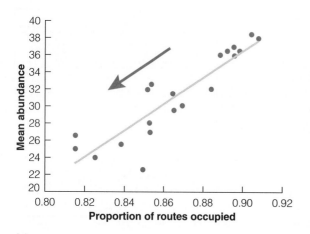

(a)

Eastern meadowlark

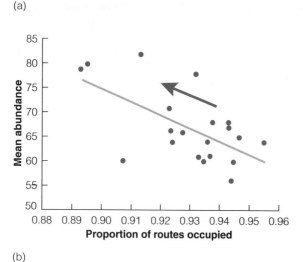

(b)

Common grackle

FIGURE 8.7

Changes in geographic range size for two North American birds that have been declining since 1970 in the Breeding Bird Surveys. Breeding Bird Surveys are carried out each June in a standard manner (50 stops over 24.5 miles) on more than 3700 specified routes in North America. (a) The eastern meadowlark has declined in abundance and its geographical range has also shrunk, so that there is a positive relationship between distribution and abundance in this species over the 20 years of data. (b) The common grackle has declined in abundance, but during this decline in abundance its geographical range has been increasing, so that there is a negative relationship between distribution and abundance for this bird species. The reasons for these differences are not known. (From Gaston and Curnutt 1998.)

of counting rare organisms. It cannot, however, be the explanation for this pattern in birds, butterflies, and mammals, which have been very well sampled and studied.

The second explanation is the *ecological specialization model*, or *Brown's model*, because Jim Brown first suggested it in 1984. This model argues that species

that can exploit a wide range of resources become both widespread and common. These species are called *generalists* and are to be distinguished from *specialists*, which exploit only a few resources. Provided that one can determine which species are generalists and which are specialists, one should be able to test this model. A corollary of this model is that widespread generalist species should use food and habitat resources that are themselves abundant.

The third explanation is the *local population model*. In this model a population is subdivided into a series of discrete patches, or local populations[1], that interact because animals or plants move between the patches. Since species differ in their capacity to disperse, some will occupy more local patches than others (Hanski et al. 1993). If this model is correct, we would expect species that disperse more to be more common and more widespread, when compared with less migratory species.

The relationship between distribution and abundance is usually discussed as a pattern among many species, but it is interesting to ask if the same pattern occurs within one species. If a species is declining in abundance, does its geographic distribution also become smaller? This could be an important question for conservation biologists studying a declining species. Conversely, are species that are increasing in abundance also expanding their geographic ranges? The best data come from bird surveys, and Figure 8.7 shows two cases of birds that have declined over 20 years. Both eastern meadowlarks and common grackles declined in North America from 1970 to 1989. Meadowlarks reduced their geographic range, as pre-dicted, but grackles increased their geographic range, contrary to prediction.

Available data at present support all of these explanations for the positive relationship between distribution and abundance, but many of the critical predictions have not yet been tested (Gaston et al. 1997). It may be that some species follow one model and some the other, or that different kinds of organisms are more likely to fit one model than another model. Since all studies of distribution and abundance involve sampling, in an imperfect world the sampling model will always be part of the explanation for these positive relationships between distribution and abundance. More data are needed to determine the ecological attributes of successful species that are widespread and abundant. These attributes may help us to understand the reasons for species being rare or endangered, and could assist in conservation biology (see Chapter 19). We are led in this way out of the world of geographic distributions and into the larger and more complex world of abundance. We turn next to exploring what happens within the zone of distribution in which populations of animals and plants increase or decrease in size in response to many of the same environmental factors we have just considered.

Summary

The geographic distribution of a species is more complex than one first suspects. The scale at which a distribution is mapped can affect the answer to the simple question: What limits geographic distributions? For many species we do not know the detailed geographic range because too few data have been collected. Even for larger plants and animals in developed countries we have few details about local distributions.

Geographic ranges measured on large-scale maps show a common pattern in all groups studied—most species have small geographic ranges, and only a few species are very widespread. This relationship holds for data on fishes, birds, and mammals, as well as for plant groups. In addition, in many taxonomic groups, geographic ranges follow Rapoport's Rule, which states that polar species have larger geographical ranges than tropical species. Climatic variability, glaciation history, or competition are cited as the main causes of this latitudinal pattern, which shows up most clearly in temperate-polar comparisons.

There is a broad positive correlation between distribution and abundance for all kinds of animals and plants, a pattern called Hanski's Rule: Widespread species are more abundant than species that have small geographic ranges. There is considerable variability in this relationship. Three explanations are currently proposed for this pattern. The pattern could be an artifact arising from the problem of sampling rare species in nature, but this is unlikely for well-known groups such as birds. Ecological specialization on rare resources may be an important factor reducing abundance, and to test this second explanation we need data on resource use by abundant species and rare species. Migration among suitable local patches may affect overall abundance and distribution, and we need movement data to test this third model. We are led to ask the question: What makes a species successful? To answer this we turn to consider the problem of abundance in more detail in Part Three.

---

[1] Also called *metapopulations*. See Chapter 16.

Key Concepts

1. Geographical distributions can be mapped at spatial scales from the continental to the local.

2. Most species occupy small geographical areas; few are widespread.

3. Polar species have larger geographical ranges than tropical species in many taxonomic groups (Rapoport's Rule).

4. Distribution and abundance are usually positively related such that widespread species are more abundant than species with small geographic ranges.

Selected References

Brown, J. H. 1984. On the relationship between abundance and distribution of species. *American Naturalist* 124:255–279.

Carter, R. N. and S. D. Prince. 1988. Distribution limits from a demographic viewpoint. In *Plant Population Ecology*, ed. A. J. Davy, M. J. Hutchings, and A. R. Watkinson, pp. 165–184. Blackwell, Oxford.

Caughley, G., D. Grice, R. Barker, and B. Brown. 1988. The edge of the range. *Journal of Animal Ecology* 57:771–785.

Fleishman, E., G. T. Austin, and A. D. Weiss. 1998. An empirical test of Rapoport's Rule: Elevational gradients in montane butterfly communities. *Ecology* 79:2482–2493.

Gaston, K. J., T. M. Blackburn, and J. I. Spicer. 1998. Rapoport's Rule: Time for an epitaph? *Trends in Ecology and Evolution* 13 (2): 70–74.

Gaston, K. J., R. M. Quinn, T. M. Blackburn, and B. C. Eversham. 1998. Species-range size distributions in Britain. *Ecography* 21:361–370.

Gotelli, N. J. and D. Simberloff. 1987. The distribution and abundance of tallgrass prairie plants: A test of the core-satellite hypothesis. *American Naturalist* 130:18–35.

Johnson, C. N. 1998. Rarity in the tropics: Latitudinal gradients in distribution and abundance in Australian mammals. *Journal of Animal Ecology* 67 (5): 689–698.

Questions and Problems

8.1 In Neotropical forest mammals a *negative* relation between geographic distribution and abundance is observed (Arita et al. 1990). Suggest two ecological explanations for this pattern.

8.2 For terrestrial birds in Australia, Schoener (1987) suggested that the greater the range size, the less the average abundance, opposite to the trends reported in this chapter. Gaston (1990, p. 111) rejected Schoener's results. Read Gaston's comments and discuss whether or not you accept Schoener's interpretation.

8.3 Discuss the application of Rapoport's Rule to the altitudinal distribution of species on mountains. What predictions does this hypothesis make for mountain species? Read Fleishman et al. (1998) and compare your analysis with theirs.

8.4 Discuss the implications of the relation between distribution and abundance for conservation biology.

8.5 The relationship between distribution and abundance is often very loose with much variability, as shown in Figure 8.6 on page 111. How does this variability affect the interpretation of these data?

Overview Question

Discuss the classification of species of plants and animals into specialists and generalists. Choose a species and describe what you would measure to convince someone that this species is either a generalist or a specialist. How does this classification assist in understanding the observed positive relationship between distribution and abundance?

ART THREE

The Problem of Abundance:
Populations

CHAPTER 9   POPULATION PARAMETERS

CHAPTER 10   DEMOGRAPHIC TECHNIQUES: VITAL STATISTICS

CHAPTER 11   POPULATION GROWTH

CHAPTER 12   SPECIES INTERACTION: COMPETITION

CHAPTER 13   SPECIES INTERACTION: PREDATION

CHAPTER 14   SPECIES INTERACTION: HERBIVORY AND MUTUALISM

CHAPTER 15   SPECIES INTERACTION: DISEASE AND PARASITISM

CHAPTER 16   POPULATION REGULATION

CHAPTER 17   APPLIED PROBLEMS I: HARVESTING POPULATIONS

CHAPTER 18   APPLIED PROBLEMS II: PEST CONTROL

CHAPTER 19   APPLIED PROBLEMS III: CONSERVATION BIOLOGY

CHAPTER 9

Population Parameters

WITHIN THEIR AREAS OF DISTRIBUTION, animals and plants occur at varying densities. We recognize this variation when we say, for example, that sugar maples are common in one woodlot and rare in another. If we are to make these statements more precise, we must quantify density. This chapter discusses some techniques used to estimate densities of animals and plants.

The Population as a Unit of Study

A population may be defined as a *group of organisms of the same species occupying a particular space at a particular time*. Thus we may speak of the deer population of Glacier National Park, the deer population of Montana, or the human population of the United States. The ultimate constituents of the population are individual organisms. For sexually reproducing organisms, the population may be subdivided into local populations called *demes*, which are groups of interbreeding organisms, the smallest collective unit of a plant or animal population. Individuals in local populations share a common gene pool. The boundaries of a population both in space and in time are vague and in practice are usually fixed arbitrarily by the investigator.

Populations as units of study, have received a good deal of interest from both ecologists and geneticists. Among the principles of modern evolutionary theory are that natural selection acts on the individual organism, and that through natural selection populations evolve. The fields of population ecology and population genetics have much in common.

The population has group characteristics—statistical measures— that cannot be applied to individuals. The basic characteristic of a population that we are interested in is its *density* and this chapter will discuss how to estimate density. The four population parameters that change density are *natality* (egg, seed, or spore production; births), *mortality* (deaths), *immigration*, and *emigration* (see Box 9.1). In addition to these attributes, one can derive secondary characteristics of a population, such as its age distribution, genetic composition, and pattern of distribution of individuals in space. Note that these population parameters result from a summation of individual characteristics.

B O X   9 . 1

DEFINITIONS OF POPULATION PARAMETERS

Every population has a spatial context and a temporal context. So, we can talk about a population of rainbow trout in a particular lake in the summer of 2000. We can describe this population by its attributes:

Density: number of organisms per unit area or per unit volume.

Natality: the reproductive output of a population.

Mortality: the death of organisms in a population.

Immigration: the number of organisms moving into the area occupied by the population.

Emigration: the number of organisms moving out of the area occupied by the population.

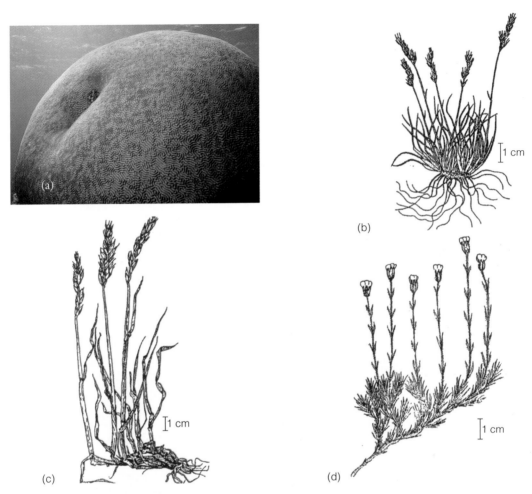

FIGURE 9.1 *Some examples of modular organisms. (a) Brain coral from the Great Barrier Reef. This large structure (1 m in diameter) is actually a colony consisting of thousands of individual coral polyps, each a module budded from the original coral polyp. (b) Fescue grass* (Festuca brachyphylla), *which forms a tussock of tightly packed modules. (c) Wheatgrass* (Agropyron boreale), *in which tillers spread laterally. (d) Sandwort* (Arenaria laricifolia) *with many tillers coming off a spreading stolon. ((a) ©Stuart Westmorland/Corbis.)*

Unitary and Modular Organisms

We are used to organisms in populations that come in individual units such as humans and birds. We call these *unitary* organisms. Deer, mice, humans, and oak trees are easy to identify as individuals. But some organisms do not come in simple units of individuals, and we call these *modular* organisms (Figure 9.1). Many plants are also difficult to categorize because they show great variation in size and structure.

Grasses are particularly difficult to fit into anyone's definition of a single individual, and many other plants have underground connections so that what appear to be separate plants are the same genetic individual (a clone). For example, aspen trees (*Populus tremuloides*) can form clones, so that a whole stand of trees may in fact be a single genetic individual.

Organisms may thus be classified as *unitary* or *modular* organisms. Most higher animals are unitary organisms in which form is determinate. A kitten is recognizable as a cat, and a young giraffe is instantly recognized as a giraffe. In unitary organisms each easily

FIGURE 9.2

Gorgonid corals, an example of a modular organism. (a) Polyps of a black flower coral, an organ-pipe coral from the Philipines, as viewed by a diver. (Photo courtesy of Robert Lin.) (b) Expanded white polyps of an organ-pipe gorgonian coral in the Philippines. Each polyp is a module, and in this group of corals there are 8 pinnate tentacles on each polyp. (©Robert Lin/Corbis.)

recognizable individual is usually a separate genetic individual. By contrast, most plants are modular organisms in which the zygote or spore develops into a unit of construction (a module) that then produces additional similar modules. Modular organisms are often branched, and individuals are composed of variable numbers of modules. Modules that can exist separately are known as *ramets*. Most plants are modular, but many of the lower animals such as hydrozoans, corals, and bryozoans, are also modular organisms (Figure 9.2). Modular organisms can have individuals that are extremely different in size.

To study populations of modular organisms, we must recognize two levels of population structure. In addition to the number of modular units, there is the number of individuals that are represented by original zygotes. These individuals are called *genets*, or genetic individuals (Harper 1977). In plants, an individual genet can be a single tree or a clone extending over a square kilometer. Each genet is composed of one or more modular units of construction, which vary with the type of organism. For example, grasses grow above ground as *tillers* (=ramets), the modular unit of construction, and each genetic individual of a grass may have many tillers. Thus to describe a population of modular organisms, we must specify both the number of genets and the number of modular ramets, in contrast to most populations of unitary organisms, in which the individual is simultaneously the genetic unit and the modular unit. Modular organisms have added another dimension to the problem of population changes-in addition to studying whole organisms, we must measure changes in modules and measure modular "birth" and "death" rates (Harper et al. 1986).

In some of these cases we can circumvent the problem by measuring *biomass* (weight) instead of counting numbers. Foresters are not interested in the number of oak trees in a woodlot but rather the sizes of the trees. In some species we will be able to apply population methodology only by making some arbitrary decision about what units to measure or count. Fortunately, in many cases, individuals and populations are easy to recognize and study.

Estimation of Population Parameters

The population attributes concerned with changes in abundance are interrelated as follows:

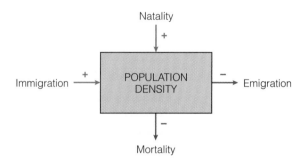

These four processes—natality, mortality, immigration, and emigration—are the *primary population parameters*. When we ask why population density has gone up or gone down in a particular species, we are asking which one (or more) of these parameters has changed. In this section we examine the methods employed in estimating these vital statistics. We can appreciate the problems involved in estimating density by considering the following approximate densities of organisms in nature:

	Density in Conventional Units	Density per m² (or m³)
Diatoms	5,000,000/m³	5,000,000
Soil arthropods	500,000/m²	500,000
Barnacles (adult)	20/100cm²	2,000
Trees	500/ha	0.0500000
Field mice	250/ha	0.0250000
Woodland mice	10/ha	0.0010000
Deer	4/km²	0.0000040
Human beings		
Netherlands	346/km²	0.0003460
Canada	2/km²	0.0000020

1 hectare = 10,000 m² = 2.47 acres
1 sq. kilometer = 100 hectares = 0.386 sq. mile

Given such a wide range of figures, covering more than a dozen orders of magnitude, it is clear that techniques for estimating density that work nicely with deer cannot be applied to bacteria or protozoa. The two fundamental attributes that affect our choice of techniques for population estimation are the *size* and *mobility* of the organism with respect to humans.

Small animals are usually more abundant than large animals. Figure 9.3 shows this trend for 350 species of mammals and 552 species of birds and allows us to predict the approximate density for a species of bird or mammal of given size. Similar plots can be constructed for other species groups (Peters 1983). Table 9.1 presents the regression estimates for several groups of animals, and Box 9.2 illustrates the use of those regressions to estimate average population density for animals of a particular size.

The systematic differences that exist among groups are clear in Figure 9.3. Birds, for example, are less abundant than mammals of equivalent size. If we were to study a 1-kg bird, we should expect a density of approximately 1 per sq. km. For a similar sized mammal, we should to expect about 100 per sq. km.

In most cases we cannot rely on these estimates of average abundance because we need to know if, for example, a population of fish is declining from over-fishing, or if a population of endangered plants is increasing or decreasing over time. How can we estimate population density? Ecologists have developed an array of techniques to estimate population density. Here we will only scratch the surface of the methods used, but it is important to understand in general how

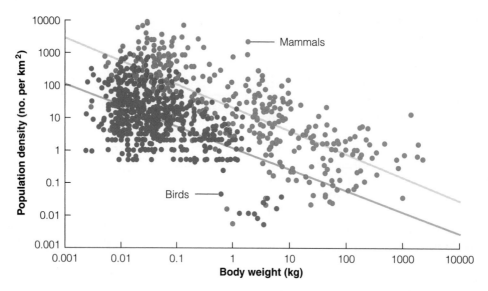

FIGURE 9.3
The relationship between body size and average abundance for 350 species of mammals (red) and 552 species of birds (blue) from around the world. Average trend lines are shown for each group. Note that the scales are logarithmic. (Data from Silva et al. 1997.)

these estimates of populations are made for two reasons: we can begin to see how ecologists quantify nature, and we will come to appreciate how difficult it is to obtain reliable estimates of populations. Reliable quantitative methods are the backbone of science, and one of the great triumphs of ecologists in the past 60 years has been the development of these methods to a high degree of precision (Sutherland 1996, Krebs 1999).

There are two broad approaches to estimating population density. In many cases we need to know the *absolute density* of a population (for example, number of individuals per hectare or per square meter) to make management decisions or conservation recommendations. In other cases we may find it adequate to know the *relative density* of the population (that is, for two areas of equal size, area *x* has more organisms than area *y*). This division of approaches is reflected in the techniques developed for measuring density.

Measurements of Absolute Density

Ecologists go about determining absolute density in two ways: by making total counts and by using sampling methods.

BOX   9.2

CALCULATION OF EXPECTED POPULATION DENSITY FROM THE REGRESSION DATA GIVEN IN TABLE 9.1

For a herbivorous mammal weighing 85 g, its expected average population density would be calculated from the coefficients given in Table 9.1 as follows:

$$\log (\text{population density}) = a + b \, (\log [\text{body mass}])$$
$$= 1.30 - 0.66 \, (\log[0.085])$$
$$= 1.30 + 0.707$$
$$= 2.007$$

- Population density is estimated as the antilog (2.007) or $10^{2.007}$, which is 102 individuals per km².

- For a seed-eating bird that also weigh's 85 g, the expected average population density would be calculated as follows:

$$\log (\text{population density}) = a + b \, (\log [\text{body mass}])$$
$$= 0.22 - 0.54 \, (\log [0.085])$$
$$= 0.22 + 0.578$$
$$= 0.798$$

- Population density for this bird is estimated as the antilog (0.798) or 6.3 individuals per km².

- As shown in Figure 9.3, birds have on average smaller populations than mammals of the same mass.

TABLE 9.1 **Predictive equations for the relationship of average population density to body size for various animal groups.**

The equation for each group has the following general form:

log (population density) = $a + b$ (log [body mass])

where all logs are base 10, population density is the number per km², and body mass is in kilograms.

Group	Intercept of regression line (a)	Slope of regression line (b)	Sample size
Mammals	1.316	−0.688	364
Herbivores	1.30	−0.66	98
Carnivores	1.69	−1.02	25
Birds	−0.045	−0.604	564
Insect eaters	−0.05	−0.64	277
Seed eaters	0.22	−0.54	80
Fishes	1.81	−0.77	11
Aquatic invertebrates	5.37	−0.58	56
Terrestrial invertebrates	3.48	−0.69	106

Source: Data are from Silva et al. (1997) and Peters (1983).

Total Counts

The most direct way to find out how many organisms are living in an area is to count them. One good example of this is a human population census. Other examples come from populations of plants and from vertebrate animals. With trees one can easily count all the individuals in a given area. With territorial birds one can count all the singing males in an area, or with bobwhite quail one can count the number of birds in each covey. Other animals, such as the northern fur seal, may be counted when they are all gathered in breeding colonies. Few invertebrates however, can be counted in total, the exceptions being barnacles and other sessile invertebrates such as some rotifers. Large animals on small areas can sometimes be counted in total or photographed and counted in the photos, but in general direct counts are possible for very few organisms.

Sampling Methods

Usually investigators must be content to count only a small proportion of the population and to use this sample to estimate the total. There are two general sampling techniques: the use of quadrats and the capture-recapture method.

Use of quadrats

The general procedure in this technique is to count all the individuals on several quadrats of known size and then to extrapolate the average count to the whole area. A *quadrat* is a sampling area of any shape. Although the word literally describes a four-sided figure, it has been used in ecology for areas of all shapes, including circles. An example will illustrate this estimation procedure: If you counted 19, 21, 17 and 19 individuals of a beetle species in four soil samples of 10 cm by 10 cm, you could extrapolate this to 1900 beetles per square meter of soil surface.

Achieving reliable estimates using this technique requires three things: (1) The population of each quadrat examined must be determined accurately, (2) the area of each quadrat must be known, and (3) the quadrats counted must be representative of the whole area. This last condition is usually achieved by random sampling procedures; students acquainted with statistics will find a good discussion of this problem in Zar (1996). The population of each quadrat may be counted without error in some organisms but only estimated in other species. Many special techniques have been developed for applying quadrat-sampling techniques to different kinds of animals and plants in terrestrial and aquatic systems. Next we examine two examples of the use of quadrats, one for animals and one for plants.

Wireworms are click beetle larvae (*Elateridae*) that live in the soil. Some species feed on and damage the roots of agricultural crops. To estimate populations of a wireworm root pest (*Agriotes* sp.), Salt and Hollick

(1944) devised a technique of extracting larvae from soil samples. This technique involved breaking the lumps of soil, separating the very coarse and very fine material using sieves, and separating the wireworms from other organic material by benzene flotation (insects accumulate at the benzene-water interface; the plant matter stays in the water). Exhaustive tests were made at each step in this process to see if larvae were lost. The investigators sampled soil by using a corer that removed a cylinder of soil 10 cm in diameter and 15 cm deep. In one pasture near Cambridge, England, they collected 240 random samples that contained a total of 3742 larvae of the wireworm *Agriotes* sp., an average of 15.6 larvae per 10 cm core, or an infestation of 19.3 million larvae per hectare. Variation among the samples can be used to construct confidence limits for this density estimate. Salt and Hollick were able to show by this careful work that wireworm populations were about three times higher in English pastureland than people had previously supposed.

Quadrats have been used extensively in plant ecology; indeed it is the most common method for sampling plants. There is an immense literature on the problems of sampling plants with quadrats; some articles for example, deal with the relative efficiency of round, square, and rectangular quadrats. We will not go into these detailed problems of methodology; interested students should refer to Krebs (1999, Chapter. 4).

To experience quadrat sampling in plant populations, my ecology classes set out a transect in an upland hardwood forest in southern Indiana. We laid out three lines, each 110 meters long, and counted all trees taller than 25 cm within a swath of 1 meter on either side of the line. Each transect line is, in effect, a very long, thin quadrat with an area of 220 m² (0.022 hectare). We obtained the following results:

	No. counted			Estimated no. per hectare		
	Line A	**Line B**	**Line C**	**Line A**	**Line B**	**Line C**
Chestnut oak	20	28	18	909	1273	818
Sugar maple	5	4	7	227	182	318
American beech	13	15	16	591	682	727

By doing this type of quadrat sampling for old trees and then again for seedlings, we could determine if populations were likely to change over time. Foresters have devised a series of ingenious techniques for estimating the abundance of forest trees, and these are reviewed by Buckland et al. (1993) and by Krebs (1999).

Capture-recapture method. The technique of capture, marking, release, and recapture is an important one for mobile animals, because it allows not only an estimate of density but also estimates of birth rate and death rate for the population being studied.

Several models can be used for capture-recapture estimation. Basically, they all depend on the following line of reasoning: If you capture animals, mark them and then release them, the proportion of animals marked in subsequent samples taken from this population should be representative of the proportion marked in the entire population. This is illustrated here for a simple example that includes two capture sessions (shown in the diagram on the facing page).

This simple type of population estimation is known as the *Petersen method*[1] because it was developed by the Danish fisheries scientist C.G.J. Petersen in 1898. It involves only two sampling periods: capture, mark, and release at time 1 and capture and check for marked animals at time 2. The time interval between the two samples must be short because this method assumes a closed population with no recruitment of new animals into the population between times 1 and 2 and no losses of marked animals. We assume that a sample, if random, will contain the same proportion of marked animals as that in the whole population:

$$\frac{\text{Marked animals in second sample}}{\text{Total caught in second sample}} = \frac{\text{Marked animals in first sample}}{\text{Total population size}}$$

For this simple example, using N for total population size:

$$\frac{5}{20} = \frac{16}{N}$$

or $N = 64$.

One of the first to use the Petersen method was Dahl (1919) who marked trout (*Salmo fario*) in small Norwegian lakes to estimate the size of the population

---

[1]Also called the *Lincoln method* by wildlife ecologists, because it was first used by F. C. Lincoln on ducks in 1930.

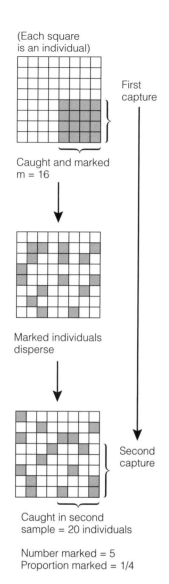

(Each square is an individual)

First capture

Caught and marked
m = 16

Marked individuals disperse

Second capture

Caught in second sample = 20 individuals

Number marked = 5
Proportion marked = 1/4

$$\begin{array}{l}\text{Total population} \\ \text{size}\end{array} = \dfrac{\begin{array}{c}\text{Size of marked} \\ \text{population}\end{array}}{\begin{array}{c}\text{Proportion of} \\ \text{population marked}\end{array}} \quad (9.2)$$

$$= \frac{109}{0.322} = 338 \text{ trout}$$

To estimate density, two situations must be considered: closed and open populations. For population estimation a population is defined as *closed* if it is not changing in size during the period of capture, marking, and recapturing; a population is defined as *open* if it is changing in size during the study period. Real populations are clearly open, unless we sample them over a very brief period.

For a *closed* population, as we have just seen, we know the size of the marked population because it is equal to the number of individuals we have actually marked in the first sample. For an open population, the estimation problem is more difficult. The marked population will decrease from one sampling period to the next because of death and emigration of marked individuals, and it will increase during the sampling as we mark new individuals.

Consider now only the marked population (M) for an open population. This segment contains two kinds of marked animals for an arbitrary time period 7:

	Time period 7		Marked population only
Total rectangle represents the entire marked population	M_C	M_C	Marked animals actually caught at time 7
	M_m	M_m	Marked animals present but not caught at time 7

Some marked animals always miss capture, and a portion of them turn up at later sampling times. In this hypothetical example, individuals could have been marked at time 6, missed at time 7, and caught again at time 8. Such individuals are included in the lower part of the rectangle.

We need to know two things to estimate the size of the marked population: (1) the number of marked animals actually caught (M_C) and (2) the proportion of marked animals actually caught (the ratio of M_C to [$M_C + M_M$]). We already know the first because it is simply a count of marked animals in the catch. There are a number of ways in which the ratio of M_C to [$M_C + M_M$] can be estimated, and

that was subject to fishing. He marked and released 109 trout, and in a second sample a few days later caught 177 trout, of which 57 were marked. From these data we estimate:

$$\begin{array}{l}\text{Proportion of total} \\ \text{population marked}\end{array} = \dfrac{\begin{array}{c}\text{Number marked} \\ \text{in sample}\end{array}}{\begin{array}{c}\text{Total number} \\ \text{caught}\end{array}} \quad (9.1)$$

$$= \frac{57}{177} = 0.322$$

The number of fish marked in the first sample was 109, and therefore

this gives rise to several models of estimation, which are discussed in detail by Seber (1982) and Krebs (1999). Appendix II gives details for one model of estimation.

All capture-recapture models make three crucial assumptions:

- Marked and unmarked animals are captured randomly.
- Marked animals are subject to the same mortality rate as unmarked animals. The Petersen method assumes there is no mortality during the sampling interval.
- Marked animals are neither lost nor overlooked.

All these assumptions have caused trouble at one time or another. For example, field mice may become trap-happy or trap-shy and thus violate the first assumption. Fish tagged on the high seas may be weakened by the nets and the tagging procedure (being held out of water) such that they suffer increased mortality just after release. In some cases, fishermen have not returned tags recovered from marked fish because they considered them good-luck charms. Leg bands may wear thin and be lost from long-lived birds. Numerous variations of the techniques of marking and recapture analysis have been designed to very cleverly circumvent some of these problems. Fisheries scientists in particular have been very active in this field (see Ricker 1975).

Two additional points should be mentioned. First, under some conditions it is possible to test from field data the crucial assumption that animals are captured randomly. Second, the reliability of these population estimates can be determined in standard statistical fashion. Both these points are reviewed and discussed by Seber (1982) and Krebs (1999).

Usually we are not interested in just a single population estimate, and we operate a mark-and-recapture scheme for several months or years, a technique called a *multiple census*. When we do this, we can begin to study the dynamics of a population. The text noted previously that it is possible to get an estimate of the natality rate and mortality rate of the population at the same time as we estimate its size. The gist of this procedure is as follows: Consider just two samples of the population, obtained by the general estimation technique outlined in Appendix II. We have estimates of the marked population at the end of time 7 (M_7) and at the start of time 8 (M_8),

and we also know the total population size at each of these times:

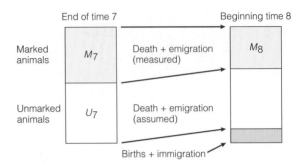

Between times 7 and 8, the marked population can decrease only, owing to death and emigration. (The marked population can *increase* only during a sampling period when we are marking animals that were formerly unmarked.) We define the survival rate as the percentage of animals that survive the time interval:

$$\text{Survival rate } (\%) = \frac{M_8}{M_7} \times 100 \qquad (9.3)$$

Suppose, for example, that we have estimated 500 marked animals at the end of time 7 and 400 at the start of time 8. Then

$$\text{Survival rate } (\%) = \frac{400}{500} \times 100 = 80\% \text{ per unit of time}$$

The mortality rate (or, more properly, *loss rate*, since it includes emigration) is defined simply as:

$$\text{Loss rate} = 100\% - \text{survival rate}$$

These rates apply to the time interval between sample times 7 and 8; if this interval is one year, the estimated survival rate is 80% per year.

The natality or "birth" rate, which is more commonly called the *dilution rate* because it includes both immigration and births, is obtained indirectly. We assume that the survival rate estimated for the marked animals also applies to the unmarked animals. This is shown in the diagram as a parallel arrow, and a numerical example will illustrate how this assumption operates.

Using the techniques just described, we obtained a population estimate for time 7 (1500) and, repeating the whole procedure, for time 8 (1400). We also obtained an estimate for the size of the marked population at the end of time 7 (500) and at the start of time 8 (400). These values are shown in bold type in the following table:

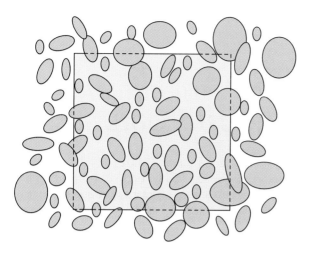

FIGURE 9.4
Schematic illustration of the problem of estimating density in mobile animals. The physical trapping area (yellow square) may be smaller than the effective grid size because of edge effects. The home ranges of mice in a forest are shown schematically as ellipses. If trapping is done throughout the square area, some animals that live on the edge of the trapping grid will be caught, so the effective grid size will be larger than the physical trapping area. Density must be estimated by dividing the estimated population size by the effective grid size. (Modified from White et al. 1982.)

	Time 7	Time 8	Comments
Marked animals	**500**	**400**	Thus survival rate is 80%
Unmarked animals	1000	800	Estimated 80% survival from 7 to time 8
Total population estimate	**1500**	**1400**	Therefore, since 400 + 800 ↑ 1400, dilution must be 200 new animals

If we project the observed survival rate, we find that we can account for $1500 \times 0.80 = 1200$ animals at time 8 as survivors of those present at time 7. Clearly, 200 new animals have appeared through natality or immigration, and this is the *dilution*.

The principle here is summed up in the equation

$$\text{Total population at time } t + 1 = \text{total population at time } t + \text{dilution} - \text{losses}$$

If we know any three of the variables in the equation, we can find the fourth by subtraction.

Developments in statistics have been very important for the study of ecology. It was not until 1936 that this technique of capture-recapture analysis was worked out in sufficient detail that it could become an important ecological tool. In the past 25 years, several important developments occurred in this field of analysis (Pollock et al. 1990, Seber 1982). George Seber, George Jolly, and Richard Cormack have made major statistical contributions to developing and refining mark-recapture methods for animal populations (see Essay 9.1). Other statistical concepts—for example, the problems of sampling—are still being improved after much work in the past 45 years (Thompson 1992). This is a good example of how developments in a science are dependent on progress in both pure and applied mathematics.

The capture-recapture technique has been used mainly on larger forms, such as butterflies, snails, beetles, and many vertebrates, that can be readily marked.

Note that when we have estimated the population size for a mobile species we still have the problem of estimating *density* because we must determine the area occupied by the population. When we are studying an island, or an isolated patch of forest, the area occupied is easy to determine. When habitats are continuous, the extent of the area occupied is less clear. Consider Figure 9.4, which illustrates the home ranges occupied by a population of mice in a forest. The trapping grid includes all of the home range of some mice in the forest, part of the home range of other mice in the forest, and none of the home range of still other mice in the forest. Because some of the animals will be caught, marked, and released in traps set on the edge of the trapping grid, the effective size of the trapping grid is larger than its physical area. To convert population estimates to density we must be able to measure the effective grid size (Anderson et al. 1983). This edge effect can be serious when the study area is small relative to the animal's scale of movement.

Indices of Relative Density

The characteristic feature of all methods for measuring relative density is that they depend on the collection of samples that represent some relatively constant but unknown relationship to total population size. These methods provide an index of abundance that is more or less accurate. When an index of abundance

E S S A Y   9 . 1

A HISTORICAL ESSAY IN ECOLOGY - THE CORMACK-JOLLY-SEBER MARK-RECAPTURE MODEL

In open populations in which births, deaths, and movements continually occur, the population estimation problem is most difficult. Beginning in the early 1900s applied ecologists developed methods that would estimate population size for a closed population. But the holy grail of a method for open populations was elusive. George Leslie, working at Oxford just after World War II, developed the first open mark-recapture model that could estimate population size as well as birth and death rates. Leslie worked with Dennis and Helen Chitty at the Bureau of Animal Population in Oxford to apply this model to field mouse populations in Britain. George Leslie was a talented applied mathematician who had trained in physiology and had no formal training in advanced mathematics but was self-taught. His papers are now classics in quantitative ecology. His collaboration with field ecologists working in Charles Elton's Bureau of Animal Population at Oxford (Crowcroft 1991) produced a series of innovative techniques that will appear throughout this part of the book. He had no computers or even calculators, and his simulated populations consisted of cardboard disks in a box. His 1940s mark-recapture model was improved by the statisticians George Seber working in New Zealand and George Jolly working independently in Edinburgh in 1965, and by Richard Cormack working at Aberdeen on survival estimation. In each case these scientists recognized that field ecologists needed ways of determining population changes accurately. This recognition had a long history in applied ecology as fisheries scientists and entomologists had developed

simpler estimation methods that could be applied to closed populations. This synergy of efforts of applied mathematicians and statisticians is a good example of how the development of rigorous statistical methods

Richard M. Cormack *(1935–) Professor of Statistics, University of St. Andrews, Scotland*

are necessary for science to proceed and how teamwork between field ecologists and applied statisticians can produce results that increase the precision of ecological data. The Cormack-Jolly-Seber model is now the cornerstone of much population estimation by mark-recapture.

(such as tracks in sand plots) is 4.0 on area x and 8.0 on area y, we can conclude that area y has a higher density of animals than area x. You *cannot* conclude that area y has twice the density of area x, because it

may be that there are only 40% more animals on area y, but they are much more active. There are a great many indices of relative density, and here we will list only a few:

1. *Traps.* We previous noted that traps are often used in capture-recapture studies to estimate absolute density. The number of individuals caught per day per trap may also be used as an index of relative density. The traps could include mousetraps spread across a field, light traps for night-flying insects, pitfall traps in the ground for beetles, suction traps for aerial insects, and plankton nets. The number of organisms trapped depends not only on the population density but also the animals' activity and range of movement, and on the researcher's skill in placing traps, so these techniques provide only a rough idea of abundance.

2. *Number of fecal pellets.* This technique has been used for snowshoe hares, deer, field mice, and rabbits. If we know the average rate of defecation the number of fecal pellets in an area can provide an

in several mammals such as Canada lynx; some records extend back 300 years.

5. *Catch per unit fishing effort.* This measure can be used as an index of fish abundance, for example, number of fish per 100 hours of trawling.

6. *Number of artifacts.* This count can be used for organisms that leave evidence of their activities, for example, mud chimneys for burrowing crayfish, tree-squirrel nests, and pupal cases from emerged insects.

7. *Questionnaires.* Questionnaires can be sent to sportsmen or trappers to get a subjective estimate of population changes. This technique is useful only for detecting large changes in population among animals large enough to be noticed.

8. *Cover.* The percentage of the ground surface covered by a plant as a measure of relative density

George M. Jolly *(1922–1998) Research Scientist Agricultural Research Council U.K.*

George A. F. Seber *(1938–) Professor of Statistics, Auckland University*

index of population size. Similar methods are used for defoliating caterpillars by estimating the amount of frass falling from trees.

3. *Vocalization frequency.* The number of bird calls heard per 10 minutes in the early morning has been used as an index of the size of bird populations. The same method can be used for frogs, crickets, and cicadas.

4. *Pelt records.* The number of animals caught by trappers has been used to estimate population changes

has been used by botanists and by invertebrate ecologists studying the rocky intertidal zone. This is an especially important method for modular organisms.

9. *Feeding capacity.* The amount of bait taken by rats and mice has been measured to obtain an index of density.

10. *Roadside counts.* The number of birds observed while driving a standard distance has been used as an index of abundance, and the same technique can be used for other highly visible organisms.

These methods for measuring relative density all need to be viewed skeptically until they have been carefully evaluated. They are most useful as a supplement to more direct census techniques and for detecting large changes in population density.

We conclude our discussion of techniques for measuring density by noting that detailed, accurate census information is obtainable for few animals. In many cases we must be content with an order-of-magnitude estimate. Because of this, a disproportionate amount of work has been done on the more easily censused forms, particularly birds, mammals, and species of economic importance. This fact introduces an obvious bias into the following discussions of the mechanisms determining population density.

Natality

A major factor in population increase is *natality*, a broad term covering the production of new individuals by birth, hatching, germination, or fission.

Two aspects of reproduction must be distinguished. The concept of *fecundity* is a physiological notion that refers to an organism's potential reproductive capacity. *Fertility* is an ecological concept that is based on the number of viable offspring produced during a period of time. We must distinguish between *realized fertility* and *potential fecundity*. For example, the realized fertility rate for a human population may be only one birth per 15 years per female in the childbearing ages, whereas the potential fecundity rate for humans is one birth per 10 to 11 months per female in the childbearing ages.

Natality rate may be expressed as the number of organisms produced per female[2] per unit time, and is synonymous with the realized fertility rate. The magnitude of the natality rate is highly dependent on the type of organism being studied. Some species breed once a year, some breed several times a year, and others breed continually. Some produce many seeds or eggs, others few. For example, a single oyster can produce 55–114 million eggs; fish commonly lay eggs

[2]For asexual organisms that reproduce by fission or budding, all individuals would be included in estimating natality rates. The same would apply to bisexual (monoecious) plants.

in the thousands; frogs produce eggs in the hundreds; birds usually lay between 1 and 20 eggs; and mammals rarely have litters of more than 10 offspring and often have only one or two. Fecundity is usually inversely related to the amount of parental care given to the young.

Mortality

Biologists are interested not only in why organisms die but also why they die at a given age. Mortality—or its converse, survival—can be looked at from several perspectives. Longevity focuses on the age of death of individuals in a population. Two types of longevity can be recognized: *potential longevity* and *realized longevity*. *Potential longevity*, the maximum life span attainable by an individual of a particular species, is a limit set by the physiology of the organism, such that it simply dies of old age. Another way of describing potential longevity is the average longevity of individuals living under optimum conditions. But organisms in nature rarely live in optimum conditions; most animals and plants die from disease, or are eaten by predators, or succumb to any one of a number of natural hazards. *Realized longevity* is the actual life span of an organism. Realized longevity can be averaged for all the individuals in a population living under real environmental conditions, and this average longevity can be measured in the field, whereas potential longevity can be measured only in the laboratory or in a zoo or botanical garden.

Two examples will illustrate these distinctions. The European robin has an average life expectation of one year in the wild, whereas it can live at least 11 years in captivity (Lack 1954). In ancient Rome, the average life expectation at birth for human females was about 21 years, and in England in the 1780s it was about 39 years (Pearl 1922). Realized longevity in humans has risen dramatically during the twentieth century. In the United States in 1999, females at birth could expect to live 81 years on average. Potential longevity in humans is around 100 years. Some individuals in Rome and pre-industrial England did live to be 80 or more, but very few. Low longevity in human populations is due to high mortality in infants and children.

The measurement of mortality may be done directly or indirectly. The direct measure is achieved by marking a series of organisms and observing how

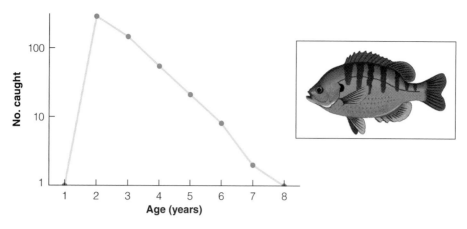

FIGURE 9.5
Catch curve for bluegill sunfish (Lepomis macrochirus) *from Muskellunge Lake, Indiana, 1942. The descending curve after age II can be used to estimate the adult mortality rate. (Data from Ricker 1958.)*

many survive from time t to time $t + 1$. We discussed this technique previously in the section on capture-recapture methods.

An indirect measure of survival may be gotten in several ways. For example, if one knows the abundance of successive age groups in the population, one can estimate the mortality between these ages. Data of this type are widely used in fisheries work, in the analysis of *catch curves*. Figure 9.5 illustrates a catch curve for bluegill sunfish in an Indiana lake. The survival rate can be estimated from the decline in relative abundance from age group to age group (except in the case of 1 year old fish, which are usually too small to be caught in the nets used). For example,

$$\begin{array}{c}\text{Survival rate between} \\ \text{II and III years}\end{array} = \dfrac{\begin{array}{c}\text{Relative abundance} \\ \text{of III fish}\end{array}}{\begin{array}{c}\text{Relative abundance} \\ \text{of II fish}\end{array}}$$

$$= \frac{147}{292} = 0.50$$

Similarly, a drop of 147 to 54 between ages III and IV gives a survival estimate of 0.37 per year between these ages. In making these calculations, we are assuming that the initial number of fish in each of the two age groups was the same and that the survival rate has been constant over time for each age group, so that population size is constant over time. Ricker (1975, Chapter. 2) discusses these problems in detail for fish populations. All indirect measures of survival involve some assumptions and should be used only after careful evaluation.

Immigration and Emigration

Dispersal-immigration and emigration-is seldom measured in a population study. In most cases it is either assumed that the two components are equal, or else work is done in an island type of habitat, where dispersal is presumably of reduced importance. Both assumptions are highly questionable. The capacity to disperse is an essential part of the life cycle of most organisms; it is the ecological process that produces gene flow between local populations and thus helps to prevent inbreeding. Dispersal can set limits on geographic distributions (as we saw in Chapter 4), and it affects community composition (as we will see in Chapter 24). Some populations sustain a net emigration and thus export individuals; others are sustained only by a net immigration. One example is small songbirds in woodlots in the eastern United States. Small woodlots are not productive for birds because of heavy nest predation, and these populations can be sustained only by immigration (Wilcove and

Robinson 1990). Dispersal may be a critical parameter in population changes.

Dispersal can be measured if individuals can be marked in a population. The use of radio-telemetry has revolutionized the study of animal movements, particularly for larger organisms (White and Garrott 1990). The major technical problem in studying dispersal is the scale of the movements involved. Because animals move distances greater than the size of study areas, information on "long distance" dispersal can be lost. One of the major unsolved problems of conservation biology is how to facilitate immigration and emigration from populations in isolated parks or refuges in a fragmented landscape (see Chapter 19).

Limitations of the Population Approach

Two fundamental limitations restrict the methods used for studying populations. First, how can we determine what constitutes a *population* for any given species? What are the boundaries of a population in space? In some situations the boundaries are clear. Wildebeest populations in the Serengeti area of East Africa form five herds that rarely exchange members (Sinclair 1977). The largest herd is highly migratory and moves seasonally between plains and woodlands following the rainy season. The four smaller herds are less migratory and breed in different areas and at slightly different times of the year.

But in many other cases, organisms are distributed in a continuum, and no boundaries are evident. White spruce trees grow in northern coniferous forests from Newfoundland to Alaska. Do all these white spruce trees belong to one population? Most population biologists would answer no to this question, but the reasons for their answer would likely differ. Part of the definition of a population should involve the probability of genetic exchange between members of a given population, but no one is able to specify this probability in a rigorous manner. Moreover, some species are asexual, and we are left with the same general problem that troubles systematists when they seek to determine what constitutes a *species*. One pragmatic answer we can give is: A population is a group of individuals that a population biologist chooses to study. To say this is only to say that we may have to start our study by making a completely arbitrary decision on what to call a population. To remember this decision after the population study is completed is a mark of ecological wisdom.

The second limitation is that populations do not exist as isolates but are imbedded in a community matrix of associated species. When we study a population, we assume that we can abstract from the whole community a single species and the small number of species that interact with it. Whether this abstraction can be done effectively is controversial, as we will see when we discuss community organization in Chapter 24, but for the moment we will proceed with the assumption that a population can be isolated from the complex tapestry of a biological community. We must keep in mind that just as communities are made of populations, populations are made of individuals.

Composition of Populations

Populations are indeed composed of individuals, but not of a series of *identical* individuals, yet we tend to forget this heterogeneity when considering population density. Three major variables distinguish individuals in many populations: *sex, age,* and *size*. The composition of many populations deviates from the expected sex ratio of 50% of each gender, so we cannot assume a constant 1:1 (50%) sex ratio. Populations of the wood lemming (*Myopus schisticolor*) in Finland contain only about 25% males (Fredga et al. 1977). Adult populations of the great-tailed grackle (*Quiscalus mexicanus*) in Texas contain about 29% males, although 50% of the nestlings are males (Selander 1965). The sex ratio of a population clearly affects the potential reproductive rate and the level of inbreeding, and it may affect social interactions in many vertebrates (Gomendio et al. 1990).

Age is a significant variable in human populations, and age effects are common to many species. Older individuals are frequently larger, and changes in size may be the main mechanism by which age effects occur. Larger fish lay many more eggs than smaller fish, and larger plants produce more seeds. Young mammals may be prone to diseases that older animals can resist. Age and size are very significant individual attributes in all animals that have social organization because they help to specify an individual's social position. Old individuals in some species may be post-reproductive, as they are in humans.

Size is a particularly significant variable in modular organisms. Size and age may be correlated in plants and animals with indeterminate growth, but much individual variation in size is independent of age. For modular plants and animals size is usually the ecologically relevant variable that defines their importance in ecological processes.

Other, secondary variables may distinguish individuals in some populations. Color is one obvious trait. Social insects such as ants have castes with distinctive individual morphology. Many phenotypic traits can affect survival, reproduction, or growth and be important to a population.

Because of these individual differences, population ecology must be seen to have a split personality. While we view populations as aggregates of individuals and calculate population density and other population parameters as averages over all individuals, we also recognize that to explain population processes we must understand the individuals that make up the population, and the mechanisms by which they reproduce, move, and die. Not all individuals are equal in a population, and individual variability affects the mechanisms behind population processes. In the next nine chapters we will flip back and forth between this dual view of populations as statistical averages and as mixtures of heterogeneous individuals as we try to understand population dynamics.

Summary

In this chapter we have looked briefly at methods of estimating population size for plants and animals. Every aspect of the problem of abundance comes down to one crucial question: *How can we estimate population size?* Absolute density can be estimated by total counts or by sampling methods using quadrats or capture-recapture methods. Relative density can be estimated by many techniques, depending on the species studied. Once we obtain estimates of population size, we can investigate changes in numbers by analyzing the four primary demographic parameters of natality, mortality, immigration, and emigration.

Organisms may be *unitary* or *modular*. Most animals are unitary with determinate form—cats all have four legs. In most unitary organisms the individual is easily recognized, and each individual is an independent genetic unit. Most plants are modular with repeated units of construction (modules)—oak trees can have any number of leaves. Individuals may be difficult to recognize in modular organisms, and a genetic individual (*genet*) may be a clone with many separate modules (*ramets*). Population changes in modular organisms involve measuring the "birth" rate of new modules and the "death" rate of old modules.

Key Concepts

1. Population density is defined as the number of organisms per unit area or per unit volume.

2. Population density is an integration of natality, mortality, immigration and emigration.

3. Individuals are clearly defined in *unitary* organisms such as deer, but less clearly defined in *modular* organisms of variable size such as grasses.

4. Population size, natality rates and mortality rates can be estimated accurately if one pays attention to the statistical assumptions of the methods employed.

Selected References

Begon, M. 1979. *Investigating animal abundance.* Edward Arnold, London. 97pp.

Blackburn, T. M. and K. J. Gaston. 1997. A critical assessment of the form of the interspecific relationship between abundance and body size in animals. *Journal of Animal Ecology* 66:233–249.

Harper, J. L., R. B. Rosen, and J. White (eds.). 1986. The growth and form of modular organisms. *Philosophical Transactions of the Royal Society of London*, Series B, 313:1–250.

Jackson, J. B. C., L. W. Buss, and R. E. Cooke (eds.). 1985. *Population Biology and Evolution of Clonal Organisms.* Yale University Press, New Haven, Conn.

Krebs, C. J. 1999. *Ecological Methodology*, Chapters 2–4. Addison, Wesley, Longman, Menlo Park, Calif.

Mueller-Dombois, D. and H. Ellenberg. 1974. Measuring species quantities. In *Aims and Methods of Vegetation Ecology*, Chapter 6. Wiley, New York.

Olshansky, S. J., B.A. Carnes, B. A. and C. Cassel. 1990. In search of Methuselah: estimating the upper limits to human longevity. *Science* 250:634–640.

Seber, G. A. F. 1982. *The Estimation of Animal Abundance and Related Parameters.* Griffin, London. 654 pp.

Sutherland, W. J. 1996. *Ecological Census Techniques: A Handbook.* Cambridge University Press, Cambridge. 336 pp.

Questions and Problems

9.1 Canada lynx are now listed as a threatened species in the contiguous 48 United states. Given an average body mass of 9.7 kg, calculate what population density you would expect for lynx from the global relationship for mammalian carnivores in Table 9.1.

9.2 The following table presents the catch (per 100 hours of trawling) of plaice (*Pleuronectes platessa*) in the southern North Sea for three fishing seasons. Plot catch curves (like the one in Figure 9.5) from these data, and use the data to calculate survival rates for the various age classes.

Catch

Age of Plaice	1950–1951	1951–1952	1952–1953
II	39	91	142
III	929	559	999
IV	2320	2576	1424
V	1722	2055	2828
VI	389	982	1309
VII	198	261	519
VIII	93	152	123
IX	95	71	106
X	81	57	61
XI	57	60	40
XII and older	94	87	99

Source: After Gulland (1955).

9.3 An ecology class trapped, marked, released, and retrapped grasshoppers and obtained the following data: Morning sample: 432 marked and released, Afternoon sample: 567 caught, of which 47 were marked. Apply the Petersen method of population estimation to these data, and discuss the necessary assumptions with particular reference to these animals. What are the implications of accidentally killing a grasshopper in the morning sample? In the afternoon sample?

9.4 One technique for estimating the springtime abundance of sheep ticks in Scotland is by dragging a wool blanket over the grass. (Ticks will cling to anything that brushes against them during the spring.) Does this technique measure absolute density or relative density? How might you determine this?

9.5 Populations of some organisms, such as ground-dwelling carabid beetles, can be estimated either by quadrat methods or by capture-recapture techniques. List some of the advantages and disadvantages of each approach.

9.6 Compare the definition of *population* presented in this text with that used in statistics texts (see Sokal and Rohlf 1995, p. 9; Zar 1996, p. 15) and in evolution texts (see Futuyma 1998).

9.7 Aerial surveys for large mammals and birds are simply a type of quadrat sampling applied on a large geographic scale. Discuss the assumptions of quadrat sampling for population estimation (see page 121) for the special case of aerial surveys. Krebs (1999) gives a detailed discussion of these problems.

9.8 Some modular organisms such as trees have a tightly controlled growth form. Others, such as clover or corals, have a very opportunistic and variable pattern of growth. Discuss what environmental factors would select for these two extreme patterns of growth form. Harper (1988) discusses this question.

9.9 Suggest three hypotheses to explain why, in Figure 9.3, birds should in general exist at a lower density than mammals of the same size. Make predictions from each hypothesis and discuss how the predictions could be tested. Silva et al. (1997) discuss this general problem.

9.10 What are the problems of using the oldest living human to estimate the upper limit to human longevity? What other approaches might you use to estimate potential longevity? Is there much scope for increasing human longevity in the developed countries? Read Olshansky et al. (1990) for an analysis of this question.

Overview Question

Roadside counts of the numbers of a particular bird species are available for many years. Design a research program that will allow you to convert these indices of abundance to absolute densities. Identify the assumptions you must make to achieve this goal.

CHAPTER 10

Demographic Techniques:
Vital Statistics

ONE OF THE GREAT STRENGTHS of population ecology is that it is quantitative. If the survival rate of adult bald eagles decreased 2% per year, would their populations decline? If we could increase the survival of juvenile salmon 0.5% in their first year, how many more adults would reach maturity and be available for fishermen? It is possible to answer these questions precisely with some simple mathematics. Population mathematics is not difficult, but it is sufficiently different to merit some of your attention if you wish to achieve a more precise understanding of how and why populations change. The next two chapters provide this quantitative background for population ecology.

Life Tables

Mortality is one of the four key parameters that drive population changes, as we saw in Chapter 9. We need a technique to summarize how mortality is occurring in a population. Is mortality high among juvenile organisms? Do older organisms have a higher mortality rate than younger organisms? We can answer these kinds of questions by constructing a *life table*, a convenient format for describing the mortality schedule of a population. Life tables were developed by human demographers, particularly those working for life insurance companies, which have a vested interest in knowing how long people can be expected to live. There is a correspondingly immense literature on human life tables, but few data are available on other animals or on plants.

Plant and animal populations may be composed of several types of individuals, and in any given analysis a demographer may group them together or may keep them separate. A life insurance company offers to males a policy different from the one they give to females for good demographic reasons, and thus it may be useful for some purposes to classify individuals by sex or age.

A life table is an age-specific summary of the mortality rates operating on a *cohort* of individuals. A cohort may include the entire population, or it may include only males, or only individuals born in a given year. An example of a cohort life table for song sparrows is given in Table 10.1. The columns of this life table are assigned the following symbols, which are consistently used in ecology:

x = age
n_x = number alive at age x
l_x = proportion of organisms surviving from the start of the life table to age x
d_x = number dying during the age interval x to $x + 1$
q_x = per capita rate of mortality during the age interval x to $x + 1$

To set up a life table, we must decide on age intervals in which to group the data. For humans or trees the age interval may be five years; for deer, birds, or perennial plants one year, and for annual plants or field mice one month. By making the age interval shorter, we increase the detail of the mortality picture shown by the life table.

TABLE 10.1 Cohort life table for the song sparrow on Mandarte Island, British Columbia.[a]

Age in years (x)	Observed no. of birds alive (n_x)	Proportion surviving at start of age interval x (l_x)	No. dying within age interval x to x + 1 (d_x)	Rate of mortality (q_x)
0	115	1.0	90	0.78
1	25	0.217	6	0.24
2	19	0.165	7	0.37
3	12	0.104	10	0.83
4	2	0.017	1	0.50
5	1	0.009	1	1.0
6	0	0.0	—	—

[a] Males hatched in 1976 were followed from hatching until all had died six years later. *Source:* From Smith (1988).

Note that if you are given any one of the columns of the life table, you can calculate the rest. Put another way, there is nothing "new" in each of the three columns l_x, d_x and q_x they are just different ways of summarizing one set of data. The columns are related as follows:

$$n_{x+1} = n_x - d_x \qquad (10.1)$$

$$q_x = \frac{d_x}{n_x} \qquad (10.2)$$

$$l_x = \frac{n_x}{n_0} \qquad (10.3)$$

For example, from Table 10.1,

$$n_3 = n_2 - d_2 \qquad q_2 = \frac{d_2}{n_2} \qquad l_4 = \frac{n_4}{n_0}$$

$$= 19 - 7 = 12 \qquad = \frac{7}{19} = 0.37 \qquad = \frac{2}{115} = 0.017$$

The rate of mortality q_x is expressed as a rate for the time interval between successive census stages of the life table. For example, q_x for the song sparrows in Table 10.1 is 0.78 for the interval between egg and one year, or per year. Thus 78% of the birds are lost in the nest or during their first year of life.

The most frequently used part of the life table (see Table 10.1) is the n_x column, the number of survivors at age x. This is often expressed from a starting cohort of 1000, but some human demographers prefer a starting cohort of 100,000. Other workers prefer to plot the l_x column to show the proportion surviving. The n_x (or l_x) data are plotted as a survivorship curve; Figure 10.1 presents the survivorship curves for the human population of the United States in 1998. Note that the n_x values are plotted on a logarithmic scale.

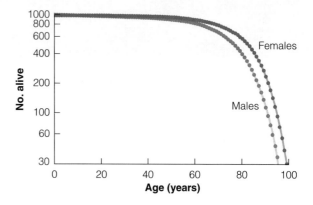

FIGURE 10.1
Survivorship curves for all males (red) and females (blue) in the United States, 1998, for a starting cohort of 1000 individuals. Life expectancy from birth for males is 73 years and for females 80 years. (Data from the U.S. National Center for Health Statistics 1999.)

Population data should be plotted this way when one is interested in per capita *rates* of change rather than absolute numerical changes (Box 10.1).

The life table was introduced to ecologists in 1921 by Raymond Pearl, one of the most important population ecologists in the United States during the first four decades of the twentieth century. Pearl (1928) described three general types of survivorship curves (Figure 10.2). Type 1 curves are characteristic of populations with low per capita mortality for most of the life span and then high losses of older organisms. The linear survivorship curve (type 2) implies a constant per capita rate of mortality independent of age. Type 3 curves indicate high per capita mortality early in life, followed by a period of much lower and relatively constant loss.

No population has a survivorship curve exactly like these idealized ones, and real curves are compos-

BOX 10.1

CALCULATION OF PER CAPITA RATES

Demographers are usually interested in per capita rates of birth and death. You can see why with a simple example. If you were told that 400 ducks had been killed in a disease outbreak, your reaction would be that this number is difficult to evaluate without knowing the size of the duck population. If you were then told that the duck population of this region was 250,000 individuals, you would be able to evaluate this mortality as a per capita rate:

$$\text{Per capita death rate} = \frac{\text{No. of deaths}}{\text{Population at risk}}$$

$$q_x = \frac{d_x}{n_x} = \frac{400}{250,000} = 0.0016$$

or a death rate of 0.16%, a tiny figure. Similarly, if you were told that nine marmots died in a severe storm, you would need to know that the total marmot population was 26 individuals to know that this is a high per capita mortality rate:

$$q_x = \frac{d_x}{n_x} = \frac{9}{26} = 0.346$$

and that more than one-third of the population died.

Natality rates should also be expressed as per capita rates for the same reasons:

$$\text{Per capita birth rate} = \frac{\text{No. of births}}{\text{Size of the reproductive population}}$$

Plotting population data on a logarithmic scale is a simple way to emphasize per capita rates. A numerical example shows this. If half of a population dies, we obtain:

Starting population size	No. dying	Per capita death rate	Final population size
1000	500	0.5	500
500	250	0.5	250
250	125	0.5	125

On a logarithmic scale, all these decreases are equal (base 10 logs):

$$\log(1000) - \log(500) = \log(500) - \log(250)$$
$$3.00 - 2.70 = 2.70 - 2.40$$

Even though *numbers* lost are greatly different, the *per capita death rates* are the same.

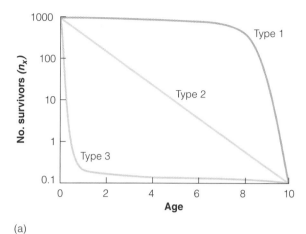

(a)

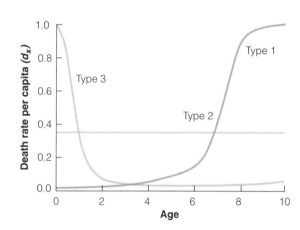

(b)

FIGURE 10.2

(a) Hypothethical survivorship curves (n_x). *(b) Mortality rate* (d_x) *curves corresponding to these hypothetical survivorship curves. Type 2 curves show constant survival rate with respect to age. Type 1 curves show increasing mortality late in life, and Type 3 curves show the highest mortality early in life. (After Pearl 1928)*

ites of the three types. In developed nations, for example, humans tend to have a type 1 survivorship curve (except for the first few days of life). Many birds have a type 2 survivorship curve, and a large number of

populations would fall in the area intermediate between types 1 and 2. Often a period of high loss in the early juvenile stages alters these ideal type 1 and 2 curves. Type 3 curves occur in many fishes, marine invertebrates, and parasites.

Now that we have seen what a life table looks like, how do we get the data to construct one? The answer is: it depends, because there are two very different ways of gathering data for life tables, and they produce two different types of life tables: the cohort life table (which we have already seen in table 10.1) and the static life table. These two life tables are different in form, except under unusual circumstances, and are always quite different in meaning (Caughley 1977).

TABLE 10.2   Static life table for the Human Female Population of Canada, 1996.

Age group (yr)	No. in each age group	Deaths in each age group	Mortality rate per 1000 persons (1000 q_x)
0–4	955,000	1211	1.28
5–9	984,500	159	0.16
10–14	987,700	162	0.16
15–19	976,500	179	0.18
20–24	1,002,900	377	0.38
25–29	1,102,100	525	0.48
30–34	1,297,200	633	0.49
35–39	1,322,500	804	0.61
40–44	1,195,700	1031	0.86
45–49	1,074,700	1422	1.32
50–54	834,000	1899	2.28
55–59	670,700	2444	3.64
60–64	616,900	3820	6.19
65–69	593,100	5900	9.95
70–74	547,100	8642	15.80
75–79	415,100	10,789	25.99
80–84	292,700	14,226	48.60
85–90	162,300	15,739	96.97
90 and above	88,000	28,763	326.85

Source: Statistics Canada (1999).

The *static life table* (also called a stationary, time-specific, current, or vertical life table) is calculated on the basis of a cross section of a population at a specific time. Table 10.2 is a static life table composed from the census data and mortality data for human females in Canada in 1996. A cross section of the female population in 1996 provides the number of deaths (d_x) in each age group and the number of individuals in that age group. This allows us to estimate a set of mortality rates (q_x) for each age group, and the q_x values can be used to calculate a complete life table in the way outlined previously, if we assume that the population is stationary.

The *cohort life life table* (also called a generation or horizontal life table) is calculated on the basis of a cohort of organisms followed throughout life. For example, we could, in principle, get all the birth records from New York City for 1931 and trace the history of all these people throughout their lives, following those that move out of town a very tedious task.

Raymond Pearl *(1879–1940) Professor of Biometry, Johns Hopkins University*

We could then tabulate the number surviving at each age interval. Very few data like these are available for human populations[1]. This procedure would give us the survivorship curve directly, and we could calculate the other life-table functions, as previously described.

---

[1]For human populations, unlike those of other animals and plants, it is possible to construct cohort life tables indirectly from mortality rate (q_x) data. To construct a cohort life table for the 1931 New York City cohort, we can obtain the mortality statistics for the 0- to 1-year-olds for 1931, the 1- to 5-year-olds for 1932–1935, the 6-to-10-year-olds for 1936–1940, and so on, and use these q_x rates to estimate the life-table functions.

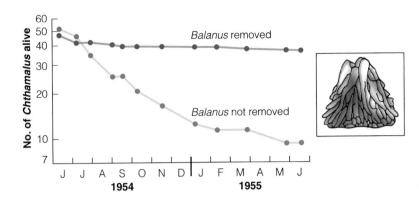

FIGURE 10.3
Survivorship curves of the barnacle Chthamalus stellatus, *which had settled naturally on the shore at Millport, Scotland, in the autumn of 1953. The survival of* Chthamalus *growing without contact with* Balanus *is compared with survival in an area with both species.* Balanus *crowds out* Chthamalus *when the two species are side by side. (Data from Connell 1961a and personal communication.)*

These two types of life table will be identical if and only if the environment does not change from year to year and the population is at equilibrium. But normally birth rates and death rates do vary from year to year, and consequently large differences exist between the two forms of life table. These differences can be illustrated most easily for human populations. For example, a static life table for humans born in 1900 in the United States would show what the survivorship curve would have been if the population had continued surviving at the rates observed in 1900. But of course the human population did not retain these same 1900 rates. The continual improvement in medicine and sanitation in the past 100 years has increased survival rates and life expectancy by more than 15 years, and the people born in 1900 had a cohort or generation survivorship curve unlike that of any of the years through which they lived. Static life tables assume static (stationary) populations.

Insurance companies would like to have data from cohort life tables covering the future, but these data are obviously impossible to get. Insurers are definitely not interested in cohort life tables covering the past—the life table for the 1900 cohort would be of little use for predicting mortality patterns today. So insurers use static life tables and correct them at each census. These predictions will never be completely accurate but will be close enough for their purposes.

Life tables from nonhuman populations are more difficult to come by. In general, ecologists use three types of data to construct life tables:

- *Survivorship directly observed.* The information on survival (l_x) of a large cohort born at the same time, followed at close intervals throughout its existence, is the best to have, since it generates a cohort life table directly and does

not involve the assumption that the population is stable over time. A good example of data of this type is that of Connell (1961a) on the barnacle *Chthamalus stellatus* in Scotland. This barnacle settles on rocks during the autumn. Connell did several experiments in which he removed a competing barnacle, *Balanus balanoides*, from some rocks but not from others, and then about once a month counted the *Chthamalus* surviving on these defined areas (Figure 10.3). Barnacles that disappeared had certainly died; they could not emigrate.

- *Age at death observed.* Data on age at death may be used to estimate the life-table functions for a static life table. In such cases we must assume that the population size is constant over time and that the birth and death rates of each age group remain constant. A good example of this type of data comes from the work of Sinclair (1977) on the African buffalo (*Syncerus caffer*) in the Serengeti area of east Africa. On his study area Sinclair collected 584 skulls of buffalo that had died from natural causes and classified them by age and sex. The age at death was determined by examining the annular rings on the horns. Young animals were difficult to sample properly because their fragile skulls were more susceptible to damage by weather and carnivores. Sinclair estimated the losses during the first two years of life by direct observations on the herd and obtained the mortality estimates shown in Figure 10.4.

- *Age structure directly observed.* Ecological information on age structure, particularly of trees, birds, and fishes, is considerable and in some cases can be used to construct a static life table.

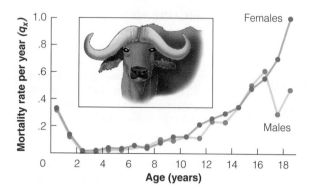

FIGURE 10.4
Mortality rate per year (q_x) *for each one year age interval for the African buffalo. Age at death was determined from skulls of dead buffalo collected during a period of steady population increase. (Data from Sinclair 1977.)*

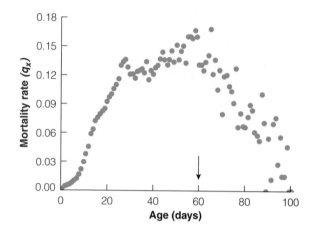

FIGURE 10.5
Age-specific mortality rates in a cohort of 1.2 million Mediterranean fruit flies (Ceratitis capitata) *raised in laboratory cages. A large initial cohort size was needed in order to insure that adequate sample sizes would be available for old flies. Note that the mortality rate does not increase continuously with age, as most models of aging have assumed, but declines after age 60 days (arrow). (Data from Carey et al. 1992.)*

In these cases, we can often determine how many individuals of each age are living in the population. For example, if we fish a lake, we can get a sample of fish and determine the age of each from annular rings on the scales. (The same type of data can be obtained from tree rings.) The difficulty is that to produce a life table from such data, we must assume a constant age distribution, something that is rare for many populations. Consequently, data of this type are not always suitable for constructing a life table.

Attempts to gather life-table data on organisms other than humans and to establish a general theory of senescence have suggested that, except for early ages when mortality is high, mortality rates (q_x) increase inexorably with age, so that for all organisms the mortality curve is roughly U-shaped, as illustrated in Figure 10.4. But this commonly accepted idea of senescence of mortality rates increasing inexorably with age, has been challenged in recent years by large-scale experiments on Mediterranean fruit flies (*Ceratitis capitata*) conducted by Carey et al. (1992) and by extensive summaries of human data by Vaupel et al. (1998). Figure 10.5 plots mortality rates for a laboratory cohort of 1.2 million Mediterranean fruit flies. The death rate for flies that survive beyond 60 days falls instead of rising, as senescence theory would predict. The implication is that our simple ideas of senescence of organisms are not correct, even for humans (Vaupel et al. 1998).

Intrinsic Capacity for Increase in Numbers

A life table summarizes the mortality schedule of a population, and we have just seen several examples. We must now consider the reproductive rate of a population and techniques by which we can combine reproduction and mortality estimates to determine net population changes. Students of human populations were the first to appreciate and solve these problems. One way of combining reproduction and mortality data for populations utilizes a demographic parameter called the *intrinsic capacity for increase* derived by Alfred Lotka in 1925.

Any population in a particular environment will have a mean longevity or survival rate, a mean natality rate, and a mean growth rate or speed of development of individuals. The values of these means are determined in part by the environment and in part by the innate qualities of the organisms themselves. These qualities of an organism cannot be measured simply because they are not a constant, but by measuring their expression under specified conditions we can define for each population its *intrinsic capacity for increase* (also called the *Malthusian parameter*), a sta-

TABLE 10.3 Survivorship schedule (l_x) and fertility schedule (b_x) for women in the United States, 1996,

Age group	Midpoint or pivotal age x	Proportion surviving to pivotal age l_x	No. female offspring per female aged x per 5-year period (b_x)	Product of l_x and b_x
0–9	5.0	0.9932	0.0	0.00
10–14	12.5	0.9921	0.003	0.0030
15–19	17.5	0.9905	0.137	0.1357
20–24	22.5	0.9883	0.278	0.2747
25–29	27.5	0.9860	0.285	0.2810
30–64	32.5	0.9829	0.211	0.2074
35–39	37.5	0.9785	0.089	0.0871
40–44	42.5	0.9725	0.017	0.0165
45–49	47.5	0.9636	0.0007	0.0007
50 +	—	—	0.0	0.00

$$R_0 = \sum_0^\infty l_x b_x = 1.0061$$

Source: Statistical Abstract of the United States 1998.

tistical population characteristic that depends on environmental conditions.

Environments in nature vary continually. They are never consistently favorable or consistently unfavorable but fluctuate between these two extremes for example, from winter to summer. When conditions are favorable, numbers increase; when conditions are unfavorable, numbers decrease. It is clear that no population goes on increasing forever. Darwin (1859, Chapter 3) recognized the contrast between a high potential rate of increase and an observed approximate balance in nature. He illustrated this problem by asking why there were not more elephants, given his estimate that two elephants could give rise to 19 million elephants in 750 years.

Therefore, in nature we observe an actual rate of population change that is continually varying from positive to negative in response to changes within the population in age distribution, social structure, and genetic composition, and in response to changes in environmental factors. We can, however, ask what would happen to a population if it persisted in its current configuration of births and deaths. This abstraction is the ecologist's version of the perfect vacuum of introductory physics: we ask what would happen in terms of population increase if conditions remained unchanged for a long time in a particular environment.

An organism's innate or intrinsic capacity for increase depends on its fertility, longevity, and speed of development. For any population, these processes are integrated and measured by the natality rate and

the death rate. When the natality rate exceeds the death rate, the population will increase. If we wish to estimate quantitatively the rate at which the population increases or decreases, we need to describe how both the natality rate and the death rate vary with age.

How can we express the variations of natality and mortality rates with age? We have just discussed the method of expressing survival rates as a function of age. The life table includes a table of age-specific survival rates. The portion of the life table needed to compute the capacity for increase is the l_x column, the proportion of the population surviving to age x. Similarly, the natality rate of a population is best described by an age schedule of births, seed production, egg production, or fission. This is a table that gives (for sexual species) the number of female offspring produced per female aged x to $x+1$ and is called a *fertility schedule*, or b_x function. Usually only females are counted, and the demographer typically views populations as females giving rise to more females. Table 10.3 gives the survivorship table the l_x schedule with which we are familiar and the *fertility schedule* for women in the United States in 1996. In this case, the great majority of women live through the childbearing ages. The fertility schedule gives the expected number of *female* offspring for each woman living through the five years of each age group. For example, slightly fewer than three women in 10 between the ages of 25 and 29 will, on average, have a female baby.

Given these data, we can obtain a useful statistic, the *net reproductive rate* (R_0). If a cohort of females

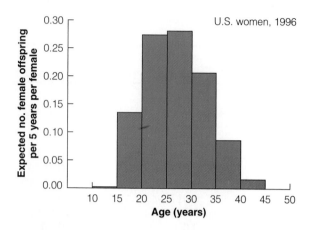

FIGURE 10.6
Expected number of female offspring per five year period for each female in the United States in 1996. Data are from the final column in Table 10.3. The area under the histogram is the net reproductive rate R_0. *(Data from the* Statistical Abstract of the United States 1998.*)*

lives its entire reproductive life at the survival and fertility rates given in Table 10.3, what will this cohort or generation leave as its female offspring? We define as the net reproductive rate as follows:

$$\text{Net reproductive rate} = R_0 \qquad (10.4)$$

$$= \frac{\begin{array}{c}\text{Number of daughters}\\\text{produced in}\\\text{generation } t+1\end{array}}{\begin{array}{c}\text{Number of daughters}\\\text{produced in}\\\text{generation } t\end{array}}$$

R_0 is thus the multiplication rate per generation[2] and is obtained by multiplying together the l_x and b_x schedules and summing over all age groups, as shown in Table 10.3:

$$R_0 = \sum_0^\infty l_x\, b_x \qquad (10.5)$$

Thus we temper the natality rate by the fraction of expected survivors to each age. If survival were 100%, R_0 would just be the sum of the b_x column. In this example (see Table 10.3), if the human population of the United States continued at these 1996 rates, it

would multiply 1.006 times in each generation. If the net reproductive rate is 1.0, the population is replacing itself exactly; when the net reproductive rate is below 1.0, the population is not replacing itself; and if the rates in the example continue for a long time, the population will increase about 0.6 percent each generation in the absence of immigration or emigration. The net reproductive rate is illustrated in Figure 10.6.

Given these two schedules expressing the age-specific rates of survival and fertility, we may inquire at what rate a population subject to these rates would increase, assuming (1) that these rates remain constant and (2) that no limit is placed on population growth. Because these survival and fertility rates vary with age, the actual natality and mortality rates of the population will depend on the existing age distribution. If the whole population were over 50 years of age, it would not increase. Similarly, if all females were between 20 and 25, the rate of increase would be much higher than if they were all between 35 and 39. Before we can calculate the population's rate of increase, it would seem that we must specify (1) age-specific survival rates (l_x), (2) age-specific natality rates (b_x), and (3) age distribution.

This intuitive conclusion is not correct. Contrary to intuition, we do not need to know the age structure of the population. Lotka (1922) showed that a population that is subject to a constant schedule of natality and mortality rates will gradually approach a fixed or *stable age distribution*, whatever the initial age distribution may have been, and will then maintain this age distribution indefinitely. This theorem is one of the most important discoveries in mathematical demography. When the population has reached this stable age distribution, it will increase in numbers according to the differential equation

$$\frac{dN}{dt} = rN \qquad (10.6)$$

or, as rewritten in integral form:

$$N_t = N_0 e^{rt} \qquad (10.7)$$

where N_0 = number of individuals at time 0

N_t = number of individuals at time t

$e = 2.71828$ (a constant)

r = intrinsic capacity for increase for the particular environmental conditions

t = time

---

[2]A generation is defined as the mean period elapsing between the birth of parents and the birth of offspring; see page 141 and Figure 10.9.

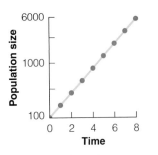

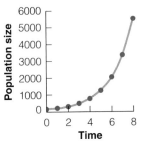

(a) Logarithmic scale (b) Arithmetic scale

FIGURE 10.7
Geometric growth of a hypothetical population when
$N_0 = 100$ *and* r = 0.5, *according to Equation (10.7).*
(a) On a logarithmic scale, geometric population growth
appears as a straight line. (b) On an arithmetic scale,
geometric population growth is a curve that rises more
rapidly with time.

x	l_x	b_x	$l_x b_x$	$(x)(l_x)(b_x)$
0	1	0	0	0
1	1	2	2	2
2	1	1	1	2
3	1	0	0	0
4	0	–	–	–

$$R_0 = \sum_0^4 l_x b_x = 3$$

This equation describes the curve of geometric increase in an expanding population (or geometric decrease to zero if *r* is negative).

A simple example illustrates this equation. Let the starting population (N_0) be 100 and let *r* = 0.5 per female per year. The successive populations would be:

Year	Population size
0	100
1	$(100)(e^{0.5}) = 165$
2	$(100)(e^{1.0}) = 272$
3	$(100)(e^{1.5}) = 448$
4	$(100)(e^{2.0}) = 739$
5	$(100)(e^{2.5}) = 1218$

This hypothetical population growth is plotted in Figure 10.7. Note that on a logarithmic scale the increase is linear, but on an arithmetic scale the curve swings upward at an accelerating rate.

To summarize to this point: (1) Any population subject to a fixed age schedule of natality and mortality will increase in a geometric way, and (2) this geometric increase will dictate a fixed and unchanging age distribution called the *stable age distribution*.

Let us invent a simple hypothetical organism to illustrate these points. Suppose that we have a parthenogenetic animal that lives three years and then dies. It produces two young at exactly one year of age, one young at exactly two years of age, and no young at year 3. The life table and fertility table for this hypothetical animal are thus extremely simple:

If a population of this organism starts with one individual at age 0, the population growth will be as shown in Figure 10.8 (page 142), or, in tabular form, as follows:

	Number at Ages				Total population	% Age 0 in total
Year	0	1	2	3	size	population
0	1	0	0	0	1	100.0
1	2	1	0	0	3	66.7
2	5	2	1	0	8	62.50
3	12	5	2	1	20	60.00
4	29	12	5	2	48	60.42
5	70	29	12	5	116	60.34
6	169	70	29	12	280	60.36
7	408	169	70	29	676	60.36
8	985	408	169	70	1632	60.36

Note that the age distribution quickly becomes fixed or stable with about 60% at age 0, 25% at age 1, 10% at age 2, and 4 % at age 3. This demonstrates Lotka's (1922) conclusion that a population growing geometrically develops a stable age distribution.

We may also use our hypothetical animal to illustrate how the intrinsic capacity for increase *r* can be calculated from biological data. The data of the l_x and b_x tables are sufficient to allow the calculation of *r*, the intrinsic capacity for increase in numbers. To do this, we first need to calculate the net reproductive rate (R_0), explained earlier. For our hypothetical animal, $R_0 = 3.0$, which means that the population can triple its size each generation. But how long is a generation? The *mean length of a generation (G)* is the mean period elapsing between the production or "birth" of parents and the production or "birth" of offspring. This is only an approximate definition, because offspring are produced over a period of time and not all at once. The mean length of a generation is defined approximately as follows (Dublin and Lotka 1925):

$$G_c = \frac{\sum l_x b_x x}{\sum l_x b_x} = \frac{\sum l_x b_x x}{R_0} \qquad (10.8)$$

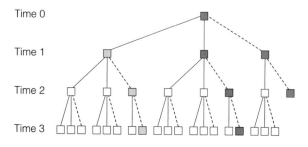

FIGURE 10.8

Population growth of a simple hypothetical organism that is parthenogenetic. Start at the top of the diagram with one green individual (each box represents one individual). At time 1 this individual gives birth to two young (yellow, red), so that there are now three individuals at time 1. At time 2 the two young individuals (red and yellow) give birth to two young each and the old green individual gives birth to one young, so at time 2 there are now eight individuals. The green individual then dies and the others reproduce, so that at time 3 there are 19 individuals. Solid lines indicate reproduction and dashed lines indicate the aging of individuals from one time to the next. Three of the individuals are color-coded to show their presence through time.

For our model organism, $G = 4.0/3.0 = 1.33$ years. Figure 10.9 uses the metaphor of a balance to illustrate the approximate meaning of generation time for a human population. Leslie (1966) has discussed some of the difficulties of applying the concept of generation time to a continuously breeding population with overlapping generations. For organisms such as annual plants and many insects with a fixed length of life cycle, the mean length of a generation is simple to measure and to understand.

Knowing the multiplication rate per generation (R_0) and the length of a generation (G), we can now determine r directly as an instantaneous rate:

$$r = \frac{\log_e(R_0)}{G} \qquad (10.9)$$

For our hypothetical organism,

$$r = \frac{\log_e(3.0)}{1.33} = 0.824 \text{ per individual per year}$$

Because the generation time G is an approximate estimate[3], this value of r is only an approximate estimate when generations overlap.

The capacity for increase is an instantaneous rate and can be converted to the more familiar finite rate[4] by the formula

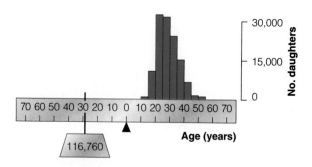

FIGURE 10.9

A mechanical balance to illustrate the idea of the mean length of one generation. Histogram of daughters from a cohort of 100,000 mothers starting life together (right side) is balanced by sum of total daughters (116,760) at exactly 28.46 years from the fulcrum. The mean length of a generation (G_c) is thus 28.46 years for these data. Data from the U.S. population of 1920, $R_0 = 1.168$.

$$\text{Finite rate of increase} = \lambda = e^r \qquad (10.11)$$

Box 10.2 illustrates how to calculate the intrinsic capacity for increase from survivorship and fertility schedules.

It should now be clear why the intrinsic capacity for increase in numbers cannot be expressed quantitatively except for a particular environment. Any component of the environment, such as temperature, humidity, or rainfall, might affect the natality and mortality rates and hence r.

Charles Birch, working at the University of Sydney, did some of the classic early research applying these quantitative demographic techniques to insects. One illustration of the effect of the environment on the capacity for increase was developed by Birch (1953a) in his work on *Calandra oryzae*, a beetle pest that lives in stored grain. The capacity for increase in this species varied with the temperature and with the moisture content of the wheat, as shown in Figure 10.10. The practical implications of these results are that wheat should be stored where it is cool and dry to prevent losses from *C. oryzae*.

---

[3] Generation time has also been defined by Caughley (1977) as:

$$G_M = \frac{\sum\left(l_x b_x x e^{-rx}\right)}{\sum\left(l_x b_x e^{-rx}\right)}$$

This will not give exactly the same value for generation time as defined in Equation (10.8); see Gregory (1997).

[4]Appendix III gives a general discussion of instantaneous and finite rates.

E S S A Y   1 0 . 1

DEMOGRAPHIC PROJECTIONS AND PREDICTIONS

How much will the whooping crane population grow in the next two years? What will the AIDS epidemic do to the population of Africa between now and 2050? To answer questions such as these, we can use the demographic methods outlined in this chapter, but in doing so it is crucial that we make one subtle but important distinction: These methods can provide *projections*—that something will happen *if* conditions a and b are met—but not *predictions* that something will happen, period. (Scientists cannot predict the future; if you want a prediction, consult an astrologer or a Ouija board.) A *demographic projection* is a statement of what will happen to a population if certain assumptions are met, and demographic projections are correct only under very specific assumptions. A demographer can project population changes into the future on the assumption that, say the age-specific birth and death rates will remain constant. But in the real world the simple assumption that things will remain as they are now is rarely a correct one. Thus projections on the effects of AIDS on a population are most difficult because they require some uncertain assumptions about future death rates. Moreover, unpredictable changes such as catastrophic environmental events are especially damaging to demographic projections. No demographer can foresee mortality to half of the whooping crane population caught in an episode of severe weather.

In spite of the fact that they cannot predict the future, it is still useful for conservationists and resource managers to make projections of what will happen if specific assumptions are fulfilled. Such projections, many of which we will examine in the next several chapter, can limit our optimism and pessimism alike.

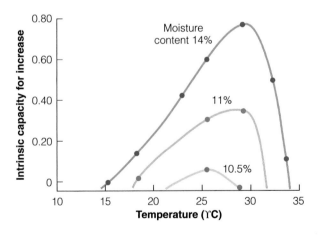

Calandra oryzae

FIGURE 10.10
Intrinsic capacity for increase (r) *of the grain beetle* Calandra oryzae *living in wheat of different moisture contents and at different temperatures. The higher the moisture content of the wheat, the more rapidly these beetles can increase in numbers. (After Birch 1953a.)*

Grain beetles live in an almost ideal habitat, surrounded by food, protected from most enemies, and with relatively constant physical conditions. They are also easy to deal with in the laboratory and are thus used extensively in ecology lab experiments. When Birch (1953a) studied two species, *Calandra oryzae* (a temperate species) and *Rhizopertha dominica* (a tropical species), he found that in both species, *r* varied with temperature and moisture (Figure 10.11). The lines *r* = 0 mark the limits of the possible ecological range for each species with respect to temperature and moisture. *Calandra* is more cold-resistant; *Rhizopertha* can increase at higher temperatures and lower humidities. The distribution of the two species in Australia agrees with these results: *Rhizopertha* is a pest only in the warmer parts of the

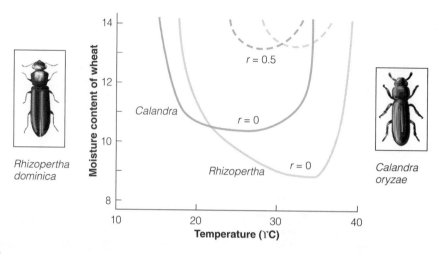

FIGURE 10.11

Intrinsic capacity for increase (r) *of the two grain beetles* Calandra oryzae *and* Rhizopertha dominica *living in wheat of different moisture content and at different temperatures. The higher the temperature and the dryer the wheat, the more* Rhizopertha *is favored. (After Birch 1953a.)*

Charles Birch *(1918–) Professor of Zoology, University of Sydney*

country and is absent from Tasmania, where *Calandra* occurs as a pest.

In general, the intrinsic capacity for increase is not correlated with the abundance of species: Species with a high *r* are not always common, and species with a low *r* are not always rare. Some species, such as the bison in North America, the elephant in central Africa, and the periodical cicadas, are (or were) quite common and yet have a low *r* value. Many parasites

and other invertebrates with a high capacity for increase are nevertheless quite rare. Darwin (1859) pointed this out in *The Origin of Species.* From a conservation viewpoint species with a high *r* can recover more quickly from disturbances, and these calculations will permit us to calculate exactly how fast they might recover.

We can calculate how certain changes in the life history of a species would affect its capacity for increase in numbers. In general, three factors will increase *r*: (1) reduction in age at first reproduction, (2) increase in number of progeny in each reproductive event, and (3) increase in number of reproductive events (increased longevity). In many cases when *r* is large, the most profound effects are achieved by changing the age at first reproduction. For example, Birch (1948) calculated for the grain beetle *C. oryzae* the number of eggs needed to obtain *r*= 0.76 according to the age at first reproduction:

Age at which breeding begins (weeks)	Total no. eggs that must be laid to produce r = 0.76
1	15
2	32
3	67
4	141 (actual life history)
5	297
6	564

BOX 10.2

CALCULATION OF THE INTRINSIC CAPACITY FOR INCREASE FROM LOTKA'S CHARACTERISTIC EQUATION

The intrinsic capacity for increase can be determined more accurately by solving the characteristic equation, a formula derived by Lotka (1907, 1913):

$$\sum_0^\infty e^{-rx} l_x b_x = 1$$

This equation cannot be solved explicitly for r because it cannot be rearranged to have r on one side and all else on the other. By substituting trial values of r, we can solve this equation iteratively, by trial-and-error. Our hypothetical animal (see Figure 10.8) can be used as an example. For our estimate of $r = 0.824$, we get

x	$l_x b_x$	$e^{-0.824x}$	$e^{-8.24x} l_x b_x$
0	0.0	1.00	0.000
1	2.0	0.44	0.880
2	1.0	0.19	0.190
3	0.0	0.08	0.000
4	0.0	0.04	0.000

$$\sum_0^\infty e^{-rx} l_x b_x = 1.070$$

If the sum is too large (as it is here), then the estimate of $r = 0.824$ is too low. We repeat with $r = 0.85$, and after several trials we find that for this hypothetical organism, $r = 0.881$ provides

$$\sum_0^\infty e^{-rx} l_x b_x = 1.004$$

which is a close enough approximation. Carey (1995) works out another example in detail.

The intrinsic capacity for increase is an instantaneous rate (see Appendix III) and can be converted to the more familiar finite rate by the formula

$$\text{Finite rate of increase} = \lambda = e^r \quad (10.11)$$

For example, if $r = 0.881$, then $\lambda = 2.413$ per individual per year in our hypothetical organism. Thus for every individual present this year, 2.413 individuals will be present next year.

The earlier the peak in reproductive output, the larger the r value, as a rule. Lewontin (1965) provides an excellent example to illustrate this in *Drosophila serrata* (Figure 10.12). The Rabaul race of this fruit fly survives poorly and lays fewer eggs than the Brisbane race, but because it begins to reproduce at an earlier age (11.7 days compared with 16.0 days) and has a shorter generation length, its capacity for increase is equal to that of the longer-living, more fertile Brisbane race.

Demographic analyses aggregate individuals into statistical population measures such as r, the intrinsic capacity for increase. One of the difficulties of this approach is that it ignores individual variation in performance. Carey et al. (1996) developed a simple graphical technique for illustrating individual variation in demographic performance. Figure 10.13 shows the survival and lifetime reproduction for 1000 individual Mediterranean fruit fly females. By color coding the rate of egg laying and rank-ordering the individual lifetimes, these individual graphs allow us to see the detailed life history pattern of the population. The details of individual life histories can shed light on how population changes originate in the properties of individual organisms.

To conclude: The concept of an intrinsic capacity for increase in numbers, which we have just discussed, is an oversimplification of nature. In nature, we do not find populations with stable age distributions or with constant age-specific mortality and fertility rates. The actual rate of increase we observe in natural populations varies in more complex ways than the theoretical constant r. The importance of r lies mostly in its use as a model for comparison with the actual rates of increase we see in nature. The actual rate of increase along with its components in the life table and fertility table, can be used in the diagnosis of environmental quality because they are sensitive to environmental conditions.

Reproductive Value

We can use life tables and fertility tables to determine the contribution to the future population that an individual female will make. We call this the *reproductive value* of a female aged x (Williams 1966), and this is most easily expressed for a population that is stable in size as follows:

$$\text{Reproductive value at age } x = V_x = \sum_{t=x}^{w} \frac{l_t b_t}{l_x} \quad (10.12)$$

FIGURE 10.12

Observed $l_x b_x$ *functions for two races of* Drosophila serrata. *Both* $l_x b_x$ *functions give the same value of the innate capacity for increase* (r) *because of the overriding importance of earlier reproduction and shorter generation length of the Rabaul race. Brisbane females lay an average of 546 eggs at 20°C, while Rabaul females lay only 151 eggs during their life span. (After Lewontin 1965.)*

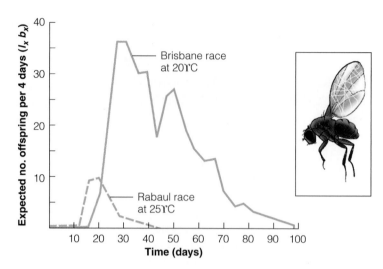

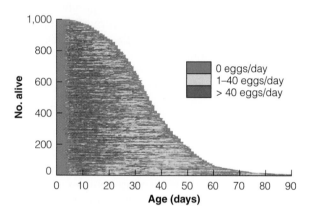

FIGURE 10.13

Graphical technique for displaying individual fertility and survival data. The life spans of 1000 Mediterranean fruit fly females is shown. Each horizontal colored line represents one individual. Individuals are ranked from the shortest life span at the top to the longest- lived individual at the bottom. (From Carey et al. 1998.)

where t and x are age and w is the age of last reproduction. Note that as defined here, reproductive value at age 0 is the same as net reproductive rate (R_0) defined on page 140.

Reproductive value can be partitioned into two components (Pianka and Parker 1975):

Reproductive value at age x = present progeny + expected future progeny

$$V_x = b_x + \sum_{t=x+1}^{w} \frac{l_t b_t}{l_x} \qquad (10.13)$$

We call the second term residual reproductive value, because it measures the number of progeny on average that will be produced in the rest of an individual's lifespan.

Reproductive value is more difficult to define if the population is not stable (Roff 1992, Stearns 1992). In this case we must discount future reproduction if population growth is occurring because the value of one progeny is less in a larger population. Figure 10.14 illustrates the change of reproductive value with age in a red deer population in Scotland. Red deer stags defend harems, and their effective breeding span is three to five years between the ages of six and 11 years. By contrast, red deer hinds start to produce calves at age 3 and breed until they are 15 years old or older. These differences in reproductive biology explain the shapes of the reproductive value curves in Figure 10.14.

Reproductive value is important in the evolution of life-history traits. Natural selection acts more strongly on age classes with high reproductive values and very weakly on age classes with low reproductive values. Predators will have a greater effect on a population if they prefer individuals of high reproductive value.

Age Distributions

We have already discussed the idea of age distribution in connection with the intrinsic capacity for increase. We noted that a population growing geometrically with constant age-specific mortality and fertility rates

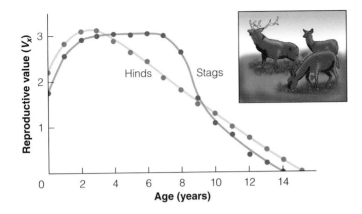

FIGURE 10.14
*Reproductive value for red deer stags (males) of
different ages, compared with that of hinds (females)
on the island of Rhum, Scotland. Reproductive value
is calculated in terms of the number of female
offspring surviving to one year of age that parents of
different ages can expect to produce in the future.
(From Clutton-Brock et al. 1982, p. 154.)*

would assume and maintain a stable age distribution.
The stable age distribution can be calculated for any
set of life tables and fertility tables. The stable age distribution is defined as follows:

C_x = proportion of organisms in the age
category x to $x + 1$ in a population
increasing geometrically

Mertz (1970) has shown that:

$$C_x = \frac{\lambda^{-x} l_x}{\sum_{i=0}^{\infty} \lambda^{-i} l_i} \quad (10.14)$$

where $\lambda = e^r$ = finite rate of increase

l_x = survivorship function from life table

x, i = subscripts indicating age

Let us go through these calculations with our
hypothetical organism:

$\lambda = e^r = e^{0.0881} = 2.413$

Age (x)	l_x	λ^{-x}	$\lambda^{-x} l_x$
0	1.0	1.0000	1.00001
1	1.0	0.4144	0.4144
2	1.0	0.1717	0.1717
3	1.0	0.0711	0.0711
4	0.0	0.0295	0.0000

$$\sum_{x=0}^{4} \lambda^{-x} l_x = 1.6572$$

Thus to calculate C_0, the proportion of organisms
in the age category 0 to 1 in the stable age distribution, we have

$$C_x = \frac{\lambda^{-0} l_0}{\sum_{i=0}^{4} \lambda^{-i} l_i} = \frac{(1.0)(1.0)}{1.6572} = 0.6035$$

For C_1, we have

$$C_1 = \frac{\lambda^{-1} l_1}{\sum_{i=0}^{4} \lambda^{-i} l_i} = \frac{(0.4144)(1.0)}{1.6572} = 0.250$$

In a similar way,

$$C_2 = 0.104$$
$$C_3 = 0.043$$

Compare these calculated values with those
obtained empirically earlier (page 141). Carey (1993)
illustrates another method of calculating the stable
age distribution for a set of l_x and b_x schedules.

Populations that have reached a constant size, in
which the fertility rate equals the mortality rate, will
also assume a fixed age distribution, called a *stationary
age distribution* (or *life-table age distribution*) and will
maintain this distribution. The stationary age distribution is a hypothetical one and illustrates what the
age composition of the population would be at a particular set of mortality rates (q_x) if the fertility rate
were exactly equal to the mortality rate. Figure 10.15
contrasts the stable and stationary age distributions
for the short-tailed vole in a laboratory colony.

A constant age structure in a population is
attained only if the l_x and b_x distributions are fixed and

FIGURE 10.15

(a) Stable age distribution and (b) stationary age distribution for the vole Microtus agrestis *in the laboratory. The stable age distribution should be observed when populations are growing rapidly, and the stationary age distribution when populations are constant in size. (After Leslie and Ranson 1940.)*

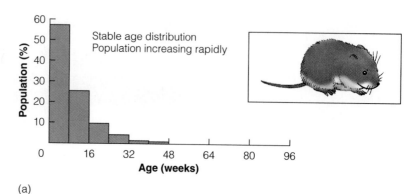

(a)

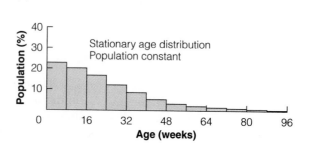

unchanging. This typically occurs in only two situations: (1) When the age-specific fertility and mortality rates are fixed and unchanging and the population grows exponentially, the population assumes a constant age structure called the *stable age distribution;* and (2) when the fertility rate exactly equals the mortality rate and the population does not change in size over time, the population assumes a constant age structure called the *stationary age distribution* which has the same form as the l_x distribution. Under any other circumstances, the populations age structure is not a constant but the changes over time. In natural populations, the age structure is thus almost constantly changing. We rarely find a natural population that has a stable age structure because populations do not increase for long in an unlimited fashion. Nor do we often find a stationary age distribution, because populations are rarely in a stationary phase for long. We can illustrate these relationships as follows:

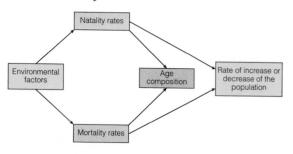

With proper care, information on age composition can be used to judge the status of a population. Increasing populations typically have a predominance of young organisms, whereas constant or declining populations do not (see Figure 10.15). Figure 10.16 illustrates this contrast among the human population of Kenya, which was increasing at 2.1% per year in 1995 and had an average life expectation at birth of 49 years; that of the United States, which was increasing at 0.6% per year in 1995 and had an average life expectation of about 77 years; and that of Italy, which had a zero rate of increase with an average life expectation of 78 years. The age structure of human populations has been analyzed in detail because of its economic and sociological implications (Weeks 1996). A country with a high fertility rate and a large proportion of children such as Kenya, with 46% under age 15 has a much greater demand for schools and other child services than do countries such as the United States, with 21% under age 15.

In populations of plants and animals, even more variation in age composition is apparent. In long-lived species such as trees and fishes, one may find *dominant year-classes*. Figure 10.17 illustrates this for Engelmann spruce and subalpine fir trees of the Rocky Mountains, in which some year-classes may be 100 times as numerous as others. In these situations, the age composition can change greatly from one year to

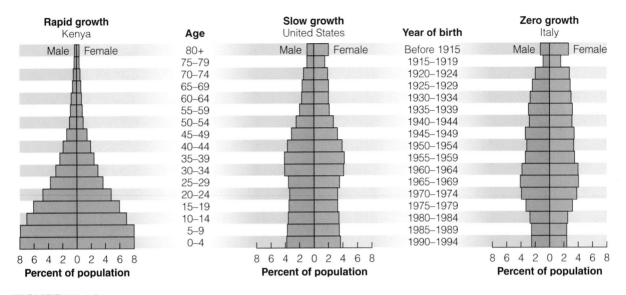

FIGURE 10.16
Age structure pyramids for the human population of Kenya (growing at 2.1% per year), the United States (growing at 0.6% per year), and Italy (zero growth) for 1995. (From McFalls 1998.)

the next. Eberhardt (1988) discusses the use of age composition information in the management of wildlife populations, and Ricker (1975, Chapter 2) discusses this problem in exploited fish populations.

Evolution of Demographic Traits

We can use the demographic techniques just described to investigate one of the most interesting questions of evolutionary ecology: Why do organisms evolve one type of life cycle rather than another? Only certain kinds of l_x and b_x schedules are permissible if a population is to avoid extinction. How does evolution act, within the framework of permissible demographic schedules, to determine the life cycle of a population?

Pacific salmon grow to adult size in the ocean and return to fresh water to spawn once and die. We may call this *big-bang reproduction*.[5] Oak trees may become

mature after 10 or 20 years and drop thousands of acorns for 200 years or more. We call this *repeated reproduction*. How have these life cycles evolved? What advantage might be gained by salmon that breed more than once, or by oak trees that drop only one set of seeds and then die?

The population consequences of life cycles were first explored by Cole (1954), who asked a simple question: What effect does repeated reproduction have on the intrinsic capacity for increase (r)? Assume that we have an annual species that produces offspring at the end of the year and then dies, has a simple survivorship of 0.5 per year, and has a fertility rate of 20 offspring. The life table for this species is as follows:

Age (x)	Proportion surviving (l_x)	Fertility (b_x)	Product ($l_x b_x$)
0	1.0	0	0
1	0.5	20	10
2	0.0	–	0
			$R_0 = 10.0$

[5]Big-bang reproduction = semelparity, and repeated reproduction = iteroparity, for those who prefer the more classical terms derived from Greek roots.

FIGURE 10.17
Age structure of (a) Engelmann spruce and (b) subalpine fir in a forest stand at 3150 m elevation in northern Colorado. Neither of these tree species has an age distribution like those shown in Figure 10.15 for stable or stationary age distributions. (Data from Aplet et al. 1988, Table 2.)

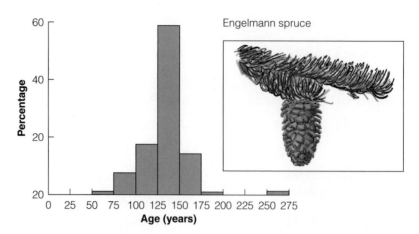

(a)

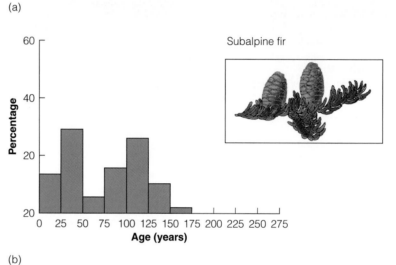

(b)

The net reproductive rate (R_0) is 10.0, which means that the species could increase 10-fold in one generation (= 1 year). We can determine r from the characteristic equation of Lotka:

Reproductive value at age $x = V_x = \sum_{t=x}^{w} \frac{l_t b_t}{l_x}$ (10.12)

from which we determine that r=2.303 for the annual species with big-bang reproduction. What advantage could this species gain by continuing to live and reproduce at years 2, 3,...° . Let us assume the most favorable condition, no mortality after age 1 and survival to age 100. The life table now becomes the following:

Age (x)	Proportion Surviving (l_x)	Fertility (b_x)	Product ($l_x b_x$)
0	1.0	0.0	0.0
2	0.5	20	10.0
3	0.5	20	10.0
4	0.5	20	10.0
5	0.5	20	10.0
-	-	-	-
-	-	-	-
-	-	-	-
99	0.5	20.0	10.0
100	0.0	0.0	0.0

$$R_0 = \sum l_x b_x = 990.0$$

In the manner outlined above, we determine that $r = 2.398$ for the perennial species with repeated reproduction. If we adopt repeated reproduction in our hypothetical organism, we raise the intrinsic capacity for increase only about 4%:

$$\frac{2.398}{2.303} = 1.04$$

Now let us work backward. What fertility rate at year 1 would equal the r of the perennial (2.398)? We can solve this problem algebraically (Cole 1954) or by trial and error. Suppose we increase the birth rate by one individual. The annual life table is now:

Age (x)	Proportion Surviving (l_x)	Fertility (b_x)	Product ($l_x b_x$)
0	1.0	0.0	0.0
1	0.5	21.0	10.5
2	0.0	-	0
			$R_0 = 10.5$

This is almost the gain achieved by repeated reproduction. If we increase the fertility rate by two individuals, we get $r = 2.398$, equal to the r for the perennial. This is obviously an ideal case, because we assume no mortality after age 1 in the perennial form. Cole (1954) generalized this ideal case to a surprising conclusion: For an annual species, the maximum gain in the intrinsic capacity for increase (r) that could be achieved by changing to the perennial reproductive habit would be equivalent to adding one individual to the effective litter size $(l_x b_x$ for age 1). Cole assumed for his ideal case perfect survival to reproductive age (Charnov and Schaffer 1973). In our hypothetical example we assumed that half of the organisms die before reaching reproductive age.

This simple model for the evolution of big-bang reproduction is unrealistic because it is a "cost-free" model: present reproduction is assumed to have no effect on future reproduction or future survival (Roff 1992, Bell 1980). Let us assume that an organism can "decide" how much of its resources it will devote to reproduction. If it uses all its resources to reproduce, it will die and thus be a big-bang reproducer. Big-bang reproduction will be favored if the greater benefits of reproduction come only at high levels of reproductive effort; conversely, if good reproductive success can be achieved at low

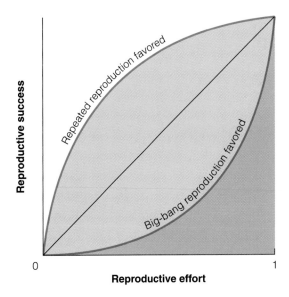

FIGURE 10.18
The reproductive effort model for the evolution of big-bang and repeated reproduction in plants and animals. The trade-off for an organism involves how much it will gain in reproductive success as it expends more and more effort on reproduction. When reproductive effort is 1, the organism breeds once and then dies (big-bang or semelparous reproduction). (Modified from Young 1990.)

levels of effort, organisms will be selected to be repeat reproducers. Figure 10.18 shows this trade-off between reproductive effort and reproductive success as implied in the reproductive effort model. The key demographic effect of big-bang reproduction is higher reproductive rates. Plants that reproduce only once typically produce 2-5 times as many seeds as closely related species that reproduce repeatedly (Young 1990). Repeated reproduction can also be favored when adult survival rates are high and juvenile survival is highly variable. The critical division between big-bang reproduction and repeated reproduction is set by the survival rate of the juvenile stages. If survival of juveniles is very poor or unpredictable, selection will usually favor repeated reproduction (Roff 1992). Let us look at one example to illustrate this theory.

Two species of giant rosette plants occur abundantly above treeline on Mount Kenya in Africa. *Lobelia telekii* is a big-bang reproducer that lives on

FIGURE 10.19
*Giant lobelias (*Lobelia telekii*) in the alpine zone of Mount Kenya, at 4200 m. The inflorescences of this semelparous plant are 1.5-3 m tall, and all plants die after flowering. The other giant rosette plant to the left and top of the photo is* Dendrosenecio keniodendron. *(Photo courtesy of Truman R. Young.)*

relatively dry, less productive slopes, whereas *Lobelia deckenii keniensis* is a repeated reproducer that lives in moist, more productive sites (Young 1990). Rosettes grow slowly from germination to reproductive size over 40-60 years for both species (Figure 10.19). In *Lobelia telekii* the resources of the entire plant go into reproduction, and the inflorescence may exceed 3 m in height and contain on average 500,000 seeds. After reproduction the entire plant dies. In *Lobelia deckenii keniensis* only a portion of the plant's resources goes into reproduction, and the inflorescence rarely exceeds 1 m tall and contains on average about 200,000 seeds. Big-bang reproduction in *Lobelia telekii* is favored by high mortality of adult plants in between flowering episodes the probability of future repro-

duction is outweighed by the greater fecundity of big-bang reproduction.

Some of the best examples of the evolution of life history strategies come from studies within a single species. Capelin are a good example because males are big-bang reproducers while females are repeated reproducers. Capelin, small (15-25g) sardine-like, pelagic fish with a circumpolar arctic distribution, form an important part of the food chain for seals, seabirds, and other fish such as cod. Males have adopted the big-bang strategy because each male can mate with several females during a spawning season and because male mortality is very high after spawning (Huse 1998). Female capelin are limited by the number of eggs they can carry, and they can improve their reproductive success only by spawning several times at yearly intervals.

Much interest in life history evolution has centered on determining the costs of reproduction. Reproductive effort at any given age can be associated with a biological *cost* and a biological *profit*. The biological cost derives from the reduction in growth or survival that occurs as a consequence of using energy to reproduce. For example, the more seeds a meadow grass (*Poa annua*) plant produces in one year, the less it grows the following year (Law 1979). Fruit fly (*Drosophila melanogaster*) females that mate often typically live shorter lives than females that mate less often (Fowler and Partridge 1989). The biological profit associated with reproduction is measured in the number of descendants left to future generations, which will be affected by the survival rate and the growth rate. The hypothetical organism must in effect ask at each age: Should I reproduce this year, or would I profit more by waiting until next year? Obviously, if the mortality rate of adults is high, it would be best to reproduce as soon as possible. But if adult mortality is low, it may pay for an organism to put its energy into growth and wait until the next year to reproduce.

Many organisms do not reproduce as soon as they are physiologically capable of doing so. The key quantity that we must measure to predict the optimal age at maturity is the *potential fecundity cost* (Bell 1980). Individuals that reproduce in a given year will often be smaller and less fecund in the following year than an individual that has previously abstained from reproduction. This is best established in poikilotherms, such as fishes, that show a reduction in growth associated with spawning. Potential fecundity costs also occur in homeotherms (Clutton-Brock et al. 1989), and the period of lactation in mammals is energetically very expensive for females (Figure 10.20). Social behavior associated with reproduction can produce great differences in the costs associated with breeding in the two sexes and thus cause differences in the optimal age at maturity for males and females. Red deer stags, for example, defend harems and attain a breeding peak after seven years of age through their fighting ability. Females mature at three years and live longer than males.

Repeated reproducers must "decide" in an evolutionary sense to increase, decrease, or hold constant

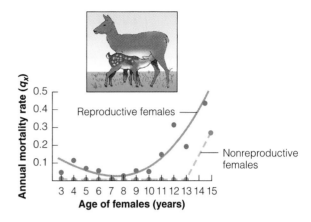

FIGURE 10.20
Cost of reproduction in female red deer on the island of Rhum in Scotland. Mortality in winter is always higher in females that reproduced during the previous summer, no matter the age of the female. (After Clutton-Brock et al. 1982.)

their reproductive effort with age. In every case analyzed so far, reproductive effort increases with age (Sydeman et al. 1991), and this may be a general evolutionary trend in organisms.

Why do species expend the effort to have repeated reproduction? The answer seems to be that repeated reproduction is an adaptation to something other than achieving maximum reproductive output. Repeated reproduction may be an evolutionary response to uncertain survival from zygote to adult stages (Roff 1992). The greater the uncertainty, the higher the selection for a longer reproductive life. This may involve channeling more energy into growth and maintenance, and less into reproduction. Thus we can recognize a simple scheme of possibilities:

	Long life span	*Short life span*
Steady reproductive success	?	Possible
Variable reproductive success	Possible	Not possible

We now believe that the advantage of repeated reproduction is that it spreads the risk of reproducing over a longer time period and thus acts as an adaptation that thwarts environmental fluctuations.

Summary

Population changes can be analyzed with a set of quantitative techniques first developed for human population analysis. A life table is an age-specific summary of the mortality rates operating on a population. Life tables are necessary because mortality does not fall equally on all ages, and in most species the very young and the old suffer high mortality.

A fertility schedule that summarizes reproduction with respect to age can describe the reproductive component of population increase. The intrinsic capacity for increase of a population is obtained by combining the life table and the fertility schedule for specified environmental conditions. This concept leads to an important demographic principle: A population that is subject to a constant schedule of mortality and natality rates will (1) increase in numbers geometrically at a rate equal to the capacity for increase (r), (2) assume a fixed or *stable age distribution*, and (3) maintain this age distribution indefinitely. The age distribution of a population is constant and unchanging only as long as the life table and the fertility table remain constant. Populations undergoing geometric increase reach a stable age distribution, and those showing a period of constant population size reach a *stationary age distribution*. Under other circumstances the age distribution will shift over time, which is the usual condition in natural populations. Demographic techniques are useful for comparing quantitatively the consequences of adopting an annual life cycle versus a perennial one. Very little gain in potential for population increase occurs in species that reproduce many times in each generation, and repeated reproduction seems to be an evolutionary response to conditions in which survival from zygote to adult varies unpredictably from good to poor. An organism thus "hedges its bets" by reproducing several times.

Key Concepts

1. Age-specific natality and mortality rates for any population can be summarized quantitatively in fertility schedules and in life tables.

2. The intrinsic capacity for increase (r) summarizes the natality and mortality schedules and forecasts the rate of population growth implicit in these schedules.

3. The age structure of a population is determined by these rates of natality and mortality.

4. These quantitative methods are useful for comparing the life history consequences of particular natality and mortality schedules in populations.

Selected References

Caughley, G. 1977. *Analysis of Vertebrate Populations.* Chapters 7-9. Wiley, New York.

Crooks, K. R., M.A. Sanjayan, and Doak D. F. 1998. New insights on cheetah conservation through demographic modeling. *Conservation Biology* 12: 889-895.

Eberhardt, L. L. 1988. Using age structure data from changing populations. *Journal of Applied Ecology* 25:373-378.   Gregory, P. T. 1997. What good is R_0 in demographic analysis? (And what is a generation anyway?). *Wildlife Society Bulletin* 25:709–713.

Hutchings, J. A. 1994. Age- and size-specific cost of reproduction within populations of brook trout, *Salvelinus fontinalis. Oikos* 40:12–20.

Reznick, D. 1985. Costs of reproduction: an evaluation of the empirical evidence. *Oikos* 44:257–267.

Roff, D. A. 1992. *The Evolution of Life Histories: Theory and Analysis.* Chapman and Hall, New York.

Stearns, S.C. 1989. Trade-offs in life history evolution. *Functional Ecology* 3:259–268.

Vaupel, J.W., J.R. Carey, K. Christensen, T.E. Johnson, A. I. Yashin, N.V. Holm, I.A. Iachine, V. Kannisto, A. A. Khazaeli, P Liedo, V.D. Longo, Y. Zeng, K.G. Manton, and J.W. Curtsinger. 1998. Biodemographic trajectories of longevity. Science 280:855–859.

Ward, J. and G. P. Adams 1998. Body condition and adjustments to reproductive effort in female moose (*Alces alces*). *Journal of Mammology* 79:1345–1354.

Questions and Problems

10.1 Reproductive success declines in old animals. How could one determine if this decline is caused by aging (senescence) or is a reproductive cost associated with previous breeding activities? Sydeman et al. (1991) discuss this question for the northern elephant seal.

10.2 In human populations in developed countries women generally outlive men by a margin of 5-10 years. Is this advantage in female longevity a general characteristic of nonhuman animal species as well? What might explain such a pattern in humans and other animals? Smith (1989) reviews this question, and Carey et al. (1995) provide data on fruit flies to test this generalization.

10.3 Connell (1970) gives the following data for the barnacle *Balanus glandula*:

Age (yr)	1959 Settlement l_x	1959 Settlement b_x	1960 Settlement l_x	1960 Settlement b_x
0	1.0000000	0	1.0000000	0
1	0.0000620	4,600	0.0000640	4,600
2	0.0000340	8,700	0.0000290	8,700
3	0.0000200	11,600	0.0000190	11,600
4	0.0000155	12,700	0.0000090	12,700
5	0.0000110	12,700	0.0000045	12,70
6	0.0000065	12,700	0.0000000	-
7	0.0000020	12,700	-	-
8	0.0000020	12,700	-	-

Calculate the net reproductive rate (R_0) for these two year-classes. What does this tell you about these populations? Estimate the capacity for increase from these data, and calculate the stable age distribution (C_x). Explain any difficulties and interpret the results in biological terms.

10.4 What additional data, if any, are required to determine the stable age distribution for the human population described in Table 10.3?

10.5 "A woman who gives birth to a set of twins at the age of 19, and subsequently gives birth to one other child, contributes as much to the future population of America as does a woman who produces five children, but whose age at birth of the initial child was 30" (Slobodkin 1961, p. 54). Under what conditions is this not true? Estimate the r value for populations with these two types of reproductive schedules. In making these estimates, assume for simplicity that the female survivorship is constant at $l_x = 0.980$ for the reproductive ages (approximately true for U.S. females in 1996).

10.6 The life table and the seed production of the winter annual plant *Collinsia verna* for 1983-84 was as follows (Kalisz 1991):

Life cycle	Age interval (months)	Number alive n_x	Average no. seeds produced per plant b_x
Seed	0-5	23061	0
Seedling	5-7	6019	0
Overwintering plants	7-12	4617	0
Flowering plants	12-13	2612	0
Fruiting plants	13-14	692	10.754

Calculate the net reproductive rate for these plants and discuss the biological interpretation of this rate.

10.7 Calculate a complete life table for the data in Table 10.2.

10.8 Forest ecologists usually measure the *size* structure of a forest and less often make use of the annual rings of temperate-zone trees to get the age structure of the forest. What might one learn from determining age structure in addition to size structure in a forest stand?

10.9 The population of the United States was growing in 1996 at 0.6 % per year (not counting immigration), yet Table 10.3 shows a net reproductive rate of 1.0. How can this happen?

10.10 Can the reproductive value of males and females at a given age differ? Discuss the data presented on red deer in Clutton-Brock et al. (1982, p. 154) as an example.

10.11 What effects might herbivores have on a plant species that would push the species toward the evolution of big-bang reproduction? Klinkhamer et al. (1997) discuss this issue.

Overview Question

A life table and a fertility schedule are available for a species of threatened plant. If you were in charge of a management plan for this plant species, what could you conclude from these two tables, and what further demographic information would you want to have?

CHAPTER 11

Population Growth

POPULATION GROWTH IS A CENTRAL PROCESS of ecology. But no population goes on growing forever, and this leads us to the problem of *population regulation* (see Chapter 16). Because species interactions such as *predation*, *competition*, *herbivory*, and *disease* affect population growth (see Chapters 12–15), and population growth produces changes in *community structure* (see Chapters 21–24), it is important to understand how population growth occurs.

The demographic techniques described in Chapter 10 are useful because they permit us to project future changes in population density in a precise manner. In this chapter we will apply these demographic parameters to the description of population growth and explore some of the difficulties of analyzing the growth of natural populations. To illustrate these methods' utility, we will use them to address a practical problem in conservation biology.

Mathematical Theory

A population that has been released into a favorable environment will begin to increase in numbers. What form will this increase take, and how can we describe it mathematically? We start by considering a simple case in which generations are separate, as in univoltine insects (one generation per year) or annual plants.

Growth in Populations with Discrete Generations

Consider a species with a single annual breeding season and a life span of one year. Let each female on average produce R_0 female offspring that survive to breed in the following year. Then

$$N_{t+1} = R_0 N_t \qquad (11.1)$$

where N_t = population size of females at generation t

N_{t+1} = population size of females at generation $t + 1$

R_0 = net reproductive rate, or number of female offspring produced per female per generation

What happens to this population will very much depend on the value of R_0. Consider two cases:

1. *Multiplication rate constant.* Let R_0 be a constant. If $R_0 > 1$, the population increases geometrically

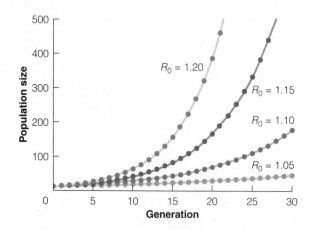

FIGURE 11.1
Geometric or exponential population growth, discrete generations, reproductive rate constant. Starting population size = 10. Equation (11.1).

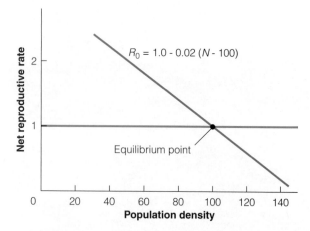

FIGURE 11.2
Net reproductive rate (R_0) as a linear function of population density (N) at time t. *In this hypothetical example, equilibrium density is 100 (black dot). The red line marks the equilibrium line at which $R_0 = 1.0$ and the population remains constant. The blue line is an example of the relationship given in Equation (11.3).*

without limit; if $R_0 < 1$, the population decreases to extinction. For example, let $R_0 = 1.5$ and $N_t = 10$ when $t = 0$:

Generation	Population Size (N_t)
0	10
1	15 = (1.5)(10)
2	22.5 = (1.5)(15)
3	33.75 = (1.5)(22.5)

Figure 11.1 shows some examples of geometric population growth with different R_0 values.

2. *Multiplication rate dependent on population size.* Populations do not normally grow with a constant multiplication rate as in Figure 11.1. If we look at the trajectory of a species population through time, we observe a variety of dynamics, including populations that fluctuate very little, others that fluctuate chaotically, and still others that fluctuate in cycles. How can we explain this variety of dynamic behavior?

The simplest way is to assume that the multiplication rate changes as population density rises and falls. At high densities, birth rates will decrease or death rates will increase from a variety of causes, such as food shortage or epidemic disease. At low densities birth rates will be high and losses from diseases

and natural enemies low. We need to express the way in which the multiplication rate slows down as density increases. The simplest mathematical model is linear: Assume that there is a straight-line[1] relationship between the density and multiplication rates such that the higher the density, the lower the multiplication rate (Figure 11.2). The point where the line crosses $R_0 = 1.0$ is a point of equilibrium in population density at which the birth rate equals the death rate. It is convenient to measure population density in terms of deviations from this equilibrium density, expressed as

$$z = N - N_{eq} \qquad (11.2)$$

where z = deviation from equilibrium density

N = observed population size

N_{eq} = equilibrium population size (where $R_0 = 1.0$)

The equation of the straight line shown in Figure 11.2 is thus

$$R_0 = 1.0 - B\left(N - N_{eq}\right) \qquad (11.3)$$

where R_0 = net reproductive rate

$(-)B$ = slope of line

---

[1] A straight line is described by the equation $y = a + bx$, where b is the slope and a is the y-intercept (the y value when $x = 0$).

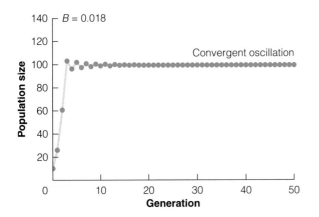

(a)

FIGURE 11.3

Examples of population growth with discrete generations and multiplication rate as a linear function of population density as in Figure 11.2. Starting density, is 10 and equilibrium density is 100. Three examples with different slopes are shown (a) for B = 0.018, the population shows convergent oscillations to equilibrium density at 100. (b) for B = 0.025, the population oscillates in a two-generation limit cycle. (c) for B = 0.029, the population fluctuates chaotically in an irregular pattern that never repeats itself.

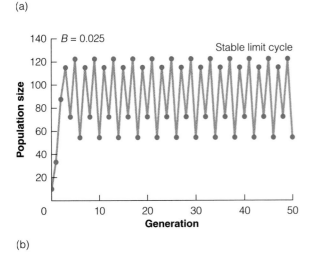

(b)

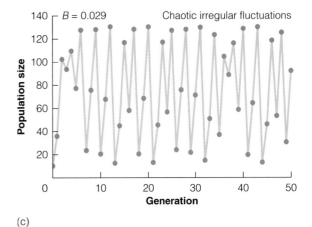

(c)

In Figure 11.2, $B = 0.02$ and $N_{eq} = 100$. Equation (11.1) can now be written

$$N_{t+1} = R_0 N_t = (1.0 - B z_t) N_t \quad (11.4)$$

The properties of this equation depend on the equilibrium density and the slope of the line. Let us work out a few examples to illustrate this. Consider first a simple example in which $B = 0.011$ and $N_{eq} = 100$. Start the population at $N_0 = 10$:

$N_1 = [1.0-0.011(10-100)]10$
$= (1.99)(10) = 19.9$
$N_2 = [1.0-0.011(19.9-100)]19.9$
$= (1.881)(19.9) = 37.4$

Similarly,

$N_3 = 63.2$
$N_4 = 88.8$
$N_5 = 99.7$

and the population density converges smoothly toward the equilibrium point of 100. A second example is worked out in Table 11.1, and three additional examples are plotted in Figure 11.3.

The behavior of this simple population model is very surprising because it generates many different patterns of population changes. If we define $L = BN_{eq}$, then:

- If L is between 0 and 1, the population approaches the equilibrium without oscillations.

- If L is between 1 and 2, the population undergoes oscillations of decreasing amplitude to the equilibrium point (*convergent oscillations*). (see Figure 11.3a).

- If L is between 2 and 2.57, the population exhibits stable limit cycles that continue indefinitely (see Figure 11.3b).

- If L is above 2.57, the population fluctuates chaotically showing what appear to be random

TABLE 11.1 Growth of a hypothetical population with discrete generations and net reproductive rate that is a linear function of density.

Population sizes are calculated from Equation (11.4) using $B = 0.025$, $N_{eq} = 100$, and a starting density of 50 individuals.

General formula: $N_{t+1} = [1.0 - 0.025(N_t - 100)]N_t$

$$N_1 = [1.0 - 0.025(50 - 100)]50$$
$$= (2.25)(50) = 112.5$$

$$N_2 = [1.0 - 0.025(112.5 - 100)]112.5$$
$$= (0.6875)(112.5) = 77.34$$

$$N_3 = [1.0 - 0.025(77.34 - 100)]77.34$$
$$= (1.5665)(77.34) = 121.15$$

$$N_4 = [1.0 - 0.025(121.15 - 100)]121.15$$
$$= (0.4712)(121.15) = 57.09$$

Similarly,

$$N_5 = 118.33$$
$$N_6 = 64.10$$
$$N_7 = 121.63$$
$$N_8 = 55.86$$

The population continues to oscillate in a stable two-generation cycle.

changes, depending on the starting conditions (May 1974a, Maynard Smith 1968) (Figure 11.3c).

Much of this mathematical theory of population growth was clarified and elaborated by the mathematical ecologist Robert May (see page 163) working at Princeton University and later at Oxford University. The fact that such a simple population model can produce such a diversity of population growth trajectories is one of the most surprising results found by twentieth-century mathematical ecologists. This model, in which the net reproductive rate decreases in a linear way with density, is the discrete-generation version of the *logistic equation* described in the next section, in which we consider populations with overlapping generations.

Growth in Populations with Overlapping Generations

In populations that have overlapping generations and a prolonged or continuous breeding season, we can

describe population growth more easily by using differential equations. As earlier, we will assume for the moment that the growth of the population at time t depends only on conditions at that time and not on past events of any kind.

1. *Multiplication rate constant.* Assume that, in any short time interval $\delta\Delta t$ (usually written as dt), an individual has the probability $b\,dt$ of giving rise to another individual. In the same time interval, it has the probability $d\,dt$ of dying. If b and d are instantaneous rates[2] of birth and death, the instantaneous rate of population growth per capita will be

instantaneous rate of population
growth $= r = b - d$ (11.5)

and the form of the population increase is given by

$$\frac{dN}{dt} = rN = (b - d)N \qquad (11.6)$$

where N = population size

t = time

r = per-capita rate of population growth

b = instantaneous birth rate

d = instantaneous death rate

This is the curve of geometric increase in an unlimited environment that we discussed in Chapter 10 with regard to the intrinsic capacity for increase.

Note that we can use the geometric growth model to estimate the doubling time for a population growing at a certain rate:

$$\frac{N_t}{N_0} = e^{rt} \qquad (11.7)$$

But if the population doubles, $N_t/N_0 = 2$. Thus

$$\frac{N_t}{N_0} = 2 = e^{rt}$$

or

$$\log_e(2) = rt \quad \text{or,} \quad \frac{0.69315}{r} = t \qquad (11.8)$$

where t = time for population to double its size

r = realized rate of population growth per capita

---

[2] See Appendix III.

A few values for this relationship are given for illustration:

r	t
0.01	69.3
0.02	34.7
0.03	23.1
0.04	17.3
0.05	13.9
0.06	11.6

Thus if a human population is increasing at an instantaneous rate of 0.0300 per year (finite rate = 1.0305), its doubling time would be about 23 years, if geometric increase prevails.

   2. *Multiplication rate dependent on population size.* Populations, however, do not show continuous geometric increase. When a population is growing in a limited space, the density gradually rises until eventually the presence of other organisms reduces the fertility and longevity of the individuals in the population. This reduces the rate of increase of the population until eventually the population ceases to grow. The growth curve defined by such a population is *sigmoid*, or S-shaped (Figure 11.4). The S-shaped curve differs from the geometric curve in two ways: It has an upper asymptote (that is, the curve does not exceed a certain maximal level), and it approaches this asymptote smoothly, not abruptly.

   The simplest way to produce an S-shaped curve is to introduce into our geometric equation a term that will smoothly reduce the rate of increase as the population builds up. We can do this by making each individual added to the population reduce the rate of increase an equal amount. This produces the equation

$$\frac{dN}{dt} = rN\left(\frac{K - N}{K}\right) \tag{11.9}$$

where N = population size

   t = time

   r = intrinsic capacity for increase

   K = upper asymptote or maximal value of N ("carrying capacity")

This equation states that

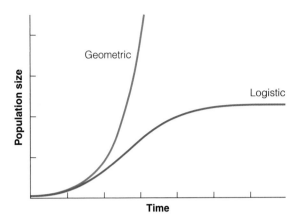

FIGURE 11.4
Population growth: geometric growth in an unlimited environment, and logistic (sigmoid) growth in a limited environment.

$$\begin{pmatrix}\text{Realized rate} \\ \text{of population} \\ \text{increase per} \\ \text{unit time}\end{pmatrix} = \begin{pmatrix}\text{Potential rate} \\ \text{of population} \\ \text{growth per} \\ \text{capita}\end{pmatrix} \times \begin{pmatrix}\text{Population} \\ \text{size}\end{pmatrix}$$

and is the differential form of the equation for the *logistic curve*. Verhulst first suggested this curve to describe the growth of human populations in 1838. Pearl and Reed (1920) independently derived the same equation as a description of the growth of the population of the United States.

   Note that r is the potential rate of population growth per individual in the population (Box 11.1). It is the same as the intrinsic capacity for increase discussed in Chapter 10.

   The integral form of the logistic equation can be written as follows:

$$N_t = \frac{K}{1 + e^{a - rt}} \tag{11.10}$$

where N_t = population size at time t

   t = time

   K = maximal value

   e = 2.71828 (base of natural logarithms)

   a = a constant of integration defining the position of the curve relative to the origin

   r = intrinsic capacity for increase

B O X   1 1 . 1

WHAT IS LITTLE-*r*, AND WHY IS IT SO CONFUSING?

Unfortunately ecologists have used *r* to mean two quite different things: *r* the intrinsic capacity for increase (Chapter 10), and *r* the realized population growth rate per capita.

When populations are growing geometrically, these two meanings are identical:

$$\frac{dN}{dt} = rN \quad \text{or,}$$

$$\frac{dN}{dt\,N} = r = \text{per capita rate of population growth}$$

This is good mathematics, but it becomes confusing when we deal with population growth that is not geometric. To keep matters clear we define two concepts:

r = potential per capita population growth rate = intrinsic capacity for increase

dN/dtN = realized per capita population growth rate

The distinction between these two concepts is easily seen in the following logistic equation:

$$\frac{dN}{dt} = rN \left(\frac{K-N}{K} \right)$$

$$\frac{dN}{dt\,N} = r \left(\frac{K-N}{K} \right)$$

The realized population growth rate (*dN/dtN*) is not equal to the potential growth rate (*r*). Ecologists in the field measure the realized growth rate, which depends (in the logistic model) on the intrinsic capacity for increase *r*, the carrying capacity *K*, and the existing population size *N*.

It is important to keep these two concepts clear. The intrinsic capacity for increase can be considered a constant for a particular population and thus is always a positive number. The realized population growth rate can be negative when a population is declining and then becomes positive when the population grows. Even though in ideal situations the population grows geometrically, and the potential growth rate and the realized rate are the same, but this is rarely the case in the real world.

$K = 100$
$r = 1.0$
$N_0 = 1.0$ (starting density)

Very early in population growth, there is little difference between the curves for the logistic and the geometric equations (see Figure 11.4). As we approach the middle segments of the curves, they diverge more. As we approach the upper limit of the logistic curve, the curves diverge much farther, and when we reach the upper limit, the population stops growing because $(K - N)/K$ becomes zero. The following calculations demonstrate this:

r	Population size	Unutilized opportunity for population growth $[(K-N)/K]$	Rate of populaton growth
1.0	1	99/100	0.99
1.0	50	50/100	25.00
1.0	75	25/100	18.75
1.0	95	5/100	4.75
1.0	99	1/100	.99
1.0	100	0/100	0.00

Note that the addition of one animal has the same effect on the rate of population growth at both the low and high ends of the curve (in this example, 1/100).

Two attributes of the logistic curve make it attractive: its mathematical simplicity and its apparent reality. The differential form of the logistic curve contains only two constants, *r* and *K*. Both these mathematical symbols can be translated into biological terms. The constant *r* is the per capita (or per individual) potential rate of population increase (the intrinsic capacity for increase of Chapter 10). It seems reasonable to attribute to *K* a biological meaning—the density at which the space being studied becomes "saturated" with organisms, the "carrying capacity" of the environment.

There are two ways of viewing the logistic curve. The more general, more flexible viewpoint is to consider it an empirical description of how populations tend to grow in numbers when conditions are initially favorable. The other way is to view the logistic curve as an implicit strict theory of population growth, as a "law" of population growth. Because the logistic curve was proposed as a strict theory of population growth, we will examine it from this point of view.

Does the logistic curve fit the facts? One way to find out is to rear a colony of organisms in a constant space with a constant supply of food. From this infor-

Let us look for a minute at the factor (K - N)/K, also called the "unutilized opportunity for population growth." To demonstrate that this factor does in fact put the brakes on the basic geometric growth pattern, we consider a situation like the following:

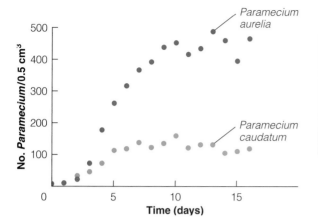

FIGURE 11.5
Population growth in the protozoans Paramecium aurelia *and* P. caudatum *at 26°C in buffered Osterhout's medium, pH 8.0, with "one loop" of bacteria added as food. (Data from Gause 1934.)*

Sir Robert May *(1936–) Professor of Zoology, Oxford University*

mation we can calculate a logistic curve. If the data fit the sigmoid pattern of the logistic, we can confirm this model of population growth. We look into this approach next.

Laboratory Tests of the Logistic Theory

Many populations have been observed in the laboratory as they increase in size. Let us consider a few relatively simple organisms first. Gause (1934) studied the growth of populations of *Paramecium aurelia* and *P. caudatum*. He began his experiments with 20 *Paramecium* in a tube with 5 ml of a salt solution buffered

to pH 8. Each day Gause added a constant quantity of bacteria, which served as food and could not multiply in the salt solution. The cultures were incubated at 26°C, and every second day they were washed with fresh salt solution to remove any waste products. Thus Gause had a *constant environment in a limited space*; the temperature, volume, and chemical composition of the medium were constant, waste products were removed frequently, and food was added in uniform amounts each day. The growth of some of Gause's *Paramecium* populations is shown in Figure 11.5. In general, the fit of these data to the logistic curve was quite good. Under these conditions the asymptotic density (*K*) was approximately 448 individuals per ml for *P. aurelia* and 128 individuals per ml for *P. caudatum*.

Populations of organisms with more complex life cycles may also increase in an S-shaped curve. Pearl (1927) fitted a logistic curve to the growth of *Drosophila melanogaster* populations he maintained in bottles with yeast as food. The fit of the data was fairly good (Figure 11.6), and Pearl ushered in the "logistic era" when he proclaimed the logistic curve to be the universal law of population growth. But Sang (1950) criticized the application of the logistic curve to *Drosophila* populations by identifying complexities in the *Drosophila* cultures that Pearl did not recognize. First, the flies did not receive a constant amount of food because the yeast that was the source of food was itself a growing population. Also, the composition of the yeast varied as the cultures aged. Second, because the fruit fly has several stages in its life cycle, it is not clear just which stage should be used in measuring "population size." Pearl counted only the adult flies, but to some extent adults and larvae feed on the same thing.

Beetles that live in flour (*Tribolium*) and wheat (*Calandra*) have been also used very often for experimental population studies. These beetles are preferable to

FIGURE 11.6
Growth of an experimental laboratory population of the fruit fly Drosophila melanogaster. *The circles are observed census counts, and the smooth curve is the fitted logistic equation. (After Pearl 1927.)*

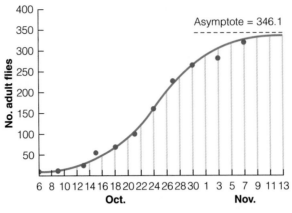

FIGURE 11.7
Population growth of the flour beetle Tribolium castaneum *at 29°C and 70% relative humidity in 8 grams of flour. Considerable variation in population growth occurs among different genetic strains of this flour beetle. (Data from Park et al. 1964.)*

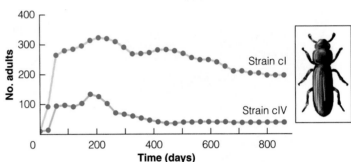

Drosophila because, even though they have as complex a life cycle (involving eggs , larvae , pupae and adults), their food source is nonliving, so their medium can be precisely controlled. Chapman (1928), one of the first to use *Tribolium* for laboratory studies in ecology, found that colonies of these beetles grew in a logistic fashion. Most workers stopped their cultures as soon as they reached the upper asymptote. Thomas Park, however, reared populations of *Tribolium* for several years and obtained the results shown in Figure 11.7. The upper asymptote of the logistic is imaginary the density does not stabilize after the initial sigmoid increase but rather shows a long-term decline. When Birch (1953b) did similar studies on *Calandra oryzae*, he found logistic growth initially, followed by large fluctuations in density with no indication of stabilization around an asymptote.

It is important to note that these populations of a single species of beetle living in a constant climate with constant food supply show wide fluctuations in numbers. These fluctuations are brought about by the influence of the animals on each other completely independent of any fluctuations in temperature, food, predators, or disease. No cases have as yet been demonstrated in which the population of any organism with a complex life history comes to a steady state

at the upper asymptote of the logistic curve. For these reasons the logistic "law" of population growth has been rejected as a general model of how populations increase in size (Kingsland 1995).

Interestingly population ecologists have historically focused on the logistic model for overlapping generations and have largely overlooked the simpler discrete models, with their much richer dynamic behavior (May 1981). Insects constitute a large fraction of animal species, and many insects have nonoverlapping generations that are described well by the simpler discrete models. This change of focus away from the logistic equation as a model for population growth has been highlighted by data on laboratory populations and has been demonstrated even more graphically by data on field populations.

Field Data on Population Growth

Population growth does not occur continuously in field populations. Many species living in seasonal environments show population growth during the favorable season each year. Long-lived organisms may show population growth only rarely, and few populations in nature fill up a vacant habitat the way they do in the laboratory. Some populations have been released from hunting pressure, and we have good records of how they subsequently increased in numbers. Some of the best examples are goose populations in Europe.

For the past 30 years many goose populations in both Europe and North America have been increasing in abundance (Ebbinge 1992; Cooch 1991). Brant geese wintering in western Europe have increased from about 20,000 birds in 1960 to over 250,000 in 1996 (Figure 11.8). Barnacle geese and white-fronted geese from western Europe have also increased in recent years. Ebbinge (1992) showed that reduction in hunting pressure is the most likely cause of these population increases. Population growth in these geese has not followed a smooth sigmoid pattern, but an upper limit of about 270,000 geese appears to have been reached in the mid-1990s. Whether this upper limit will be stable is not yet known.

Other good examples of population growth have been observed when animals introduced onto islands or into other new habitats were studied as they increased in numbers. Reindeer have been introduced into many parts of Alaska since 1891 to replace the dwindling caribou herds in the economy of the Eskimo. In 1911 reindeer were introduced onto two of the Pribilof Islands in the Bering Sea off Alaska.

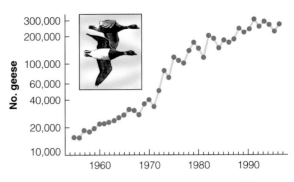

FIGURE 11.8
*Growth in Brent geese (*Branta bernicla bernicla*) populations that winter along the coasts of France, Britain, and the Netherlands. Year refers to winter (1960 = 1960-61 winter). (Data from Ebbinge 1992 and personal communication.)*

Four males and 21 females were released on St. Paul Island (106 km²) and three males and 12 females were introduced on St. George Island (90 km²). These introductions were immediate successes. The subsequent history of these herds is of interest because the islands were completely undisturbed environments there was little hunting pressure, and there were no predators. The two herds have had quite different histories on the two islands (Figure 11.9). The St. George

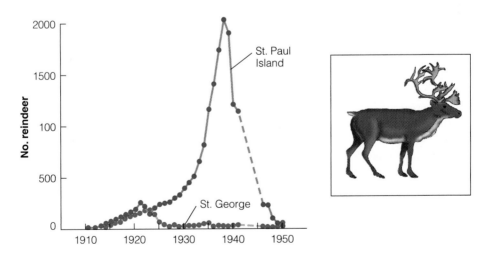

FIGURE 11.9
Reindeer population growth on two of the Pribilof Islands, from their introduction in from 1911 until 1950. The two islands are 60 km apart and have the same vegetation and climate. The St. Paul Island population shows a classic ungulate irruption and crash associated with overgrazing and starvation. (Data from Scheffer 1951.)

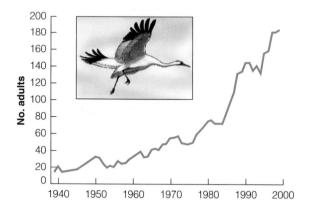

FIGURE 11.10
Population growth of the whooping crane, an endangered species that has recovered from near extinction in 1941. Counts of adults are made annually on the wintering grounds at Aransas, Texas. (Data from Binkley and Miller 1983, and Cannon 1996.)

herd reached a low ceiling of 222 reindeer in 1922 and then subsided to a small herd of 40 to 60 animals. The St. Paul herd grew continuously to about 2000 reindeer in 1938 and then abruptly declined to only eight animals in 1950. The ecological differences between these two islands appear to have been very slight (they had the same type of vegetation and the same climate), and there was no obvious environmental reason why the two populations behaved so differently (Scheffer 1951), although illegal hunting on St. George Island may have contributed to the differences shown in Figure 11.9

Reindeer were also introduced onto St. Matthew Island (332 km²) in the Bering Sea in 1944. They increased from an initial 29 animals (24 females, five males) in 1944 to 1350 in 1957, to 6000 in 1963, and then crashed to 42 animals in 1966 (Klein 1968), thus repeating the St. Paul Island sequence in a slightly shorter time. In both these cases, winter food shortage,

E S S A Y   1 1 . 1

WHAT IS A "GOOD" POPULATION GROWTH MODEL?

Raymond Pearl in the 1920s had a vision of the logistic equation as a universal model of population growth. We now know that this was too optimistic and that many if not most episodes of population growth do not fit this model. In the real world most population growth patterns are so highly variable that they defy description by a simple model. Nevertheless, applied ecologists are often faced with a need to forecast how the population of an endangered species might increase if it were protected, or how a fish population might recover from overharvesting. So, what can we do? Our choice of approach depends very much on how much we already know about the species in question:

1. *Considerable background knowledge.* For some species we know the approximate birth rate, the number of eggs they lay, the approximate generation time, and their life expectancy. For these species we can use the Leslie matrix models or stage-based matrix models to make a simple forecast of short-term changes in population size.

2. *Little background knowledge.* For many species we know almost nothing about the vital demographic parameters. These species are probably best treated by the use of simple models such as the logistic equation or the geometric growth equation for short-term forecasts. We know these simple models are not precise but they are better than nothing, and ecologists must often follow the old adage that a poor model is better than no model.

We must keep in mind that mathematical models should not be classified as *right* or *wrong*, or as *valid* or *invalid*. All models are wrong but some are useful. Models must be evaluated primarily by their *utility* in helping to answer a question. All models simplify reality to help us understand it, and to help us explore *what if* questions in population dynamics. So even though we well conclude that the geometric model of population growth is a poor general model for describing population growth, we may still decide to use it to forecast the short-term path of recovery of an endangered frog species. *Utility* is the key to deciding which models are valuable to ecologists.

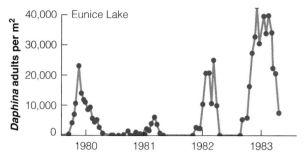

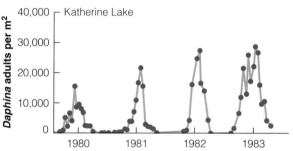

FIGURE 11.11
Density of the cladoceran Daphnia rosea *in Eunice Lake and Katherine Lake, British Columbia, from 1980 to 1983. Because these temperate lakes show strong seasonal dynamics that vary from year to year, the population growth curve cannot be described by a simple equation like the logistic equation. (Data from Walters et al. 1990.)*

particularly due to overgrazing of lichens, was the major cause of the dramatic population crashes. Reindeer populations on islands seem to be a dramatic example of population growth according to the discrete population model when L is large (see Figure 11.3). Rapid increase in population size is followed by a dramatic crash in numbers such that no stable state is seen.

The whooping crane (*Grus americana*) is a good example of an endangered species now recovering from the brink of extinction. Only 47 whooping cranes existed when this species was first protected in 1916, and only 15 birds were still alive in 1941. The whooping crane breeds in the Northwest Territories of Canada and migrates to overwinter on the Texas coast at the Aransas National Wildlife Refuge. Counts of the entire population on the wintering grounds in Texas since 1938, have yielded the population growth curve shown in Figure 11.10. Population growth has been irregular in the whooping crane. Binkley and Miller (1983, 1988) found that the rate of increase (r) changed around 1956, at which time the population began to recover more rapidly than before. Moreover, a ten-year cycle, possibly a spin-off of predation from the 10-year cycle of snowshoe hares in the breeding areas, is superimposed on the population growth curve (Nedelman et al. 1987). Hare predators like coyote and lynx may turn to whooping crane nests and chicks once hares begin to decline

Many organisms show strong annual fluctuations in density, and thus the pattern of population growth can be observed once a year. The cladoceran *Daphnia*, common in the plankton of many temperate lakes and ponds, shows a spring increase in numbers that varies dramatically from year to year (Walters et al. 1990). These cladocerans increase in numbers in an almost exponential manner (Figure 11.11), remain abundant for a variable amount of time in midsummer, and then decline in autumn, possibly because of reduced algal density in the lake water. The maximum density reached varies greatly in different years, so there is no constant carrying capacity (K) for these planktonic organisms.

Very often, field data on population growth are too crude to show definitely whether or not the logistic curve is a good representation of the data. The cases we illustrated here suggest that the logistic curve only approximately describes field population increases.

Pearl and Reed (1920) used the logistic curve to predict the future growth of the U.S. population. They fitted the logistic curve to the census data from 1790 to 1910 and projected it to reach asymptotic density, a value of 197 million, around the year 2060 (Figure 11.12). The census data for 1920 to 1940 fit the curve very well (Pearl et al. 1940), but subsequent census data show a nearly geometric increase rather than a logistic one. The predicted asymptote of 197

B O X   1 1 . 2

A SIMPLE TIME-LAG MODEL OF POPULATION GROWTH

Consider a simple population growth model with discrete generations. Assume that the reproductive rate at generation t depends on density in a linear manner but that, instead of depending on density at generation t (as in Figure 11.2), it depends on density at the previous generation $(t-1)$. We measure density as a deviation from the equilibrium point:

$$z = N - N_{eq} \quad (11.11)$$

where $z =$ deviation from equilibrium density

$\qquad N =$ observed population size

$\qquad N_{eq} =$ equilibrium population size (where $R_0 = 1.0$)

The reproductive rate is described in Figure 11.2 as a straight line, $R_0 = 1.0 - Bz$. The population growth model can thus be written as:

$$N_{t+1} = R_0 N_t \qquad (11.12)$$

$$= \left(1 - Bz_{t-1}\right)N_t$$

which is similar to the preceding treatment except that the reproductive rate is now defined by the density of the previous generation. The properties of this equation depend on the equilibrium density and the slope of the line.

Let us work out a hypothetical case with a time lag to illustrate a simple model of time-lag population growth:

$B = 0.011 \qquad\qquad N_{eq} = 100$

Start a population at $N_0 = 10$ (and use $N = 10$ for first-generation calculation of the time-lag term). From Equation (11.12):

$$N_1 = \left[1.0 - 0.011(10 - 100)\right]10 = 19.9$$

$$N_2 = \left[1.0 - 0.011(10 - 100)\right]19.9 = 39.6$$

$$N_3 = \left[1.0 - 0.011(19.9 - 100)\right]39.6 = 74.4$$

These results are plotted in Figure 11.13. This population oscillates more or less regularly, with a period of six or seven generations between peaks in numbers, in contrast to the smooth approach to equilibrium density that occurred in the absence of a time lag.

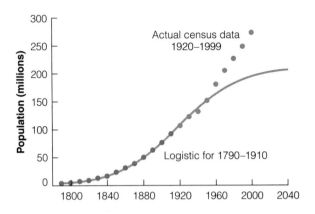

FIGURE 11.12
Census counts of the human population of the United States from 1790 to 1990. The smooth curve is the logistic equation fitted to the census counts from 1790 to 1910, inclusive (Pearl and Reed 1920). extrapolated beyond the 1910 data to its predicted asymptote around 200 million. In mid-1999 the U.S. population was an estimated 273 million. Human population growth in the United States has not followed the logistic equation. (Modified from Pearl et al. 1940; additional data from the Population Reference Bureau and U.S. Bureau of Census 1999.)

million was, in fact, reached in 1968, and current estimates for the U.S. population in 2025 range from 260 million to 357 million (*Statistical Abstract of the United States, 1998*). The lesson is to avoid trying to predict population growth from extrapolation of simple models.

We conclude from this analysis that population growth may sometimes be sigmoid in natural populations and thus fit the logistic model, but often it is not. Natural populations almost never achieve the asymptotic stable density of the logistic curve, and hence the logistic model has serious drawbacks as a general model of population growth. What can be done about this? Work on population growth models has proceeded along three lines. One has been to analyze the effect of *time lags* on the logistic model, because the assumption of no time lags in the logistic model is most clearly at odds with the biological realities of complex organisms. A second approach has been to construct *probabilistic (stochastic) models* of population growth. The third approach has been to use more specific models based on age or size to project population changes (*Leslie matrix models*). Next we look briefly at these three approaches.

Time-Lag Models of Population Growth

Animals and plants do not respond instantly to changes in their environment, and this leads us to consider what effect time lags might have on population growth models. We can look at this problem in the simplest way by changing our assumption that the reproductive rate at generation t depends on density not in the same generation but instead on density in the last generation ($t-1$) (see Figure 11.2). Box 11.2 gives the details of how to do these calculations, and Figure 11.13 shows the results. A delay in feedback of only one generation can change a stable population growth pattern into an unstable one. Maynard Smith (1968, p. 25) has shown that, defining $L = BN_{eq}$

If $0 < L < 0.25$, then stable equilibrium with no oscillation

If $0.25 < L < 1.0$, then convergent oscillation

If $L > 1.0$, then stable limit cycles or divergent oscillation to extinction

Compare the results of this time-lag model with those obtained without any time lags (see page 159).

Laboratory populations of *Daphnia* are a good example of the effect of time lags on population growth. Pratt (1943) followed the development of *Daphnia* populations in the laboratory at two temperatures. The populations, in 50 ml of filtered pond water, started with two parthenogenetic females each. *Daphnia* were counted every two days and transferred to a fresh culture. The only food used was a green alga, *Chlorella*. Populations at 25°C showed oscillations in numbers, whereas those at 18°C were approximately stable (Figure 11.14). Oscillations that occurred at 25°C resulted from a delay in the depressing effect of population density on birth rates and death rates. At 25°C, the birth rate is affected first by rising density, and only later is the death rate increased. This causes the *Daphnia* population to continually "overshoot" and then "undershoot" its equilibrium density. Note that these oscillations are intrinsic to the biological system and are not caused by external environmental changes.

The biological mechanisms in *Daphnia* that account for these time lags are now well understood (Goulden and Hornig 1980). *Daphnia* store energy in the form of oil droplets, mainly as triacylglycerols, when food is superabundant. They use these energy reserves once the food supply has collapsed, so the effects of low food supply are not instantaneous but

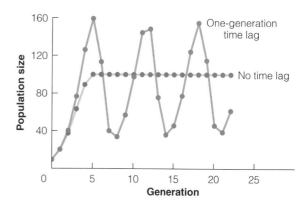

FIGURE 11.13

Hypothetical population growth with and without a time lag, with discrete generations, and with reproductive rate a linear function of density. Starting density = 10, slope of reproductive curve = 0.011, equilibrium density = 100.

delayed. Females can thus continue to produce offspring even after food has become scarce. After these energy reserves are exhausted, *Daphnia* starve and die, producing the oscillations shown in Figure 11.14.

So we see that the introduction of time lags into simple models of population growth permits three possible alternatives: (1) a converging oscillation toward equilibrium, (2) a stable oscillation around the equilibrium level, or (3) a smooth approach to equilibrium density. In addition, some configurations of time lags will produce a divergent oscillation that is unstable and leads to extinction of the population. These outcomes are clearly more realistic models of what seems to occur in natural populations.

Stochastic Models of Population Growth

The models we have discussed so far are *deterministic* models, which means that given certain initial conditions, each model predicts one exact outcome. But biological systems are probabilistic, not deterministic. Thus we speak of the *probability* that a female will have a litter in the next unit of time, or the probability that there will be a cone crop in a given year, or the probability that a predator will kill a certain number of animals within the next month. The realization that population trends are thus the joint outcome of many individual probabilities has led to the development of probabilistic, or *stochastic*, models.

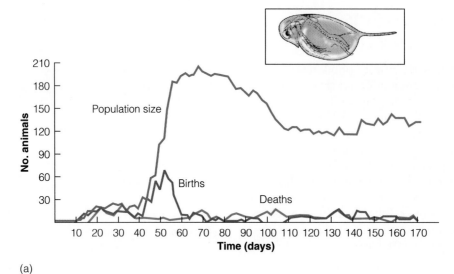

(a)

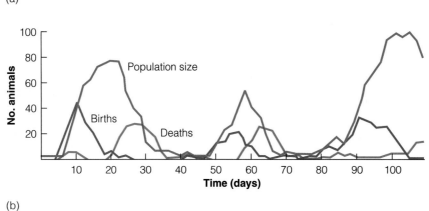

(b)

FIGURE 11.14
Population growth in the water flea (Daphnia magna) *in 50 ml of pond water (a) at 18°C and (b) at 25°C. The numbers of births and deaths have been doubled to make them visible at this scale. (Modified from Pratt 1943.)*

We can illustrate the basic nature of stochastic models very simply. Recall the geometric growth equation, a deterministic model we previously developed for discrete generations (Equation 11.1):

$$N_{t+1} = R_0 N_t$$

Consider an example in which the net reproductive rate (R_0) is 2.0 and the starting density is 6:

$$N_{t+1} = (2.0)(6) = 12$$

The deterministic model thus predicts a population size of 12 at generation 1. In constructing stochastic model for this, we might assume that the probabilities of reproduction are as follows:

	Probability
One female offspring	0.50
Three female offspring	0.50

Clearly, on the average, two female parents will leave four female offspring, so $R_0 = 2.0$. Let us use coin tosses to construct some numerical examples. If the coin comes up heads, one offspring is produced; if tails, three offspring.

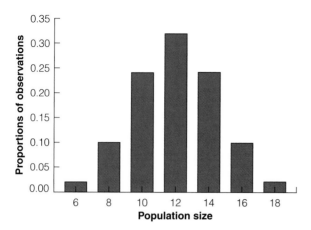

FIGURE 11.15
Frequency distribution size of female population after one generation for the example discussed in the text. $N_0=6$, $R_0=2.0$, probability of having one female offspring=0.5, probability of having three female offspring = 0.5.

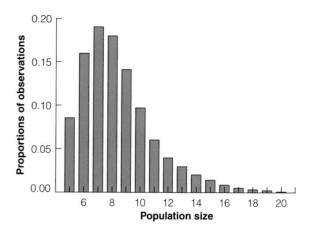

FIGURE 11.16
Frequency distribution at time 1 of the size of a population undergoing a pure birth process, $b = 0.5$, $N_0 = 5$. (After Pielou 1969.)

| | Outcome | | | |
Parent	Trial 1	Trial 2	Trial 3	Trial 4
1	(h)1	(t)3	(h)1	(t)3
2	(t)3	(h)1	(t)3	(h)1
3	(h)1	(t)3	(h)1	(h)1
4	(t)3	(t)3	(t)3	(t)3
5	(t)3	(t)3	(t)3	(h)1
6	(t)3	(t)3	(h)1	(h)1
Total population in next generation	14	16	12	10

Some of the outcomes are above the expected value of 12, and some are below it. If we continued doing this many times, we could generate a frequency distribution of population sizes for this simple problem; an example is shown in Figure 11.15. Note that populations starting from exactly the same point with exactly the same biological parameters could, in fact, finish one generation later with either as few as six or as many as 18 members.

The population growth of species with overlapping generations can also be described by stochastic models. Geometric growth in this case follows the differential equation

$$\frac{dN}{dt} = r N = (b - d)N \qquad (11.13)$$

where b = instantaneous birth rate

d = instantaneous death rate

In the simplest case (the pure birth process), we assume that $d = 0$, so no organisms can die. If we assume a simple binary fission type of reproduction, the probability that an organism will reproduce in the next short time interval dt is $b\,dt$, in which b is the instantaneous birth rate. Consider an example where $b = 0.5$ and $N_0 = 5$ (starting population). In one time interval, according to the deterministic model (Equation 11.7)

$$N_t = N_0 e^{rt}$$

$$N_1 = (5)e^{(0.5)(1)} = 8.244$$

For the stochastic equivalent of this simple model, we must determine two things from the instantaneous rate of birth:

Probability of not reproducing in one time interval = e^{-b} = 0.6065
Probability of reproducing at least once in one time unit = 1.0 - e^{-b} = 0.3935
Thus for five organisms, the chance that none of the five will reproduce in the next unit of time is
(0.6065)(0.6065)(0.6065)(0.6065)(0.6065) = 0.082

so, in approximately one trial out of 12, no population change will occur in the unit of time ($N_1 = 5$). We could laboriously count up all the other possibilities,

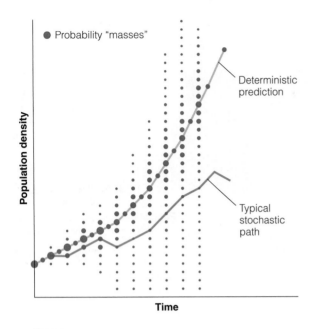

FIGURE 11.17
Stochastic model of geometric population growth for continuous overlapping generations. Population predictions cannot be represented by a single value in stochastic models, as they can with deterministic models, and the uncertainty of the prediction increases over time. (After Skellam 1955.)

remembering that each individual may undergo fission more than once in each unit of time. Or we may follow a mathematician's application of probability theory to this problem (Pielou 1969, p. 9) to reach the conclusions shown in Figure 11.16. Note again the possible variation in final population size when births are considered in a probabilistic manner. Figure 11.17 illustrates the principles of stochastic models of population growth.

If we use probabilistic models and allow both births and deaths to occur randomly, there is a chance that a population will become extinct. What is the chance of extinction for a population starting with N_0 organisms and undergoing stochastic changes in size with average instantaneous birth rate b and death rate d, as in Figure 11.17? Pielou (1969, p. 17) discussed two cases:

1. *Average birth rate greater than average death rate.* These populations should increase geometrically but may by chance drift to extinction, particularly during the first few time periods if population size is small. The probability of extinction at some time is given by

$$\text{Probability of extinction} = \left(\frac{d}{b}\right)^{N_0} \quad (11.14)$$

For example, if $b = 0.75$ and $d = 0.25$ for $N_0 = 5$, we have:

$$\text{Probability of extinction} = \left(\frac{0.25}{0.75}\right)^5 = 0.0041$$

But if $b = 0.55$, $d = 0.45$, and $N_0 = 5$, then

$$\text{Probability of extinction} = \left(\frac{0.45}{0.55}\right)^5 = 0.367$$

Thus the larger the initial population size and the greater the difference between birth and death rates, the greater chance a population has of staying in existence. The effects of random fluctuations in birth and death rates on individuals is called demographic stochasticity, or demographic uncertainty (Soulè 1987). The important principle is that, even when the birth rate exceeds the death rate on average, there is a finite probability of a population going extinct.

2. *Average birth rate equals average death rate.* These populations are stationary in numbers, fluctuating around constant densities, as is typical of the real world on the average, and by Equation 11.14:

$$\text{Probability of extinction} = \left(\frac{d}{b}\right)^{N_0} = \left(1.0\right)^{N_0} = 1.0$$

as time approaches infinity. Thus, when births equals deaths on average, extinction is a certainty for any population subject to stochastic variations in births and deaths, if we allow a long enough time span.

Stochastic models of population growth thus introduce the important idea of biological variation into the consideration of population changes. The probability approach to these ecological problems is consequently more realistic. The price we must pay for the greater realism of stochastic models is the greater difficulty of the mathematics, but part of this difficulty can be resolved by using computers. The variation inherent in stochastic models becomes more important as population size becomes smaller just as predictions about the change in size of an individual family from one year to the next are much less certain than predictions about the change in size of the world's population. If all populations were in the millions, stochastic models could be eliminated, and deterministic models would be adequate.

Population Projection Matrices

One realistic way of estimating population growth was pioneered by P. H. Leslie (1945), who calculated population changes from age-specific birth and survival rates. Such an age-classified model is called a *Leslie matrix*. Leslie, who worked closely with Charles Elton's ecologists at the Bureau of Animal Population in Oxford, was responsible for many applications of mathematics to ecological questions (Crowcroft 1991). The essential feature of Leslie matrix models is that the organism's life cycle is broken down into a series of stages. Each age class is one stage in a simple Leslie matrix (Figure 11.18a), and this is exactly the same approach we used in Chapter 10 for life tables. Organisms survive from one stage to the next with probability P_x, and they produce a number of offspring F_x. In the conventional life table notation

$$P_x = \frac{l_{x+1}}{l_x} = (1 - q_x) = \left\{ \begin{array}{l} \text{Probability that an} \\ \text{individual of age group} \\ x \text{ will survive to enter} \\ \text{age group } x + 1 \text{ at the} \\ \text{next time interval of} \\ \text{the life table} \end{array} \right\} \quad (11.15)$$

$$F_x = b_x s_x = \left\{ \begin{array}{l} \text{Number of female offspring born} \\ \text{in one time interval per female} \\ \text{aged } x \text{ to } x + 1; \text{ these offspring} \\ \text{must survive to enter age group} \\ 0 \text{ at the next time interval} \end{array} \right\} \quad (11.16)$$

where l_x = number of individuals alive at start of age interval x

b_x = number of births in one time interval per adult female aged x to $x + 1$

s_x = proportion of the b_x offspring that are alive at the start of the next time interval

q_x = probability that an individual of age group x will not survive to enter age group $x + 1$ at the next time interval

Begin with a population having specified age structure at time t:

N_0 = number of organisms between ages 0 and 1
N_1 = number of organisms between ages 1 and 2 (and so on to the oldest age class)
N_k = number of organisms between ages k and $k+1$ (oldest organisms)

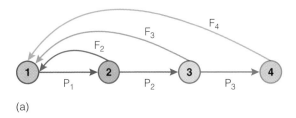

(a)

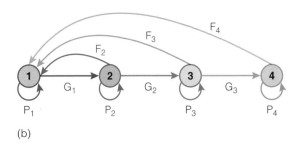

(b)

FIGURE 11.18
Population projection matrices. (a) The Leslie matrix, or age-classified life cycle. Four age classes are shown in this example, with different fecundities (F_x) for each age class and different probabilities P_x of surviving from one age class to the next. (b) A size- or stage-based life cycle, in which the only added complication is that an individual has probability P_x of remaining in the same life cycle stage in one time period and a probability G_x of surviving and moving on into the next stage of the life cycle. (After Caswell 1989.)

Patrick H. (George) Leslie *(1901–1972) Bureau of Animal Population, Oxford University*

TABLE 11.2 Stage-based life table and fecundity table for the loggerhead sea turtle.[a]

Stage number	Class	Size (carapace length) (cm)	Approximate age (yr)	Annual survivorship	Fecundity (eggs/yr)
1	Eggs, hatchlings	<10	<1	0.6747	0
2	Small juveniles	10.1–58.0	1–7	0.7857	0
3	Large juveniles	58.1–80.0	8–15	0.6758	0
4	Subadults	80.1–87.0	16–21	0.7425	0
5	Novice breeders	>87.0	22	0.8091	127
6	First-year remigrants	>87.0	23	0.8091	4
7	Mature breeders	>87.0	24–54	0.8091	80

[a]These values assume a population declining at 3% per year.
Source: Data from Crouse et al. (1987).

TABLE 11.3 Stage-class population matrix for the loggerhead sea turtles.[a]

0	0	0	0	127	4	80
0.6747	0.7370	0	0	0	0	0
0	0.0486	0.6610	0	0	0	0
0	0	0.0147	0.6907	0	0	0
0	0	0	0.0518	0	0	0
0	0	0	0	0.8091	0	0
0	0	0	0	0	0.8091	0.8089

[a]Estimates based on the life table presented in Table 11.2, with the survival estimates broken down into survival within the same stage and survival and movement into the next stage.
Source: Data from Crouse et al. (1987).

Time units for age in a Leslie matrix are often one year but can be any fixed time unit, depending on the organism. Usually only the female population is considered for sexually reproducing species.

If we assume no emigration and no immigration, the population's age structure at the next time interval is defined as follows:

New age structure   (11.17)
↓

$$\left\{ \begin{array}{l} \text{Number of} \\ \text{new organisms} \\ \text{at time } t+1 \end{array} \right\} = F_0 N_0 + F_1 N_1 + F_2 N_2 + F_3 N_3 + \ldots + F_k N_k$$

$$= \sum_{x=0}^{k} F_x N_x$$

Number of age 1 organisms at time $t + 1 = P_0 N_0$
Number of age 2 organisms at time $t + 1 = P_1 N_1$
Number of age 3 organisms at time $t + 1 = P_2 N_2$

and so on.

Leslie (1945) recognized that this problem could be cast as a simple matrix problem if one defined a transition matrix **M** as follows:

$$\mathbf{M} = \begin{bmatrix} F_0 & F_1 & F_2 & F_3 & F_4 & F_5 & \ldots & F_{k-1} & F_k \\ P_0 & 0 & 0 & 0 & 0 & 0 & \ldots & 0 & 0 \\ 0 & P_1 & 0 & 0 & 0 & 0 & \ldots & 0 & 0 \\ 0 & 0 & P_2 & 0 & 0 & 0 & \ldots & 0 & 0 \\ 0 & 0 & 0 & P_3 & 0 & 0 & \ldots & 0 & 0 \\ 0 & 0 & 0 & 0 & P_4 & 0 & \ldots & 0 & 0 \\ & & \ldots & & & & & & \\ & & \ldots & & & & & & \\ 0 & 0 & 0 & 0 & 0 & 0 & \ldots & P_{k-1} & 0 \end{bmatrix} \quad (11.18)$$

where $F_x \geq 0$ and P_x ranges from 0 to 1. By casting the present age structure as a column vector, we get

$$\vec{N}_t = \begin{bmatrix} N_0 \\ N_1 \\ N_2 \\ N_3 \\ N_4 \\ \cdot \\ \cdot \\ N_k \end{bmatrix} \quad (11.19)$$

Leslie showed that the age distribution at any future time could be found by premultiplying the column vector of age structure by the transition matrix **M**:

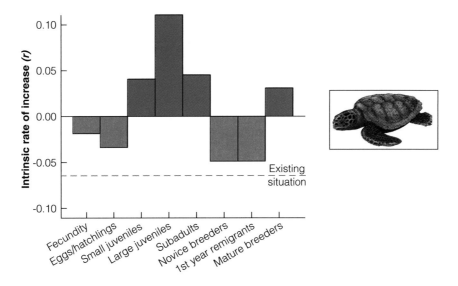

FIGURE 11.19
Hypothetical changes in the rate of population increase of loggerhead sea turtle populations off the southeastern United States resulting from either simulated increases of 50% in fecundity or simulated increases in survival to 100% for the stages of the life cycle. For four stages of the life cycle, improving survival or fecundity would still leave the population in decline (red). The greatest improvement for this endangered turtle would occur by improving the survival of the large juveniles (blue). (From Crouse et al. 1987.)

$$\mathbf{M}\vec{N}_t = \vec{N}_{t+1}$$ (11.20)
$$\mathbf{M}\vec{N}_{t+1} = \vec{N}_{t+2}$$

Students who are familiar with matrix algebra will benefit from the discussion of the properties of this matrix in Leslie (1945) and in Caswell (1989).

Lefkovitch (1965) realized that the Leslie matrix was a special case of a more general *stage-based matrix*, in which life history stages replace ages. Such a stage-based or size-based model is illustrated in Figure 11.18b. One new complexity is added to the age-based model: Whereas all individuals of age x move to age $x + 1$ after 1 unit of time, in a stage- or size-based model some individuals will remain in the same life cycle stage. We thus have two probabilities associated with each stage:

P_x = probability that an individual will survive and remain in stage- or size-class x in the next time unit
G_x = probability that an individual will survive and move up to the next stage- or size-class $x+1$ in the next time unit

Note that in stage-based matrices we set the time unit such that it is impossible for the organism to jump up two or more stages in one time step.

All of this seems uncomfortably abstract, so let us look at an example of a size-based matrix model. Crouse et al. (1987) analyzed the dynamics of the loggerhead sea turtle (*Caretta caretta*), an endangered species from the Atlantic Ocean off the southeastern United States. Sea turtles have a long life-span that can be broken down into seven stages based on size. Table 11.2 lists these stages along with the size and approximate age of turtles in each stage. Survivorship varies with size, and only individuals over 87 cm long are sexually mature.

The population projection matrix based on this life history takes the following form:

$$\mathbf{M} = \begin{bmatrix} P_1 & F_2 & F_3 & F_4 & F_5 & F_6 & F_7 \\ G_1 & P_2 & 0 & 0 & 0 & 0 & 0 \\ 0 & G_2 & P_3 & 0 & 0 & 0 & 0 \\ 0 & 0 & G_3 & P_4 & 0 & 0 & 0 \\ 0 & 0 & 0 & G_4 & P_5 & 0 & 0 \\ 0 & 0 & 0 & 0 & G_5 & P_6 & 0 \\ 0 & 0 & 0 & 0 & 0 & G_6 & P_7 \end{bmatrix}$$ (11.21)

The best estimates of the parameters of this matrix are given in Table 11.3.

Given this model of population growth for the loggerhead sea turtle, we can ask some interesting questions about how to reverse the population decline of this endangered species. By holding all but one of the life history parameters constant, we can investigate quantitatively the impact of conservation efforts. Figure 11.19 shows the results of either increasing fecundity 50% or improving survival in each stage of the life cycle. Improving fecundity 50% still leaves the population declining. Maximum improvement is achieved by improving the survival of juvenile turtles. At the present time most sea turtle conservation efforts are focused on protecting the eggs on beaches, even though 20-30 years of protecting nests on beaches has produced no increase in sea turtle abundance (Crouse et al. 1987). In fact this is exactly what the model in Figure 11.19 would predict. What is needed for conservation is an improvement of juvenile turtle survival at sea. Much juvenile loss is caused when turtles are caught in shrimp nets and

drown, so shrimp trawlers are now being fitted with a device to prevent the capture and drowning of sea turtles (Crowder et al. 1994).

Stage-based or size-based matrix models have been used extensively for plant populations in which size is a more useful measure of an individual than is age (Caswell 1989). Matrix models also permit plants to grow or decrease in size, a biologically useful feature. The solution to matrix population growth models based on age- or size stages is just as complex as those previously illustrated for simple population growth models. Populations may increase or decrease geometrically or may show oscillations. Because these models assume a constant schedule of survival and reproduction, they can be applied to natural populations only for the short time periods for which this assumption is valid.

Summary

The growth of a population can be described with simple mathematical models for organisms with discrete generations and for those with overlapping generations. If the multiplication rate is constant, geometric population growth occurs. Populations stabilize at finite sizes only if the multiplication rate depends on population size, and large populations have lower multiplication rates than small populations. Simple models for discrete generations can lead to complex dynamics, from stable equilibria to cycles and chaos. For species with overlapping generations, the logistic equation is a simple mathematical description of population growth to an asymptotic limit.

The S-shaped logistic curve is an adequate description for the laboratory population growth of *Paramecium*, yeast, and other organisms with simple life cycles. Population growth in organisms with more complex life cycles seldom follows the logistic curve very closely. In particular, the stable asymptote of the logistic is not achieved in natural populations and numbers fluctuate.

Three different types of population growth models have been developed to improve on the simple logistic model. *Time-lag models* have been used to analyze the effects of different time lags on the population growth curve. The introduction of time lags into the simple models of population growth can produce oscillations in population size instead of a stable asymptotic density. *Stochastic models* of population growth introduce the effects of chance events on populations. Populations starting from the same density and having the same average birth and death rates may increase at different rates because of chance events, which can lead to extinction and are particularly important in small populations. *Matrix models* of population growth can be age- or size-based and thus can be used for plants and animals alike. Matrix models are ideally suited to asking hypothetical questions about the contribution of specific life table parameters to population growth and to exploring the consequences of alternative management plans for endangered species or pests.

Key Concepts

1. Population growth can be described with simple mathematical models both for organisms with discrete generations and for organisms with overlapping generations.

2. Simple models for discrete generations can lead to complex dynamics from a stable equilibrium to cycles and chaotic fluctuations in numbers.

3. For species with overlapping generations, the logistic equation is a simple mathematical description of population growth to an asymptotic limit.

4. Natural populations often grow rapidly, but then density fluctuates widely rather than remaining at an equilibrium density.

5. More complex and more realistic models of population growth incorporate time lags and chance into population growth models.

Overview Question

Most analyses of population growth describe processes applicable to unitary organisms. Plants and other modular organisms also undergo population growth. Discuss the application of the models discussed in this chapter to population growth in modular organisms.

Selected References

Berryman, A. A. 1981. *Population Systems: A General Introduction*, Chapter 2. Plenum Press, New York.

Binkley, C. S. and R. S. Miller. 1983. Population characteristics of the whooping crane, *Grus americana*. *Canadian Journal of Zoology*:61 2768-2776.

Caswell, H. 1989. *Matrix Population Models*. Chapters 1-4. Sinauer Associates, Sunderland, Mass.

Clutton-Brock, T.H., A. W. Illius, B.T. Grenfell, A. D.C. MacColl, and S.D. Albon. 1997. Stability and instability in ungulate populations: an empirical analysis. *American Naturalist* 149:195–219.

Dennis, B., R. A. Desharnis, J.M. Cushing, and R.F. Costantino. 1997. Transitions in population dynamics: equilibria to periodic cycles to aperiodic cycles. *Journal of Animal Ecology* 66:704–729.

Gaillard, J. M., M. Festa-Bianchet, and N. G. Yoccoz. 1998. Population dynamics of large herbivores: Variable recruitment with constant adult survival. *Trends in Ecology and Evolution* 13:58–63.

Pearl, R. 1927. The growth of populations. *Quarterly Review of Biology* 2:532–548.

Pielou, E.C. 1969. *An Introduction to Mathematical Ecology*, Chapters 1–2. Wiley-Interscience, New York.

Silvertown, J. W. and J. Lovett-Doust. 1994. *Introduction to Plant Population Ecology*, Chapters 3 and 5. Wiley, New York.

Watkinson, A. R. 1997. Plant population dynamics. Pages 359–400 in M. J. Crawley, ed. *Plant Ecology*, 2nd ed. Blackwell Science, Oxford.

Zeng, Z., R. M. Nowierski, M. L. Taper, B. Dennis and W. P. Kemp. 1998. Complex population dynamics in the real world: modeling the influence of time-varying parameters and time lags. *Ecology* 79:2193–2209.

Questions and Problems

11.1 List for plants and animals six reasons why the assumption that population growth at a given point in time depends only on conditions at that time and not on past events (see p. 160) might be incorrect.

11.2 An isolated elk population in Washington State increased as follows from 1975 to 1993 as follows (Eberhardt et al. 1996):

Year	Total No. Elk
1975	8
1976	13
1977	14
1979	15
1981	20
1982	21
1983	27
1984	40
1985	54
1986	68
1987	67
1988	72
1989	79
1990	94
1991	110
1992	146
1993	179

What shape of population growth is shown by these data? Could one fit a logistic equation to these data? Why or why not? Calculate the rate of increase (r) for this population.

11.3 Determine the population growth curve for ten generations for an annual plant with a net reproductive rate of 25 and a starting density of 2. Assume a constant reproductive rate (Equation 11.1).

11.4 Determine the population growth curve for the annual plant in question 11.3 if reproductive rate is a linear function of density of the form $R_0 = 1.0 - 0.01z$ (where z = deviation from equilibrium density), equilibrium density is 1000, and starting density is 2 (Equation 11.4). Repeat under the assumption that there is a one-generation time lag in changing reproductive rate (Equation 11.12).

11.5   Determine the doubling time for the following human populations (1999 data) from Equation 11.8:

Country	Realized instantaneous rate of population growth
Ghana	0.029
South Africa	0.016
Canada	0.004
Argentina	0.012
United Kingdom	0.002
Ireland	0.006
Russia	-0.005

What assumptions must one make to predict these doubling times?

11.6   The snapping turtle (*Chelydra serpentina*) has the following life history parameters (Cunnington and Brooks 1996):

Stage of life cycle	Annual probability of survival for this stage	Probability of remaining in this stage for next year	Fecundity (No. of eggs per year)
Eggs	0.0635	0.0	0
Small juveniles	0.0554	0.6985	0
Large juveniles	0.0554	0.6985	0
Subadults	0.0554	0.6985	0
Novice breeders	0.9660	0.0	15.63
Second-year breeders	0.9660	0.0	15.63
Mature breeders	0.9660	0.9638	15.63

Calculate the course of population growth for a snapping turtle population that begins with eight adults. Compare the demographic rates of this turtle with those of the loggerhead sea turtle (see Table 11.3).

11.7   The whooping crane may be removed from the endangered species list once it reaches a total population size of 200 birds, which should provide 50 nesting pairs. If this population has been growing geometrically since 1957 at $r = 0.0416$ per year (Binkley and Miller 1983), calculate when it will be removed from the endangered list, given a population of 183 birds in 1998.

11.8   Discuss how the logistic pattern of population growth might be changed if K and r are not constant but vary over time. May (1981, pp. 24–27) discusses some simple examples.

11.9   The giant lobelia *Lobelia deckenii keniensis* on Mount Kenya produces on average 250,000 seeds and flowers every eight years. Average adult survival is 0.984 per year. Plants do not begin setting seed until they are 50 years old. Assuming for simplicity a two-stage life cycle (seeds, adult plants), calculate what survival rate of seeds would produce a stable population ($\lambda = 1.0$). How would this survival rate change if the plants flowered every year instead of only once every eight years?

11.10   A feral house mouse population can increase at $r = 0.0246$ per day. At this rate of increase, how many days are needed for the population to double?

11.11   What factors determine the carrying capacity of the earth for humans and how might you determine it? Discuss the historical estimates of the Earth's carrying capacity given in Cohen (1995, 1997) in light of the current population (6 billion in 2000).

11.12   If the human population instantly adopted zero population growth ($R_0 = 1.0$), the population would continue to grow until it reached the stationary age structure. Keyfitz (1971) showed that such a population would increase by demographic momentum, as follows:

$$Q = \frac{be}{ra}\left[\frac{R_0 - 1.0}{R_0}\right]$$

where Q = finite rate of population change (1.0 = no change)

   b = crude birth rate per 1000 persons

   e = life expectancy at birth in years

   r = current rate natural increase per 1000 persons

   a = average age at first reproduction in years

   R_0 = current net reproductive rate

A human population growing at these rates would increase Q times before it reached equilibrium, if zero population growth was instantly adopted.

Calculate how much the human population of the Earth would increase from current levels if zero population growth happened overnight. In 1998 the human population parameters were: $b = 28$, $e = 61$ years, $r = 17$, $a = 25$ years, and $R_0 = 1.77$.

How sensitive is this estimate to changes in the birth rate? To changes in average age at reproduction?

CHAPTER 12

Species Interactions:
Competition

ORGANISMS DO NOT EXIST ALONE IN NATURE but instead in a matrix of other organisms of many species. Many species will be unaffected by the presence of one another in an area, but in some cases two or more species will interact. The evidence for this interaction is quite direct: Populations of one species change in the presence of a second species.

Classification of Species Interactions

Interactions between populations can be classified on the basis of either the *mechanism* of the interaction or the *effects* of the interaction (Abrams 1987). Ecologists use both of these classifications and often combine them. In categorizing interactions on the basis of mechanism, we can identify six interactions between individuals of different species:

- *Competition:* two species use the same limited resource, or seek that resource, to the detriment of both.
- *Predation:* one animal species eats all or part of a second animal species.
- *Herbivory:* one animal species eats part or all of a plant species.
- *Parasitism:* two species live in a physically close, obligatory association in which the parasite depends metabolically on the host.

- *Disease:* an association between a pathogenic microorganism and a host species in which the host suffers physiologically.
- *Mutualism:* two species live in close association with one another to the benefit of both.

Some authors do not distinguish parasitism from disease, or predation from herbivory, and there is great variability in how loosely these terms are used in the ecology literature.

In categorizing interactions on the basis of *effects*, the most common effects studied are on population growth. Odum (1983) categorized effects as 0, +, and –. A zero indicates no effect of one species on the other, a plus indicates that the population has benefited at the expense of the other, and a minus indicates that the population has been adversely affected by the other. This system has fatal flaws for classifying interactions because it does not specify a time frame for recognizing effects, and more importantly it cannot describe many indirect interactions (Abrams 1987) that result when one species affects a second species which in turn affects a third species. Ecological communities are composed of many species linked in complex food webs, and a simple two-species interaction such as (+,–) cannot adequately summarize the possible interactions in a web. For this reason, we will define species interactions based on their mechanism and then explore the variety of effects these mechanisms can produce in populations of plants and animals.

In this chapter we discuss the interactions between two species that result from competition. There are

179

two different types of competition, defined as follows (Birch 1957):

- *Resource competition* (also called *scramble* or *exploitative competition*) occurs when a number of organisms (of the same or of different species) utilize common resources that are in short supply.

- *Interference competition* (also called *contest competition*) occurs when the organisms seeking a resource harm one another in the process, even if the resource is not in short supply.

In scramble competition, all individuals are equally affected; there are no "winners" or "losers." In contest competition, some individuals acquire resources at the expense of other individuals, so there are "winners" and "losers." Note that competition may be *interspecific* (between two or more different species) or *intraspecific* (between members of the same species). In this chapter, we discuss interspecific competition only.

Competition occurs for resources, and a variety of resources may become the center of competitive interactions. For plants, light, nutrients, and water may be important resources, but plants may also compete for pollinators or for space. Water, food, and mates are possible sources of competition for animals. In some animals, competition for space may involve many types of specific requirements, such as nesting sites, wintering sites, or resting sites that are safe from predators. Species must share a common interest in one or more resources before they can be potential competitors.

Several aspects of the process of competition must be kept clear. First, animals need not see or hear their competitors. A species that feeds by day on a plant may compete with a species that feeds at night on the same plant if the plant is in short supply. Second, many or most of the organisms that an animal sees or hears will not be its competitors. This is true even if resources are shared by the organisms. Thus even though oxygen is a resource shared by most terrestrial organisms, there is no competition for it among these organisms because this resource is superabundant. Third, competition in plants usually occurs among individuals rooted in position and therefore differs from competition among mobile animals. The spacing of individuals is thus more important in plant competition.

Theories on Competition for Resources

Mathematical models have been used extensively to build hypotheses about what happens when two species live together, either sharing the same food, occupying the same space, or preying on or parasitizing the other. The classical models of these phenomena are the *Lotka-Volterra equations*, which were derived independently by Lotka (1925) in the United States and Volterra (1926) in Italy. More recent models by Tilman (1982) have provided another important perspective on competition theory.

Mathematical Model of Lotka and Volterra

Lotka and Volterra each derived two different sets of equations: One set applies to *predator-prey* interactions, the other set to *nonpredatory* situations involving competition for food or space. We are concerned here only with their second set of equations for nonpredatory competition.

The Lotka-Volterra equations, which describe competition between organisms for food or space, are based on the logistic curve. We have seen that the logistic curve is described by the following simple logistic equations: for species 1,

$$\frac{dN_1}{dt} = r_1 N_1 \left(\frac{K_1 - N_1}{K_1} \right) \qquad (12.1)$$

and for species 2,

$$\frac{dN_2}{dt} = r_2 N_2 \left(\frac{K_2 - N_2}{K_2} \right)$$

where N_1 = population size of species 1

t = time

r_1 = intrinsic capacity for increase of species 1

K_1 = asymptotic density or "carrying capacity" for species 1

and these variables are similarly defined for species 2. We can visualize two species interacting—that is, affecting the population growth of each other—with the following simple analogy. Consider the environment to contain a certain amount of a limiting resource, such as nitrogen in the soil. Species 1 uses this resource, and the environment will hold K_1 individuals of this species (shown in green) when all the resource is being monopolized. But some of this resource can also be used by a competitor, species 2 (shown in yellow), which in this example needs much less of the resource to support one individual:

In most cases, the amount of resource used by one individual of species 2 is not exactly the same as that used by one individual of species 1. For example, species 2 may be smaller and require less of the critical resource that is contained in the environment. For this reason, we need a factor to convert species 2 individuals into an equivalent number of species 1 individuals. For this competitive situation, we define

$$\alpha N_2 = \text{equivalent number of} \atop \text{species 1 individuals} \quad (12.3)$$

where α is the conversion factor for expressing species 2 in units of species 1. This is a very simple assumption, which states that under all conditions of density there is a constant conversion factor between the competitors. We can now write the competition equation for species 1 as

$$\frac{dN_1}{dt} = r_1 N_1 \left(\frac{K_1 - N_1 - \alpha N_2}{K_1} \right) \quad (12.4)$$

This equation is mathematically equivalent to the simple analogy we just developed. Figure 12.1 shows this graphically for the equilibrium conditions, when dN_1/dt is zero. The two extreme cases are shown at the ends of the diagonal line in Figure 12.1. All the "space" for species 1 is used (1) when there are K_1 individuals of species 1, or (2) when there are K_1/α individuals of species 2. Populations of species 1

below this line will increase in size until they reach the diagonal line, which represents all points of equilibrium and is called the *isocline*. Note that we do not yet know where along this diagonal we will finish, but it must be somewhere at or between the points $N_1 = K_1$ and $N_1 = 0$.

Now we can retrace our steps and apply the same line of argument to species 2. We now have a volume of K_2 spaces to be filled by N_2 individuals but also by N_1 individuals. Again we must convert N_1 into equivalent numbers of N_2, and we define

$$\beta N_1 = \text{equivalent number of} \atop \text{species 2 individuals} \quad (12.5)$$

where β is the conversion factor for expressing species 1 in species 2 units. [1] We can now write the competition equations for the second species, as follows:

$$\frac{dN_2}{dt} = r_2 N_2 \left(\frac{K_2 - N_2 - \beta N_1}{K_2} \right) \quad (12.6)$$

Figure 12.2 shows this equation graphically for the equilibrium conditions when dN_2/dt is zero.

Now if we put these two species together, what might be the outcome of this competition? Only three outcomes are possible: (1) Both species coexist, (2) species 1 becomes extinct, or (3) species 2 becomes extinct. Intuitively, we would expect that species 1, if it had a very strong depressing effect on species 2, would win out and force species 2 to become extinct. The converse would apply for the situation in which species 2 strongly affected species 1. In a situation in which neither species has a very strong effect on the other, we might expect them to coexist. These intuitive ideas can be evaluated mathematically in the following way.

Solve the following simultaneous equations at equilibrium:

$$\frac{dN_1}{dt} = 0 = \frac{dN_2}{dt} \quad (12.7)$$

This can be done by superimposing figures (such as Figures 12.1 and 12.2) and adding the arrows by vector addition. Figure 12.3 shows the four possible geometric configurations. In each of these, the vector arrows have been abstracted, and the results can be traced by following the arrows. Species 1 will increase in yellow and green areas, and species 2 will increase

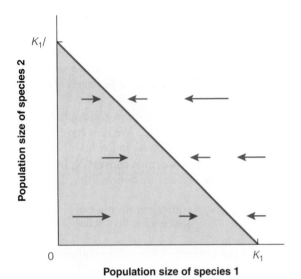

FIGURE 12.1

Changes in population size of species 1 when competing with species 2. Populations in the yellow area will increase in size and will come to equilibrium at some point on the blue diagonal line. The size of the arrows indicate the approximate rate at which the population will move toward the blue diagonal line. The blue diagonal line represents the zero growth isocline, all those points at which $dN_1/dt = 0$.

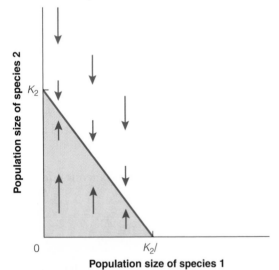

FIGURE 12.2

Changes in population size of species 2 when competing with species 1. Populations in the yellow area will increase in size and will come to equilibrium at some point on the blue zero growth isocline, all those points at which $dN_2/dt = 0$. The sizes of the arrows indicate the approximate rates at which the population will move toward the isocline.

in yellow and brown areas. There are a number of principles to keep in mind in viewing these kinds of curves. First, there can be no equilibrium of the two species unless the diagonal curves cross each other. Thus in cases 1 and 2, there can be no equilibrium, because one species is able to increase in a zone in which the second species must decrease. These cases lead to the extinction of one competitor. Second, if the diagonal lines cross, the equilibrium point represented by their crossing may be either a *stable* point or an *unstable* point. It is stable if the vectors about the point are directed *toward* the point, and unstable if the vectors are directed *away* from it. In case 4, the point where the two lines cross is unstable because if in response to some small disturbance the populations move slightly downward, they reach a zone in which N_1 can increase but N_2 can only decrease, which results in species 1 coming to an equilibrium by itself at K_1. Similarly, slight movement upward will lead to an equilibrium of only species 2 at K_2.

Tilman's Model

The Lotka-Volterra equations describe competition only by its results—that is, according to changes in the population sizes of the two competing species. In these models, no mechanisms are specified by which the effects of competition are produced. Tilman (1987) criticized this approach to competition and emphasized that we need to study the mechanisms by which competition occurs.

Tilman (1977, 1982) presented a mathematical model of competition based on resource use. We begin our examination of the essential features of Tilman's model by considering Figure 12.4, which illustrates the response of an organism to two essential resources; for terrestrial plants these might be nitrogen and light, for example. If the abundance of either resource 1 or resource 2 is too low, the population declines, conversely, if both resources are abundant, the population increases. The boundary between population growth and decline is the zero growth isocline of this species. A second key parameter for Tilman's model is the rate of consumption of the two essential resources. Each species will consume resources at different rates. For example, a plant might utilize water more rapidly than it utilizes nitrogen. These rates of consumption will

---

[1]α and β can be written more generally as a_{ij}, the effect of species *j* on species *i*. Thus $\alpha = a_{12}$ and $\beta = a_{21}$.

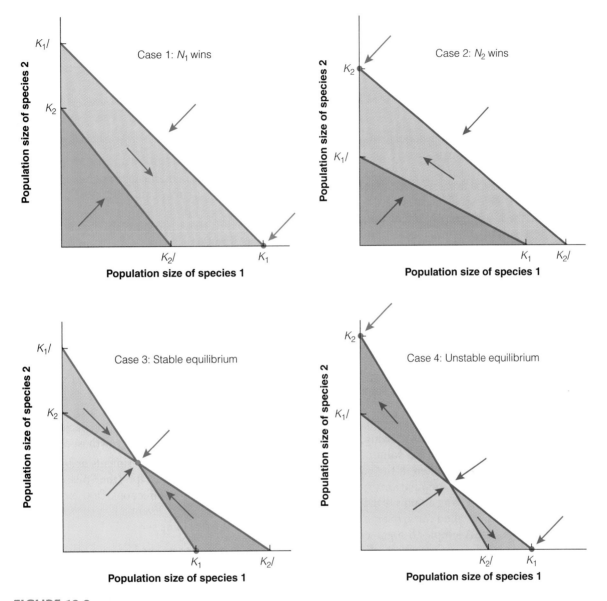

FIGURE 12.3

Four possible outcomes of competition between two species. Blue arrows indicate direction of change in populations, and red dots and red arrows indicate the final equilibrium points. In the yellow zone, both species can increase; in the green zone, only species 1 can increase; in the brown zone, only species 2 can increase; and in the white zone both species must decline.

determine the slope of the consumption vectors illustrated in Figure 12.4.

If we repeat this analysis for a second species, we can superimpose the two zero growth isoclines. Figure 12.5 shows the possible outcomes of competition for the two competing species. In the first case (Figure 12.5a), species B needs more of both resources than species A. Thus species A will win out in competition, and species B will go extinct. The second case (Figure 12.5b) is the mirror image of the first case, and species A goes extinct. In the remaining case (Figure 12.5c) the zero growth isoclines cross, so there is an equilibrium

ESSAY 12.1

WHAT IS A PHASE PLANE, AND WHAT IS AN ISOCLINE?

The dynamics of two interacting populations can be illustrated graphically in two ways:

1. *Time series*: This is the usual way that population data are plotted, with population size on the Y-axis and time on the X-axis. For two interacting populations there will be two lines on the graph (see, e.g., Figure 11.7, page 164).

2. *Phase plane*: By taking cross sections of the time series plot, one can abstract time from the diagram and plot species 1 numbers (X-axis) directly against species 2 numbers (Y-axis). This type of diagram is called a phase plane. Figure 12.1 is an example, and phase-plane diagrams will be used in this and subsequent chapters to illustrate the changes of one species population relative to another.

Phase planes are most useful for plotting models that have an equilibrium as a final solution. For population growth we are interested in the equilibrium that represents zero population growth $dN_1/dt = 0$. The line connecting all points that have zero population growth is called the *zero growth isocline* for that population. A simple thought-experiment will illustrate how the zero growth isocline can be constructed.

Consider Figure 12.1 with respect to species 1 numbers:

1. Start a population in the upper right corner of the graph, with high numbers of species 1 and species 2 individuals. Species 1 will decrease in numbers because it is above carrying capacity in this simple model. From your starting point put an arrow on the graph pointing to the left for species 1.

2. Now start another population in the upper left side of the graph, with low numbers of species 1 and high numbers of species 2. Species 1 will now increase in numbers because it is below carrying capacity. Put another arrow on the graph pointing to the right from this starting point.

3. Somewhere between these extremes you could start a population that would not change in numbers for species 1 because it was exactly at carrying capacity for the fixed number of species 2. This point would be on the zero growth isocline.

4. Repeat these thought-experiments many times all over the area of this graph (the "phase plane"), placing arrows in the direction of movement of the species 1 population. Eventually you would define the blue diagonal line shown in Figure 12.1.

You have now constructed a phase plane diagram with a zero growth isocline.

point. To determine whether this equilibrium is stable or unstable, we need additional information on the consumption curves for each species. At the equilibrium point in Figure 12.5c, species A is limited by resource 2, and species B is limited by resource 1. If species A consumes relatively more of resource 1 than does species B, the equilibrium point is unstable, and one species or the other will go extinct. To apply Tilman's model to a particular environment, we must know the rate of supply of the limiting resources to the populations (a function of the habitat), and the rates of

consumption of these resources by each species (represented by the vectors in Figure 12.4).

Figure 12.6 illustrates five examples of competition in Tilman's model and shows that the rate of supply of the limiting nutrients is critical in determining the outcome of a competitive situation. Given a resource supply rate for a habitat, Tilman's model can predict the outcome of competition if we know for each species the concentration to which the limiting nutrient would be reduced by that species (called R^*; Tilman 1990). This is the key point for Tilman's

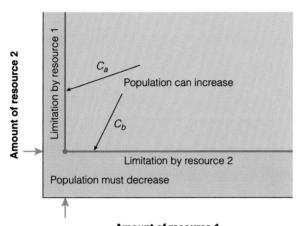

FIGURE 12.4

The response of an organism to variations in two essential resources (such as nitrogen and water for plants). The blue lines represent the zero growth isoclines, the lower one set by resource 2 and the left one set by resource 1 (arrows). Above these isoclines in the green shaded area, the population can increase in size; below these isoclines in the purple area, the population will decline. In the left side of the purple area, resource 1 is limiting; in the bottom side of the purple area, resource 2 is limiting. Only at the intersection point (blue dot) are both resources simultaneously limiting. At the hypothetical consumption vectors C_a the organism uses resource 1 more rapidly and resource 2 more slowly; C_b represents the opposite case. (Modified from Tilman 1982.)

model of competition, and we can measure this concentration R^* by growing the organism in monoculture. For example, for a grass we could plant the species in single-species plots in a greenhouse and compare soil nutrient levels to those in unvegetated plots. Superior competitors are able to consume limiting resources to low levels, in the process eliminating competitors. Tilman's model represents an important advance in models of competition.

Tilman's model provides the same final predictions as the Lotka-Volterra model (compare Figure 12.5 with Figure 12.3), but Tilman's model can be extended to make community-level predictions about species diversity and succession (Tilman 1986, 1990). The strength of Tilman's model is in its emphasis on mechanism, and because of this it can help us understand more precisely how species interact over limited resources.

Now that we have these mathematical formulations, we must see if they are an adequate representation of what happens in actual biological systems.

Competition in Experimental Laboratory Populations

One of the first and most important investigations of competitive systems was conducted by a Russian

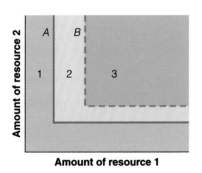
(a) Exclusion: species *A* wins

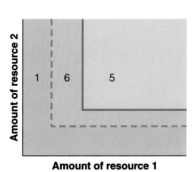
(b) Exclusion: species *B* wins

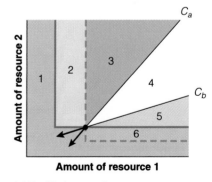
(c) Equilibrium coexistence

FIGURE 12.5

Tilman's model of competition for two essential resources. The zero isoclines for Species A (blue) and species B (red, dashed line) are shown, along with the consumption rate vectors for each species (C_a and C_b). For all three cases the regions are labeled and colored as follows: 1 (purple) = neither species can live; 2 (yellow) = only species A can live; 3 (blue) = species A wins out in competition; 4 (white) = stable coexistence; 5 (pink) = species B wins out in competition; 6 (green) = only species B can live. • = stable equilibrium point. (From Tilman 1982.)

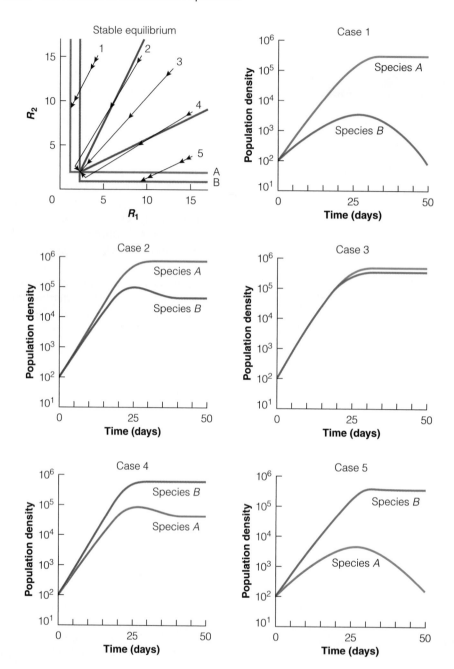

FIGURE 12.6

The outcome of competition for Tilman's model for five different resource supply rates for equilibrium coexistence. The resulting five population curves for the two species show the time course of competition. Note that even though this model permits coexistence, in some cases one species will go extinct because the resource supply rates of the habitat are not in the zone of stable coexistence show in Figure 12.5c. (From Tilman 1982.)

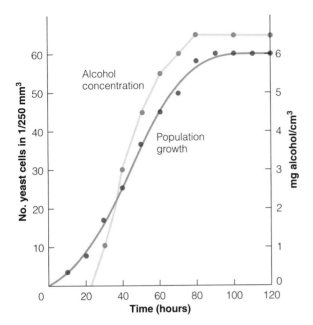

FIGURE 12.7
Population growth and ethyl alcohol accumulation in a population of yeast (Saccharomyces). *(After Richards 1928.)*

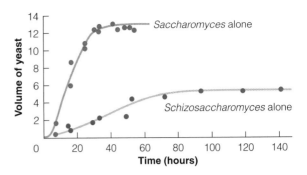

FIGURE 12.8
Population growth of pure cultures of two yeasts, Saccharomyces *and* Schizosaccharomyces. *(After Gause 1932.)*

microbiologist named Georgyi Frantsevich Gause working at Moscow University. Gause (1932) studied in detail the mechanism of competition between two species of yeast, *Saccharomyces cervisiae* and *Schizosaccharomyces kephir*.[2] In the first aspect of his investigations, concerning the growth of these two species in isolation, he found that the population growth of both species of yeast was sigmoid and could reasonably be fitted by the logistic curve.

Gause then asked: What are the factors in the environment that depress and stop the growth of the yeast population? Richards (1928) had previously shown that when the growth of yeast stops under anaerobic conditions, a considerable amount of sugar and other necessary growth substances remain. Because growth ceases before the reserves of food and energy are exhausted, something else in the environment must be responsible for the restriction of population increase. The decisive factor seems to be the accumulation of ethyl alcohol, which is produced by the breakdown of sugar for energy under anaerobic conditions (Figure 12.7). High con-

centrations of alcohol kill the new yeast buds just after they separate from the mother cell. Richards showed that the yeast growth could be reduced by artificially adding alcohol to cultures and that changes in the pH of the medium were of secondary importance. Thus with yeast we apparently have a quite simple relationship, with the population being limited principally by one factor: ethyl alcohol concentration.

When grown separately, the two yeast species reacted as shown in Figure 12.8. From these curves, Gause calculated logistic curves (calculated in units of volume):

	Saccharomyces	Schizosaccharomyces
K	13.00	5.80
r	0.22	0.06

Gause then investigated what would happen when the two yeast species were grown together, and he obtained the results shown in Figures 12.9 and 12.10. Gause assumed that these data fit the Lotka-Volterra equations, and using the equations on the data from the mixed cultures, he obtained the following data:

Competition Coefficients

Age of Culture (hr)	α Saccharomyces	β Schizosaccharomyces
20	4.79	0.501
30	2.81	0.349
40	1.85	0.467
Mean value	3.15	0.439

The influence of *Schizosaccharomyces* on *Saccharomyces* is measured by α, and this means that, in terms of

[2]These organisms' scientific names have changed since Gause's studies.

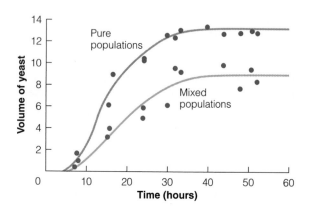

FIGURE 12.9
Growth of populations of the yeast Saccharomyces *in pure cultures and in mixed cultures with* Schizosaccharomyces. *(After Gause 1932.)*

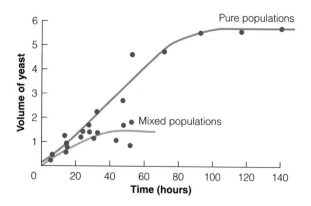

FIGURE 12.10
Growth of populations of the yeast Schizosaccharomyces *in pure cultures and in mixed cultures with* Saccharomyces. *(After Gause 1932.)*

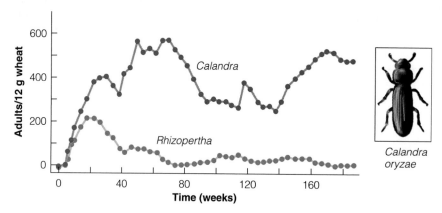

FIGURE 12.11
Population trends of adult grain beetles (Calandra oryzae and Rhizophertha dominica) *living together in wheat of 14% moisture content at 29.1°C. (After Birch 1953b.)*

competition, *Saccharomyces* can fill its K_1 spaces according to the equivalence

1 volume of *Schizosaccharomyces* = 3.15 volumes of *Saccharomyces*

Note that the α and β values tend to change with the age of the culture, but as a first approximation we can assume α and β to be constants.

If alcohol concentration is the critical limiting factor in these anaerobic yeast populations, Gause argued, then we should be able to determine the competition coefficients α and β by measuring the alcohol production rate of the two yeasts. He found:

	Alcohol Production (% EtOH/ml yeast)
Saccharomyces	0.113
Schizosaccharomyces	0.247

Gause then argued that since alcohol was the limiting factor of population growth, the competition coefficients, α and β, should be determined by a direct ratio of these alcohol production figures:

$$\alpha = \frac{0.247}{0.113} = 2.18$$

$$\beta = \frac{0.113}{0.247} = 0.46$$

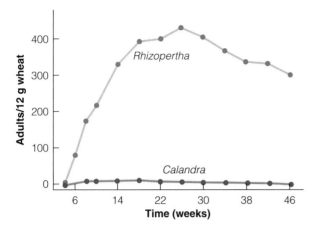

*Rhizopertha
dominica*

FIGURE 12.12
Population trends of adult grain beeltes (Calandra oryzae and Rhizopertha dominica) *living together in wheat of 14% moisture content at 32.3°C. (After Birch 1953b.)*

These independent physiological measurements agree in general with those obtained from the population data given previously. Gause attributed the differences in the α values to the presence of other waste products affecting *Saccharomyces*. Gause assumed that the competition coefficients would be the reciprocals of each other, but this assumption need not apply to all cases of competition.

Georgyi F. Gause *(1910–1986) Russian ecologist and microbiologist*

In many laboratory experiments, a species can do well when raised alone but can be driven to extinction when raised in competition with another species. When Birch (1953b) raised the grain beetles *Calandra*

oryzae and *Rhizopertha dominica* at several different temperatures, he found that *Calandra* would invariably eliminate *Rhizopertha* at 29°C (Figure 12.11) and that *Rhizopertha* would always eliminate *Calandra* at 32°C (Figure12.12). Birch found that he could predict these results from the intrinsic capacity for increase; for example,

	r	Temperature	Winner
Calandra	0.77	29.1°C	Calandra
Rhizopertha	0.58		
Rhizopertha	0.69	32.3°C	Rhizopertha
Calandra	0.50		

Thus we could change the outcome of competition by changing only one component of the environment, temperature, by only 3°C.

In all the grain beetle experiments just discussed, one species or the other died out completely. All these situations fall under cases 1 or 2 in our treatment of the Lotka-Volterra equations. What about case 3, in which the species coexist? Yeasts coexisted in Gause's experiments; does coexistence ever occur in grain beetles?

Under the conditions of extreme crowding in laboratory experiments, it is possible for two species to live together indefinitely if they differ even slightly in their requirements. For example, Crombie (1945) reared the grain beetles *Rhizopertha* and *Oryzaephilus* in wheat and found that they would coexist indefinitely. The larvae of *Rhizopertha* live and feed inside the grain of wheat; the larvae of *Oryzaephilus* live and feed outside the grain. (The adults of both species have the same feeding behavior, feeding outside the wheat grain.) Apparently these larval differences were sufficient to allow coexistence.

Gause (1934) found that *Paramecium aurelia* and *P. bursaria* would coexist in a tube containing yeast. *P. aurelia* would feed on the yeast suspension in the upper layers of the fluid, whereas *P. bursaria* would feed on the bottom layers. This difference in feeding behavior allowed these species to coexist.

Thus by introducing only very slight differences in the environment, or given very slight differences in species habits, coexistence can occur between competing animal species under laboratory conditions.

Competition in Natural Populations

We now come to the question of how these theoretical and laboratory results apply to nature. In asking this question, we come up against a controversy of modern ecology, the problem of Gause's hypothesis.

Gause (1934) wrote: "As a result of competition two similar species scarcely ever occupy similar niches, but displace each other in such a manner that each takes possession of certain peculiar kinds of food and modes of life in which it has an advantage over its competitor" (p. 19). Gause referred to Elton (1927), who had defined *niche* as follows: "The niche of an animal means its place in the biotic environment, its relations to food and enemies" (p. 64). Thus Elton used the term *niche* to describe the *role* of an animal in its community, so one could speak (for example) of a broad herbivore niche, which could be further subdivided.

Gause went on to say that the Lotka-Volterra equations do "not permit any equilibrium between the competing species occupying the same 'niche,' and [lead] to the entire displacing of one of them by another." . . . Both species survive indefinitely only when they occupy different niches in the microcosm in which they have an advantage over their competitors" (p. 48). Gause identifies case 3 (stable coexistence) with the situation of "different niches" and cases 1, 2, and 4 with the situation of "same niche."

Gause himself never formally defined what is now called *Gause's hypothesis*. In 1944 the British Ecological Society held a symposium on the ecology of closely related species. An anonymous reporter (who turned out to be David Lack) wrote that year in the *Journal of Animal Ecology* that "the symposium centered about Gause's contention (1934) that two

species with similar ecology cannot live together in the same place . . ." (p. 176).

As is usual, several workers immediately searched out and found earlier statements of "Gause's hypothesis." Monard, a French freshwater biologist, had expressed the same idea in 1920, and Grinnell, a California biologist, had written much the same thing in 1904. Darwin apparently had the same idea but never clearly expressed it. Perhaps we should drop the use of names and call this idea the *competitive exclusion principle*, which Hardin (1960) states succinctly: "Complete competitors cannot coexist." The competitive exclusion principle encapsulates the conclusions of the Lotka-Volterra models for competition.

The concept of the niche is intimately involved with the competitive exclusion principle, and so we must clarify this concept first. The term *niche* was almost simultaneously defined to mean two different things. Joseph Grinnell, who in 1917 was one of the first to use the term *niche*, viewed it as a subdivision of the habitat: Each niche was occupied by only one species. Elton in 1927 independently defined the niche as the "role" of a species in the community. These vague concepts were incorporated into Hutchinson's redefinition of the niche in 1958. If we consider just two environmental variables, such as temperature and humidity, and determine for each species the range of values that allows the species to survive and multiply, we can produce a graph like that in Figure 12.13a. This area in which the species can survive is part of its niche. We would then introduce other environmental variables, such as pH or size of food, until all the ecological factors relative to the species have been measured. The addition of the third variable produces a volume, and ultimately we arrive at an *n*-dimensional hypervolume, which we call the *fundamental niche* of the species—the set of resources it can utilize in the absence of competition and other biotic interactions.

This idea of a fundamental niche has some practical difficulties. First, it has an infinite number of dimensions, and thus we cannot completely determine the niche of any organism. Second, we assume that all environmental variables can be linearly ordered and measured, an assumption that is particularly difficult for the biotic dimensions of the niche. Third, the model refers to a single instant in time, and yet competition is a dynamic process. One way to escape these problems is to examine in only one or two dimensions the *realized niche*, the observed resource use of a species

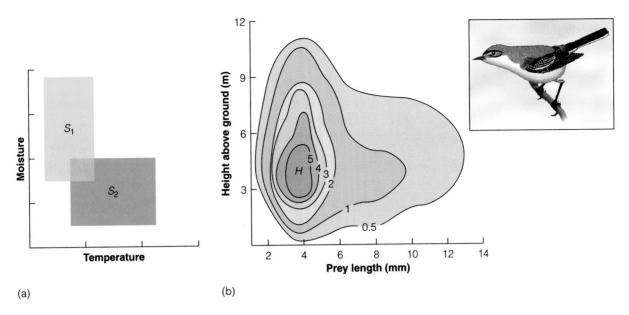

FIGURE 12.13
(a) Hypothetical diagram of part of the niche space of two species, S₁ and S₂. Only two environmental variables are used for illustration, but this approach could be extended to three variables to define a volume, or to four or more variables to define a hypervolume which Hutchinson (1958) called the fundamental niche *of a species. Note that the fundamental niche space of species may overlap to a greater or lesser extent. (b) The realized feeding niche of the blue-grey gnatcatcher (Polioptila caerulea) in California, represented by capture of insect prey of different sizes taken at different heights above the ground. The contour lines map the feeding frequencies (in terms of percentage of total diet) to these two niche axes for adult gnatcatchers during the incubation period in July and August in oak woodlands in California. (Data of Root 1967, in Whittaker et al. 1973.)*

in the presence of competition (Figure 12.13b). Thus we can discuss the differences in the *feeding niches* of two closely related birds, and can avoid discussing the attributes of unmeasurable factors that are part of the entire fundamental niches of the two species.

The *fundamental niche*, which describes a species's role in the absence of competition and other interactions, can differ dramatically from the *realized niche*, a species's actual usage of resources after competition and other biotic interactions (Figure 12.14). Species do not always occupy the best part of their fundamental niche, but may be restricted by competition or predation to the peripheral areas of their niche. Thus, for example, when a plant species is grown alone so that it can express its fundamental niche, it may grow best in high-nitrogen soils but be restricted to poorer soils by a superior competitor, as shown in Figure 12.14b (Austin 1999).

Given that we have now defined a niche, we can next ask whether two species in the same community

can exist in a single niche. Does competitive exclusion occur in natural communities? Before answering this question, we must realize that every hypothesis has its limits, and thus we should be careful to set down at the start some situations in which competitive exclusion would *not* be expected to occur. These situations are (1) unstable environments that never reach equilibrium and are occupied by colonizing species, (2) environments in which species do not compete for resources, and (3) fluctuating environments that reverse the direction of competition before extinction is possible (Hutchinson 1958).

Field naturalists were the first to question Gause's hypothesis. They pointed out that one might see in the field many examples of closely related species living together and apparently in the same habitat. Anyone who has made field collections of plants or insects will attest to the great number of species living in close association. This observation brings us to the ecological paradox of competition: How can we reconcile the

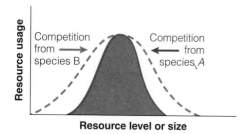

(a)

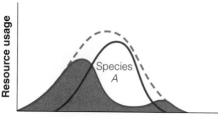

(b)

FIGURE 12.14

Two possible relationships between fundamental and realized niches for a hypothetical species C. *The fundamental niche is shown as a dashed red line and the realized niche is shaded red. (a) Species* A *and* B *compete for the resource and cause the realized niche of species* C *to shrink to the central optimum. Note that the curves are symmetrical and bell shaped. (b) A dominant species* A *(blue line) forces species* C *out of its optimum into the peripheral part of its fundamental niche, causing a bimodal, asymmetrical realized niche shown in red. Plants competing for nitrogen could show either of these patterns. (Modified from Austin 1999.)*

frequent extinction of closely related species in laboratory cultures with the apparent coexistence of large numbers of species in field communities?

Ecologists have developed two simple views in attempting to answer this question. One holds that competition is rare in nature, and since species are not competing for limited resources, there is no need to expect evidence of competitive exclusion in natural communities. The other view holds that competition has been very common throughout the evolutionary history of communities and has resulted in adaptations that serve to minimize competitive effects.

How common is competition in nature? Much investigation has centered on closely related species on the assumption that taxonomic similarity should promote competition. Lack (1944a, 1945), for example, studied the ecology of closely related species of

Cormorant (Phalacrocorax carbo) (©CK Lorenz/Photo Researchers, Inc.)

Shag (Phalacrocorax aristotelis)

birds in an attempt to test Gause's hypothesis. One example of his work involved the cormorant (*Phalacrocorax carbo*) and the shag (*P. aristotelis*). These species are very similar in habits and appeared to overlap widely in their ecological requirements; both are cliff nesters and feed on fish. Lack showed that the cormorant nests chiefly on flat, broad cliff ledges and feeds chiefly in shallow estuaries and harbors; the shag nests on narrow cliff ledges and feeds mainly out at sea. Thus Lack identified significant ecological dif-

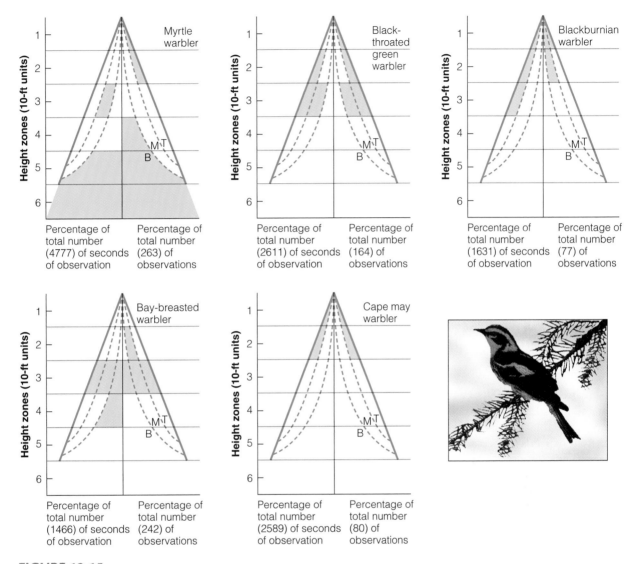

FIGURE 12.15

Feeding positions of five species of warblers in the coniferous forests of the northeastern United States. The zones of most concentrated feeding activity are shaded. B = base of branches, M = middle of branches, T = terminal portions of branches. The blackburnian warbler is illustrated. (After MacArthur 1958.)

ferences between these closely related species and assumed that competition was minimized.

Robert MacArthur of Princeton University was instrumental in bringing the study of competition to the fore in North America because of his theoretical and empirical work on birds. His thesis work involved a classic study of potential competition among a group of closely related birds. The boreal forests of New England are inhabited by five warbler species of the genus *Dendroica*, all of which are insect eaters and about the same size. Why does one species not exterminate the others by competitive exclusion? MacArthur (1958) showed that these warblers feed in different positions in the canopy (Figure 12.15), feed in different manners, move in different directions through the trees, and have slightly different nesting dates. The feeding-zone differences seem sufficiently large to explain the coexistence of the blackburnian, black-throated green, and

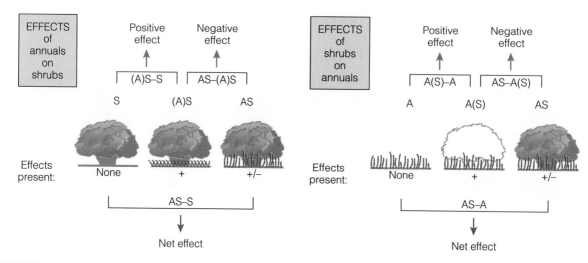

FIGURE 12.16

Experimental design for separating competition and facilitation between desert shrubs and annual grasses and herbs in the Mojave Desert. S = annual removals, shrubs alone; (A)S = artificial annuals (straw) and shrubs; AS = controls, both species present; A = shrub removals, annuals alone; A(S) = artificial shrubs, annuals under artificial canopy. + = potential positive effects (facilitation); − = potential negative effects (competition). (From Holzapfel and Mahall 1999.)

bay-breasted warblers. The myrtle warbler is uncommon and less specialized than the other species. The Cape May warbler is different from these other species because it depends on occasional irruptions of forest insects to provide a superabundant food source for its continued existence. During irruptions of insects, the Cape May warbler increases rapidly in numbers and obtains a temporary advantage over the others. During years between irruptions, they are reduced in numbers to low levels.

Thus closely related species of birds either live in different sorts of places or else use different sorts of foods. Lack suggested that these differences arose because of competition in the past between these closely related species. Therefore, because of Gause's hypothesis and its associated selection pressure, species either "moved" to different places and so avoided competition, or they changed their feeding behavior to avoid competition. There is, however, a great logical difficulty in testing Lack's hypothesis that competition has caused two species to differ (Simberloff and Boecklen 1981). Even though two species are always somewhat different from each other, as a by-product of speciation, observing differences between species does not necessarily mean that competition caused the differ-

ences. This fundamental difficulty means that descriptive studies of species differences by themselves are not useful for understanding the importance of competition in natural populations. What is needed is experimental work on populations that are possible competitors.

Animal ecologists have attributed the coexistence of many different species to these species' abilities to specialize in their diet: Herbivores feed on different plant species or different parts of plants, and carnivores feed on different animal species, or eat both plants and animals. Thus many food resources are available to animals. But plants all need only a few resources, and it has never been clear how all the plant species we see can coexist in natural communities (Grace 1995). One possible explanation is that plant communities are not in equilibrium, so competition can never reach the end point of competitive exclusion. How might this work? One way to approach this involves the *paradox of the plankton.*

Many natural plant communities contain many species that should be potential competitors. The phytoplankton of marine and freshwater environments consists of a large number of autotrophic species that utilize a common pool of nutrients and undergo photosynthesis in a relatively unstructured environment.

All vascular plants require water, light, and nutrients, and consequently competition between plants over essential resources ought to be common. But plant interactions can also have a positive effect on other plants, a phenomenon called *facilitation*. The focus of ecologists on negative effects has tended to obscure positive effects that might be common. In arid environments, competition among roots for water should be expected, and range managers have long been interested in improving arid zone grazing by replacing shrubs such as sagebrush with grasses or legumes. Holzapfel and Mahall (1999) tested for the effects of competition and facilitation between a desert shrub (*Ambrosia dumosa*, burroweed) and annual grasses and herbs in the Mojave Desert of California. By eliminating shrubs and replacing them with an artificial canopy, the investigators could measure both the positive and negative effects in this desert system, where water is the limiting factor. Figure 12.16 shows the experimental design and the means by which the investigators could separate positive and negative effects, and Figure 12.17 shows the results of this

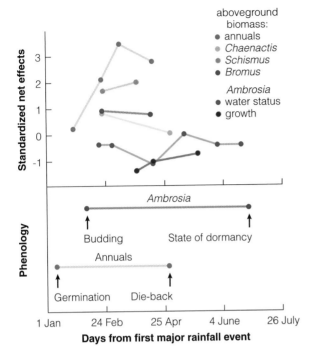

FIGURE 12.17
Changes in the net effects of shrubs (Ambrosia) *on annuals, and of annuals on shrubs, in relation to phenology, Mojave Desert of California. The net positive effects of shrubs on annual plants are show in green on the top panel. The net negative effects of annuals on shrub growth and on water balance are shown in brown. Annuals includes the entire annual community. Seasonal shifts of positive and negative effects are clear for this desert system. (Holzapfel and Mahall 1999.)*

Robert H. MacArthur *(1930–72) Princeton University*

How can all these species coexist especially given that, because natural waters are often deficient in nutrients, competition should be strong? This dilemma, the *paradox of the plankton*, has been aptly described by Hutchinson (1961) as a possible exception to the competitive exclusion principle. Hutchinson suggested that these species could coexist because of environmental instability; before competitive displacement could have time to occur, seasonal changes in the lake or the sea would occur (see Figure 11.11, page 167). The phytoplankton may thus be viewed as a non-equilibrium community of competing species and thus are not an exception to the principle of competitive exclusion.

analysis. Annual grasses and herbs had strong negative effects on *Ambrosia*, measured by water balance and by shrub growth. By contrast, *Ambrosia* had strong positive effects on all the annual plant species, improving biomass, seed production, and survival. Shading by shrubs lowers ambient temperature and improves

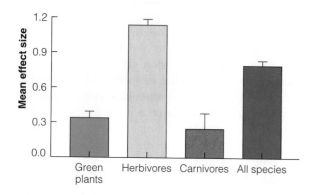

FIGURE 12.18
Relative strength of competition between species as measured by mean effect size (Equation 12.8) on biomass in 93 species of organisms in diverse groups. 95% confidence limits are shown for each category. (Data from Gurevitch et al. 1992, Table 1.)

water availability to annuals, and this is the mechanism behind facilitation. Plant-plant interactions are not always negative.

Interspecific competition has been analyzed in a wide variety of plants and animals during the past 40 years, and we can now ask how frequently competition occurs between species in nature, and how strong its effects are. Gurevitch et al. (1992) have tabulated the results of 218 competition experiments. To compare different groups of plants and animals they defined effect size in the usual statistical manner:

$$\text{Effect size} = \frac{\overline{X}_e - \overline{X}_c}{s} \qquad (12.8)$$

where $\overline{X}_c$ = mean biomass of the control group

$\overline{X}_e$ = mean biomass of the experimental group

s = standard deviation of both groups pooled

Since the experimental treatments involve the removal of potential competitors, a positive effect size means that competition is reducing the density or biomass of the species. A negative effect size implies facilitation, a higher density or biomass under conditions of interaction.

Competition had a strong overall effect in 218 studies covering 93 species (average effect size = 0.8;

Gurevitch et al. 1992). Figure 12.18 shows the average effect sizes in four categories of organisms. Several general trends are apparent in this figure. Plants and carnivores showed relatively small effects of interspecific competition, compared with herbivores. Considerable variation in competitive effects is apparent within herbivores. Some herbivores like frogs and toads show strong effects of interspecific competition, while other herbivores like marine molluscs show only moderate effects. The overall conclusion of this survey of interspecific competition is that it occurs frequently (but not always) in natural populations, and that more information is needed on the mechanisms by which competition operates in nature and how large the effects of competition might be in different species groups.

Gause's hypothesis would seem to predict that if two competitors are very similar, competition would lead to the rapid extinction of one species or the other because of very strong competition. One way to test this using laboratory populations is to use strains of microorganisms with known genetic differences. Kashiwagi et al. (1998) used mutants of the bacterium *Escherichia coli* to test for competition between two strains that differ at a single genetic locus, the smallest possible difference between competitors. They found that these two mutants would coexist in a chemostat, even when the starting population sizes of the two strains were varied. These results suggest that Gause's hypothesis should be rejected as a general ecological model for competition, since even the tiniest differences can permit coexistence of closely related forms.

The fact that strong competitive interactions are found in some natural populations led MacArthur (1972) to explore the consequences of competition for the evolution of species differences. Species that come into competition may evolve differences to minimize the impact of competition. The process can be illustrated with a simple example. Let us assume that we have identified two species that may compete for food at a certain time of the year. When we measure their food consumption for items of different size and obtain the *resource-utilization* curves shown in Figure 12.19, Three cases become apparent: (a) If the curves are completely separate, some food resources are not being utilized, and (unless some other constraints are imposed) one or both species will benefit by feeding on the unutilized food items. This change in feeding

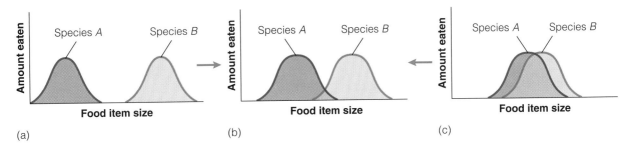

FIGURE 12.19
Hypothetical resource-utilization curves for two species that are potential competitors. Food size is the resource for which competition may occur in this situation. Vertical arrows indicate the direction of evolution pressures toward case (b).

will lead to one of the next two cases. (b) If the curves overlap only slightly so that each species has a set of food items largely to itself, the two species will each be able to survive and reproduce in the same habitat. (c) If the curves overlap a great deal so that both species eat most of the same foods, competition will be so severe that one species may be driven to extinction, or may move to a different habitat to avoid competition, or, over many generations, may evolve feeding differences, as in case (b).

The result of this simple graphic analysis is that competing species should evolve toward case (b) as the limiting similarity possible for coexistence. This is the classical theory of niche evolution under competition. A great deal of theoretical work has been devoted to attempts to quantify the limiting similarity or maximum possible overlap illustrated in Figure 12.19b (May 1974b), but there appears to be no "ecological constant" that can describe limiting similarity for all possible competitive situations (Abrams 1975, 1996). Nevertheless it is interesting to ask how much species overlap in their resource use, and resource partitioning has been an important focus of study for the analysis of species interactions.

Other difficulties confound the simple situation depicted in Figure 12.19. Niches may have more than one dimension in which competition is possible. Most studies have emphasized size of food as one dimension, but type of food, habitat usage, and feeding times could also be important dimensions along which competitors segregate to avoid competition. Cases of nearly complete overlap in one niche dimension (as depicted in Figure 12.19c) may be misleading if two or more niche dimensions are important to the competing organisms.

The role of competition in natural populations can be analyzed in several ways (Wiens 1989). We can search for patterns in resource utilization like those shown in Figure 12.19. One good example comes from the study of the diets of five species of terns on Christmas Island in the Pacific Ocean (Figure 12.20). Terns on Chirstmas Island are ecologically segregated according to their diets, and these data are consistent with the idea that competition has favored ecological divergence in diets. Schoener (1986b) compiled data from many studies of this sort that show ecological segregation. Even if such segregation occurred by evolutionary changes in the past, it is still an open question whether interspecific competition is operating today in these populations.

Wiens (1989) pointed out that in order to show that interspecific competition occurs, one must demonstrate that the species involved overlap in resource use and that competition over these resources has negative effects. Table 12.1 lists criteria that ecologists use to become more convinced that interspecific competition is producing negative effects in modern populations. Much of the data on resource utilization, such as Figure 12.20 satisfy criteria 1 and 2 only. A second, better way to analyze competition is thus to conduct experiments of the type described in Chapter 6. Humans have inadvertently conducted some of the best experiments in competition by introducing species into new areas.

Some introduced ants have extended their distributions with the help of human beings and in the

TABLE 12.1 **Criteria for establishing the occurrence of interspecific competition, listed according to the strength of the evidence for its occurrence in natural populations.**

Criteria	Strength of evidence
1. Observed checkerboard patterns of distribution consistent with predictions	Weak
2. Species overlap in resource use	↓
3. Intraspecific competition occurs	Suggestive
4. Resource use by one species reduces availability to another species	↓
5. One or more species is negatively affected	Convincing
6. Alternative process hypotheses are not consistent with patterns	

Source: Wiens (1989), p. 17.

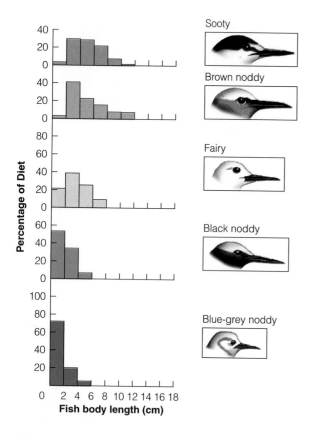

FIGURE 12.20

Resource partitioning in five species of terns on Christmas Island, Pacific Ocean, as seen in frequency distributions of prey size. Terns are arranged in order of size from the largest species at the top to the smallest at the bottom. The two largest terns are nearly the same size and eat very similar sizes of fish, but the sooty tern feeds at sea several hundred kilometers from land and the brown noddy tern feeds within 100 km of land. (From Ashmole 1968.)

process have eliminated the native ant fauna through competition. Relatively few species of ants have shown a striking ability to displace resident species. The Argentine ant (*Linepithema humile*), which was first discovered in California in 1907 and has been spreading ever since, is displacing native ants in many temperate and subtropical areas (Holway 1999). Holway (1999) has analyzed the reasons for the strong competitive ability of Argentine ants. He found that Argentine ants were more effective at exploitative competition than native ants in northern California. Baits set out at fixed distances from ant colonies were found within four minutes by Argentine ants, in contrast to the 10–35 minutes required by native ants. Argentine ants were not more successful in direct aggression—sometimes native ants won, and sometimes Argentine ants won. The situation concerning chemical defensive compounds is similar: Both native ants and Argentine ants produced chemical repellents that worked equally well against other species. The key to Argentine ant success seems to be in pure numbers—they are more numerous than native ants, because they form supercolonies. Whereas native ants form individual colonies and defend the territory around their colony from other ants, Argentine queens and workers move freely between different nests without any territorial defense. Worker numbers are much larger in supercolonies, and Argentine ants thus overwhelm their competitors by force of numbers. By securing most of the food resources in an area, Argentine ants and drive competing native ant species extinct.

Plant ecologists have repeatedly demonstrated the effect of one plant on another for pasture species and agricultural crops; one example will illustrate this. Two annual plants, *Bromus mollis* and *Erodium botrys*, dominate the grasslands of the Central Valley of California. They are important species to grazers, and unfertilized land is dominated by grasses like *Bromus*. *Erodium* has a low growth form but rapid root penetration in the seedling stage, so it is favored in competition for soil nutrients. *Bromus* grows taller, and if adequate nutrients are available it becomes the superior competitor for light and shades out *Erodium*. Thus a balanced competition seems to exist in which *Erodium* is favored in poor soils and drought years, and *Bromus* is favored in better soils when adequate moisture is available (McCown and Williams 1968).

Potential competitors can avoid competition by ecological segregation in time, in space, or in food resources. Animals most often segregate in space (by using different habitats), segregate in diet less often, and rarely segregate by time (Schoener 1986b). Plants all require the same resources of light, water, and nutrients, and consequently they can usually partition resources only in space (Grace 1995). Thus we might expect plants to compete more often than animals, but as Figure 12.18 indicates, the available data do not support this generalization.

Much of competition theory has been developed with birds in mind (Wiens 1989), and yet other groups

Argentine ant (©Georg D. Lepp/Photo Researchers, Inc.)

of organisms are not as specialized as birds. Many fishes, for example, have wide tolerances to habitat conditions and feeding conditions, and very plastic growth rates. This plasticity or flexibility is also found in many invertebrates and in many plants, and contrasts with the specialized and often fixed patterns of growth and reproduction in birds and mammals. Thus poor environmental conditions that might cause a mammal population to die out may cause a fish population to stop growing or an invertebrate population to enter a dormant phase. Ecological segregation might be less important for these kinds of plants and animals.

We should not assume that competition in natural populations is always occurring. Wiens (1977) has argued that competition may be rare in some populations because of high environmental fluctuation. According to this argument, populations are typically below the carrying capacity of their environment, and thus resources are plentiful. Occasionally a "crunch" occurs, a period of scarcity in which competition does occur. This may happen only once every five or ten years, or even less frequently, implying that competition may be difficult to detect in most short-term studies.

Evolution of Competitive Ability

If two species are competing for a resource that is in short supply, both would benefit by evolving differences that reduce competition, as indicated in Figure 12.19. The benefit involved is a higher average population size for each species, and presumably a reduced possibility of extinction. But in many cases it will be impossible to evolve differences that reduce competition. Consider, for example, food size as a limiting resource, as shown in Figure 12.19. If species A evolves such that it uses smaller food items than species B, it still may encounter a third species, C, that also feeds on small-sized food. Thus species may be constrained by a web of other possible competitors, such that the option of evolving to avoid competition is not always feasible. If a species cannot avoid competition, it must evolve competitive ability. Competitive ability is one element of the more general problem of the evolution of life history strategies.

Ecologists have used two general approaches to the general question of the evolution of competitive ability as an element in life history strategies. Animal ecologists have utilized the theory of *r*-selection and *K*-selection first proposed by MacArthur and Wilson (1967), while plant ecologists have utilized a related theory of plant strategies, the *C-S-R* model developed by Grime (1979).

Theory of *r*-Selection and *K*-Selection

The idea of competitive ability in animals is an ecological concept that is intuitively clear but difficult to define, and to understand how competitive ability might evolve we need to look at life history strategies more broadly. To understand life history evolution, we can begin with the Lotka-Volterra equations for competition, which are based on the logistic curve for each competing species. Two parameters characterize the logistic curve: r (rate of increase) and K (saturation density). We can characterize organisms by the relative importance of r and K in their life cycles.

In some environments, organisms exist near the asymptotic density (K) for much of the year, and these organisms are subject to K-selection. In other habitats the same organisms may rarely approach the asymptotic density but instead remain on the rising portion of the curve for most of the year; these organisms are subjected to r-selection. MacArthur and Wilson (1967) defined r-selection and K-selection to be density-dependent natural selection. As a population initially colonized an empty habitat, r-selection would predominate for a time, but ultimately the population would come under K-selection.

Species that are r-selected seldom suffer much pressure from interspecific competition, and hence they evolve no mechanisms for strong competitive ability. Species that are K-selected exist under both intraspecific and interspecific competitive pressures. The pressures of K-selection thus push organisms to use their resources more efficiently. The individual that can most quickly convert limiting resources into reproductive adults is usually declared the superior competitor (Gill 1974).

If K-selection is a complete description of competitive ability, we should be able to predict the outcome of competition in laboratory situations by knowing the K values for the two competing species. We cannot do this, however, because of the third parameter in the Lotka-Volterra equations for competition—the competition coefficients α and β. Species can evolve competitive ability by the process of α-selection (Gill 1974). Any mechanism that prevents a competitor from gaining access to limiting resources will increase α (or β) and thereby improve competitive ability. Most types of interference phenomena fall into this category. Territorial behavior in mammals and birds, and allelopathic chemicals in plants, are two examples of interference attributes that keep competing species from using resources.

One major evolutionary problem with α-selection is that the strategy of interference often affects members of the same species as well as members of competing species, such that competitive ability is achieved

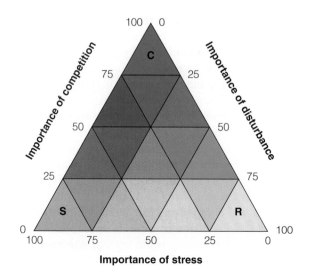

FIGURE 12.21
Grime's triangle model of plant life history strategies. Plant attributes evolve within this life history space depending on the relative importance of three factors: competition (C), stress (S), and disturbance (represented by R, for ruderal or weed strategy). It is for these factors that this model is also called the C-S-R model. (Modified from Grime 1979.)

only at the expense of a reduction in the species's own values of r and K. An example is a shrub that produces chemicals that retard the germination and growth of competing plants but that also induces autointoxication after several years (Rice 1984). An individual's negative effects on members of its own species present no evolutionary problem so long as the affected individuals do not include the individual itself or its kin.

Alpha-selection for interference attributes can also operate when organisms are at low density. In animals, the evolution of a broad array of aggressive behaviors has been crucial in substituting ability in mock combat for ability to utilize resources in competition in many situations (MacArthur 1972), and we can recognize an idealized evolutionary gradient:

Low density – colonization and growth (r selection)
↓
High density – resource competition   (K selection)
↓
High density – interference mechanisms (α selection)
prevent resource competition

Populations may exist at all points along this evolutionary gradient because competition for limiting resources is only one source of evolutionary pressure that molds the life cycles of plants and animals (Roff 1992).

Grime's Theory of Plant Strategies

Vascular plants face two broad categories of factors that affect their growth and reproduction. One category includes shortages of resources such as light, water, or nitrogen; temperature stresses; and other physical-chemical limitations. This category Grime (1979) called *stress*. A second category includes all the factors classified as *disturbances*, including grazing, diseases, wind storms, frost, erosion, and fire. Grime examined the four possible combinations of these two categories and recognized that for one combination, no strategy was possible:

	Intensity of Stress	
Intensity of disturbance	*Low*	*High*
Low	Competitive (*K*) strategy	Stress-tolerant strategy
High	Ruderal (weed or *r*) strategy	None possible

If stress and disturbance are too severe, no plant can survive. Grime (1979) suggested that the other three strategies formed the primary focus of plant evolution, and that individual plant species have tended to adopt one of these three life history models.

The three strategies can be diagrammed as a triangle (Figure 12.21), which emphasizes that these three strategies represent tradeoffs in life history traits. A plant cannot be good at all three strategies but must trade off one set of traits against another (Loehle 1988). Competitive plants show characteristics of *K*-selection—dense leaf canopies, rapid growth rates, low levels of seed production, and relatively short life-spans—and many perennial herbs, shrubs, and trees show these characteristics. Stress-tolerant plants often have small leaves, slow growth rates, evergreen leaves, low seed production, and long life spans. Ruderals are weeds that thrive on disturbance; they exhibit small size, rapid growth, are often annual plants, and devote much of their resources to seed production. Ruderals are the *r*-strategists of the plant world.

The evolution of competitive ability, although viewed differently by botanists and zoologists, has achieved a convergence of ideas. Both *r*- and *K*-selection theory and Grime's theory describe well the tradeoffs organisms must face in evolutionary time. Organisms cannot become good at everything, and adaptations are always a compromise between conflicting goals.

Character Displacement

One evolutionary consequence of competition between two species has been the divergence of the species in areas where they occur together. This sort of divergence is called *character displacement* (Figure 12.22) and can arise for two reasons. Because two closely related species must maintain reproductive isolation, some differences between them may evolve that reinforce reproductive barriers (see Chapter 2). In other cases, interspecific competition causes divergence, as illustrated by evolution from part c to part b in Figure 12.19.

In most cases, character displacement is inferred from studies in areas where the two species occur together and where they occur alone. Figure 12.23 gives a classic example of character displacement from Darwin's finches on the Galapagos Islands. Before we conclude that this example is a good illustration of evolutionary changes in competing populations, we must satisfy four criteria (Arthur 1982):

1. The change in mean value of the character in areas of overlap should not be predictable from variation within areas of overlap or areas of isolation.

2. Sampling should be done at more than one set of locations so that local effects can be eliminated.

3. Heritability of the character must be estimated to be high so that genetic variation may be assumed to underlie the differences observed.

4. Evidence must be presented to show that the species are indeed competing and that the character measured has some relevance to the competitive process.

For Darwin's finches, criteria 1, 3, and 4 are satisfied. The change in beak size in *Geospiza fortis* in isolation on Daphne Island, for example, is much greater than one would predict from observed variation on any of the other islands. Beak characters in *G. fortis* have a very high heritability (offspring resemble parents),

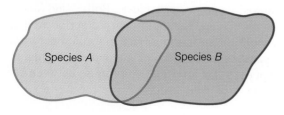

(a) Geographic distribution

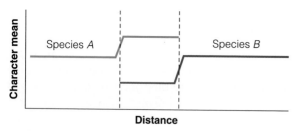

(b) Character changes

FIGURE 12.22
Schematic view of character displacement arising from interspecific competition in the zone of overlap of two species. This scheme is inferred as an explanation of the observations illustrated in Figure 12.23.

which suggests that the variation in beak depth in *Geospiza* shown in Figure 12.23 is largely genetic in origin. There is good observational evidence of competition for food in Darwin's finches, as discussed by Grant (1999). Unfortunately, because of the finite size of the Galapagos Islands, we cannot replicate these findings in another set of the islands (criterion 2). There are at present few cases in which character displacement has been conclusively demonstrated (Arthur 1982). We do not know whether this means that character displacement rarely occurs, or that it is common but very difficult to demonstrate in natural populations.

Diffuse Competition and Indirect Effects

Competition between species is usually thought of in terms of two species interacting over limited resources. MacArthur (1972) recognized that competition could also occur among many species. He

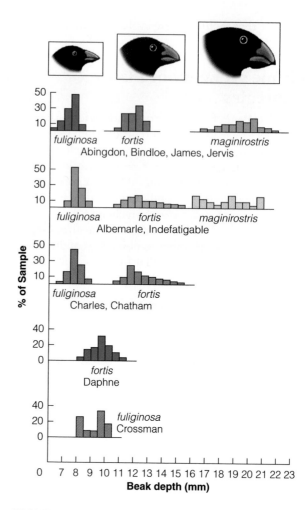

FIGURE 12.23
Character displacement in beak size in Darwin's finches from the Galapagos Islands. Beak depths are given for Geospiza fortis *and* G. fuliginosa *on islands where these two species occur together (upper three sets of islands) and alone (lower two islands).* Geospiza magnirostris *is a large finch that occurs on some islands. (After Lack 1947.)*

defined *diffuse competition* as the combined effects of many species upon a given species. In diffuse competition any single pair of species has very weak interactions, so it may be difficult to measure any effect of one species on another. The principal ideas about competition also apply to cases of diffuse competition, and the main reason for emphasizing this idea is to remind us that species in nature rarely interact as pairs but are instead involved in many types of interactions with many other species (Moen 1989).

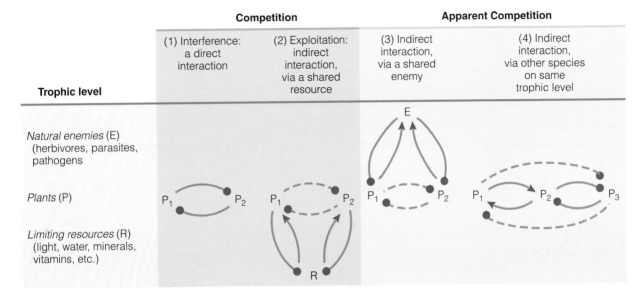

FIGURE 12.24
Schematic illustration of possible pathways of interspecific competition, in this case for plants. Solid lines are direct interactions, dashed lines are indirect ones. An arrowhead indicates a positive effect, a circle indicates a negative effect. A similar type of interaction scheme can be applied to animals. (After Connell 1990.)

Organisms of different species may interact directly or indirectly (Figure 12.24). Interference competition occurs by direct effects in which, for example, two species of birds vie for access to tree holes for nesting. Exploitative competition involves indirect effects because the two species have no interactions with each other but interact only through a third species or a shared resource. For example, if buffalo and grasshoppers eat the same grass, exploitative competition may occur even though buffalo have nothing directly to do with grasshoppers. Indirect effects are often surprising and can take on a variety of forms (Abrams 1987). For example, Holt (1977) recognized how indirect effects could produce apparent competition. Con-

sider two herbivores, such as rabbits and pheasants, that do not eat any of the same foods or compete for any essential resources. If these two species have a common predator, an increase in the abundance of rabbits could increase the abundance of the predator, which might then eat more pheasants and reduce their numbers. In systems like this, one could easily be fooled into thinking that two species were competing because when one increased in numbers, the other decreased, and vice versa. The important idea here is that we should try to understand the mechanisms behind interactions between species and not simply describe how numbers may go up or down without knowing why.

Summary

Competition between species occurs when both species strive to obtain resources that each needs. Theoretical models of competition indicate that, in cases of competition between two similar species, one species may be displaced, or both may reach a stable equilibrium. The possibility of

displacement has given rise to the *competitive exclusion principle*, which states that complete competitors cannot coexist. Under simple laboratory conditions, one species often becomes extinct but sometimes coexists with another species. Natural communities show many examples of the coexistence of similar species, and this must be reconciled with the principle of competitive exclusion. One approach to solving this paradox is to suggest that competition is rare in nature, and hence ecological displacement is not to be expected. Another approach is to suggest that competition has occurred and that the interrelations we now see are the outcome of competition, displacement, and subsequent evolution in the past. Organisms evolve competitive ability by becoming more efficient resource users and by developing interference mechanisms that keep competing species from using scarce resources.

Interspecific competition is common and can exert a major influence on population size in many natural populations. Experimental work suggests that the effects of competitive interactions in field populations are greater in herbivores than in plants or carnivores. Detailed studies of the mechanisms of competition between species are needed to understand multispecies systems and to predict patterns in natural and agricultural communities.

The evolution of competitive ability can be evaluated within a broad framework of the evolution of life history traits. Weedy species colonize quickly and avoid competition, while species in stable communities are under evolutionary pressure to minimize competition by niche differentiation and specialization.

Key Concepts

1. Competition between species can result from exploitation of resources that are in short supply or from interference in gaining access to needed resources.

2. Competition between species can be analyzed with simple mathematical models based on the logistic growth equation.

3. Competition is common in natural populations of plants and animals, and is particularly strong among herbivores.

4. In natural populations, competition over evolutionary time leads to niche differentiation, which acts to minimize competition between species.

5. To understand the effects of competition we need to study the mechanisms by which it operates and the resources that are being utilized.

Selected References

Abrams, P. A. 1998. High competition with low similarity and low competition with high similarity: Exploitative and apparent competition in consumer-resource systems. *American Naturalist* 152:114–128.

Bergelson, J. 1996. Competition between two weeds. *American Scientist* 84:579–584.

Grace, J. B. 1995. In search of the Holy Grail: Explanations for the coexistence of plant species. *Trends in Ecology & Evolution* 10:263–264.

Gurevitch, J., L. L. Morrow, A. Wallace, and J. S. Walsh. 1992. A meta-analysis of competition in field experiments. *American Naturalist* 140:539–572.

Holway, D. A. 1999. Competitive mechanisms underlying the displacement of native ants by the invasive Argentine ant. *Ecology* 80:238–251.

Kashiwagi, A. T., T. Kanaya, T. Yomo, and I. Urabe. 1998. How small can the difference among competitors be for coexistence to occur? *Researches on Population Ecology* 40:223–226.

Liebold, M. A. 1995. The niche concept revisited: mechanistic models and community context. *Ecology* 76:1371–1382.

Paine, R. T. 1984. Ecological determinism in the competition for space. *Ecology* 65:1339–1348.

Wiens, J. A. 1989. *The Ecology of Bird Communities, Volume 2. Processes and variations.* Chapter 1, pp. 3–63. Cambridge University Press, Cambridge.

Winston, M. R. 1995. Co-occurrence of morphologically similar species of stream fishes. *American Naturalist* 145:527–545.

Questions and Problems

12.1 The introduced house sparrow (*Passer domesticus*) competes with the native house finch (*Carpodacus mexicanus*) in the western United States for nesting sites, and the house finch seems to lose out more frequently in interference competition. In 1940 the house finch was introduced into the eastern United States. Discuss the potential impact of this eastern introduction of the house finch on the house sparrow, and list the observations and experiments you would like to do to investigate this species interaction. Bennett (1990) summarizes the data currently available on these species.

12.2 In the Lotka-Volterra competition model, what is the meaning of a situation in which $\alpha = \beta$? In which $\alpha = \beta = 1$? What outcome is predicted when $\alpha = \beta = 1$ and $K_1 = K_2$? What is implied if $\alpha = 1/\beta$ and if $\alpha \neq 1/\beta$?

12.3 Charles Darwin in *The Origin of Species* (1859, Chapter 3) states:

> As the species of the same genus usually have, though by no means invariably, much similarity in habits and constitution, and always in structure, the struggle will generally be more severe between them, if they come into competition with each other, than between the species of distinct genera.

Discuss.

12.4 This chapter has discussed interspecific competition. What should be the relationships between interspecific competition and intraspecific competition? How could one measure the relative strengths of these two types of competition for a plant or animal species?

12.5 What is the evidence for plants that realized niches are bell-shaped, as in Figure 12.19 (page 197)? What are the implications for competition theory if niches are not symmetrical and bell-shaped? Austin et al. (1990) discuss this problem for plants.

12.6 Plant populations respond to crowding by reducing the biomass of individual plants. The relationship between plant density and individual biomass has been called the -3/2 power law of plant thinning. Review the evidence for and against this law of plant growth as summarized in Weller (1987, 1991).

12.7 Analyze the yeast results of Gause (1932) by the use of Lotka-Volterra plots (as in Figure 12.3), and predict the outcome of this competition from the estimates of α, β, K_1, and K_2.

12.8 Review the work of Connell (1961a), discussed in Chapter 6, and discuss the role of competitive exclusion in affecting distributions of organisms.

12.9 Competition for light in trees should produce an immediate benefit for individuals that are taller than their neighbors. Discuss the factors that may affect the height to which trees grow in terms of the costs and benefits of being tall. King (1990) discusses this problem in detail.

12.10 Where in Grime's triangle (see Figure 12.21, page 200) would one expect to find annual plants? Trees? Cacti? What characteristics of plants might one use to quantify these three axes?

Overview Question

Sheep, rabbits, eastern grey kangaroos, and red kangaroos are possible competitors for food in the rangelands of eastern Australia. Design an interactive flow-chart for testing the presence and intensity of competition among these herbivores.

CHAPTER 13

Species Interactions:
Predation

I̶N ADDITION TO COMPETING FOR FOOD or space, species may interact directly via predation. Predation in the broad sense occurs when members of one species eat those of another species. Often, but not always, this involves the killing of the prey. Humans are now one of the major predators of the Earth's ecosystem. We prey on fishes in the oceans, and hunt grouse, geese, and deer for sport. In this chapter we explore how predation operates and what we need to know to understand its effects.

Five specific types of predation may be distinguished. *Herbivores* are animals that prey on green plants or their seeds and fruits; often the plants eaten are not killed but may be damaged. Typical predation occurs when *carnivores* prey on herbivores or on other carnivores. *Insect parasitoids* are a type of predator that lay eggs on or near the host insect, which is subsequently killed and eaten. *Parasites* are plants or animals that live on or in their hosts and depend on the host for nutrition. They do not consume their hosts and thus differ little in their effects from herbivores. Finally, *cannibalism* is a special form of predation in which the predator and the prey are members of the same species. All these processes can be described initially with the same kind of mathematical models, and we will begin by considering them together as "predation" in the broad sense.

Predators do not interact only with their prey species, they can also interact with one another via competition (Figure 13.1). Competition between predators may be indirect when both predator species eat the same prey species that is in short supply, or it may be indirect via prey species that themselves compete for space or food. The important point is that predation in nature goes on within a context of other biotic interactions, including competition.

Predation is an important process from three points of view. First, predation on a population may restrict distribution (see Chapter 6) or reduce abundance of the prey. If the affected animal is a pest, we may consider predation useful. If the affected animal is a valuable resource like caribou or domestic sheep, we may consider the predation undesirable. Second, along with competition, predation is another major type of interaction that can influence the organization of communities, and we will discuss the role of predation in affecting community structure in Chapters 23 and 24. Third, predation is a major selective force, and many adaptations we see in organisms, such as warning coloration, have their explanation in predator-prey coevolution.

We begin our analysis of the predation process by constructing some simple models. All these models have the underlying assumption that we can isolate in nature a system consisting of one predator species and one prey species.

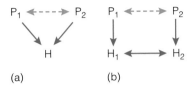

FIGURE 13.1
Schematic diagram of possible indirect effects (dotted arrows) between two predator species P_1 and P_2 that eat herbivores H_1 and H_2. (a) Indirect effects via exploitation. Two predators that share a common prey species may interact indirectly through exploitative competition of the common prey species such that there could be an indirect effect of P_1 on P_2. (b) Indirect effects without competition. Two predators eat two different prey species and do not interfere with one another and do not compete for food. But competition between the two prey species can cause effects on either or both predators indirectly. For example, if H_1 increases, P_1 will increase, H_2 will decrease because of competition with H_1, and because H_2 decreases, P_2 will also decrease. This indirect effect between the two predators is called apparent competition because an examination of predator numbers alone suggests that P_1 increases and P_2 decreases as a result of these interactions. The important point is that food web linkages can produce effects between two predators that ecologically do not interact directly.

Mathematical Models of Predation

The models we discuss in this section are of two types: those for organisms with discrete generations and those for organisms with continuous generations.

Discrete Generations

First we explore a simple model of predator-prey interactions using a discrete generation system. In seasonal environments, many insect parasitoids (predator) and their insect hosts (prey) have one generation per year and can be described by a model of the following type.

Assume that a small prey population will increase in the absence of predation, and this increase that can be described by the logistic equation (see Chapter 11):

$$N_{t+1} = (1.0 - B z_t)N_t \qquad (13.1)$$

where N_1 = population size

t = generation number

B = slope of reproductive curve (of Figure 11.2)

$z_t = (N_t - N_{eq})$ = deviation of present population size from equilibrium population size in the absence of the predator

In the presence of a predator, we must subtract from this equation a term accounting for the individuals eaten by predators, and this could be done in a number of ways. All the prey above a certain number (the number of *safe sites*) might be killed by predators, or each predator might eat a constant number of prey. If, however, the abundance of the prey is determined by the abundance of predators, the whole predator population must eat proportionately more prey when prey are abundant and proportionately less prey when prey are scarce. They could do this by becoming more abundant when prey are abundant or by being very flexible in their food requirements. We subtract a term from the prey's logistic equation:

$$N_{t+1} = (1.0 - B z_t)N_t - CN_tP_t \qquad (13.2)$$

where P_t = population size of predators in generation t

C = a constant measuring the efficiency of the predator

What about the predator population? We assume that the reproductive rate of the predators depends on the number of prey available. We can write this simply as

$$P_{t+1} = QN_tP_t \qquad (13.3)$$

where P_t = population size of predator

N = population size of prey

t = generation number

Q = a constant measuring the efficiency of utilization of prey for reproduction by predators

Note that if the prey population (N) were constant, this equation would describe geometric population growth (Chapter 11) for the predator.

To put these two equations together and interpret them, we must first obtain the maximum reproductive rates of both predator and prey. When predators are absent and prey are scarce, the net reproductive rate of the prey will be, approximately,

$$N_{t+1} = (1.0 - BN_{eq})N_t \qquad (13.4)$$

or:

$$R = \frac{N_{t+1}}{N_t} = 1.0 - BN_{eq} \qquad (13.5)$$

where R = maximum finite rate of population
increase of the prey

For the predator, when the prey population is at equilibrium and predators are scarce, predators will increase at

$$P_{t+1} = QN_{eq}P_t \qquad (13.6)$$

or

$$S = \frac{P_{t+1}}{P_t} = QN_{eq} \qquad (13.7)$$

where S is the maximum finite rate of population increase of the predator.

Let us now work out an example. Let the maximum rate of increase of the prey (R) = 1.5 and N_{eq} = 100, so that the absolute value of the slope of the reproductive curve B = 0.005 (see Figure 11.2). Assume that the constant C measuring the efficiency of the predator is 0.5. Thus

$$N_{t+1} = (1.0 - 0.005z_t)N_t - 0.5N_tP_t \qquad (13.8)$$

Assume that under the best conditions, the predators can double their numbers each generation (S = 2.0), so that the constant Q is

$$S = QN_{eq} \qquad (13.9)$$

$$2.0 = Q(100)$$

or

$$Q = 0.02 \qquad (13.10)$$

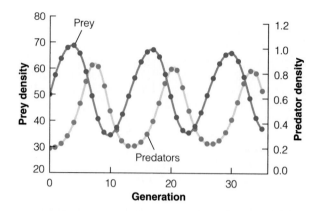

FIGURE 13.2
Population changes in a hypothetical predator-prey system with discrete generations. For the prey population N_{eq} = 100, B = 0.005, and C = 0.5. For the predator, Q = 0.02.

Consequently, the second equation is

$$P_{t+1} = 0.02N_tP_t \qquad (13.11)$$

Start a population at N_0 = 50 and P_0 = 0.2:

$$N_1 = ([1.0 - 0.005(50 - 100)]50) - [(0.02)(50)(0.2)]$$

$$= 62.5 - 5.0 = 57.5$$

$$P_1 = (0.02)(50)(0.2)$$

$$= 0.2$$

For the second generation,

$$N_1 = ([1.0 - 0.005(57.5 - 100)]57.5)$$

$$- [(0.5)(57.5)(0.2)]$$

$$= 69.72 - 5.75 = 63.97$$

$$P_1 = (0.02)(57.5)(0.2)$$

$$= 0.23$$

These calculations can be carried over many generations to produce the results shown in Figure 13.2 — a cycle of predator and prey numbers.

A stable oscillation in the numbers of predators and prey is only one of four possible outcomes; the others are stable equilibrium with no oscillation, convergent oscillation, and divergent oscillation leading to the extinction of either predator or prey. Maynard Smith

(1968) has shown that the range of variables for a stable equilibrium without oscillation is very restricted. An example will illustrate this solution. Let $N_{eq} = 100$, $B = 0.005$, and $C = 0.5$ for the prey, while $Q = 0.0105$ ($S = 1.05$) for the predator. For the first generation, from a starting population of 50 prey and 0.2 predators:

$$N_1 = \left(\left[1.0 - 0.005(50 - 100)\right]50\right) - \left[(0.5)(50)(0.2)\right]$$

$$= 69.72 - 5.75 = 57.50$$

$$P_1 = (0.0105)(50)(0.2)$$

$$= 0.105$$

Similarly,

	N	P
Second generation	66.70	0.063
Third generation	75.70	0.044
Fourth generation	83.20	0.035
Fifth generation	88.70	0.031.

The populations show decreasing small oscillations and gradually stabilize around a level of 95.2 for the prey and 0.048 for the predator.

Discrete generation predator-prey models show a variety of dynamic behaviors much like those seen in discrete population growth models (see page 159).

Continuous Generations

Many predators and prey have overlapping generations, with births and deaths occurring continuously; vertebrate predators provide many examples. For the continuous-generation case, Lotka (1925) and Volterra (1926) independently derived a set of equations to describe the interaction between populations of predators and prey. Vito Volterra, a professor of physics in Rome, became interested in population fluctuations in 1925 when his daughter became engaged to a young marine biologist who was studying the effects of World War I on fish catches in the Adriatic. The early models of Lotka and Volterra were unrealistic, and other models that are capable of greater biological realism have replaced them (Berryman 1992). The best general models were developed by Rosenzweig and MacArthur (1963) as graphic models.

Consider first the population growth of a prey species in relation to predator and prey abundance (Figure 13.3) . Now do a hypothetical experiment: Construct a series of populations at different predator and prey densities, and at each point measure whether the prey increase or decline. For example, at point A in Figure 13.3 there are many predators, and prey will certainly decline. At point B, there are few predators, and prey will increase. At point C there are many predators and many prey, and excessive predation will drive prey numbers down. By following this process for a series of points we can divide the area of the graph into a zone of prey increase and a zone of prey decrease. This fixes the *prey isocline*, the boundary between these two zones at which the rate of increase of the prey population is zero. At equilibrium the prey population must exist somewhere on this line. Lotka and Volterra made the simple assumption that the prey isocline was a horizontal line (see Figure 13.3), but in the more realistic Rosenzweig-MacArthur model the prey isocline always has a "hump." What ecological factors cause a hump-shaped prey curve? The key process is that, as prey numbers build up, prey begin to limit their own rate of increase because of food shortage, disease, or social interactions. To the left of the hump the dominant limitation on the prey is from the predators. Above the isocline hump, at higher prey numbers, the prey isocline curve falls off because the dominant limitation on the prey rate of increase comes from prey

Vito Volterra *(1860–1940) Professor of Mathematical Physics, University of Rome*

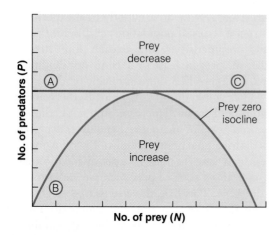

FIGURE 13.3

The prey isocline of the Rosenzweig-MacArthur model for a predatory-prey interaction. In the purple zone the prey can increase in abundance. The simple model of a horizontal prey isocline assumed by Lotka and Volterra is shown in blue. The hump-shaped Rosenzweig-MacArthur prey isocline is more realistic than the Lotka-Volterra isocline because as prey numbers increase more predators can be supported but at a diminishing rate. As prey numbers build up, prey begin to limit their own increase because of food shortage, disease or social interactions. At the hump of the isocline a maximum number of predators can be supported. Above the hump, at higher prey numbers, the prey isocline curve falls off because the dominant limitation on the prey rate of increase comes from prey intraspecific competition, and fewer predators are needed to hold down the prey rate of increase. (After Rosenzweig and MacArthur 1963.)

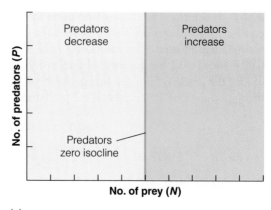

(a)

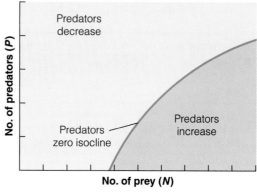

(b)

FIGURE 13.4

The predator isocline of the Rosenzweig-MacArthur Model for predator-prey interaction. In the pink colored zones the predator can increase in abundance. (a) In the simple model of a vertical isocline assumed by Lotka and Volterra, there is a single prey density above which predator populations can grow, and below which they decrease. (b) A more realistic predator isocline, which bends to the right because as predators increase in number they compete with one another for breeding sites and other resources. Not all predators will have the same shape of incline.

intraspecific competition, and predator limitation on the prey becomes less and less significant. The exact shape of the prey curve will depend on the demographic characteristics of the prey and the carrying capacity of the environment, which sets an upper limit to prey abundance.

Now consider the population changes of a predator that is food-limited at low prey densities and eats only a single prey species. When prey numbers are high, predator numbers should increase. But at high predator density, predators stop increasing because of other limitations, such as territorial behavior in wolves or a shortage of burrow sites for predatory crabs. The resulting predator isocline is shown in Figure 13.4. The predator isocline will not always be this shape, and not all predators will have exactly the same shape of isocline. A key point to note is that the more efficient the predator, the more the predator isocline is positioned to the left in Figure 13.4.

By superimposing the two isoclines in Figures 13.3 and 13.4, we get a graphic model of a predator-prey interaction. In this case, by examining the vectors around the equilibrium point, we can see that this is a stable equilibrium for both predator and prey (Figure 13.5a). In the lower right quadrant (C in Figure 13.5), the prey is decreasing but the predator is increasing, so the vector points upward and inward. The lower left quadrant (B) represents increasing prey and decreasing predators. The upper left quadrant represents both species decreasing. This model, the

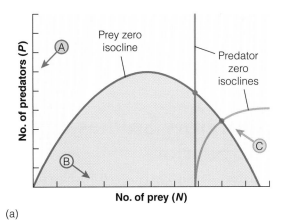

(a)

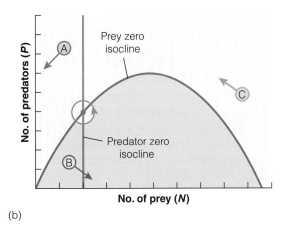

(b)

FIGURE 13.5

The predator and prey isoclines superimposed for the Rosenzweig-MacArthur model for predator-prey interaction. The equilibrium points are indicated by red dots, and the vectors from points A, B and C indicate the direction of movement in the phase plane. (a) When the predator isocline intersects the prey isocline to the right of the hump, there is a stable equilibrium point, no matter what the exact shape of the predator isocline is. (b) When the predator isocline intersects the prey isocline to the left of the hump, limit cycles like those in Figure 13.2 arise. Depending on the exact slopes of the lines, these cycles may be large enough to lead to the extinction of the predator, the prey, or both. The key point is that predator-prey systems that intersect to the left of the hump in the prey-zero isocline are unstable compared with those that intersect to the right of the hump.

Rosenzweig-MacArthur model of predator-prey interactions, is useful because we can explore in a graphic manner the effects of simple changes to the predator-prey system.

Consider the situation in which the predator is not restricted by any limitations other than its food supplies (the assumption of Lotka and Volterra). In

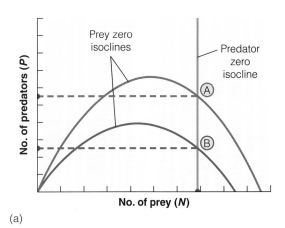

(a)

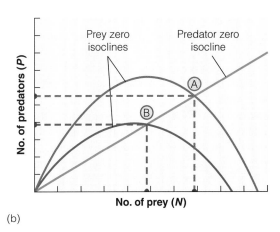

(b)

FIGURE 13.6

Predator-prey isoclines in (a) the classical Rosenzweig-MacArthur model and (b) the ratio-dependent model. Two prey isoclines are shown for less productive (blue) and more productive (purple) habitat. Increasing prey productivity changes only the equilibrium predator abundance from A to B in the classical model, but changes both predator and prey abundance in the ratio-dependent model. The equilibrium intersection points are shown by dotted lines, and the resulting equilibrium numbers of predators and prey by the dots on each axis.

this case the predator isocline is vertical and remains linear (Figure 13.5a). This system is stable, and if disturbed from equilibrium it will show convergent oscillations back to the equilibrium point. Now consider this same system with a more efficient predator. Predator efficiency in this graphic presentation means that the predators can subsist on lower prey numbers, so the predator isocline is moved to the left on the graph (Figure 13.5b). When the predator isocline intersects the prey isocline to the left of the hump, there is no point equilibrium for the system, and populations endlessly follow a stable cycle around the hypothetical equilibrium point. The farther the equilibrium point is from the hump of the prey isocline, the larger will be the amplitude of the resulting cycles and the greater the possibility of extinction.

The Rosenzweig-MacArthur model of predator-prey interactions thus reveals a wide variety of dynamic behavior, from stability to strong oscillations. This model provides a focus for asking simple questions about predator-prey systems, such as What would happen if prey became less abundant? The model also serves as an entry point into understanding the more complex real world.

All these predator-prey models make a series of simplifying assumptions about the world, including a homogeneous world in which there are no refuges for the prey or different habitats, and that the system is one predator eating one prey. Relaxing these assumptions leads to more complex and more realistic Rosenzweig-MacArthur models (Taylor 1984).

The classical form of the Rosenzweig-MacArthur model uses a vertical predator isocline, as in Figure 13.6a, which implies that the rate of increase of the predator population is controlled completely by the density of the prey. In this model the exact equilibrium for the prey species depends only upon the predator's characteristics. In particular, if the productivity of the prey population increases, the equilibrium density of the prey does not change (see Figure 13.6a). All the gain in prey productivity goes to the predators, which increase in abundance. An alternative model, the ratio-dependent model suggested by Arditi et al. (1991a, 1991b), postulates a predator isocline that runs diagonally upward (see Figure 13.6b). The ratio-dependent model assumes that the predation rate depends on the ratio of predators to prey, rather than just on prey numbers alone (Arditi et al. 1991a, Akcakaya et al. 1995). These two models make quite different predictions about the relationship between prey abundance and predator abundance. In the ratio-dependent model, as prey productivity is increased, predator and prey equilibria both rise. In some biological systems the classical theory may be adequate, but in other systems the ratio-dependent theory fits better.

The simple models of predation that we have just discussed are interesting in that they indicate that oscillations may be an outcome of a simple interaction between one predator species and one prey species in an idealized environment. In discrete generation systems, the outcome of a simple predation process may be stable equilibrium, oscillations, or extinction. Discrete

systems are more likely to lead to extinction in a fluctuating environment (May 1976). We next consider evidence from laboratory and field populations to see how well these simple models fit real predator-prey systems.

Laboratory Studies of Predation

Laboratory systems can be set up in which the major assumptions of predator-prey models can be met, and then we can investigate how these simple laboratory systems work before we tackle the more complex natural world.

Gause (1934) was the first to make an empirical test of the models for predator-prey relations. He reared the protozoans *Paramecium caudatum* (prey) and *Didinium nasutum* (predator) together in an oat medium. In his initial experiments, *Didinium* always exterminated *Paramecium* and then died of starvation—that is, the system went to extinction (Figure 13.7a)—and this is not very interesting biologically. Extinction occurred under all the circumstances Gause used for this system—making the culture vessel very large, introducing only a few *Didinium*, and so on. The conclusion was that the *Paramecium-Didinium* system did not show either a stable equilibrium or a stable limit cycle. Gause thought that stability could not be achieved because of a biological peculiarity of *Didinium:* It was able to multiply very rapidly even when prey were scarce, the individual *Didinium* becoming smaller and smaller in the process.

Gause then introduced a complication into the system: To the oat medium he added sediment, which constituted a refuge for the prey. *Paramecium* in the sediment were safe from *Didinium*, which never entered it. In this system, the *Didinium* again eliminated the *Paramecium*, but only from the clear-fluid medium; *Didinium* then starved to death, and the *Paramecium* hiding in the sediment emerged to increase in numbers (Figure 13.7b). The experiment ended with many prey and no predators. The system had reached a stable point predicted by the mathematical model, but it was a biologically uninteresting system with the predators extinct. In doing these experiments Gause added an important idea to our understanding of predation: the potential importance of refuges for prey species.

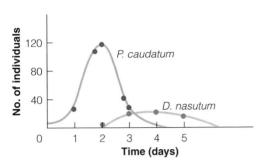
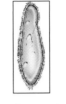

Paramecium

(a)

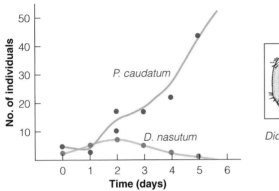

Didinium

(b)

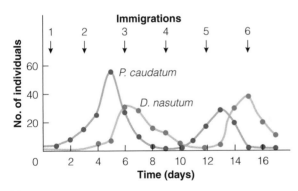

(c)

FIGURE 13.7
Predator-prey interactions between the protozoans Paramecium caudatum *and* Didinium nasutum *in three microcosms: (a) oat medium without sediment, (b) oat medium with sediment, and (c) oat medium without sediment and with immigration. (After Gause 1934.)*

Gause, quite determined, tried yet another system, introducing *immigration* into the experimental setup. Every third day he added one *Paramecium* and one *Didinium*, which produced the results shown in Figure 13.7c. Gause concluded that in *Paramecium* and *Didinium* stable oscillations in predator and prey numbers are not a property of the predator-prey interaction itself, as some models predict, but apparently are a result of constant interference from outside the system.

Carl Huffaker, working at Berkeley on the biological control of insect pests, completed a classic set

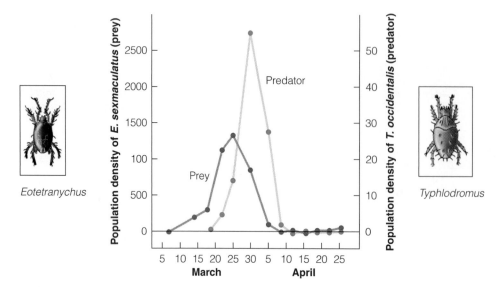

FIGURE 13.8
Densities (per unit area of orange) for the prey mite Eotetranychus sexmaculatus *and the predator mite* Typhlodromus occidentalis, *with 40 oranges, 20 of which provided food for the prey alternating with 20 foodless (covered) oranges. (After Huffaker 1958.)*

of experiments on predator-prey dynamics that had important implications for predator-prey theory. Huffaker (1958) questioned Gause's conclusions that the predator-prey system was inherently self-annihilating without some outside interference such as immigration. He claimed that Gause had used too simple a microcosm. Huffaker studied a laboratory system containing a phytophagous mite, *Eotetranychus sexmaculatus*, as prey, and a predatory mite, *Typhlodromus occidentatis*, as predator. The prey mite infests oranges, so Huffaker used these fruits for his experiments. When the predator was introduced onto a single prey-infested orange, it completely eliminated the prey and died of starvation (like Gause's *Didinium*). Huffaker gradually introduced more and more spatial heterogeneity into his experiments. In some cases he placed 40 oranges on rectangular trays similar to egg cartons and partly covered some oranges with paraffin or paper to limit the available feeding area; in other cases he used rubber balls as "substitute oranges" so that he could either disperse the oranges among the rubber balls or place all the oranges together. In still other cases, he added whole new trays that included artificial barriers of petroleum jelly, which the mites could not cross.

All of Huffaker's simple systems eventually resulted in extermination of the populations. Figure 13.8 illustrates a population that became extinct in a moderately complex environment containing 40 oranges. Finally, Huffaker produced the desired oscillation in a 252-orange universe with a complex series of petroleum-jelly barriers; in this system, the prey were able to colonize oranges in "hop, skip, and jump" fashion and keep one step ahead of the predator, which exterminated each little colony of the prey it found (Figure 13.9). The predators died out after 70 weeks, and the experiment was terminated.

Huffaker concluded that he could establish an experimental system in which the predator-prey relationship would not be inherently self-destructive. He admitted, however, that his system was dependent on local emigration and immigration, and that a great deal of environmental heterogeneity was necessary to prevent immediate annihilation of the system. The important idea that Huffaker's work added to our perspective on predator-prey theory is the concept of *environmental heterogeneity*. The world is not a uniform environment but consists of a variety of patches that are either good or bad for predator and prey alike. The addition of the simple idea of environ-

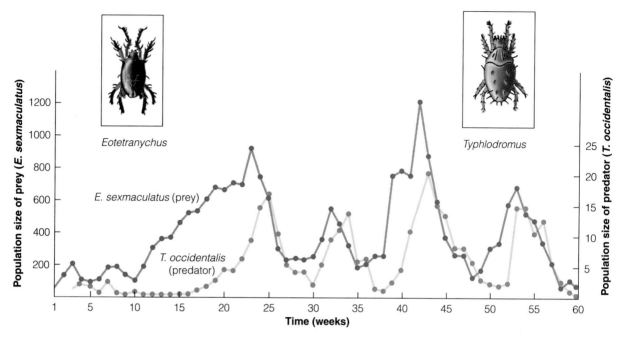

FIGURE 13.9
Predator-prey interaction between the prey mite Eotetranychus sexmaculatus *and the predator mite* Typhlodromus occidentalis *in a complex laboratory environment consisting of a 252-orange system in which one-twentieth of each orange was exposed for possible feeding by the prey. (After Huffaker et al. 1963.)*

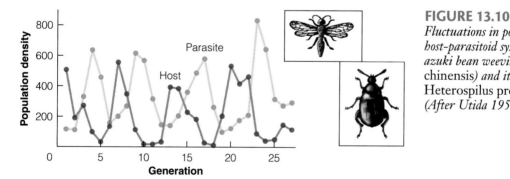

FIGURE 13.10
Fluctuations in population density in a host-parasitoid system containing the azuki bean weevil (Callosobruchus chinensis) *and its parasitoid* Heterospilus prosopidis *(a wasp). (After Utida 1957.)*

mental heterogeneity into our thinking about ecological systems has had a revolutionary effect on our thinking about ecological communities, as we will see in our discussions of community dynamics.

Stable oscillations of predator-prey interactions have been obtained in several laboratory systems. Utida (1957) maintained a system containing the azuki bean weevil as a host (prey) and a wasp that is parasitoid on the larvae of the weevil as a parasitoid (predator) in a petri dish. Systems of this type show

oscillations (Figure 13.10), which Utida followed for a maximum of 112 generations (14 complete oscillations). The oscillations were gradually damped in amplitude (convergent oscillations), and Utida noted that a long-term trend was imposed on the cycle—the host population gradually increased in density, and the parasitoid population gradually declined. This raised an interesting question: Can evolutionary changes in laboratory predator-prey systems occur during such brief experiments? Lotka and Volterra,

Rosenzweig and MacArthur, and other modelers of predator-prey systems have assumed a constant and unchanging prey species and a constant and unchanging predator species, and other ecologists have often followed their lead in assuming that evolution cannot occur on an ecological time scale.

Laboratory studies of predator-prey systems have carried us a long way from our starting point. What might we look for in predator-prey systems in the field? We must consider four aspects of predator-prey dynamics that have been simplified in both theoretical and laboratory studies:

- Multiple prey species being eaten by multiple predator species
- Refuges for the prey
- Spatial heterogeneity in habitat suitability for both the predator and the prey
- Evolutionary changes in predator and prey characteristics

We have assumed so far that predators determine the abundance of their prey and vice versa, and we should consider whether this generalization holds for field situations. If it does, we might expect to see evidence of stability in predator-prey associations in some natural systems. Associations of predator and prey might show evolutionary changes, and these evolutionary changes could be looked for in species that have recently come into contact in the field. Highly efficient predators introduced into a new ecosystem might cause the extinction of vulnerable prey.

Field Studies of Predation

How can we find out whether predators determine the abundance of their prey? The obvious experiment is to remove predators from the system and to observe its response. Few such direct experiments have been properly conducted with adequate controls, but let us examine some case studies.

Atlantic salmon (*Salmo salar*) are important in both commercial and sport fishing along the east coast of Canada, and declining stocks have been a serious problem. One attempt to increase production arose out of the observation that avian predators, particularly kingfishers and mergansers, were eating a large proportion of the young salmon population. Atlantic salmon lay their eggs in fresh water, and the young salmon live two or three years in fresh water before they emigrate to sea as smolts. After White (1939) removed 154 kingfishers and 56 mergansers from the Margaree River in Nova Scotia in 1937–1938, he recorded increased numbers of young salmon:

	No. Salmon Smolts Going to Sea
1937, before bird control	1834
1938, after bird control	4065

Clearly, bird control helped salmon to survive, but two objections can be raised to this experiment. (1) Production of salmon smolts varies greatly from year to year normally, and no control (unmanipulated) data are available to counter the possibility that 1938 was just a "good" year for salmon; and (2) an increased number of young salmon does not necessarily mean that more adult salmon will return one or two years later. Perhaps if more salmon smolts survive to reach the ocean, predators there will compensate by eating more of them.

Elson (1962) repeated this experiment on a longer time scale (six years) by removing an average of 54 mergansers and 164 kingfishers a year from a 10-mile (16-km) stretch of the Pollett River in New Brunswick. He obtained the following results:

Year	No. Salmon Fry Released	Surviving to smolts (%) (2 yr)
No Bird Control		
1942	16,000	12
1943	16,000	6
1945	249,000	2
Bird-control Program		
1947	273,000	8
1948	235,000	6
1949	243,000	8
1950	246,000	10

Some side effects of bird control were observed. Most species of fish in the river approximately doubled their numbers as a result of the bird control. Elson concluded that intensive bird control could increase the production of Atlantic salmon in streams.

Wildlife managers have a limited range of possible treatments to apply to increase desirable wildlife populations, and predator control has been among the most popular of treatments. In some cases predator control is highly effective. Duck populations in the

ESSAY 1 3 . 1

LABORATORY STUDIES AND FIELD STUDIES

Predator-prey dynamics can be studied either in the laboratory or in the field. What are the advantages and disadvantages of these two kinds of studies for analyzing biological interactions? Can we directly apply the results of laboratory studies to field situations? These critical questions are not easy to answer.

Ecological laboratory studies are done in model systems or *microcosms*—small ecosystems housed in containers. Microcosms can range from simple two species systems to complex communities of many different species. Although most microcosms are small, some—such as Biosphere 2 in Arizona or the Ecotron in England—are very large. We have already seen good examples of microcosms in Gause's work on competition in Chapter 12 and in his work on predation in this chapter. Many of our ideas about competition and predation have come from microcosm research.

Laboratory studies of microcosms are controlled, and in the classical laboratory study only one or two factors are manipulated. In his predator-prey studies, Gause could vary the number of prey and predators introduced to start the cultures. Other factors that may effect the system, such as temperature or the size of the containers, are held constant. Replication is relatively easy to achieve, particularly with small organisms. In some but not all cases, results are obtained in a short time period. Costs of doing experiments are relatively low. Small-sized containers are typically used in laboratory microcosms.

Field studies are uncontrolled, and even in experimental field studies in which one or two factors are manipulated, all other factors are left to vary naturally.

Consequently because there are warm years and cold years, wet years and dry years, all this natural variation can impinge on the results obtained. At first sight this would seem to be a great disadvantage of field studies. But in fact this variability is part of the real world, and the results of field studies are thus robust with natural variation in uncontrolled environmental factors. The greatest advantages of field studies relate to scale; some processes are too large spatially to study in the laboratory. An example would be turbulence in lakes or the oceans as it affects predation of larval fish, or dispersal and territorial social organization of wolves as it affects prey consumption rates. Another difference between field and microcosm studies is duration. Microcosm studies are typically of short duration, and many mistakes in ecology have been prompted by microcosm studies of too short duration (Tilman 1989, Carpenter 1996). Field studies of longer duration often uncover more of the complexity of ecological relationships that require additional study. Of course, field experiments are usually more expensive and require a long time commitment to obtain the results.

The best model for ecological studies is to use laboratory and field experiments together. Microcosms can suggest hypotheses and mechanisms that can be tested in longer-term field manipulations. Microcosms are not suited to studies of more than a few years in duration, but by combining their statistical power and experimental rigor with long-term field experiments, ecologists can have the best of both worlds (Fraser and Keddy 1997).

northern prairies of North America are heavily affected by predation on eggs by a variety of vertebrate predators, including foxes, coyotes, skunks, ground squirrels, badgers, and raccoons. Although predator control could be implemented on all of these predators, it is important to identify the most significant predator species. Greenwood (1986) reported an experiment in which striped skunks were removed

from waterfowl nesting areas over three years. Figure 13.11 illustrates the increases in duck nesting success after skunk removal.

Other examples of duck egg predation illustrate the interactions that can occur between predators. In the prairie pothole areas of North America the red fox is a major predator of duck eggs, but fox abundance is depressed by coyotes (Sovada et al. 1995),

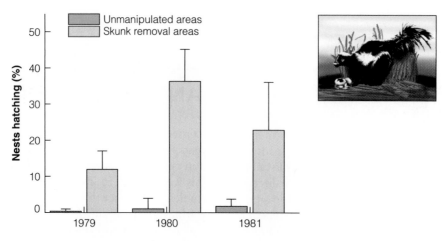

FIGURE 13.11
*Mean hatching rates of upland duck nests in waterfowl areas of North Dakota from which striped skunks (*Mephitis mephitis*) were removed during the nesting season, April–July 1979–1981. Skunk removal dramatically improved duck nesting success. (Data from Greenwood 1986, Table 3.)*

which are territorial and kill or exclude foxes from areas. Nest success in North Dakota averaged 32% on areas where coyotes were common and 17% on areas where the red fox predominated. The key point is that the composition of the predator community and the interactions between the predators can affect the level of predation in complex predator-prey systems.

Paul Errington studied muskrats (*Ondatra zibethicus*) in the marshes of Iowa for 25 years to determine the effects of predation on muskrat populations. He questioned the common assumption that if a predator kills a prey animal, the prey population must then be one animal lower than it would have been without predation. You cannot study the effects of predation, Errington argued, by counting the numbers of prey killed; one must determine the factors that condition predation, the factors that make certain individuals vulnerable to predation while others are protected. Mink predation on muskrats was indeed a primary cause of death in Iowa marshes, but Errington contended that mink were removing only surplus muskrats that were doomed to die for other reasons. The numbers of muskrats were determined by the territorial hostility of muskrats toward one another, and the muskrats driven out by this hostility over space were doomed to die—if not from predators, then from disease or exposure; predators were merely acting as the "executioners" for animals

excluded by the social system (Errington 1963). Errington introduced the important idea that in some systems predation may remove from populations only the doomed surplus, and that predator effects should be inferred only from proper experiments involving both predator reduction areas and unmanipulated control [1] areas.

The role of predators in limiting the abundance of mammals is controversial. When a proper experimental design is used, the question can be clearly answered. The best example involves dingo predation on kangaroos in Australia (Caughley et al. 1980). The dingo (*Canis familiaris dingo*), a dog, is the largest carnivore in Australia. Because dingoes eat sheep, some very long fences (9660 km) have been built in southern and eastern Australia to prevent dingoes from moving into sheep country. Intensive poisoning and shooting of dingoes in sheep country, coupled with the dingo fence to prevent recolonization, has produced a classic experiment in predator control. A spectacular increase in the abundance of red kangaroos (*Macropus rufus*) resulted when dingoes were eliminated (Figure 13.12); densities of kangaroos are 166 times higher in New South Wales than in South Australia. Emus (*Dromaius novaehollandiae*), large

---

[1]*Control* here is used in the experimental design sense to mean an unmanipulated area. It should not be confused with the use of *control* to mean animal or plant removal, as in "pest control."

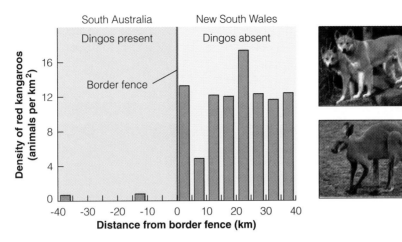

FIGURE 13.12
Density of red kangaroos on a transect across the New South Wales-South Australia border in 1976. The border is coincident with a dingo fence that prevents dingos from moving from South Australia into the sheep country of New South Wales. (After Caughley et al. 1980.)

Carl B. Huffaker *(1914–1995) Professor of Entomology, University of California, Berkley*

flightless birds, are also over 20 times more abundant in the dingo-free areas. There is no alternative explanation of these dramatic differences across the dingo fence (Caughley et al. 1980), and we must conclude that dingo predation limits the density of red kangaroos. Dingoes are able to hold kangaroo numbers low without going extinct themselves because they have alternate prey such as rabbits and rodents to sustain them when kangaroos are in short supply.

In contrast, the Serengeti Plains of eastern Africa contain a suite of large mammals and their predators, but the predators—lions, leopards, cheetahs, wild dogs, and spotted hyenas—seem to have little effect on their large mammal prey (Sinclair and Arcese 1995). Most of the prey individuals taken by predators are doomed surplus—older, injured, or diseased animals. Also, the vast majority of the prey species are migratory, whereas most of the predators are resident. Lions, for example, seem to be limited in numbers by the resident prey species available in the dry season, when the migratory ungulates are elsewhere.

Wildlife managers have instituted many predator control programs on the assumption that predation was limiting the abundance of prey populations. There has been growing opposition from the conservation movement to the idea of predator control, and there is an urgent need to resolve these issues so that proper management can be implemented (Haber 1996). One of the classic controversies in North America has concerned wolf predation on moose. If wolf predation affects the abundance of moose, then removing wolves should increase moose numbers. Of the five experiments that have been conducted in Canada and Alaska (Boutin 1992), an improvement in moose calf survival occurred in three of the five studies, but only in one study was a significant increase in the moose population observed.

Larsen et al. (1989) put radio-collars on moose calves in the Yukon and found that 58% of all calf deaths were caused by grizzly bears, and 25% by wolves. Predation may affect moose numbers, but it is the combined predation by grizzly bears, black bears, and wolves that is significant (Gasaway et al. 1992), and not predation by wolves alone.

Caribou herds in North America are also preyed on by wolves, and there is considerable argument about whether predation limits caribou density or whether food resources are limiting (Bergerud and Elliot 1998). This controversy has important practical consequences because of pressure for wolf-control programs. Unfortunately, although many wolf-control programs have been implemented, few of them have used a proper experimental design that included unmanipulated populations for comparison (Boutin 1992). Bergerud (1980) used circumstantial evidence to suggest that a combination of predation on young caribou and human hunting of adult caribou were jointly holding caribou populations in check. He compared the realized rates of increase

$$\frac{dN}{dt}\frac{1}{N}$$

of 40 different caribou herds and got the following average values for annual rates of population change:

| Hunting Mortality | Predators | | |
	None	Few	Normal
None	0.28	1.11	2.02
< 5% per year	0.08	0.05	-0.01
> 5% per year	-0.11	-0.19	-0.13

Predators alone, according to Bergerud (1980), would hold caribou populations around 0.4 per km². Herds without wolves or other predators increase up to 20 or more individuals per km² and then suffer starvation (see, e.g., Figure 11.9, page 165).

Caribou, moose, and their predators may represent an interesting example of net reproduction curves with two stable points (Figure 13.13) , instead of the one stable point we previously discussed (see Figure 11.2, page 158). Populations may escape from the factor setting the lower equilibrium (such as predation) and then increase further to an upper equilibrium set by a second factor such as food shortage (Sinclair 1989). When the lower equilibrium is set by predation, this lower region is called a *predator pit*.

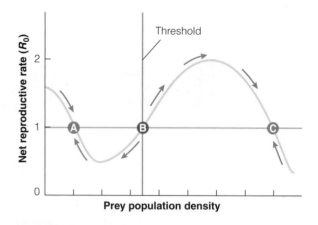

FIGURE 13.13
Simple population model with two stable points (red). The blue line marks the equilibrium line at which the net reproductive rate of the prey is 1.0. Points A and C are stable equilibria; point B is an unstable equilibrium that marks a boundary or threshold from which populations always diverge. This type of model might be appropriate for caribou populations in which the lower equilibrium is set by predators, and the upper equilibrium by food shortage.

Spectacular examples of the influence of predators have occurred where humans have accidentally introduced a new predator. A striking example is the virtual elimination of the lake trout fishery in the Great Lakes by the sea lamprey (*Petromyzon marinus*). The marine lamprey lives on the Atlantic coast of North America and migrates into fresh water to spawn. Adult lampreys have a sucking, rasping mouth by which they attach themselves to the sides of fish, rasp a hole, and suck out body fluids. Only a few fish attacked by lampreys survive. Niagara Falls presumably blocked the passage of the lamprey to the upper Great Lakes before the Welland Canal was built in 1829. The first sea lamprey was found in Lake Erie in 1921, in Lake Michigan in 1936, in Lake Huron in 1937, and in Lake Superior in 1938 (Applegate 1950). Lake trout catches decreased to virtually zero within about 20 years of the lamprey invasion (Figure 13.14). Control efforts to reduce the lamprey population have been implemented since 1951 and lamprey are now rare, and attempts to rebuild the Great Lakes fishery have been made by releasing trout bred in hatcheries. Lake trout have increased reasonably well in Lake Superior but are still rare in all the other Great Lakes (Krueger et al. 1995). Restoration of lake trout in these lakes has been hampered by a loss of genetic

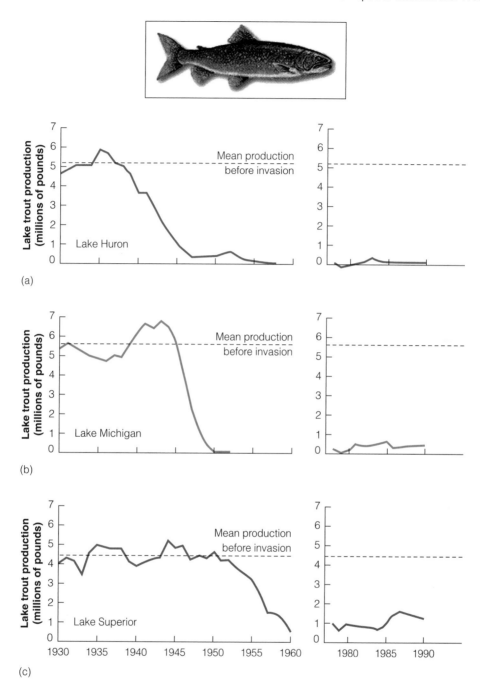

FIGURE 13.14

Effect of sea lamprey introduction on the lake trout fishery of the upper Great Lakes of North America. Lampreys were first seen in (a) Lake Huron in 1937, (b) Lake Michigan in 1936, and (c) Lake Superior in 1938. Commercial fish production from 1978 to 1990 is shown in the right panel. Lake trout have not recovered in the Great Lakes in spite of sea lamprey control. (Data from the Great Lakes Fishery Commission.)

FIGURE 13.15

Numerical response of wolves to moose in North America. The density of wolves increases with moose density up to about 1 moose per km² and then may reach a plateau. Wolves defend territories, which may restrict their numerical response to prey abundance. (From Messier 1994.)

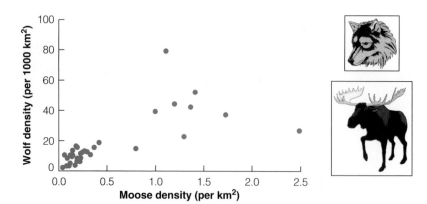

diversity, loss of spawning areas, chemical contaminants, and the introduction of new exotic species such as Pacific salmon (Holey et al. 1995).

We conclude that in some but not all cases, the abundance of predators does influence the abundance of their prey in field populations. This raises an important question: *What is it about certain predators that makes them effective in controlling populations of their prey?* Can we find some type of system by which we can effectively classify predators? This question has great economic implications both in the management of fish and wildlife populations and in agricultural pest control. It is, of course, possible to proceed in a case-by-case manner and to investigate each individual predator-prey system on its own, but this is clearly inefficient, and we would rather attempt to reach some generalizations that apply to many individual cases.

Buzz Holling was instrumental in focusing attention on the components of predation. Working at the Canadian Forest Research Laboratory at Sault Ste. Marie, Ontario, in the late 1950s, he clarified this alternative approach to understanding how predation operates. We should begin, Holling argued, by assuming a simple one predator–one prey system and ask how predators respond to an increase in prey population density. Four possible responses are (1) a *numerical* response, in which the density of predators in a given area increases by reproduction; (2) a *functional* response, in which the number of prey eaten by individual predators changes; (3) an *aggregative* response, in which individual predators move into and concentrate in certain areas within the study area; and (4) a *developmental* response, in which individual predators eat more or fewer prey as predators grow toward

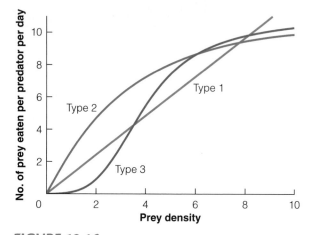

FIGURE 13.16

Three types of possible functional responses for predators to changes in prey abundance. Type 1 responses show a constant consumption of prey, with no satiation; Type 2 and Type 3 responses reach saturation at high prey densities.

maturity (Murdoch 1971). Considerable theoretical and practical work has been done on the numerical, functional, aggregative and developmental responses since the early analyses by Holling (1959).

A *numerical response* of predators can occur because of reproduction by the predator, and an *aggregative response* results from the movements or concentration of predators in areas of high prey density. Predators are usually mobile, and they do not search at random but instead concentrate on patches of high prey density. Figure 13.15 illustrates the numerical response of wolves to moose in North America. Where there are more moose, there are more wolves, and this is a common observation for many predator-prey systems. The

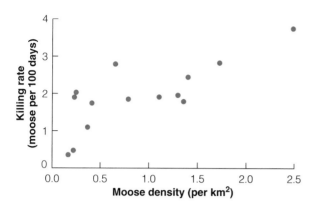

FIGURE 13.17
Functional response of wolves to moose in North America. The predation rate of wolves increases with moose density; this pattern most closely resembles a Type 2 functional response. Additional data at higher moose densities could resolve the exact shape of this relationship. (From Messier 1994.)

ability of predators to aggregate in patches of high prey density is a critical element in determining how effective the predator can be at limiting prey populations.

The *functional response* measures for each individual predator how many prey it eats in a given time period. Three general types of functional responses are recognized (Figure 13.16). The functional response of many predators rises to a plateau as prey density increases, so that over some range of prey density each individual predator eats more prey, but at some high prey density the predator becomes satiated and will not eat more. Figure 13.17 shows one example of a Type 2 functional response for wolves preying on moose in North America. The upper plateau of these functional responses is fixed by *handling time*, the time it takes for a predator to catch, kill, and eat a prey organism. The curve rises rapidly when the searching capacity of the predator is high. Note that the exact shape of the functional response curve observed for field populations will depend on the range of prey densities observed. If only low prey densities occur, the functional response may be a rising straight line; if only high prey densities occur, the functional response may be a horizontal line with no relationship between prey density and the number of prey eaten per predator per day.

The *developmental response* occurs because predators are often growing and maturing during laboratory and field studies of predation. Figure 13.18 illustrates the effects of a functional response and the additional effects of the developmental response on the number of mosquito larvae eaten by backswimmers (*Notonecta hoffmanni*) in the laboratory. Backswimmers grow more rapidly at higher food levels, and this explains the rise in the curve for total consumption in Figure 13.18 (Murdoch and Sih 1978).

Much of the work on predator-prey models has been conducted on laboratory populations and has proved difficult to translate to field populations (Sih et al. 1998). Thus we cannot give more than a vague answer to our general question about what makes some predators effective in controlling their prey. Because much of the theoretical work has concen-

C.S. (Buzz) Holling *(1930–) Arthur R. Marshall Jr. Chair in Ecological Sciences, University of Florida*

trated on single-species systems, there is a need to consider the more complex cases in which predators feed on several prey species (Pech et al. 1995).

If we can measure the functional, numerical, developmental, and aggregative responses for a predator-prey

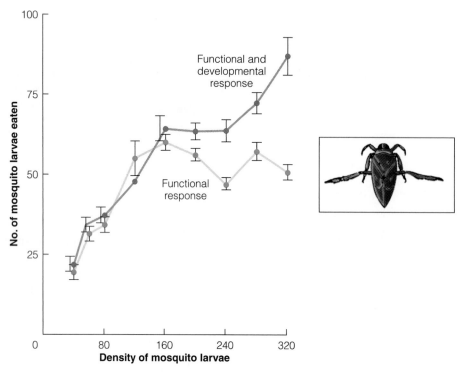

FIGURE 13.18
Functional and developmental responses of the predatory insect Notonecta *feeding on mosquito larvae in the laboratory. The prey consumption rate is measured by the number of mosquito larvae eaten per day. At high food levels* Notonecta *grow larger faster, and thus the combined functional and developmental response curve accelerates upward. (From Murdoch and Sih 1978.)*

system, we can determine the *total response* of the predators (Figure 13.19). The total response of the predators is illustrated by the following simple flow-chart:

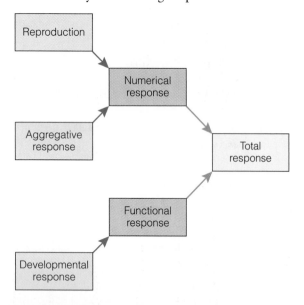

The total response gives the percent of prey organisms eaten per unit time by the entire predator population, plotted against prey density, as illustrated in Figure 13.19. If the total response increases as prey density increases, the predator may limit the density of the prey. By contrast, if the total response remains constant or falls as prey density increases, the predator cannot limit prey numbers. The key question is always whether percent mortality imposed by the predator on the prey increases as prey density increases. At high prey densities, some predators will exert no controlling influence on the prey because they will be swamped by prey numbers. For example, the bay-breasted warbler (*Dendroica castanea*) increased 12-fold during an irruption of the spruce budworm (*Choristoneura fumiferana*) in eastern Canada. Both a numerical and functional response occurred in this warbler (see Figure 13.19a and b), but an 8000-fold increase in the budworm reduced this predator to an insignificant agent of loss (Morris 1963). This threshold of prey density above which

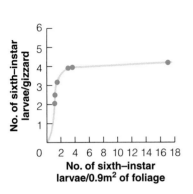
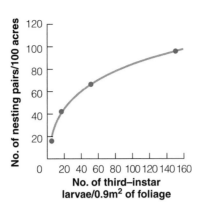
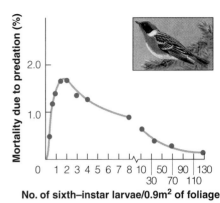

FIGURE 13.19
(a) Functional and (b) numerical responses of the bay-breasted warbler to changes in the abundance of spruce budworm in New Brunswick, Canada. (c) The total (combined) response, based on the assumption that larvae are available for 30 days, the average feeding day is 16 hours, and the digestive period is two hours. (From Mook 1963.)

prey escape from being limited by predators may be important in the conservation of endangered species (Sinclair et al. 1998).

One important general implication of our analysis is that predators may have effects on prey abundance that are important when prey populations are low but become unimportant when prey densities are high. Populations of this sort can exist in two different phases, a low-density *endemic* phase and a high-density *epidemic* phase (illustrated in Figure 13.13). Some forest insect pests, such as the spruce budworm (Swetnam and Lynch 1993), show such biphasic densities, and the key to the endemic phase may be the action of predators at low budworm densities.

Optimal Foraging Theory

The behavior of predators in choosing prey is part of *optimal foraging theory*. When a predator has a choice of two or more different foods, the situation becomes more complex than a simple functional response. How should a predator decide what items to eat? What is an optimal diet for an animal faced with many different prey items? One way to answer this question is to assume that the predator is completely wise so that it can choose the optimal diet (Stephens and

Krebs 1986). What is optimal depends on what the predator is trying to achieve.

Let us begin by assuming that predators will maximize the net rate of energy gain while foraging. More energy is assumed to be better for a predator, because if it has more energy, it will be able to meet its metabolic requirements and still have energy for other activities such as defending a territory, fighting, reproducing, and moving. Predators may be either energy-maximizers or time-minimizers. They are energy-maximizers if they maximize the amount of energy gained in a fixed time unit; they are time-minimizers if they minimize the time needed to gain a specified amount of energy. Diet choice can be determined only after the overall strategy is fixed.

Assume for simplicity that the predator "wishes" to maximize its daily energy intake. We will assume that for most predators it is not possible to handle and eat a prey while searching for another prey, and thus we can separate searching time from handling time which includes the time it takes to kill a prey and eat it. Second, we assume that prey are encountered sequentially, one by one. We define the *profitability* of a prey type as

$$\text{Profitability} = \frac{\text{energy value}}{\text{handling time}} = \frac{E}{h} \quad (13.12)$$

The optimal diet is then determined by a series of choices, and we can illustrate it with a simple example in which a predator has two prey types from which to choose. Our predator may eat large prey 1 with energy value E_1 and handling time b_1, or it can eat smaller prey 2 with energy value E_2 and handling time b_2. Assume that the profitability of prey 1 is higher, or in mathematical terms:

$$\frac{E_1}{b_1} > \frac{E_2}{b_2} \qquad (13.13)$$

where E_1 = energy value of prey 1

b_1 = handling time for prey 1

E_2 = energy value of prey 2

b_2 = handling time for prey 2

If a predator encounters a prey, it must decide whether to eat it or ignore it. The predator will have two rules:

1. If the predator encounters prey 1, it should always eat it because it is the most profitable prey.

2. If it encounters prey 2, it should eat it if the gain from eating it exceeds the gain from rejecting it and searching for a more profitable prey 1.

If we define S_1 as the average search time to find a prey 1 individual, this rule can be translated to

$$\frac{E_2}{b_2} > \frac{E_1}{S_1 + b_1} \qquad (13.14)$$

or if we rearrange terms,

$$S_1 > \frac{E_1 b_2}{E_2} - b_1$$

This optimal foraging model makes some simple predictions. A predator will specialize on one prey only if the average search time S_1 for that prey is relatively low. Note that this decision does not depend on the abundance of the alternative prey (the search time S_2). The predator who is optimally foraging in a simple two-prey system will switch from being a specialist that eats only prey 1 to a generalist that eats both prey items as the search time for prey 1 increases, according to Equation (13.14). In natural foraging situations the predator switches from eating mainly one prey type, to eating some of both, as the abundance of prey 1 changes. Switching may be common in social animals that feed in groups (Murdoch and Oaten 1975).

Switching can be important in predators that feed on several types of prey because it could act to stabilize the density fluctuations of the prey species. As one prey species increases in abundance relative to the others, the predator would concentrate its feeding on the more abundant prey species and possibly restrict that prey's population growth. Conversely, switching to alternative foods may help a prey population to recover if it falls to a low level. Switching behavior could thus be a benefit to the predator by allowing it to maintain a stable population size.

Some predators seem to maintain stable population densities by living on alternative foods when their main prey is scarce. Tawny owl (*Strix aluco*) populations in Europe are remarkably stable despite great fluctuations in the abundance of the small rodents that serve as its main prey species (Figure 13.20). The amount of successful reproduction in tawny owls is determined by the abundance of the prey rodents, but reproductive success does not influence subsequent population size of the owls (Jedrzejewski et al. 1996). Instead, territorial behavior limits the density of tawny owls.

Much of predation theory has been directed toward understanding how predators might stabilize prey populations (Murdoch and Bence 1987, Gotelli 1998). It is clear from many field studies that predators do not necessarily stabilize prey numbers, as we saw with respect to Atlantic salmon and their bird predators. Predators can be loosely classified into generalist predators and specialist predators. Generalist predators eat a great variety of prey and do not heavily depend on one species; specialist predators, by contrast, depend on only one or two species for the majority of their diet. The effects of specialist and generalist predators on prey populations differs:

• Generalist predators tend to stabilize prey numbers

• Specialist predators tend to cause instability in prey numbers

These generalizations are not iron-clad and should be treated as hypotheses rather than facts. The Lotka-Volterra model and other simple predator-prey models all deal with specialist predators eating one prey species.

The fact that some predator-prey models result in oscillations (see Figure 13.2) appears to be strikingly applicable to some biological systems. The Canada lynx (*Lynx canadensis*) eats snowshoe hares (*Lepus americanus*), and both species show dramatic cyclic oscilla-

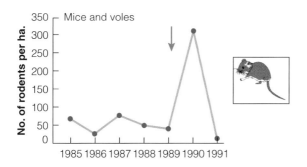

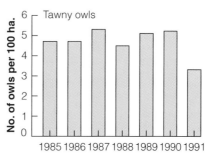

FIGURE 13.20

Autumn densities of forest rodents and tawny owls in Bialowieza National Park, Poland, 1986–1992. The red arrow marks 1989, a year in which heavy seek crops from oak, maple, and hornbeam trees triggered a large irruption in the seed-eating rodents. Tawny owls maintain stable numbers on territories regardless of large changes in prey abundance. (Data from Jedrzejewski et al. 1996.)

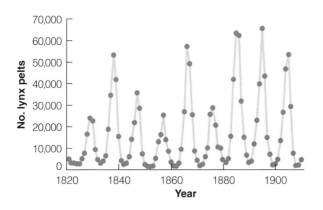

FIGURE 13.21

Canada lynx fur returns of the Northern Department, Hudson's Bay Company, 1821–1913. Canada lynx are specialist predators of snowshoe hares, and both hares and lynx oscillate in numbers in a 9–10 year cycle. (After Elton and Nicholson 1942. Photo courtesy of M. O'Donoghue.)

tions in density with peaks every 9 to 10 years (Figure 13.21) . Charles Elton analyzed the records of furs traded by the Hudson's Bay Company in Canada and showed that the cycle is a real one that has persisted unchanged for at least 200 years (Elton and Nicholson 1942). This lynx-hare cycle has been interpreted as an example of an intrinsic predator-prey oscillation, but this has been disputed by Keith (1991), who suggested that both food shortage and predation were necessary to generate cycles. Lynx depend on snowshoe hares, and are thus food-limited, whereas hares are both food- and predator-limited (Krebs et al. 1995).

Islands can provide natural experiments in predator removal to investigate predator-prey interactions. Lizards (*Anolis* spp.) and spiders are conspicuous animals on small islands in the Caribbean, and these lizards are known to eat spiders. When Schoener and

Toft (1983) surveyed 93 small islands in the Bahamas and counted all the orb-weaving spiders, they found ten times as many spiders on islands that had no lizards, compared with islands that had lizards. To analyze this system experimentally, Spiller and Schoener (1988) constructed enclosures on one island that contained both lizards and spiders, and then removed lizards from three replicate enclosures. Figure 13.22 shows the results of this experiment. Spider numbers increased dramatically in the enclosures that had no lizards. The interaction between lizards and spiders on these Caribbean islands involves both predation by lizards on spiders and competition for food. The diets of lizards and spiders overlap considerably, and lizards consume many of the larger insects that spiders could also eat. The impact of lizards on spiders is also affected by rainfall (Spiller and Schoener 1990, 1995).

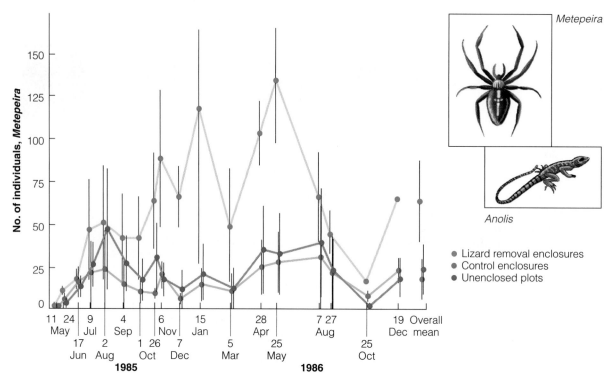

FIGURE 13.22
Density of the orb spider Metepeira datona *in lizard removal enclosures, control enclosures, and open plots on Staniel Cay, Bahamas. Spider densities increased three to five times in plots from which lizard predators were removed. (From Spiller and Schoener 1988.)*

Evolution of Predator-Prey Systems

One of the striking features of the simple models of predator-prey interactions is that these models are often unstable. Oscillations are common in many predator-prey models (see Figure 13.2) but are not common in the real world. One way in which we can explain the stability of real predator-prey systems is to postulate that natural selection has changed the characteristics of predators and prey alike such that their interactions produce population stability. Evolutionary change in two or more interacting species is called *coevolution*, and we are concerned in this case with the coevolution of predator-prey systems.

If one predator is better than another at catching prey, the first individual will probably leave more descendants to subsequent predator generations. Thus predators should be continually selected to become more efficient at catching prey. The problem,

of course, is that by becoming too efficient, the predator will exterminate its prey and then suffer starvation. The prey at the same time are being selected to be better at escaping predation. Because of the conflicting adaptive goals of predator and prey, many evolutionists have described predator-prey evolution as an "arms race" (Dawkins and Krebs 1979). Predators may have an inherent disadvantage in this arms race because of the "life-dinner" principle, which states that selection will be stronger on the prey than on the predator because a prey individual that loses the race loses its life, whereas the unsuccessful predator loses only a meal. Dawkins (1982) suggested that the inherent disadvantage of a predator could be offset if the predator is rare and the prey is common. In this case the predator will be only a minor selective agent on the prey population as a whole.

Abrams (1986) criticized the "arms race" analogy for predator-prey coevolution, citing many theoretical situations in which the arms race would not occur,

TABLE 13.1 Evaluation of prey quality in predation by a trained red-tailed hawk on individuals of three prey species in Wisconsin. Results based on 447 attacks.

Species of prey	Difficulty of prey capture	Percent of attacks that failed	Percent of substandard individuals in hawk kills—percent of substandard individuals in the population
Eastern chipmunk	Easy	72	8
Cottontail rabbit	Moderate	82	21
Gray squirrel	Difficult	88	33

Source: Modified from Temple (1987).

but he recognized that some asymmetry in the evolutionary responses of predators and prey was common. Prey should always increase their investment in escape mechanisms, if predators invest in becoming more efficient. But the reverse is often not true—predators do not always respond to prey investment—and whether or not predators will respond depends on the details of specific predator-prey system. The "arms race" analogy may not be correct in many particular cases of predator-prey coevolution.

Two obvious constraints operate in systems having several species of predators and prey. The existence of several species of predators feeding on several species of prey places limits on predator efficiency. For example, one prey species may escape by hiding under rocks, while a second species may run very fast. Clearly, a predator is constrained by conflicting pressures either to get very good at turning rocks over or to get very good at running, and it is difficult to be good at both these activities. Conversely, we can imagine that the prey population is always being selected for escape responses. Faced with several predators with different types of hunting strategy, prey will not be able to evolve a specific escape behavior suitable to all species of predators.

One persistent belief about predator-prey systems is that predators typically capture substandard individuals from prey populations, so that weak, sick, aged, and injured prey are culled from prey populations. Temple (1987) tested this idea by flying a trained red-tailed hawk (*Buteo jamaicensis*) and measuring the outcome of its attacks on eastern chipmunks, cottontail rabbits, and gray squirrels. Table 13.1 shows that the more difficult the prey is to catch, the higher the fraction of substandard individuals that are caught. Gray squirrels are particularly difficult for red-tailed hawks to catch, and squirrels taken were in markedly poorer condition than squirrels in the general population. The same generalization seems to hold for other ver-

tebrate predators—substandard individuals are captured disproportionately when the type of prey is difficult to catch but not when it is easy to capture. Thus hyenas in Africa take wildebeest in poor condition because wildebeest are difficult to capture, but take gazelles at random because they are relatively easy to catch (Kruuk 1972).

The coevolution of predator-prey systems occurs most tightly when the predators strongly affect the abundance of the prey. In some predator-prey systems the predator does not determine the abundance of the prey, so the evolutionary pressures are considerably reduced. In some cases the prey has refuges available where the predator does not occur, or the prey may have certain size classes that are not vulnerable to the predator. In other cases (see Figure 13.20) the predators have developed territorial behavior that restricts their own density so that they cannot easily respond to excessive numbers of prey animals (Taylor 1984).

Much of the stability we see in the natural world may result from the continued coevolution of predators and prey. Predators that do not have prudence forced on them by their prey may exist for only a short time in the evolutionary record, and we are left today with a residue of highly selected predator-prey systems.

Predators need not limit the density of their prey species to play an important role in the evolution of prey characteristics. Two antipredator defense strategies that are common in animals—warning coloration and group living—illustrate how evolutionary pressures can affect predator-prey systems.

Warning Coloration

Among animals there is a widespread correlation between conspicuous coloration and the presence of aversive qualities for potential predators. For example, many butterflies and other insects are brightly colored and contain poisons that are distasteful to

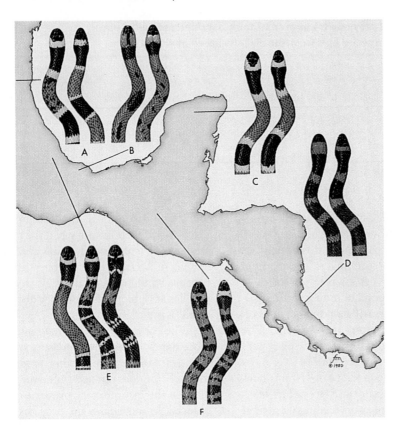

FIGURE 13.23
Geographic variation in color pattern in poisonous coral snakes and their nonvenomous mimics in Central America. The poisonous models (genus Micrurus*) are shown on the left, and the mimics (genus* Pliocercus*) on the right, for five different areas (A–D, F). In E, simultaneous mimicry of two models is shown. The colubrid snake* Pilocercus elapholdes *(center) combines elements of the patterns of* Micurus diastema *(left) and* Micurus elegans *(right). (From Greene and McDiarmid 1981. Illustration copyright by the artist Frances J. Irish; used with permission.)*

predators. The theory of warning (or aposematic) coloration is usually put forward as an explanation of this correlation (Guilford 1988).

Mechanisms of prey defense using warning coloration must evolve by increasing the chances of survival of the individuals in which they are found. But for distasteful species, the predator must first sample one individual before the predator learns to avoid other prey of similar color. If the prey are gregarious and nearby individuals are closely related, kin selection would operate to favor the warning coloration. If only a few siblings are sampled from a large brood and the predator learns to avoid other individuals of the group, an allele for distastefulness can increase in frequency by kin selection. Predators do in fact seem to learn very quickly to avoid distasteful insects (Brower 1988).

Coral snakes are brightly colored with red, yellow, and black bands. All of the 120 species of coral snakes in tropical America are extremely poisonous. Many other nonpoisonous snakes have evolved color patterns to mimic the appearance of coral snakes (Figure 13.23) . These nonpoisonous snakes are called Batesian mimics[2] because they mimic the color patterns of unrelated poisonous species. Birds that live in areas occupied by coral snakes have an innate tendency to avoid snakes with these color patterns (Smith 1980), so a predator need not have a lethal encounter to avoid the poisonous species (Pough 1988). The mimic species can profit from resembling a poisonous or unpalatable prey species, and this coevolution has been particularly well developed in tropical species groups.

Group Living

Why do birds form flocks and ungulates congregate in herds? What are the advantages of group living? First, we should be clear that not all birds flock and that some ungulates are solitary. Group living is not always advantageous, and when it is, there may be several reasons why. One possible reason for living in groups is to reduce the risk of predation (Krebs and Davies 1993).

There are three main advantages a prey organism can obtain by living in a group. First, early detection of predators may be facilitated if many eyes are looking. Figure 13.24 plots detection success versus group size in wood pigeons. Prey in groups may thus be able to spend more time feeding and less time watching out for potential predators.

Second, if the prey are not too much smaller than the predator, several of them acting together may be able to deter the predator from attacking. One possible example of this is mobbing in birds. Primate troops will also mob predators in some circumstances, and musk-oxen threatened by wolves gather into a circular defense formation.

Third, if the predator is still able to attack a group, it must select and pursue one individual in the group to capture. By fleeing in all directions, the prey

---

[2]A Batesian mimic could be likened to a sheep in wolf's clothing.

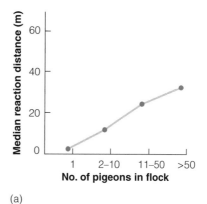

(a)

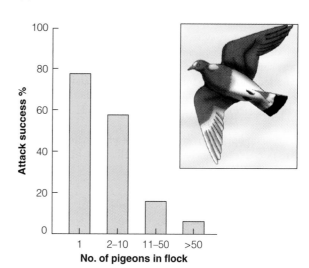

(b)

FIGURE 13.24
The value of flocking in wood pigeons (Columba palumbus) *measured by the distance at which they detected the approach of a goshawk* (Accipiter gentilis). *Goshawk attack success is much reduced against flocks. (From Kenward 1978.)*

may confuse the predator, who may not be able to concentrate on any one individual. Alternatively, by being in the center of a group, an individual may reduce its chances of being eaten.

Many of the most striking characteristics of animal social structure and behavior are adaptations related to predation.

Summary

One species interaction involves predation. Simple mathematical models can be used to describe this interaction. When generations are discrete, simple models can produce stable equilibria of predator and prey, but usually produce oscillations in the numbers of both species. When generations are continuous, graphic models developed by Rosenzweig and MacArthur can be used to evaluate the equilibrium levels and the stability of predator-prey systems. Both stable equilibria and cyclic oscillations may occur. All these simple models make the assumption that the world is homogeneous (one habitat), that there are no prey refuges, and that only one predator species eats one prey species. Relaxing these assumptions leads to more complex models.

In laboratory systems of predators and prey, cyclic oscillations are produced only in complex environments, and most simple systems do not reach stability but instead are self-annihilating. The importance of refuges and spatial heterogeneity can be illustrated readily in laboratory systems, and these factors are even more critical in field populations of predators and prey.

Field populations can be models of predator-prey systems only if predators have a strong effect on the abundance of their prey. This assumption can be tested by predator-removal experiments. In some but not all cases studied, the abundance of predators does influence the abundance of prey. The properties of effective predators can be described in a general manner, but we cannot yet predict which predators will be good agents of prey control without actually doing field tests. Both predator and prey species are affected by many other factors in the environment, and consequently the population trends predicted by simple predator-prey models are rarely found in field populations.

Predator-prey systems always involve a coevolutionary race in which prey are selected for escape and predators for hunting ability. These systems stabilize most easily when several species are involved, when prey have safe refuges from predators, and when predators take old animals of little reproductive value. Many characteristic structures and behavior patterns of animals are adaptations related to predation.

Predation is a major process in the organization of communities, and we will return to consider its effect on the structure of communities in Chapters 23 and 24.

Key Concepts

1. Predator-prey interactions can be analyzed with simple models for one predator-one prey systems.

2. Simple models of predation often lead to predator-prey cycles rather than a stable equilibrium.

3. Laboratory systems rarely lead to stable interactions between predators and prey, but they show the importance of prey refuges and spatial heterogeneity.

4. Predation can be broken down into components—numerical, functional, developmental, and aggregative responses of predators to prey—to aid our understanding of the predation process.

5. Multiple predator-multiple prey systems lead to more complex dynamics, and show the importance of predation to the evolution of escape behavior and warning coloration in animals.

Selected References

Abrams, P. A. and C. J. Walters. 1996. Invulnerable prey and the paradox of enrichment. *Ecology* 77:1125–1133.

Berryman, A. A. 1992. The origins and evolution of predator-prey theory. *Ecology* 73:1530–1535.

Boutin, S. 1995. Testing predator-prey theory by studying fluctuating populations of small mammals. *Wildlife Research* 22:89–100.

Boveng, P. L., L. M.. Hiruki, M .K. Schwartz, and J. L. Bengtson. 1998. Population growth of Antarctic fur seals: limitation by a top predator, the leopard seal? *Ecology* 79:2863–2877.

Caughley, G., G. C. Grigg, J. Caughley, and G. J. E. Hill. 1980. Does dingo predation control the densities of kangaroos and emus? *Australian Wildlife Research* 7:1–12.

Cote, I. M. and W. J. Sutherland. 1997. The effectiveness of removing predators to protect bird populations. *Conservation Biology* 11:395–405.

Lima, S. L. 1998. Nonlethal effects in the ecology of predator-prey interactions. *Bioscience* 48:25–34.

Pech, R. P., A. R. E. Sinclair, and A. E. Newsome. 1995. Predation models for primary and secondary prey species. *Wildlife Research* 22:55–64.

Sih, A., G. Englund and D. Wooster. 1998. Emergent impacts of multiple predators on prey. *Trends in Ecology and Evolution* 13:350–355.

Sovada, M. A., A. B. Sargeant, & J. W. Grier. 1995. Differential effects of coyotes and red foxes on duck nest success. *Journal of Wildlife Management* 59:1–9.

Spiller, D. A. and T. W. Schoener. 1994. Effects of top and intermediate predators in a terrestrial food web. *Ecology* 75:182–196.

Temple, S. A. 1987. Do predators always capture substandard individuals disproportionately from prey populations? *Ecology* 68:669–674.

Questions and Problems

13.1 Plot the data from Figure 13.10 on a phase plane like Figure 13.5, and sketch on the phase plane the predator isocline and the prey isocline. What can you conclude about the stability properties of this predator-prey system?

13.2 Calculate the population changes from Equations (13.2) and (13.3) for ten generations in a hypothetical predator-prey system with discrete generations in which the parameters for the prey are $B = 0.03$, $Neq = 100$, $C = 0.5$, and starting density is 50 prey, and for the predators, $Q = 0.02$ (or $S = 2.0$) and starting density is 0.2. How would the prey population change in the absence of the predators?

13.3 Assume that in the striped skunk predator-control experiments plotted in Figure 13.11, nest losses of ducks were reduced, but duckling and adult survival were unchanged, and that population density changes were unaffected by the predator removal program. Discuss the demographic mechanics of how this might be possible.

13.4 Buckner and Turnock (1965) studied bird predation on the larch sawfly in Manitoba. They obtained the following data for the chipping sparrow (*Spizella passerina*):

	Plot I		Plot II	
Year	Sparrows per acre	Sawfly larvae per acre	Sparrows per acre	Sawfly larvae per acre
1954	—	—	3.2	235,000
1956	—	—	2.9	33,400
1957	1.4	2,138,700	2.3	40,000
1958	0.5	879,400	2.5	41,200
1959	0.4	437,800	2.2	27,300
1960	0.2	354,300	2.2	54,600
1961	0.5	199,900	2.3	15,000
1962	1.1	191,800	5.0	3,200
1963	0.2	366,800	0.3	3,900

Plot the numerical response of chipping sparrows to changes in sawfly larval abundance, and discuss the differences between plots I and II for this predator-prey system.

13.5 The collapse of lake trout populations in the Great Lakes coincided with a general increase in commercial fishing of the lakes. Discuss the hypothesis that the collapse of fish stocks in the Great Lakes (see Figure 13.14) was caused more by overfishing (human predation) than by the introduction of the sea lamprey. Steedman and Regier (1987) and Coble et al. (1990) give references.

13.6 How does the predation by herbivores on green plants differ from either the predation of insect parasitoids on their hosts or the predation of carnivores on herbivores? Make a list of similarities and differences, and discuss how they affect the simple models of predation discussed in this chapter.

13.7 Introduce a time lag into the simple predator-prey model for discrete generations by making the predator density change in relation to prey density at generation $t - 1$ rather than at generation t. Repeat the calculations for one of the examples discussed in the text, and determine what effect this time lag has on the system's behavior.

13.8 One of the "textbook" examples of the effects of predator control involves the Kaibab deer herd. According to several textbooks, the removal of predators from the Kaibab mule deer population allowed the deer to increase dramatically in numbers, to overgraze their food supply, and to starve. Trace the history of this example from Rasmussen (1941) to the critique by Caughley (1970).

13.9 Wildebeest in the Serengeti area of east Africa have a very restricted calving season. All females give birth within a space of three weeks at the start of the rainy season (Sinclair and Arcese 1995). How would you test the hypothesis that this restricted calving season is an adaptation to reduce predation losses of calves?

13.10 The graphic model of Rosenzweig and MacArthur (see Figure 13.5) predicts that the predator-prey system will become unstable when nutrients are added to the prey population (the paradox of enrichment). Evaluate the evidence for the occurrence of the paradox of enrichment in laboratory and field populations. Does the same prediction follow from ratio-dependent predation theory? McCauley and Murdoch (1990), Artidi et al. (1991), and Abrams (1997) provide references and a discussion of the problem.

13.11 The birds, lizards, and mammals of Guam in the western Pacific Ocean have been driven to extinction or to low numbers by the introduced brown tree snake (*Boiga irregularis*). How could this happen? Is it adaptive for a predator to drive its prey to extinction? Read the discussion in Rodda et al. (1997) and evaluate the uniqueness of this situation.

Overview Question

When populations of moose, caribou, or deer decline in Alaska or Canada, a great public pressure to instigate wolf control programs typically ensues. What data would you collect to describe and understand the dynamics of this predator-prey system? List the alternative hypotheses you would test and their management implications.

CHAPTER 14

Species Interactions:
Herbivory and Mutualism

PLANT-ANIMAL INTERACTIONS are the focus of many population interactions. *Herbivory* is a major interaction in which animals prey on plants, and herbivory is traditionally considered a profit for the animals and a loss for the plants. Many examples of *mutualism* involve plant-animal interactions that are by contrast a gain for both species. In this chapter we discuss herbivory and mutualism to assess their effects on abundance and their evolutionary origins.

Herbivory is a special kind of predation because the herbivore does not kill the plant but eats only part of it. Over half of the macroscopic species on Earth are plants, and consequently a major part of species interactions involve plant-herbivore interactions. In this chapter we will examine some of the specific relationships between herbivores and plants. The uniqueness of these relationships is often only a reflection of the simple fact that most plants cannot move, so "escape" from herbivores can be achieved only by some clever adaptations. Herbivores can be important selective agents on plants, and the evolutionary interplay between plants and animals is a major theme in this chapter.

Defense Mechanisms in Plants

The world is green, and there are three possible explanations for this. First, some herbivore populations may evolve self-regulatory mechanisms that prevent them from destroying their food supply (see Chapter 16). Or second, other control mechanisms,

such as predation, may hold herbivore abundance down so that plants escape being totally eaten. Third, not all that is green may be edible. Plants have evolved an array of defenses against herbivores, and this has set up a coevolutionary contest between plants and herbivores in evolutionary time.

Plants may discourage herbivores by structural adaptations, as anyone who has tried to prune a rosebush will attest, but they may also use a variety of chemical weapons that we are only now starting to appreciate. Plants contain a variety of chemicals that have always puzzled plant physiologists and biochemists (Feeny 1992). These chemicals, called *secondary plant substances*, are found only in some plants and not in others and are by-products of the primary metabolic pathways in plants. Figure 14.1 gives a simplified view of the biochemical origins of some of the major chemical groups of secondary plant substances. A number of these substances are familiar to us already. One acetogenin, juglone, is produced by walnut trees (see page 78). Among the phenylpropanes found in some trees, are the spices cinnamon and cloves. Familiar terpenoids include peppermint oil and catnip. Well-known alkaloids include nicotine, morphine, and caffeine.

Two views have developed about the origins of secondary plant substances. One view is that these compounds are primarily waste products of plant metabolism, and that the evolutionary origin of secondary plant substances can be understood as an adaptation to avoid autointoxication (Muller 1970). Excretion is a necessary part of metabolism, and in plants it takes much different forms from excretion in

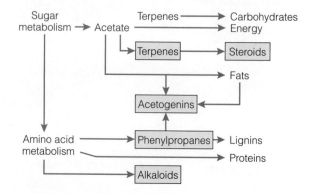

FIGURE 14.1
Relationships of the major groups of secondary plants substances (shown in boxes) to the primary metabolic pathways of plants. (After Whittaker and Feeny 1971.)

animals. According to this view, plants have evolved numerous ways of eliminating toxic organic chemicals by volatilization or leaching and other means of chemically altering toxic substances to render them harmless within the plant. This done, the plant may derive benefit in one of two ways: First, by releasing chemicals into its immediate environment, a plant may be able to suppress competitors, producing allelopathic effects (see Chapter 6); second, by accumulating some chemicals in its leaves or stem, a plant may become toxic or distasteful to herbivores.

A second view of the origin of secondary plant substances is that these chemicals were specifically evolved by plants to thwart herbivores (Frankel 1959, Ehrlich and Raven 1964). This view assumes that few secondary plant substances are really useful to the plant as excretory products, and that most are actively produced at a metabolic cost to the plant. Such a chemical variety exists in different plant groups only because plants that produce secondary substances are at a selective advantage. If all animals could be removed from the community, plants would not produce secondary substances.

This plant-defense view argues that herbivores have a strong effect on plant fitness, and that well-defended plants are fitter. If plant defense characteristics are inherited, then all the elements needed for natural selection are present.

If plant defense has a cost in terms of plant fitness, we can make four general predictions:

- Plants evolve more defenses if they are exposed to much damage, and fewer defenses if the cost of defense is high.

- Plants allocate more defenses to valuable tissues that are at risk.

- Defense mechanisms are reduced when enemies are absent, and increased when plants are attacked.

- Defense mechanisms are costly and cannot be maintained if plants are severely stressed by environmental factors.

The cost of defense is due to the diversion of energy and nutrients from other needs. Much evidence to support these four predictions has now accumulated. The autointoxication view is now thought incorrect, and the hypothesis that secondary substances have an ecological role as deterrents to herbivory has become a fruitful and exciting area of research (Karban and Baldwin 1997). Secondary substances in plants are not static end-products of metabolism but have rapid turnover rates in the metabolic pool.

Paul Feeny, working at Cornell University, was one of the first to recognize the importance of plant defenses and to suggest a theoretical basis for it. Feeny's plant apparency theory (1976) was the first general theory to attempt to explain plant-herbivore interactions. The major premise of this theory is that the type of defenses the plant uses depends on how easily a herbivore can find the plant. Plants that are easily found by herbivores evolve chemical defenses of a different type from those used by plants that are difficult for herbivores to locate.

Paul Feeny *(1940–) Professor of Ecology, Cornell University*

Short-lived plants may be able to escape the attention of herbivores by developing so quickly that their herbivores are unlikely to discover them—such plants are "unapparent" to their herbivores. In contrast, "appar-

TABLE 14.1 **Characteristics of inherently fast-growing and slow-growing plant species.**

Variable	Fast-growing species	Slow-growing species
Growth characteristics		
Maximum growth rates	High	Low
Maximum photosynthetic rates	High	Low
Dark respiration rates	High	Low
Leaf protein content	High	Low
Responses to pulses in resources	Flexible	Inflexible
Leaf lifetimes	Short	Long
Successional status	Often early	Often late
Antiherbivore characteristics		
Expected rates of herbivory	High	Low
Amount of defense metabolites	Low	High
Type of defense (in the sense used by Feeny)	Qualitative (alkaloids)	Quantitative (tannins)
Turnover rate of defense	High	Low
Flexibility of defense expression	More flexible	Less flexible

Source: Coley et al. (1985).

ent" plants are sure to be found by herbivores because they are long-lived.

Some defense mechanisms in plants are quantitative, and secondary compounds may vary in concentration. For example, tannins and resins in leaves may occupy up to 60% of the dry weight of a leaf (Feeny 1976). Quantitative plant defenses should be used by "apparent" plants. Other defense mechanisms may be qualitative or +/– defenses because the compounds involved are present in very low concentrations (less than 2% dry weight). Examples are alkaloids and cyanogenic compounds in leaves. Qualitative defenses are poisons that protect plants against generalized herbivores that are not adapted to cope with the toxic chemicals, but they do not stop specialized herbivores that have evolved detoxification mechanisms in the digestive system. Qualitative defenses should be used by "unapparent" plants (Feeny 1976).

Both the type of defense and the amount of defense plants use depend on the vulnerability of the plant tissues (Rhoades and Cates 1976). Growing shoots and young leaves are more valuable to plants than mature leaves, so plants typically invest more heavily in the defense of growing tips and young leaves. Tannins, resins, alkaloids, and other defense chemicals are concentrated at or near the surface of the plant, thereby increasing their effectiveness.

The plant apparency theory stimulated a great deal of work on plant defense during the 1970s and 1980s, and it was soon found to be inadequate as an explanation of plant defense. Some "apparent" plants have qualitative, toxic defenses, and some "unapparent" species use quantitative defenses. In some cases it is difficult for researchers to decide whether or not plants are apparent to their herbivores and not just to humans. It became clear that a more general theory was needed (Feeny 1992).

One element missing from the previous ideas was the ability of a plant to defend itself, given the resources available to it. Coley et al. (1985) proposed the *resource availability hypothesis* to take into account the plant's ability to replace tissues taken by herbivores. Both fast-growing and slow-growing plants have a suite of physiological characteristics that are summarized in Table 14.1. Herbivores prefer fast-growing plants and tend to avoid slow-growing plants. Because each leaf represents a greater investment for a slow-growing plant, slow-growing plants stand to lose more to herbivores and thus invest more in defensive chemicals. Figure 14.2 summarizes a conceptual model of the resource availability hypothesis. The higher the plant's growth rate, the lower the predicted investment in defense.

Plants can either defend themselves at all times, or only when attacked by a herbivore. Structural defenses are clearly more permanent than chemical defenses. By utilizing *inducible defenses*, plants can avoid the cost of producing defensive chemicals when they are not needed. Once a plant is attacked by a herbivore, it can then activate its defenses. Induction

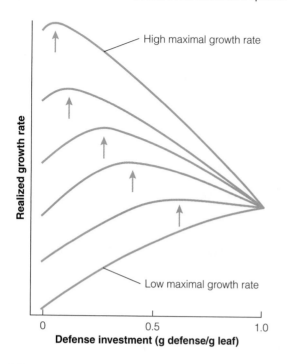

FIGURE 14.2
The resource availability hypothesis of plant defense. Each curve represents a plant species with a different maximal growth potential. Levels of defense that maximize growth rate are indicated by an arrow. Slow-growing plants should invest much more in defense because losses to herbivores are more difficult to replace. (From Coley, Bryant and Chapin 1985.)

times for defensive reactions by plants have been studied for only a few species and vary from 12 hours to one year or more (Tollrian and Harvell 1999). Rapid defensive responses in plants were unexpected and are now the subject of intensive interest in plant-herbivore research. If defenses are costly, we would expect a relaxation of defenses after a herbivore's attack, but few measurements have yet been made (Karban and Baldwin 1997). If induced responses are occurring in a plant, we need to answer two questions to determine if these responses are antiherbivore defense responses:

1. Do the induced changes affect herbivore foraging or herbivore distributions?
2. Do plants suffer less damage and have greater fitness as a result of induced changes in leaf chemistry?

Herbivores do not, of course, sit idly by while plants evolve defense systems (Strong et al. 1984). Herbivores circumvent plant defenses either by evolving enzymes to detoxify plant chemicals or by

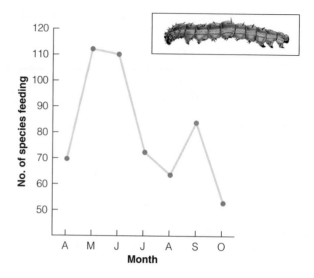

FIGURE 14.3
Number of Lepidoptera *species feeding as larvae on common oak leaves in Britain from April to October. (After Feeny 1970.)*

altering the timing of their life cycle to avoid the noxious chemicals of the plants (as in the next example of tannins in oak leaves). The coevolution of animals and plants can thus occur, and we will examine four cases to illustrate this.

Tannins in Oak Trees

The common oak (*Quercus robur*), a dominant tree in the deciduous forests of western Europe, is attacked by the larvae of over 200 species of Lepidoptera, more species of insect attackers than any other tree in Europe withstands. The attack of insects is concentrated in the spring (Figure 14.3), with a smaller peak of feeding in the fall. Among the most common of these insects is the winter moth, whose larvae feed on oak leaves in May and drop to the ground to pupate late in that month. Why is insect attack concentrated in the spring? One possibility is that oak leaves become less suitable insect food as they age (Feeny 1970).

Winter moth larvae fed "young" oak leaves grow well, but if larvae are fed slightly "older" leaves, they grow very poorly:

Winter moth larvae diet		Mean peak larval weight (mg)
May 16:	"young" oak leaves	45
May 28–June 8:	"old" oak leaves	18

No adults emerged from the larvae fed older leaves. Thus some change occurs very rapidly in oak

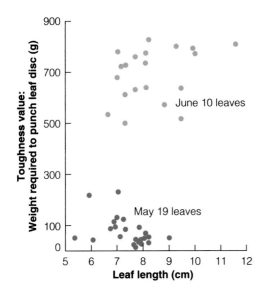

FIGURE 14.4
Toughness of "young" oak leaves collected May 19 and "old" oak leaves collected June 10. (After Feeny 1970.)

leaves in the spring to make them less suitable for winter moth larvae. The most obvious changes in oak leaves during the spring are a rapid darkening and an increase in toughness. The thin oak leaves of May become thick and more difficult to tear by early June (Figure 14.4). If leaf toughness is a sufficient explanation for the feeding pattern of oak insects in the spring, then grounds up older leaves should provide an adequate diet. But if chemical changes have occurred as well, ground-up older leaves should still be inadequate as a larval diet. Ground-up leaves seem to be an adequate diet, at least until early June:

Larvae fed ground-up leaves	Mean peak larval weight (mg)
May 13: "young" leaves	37
June 1: "old" leaves	35

If mature oak leaves can provide an adequate diet, why has natural selection not favored insect mouthparts able to cope with tough leaves? Some Lepidoptera do feed on summer oak leaves, so that it is possible to feed on tough leaves. If mature oak leaves later in summer are relatively poor nutritionally compared with young spring leaves from May and early June, this would produce natural selection toward early feeding.

Two related chemical changes in oak leaves seem to be significant for feeding insects: The amount of tannins in the leaves increases from spring to fall (especially after July), and the amount of protein decreases from spring to summer and remains low from June onward. Tannins, which are secondary plant substances that may reduce palatability, may act in oak leaves by tying up proteins in complexes that insects cannot digest and utilize. Larval weights of winter moths are significantly reduced if their diet contains as little as 1% oak-leaf tannin.

Nevertheless, some insects have evolved ways of minimizing the effect of tannins. Insects that feed on oak leaves in the summer and fall tend to grow very slowly, which may be an adaptation to a low-nitrogen diet. Table 14.2 shows that many of the late-feeding insects on oak overwinter as larvae and complete their development on the spring leaves. Many others are leaf miners, which may avoid tannins by feeding on leaf parts that contain little tannin.

Thus the oak tree has defended itself against herbivores by the use of tannins as a chemical defense and altered leaf texture (toughness) as a structural defense. Herbivores have compensated by concentrating feeding in the early spring on young leaves and by altering life cycles in the summer and fall.

Ants and Acacias

A mutualistic system of defense has coevolved in the swollen-thorn acacias and their ant inhabitants in the New World tropics. The ants depend on the acacia tree for food and a place to live, and the acacia depends on the ants for protection from herbivores and neighboring plants. Not all of the approximately 700 species of acacias (*Acacia* spp.) depend on the ants in the New World tropics, and not all the acacia ants (*Pseudomyrmex* spp.),

FIGURE 14.5

(a) Acacia collinsii *growing in open pasture in Nicaragua. This tree had a colony of about 15,000 worker ants and was about 4 m tall. (b) Area cleared over ten years around a growing* Acacia collinsii *in Panama by ants chewing on all vegetation except the acacia. Machete in photo is 70 cm long. The area was not disturbed by other animals. (c) Swollen thorns of* Acacia cornigera *on a lateral branch. Each thorn is occupied by 20–40 immature ants and 10–15 worker ants. All the thorns on the tree occupied by ants belonging to a single colony. An ant entrance hole is visible in the left tip of the fourth thorn up from the bottom. (Photos courtesy of (a) N. Smythe/National Audobon Society/Photo Researchers, Inc., (b) Gregory D. Pimijian/Photo Researchers, Inc. (c) Stephen J. Knasemann, Photo Researchers, Inc.)*

TABLE 14.2   Larval feeding habits of early-feeding and late-feeding lepidoptera species on leaves of the common oak in Britain.

Feeding habit	Early-feeding species[a] (%)	Late-feeding species[b] (%)
Larvae complete growth on oak leaves in one season	92	42
Larvae complete growth on low herbs after initial feeding on oak leaves	3	11
Larvae overwinter and complete growth in following year	4	38
Larvae bore into leaf parenchyma (leaf miners)	3	26

[a]Early-feeding larvae are in May and June; total of 111 species.
[b]Total of 90 species.
Note. Some species exhibit more than one of the feeding habits, so the columns do not add to 100%.
Source: After Feeny (1970).

150 species or more, depend completely on acacia. In a few cases a high degree of mutualism has developed, described in detail by Janzen (1966). Some of the species of ants that inhabit acacia thorns are obligate acacia ants and live nowhere else.

Swollen-thorn acacias have large, hollow thorns in which the ants live (Figure 14.5). The ants feed on modified leaflet tips called Beltian bodies, which are the primary source of protein and oil for the ants, and also on enlarged extrafloral nectaries, which supply sugars. Swollen-thorn acacias maintain year-round leaf production, even in the dry season, providing food for the ants. If all the ants are removed from swollen-thorn acacias, the trees are quickly destroyed by herbivores and crowded out by other plants. Janzen (1966) showed that acacias without ants grew less and were often killed:

	Acacias with ants removed	Acacias with ants present
Survival rate over 10 months (%)	43	72
Growth Increment		
May 25–June 16 (cm)	6.2	31.0
June 16–August 3 (cm)	10.2	72.9

Swollen-thorn acacias have apparently lost (or never had) the chemical defenses against herbivores found in other trees in the tropics.

The acacia ants continually patrol the leaves and branches of the acacia tree and immediately attack any herbivore that attempts to eat acacia leaves or bark. The ants also bite and sting any foreign vegetation that touches an acacia, and they clear all the vegetation from the ground beneath the acacia tree. As a result the swollen-thorn acacia often grows in a cylinder of space virtually free of all competing vegetation (see Figure 14.5b).

Thus the ant-acacia system is a model system of the coevolution of two species in an association of mutual benefit. By reducing herbivore destruction and competition from adjacent plants, the ants serve as a living defense mechanism for the acacias.

Spines in a Marine Bryozoan

Induced defense mechanisms are not confined to plants. Both freshwater and marine animals produce spines or thickened shells in response to chemical cues released by the species that feed on them (Harvell 1986). These spines or thickened shells are inducible defenses; they do not appear if the predator is not present in the area. In many cases these systems are predator-prey interactions, rather than plant-herbivore interactions, but the principles of defense are the same—the organisms respond morphologically to reduce their chances of being eaten (Myers and Bazely 1991).

The marine bryozoan *Membranipora membranacea* produces large, chitinous spines very rapidly within 36 hours of attack by nudibranch predators (Figure 14.6). Harvell (1986) analyzed the cues that triggered spine production in these modular organisms, and how effective it was against predators. The mere presence of nudibranchs stimulated spine production, and this

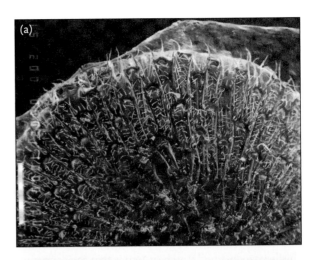

FIGURE 14.6
Inducible spines on colonies of the marine bryozoan Membranipora membranacea. *Spines are typically produced as a peripheral band of one or two zooids. (a) Colony view (microscope field: 20 mm). (b) Zooid view (field = 100 µm). These spines deter feeding by nudibranch predators. (From Harvell 1986.)*

occurred even in bryozoan colonies distant from the nudibranchs. A chemical cue released from the nudibranchs into the water is presumed to be the triggering cue for spine production. On spined colonies, nudibranchs ate only 40% as many zooids per day as those on unspined colonies. Producing spines is thus effective for reducing damage in bryozoans. There is a cost to spine production. Harvell (1986) found that bryozoan colonies producing spines grew at only 85% the rate of unspined colonies. In these colonial animals, a decrease in growth is converted directly to a decrease in reproductive output. The key feature of

FIGURE 14.7

Percentage of the cactus Opuntia stricta *having spines on three islands off the coast of Queensland, Australia. Cattle were present on one island and grazing damage was observed, but cattle were absent from the other two islands. (After Myers and Bazely 1991.)*

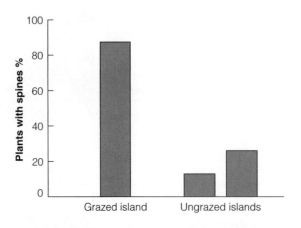

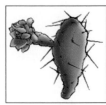

this bryozoan-nudibranch system is that the consumer does not eat the entire bryozoan colony, and induced defenses may deter further losses. But this benefit of defense has a cost in energy diverted from growth and reproduction into spine production.

Spines and Thorns in Terrestrial Plants

Thorns, spines, and prickles also occur widely on terrestrial plants, and even though everyone assumes that they act as physical defenses against large herbivores, there is remarkably little evidence that this is true (Myers and Bazely 1991). A variety of observations are consistent with this idea, but there is little experimental evidence. For example, the cactus *Opuntia stricta* has more spiny individuals on Australian islands on which cattle graze than on islands with no grazing (Figure 14.7). If thorns and spines are herbivore defense mechanisms, they could be used to test ideas about plant defenses. The resource availability hypothesis (see Figure 14.2) predicts that plants growing in nutrient-poor soils should invest more in plant defense than plants on rich soils. This is not the case for the fynbos vegetation of South Africa (Campbell 1986). Fynbos is a shrubland of sclerophyllous, evergreen plants growing on very poor soils. Only 4% of the total plant cover in fynbos has spines, compared with 13% of the plant cover in nutrient-richer areas that lack fynbos. Campbell (1986) suggests that the fynbos vegetation is so poor that no large herbivores can live on it, and consequently there is no selection for physical plant defenses like thorns.

Acacias in Africa and Israel grow spines and thorns that appear to be an adaptation against large herbivores. In Tanzania *Acacia tortilis* trees protected from grazing do not grow spines (Gowda 1996). Goats feeding on these acacias induce spines on the trees, and Gowda (1996) found that the more spines on individual plants, the fewer shoots they lost to goat browsing on branches and leaves. In the Negev Desert of Israel, Rohner and Ward (1997) compared acacias on fenced areas that excluded large herbivores for more than ten years with acacias on open areas. They found that browsed acacias increased the numbers of spines and thorns, but they did not consistently increase their chemical defenses compared with controls.

Herbivores on the Serengeti Plains

Because of the many defense mechanisms of plants, all that is green is not necessarily edible and herbivores may be more food-limited than they appear. The result is that herbivores may still compete for food plants, as we saw in Chapter 13. But in some cases herbivores may cooperate in the harvesting of plant matter. The grazing system of ungulates on the Serengeti Plains of east Africa is an excellent illustration of how herbivores may interact over their food supply. The Serengeti Plains contain the most spectacular concentrations of large mammals found anywhere in the world. A million wildebeest (Figure 14.8), 600,000 Thomson's gazelles, 200,000 zebras, and 65,000 buffaloes occupy an area of 23,000 km^2 (9000 mi^2), along with undetermined numbers of 20 other species of grazing animals (Sinclair and Arcese 1995).

The dominant grazers of the Serengeti Plains are migratory and respond to the growth of the

FIGURE 14.8
Blue wildebeest grazing on the Serengeti Plains of East Africa. An estimated 1 million migratory wildebeest inhabit the Serengeti region. (Photos courtesy of A. R. E. Sinclair and Nigel J. Dennis/Photo Researchers, Inc.)

grasses in a fixed sequence (Figure 14.9). First, zebras enter the long-grass communities and remove many of the longer stems. Zebras are followed by wildebeest, which migrate in very large herds and trample and graze the grasses to near the ground. Wildebeest are in turn followed by Thomson's gazelles, which feed on the short grass during the dry season (Bell 1971).

Different grazers in the Serengeti system do not select different species of grasses but instead select different parts of the grass plant during different seasons (Figure 14.10). Zebras eat mostly grass stems and sheaths and almost no grass leaves. Wildebeest eat more sheaths and leaves, and Thomson's gazelles eat grass sheaths and a large fraction of herbs not touched by the other two ungulates. These feeding differences have significant consequences for the ungulates because grass stems are very low in protein and high in lignin, whereas grass leaves are relatively high in protein and low in lignin, such that leaves provide more energy per gram of dry weight. Herb leaves typically contain even more protein and energy than grass leaves (Gwynne and Bell 1968). So zebras seem to have the worst diet and Thomson's gazelles the best.

How can zebras cope with grass stems as the major part of their diet during the dry season? Most of the ungulates in the Serengeti are ruminants, which have a specialized stomach containing bacteria and

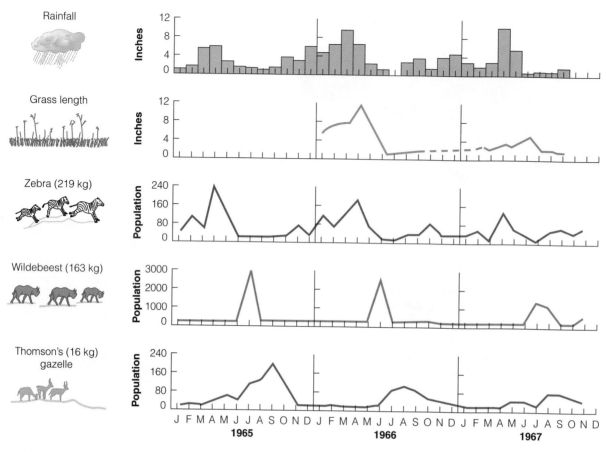

Rainfall

Grass length

Zebra (219 kg)

Wildebeest (163 kg)

Thomson's (16 kg) gazelle

FIGURE 14.9

Populations of three migrating ungulates in relation to rainfall and grass length on the Serengeti Plains of east Africa. The figures were obtained in the western Serengeti by a series of daily transects in a strip approximately 3000 m long and 800 m wide. Successive peaks during each year mark the passage of the main migratory species in the early dry season. (From Bell 1971.)

protozoa that break down the cellulose in the cell walls of plants. But the zebra is not a ruminant and is similar to the horse in having a simple stomach. Zebras survive by processing a much larger volume of plant material through their gut than ruminants do, perhaps roughly twice as much. So even though a zebra cannot extract all the protein and energy from the grass stems, it eats more and compensates by volume. Zebras also have an advantage of being larger than wildebeest and Thomson's gazelles, and larger animals need less energy and less protein per unit of weight than smaller animals. The net result of these factors is that in times of dietary stress, large animals are able to tolerate low food quality better than small animals can.

Competition for food may occur between wildebeest and Thomson's gazelles because they eat the same parts of the grass. Wildebeest have what appears to be a devastating effect on the grassland as they pass through in migration. Green biomass was reduced by 85% and average plant height by 56% on sample plots. By establishing fenced areas as grazing exclosures, McNaughton (1976) was able to follow the subsequent changes both in grassland areas subject to wildebeest grazing and in areas protected from all grazing (Figure 14.11). Grazed areas recovered after the wildebeest migration had passed and produced a short, dense lawn of green grass leaves. As gazelles entered the area during the dry season, they concentrated their feeding on areas where wildebeest had previously grazed and avoided areas of grassland that the wildebeest herd had missed.

Grass production was reduced by both wildebeest and gazelles, but no signs of competition were found.

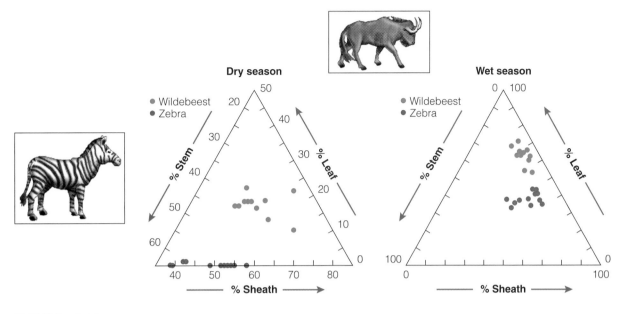

FIGURE 14.10
Frequency of the three structural parts of grass in the diets of wildebeest and zebras during the dry season and the wet season, Serengeti Plains, East Africa. (After Gwynne and Bell 1968).

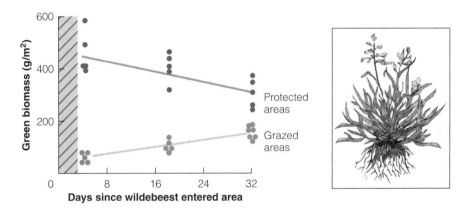

FIGURE 14.11
Vegetation recovery after a wildebeest migration passed through the western Serengeti Plains of east Africa. The three to five days of wildebeest passage is marked by the hatched area at the left of the graph. Some areas were protected from wildebeest grazing by fencing before the migration arrived. (Modified from McNaughton 1976.)

The Serengeti ungulate populations show evidence of *grazing facilitation*, in which the feeding activity of one herbivore species improves the food supply available to a second species. Heavy grazing by wildebeest prepares the grass community for subsequent exploitation by Thomson's gazelles in the same general way that zebra feeding improves wildebeest grazing. Thus potential competition may be replaced by mutualism.

Feeding systems of this type may be severely upset by the selective removal of one herbivore in the sequence.

The grazing-facilitation hypothesis was tested by comparing population trends of wildebeest, zebras, and Thomson's gazelles in the Serengeti (Sinclair and Norton-Griffiths 1982). If grazing facilitation is mutualistic and obligatory, wildebeest numbers should not increase if zebra numbers do not

FIGURE 14.12

Population changes in migratory ungulates in the Serengeti Plains of East Africa. Estimates were obtained by aerial census. Vertical lines are error estimates (1 standard error). (Data from Sinclair and Norton-Griffiths 1982, and Dublin et al. 1990.)

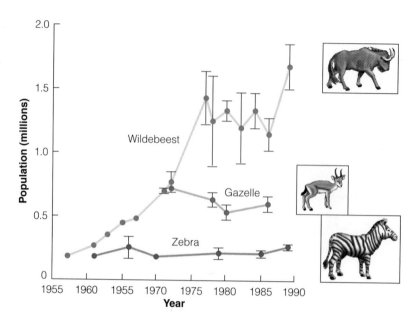

increase, and if wildebeest numbers increase, gazelle numbers should also increase. Figure 14.12 shows that this has not happened. Wildebeest numbers more than doubled during the 1970s while gazelle numbers fell slightly and zebra numbers remained constant. Predation may hold zebra numbers down, and these three ungulates apparently are not as closely linked as Bell (1971) suggested and may not form a mutualistic association.

Competition for grass in the Serengeti region may occur between very different types of herbivores (Sinclair and Arcese 1995). In addition to the large ungulates, 38 species of grasshoppers and 36 species of rodents consume parts of the grasses and herbs. In the Serengeti Plains, most of the plant material consumed by herbivores is consumed by the large ungulates, but in some plant communities within the Serengeti, grasshoppers consumed nearly half as much grass as did the ungulates. The grazing system of the Serengeti is thus even more complex than is suggested in Figure 14.9. In any grazing system, herbivores of greatly differing size and taxonomy may be affecting one another positively or negatively.

Can Grazing Benefit Plants?

Herbivores eat parts of plants, and at first view this action would appear to be detrimental to the individual plant. But could grazing or browsing in fact be beneficial to a plant? On a more practical level of public policy, should public grazing land be protected from sheep and cattle grazing, or should we encourage cattle and sheep production? If cattle and sheep grazing is good for plants, then we would have a clear, ecologically based reason to support current land management policies in the western United States and Australia. What is the ecological evidence?

The idea that grazing is good for individual grasses, good for cattle, good for plant communities (to retard succession to shrubs), and good for ecosystems (to speed up decomposition) have been promoted by both range managers (Savory 1988) and ecologists (Owen and Wiegert 1981). This idea in the broad sense postulates that grazers and grasses are in a mutualistic relationship in which both gain. But it is clear that too much grazing is detrimental—everyone agrees with that. The question is whether or not some moderate level of grazing will stimulate plants to produce more biomass, a proposal called the *overcompensation hypothesis* (Figure 14.13).

To evaluate the idea that grazing could improve plant production, it is important to measure both aboveground and belowground biomass. There is little evidence that grazing ever increases plant production or improves the fitness of plants (Belsky 1986, 1987; Painter and Belsky 1993). Plants respond to grazing by regrowth (see Figure 14.11), but they never recover completely from the losses caused by grazing. The prevailing view of the plant-herbivore interaction for grazing systems is that it is a predator-prey type of interaction in which the herbivore gains

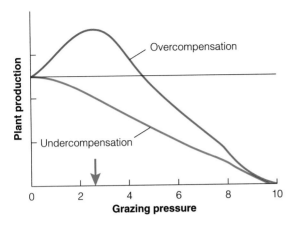

FIGURE 14.13
The overcompensation hypothesis of grazing. Grazing is postulated to be favorable for plant production up to some optimum level of grazing pressure (green arrow). This hypothesis predicts that plant production will be lower in the absence of grazing, and that moderate grazing will increase plant growth and plant fitness. The classical view of grazing is shown by the red line, in which plants cannot completely replace grazed tissues. The horizontal black line indicates exact compensation of lost tissues by plant. (Modified after Belsky 1986.)

E S S A Y   1 4 . 1

HERBIVORY, ECONOMICS, AND LAND USE

Ecological ideas about herbivory meet economic ideas about land use in the grazing lands of the western United States. About 70% of the land area in the western states is grazed by livestock, including wilderness areas, wildlife refuges, national forests, and some national parks (Fleischner 1994). Ecologists ask two questions about the effects of grazing: (1) What are the ecological costs of grazing these areas? And (2) is grazing in its current form sustainable? Economists ask about the balance of costs and benefits of grazing, but in doing so rarely consider the ecological costs, which almost never have dollar values attached to them. The result has been an ongoing and acrimonious controversy over land use in the West, a controversy with multiple dimensions.

The experimental ecologist would like to look at comparable grazed and ungrazed land in order to measure the ecological effects of grazing on populations of plants and animals. But almost no ungrazed land is available for such comparisons. Much of the land left ungrazed is on steep slopes or in rocky areas that differ dramatically from the surrounding habitats. One solution to this problem is to use livestock exclosures to study effects. But this approach also has problems because most exclosures are small in area and were previously grazed. Small exclosures do not include all the species in a community, especially the rare ones. And if the initial grazing effects are the most severe ones, historical carryover will affect even long-term exclosure studies on sites that were previously grazed. As such, exclosures will underestimate the true effects of grazing on plants and animals.

Some ecologists and land managers argue that grass needs grazing and that livestock are thus essential for the ecological health of western grazing lands. Some of the justification for this has come from the *overcompensation* or herbivore optimization hypothesis, which suggests that plant productivity may increase if plants are grazed. There is little ecological evidence for this hypothesis in western rangelands, but the idea keeps coming up as but one justification for the current grazing system.

The use of public lands for grazing must be balanced with the needs of conservation and recreation. In particular, all those concerned need to work out sustainable land-use practices that will achieve these diverse economic and ecological goals. The western rangelands should not be all national parks, nor should they be all overgrazed plant communities. There must be cooperation among all interested groups to achieve the goal of sustainable land use, and good science conducted to show us what policy goals can be achieved by good land management (Brown and McDonald 1995).

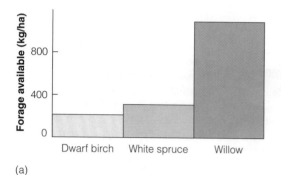

(a)

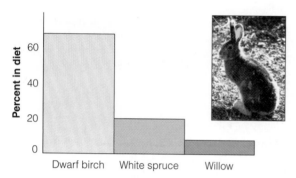

(b)

FIGURE 14.14
Selective feeding of snowshoe hares on woody plants during winter. (a) Biomass of available forage. (b) Percent of forage type in diet. Hares do not eat a random sample of the plants available but are highly selective and prefer dwarf birch. Data from Kluane Lake, Yukon, winter 1979–1980 (Unpublished data, A. R. E. Sinclair and J. N. M. Smith.)

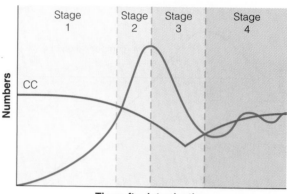

FIGURE 14.15
The general pattern of ungulate irruptions that is often characteristic for populations introduced into new areas. Four general stages of the irruption can be recognized. cc = carrying capacity of the habitat based on food supplies. (After Riney 1964.)

and the plant loses. But not all plant-herbivore interactions can be classified as +/ − or negative—some plant-animal interactions are mutualistic, as we saw for the ant-acacia system, in which both parties gain. Pollination and fruit dispersal are two additional interactions that can be beneficial for both the plant and the herbivore.

Herbivores are commonly thought to be "lawn mowers," but it is important to recognize that they are highly selective in their feeding. This selectivity is a major reason why the world is not completely green for a herbivore. Figure 14.14 illustrates selective feeding in the snowshoe hare (*Lepus americanus*). Snowshoe hares feed in winter on the small twigs of woody shrubs and trees. In the southwestern Yukon, only three main plant species are available above the snow, and hares clearly prefer dwarf birch (*Betula*

glandulosa) over willow (*Salix glauca*). These preferences may be caused by plant secondary substances such as phenols (Sinclair and Smith 1984).

Dynamics of Herbivore Populations

There are two basic types of plant-herbivore systems. We have discussed one type, called an *interactive herbivore system* because the herbivores influence the rate of growth and the subsequent fate of the vegetation: This feedback is critical for the dynamics of the plant-herbivore system. Other herbivore systems, called *noninteractive herbivore systems*, show no relationship between herbivore population density and the subsequent condition of the vegetation.

Many herbivore systems are interactive, with feedback occurring between the herbivores and the plants. Serengeti ungulates provide many examples, and most grazing systems are of this type. Next we look at an example of each type to compare and contrast the two ways in which animal populations react to their food plants.

Interactive Grazing: Ungulate Irruptions

Many ungulates introduced into a new region increase dramatically to high densities and then col-

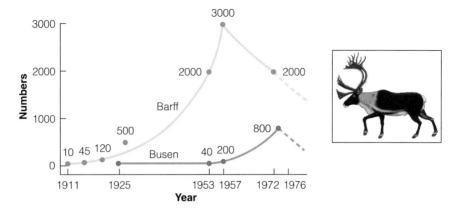

FIGURE 14.16
Population irruption of reindeer on South Georgia. Two separate herds were introduced on different parts of the island, and estimates of their numbers have been made at various times. (After Leader-Williams 1988.)

lapse to lower levels. The increase and subsequent collapse is called an *irruption*. Introduced reindeer populations have provided several examples. Figure 14.15 illustrates the stages of an irruption. Irruptions commonly occur when the introduced ungulate has an excellent food supply and no natural predators. As the population increases during stage 1 of an irruption, the food resources are reduced. During stage 2 the population exceeds the carrying capacity of the habitat, and food plants are overutilized and damaged. In stage 3 the population collapses because of food shortage, often aggravated by severe weather. This collapse may continue to near-extinction (see Figure 11.9), or the population may stabilize at lower numbers (Leader-Williams 1988).

Graeme Caughley, working at the University of Sydney and later at CSIRO Wildlife and Ecology in Canberra, Australia, made major contributions to the study of plant-herbivore dynamics. He studied irruptions of the Himalayan thar in New Zealand (Caughley 1970). The Himalayan thar, a goat-like ungulate of Asia, was introduced into New Zealand in 1904 and has since spread over a large region of the Southern Alps. As its density increased, its birth rate fell only slightly and its death rate increased, primarily because of increased juvenile mortality. After a period of high density the population declined due to a combination of reduced adult fecundity and a further increase in juvenile losses.

What caused these population changes? Caughley (1970) suggested that grazing by the thar both

reduced its food supply and changed the character of the vegetation. The link between ungulates and their food plants is critical in these irruptions. The most conspicuous effect of thar grazing was found in the abundance of snow tussocks (*Chionochloa* spp.), evergreen perennial grasses that were the dominant vegetative cover where thar were absent but were scarce where thar had become common. Snow tussocks were believed to be important as food in late winter and cannot tolerate even moderate grazing pressures. When thar reach high densities, they begin to browse on shrubs in winter and may even kill some shrubs by their feeding activities.

Norwegian whalers introduced two separate populations of reindeer to South Georgia, a subantarctic island, in 1911 and in 1925, primarily for sport hunting. During the 1950s whales became scarce and reindeer hunting nearly stopped. Figure 14.16 shows the population history of the two reindeer herds on South Georgia (Leader-Williams 1988). The Barff herd reached a peak in 1955 and collapsed to about 2000 animals; the Busen herd grew more slowly to a peak in the early 1970s and then also began to collapse. By this time reindeer were overgrazing tussock grasslands, which are their dominant winter food. This simple island system, which lacks predators and other grazing animals, clearly illustrates the interplay between plants and herbivores in an interactive grazing system.

A general picture of an ungulate irruption emerges: A small number of animals is introduced onto a range with superabundant food, and a gradual increase in animal density and decrease in plant density occurs until the animals have reduced or eliminated their best forage. Animal numbers then decline until a new, lower density is reached, at which the herbivores and their plants may stabilize. Ungulate irruptions occur in both native and introduced species (Caughley 1970).

This sequence of events is similar to that predicted by some simple predator-prey models (Noy-Meir 1975, Caughley 1976b). The Rosenzweig-MacArthur predator-prey model discussed in Chapter 13 can also be applied to a simple grazing system. Plant growth is a simple function of plant biomass for most plants, and the logistic equation can describe plant growth in the absence of grazing. Figure 14.17 illustrates some

Graeme Caughley *(1937–1994) CSIRO Wildlife and Ecology, Canberra*

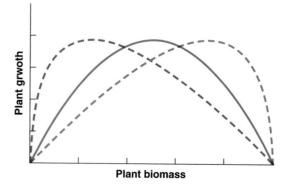

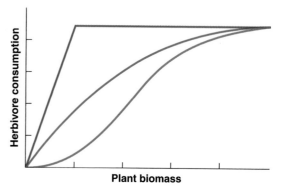

FIGURE 14.17
Simple models of a plant-herbivore grazing system. (a) Plant growth as a function of plant biomass. The logistic growth function is shown in red, and two other possible shapes of growth functions are shown in blue and green. (b) Herbivore consumption as a function of plant biomass. These are functional response curves like those for predator-prey relations. The blue curve is Type 1, the red curve is Type 2, and the green sigmoid curve is Type 3. (Modified from Noy-Meir 1975.)

possible models for plant growth and herbivore consumption in a grazing system. If we combine the logistic plant growth model with the Type 2 consumption curve, we can generate the dynamics shown in Figure 14.18, a simple herbivore-vegetation model that mimics the reindeer irruption shown in Figure 14.16 (Caughley 1976b). The behavior of simple herbivore-vegetation models is highly dependent on the rates of increase of the plants and herbivores alike, and also on the feeding rates of the herbivores. Caughley (1976b) showed that such simple model systems oscillate in cycles if the grazing pressure tends to hold the amount of vegetation below about half the

amount present in the ungrazed state. If the herbivore is a very efficient grazer, such simple systems can collapse completely (Noy-Meir 1975).

Many insect populations experience irruptions that damage their food plants (Myers 1993). The spruce budworm, for example, periodically irrupts to epidemic proportions in the coniferous forests of eastern Canada (Figure 14.19). Budworms eat the buds, flowers, and needles of balsam fir trees. Outbreaks occur every 35 to 40 years, in association with the maturing of extensive stands of balsam fir, and during budworm outbreaks many balsam fir trees are defoliated and killed. Populations of the large aspen tortrix moth irrupt in interior Alaska at intervals of 10–15 years; during these irruptions

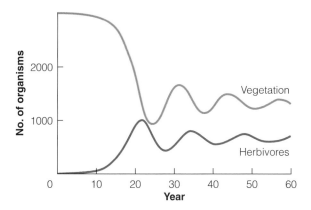

FIGURE 14.18

Simple model of an interactive herbivore-plant grazing system, which is similar to an ungulate irruption in showing a damped oscillation. The model used is a modified version of the Lotka-Volterra predator-prey model in which vegetation increases logistically instead of exponentially. (After Caughley 1976a.)

FIGURE 14.19

Spruce budworm population outbreaks since 1750 in eastern Canada. Quantitative measurements of density are available since 1946. Historical records extend back to 1878, and tree ring data permit qualitative estimates back to 1750. The adult moth (right) and the terminal branches of balsam fir trees defoliated by the larvae are shown. (After Royama 1984; photos courtesy of the U.S. Forest Service.)

quaking aspen trees are severely defoliated (80–100%) for two to four years (Clausen et al. 1991)

Many herbivorous insect populations may be held at low densities by a protein deficiency in their food plants. White (1993) has suggested that most plant material is not suitable food for insects because of nitrogen deficiencies. When plants are physiologically stressed—by water shortage, for example—they often respond by increasing the concentration of amino acids in their leaves and stems. Some larval insects may survive much better when more amino acids are available, and thus the stage is set for an insect irruption. This hypothesis, called the *plant stress hypothesis*, postulates that plants under abiotic stress become more suitable as food for herbivorous insects (Larsson 1989). This hypothesis may explain irruptions in many insect pests like the spruce budworm, which interact destructively with their food plants only occasionally (White 1993). But not all insects may respond positively to plant stress. Sucking insects such as aphids increase their reproductive rate as plants are stressed, whereas in most cases chewing insects do not improve their performance on stressed plants.

A common observation of foresters is that insect attack is often concentrated on young, vigorously growing trees. Price (1991) suggested the *plant vigor hypothesis* as an alternative to the plant stress hypothesis to explain these exceptions. Many herbivores feed preferentially on vigorously growing plants or plant modules. Moose, for example, prefer to feed on rapidly growing shoots of birch and seem immune to plant defense chemicals (Danell and Huss-Danell 1985). Different types of herbivores feed on rapidly growing plants, and others prefer stressed plants, so we should not expect all species to fit only one of these two hypotheses. Understanding patterns of herbivore attacks on plants will likely require a diversity of hypotheses.

Noninteractive Grazing: Finch Populations

European finches feed on the seeds of trees and herbs, and their feeding activities do not in any way affect the subsequent production of their food plants. These species form a good example of a noninteractive system in which controls operate in only one direction:

Food plant production → herbivore density

There is no direct feedback from the herbivores to the plants, as in the grazing systems discussed in the previous section. This lack of interaction has important consequences for plant-herbivore interactions.

Among two groups of British finches, one group feeds on the seeds of herbs, and their populations are

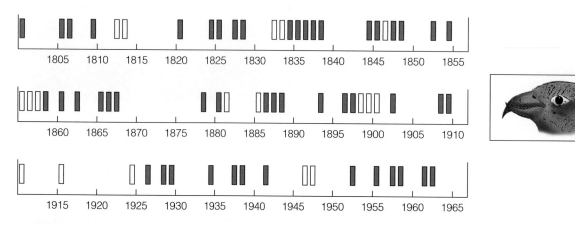

FIGURE 14.20

Years of invasions of the common crossbill into southwestern Europe. Solid blocks indicate large invasions; open blocks indicate small invasions. (Data from Newton 1972.)

quite stable (Newton 1972); a second group feeds on the seeds of trees, and their populations fluctuate greatly. Population dynamics in these finches are determined by fluctuations in seed crops from year to year. Herbs in the temperate zone produce nearly the same numbers of seeds from one year to another, but trees do not. Most trees require more than one year to accumulate the reserves necessary to produce fruit. Spruce trees in Europe, for example, have moderate to large cone crops every two to three years in central Europe, every three to four years in southern Scandinavia, and every four to five years in northern Scandinavia. Good weather is also needed when the fruit buds are forming during the year before the seed crop is produced. The net result is that trees in a given geographic region usually fruit in synchrony. Various geographic regions may or may not be in synchrony with each other, depending on local weather conditions.

Finches that depend on tree seeds experience great irruptions in population density. They exist only by being opportunistic and moving large distances to search for areas of high seed production. All the "irruptive" finches breed in northern areas and rely at some critical part of the year on seeds from one or two tree species (see Figure 6.10, page 82). Periodically these finches leave their northern breeding areas and move south in large numbers. Figure 14.20 shows the years of invasion of the common crossbill into southwestern Europe. A major invasion of crossbills into western Europe occurred most recently in 1990.

Mass emigration of crossbills and other finches is presumably an adaptation that avoids food shortages on the breeding range (Newton 1972). But crop failure alone is not sufficient to explain these mass movements. For example, in Sweden the spruce cone crop has been measured in all districts since 1900. Not all poor spruce crops in Sweden have resulted in crossbill movements. Very poor spruce crops occurred in 14 years between 1900 and 1963, but in only six of these years did crossbills move. Other evidence suggests that high population density may be necessary before larger scale movements can be triggered. In some years, crossbills began to emigrate in the spring, even before the new cone crop was available. Crossbills also put on additional fat before they emigrate, in the same way that migratory birds do. The suggestion is that high crossbill density is a prerequisite for large scale movements and that emigration occurs in response to the first inadequate cone crop once high bird densities are present.

Why emigrate? Mass emigration presumably is advantageous to the birds that stay behind, provided they find sufficient food. Emigration, by contrast, is often considered suicidal, and the question arises as to how such an adaptation could exist. Crossbill emigrants might have two potential advantages: They could colonize new habitats in the south and thereby leave descendants; more likely, however, they obtain an advantage by migrating back north again after the food crisis has passed. Newton (1972) described four

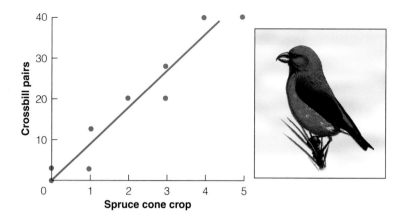

FIGURE 14.21
Relation between the early spring density of common crossbills in northern Finland and the relative size of the spruce cone crop. (Data of Reinikainen, cited in Newton 1972.)

common crossbills that were banded in Switzerland during an irruption and were recovered a year later in northern Russia. Thus some birds return north, even though many die in the south during the irruption.

A close correlation exists between crossbill breeding densities and the size of their food supply (Figure 14.21). This correspondence is obtained by having great mobility such that populations can concentrate their nesting in areas with good cone crops. How random mobility within the normal breeding becomes a unidirectional emigration in years of irruption is not understood.

Seed Dispersal: An Example of Mutualism

Not all plant-animal interactions are detrimental to plants; one good example of mutualism involves seed dispersal. Many plants depend on birds and mammals for the dispersal of their seeds. Coevolution between plants and their seed predators may help to explain some features of plant reproductive biology and animal feeding habits (Faegri and Pijl 1971, Thompson 1994). Coevolution may be seen particularly clearly in the evolution of fruits.

Fruits are basically discrete packages of seeds enclosed (typically) within a certain amount of nutritive material. Vertebrates eat these fruits and digest part of the material, but many of the seeds pass through the digestive system unharmed. Plants advertise fruits by a ripening process in which the fruits change color, taste, and odor. Once fruits are ripe, they can also be attacked by damaging agents, such as

fungi and bacteria, or by destructive feeders who will not disperse the seeds. Plants need to evolve traits that assist seed dispersal but retard seed destruction.

A plant defends its fruit against damage by herbivores in four general ways (Table 14.3). The least expensive defense is to fruit during the season of minimum pest numbers. For example, in the temperate zone, insect damage to fruits will be minimal in the autumn and winter. Second, plants can reduce their exposure to fruit damage by ripening fruit more slowly. This strategy, however, will also reduce the availability of ripe fruit to dispersers and may be counterproductive to the plant. Third, by making the fruit very unbalanced nutritionally, a plant may discourage insect and fungal attack (Herrera 1982). Fruit pulp is high in carbohydrates, low in lipids, and extremely low in protein. Fruits are among the poorest protein sources in nature, and this may be a general pest-defense mechanism.

Finally, plants may adopt a chemical or mechanical defense of their fruits, although this is the most expensive defense method. Because chemical defenses in fruits reduce herbivore attack and thus seed dispersal, it has commonly been assumed that as fruits ripen, chemical defenses should be eliminated. For example, the alkaloid tomatine in green tomatoes is degraded by an enzyme system that is activated in ripening red tomatoes (McKey 1979). But the assumption that ripe fruits will not be chemically defended is probably a biased extrapolation from cultivated fruits that have been selected for centuries. Of all the wild European plants that produce fleshy fruits, at least one-third have fruits toxic to humans (Herrera 1982). Toxic fruits are avoided by many birds and mammals, and consequently such fruits are dispersed

TABLE 14.3 Costs and effects of the main strategies plants use to protect their fruits.

Defense strategy	Cost to the plant in energy and nutrients	Effect on plant of disperser's feeding response
Fruiting during the time of lowest pest pressure	None	Variable,[a] usually positive
Reducing exposure time by decreasing ripening rate	None	Variable,[a] often negative
Unbalanced and/or poor nutritional value of fruit flesh	None	Negative
Chemical defense	Some	Negative

[a]Depends on disperser abundance and attractability.
Source: After Herrera (1982).

only by a selected subset of vertebrates who can detoxify the specific chemicals. Toxic fruits are thus one way for a plant to "choose" its dispersal agents, although the detailed reasons why this is adaptive for the plants have not been discovered.

The interaction of seed predation and seed dispersal has been analyzed particularly clearly in the case of Clark's nutcracker (*Nucifraga columbiana*) and the whitebark pine (*Pinus albicaulis*). Several soft pines in North America have large, wingless seeds that are not dispersed by wind. These seeds are frequently removed from their cones by jays and nutcrackers, transported some distance, and cached in the soil as a future food source. When the birds cache more seeds than they can subsequently eat, the surplus is available for germination. This interaction can be considered a case of mutualism only if both species increase their fitness because of the association.

Does the whitebark pine benefit from seed dispersal by Clark's nutcracker? Nutcrackers cache three to five seeds, on average, in each store, bury them an average of 2 cm, avoid damp sites, and place many of their caches in microenvironments favorable for subsequent tree growth (Tomback 1982). Each nutcracker on the average stores about 32,000 whitebark pine seeds each year, which represents three to five times the energy required by the nutcracker. Thus many caches are not fully utilized, and the survival of pine seedlings arising from unused caches is very high. Nutcrackers also disperse pine seeds to new areas within the subalpine forest zone, in the process increasing the local distribution of whitebark pine (Tomback 1982). Many other species of birds feed on whitebark pine seeds, as do mice, chipmunks, and squirrels, but none of these species caches pine seeds in ways that favor germination (Hutchins and Lanner

1982). We conclude that the Clark's nutcracker–whitebark pine system is a coevolved mutualism in which both species profit, the nutcracker by obtaining food and the pine by achieving seed dispersal.

The great variety of seed dispersal systems used by plants has important consequences for seed-eating animals, and much more exploratory work will be needed before we know how many of these systems are mutualistic and how many are exploitative.

Complex Species Interactions

Species interactions are rarely one-on-one in natural communities, and the untangling of complex sets of species interactions is an important focus in ecology today. Complex interactions illustrate how difficult it can be to determine whether a single species exerts a beneficial or a harmful influence on another species.

Interactions between homopterous insects and the ants that either tend them or prey on them have been described for nearly a century (Buckley 1987). These plant-homopteran-ant interactions are significant because many of the world's major plant pests are homopteran insects, and many of the worst crop diseases are transmitted by homopterans. The manipulation of ant assemblages to control homopteran pests has been practiced in China since at least A.D. 300 (Needham 1986).

Ants that tend homopterans provide a positive benefit for the homopterans, including protection from predators or parasitoids, sanitation in honeydew removal, and transportation to new feeding sites. Ants may also remove dead individuals from homopteran

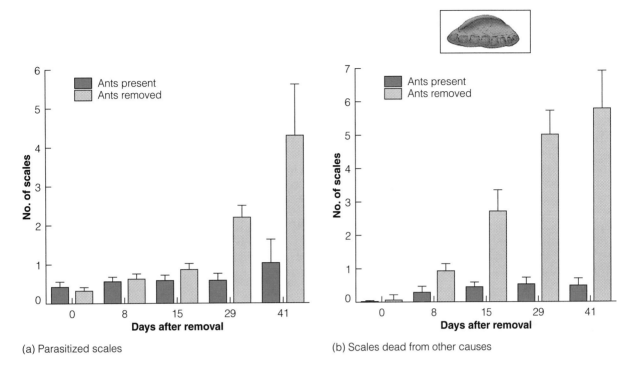

(a) Parasitized scales

(b) Scales dead from other causes

FIGURE 14.22

Number of green scales (Coccus viridis) *per leaf on Indian fleabane plants* (Pluchea indica) *with and without ants* (Pheidole megacephala)*. Mortality of the insects increased dramatically when ants were removed. (After Bach 1991.)*

populations and provide nest sites. For the plants, plant-homopteran-ant interactions impose many costs and confer few if any benefits. Homopterans consume phloem sap and tax the plant by removing metabolites, damaging plant tissues, and increasing water loss. Homopterans may also transmit plant pathogens. For the ants that tend homopterans, the main benefit is the food they obtain in the form of sugars from the honeydew secreted by homopterans (Buckley 1987), and ant colonies that feed on honeydew have higher populations than colonies with no honeydew source.

In Hawaii an insect called green scale (*Coccus viridis*) is tended by the ant *Pheidole megacephala* on the host plant Indian fleabane (*Pluchea indica*). Bach (1991) analyzed this system using removal and addition experiments to measure the strength of these interactions. When she removed ants from plants, the number of parasitized green scales increased, and the

mortality of scales to other predators and diseases also increased (Figure 14.22). Ants removed ladybird beetle larvae introduced onto plants as possible scale predators. On plants without ants, honeydew from the green scale accumulates, and a sooty mold grows on the leaves. This mold reduces the photosynthetic rate of the leaf, and leaves infested with mold were often shed by the plant. Ants thus indirectly benefited the plant by removing the honeydew. Ants may also remove other herbivorous insects such as moth larvae from plants and thus protect the plants from additional losses (Bach 1991).

Complex interactions become more difficult to unravel as the numbers of species involved grows. Interaction webs may involve herbivory, mutualism, predation, and competition, and emphasize the need to look at both direct interactions and indirect effects in ecological systems in a quantitative way (Strauss 1997).

Summary

Herbivory is a form of predation. Because plants are modular organisms, herbivores usually eat only part of the plant, and typical herbivory thus differs from typical predation. Plants have a variety of structural and chemical defenses that discourage herbivores from eating them. Many secondary plant substances stored in plant parts discourage herbivores, which have responded to these evolutionary challenges by timing their life cycle to avoid the chemical threats or by evolving enzymes to detoxify plant chemicals. Several theories have attempted to specify the strategies of plant defense. The resource availability hypothesis emphasizes the differential costs and benefits of defense for slow-growing and fast-growing plants, and predicts when plants ought to invest in either chemical or physical defense.

Some herbivores can affect the future density and productivity of their food plants. A very efficient herbivore can thus drive itself to extinction unless it has evolved some constraints that prevent overexploitation of its food plants. Most herbivore-plant systems seem to exist in a fluctuating equilibrium. Some herbivore populations track their food supply, and large fluctuations in food supply are often translated into large fluctuations in herbivore densities.

Not all herbivory is detrimental. Seed- and fruit-eating vertebrates obtain food from plants and may disperse seeds, thereby benefiting the plant in a mutualistic interaction. Fruits and seeds have a variety of mechanical and chemical defenses, and plants must time their fruiting season to assist seed dispersal but retard seed destruction.

Mutualism occurs in many plant-animal interactions. Pollination and seed dispersal are two examples of processes that benefit both plant and animal species. Ants may form mutualistic relationships with plants, particularly in tropical areas or with herbivorous insects like homopterans.

Key Concepts

1. That the world is green implies that herbivores are prevented from destroying their food sources, either by their own behavior, by their enemies, or by plant defense strategies.

2. The resource availability hypothesis predicts that plants growing slowly in poor habitats should invest most in plant defense because they have the most to lose from herbivory.

3. Herbivores may not achieve a stable interaction with their food plants, and many ungulates undergo irruptions with subsequent oscillations in numbers.

4. The MacArthur-Rosenzweig model of predator-prey dynamics can be applied to grazing systems to determine the kinds of plant-herbivore interactions that might lead to stability.

5. Not all plant-herbivore interactions are detrimental to plants. Animals disperse seeds and pollinate plants, and many mutualistic interactions have evolved in which both the plants and the herbivores gain from their association.

Selected References

Caughley, G. and A. Gunn. 1993. Dynamics of large herbivores in deserts: Kangaroos and caribou. *Oikos* 67:47–55.

Caughley, G. and J. H. Lawton. 1981. Plant-herbivore systems. In *Theoretical Ecology*, ed. R. M. May, pp. 132–166. Blackwell, Oxford, England.

Coley, P. D., J. P. Bryant, and F. S. Chapin III. 1985. Resource availability and plant antiherbivore defense. *Science* 230:895–899.

De Jong, T. J. 1995. Why fast-growing plants do not bother about defence. *Oikos* 74:545–548.

Feeny, P. 1992. The evolution of chemical ecology: Contributions from the study of herbivorous insects. In *Herbivores: Their Interactions with Secondary Plant Metabolites. Vol. II. Evolutionary and Ecological Processes*, ed. G. A. Rosenthal and M. Berenbaum, pp. 1–44. Academic Press, San Diego.

Harvell, C. D. 1998. Genetic variation and polymorphism in the inducible spines of a marine bryozoan. *Evolution* 52:80–86.

Myers, J. H. and D. Bazely. 1991. Thorns, spines, prickles, and hairs: Are they stimulated by herbivory and do they deter herbivores? In *Phytochemical Induction by Herbivores*, ed. T. W. Tallamy and M. J. Raupp, pp. 325–344. John Wiley, New York.

Painter, E. L. and A. J. Belsky. 1993. Application of herbivore optimization theory to rangelands of the western United States. *Ecological Applications* 3:2–9.

White, T. C. R. 1993. *The Inadequate Environment: Nitrogen and the Abundance of Animals*. Springer-Verlag, New York.

Wold, N. and R. J. Marquis. 1997. Induced defense in white oak: Effects on herbivores and consequences for the plant. *Ecology* 78:1356–1369.

Questions and Problems

14.1 Three species of crossbills in northern Europe (see Figure 6.10, page 80) tend to irrupt together. But two species concentrate on larch and spruce cones, which mature in one year, while the third species feeds on pine cones, which mature in two years. Poor flowering seems to occur at the same time in pine, spruce, and larch. How can you explain this puzzle? Suggest an experiment to test your hypothesis, and compare your ideas with those of Newton (1972, p. 239).

14.2 Discuss the advantages and disadvantages of physical versus chemical defenses in plants.

14.3 Plants endemic to islands without large mammalian herbivores are believed to be vulnerable to damage and possible extinction because they have no evolutionary history of being grazed. How would you test this hypothesis that island plants lack defenses against herbivores? Compare your approach with that of Bowen and Van Vuren (1997), who studied the plants of Santa Cruz Island off California.

14.4 How does herbivory on seeds and fruits differ from herbivory on leaves and stems of plants?

14.5 Wildlife managers and range ecologists both speak of the "carrying capacity" of a given habitat for a herbivore population. Write an essay on the concept of carrying capacity, how it can be measured, and how the concept can be applied to agricultural and natural situations. Dhondt (1988) and McLeod (1997) discuss the definition of this term.

14.6 Cannabin is a secondary plant substance (a terpene) produced by the hemp plant, *Cannabis sativa*. From information available in the literature, write an essay on the biological role of cannabin in this plant. Lago and Stanford (1989) and Van Der Werf et al. (1996) give some recent references.

14.7 Alkaloids are plant defense chemicals, but not all plants contain alkaloids. Among annual plants, the incidence of alkaloids is nearly twice that among perennial plants. Tropical floras also contain a much higher fraction of species with alkaloids than temperate floras, and this is true for both woody and non-woody plants. Suggest why these patterns exist, and then compare your ideas with those of Levin (1976a).

14.8 Caughley and Lawton (1981) suggest that the growth of many plant populations will be close to logistic. Review the assumptions of the logistic equation (see Chapter 11), and discuss why this suggestion might be true or false.

14.9 Eucalyptus trees in Australia have high rates of insect attack on leaves, with 10–50% of the leaves eaten every year, even though these trees also contain very high concentrations of essential oils and tannins (Morrow and Fox, 1980). Discuss how this situation could occur if eucalyptus oils and tannins are defensive chemicals.

14.10 Large herbivorous mammals are not always present in habitats dominated by spiny plants. Why might this be? Janzen (1986) reviews the vegetation of the Chihuahuan Desert of northcentral Mexico and interprets the abundance of spiny cacti as reflecting the "ghost of herbivory past." Read Janzen's analysis and discuss how one might test his ideas.

14.11 Mutualism is usually thought to be more common in tropical areas than in temperate or polar regions. Evaluate the evidence for this generalization. Boucher et al. (1982) discuss this issue.

14.12 If a plant-homopteran-ant interaction has a net negative effect on the individual plants occupied by the homopterans (see Figure 14.22), why are these plants not selectively eliminated from the population?

Overview Question
Under what conditions might herbivory benefit a plant, such that a plant-herbivore interaction could be mutualistic? How would you test the overcompensation model for a grassland grazing system like the Serengeti in Africa or the Great Plains in North America?

CHAPTER 15

Species Interactions:
Disease and Parasitism

DISEASE IS AN IMPORTANT INTERACTION between organisms, ranking with competition, predation, and herbivory as one of the four agents of population change. Disease has been one of the great preoccupations of humans through all recorded history, and our history books are replete with tales of the Black Death of the fourteenth century, the smallpox scourge of the nineteenth century, and the Great Influenza Pandemic of 1918. Today we are occupied with the AIDS epidemic, drug-resistant tuberculosis, and mad-cow disease. Generations of children come down with everyday diseases like measles, and each winter we suffer another flu epidemic, but both diseases are rarely more than inconveniences, given modern medicine.

Disease is defined as an interaction in which a disease organism lives on or within a host plant or animal, to the benefit of the disease agent and the detriment of the host. Disease agents are typically bacteria or viruses, but may be pathogenic fungi or prions (protein bodies); these agents are called *microparasites*. Parasitism has much in common with disease as a biotic interaction, and differs from disease mainly because parasites are often large, multicellular organisms such as tapeworms; these large agents are called *macroparasites*. But there is a middle ground of parasites, like the spirochetes responsible for syphilis, that are really disease organisms, and so these two interactions can be treated together. We tend to think of parasites as inflicting nonlethal harm on their hosts, but many diseases are also rarely lethal.

The effects of disease and parasitism can be either lethal or sublethal. Humans are often preoccupied with lethal diseases, but sublethal effects are probably more important to plants and animals in ecological settings. Disease can debilitate organisms and reduce fitness. Parasitized animals may reproduce less, be more subject to capture by predators, or be less tolerant of environmental extremes. Disease and parasitism can thus interact with competition and predation to affect population dynamics. Very few individuals are not affected by diseases or parasites, and the normal situation is for every individual of every species to harbor both disease agents and parasites.

We begin our analysis of disease by constructing some simple models, all of which have the underlying assumption that we can isolate in nature a system consisting of one host species and one disease organism. We will restore the system's complexities later.

Mathematical Models of Host-Disease Interacation

Human epidemiology is a focus of much disease research and has been the source of mathematical models that explore the host-disease interaction. In contrast to models of competition and predation, disease models have traditionally been continuous time models that use differential equations. These models are applicable to many ecological systems in which birth, death, and infection processes are continuous in time. Roy Anderson of Oxford University has been a world leader in bringing mathematical models of disease into ecology during the past 20 years; it is his work, and that of his colleague Robert May, that much of the following analysis is based.

Many types of models have been used in the study of disease epidemiology (Anderson and May 1978). *Compartment models* are boxes-and-arrows models that include simplified population dynamics, and they are a good starting point for learning to think about epidemics. In their simplest form they assume a constant host population, and because this fits the human situation in the short term, they have been used extensively for exploring human disease problems. More complex models can be developed that allow both host and parasite populations to vary in size and have been used to explore the dynamics of the entire ecological system of hosts and parasites. Let us explore some simple examples of these models.

Roy M. Anderson, *(1947–) Professor of Zoology, Oxford University*

Compartment Models with Constant Population Size

We begin by considering microparasites such as viruses and bacteria that are directly transmitted between hosts and reproduce within the host. Microparasites like the influenza virus typically are very small, have a short

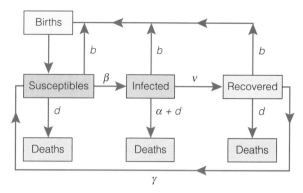

FIGURE 15.1
Compartment model for a directly transmitted microparasitic infection such as measles. Hosts are divided into susceptibles, infected, and recovered and immune. The parameters controlling this simple model are in the natural birth (b) and death (d) rates of the host, and the parameters of the disease agent: disease-induced deaths (α), recovery rate (ν), transmission rate (β) and rate of loss of immunity (γ).(From Anderson and May 1979.)

generation time, and thus have very high replication rates. Hosts that recover from infections typically acquire some immunity against reinfection, sometimes for life. In many cases the duration of the infection is short relative to the life span of the host, and we think of microparasitic infections as transient for the host.

For microparasitic infections we can divide the host population into three parts: susceptible, infected, and recovered. Figure 15.1 illustrates a simple compartment model for a microparasitic disease. The host population is characterized by the relative sizes of the three compartments and the instantaneous rates of birth (*b*) and death (*d*). The effects of the disease agent are summarized by four parameters: the per capita rate of disease mortality (α), the per capita recovery rate of hosts (ν), the transmission rate (β), and the per capita rate of loss of immunity (γ). This is a relatively simple compartment model because it does not take into account either the abundance of the disease agent in the host (individuals are either infected or not infected) or individual differences in susceptibility due to genetic or nutritional effects.

Compartment models are useful for answering questions about the stability of the host-disease interaction. Will the disease persist in a population or will it die out? How do the proportions of susceptible and infected individuals change through time as the disease goes through a population? We can answer these questions by converting the compartment model into a mathematical model, as follows (Heesterbeek and Roberts 1995).

Consider first the susceptibles in the population. We can estimate their rate of change by the differential equation

$$\begin{pmatrix} \text{Rate of change in} \\ \text{the susceptible} \\ \text{population} \end{pmatrix} = \begin{pmatrix} \text{Rate of transmission} \\ \text{of disease from} \\ \text{infected} \\ \text{to susceptibles} \end{pmatrix} \quad (15.1)$$

$$\frac{dX}{dt} = \frac{-\beta XY}{N}$$

where X = number of susceptibles

Y = number of infected individuals

N = total number of individuals = X + $Y + Z$

Z = number of recovered and immune individuals

β = transmission rate per encounter

In this simple model we assume that the population size (N) and the transmission rate (β) are constants. The number of infected individuals (Y) is given by

$$\begin{pmatrix} \text{Rate of change in} \\ \text{the infected} \\ \text{population} \end{pmatrix} = \begin{pmatrix} \text{Rate of transmission} \\ \text{of disease from} \\ \text{infected} \\ \text{to susceptibles} \end{pmatrix} -$$

$$- \begin{pmatrix} \text{Rate of recovery} \\ \text{of infected} \\ \text{individuals} \end{pmatrix} \quad (15.2)$$

$$\frac{dY}{dt} = \frac{\beta XY}{N} - \gamma Y$$

where γ = recovery rate

and all other terms are as previously defined. The term

$$\frac{\beta XY}{N}$$

is called the transmission term. In this simple model it is assumed that disease transmission is proportional to the product of the number of susceptible individuals (X) and the proportion of the population that is infected (Y/N). For simplicity we assume that the

recovery rate (γ) is a constant. Finally, the dynamics of the recovered individuals (Z) can be written as

$$\begin{pmatrix} \text{Rate of change in} \\ \text{the recovered} \\ \text{population} \end{pmatrix} = \begin{pmatrix} \text{Rate of recovery} \\ \text{of infected} \\ \text{individuals} \end{pmatrix} \quad (15.3)$$

$$\frac{dZ}{dt} = \gamma Y$$

where all terms are as previously defined above.

Compartment models are named after the types of compartments used, so this model is sometimes called an *SIR model* (susceptible, infected, recovered). If there were no recovery from the disease (as with untreated rabies), then $\gamma = 0$ and we would have an SI model.

What use can we make of this simple model? The first question we can ask is whether or not an epidemic develops when a small number of infected individuals enters a large population of susceptibles. The answer to this question depends on the value of a critical epidemiological parameter, the basic reproductive rate of the disease organism, called R_0. We define the basic reproductive rate as

R_0 = the average number of secondary infections produced by one infected individual

For this simple model,

$$R_0 = \frac{\beta}{\gamma} \quad (15.4)$$

(see Box 15.1). On average, one infected individual meets and infects β susceptible individuals per unit of time, and it does this for a time period of average length $1/\beta$ until it recovers.[1] An epidemic can develop only if $R_0 > 1$, which ensures a chain reaction of infection. In this simple model R_0 is a constant. Note that the basic reproductive rate of these disease models is analogous to the net reproductive rate of population growth models (see Chapter 11, page 152).

The course of the disease under this simple model is illustrated in Figure 15.2. The number of infected individuals rises steadily to a peak and then declines to zero, and the infection dies out. The susceptible population becomes too small after a certain time for an infected individual to encounter a susceptible one in order to cause new disease cases. In this simple model organisms become immune and the epidemic dies out.

One interesting question that we can ask about this simple model is what happens if there is a steady influx of new susceptible individuals. Populations are often growing during the breeding season, or

---

[1]For instaneous rates of death or recovery, the average duration until the event occurs is 1/rate. Thus for a death rate of d, the average will be $1/d$, and for a recovery rate of γ, the average time of being infective life span before recovery will be $1/\gamma$

ESSAY 15.1

WHAT IS THE TRANSMISSION COEFFICIENT (β), AND HOW CAN WE MEASURE IT?

All host-parasite models have within them a difficult parameters called the transmission coefficient (β), which measures the rate at which a disease or parasite moves from infected individuals to susceptibles. The transmission coefficient enters simple models as a mass-action term depending only on the numbers of susceptibles (X) and infectives (Y). For example, Equation (15.2) states that

$$\left(\begin{array}{c}\text{Change in number}\\ \text{of infected per}\\ \text{unit time}\end{array}\right) = \left(\begin{array}{c}\text{Per capita contact}\\ \text{rate between infected}\\ \text{and susceptibles}\end{array}\right)$$
$$- \left(\begin{array}{c}\text{Rate of recovery}\\ \text{of infected}\end{array}\right)$$

$$\frac{dY}{dt} = \frac{\beta XY}{N} - \gamma Y$$

In this simple model the transmission coefficient is the probability that a single contact between a susceptible host and an infected one will result in disease transmission. The transmission coefficient is thus a dimensionless number, a probability between 0 and 1.

How can we estimate β from empirical data? Hone et al. (1992) used one method for a model of swine fever in wild pigs. The critical data are the number of deaths on each day of the epidemic. For example, a swine fever epidemic in Pakistan gave the following detailed results for an initial population of 465 pigs:

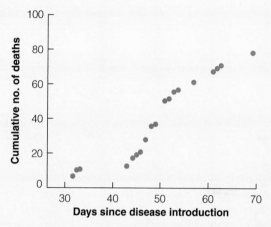

The rate at which deaths from the disease accumulate is clearly related to the transmission coefficient β as well as to the disease-induced death rate α. Given that we can get estimates of all the parameters in the disease model and that we know the starting population size, we can select an arbitrary value of β and then run the model to see if it fits the data shown in the curve. We can keep doing this until we zero in on the value of β that gives the best fit to the cumulative number of deaths curve. If we select a value of β that is too large, the deaths will happen too fast; if we select a value of β that is too small, the deaths will happen too slowly. For the previous data on swine fever, Hone et al. (1992) obtained an estimate of β of 0.001 per day, which means that a single daily contact of an infected pig with a susceptible one has a probability of only 1 in 1000 of transmitting the infection.

immigrants may move into an area. The resulting disease model becomes more realistic and more complicated if there is influx. In some cases the host-disease system reaches an equilibrium at a density of the susceptible population at which the basic reproductive rate is 1.0 (Anderson and May 1991). In other cases the populations will oscillate around an equilibrium point, or there may be no steady state for the system.

How might we control a disease described by this simple model? If we vaccinate individuals or cull susceptibles from the population, how many must we treat or remove to eradicate the disease? If we make a fraction c of the susceptible population immune, then

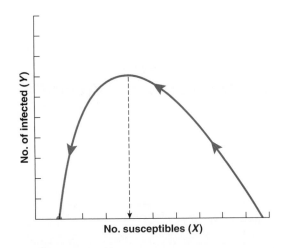

FIGURE 15.2
Trajectory for the simple epidemic described by Equations (15.1) through (15.3). Begin with a large number of susceptible individuals and no infected individuals. The number of infected individuals rises to a peak, indicated by the dotted line, when x = γ/β and then falls to zero at the equilibrium marked by the red dot. (Modified after Heesterbeek and Roberts 1995.)

a fraction $(1 - c)$ remains susceptible. Thus from Equation (15.4) we can calculate:

$$R_0 = \frac{(1 - c)\beta}{\gamma} \qquad (15.5)$$

which is equivalent to

$$c > 1 - \frac{\gamma}{\beta} = 1 - \frac{1}{R_0} \qquad (15.6)$$

Thus, for example, if R_0 is 4, then we would have to vaccinate or cull 75% of the population to control the disease.

Compartment Models with Variable Population Size

We can add more realism to this first compartment model by allowing the population size of the host to vary over time. Second, we can allow the contact rate to be a function of population size, so that disease transmission increases with population density. To keep the model simple, we assume a host-disease system in which there is no recovery from the disease, so that we construct an SI model (susceptibles and infecteds only). Infected must die in this model.

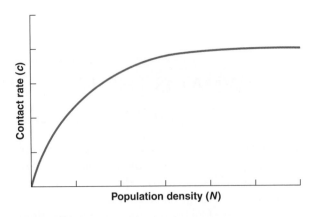

FIGURE 15.3
The expected relationship between contact rate and population density. At higher densities contact rates will increase, and disease transmission will be facilitated, but at some density contacts reach a maximum and do not increase. (Modified after Heesterbeek and Roberts 1995.)

Models of this type allow us to ask an important ecological question: Does the disease affect population size of the host? We modify the simple models described in Equations (15.1) and (15.2) to allow for changes in host numbers. For the susceptibles we have

$$\frac{dX}{dt} = bN - dX - \frac{c\beta XY}{N} \qquad (15.7)$$

where X = number of susceptibles

Y = number of infected individuals

N = total number of individuals = $X + Y$

b = instantaneous birth rate of the host (constant)

d = instantaneous death rate of the host in the absence of disease (constant)

c = contact rate, a function of population density N (Figure 15.3)

β = transmission rate per encounter

For the infected individuals we get

$$\frac{dY}{dt} = \frac{c\beta XY}{N} - (\alpha + d)Y \qquad (15.8)$$

where α = Increase in host mortality due to disease

and all other terms are as previously defined.

B O X   1 5 . 1

HOW CAN WE DETERMINE R_0?
A MATHEMATICAL EXCURSION

One of the critical parameters in simple disease models is the net reproductive rate of the disease agent, R_0. When the net reproductive rate is > 1, the disease will spread, and when it is < 1 the disease dies out. We can derive this parameter with a bit of algebra. Begin with Equation 15.2 for the simple model:

$$\frac{dY}{dt} = \frac{\beta XY}{N} - \gamma Y$$

where Y = Number of infected individuals
   X = Number of susceptible individuals
   N = Total number of individuals
   β = Transmission rate per encounter
   γ = Recovery rate per capita

By definition, at equilibrium the rate of change in the number of infected individuals is zero, so

$$\frac{dY}{dt} = 0 \quad \text{and consequently,}$$

$$0 = \frac{\beta XY}{N} - \gamma Y$$

We can divide all these terms by Y to obtain

$$0 = \frac{\beta X}{N} - \gamma \quad \text{and thus,}$$

$$\gamma = \frac{\beta X}{N}$$

By rearranging terms we can obtain

$$\frac{\gamma}{\beta} = \frac{X}{N} \quad \text{or taking reciprocals,}$$

$$\frac{\beta}{\gamma} = \frac{N}{X}$$

At equilibrium, N is carrying capacity K, and X, the number of susceptibles, is defined to be the threshold population density K_T. But at equilibrium the net reproductive rate is defined as

$$R_0 = \frac{K}{K_T} = \frac{\text{equilibrium population density}}{\text{threshold population density}}$$

Thus since X is equal to the threshold density K_T, and $N = K$, we can put these two relationships together to obtain Equation (15.4):

$$R_0 = \frac{\beta}{\gamma}$$

We can solve these equations at equilibrium

$$\left(\frac{dX}{dt} = 0 = \frac{dY}{dt} \right)$$

to get the following solutions (Heesterbeek and Roberts 1995):

$$Y^* = \left(\frac{b-d}{\alpha} \right) N^* \tag{15.9}$$

$$c_{N^*} = \frac{\alpha(\alpha + d)}{\beta(\alpha + d - b)} \tag{15.10}$$

where Y^* = number of infectives at equilibrium

   N^* = total population size at equilibrium

   c_{N^*} = contact rate at equilibrium population density

and all other terms are as previously defined.

This equilibrium will have a solution only if the contact rate (Figure 15.3) can possibly be as high as that given by Equation (15.10). If there is a solution to the equilibrium, the disease will reduce population size when

$$\alpha > b - d \tag{15.11}$$

that is, if disease-induced mortality is greater than the potential rate of population growth. Equation (15.8) can be used to investigate the possibility of eradicating a disease from a wild population by culling (or by vaccination). Assume a fixed per capita culling rate δ. Heesterbeek and Roberts (1995) showed that culling can eradicate a disease from a population if

$$\delta > \beta c - (\alpha + d) \tag{15.12}$$

where δ = culling rate per capita

   c = contact rate at the equilibrium population size after culling

   d = natural death rate at the equilibrium population size after culling

   β = transmission rate of the disease

   α = increase in host mortality due to disease

We will discuss on page 268 cases in which culling has been used to reduce wildlife diseases losses.

If infected animals can be identified in the field and killed, this action will effectively increase the

FIGURE 15.4
Clutch size of western fence lizards (Scleroporus occidentalis) *infected with lizard malaria* (Plasmodium mexicanum). *Arrows indicate the average clutch sizes for uninfected and infected females. (Data from Schall 1983.)*

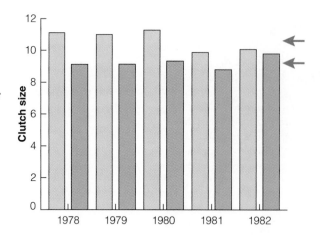

Effects of Disease on Individuals

Individual hosts are effectively islands for a disease agent, and from the viewpoint of the disease organism these islands or patches of habitat must be colonized for the disease to spread. We begin by looking at these individual hosts and ask how a disease agent might affect them as individuals.

Effects of disease and parasitism on individual organisms are relatively easy to study (Gulland 1995). We have relatively few data on wild animals and plants compared to the large amount of data on domestic animals and humans. One reason why few studies have been done in the wild is that parasites and diseases are thought to coevolve with their hosts such that they become relatively harmless (see page 276), and consequently one would not expect to find strong effects. But this may not be correct, and more studies are finding significant effects on reproduction, survival, and growth of infected organisms.

mortality rate caused by the disease (α), and reduce the amount of culling needed to achieve eradication.

These simple models can be elaborated to account for the specific details of particular diseases (Grenfell and Dobson 1995). There is a large literature on epidemiological models, and to explore it further consult Bailey (1975) or Busenberg and Cooke (1993). We next explore the effects diseases have on individuals and populations of animals and plants.

Effects on Reproduction

Because organisms have a limited amount of available energy, it is not surprising that parasite and disease infections can reduce reproductive output. A good illustration of these effects can be seen in lizards. Even though malaria parasites infect many vertebrates, including humans (four species of *Plasmodium*), a majority of the 125 malaria species attack lizards. In California, about 25% of western fence lizards (*Scleroporus occidentalis*) are infected with lizard malaria, and infected females have smaller clutch sizes than uninfected females (Figure 15.4). Clutches are about 20% smaller in malaria-infected lizards compared with uninfected individuals (Schall 1983). The cause of this reduction in reproductive effect is that individuals store less fat in a given summer and thus females have less energy available the following spring to lay eggs.

Bird chicks are often attacked by nest parasites, and if parasite infestation is severe, reproductive output can be reduced. Colonial nesting birds are particularly prone to nest ectoparasite attacks because of the ease of transmission. McKilligan (1987) studied cattle egrets nesting in heronries in Queensland and found heavy tick infestations. The argasid tick *Argas robertsi* attacks both adult birds and chicks, and many nestlings that die have heavy tick infestations (Figure 15.5). Chicks with ticks grow more slowly than those without ticks. Ticks in this colony were carrying an arbovirus, and the effects of ticks on egret chicks may arise either from direct blood loss or from subsequent arbovirus infections. McKilligan (1996) carried out two experiments comparing egret broods that were naturally tick-infested with an equal number of broods rendered

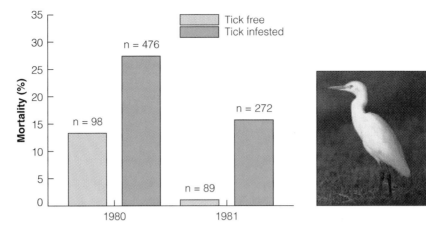

FIGURE 15.5
Mortality of cattle egret (Bubulcus ibis) *nestlings up to 11 days of age in Queensland during two breeding seasons. Numbers of chicks studied are given above each sample. (Data from McKilligan 1987.)*

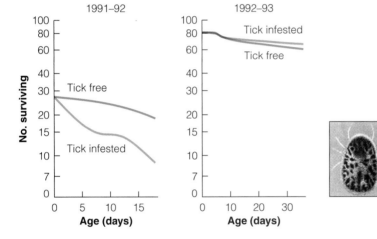

FIGURE 15.6
Survivorship curves of cattle egret chicks treated with an insecticide to remove ticks and of naturally infested chicks in Queensland. The two years of this experiment differed dramatically. Tick-infested chicks suffered high losses in 1991–1992 (n = 27). Ticks were relatively rare in 1992–1993, a dry year, and ticks had no effect on chick survival in the nest (n = 81). (Data from McKilligan 1996.)

tick-free by applying an insecticide to their nests. He did this for two years and got different results each year (Figure 15.6). In the 1991–1992 breeding season, tick infestations were severe, with an average of 24 ticks per chick per day, and tick-associated mortality was significant. In 1992–1993 the breeding season was dry and there were few ticks (a mean five per chick); very few egret chicks died from ectoparasites.

Effects on Mortality

No one doubts that diseases kill animals, and there are numerous cases in which veterinary examinations of dead animals suggest that a parasite or disease was the immediate cause of death. The population ecologist, however, needs to know more. What fraction of mortality is disease-caused? This is a more difficult ques-

tion to answer and, while we have much data of this type for humans, we have very little for natural populations of animals or plants. One example will illustrate the problems of obtaining good information.

In the spring of 1988, harbor seals in the North Sea began to die in large numbers. Dead seals were first noted in the central Baltic off Denmark, and mortality spread around the Baltic, to the Dutch coast, to Britain, and as far as Ireland by August 1988 (Figure 15.7). Harbor seals occur throughout the North Atlantic, and before the epizootic[2] approximately 50,000 harbor seals lived in European waters (Swinton et al. 1998). An estimated 60% of the total seal population in the Baltic Sea died from this epizootic, and the deaths occurred very

---

[2]An *epizootic* is a disease epidemic among wild animals.

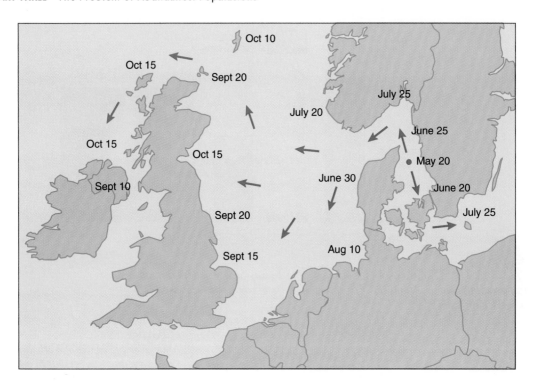

FIGURE 15.7
*Map of the spread of the phocine distemper virus epizootic among the harbor seal
populations of northern Europe in the summer of 1988. The epizootic began in a seal
colony in the central Baltic in April. This outbreak, the first well-documented epizootic
among free-ranging marine animals, had a very rapid spread and a high rate of
mortality. (Data from Swinton et al. 1998.)*

rapidly (Figure 15.8). During the outbreak the exact
cause of death was not clear, but a viral disease was suspected because the dying seals had symptoms that
resembled those of canine distemper.

One characteristic of the disease is that it caused
pregnant female seals to abort. Osterhaus and Vedder (1988) identified the infective agent as a morbil-
livirus similar to canine distemper virus, and they
named it *phocine distemper virus*. Within two weeks
of infection seals developed the symptoms and typically died of pneumonia with secondary bacterial and
viral infections.

The incidence of infection for this seal epizootic
could not be measured directly, but Heide-Jørgensen
and Härkönen (1992) estimated that 95% of the harbor seals were infected with the virus. Deaths from
phocine distemper virus seemed to be more common
in males than females, although both were infected.
There was no indication that the epizootic was
affected by the number of seals in each colony, and the
main predictor of the spread was distance between

colonies. Harbor seal colonies in northern Norway
and Iceland escaped the epizootic, presumably because
no infected seals dispersed to these distant colonies.

The key question from the harbor seal's viewpoint is whether or not this viral disease could persist
in the population. Infected individuals that recover
are immune for life, but since births occur each year
there is a continual source of susceptibles in the population. In the Baltic, seal pups are about 20–22% of
the total population in any given year. Swinton et al.
(1998) used these estimates to construct a compartment model of the 1988 epizootic. Seal colonies are
discrete population patches, and the transmission rate
(β) between individuals seems to be constant at 0.005
per day. The net reproductive rate (R_0) for this viral
disease is approximately 2.8. The critical variable
needed for the types of simple compartment models
previously discussed (one-population models) is the
rate of spread of the virus from seal colony to seal
colony. Dispersal of infected seals between colonies
must have been frequent to enable the rapid spread

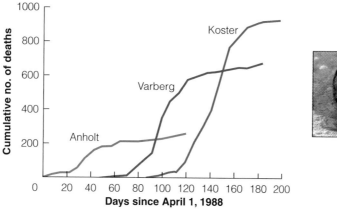

FIGURE 15.8
Cumulative number of harbor seal death recorded in the central Baltic Sea in the summer of 1998. The epizootic of phocine distemper virus started at the small Anholt seal colony in April, spread to the larger Varberg colony in mid-May, and reached the Koster colony in mid-June. On average an estimated 60% of the seals were killed at each site. (Data from Heide-Jørgensen and Härkönen 1992.)

shown in Figure 15.7. Given these estimates for a model, Swinton et al. (1998) showed that phocine distemper virus could not be maintained in harbor seal populations as a persistent infection. This conclusion relates to the origin of the disease in the first place. Phocine distemper virus is found in both grey seals and harp seals in the Atlantic and seems to be a relatively innocuous disease in harp seals (Harwood 1989). Harp seals are northern seals and are normally rare in southern waters. In 1987 and 1988 harp seals moved in large numbers from northern Norway and Spitzbergen south into the North Sea. The phocine distemper virus may have crossed species boundaries at this time to set off the 1988 epizootic among the more susceptible harbor seal population.

In spite of all the harbor seal deaths in 1988, the seal populations of western Europe were only temporarily affected and quickly recovered to their former numbers. The seal epizootic of 1988 raises the general question of how often a disease can exert a long-term effect on a population, a question we turn to next.

Effects of Disease on Populations

Few studies of plant or animal diseases have included a closely monitored population in which each individual's history is known. Most often the available data

are estimates of seroprevalence[3] from individuals of known age or size. Consequently the effect of a disease on a particular population is often not well known. Most disease studies have concentrated on effects on the humans or on agriculture, and there is a need to bring ecologists and epidemiologists together to measure population effects (Mills 1999). The following three examples illustrate the range of problems faced in trying to measure the effects of disease on populations.

Brucellosis in Ungulates

Brucellosis is a highly contagious disease of ungulates caused by a bacterium *(Brucella abortus)*. Prevalent in cattle throughout the world, it manifests itself in females by abortion, so its common name is "contagious abortion." Much effort has been expended by the livestock industry to eradicate brucellosis in cattle, but the possible transmission of infection from wild ungulates to cattle has caused much controversy in the western United States, where brucellosis is endemic in bison and elk (Aguirre and Starkey 1994). Figure 15.9 illustrates the age pattern of seroprevalence to brucellosis of bison in Yellowstone National Park. Seroprevalence increases with age in bison, so

---

[3]Seroprevalence is the percentage of individuals in the host population with antibodies to a particular disease agent. It measures how widespread a disease has been in a population.

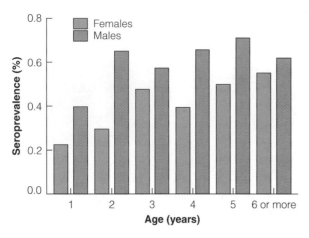

FIGURE 15.9
Seroprevalence to brucellosis for female and male bison in Yellowstone National Park in the winter of 1990–1991. Seroprevalence in males increases to two years of age, while seroprevalence in females appears to increase to age 5–6. (Data from Pac and Frey 1991.)

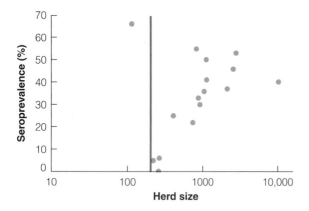

FIGURE 15.10
The relationship between population size and seroprevalence to brucellosis in bison in six national parks in Canada and the western United States. A threshold density (blue line) of about 200 bison is needed to maintain this disease in the population. (After Dobson and Meagher 1996.)

that about 60% of older adults have antibodies to the *Brucella* bacterium. There is considerable controversy over whether or not brucellosis is a native disease of bison or whether it was introduced into North America by cattle (Meagher and Meyer 1994). Most probably it was not present in bison before 1917 and was contracted from domestic cattle.

A simple model of the interaction between brucellosis and bison in Yellowstone National Park was constructed by Dobson and Meagher (1996) to deter-

mine whether brucellosis could be eliminated by a culling program. Brucellosis has a sharply defined threshold for establishment (Figure 15.10), and the proportion of bison infected rises smoothly with population density. These data illustrate one of the important principles of epidemiology: the critical threshold. Most diseases have a threshold host population density that is needed for the continued presence of the pathogen. In this case, brucellosis will persist in bison populations of 200 or larger, a low number. Bison in Yellowstone now number about 4000 animals. Whereas, it is possible to cull bison down to this low density, this action is unacceptable because it would put them in danger of extinction (and would be politically unacceptable to a variety of people). So it is unlikely that culling will be a viable strategy for eliminating brucellosis in bison in Yellowstone National Park (Dobson and Meagher 1996). Note that brucellosis could infect bison populations in very small herds, but once it passed through a small population it would fail to maintain itself and would die out.

Rabies in Wildlife

Rabies is one of the oldest known diseases, and one of the most terrifying diseases for humans. Democritus around 500 B.C. recorded a description of rabies, and Aristotle 200 years later wrote about rabies in his *Natural History of Animals*. Rabies is a directly transmitted viral infection of the central nervous system, and all mammals are susceptible. The

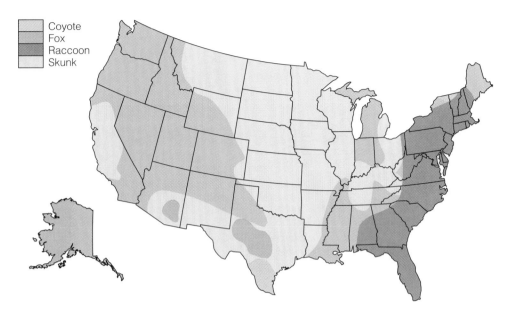

Coyote
Fox
Raccoon
Skunk

FIGURE 15.11

Map of the main rabies reservoirs in different regions of the United States. In each of these areas other mammals are also involved, but are not the major host. The fox is the red fox (Vulpes vulpes), *and the skunk is the striped skunk* (Mephitis mephitis). *The geographical range of these species is much wider than the areas in which rabies is a major problem. (From Krebs et al. 1998; figure courtesy of the Center for Disease Control and Prevention, Atlanta.)*

disease is particularly common in foxes, wolves, coyotes, skunks, raccoons, jackals, and bats, but domestic dogs most frequently transmit it to humans. Rabies virus, present in saliva, is transmitted directly by the bite of an infected animal. A few cases of aerosol transmission from bats in caves have been reported (Krebs et al. 1995). Once rabies is contracted, death is inevitable: There is no cure. Worldwide the incidence of rabies in humans is low; and about 15,000 people a year are victims, mostly in India and the Far East. The incubation period in humans is highly variable, ranging from less than ten days to more than six years. Malaria and tuberculosis are much more significant causes of human deaths globally, but no disease is as feared as rabies.

Rabies is caused by a number of different viruses belonging to the *Lyssavirus* genus in the Family Rhabdoviridae. Carnivorous mammals are the essential hosts for the virus. In Europe the red fox is the main reservoir for rabies (Anderson et al. 1981); in North America raccoons, skunks, foxes, and bats are the main reservoirs, and in 1997 wild animals represented 93% of the reported cases. The main vectors of rabies differ in different regions of the United States (Fig-

ure 15.11). These vectors carry a diverse set of rabies virus genotypes. The raccoon is a keystone host of rabies in the southeastern United States and a majority of the recorded cases in wild animals in the United States are now from raccoons (Figure 15.12). The striped skunk currently represents about 25% of rabies reports, although it was more important as a host in the 1970s. The reason for these host shifts in rabies incidence is completely unknown.

An epizootic of rabies in eastern North America began around 1970 in Virginia and has been spreading for 30 years (Figure 15.13). This epizootic probably began from diseased raccoons brought into the area by humans. Rabies in raccoons has since spread north to Ontario, crossing the border in 1998, and has also spread south to meet another epizootic moving north from Florida. Because raccoons are so common, particularly around human habitations, rabies in raccoons has been particularly targeted by control agencies in the United States and Canada in recent years.

A recent attempt has been made to reduce rabies in raccoons by vaccination of wild raccoons, using a recombinant virus vaccine approved in April 1997. A

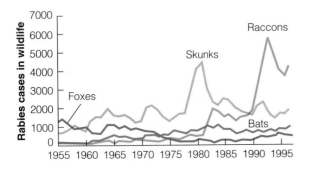

FIGURE 15.12
Number of rabies cases in wildlife in the United States reported to the Center for Disease Control and Prevention from 1955–1997. The recent rise in rabies cases in raccoons has resulted from an epizootic of rabies moving up the East Coast during the past 20 years. (Data from the Center for Disease Control and Prevention, Atlanta.)

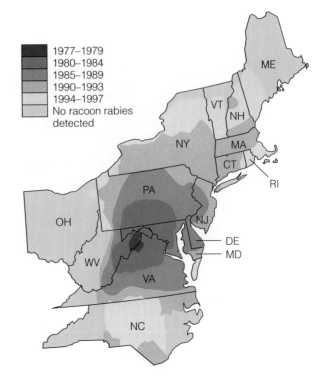

FIGURE 15.13
Spread of rabies epizootic in raccoons in the eastern United States since 1977. The epizootic began in Virginia and moved as far north as southern Canada by 1998. It has moved south as well and has met another rabies epizootic in raccoons spreading from Florida. The outbreak in Virginia was probably caused by human translocation of infected raccoons from the southeastern states during the 1970s. (From Center for Disease Control and Prevention, courtesy John W. Krebs.)

raccoon bait, a small cube of fish oil and wax polymer, contains the oral rabies vaccine. Millions of baits are distributed annually to immunize susceptible raccoons and foxes. In addition, raccoons can be easily live trapped, injected with the vaccine, and released. This vaccination program has been used in Massachusetts, New York, New Jersey, Vermont, and Ontario, but its effect on the incidence of rabies in raccoons is not yet clear.

In many parts of the world, rabies reaches humans through domestic dogs, but in North America and Europe vaccination of dogs has cut this link to humans. During the past eight years 26 people have died in the United States from rabies, and 90% of the confirmed cases have been caused by rabies virus variants carried by bats (Krebs et al. 1998). Little is known about either the incidence of rabies in bats or the impact of rabies on bat populations.

A major epidemic of rabies began in Poland in 1939 and gradually moved 1400 km westward at a rate of 20–60 km per year (Figure 15.14). The epidemic reached the Atlantic coast in northern France in the late 1980s and stopped. The main carrier has been the red fox, with over 70% of the reported cases in Europe (Anderson et al. 1981). One feature of rabies in Europe has been that it appears to be cyclic. Figure 15.15 illustrates a four-to-five-year cycle in rabies cases in red foxes in one area of eastern France. After extensive culling programs failed to stop rabies or reduce its incidence, most European countries began to use oral vaccination of foxes in baits to stop the spread of the disease. Vaccination via baits has proven highly successful in Europe. By 1999 rabies was much reduced in western Europe, and Switzerland had reached the status of rabies free as a result of this extensive vaccination program.

Figure 15.16 gives a simple compartment model for rabies. Many attempts have been made to model a rabies outbreak (Barlow 1995). Anderson et al. (1981) presented a simple model of rabies that captures much of the ecology of this disease (Box 15.2). From this model we can ask a critical management question: Can we eliminate rabies from the fox population by culling or by vaccination? Attempts to control the spread of rabies in Europe and in North America by culling have been unsuccessful despite heroic efforts. Foxes have high reproductive rates and high dispersal rates, and these two parameters combine to make culling attempts unsuccessful at controlling the disease unless the foxes are in poor habitat or the rate of culling is extremely high.

FIGURE 15.14
Rabies epizootics in Europe from 1939 to 1980. Dates with arrows indicate the year the epizootic crossed the then-existing national borders. Since 1983 the movement of rabies west has been reduced, probably due to oral vaccination of foxes using baits. (From Macdonald and Voigt 1985.)

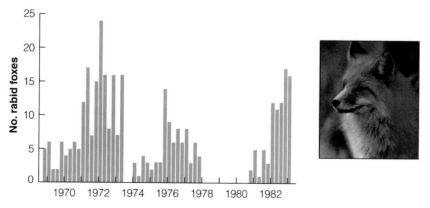

FIGURE 15.15
Number of rabies cases reported in red foxes in the Ardennes region of eastern France. Rabies spreading west entered the area in 1968. The cyclic nature of the disease is clearly evident in this local area. Since 1985 rabies has decreased rapidly in this area due to vaccination of foxes using baits. (Data from World Health Organization, Tubingen.)

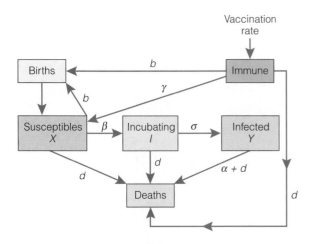

FIGURE 15.16
A compartment model for rabies. All infected animals die, so there is no recovery compartment. Because animals can be vaccinated artificially as a control measure, a vaccination rate compartment is added. (Modified from Bacon 1985.)

Vaccination directly reduces the size of the susceptible pool and is much more effective in control of rabies. Figure 15.17 shows that the proportion of foxes that would need to be vaccinated varies with the density of the fox population. If foxes are at a density of 2 per km², the model predicts that vaccinating about 50% of the foxes would break the transmission cycle and eradicate the disease. Extensive programs of vaccination of wild foxes using baits have been carried out in Switzerland (since 1978), Austria, Hungary, France, Belgium, and Germany (Pastoret and Brochier 1999). These vaccination programs have been successful in eliminating rabies from wildlife reservoirs in large areas and thus in reducing the health risk.

At present we have no data at all on the effects of rabies on mammal host populations. Most of the effort has been directed at the public health aspects of this disease, and on preventative measures to reduce damage to humans and domestic animals. The most critical issues involve the assumption that for mammalian hosts the transmission rate (β) of rabies is a constant at all host densities, and that the threshold for persistence of the disease is also a constant and thus identical in both good and poor host habitats (Barlow 1995).

Myxomatosis in the European Rabbit

The European rabbit (*Oryctolagus cuniculus*) was introduced into Australia in 1859 and increased to very high densities within 20 years. After World War II, an attempt was made to reduce rabbit numbers by releasing a viral disease called myxomatosis. Myxomatosis originated in the South American jungle rabbit *Sylvilagus brasiliensis*. In its original host, myxomatosis is a mild disease that rarely kills its host. The disease agent is the myxoma virus, a pox virus of the genus *Leporipoxvirus*. Transmission of myxomatosis occurs via biting arthropod vectors, principally mosquitoes and fleas. Transmission is passive, and the virus does not replicate in the vector.

Myxomatosis was highly lethal to European rabbits when it was introduced into Australia in 1950, killing over 99% of individuals infected. Figure 15.18 shows the precipitous crash in rabbit numbers that followed the introduction of myxomatosis in one area in southeastern Australia in 1951. Myxomatosis was also introduced to France in 1952, from where it spread throughout western Europe, reaching Britain in 1953. In Britain 99% of the entire nation's rabbit population was killed in the first epizootics from 1953 to 1955 (Ross and Tittensor 1986).

Very soon after its introduction, weaker myxoma virus strains were detected in England and in Australia (Fenner and Chapple 1965, Fenner and Ratcliffe 1965). Frank Fenner, working at the Australian National University, was instrumental in studying the changes that have occurred in the myxoma virus in Australia. Since myxomatosis was introduced into Britain and Australia, evolution has been going on in both the virus and the rabbit. The virus has become attenuated such that it kills fewer and fewer rabbits and takes longer to cause death. Because mosquitoes

Frank Fenner *(1914–) Professor of Microbiology, Australian National University, Canberra*

B O X   1 5 . 2

A SIMPLE RABIES MODEL

Anderson et al. (1981) have presented the following model as a representation of rabies in red foxes in Europe. Figure 15.16 shows the compartment model visually. The model contains susceptibles (X), incubating (I) and infected (Y) foxes, and it assumes logistic growth for the fox population without rabies. Transmission for rabies is assumed to be proportional to the product of the number of susceptible (X) foxes and the number of infected (Y) foxes. The equations for the model are as follows:

$$\frac{dX}{dt} = rX - \gamma XN - \beta XY \qquad (15.6)$$

$$\frac{dI}{dt} = \beta XY - (\sigma + d + \gamma N)I \qquad (15.7)$$

$$\frac{dY}{dt} = \sigma I - (\alpha + d + \gamma N)Y \qquad (15.8)$$

where X = number of susceptible foxes
I = number of incubating foxes
Y = number of infected foxes
N = total number of foxes = $X + I + Y$
t = transmission rate per encounter
r = population growth rate per capita in absence of disease = $b - d$
d = death rate of foxes per capita in absence of rabies (life expectancy = $1/d$)
b = birth rate of foxes per capita in absence of rabies
γ = r/K where K is the fox carrying capacity
σ = rate of incubation (incubation period = $1/\sigma$)
α = death rate of rabid foxes (life expectancy of rabid foxes = $1/\alpha$)

Anderson et al. (1981) estimated these parameters for the red fox in Europe to be as follows:

Parameter	Definition	Estimated value
b	birth rate per capita	1 per year
d	death rate per capita	0.5 per year
r	population growth rate = b-d	0.5 per year
σ	rate of incubation	13 per year
α	death rate of rabid foxes	73 per year
β	transmission coefficient	80 km^2 per year
K	fox carrying capacity	1 to 4 per km^2

This model with these parameters produces cycles in fox numbers with a three-to-five-year period, in agreement with the data shown in Figure 15.15. From this model the basic reproductive rate R_0 is given by

$$R_0 = \frac{\sigma \beta K}{(\sigma + b)(\alpha + b)} \qquad (15.9)$$

When R_0 is less than 1, rabies will die out in the fox population. The threshold density at which rabies will be maintained in the population in this model is estimated to be around 1 fox per km^2.

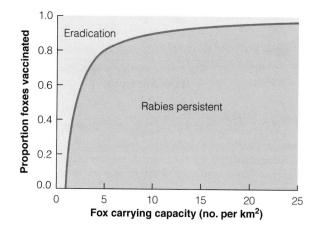

FIGURE 15.17
Proportion of red foxes that would need to be vaccinated to eliminate rabies in Europe in relation to the carrying capacity of the habitat. The simple model in Box 15.2 predicts that if fox carrying capacity is relatively low, only a small proportion of the foxes would need to be vaccinated to eradicate the disease from Europe. (Modified from Anderson et al. 1981.)

TABLE 15.1 **Virulence of field myxoma virus in laboratory rabbits in Australia, Great Britain, and France after the introduction of myxomatosis to these three countries between 1949 and 1951.**

Values in the table are the percentages of virus samples collected in the field that were classified as each virulence type. These studies measure changes in the virulence of the virus to a standardized host, the laboratory rabbit. Viruses collected in all three countries in three different time periods show the rapid change brought about by selection for less-virulent virus strains.

	Virulence Type—Grade					
	I	II	IIIA	IIIB	IV	V
Mean survival of rabbits (days)	<13	14–16	17–22	23–28	29–50	—
Mortality rate (%)	>99	95–99	90–95	70–90	50–70	<50
Australia						
1950–1951	100	—	—	—	—	—
1958–1959	0	25.0	29	27	14	5
1963–1964	0	0.3	26	33	31	9
Great Britain						
1953	100	—	—	—	—	—
1968–1970	3	15	48	23	10	1
1971–1973	0	3	37	57	3	0
France						
1953	100	—	—	—	—	—
1962	11	19	35	21	13	1
1968	2	4	14	21	59	4

Source: After Fenner and Myers (1978) and Anderson and May (1982).

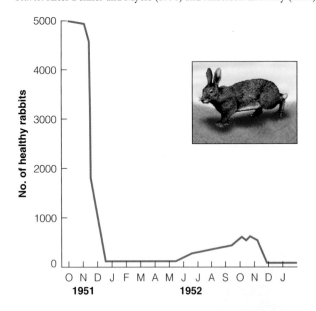

FIGURE 15.18
Population crash of the European rabbit (Oryctolagus cuniculus) *at Lake Urana, New South Wales, after the viral disease myxomatosis was introduced in 1951. Numbers of healthy rabbits were counted on standardized transects. (After Myers et al. 1954.)*

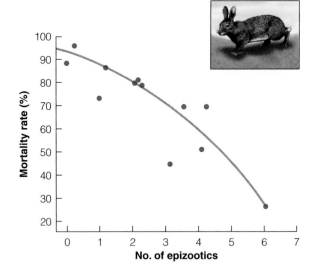

FIGURE 15.19
Mortality rates of wild European rabbits from the Lake Urana region of southeastern Australia after exposure to several epizootics of myxomatosis. Mortality was measured after a challenge infection with a strain of myxoma virus of grade III virulence. (After Fenner and Myers 1978.)

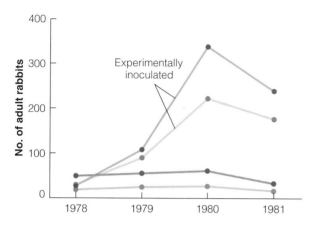

FIGURE 15.20
Numbers of adult rabbits counted in March on four fenced areas in southeastern Australia. Rabbits on two areas (blue, red) were inoculated experimentally with an attenuated myxomatosis strain that produced immunity to the more virulent natural strains. The two other populations were inoculated with a virulent strain of the virus. Rabbits on the protected areas increased eightfold and 12 fold over pretreatment levels. (Data from Parer et al. 1985.)

are a major vector of the disease, the time period between exposure and death is critical to viral spread. Table 15.1 summarizes changes that have occurred in the virus. These data were obtained by testing standard laboratory rabbits against the virus, so they measure viral changes while holding rabbit susceptibility constant. Since 1951 less-virulent grades of virus have replaced more virulent grades in field populations.

Rabbits have also become more resistant to the virus (Figure 15.19). By challenging wild rabbits with a constant laboratory virus source, we can detect that natural selection has produced a growing resistance of rabbits to this introduced disease.

What is the net effect of these changes in the virus and in the genetic resistance of the rabbits to the population dynamics of the host? Because over time myxomatosis has caused less and less mortality, it is tempting to assume that the disease was having little effect on rabbit numbers. One way to test the idea that myxomatosis was no longer effective in rabbit control is to compare rabbit populations with and without exposure to myxomatosis. This is difficult to do technically because it is impossible to find a field population of rabbits that does not already have myxomatosis. The only method possible is to reduce the effect of myxomatosis by making rabbits immune or by cutting the transmission by vectors. Two such attempts have been made. In Australia, Parer et al. (1985) compared four fenced populations of rabbits, two inoculated with an attenuated strain of the virus (to produce immunity with little mortality) and two inoculated with a virulent strain of the virus. Figure 15.20 illustrates the effects of this experiment on the numbers of rabbits. Populations protected from myxomatosis-caused mortality increased eight-fold and 12-fold over control levels. A

similar experiment in England reduced rabbit fleas (the main vector) with insecticides, and produced a twofold to threefold increase in rabbit numbers (Trout et al. 1992). These results show clearly that myxomatosis is still suppressing rabbit populations, in spite of its reduced virulence in field populations.

Myxomatosis is a good example of the strong effect that a disease can have on a wild population. The fact that this disease was transferred between species by humans, raises the broad question of how disease organisms and their hosts coevolve in evolutionary time, a critical question to which we now turn.

Evolution of Host-Parasite Systems

One of the striking features of the simple models of host-parasite interactions is that these models are often unstable. Oscillations are common in many host-parasite models, as they are in predator-prey models: diseases may explode or go to extinction in simple models. But even though in real disease systems some diseases disappear, most seem to persist. One way in which we can explain the stability of real host-parasite systems is to postulate that natural selection has changed the characteristics of both hosts and disease organisms so that their interactions produce population stability. In particular, the conventional wisdom about host-parasite evolution is that virulence is selected against, so that diseases and parasites become less harmful to their hosts and thus persist. Thus the well-adapted parasite is a benign parasite (Ewald 1995). If this traditional view of peaceful coexistence is correct, we would expect to see diseases and parasites becoming less harmful over evolutionary time. But does natural selection work that way with host-parasite systems? What can we say about the evolution of virulence?

Natural selection does not necessarily favor peaceful coexistence of hosts and parasites. To maximize fitness a parasite or a disease agent should optimize the trade-off between virulence and other fitness components such as transmissibility. If the host did not evolve, the parasite should be able to reach this optimal balance of host exploitation. But hosts do evolve, and this produces an arms race between the host and the parasite. If hosts are genetically variable, the parasite or disease agent will be on average less virulent than if the hosts are uniform (Ebert 1999). The evolutionary time scales of the host organism and the disease agent are typically greatly different. Hosts evolve slowly, bacteria and viruses evolve quickly.

One way to study the evolution of host-parasite systems involves serial passage experiments in the laboratory (Ebert 1998). In serial passage experiments, disease organisms or parasites are transferred from one host to another, holding host properties constant so that the evolutionary changes in the disease organisms can be monitored. Because the disease organisms are propagated under defined laboratory conditions, their biological attributes can be compared with those of the ancestral organism at the outset of the experiment. Although serial passage was developed for vaccine studies, it can be used very effectively in studies of the evolution of virulence. Figure 15.21 shows a serial passage experiment in laboratory mice with the mouse typhoid bacterium *Salmonella typhimurium*.

Diseases become more virulent with passage in artificial serial passage experiments in their native host species (Ebert 1998), and this appears to be a general result with many different viral, bacterial, fungal, and protozoan disease agents. One explanation of this is the *Red Queen Hypothesis*, which states that genetic variation is beneficial because it hinders parasite and disease adaptation. In laboratory serial passage experiments the host is often clonal or of limited genetic variability. What is clear is that the increase in virulence of disease agents observed in serial passage experiments in the laboratory does not occur in most natural disease systems, and host genetic variability is

E S S A Y   1 5 . 2

WHAT IS THE RED QUEEN HYPOTHESIS?

In Lewis Carroll's *Alice in Wonderland* there is a scene in which Alice and the Red Queen must run as fast as they can to get nowhere because the world is running by at the same speed. Van Valen (1973) used this metaphor to illuminate biotic evolution. Any evolutionary adjustment that a particular species makes can be countermanded by natural selection acting on all other species in the community. For example, if a prey evolves to run faster to escape its predators, the predators can also evolve to run faster to catch the prey. Thus disease-host systems, plant-herbivore systems, and predator-prey systems may show consistent evolutionary change, not to increase adaptedness but simply to maintain it. The species run, run, run but get nowhere. Increasing fitness in one species is always balanced by decreasing fitness in all other species.

Rates of evolution can be much faster in disease agents and parasites that have short generation times relative to their hosts. The Red Queen Hypothesis predicts a continuing evolutionary battle between hosts and parasites, with the important implication that because parasites evolve faster, the main selection pressures will come from the most common host genotypes. By changing genotypes over time, the host can present a moving target that the parasite or disease cannot catch. This is one possible reason for the evolution of sex, in which recombination at each generation presents a new array of host genotypes to the coevolving array of diseases and parasites. The Red Queen Hypothesis thus predicts continually changing evolutionary dynamics between parasites and hosts, not a stable equilibrium.

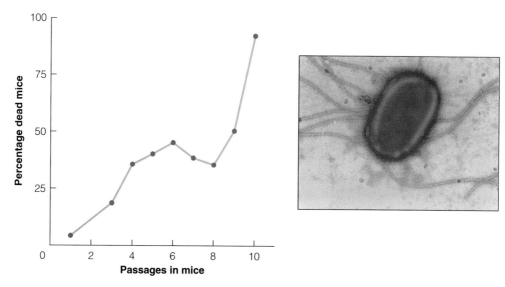

FIGURE 15.21

Change in virulence of the mouse typhoid bacterium Salmonella typhimurium *after serial passage in laboratory mice. A constant source of laboratory mice was used in these studies so that no evolution of the host could occur. Virulence increase rapidly over time as the* Salmonella *adapted to its host. (Data from Ebert 1998.)*

believed to be the principal reason that such runaway evolution does not often happen in nature.

The coevolution of rabbits and myxomatosis is one of the best empirical studies of host-parasite interactions in natural populations. The evolution of resistance to the myxoma virus in rabbits is easily explained by selection operating at the individual level—rabbits that are more resistant leave more offspring. It is more difficult to explain the evolution of reduced virulence in the virus. Virulence in a virus is related to fitness because more virulent viruses make more copies of themselves. But if more virulent viruses kill rabbits more quickly, less time will be available for transmission of the virus through mosquitoes or fleas. The result for the myxoma virus is group selection that operates to reduce virulence to a moderate level (Levin and Pimentel 1981). The basic reproductive rate (R_0) of the virus is highest at intermediate virulence. Group selection occurred because less-virulent viral strains are favored over more virulent viral strains because they take longer to kill the host rabbit (see Table 15.1). Host-parasite systems may be ideal candidates for group selection along these lines.

We do not know if the rabbit-myxoma system has reached a stable equilibrium, or whether continuing evolution will allow the rabbit population to slowly recover to its former levels. There is some evidence that the rabbit-myxomatosis interaction in Britain is changing, and the population size of rabbits in Britain seems to be slowly increasing (Trout et al. 1992). Evolutionary changes in the myxoma-rabbit system in Australia have been complicated by the accidental release in 1997 of a second viral disease, rabbit hemmorhagic disease, that has further reduced the rabbit's average density (Saunders et al. 1999).

Summary

Disease, one of the four major interactions between species, is an interaction between organisms in which the host loses and the parasite gains. Disease has been one of the major preoccupations of humans throughout history, and much of our understanding of disease dynamics has its roots in epidemiologists and medical scientists' efforts to understand the dynamics of human diseases such as malaria.

Mathematical models of host-parasite systems utilize compartment models to represent the interactions. The host population is usually broken down into susceptible, infected, and recovered individuals, and can be considered to be either constant (as in many human disease models) or variable in size (with birth and death rates). These simple models are characterized by a few parameters that define the outcome of the interaction. The most critical parameter is the *basic reproductive rate* (R_0) of the disease organism, the number of new infections produced by the average infected individual over its lifespan. If the reproductive rate is 1 or more, the disease will propagate, and if it falls below 1 the disease disappears.

Simple models of host-parasite systems all show a threshold density below which the disease or parasite will die out. The objective of much of the study of applied disease ecology is to determine how best to move the host population below threshold density. In general, culling of animals has not been very successful in achieving eradication or even control of diseases of wild animals, and vaccination may be a better general strategy for practical control.

Diseases and parasites can affect the reproductive rate or the mortality rate of their hosts. Even though many studies show effects on mortality, few either show whether a disease or parasite can reduce the average density of the host species or measure how large these effects are in nature. The chapter examined rabies to illustrate these concepts, and while we know much about the transmission of rabies to humans, we know little about its effects on the foxes, skunks, coyotes, and bats that are the main carriers. The best studies of disease in nature have been done on myxomatosis, a viral disease introduced into Australia and Europe to control European rabbits.

Our view of the evolution of virulence has progressed from the conventional wisdom, that well-adapted parasites and diseases are benign, to a more dynamic view in which diseases and their hosts are locked in an arms race, with each group evolving to maximize its fitness. Virulence will increase in evolutionary time if the parasite or disease organism can increase its fitness by harming the host more and producing more copies of itself. One of the main factors limiting disease virulence is host genetic variability, and monocultures of crops or clonal populations are particularly susceptible to virulent disease outbreaks.

Key Concepts

1. Disease is one of the major causes of debilitation and death of animals and plants, and the interactions between parasites and disease agents and their hosts are important for individuals and populations.

2. Simple host-parasite models can predict extinction, stability, or host-parasite cycles. Stable interactions of host and parasite are rather rare in most disease models.

3. Diseases and parasites can affect reproductive output or mortality rates, but in few cases do we understand the effects of disease on the host population.

4. Parasites and diseases do not necessarily coevolve to become more benign, but instead face an arms race in which each is attempting to maximize fitness in evolutionary time.

5. While human disease has been a major preoccupation of medical scientists, we know much less about the role of disease in ecological systems. Diseases introduced to new hosts have caused major effects on population dynamics.

Selected References

Anderson, R. M. 1991. Populations and infectious diseases: Ecology or epidemiology? *Journal of Animal Ecology* 60:1–50.

Anderson, R. M. and R. M. May. 1978. Regulation and stability of host-parasite population interactions. *Journal of Animal Ecology* 47:219–247.

Barlow, N. D. 1995. Critical evaluation of wildlife disease models. Pages 230–259 in B. T. Grenfell and A. P. Dobson, eds. *Ecology of Infectious Diseases in Natural Populations.* Cambridge University Press, Cambridge, England.

Bowers, R. G., M. Begon, and D. E. Hodgkinson. 1993. Host-pathogen population cycles in forest insects? Lessons from simple models reconsidered. *Oikos* 67:529–538.

Dobson, A. 1995. The ecology and epidemiology of rinderpest virus in Serengeti and Ngorongoro conservation area.

Pages 485–505 in A. R. E. Sinclair and P. Arcese, eds. *Serengeti II: Dynamics, Management, and Conservation of an Ecosystem.* University of Chicago Press, Chicago.

Dobson, A. and M. Meagher. 1996. The population dynamics of brucellosis in Yellowstone National Park. *Ecology* 77:1026–1036.

Ebert, D. 1998. Experimental evolution of parasites. Science 282:1432–1435.

Ewald, P. W. 1995. The evolution of virulence: a unifying link between parasitology and ecology. *Journal of Parasitology* 81:659–669.

Grenfell, B. T. and A. P. Dobson, eds. 1995. *Ecology of Infectious Diseases in Natural Populations.* Cambridge University Press, Cambridge, England.

Krebs, J. W., M. L. Wilson, and J. E. Childs. 1995. Rabies—epidemiology, prevention, and future research. *Journal of Mammalogy* 76: 681–694.

Laurance, W. F., K. R. McDonald, and R. Speare. 1996. Epidemic disease and the catastrophic decline of Australian rain forest frogs. *Conservation Biology* 10: 406–413.

Ostfeld, R. S. 1997. The ecology of Lyme-disease risk. *American Scientist* 85:338–346.

Questions and Problems

15.1 Serial passage experiments in which a disease organism is grown in a new host (typically in cell culture) become attenuated (less virulent) when put back in their original host, and this attenuation increases with each serial passage (Ebert 1998). Discuss what this finding tells us about the evolution of virulence in host-parasite systems.

15.2 Calculate the population changes from Equations (15.1) to (15.3) in a hypothetical host-parasite system. The parameters for the interaction are: β = 0.025 (transmission rate) and γ = 0.01 (recovery rate). Start the population with 500 susceptibles and 5 infecteds, and investigate how the dynamics would change if β increased to 0.040 or 0.060.

15.3 About 20 million waterfowl die each year in North America from avian cholera, which is caused by the bacterium *Pasteurella multocida* (Botzler 1991). Over 100 species have been known to be infected. Epizootics are typically explosive and involve hundreds and sometimes thousands of birds. There is high variation from year to year in the incidence of this disease. Plan a research program to determine the effects of avian cholera on a species of duck. What are the key questions you need to answer to be able to control this disease?

15.4 Myxomatosis has been introduced into European rabbit populations on islands in order to eradicate the rabbits for conservation purposes. Discuss what factors you would predict would affect the success of this kind of control program. Flux (1993) evaluates the success of these attempts.

15.5 Simple models of host-parasite systems do not have any spatial component. What advantages might be gained by constructing a spatial model of disease? Rabies is an example of a disease with interesting spatial spread patterns (see Figures 15.13 and 15.14). Foxes defend discrete, nonoverlapping territories. How might territorial behavior affect the spatial dynamics of rabies spread in foxes?

15.6 How does the biology of plant–plant pathogen systems differ from animal host–parasite systems? Can the simple models used in this chapter be applied to plant diseases? Swinton and Anderson (1995) discuss this question.

15.7 Barlow (1995) showed that the vaccination rate required to eliminate a disease will always be greater than the culling rate required for elimination, given the standard SIR host-parasite model. If this is correct, why might we still prefer vaccination as a strategy for disease control in wild animals?

15.8 One of the controversies in disease ecology is whether the parasitic nematode *Trichostrongylus tenuis* has a strong effect on red grouse populations in Scotland and northern England. Review this controversy and suggest experiments that could resolve the different points of view. Lawton (1990), Moss et al. (1993), and Hudson et al. (1992) discuss the differing points of view.

15.9 Anderson and May (1981) suggested that fluctuations in forest insect populations could be explained as host-parasite interactions, because simple disease models could generate population cycles or outbreaks of the host insect species. Review the subsequent history of this suggestion from the papers by Vezina and Peterman (1985), Myers (1993), and Bowers et al. (1993).

15.10 Anthrax, a bacterial disease caused by *Bacillus anthracis*, is lethal to most mammalian herbivores. Within a few months during 1983–1984 an anthrax epizootic wiped out 90% of the impala population in Lake Manyara National Park in Tanzania. How is it possible for an epizootic of this type to suddenly appear in a population and then disappear for decades? Discuss the biological mechanisms that might permit this type of phenomenon. Prins and Weyerhaeuser (1987) discuss this particular impala epizootic.

Overview Question

Snowshoe hares are hosts to many species of internal parasites (nematodes and tapeworms) and external parasites such as ticks and fleas. Outline a research program to determine the effects of parasites on individual hares and on their population dynamics.

CHAPTER 16

Population Regulation

IN THE PREVIOUS THREE CHAPTERS we have often asked the question about whether predation, disease, or competition could affect the population dynamics of a particular animal or plant. How can we decide that? If a predator kills a prey individual, does that automatically affect the population level of the prey? If we kill pests with insecticides, will they necessarily become less abundant? The answer to these questions is *no*, and in this chapter we explore why simple concepts of population arithmetic can be misleading. These questions are at the core of conservation, land management, fisheries, and pest control issues that occupy our news media daily. For that reason it is important that our understanding of population regulation is correct.

We can make two fundamental observations about populations of any plant or animal. The first observation is that abundance varies from place to place; there are some "good" habitats where the species is, on the average, common and some "poor" habitats where it is, on the average, rare. The second observation is that no population goes on increasing without limit, and the problem is to find out what prevents unlimited increase in low- and high-density populations. This is the problem of explaining fluctuations in numbers. Figure 16.1 illustrates these two problems, which are often confused in discussions of population regulation.

Prolonged controversies have arisen over the problems of the regulation of populations. Before 1900 many authors, Malthus and Darwin included, had noted that no population goes on increasing with-

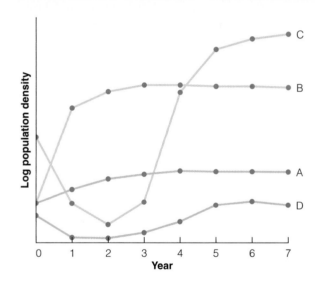

FIGURE 16.1
Hypothetical annual censuses of four populations occupying different types of habitat. Two questions may be asked about these populations: (1) Why do all populations fail to go on increasing indefinitely? (2) Why are there more organisms on the average in the good (red) habitats B and C compared with the poor (blue) habitats A and D? (After Chitty 1960.)

out limit, that there are many agents of destruction that reduce the population. It was not, however, until the twentieth century that researchers attempted to analyze these facts more formally. The stimulus for this came primarily from economic entomologists, who had to deal with both introduced and native insect pests. Most of the ideas we have on population regulation can be traced to entomologists. The basic principles of population regulation can be derived from a simple model, taken from the models of population growth presented in Chapter 11.

A Simple Model of Population Regulation

If populations do not increase without limit, what stops them? We can answer this question with a simple graphic model similar to that in Figure 12.2. A population in a closed system[1] will increase until it reaches an equilibrium point at which

Per capita birth rate = per capita death rate

Figure 16.2 illustrates three possible ways in which this equilibrium may be defined. As population density goes up, birth rates[2] may fall or death rates may rise, or both changes may occur. To determine the equilibrium population size for any field population, we need only determine the curves shown in Figure 16.2. Note that this simple model in no way depends on the shapes of the curves, provided that they rise or fall smoothly. In particular, these curves need not be straight lines.

We now introduce a few terms to describe the concepts shown in Figure 16.2. The per capita death rate is said to be *density dependent* if it increases as density increases (see Figure 16.2a and 16.2c). Similarly, the per capita birth rate is called density dependent if it falls as density rises (see Figure 16.2a and 16.2b). Another possibility is that the birth or death rates do not change as density rises; such rates are called *density-independent rates*.

Note that Figure 16.2 does not include all logical possibilities. Birth rates might, in fact, *increase* as population density rises, or death rates might *decrease*. Such rates are called *inversely density-dependent rates* because they are the opposite of directly density-

---

[1]A closed system has no immigrants and no emigrants, so the population dynamics are driven solely by births and deaths.
[2]In all discussions of population regulation, "birth rates" always refers to per capita birth rates, and death rates always refers to per capita death rates.

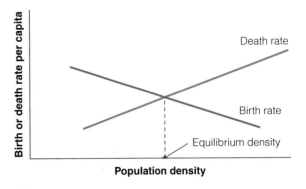

(a)

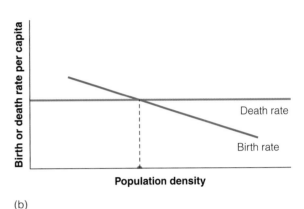

(b)

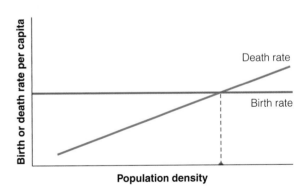

(c)

FIGURE 16.2
Simple graphic model to illustrate how equilibrium population density may be determined. Population density comes to an equilibrium only when the per capita birth equals the per capita death rate, and this is possible only if birth or death rates are density dependent. Note that these relationships need not be straight lines. (Modified from Enright 1976.)

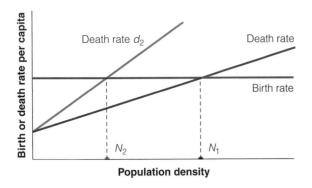

(a)

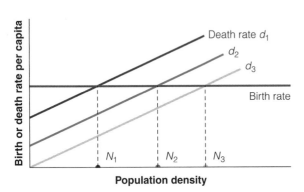

(b)

FIGURE 16.3
Simple graphic model to illustrate how two populations may differ in average abundance. In this example the birth rate is density independent and the death rate is density dependent. In (a) the two populations differ in the amount of density-dependent mortality because the slopes of the lines differ. In (b) the populations differ in the amount of density-independent mortality (the slopes of the lines do not differ). Dashed lines indicate the equilibrium population densities. (Modified from Enright 1976.)

dependent rates. Inversely density-dependent rates are not shown in Figure 16.2 because they can never lead to an equilibrium density. Figure 16.2 can be formalized into the first principle of natural regulation: *No closed population stops increasing unless either the per capita birth rate or death rate is density dependent.*

We can extend this simple model to the case of two populations that differ in equilibrium density to answer the question of why abundance varies from place to place (Figure 16.3). Consider first the simple case of populations with a constant (density-independent) birth rate. Equilibrium densities vary for two reasons: (1) Either the slope of the mortality curve changes

(see Figure 16.3a), or (2) the general position of the mortality curve is raised or lowered (see Figure 16.3b). In the first case, the density-dependent rate is changed because the slopes of the lines differ, but in the second case, only the density-independent mortality rate is changed. From this graphic model we can arrive at the second principle of natural regulation: *Differences between two populations in equilibrium density can be caused by variation in either density-dependent or density-independent per capita birth and death rates.* This principle seems simple: It states that anything that alters birth or death rates can affect equilibrium density. Yet this principle was in fact denied by many population ecologists for 40 years (Enright 1976, Sinclair 1989). We now review this historical controversy.

Historical Views of Population Regulation

The concept of the *balance of nature* has been a background assumption in natural history since the time of the early Greeks and underlies much of the thinking about natural regulation (Egerton 1973). The simple idea of early naturalists was that the numbers of plants and animals were fixed and in equilibrium, and observed deviations from equilibrium, such as the locust plagues described in the Bible, were the result of a punishment sent by divine powers. Only after Darwin's time did biologists try to specify how a balance of nature was achieved and how it might be restored in areas where it was upset.

A considerable amount of activity around the turn of the twentieth century centered on attempts to control insect pests by the introduction of parasites. L. O. Howard and W. F. Fiske (1911), two economic entomologists with the U.S. Department of Agriculture, studied the parasites of two introduced moths, the gypsy moth and the brown-tail moth, in an attempt to control the damage these defoliators inflicted on New England trees. Howard and Fiske believed that each insect species was in a state of balance such that it maintained a constant density if averaged over many years. For this balance to exist, they argued, there must be, among all the factors that restrict the insect's increase, one or more *facultative agents* that exert a relatively more severe restraint when the population increases. They argued that

E S S A Y   1 6 . 1

DEFINITIONS IN POPULATION REGULATION

The lack of clear definitions has plagued discussions of population regulation for decades. We begin by separating two problems:

1. *Population limitation*: What factors and processes can change average density?
2. *Population regulation*: What processes halt population increase?

If we keep these two problems separate, we will solve about half of the confusion in terminology. Answering the first question does not answer the second question.

Population limitation implies a before and after or experimental-control type of comparison. For example, European rabbits in Australia were at high density before myxomatosis and at low density after this disease was introduced (see Figure 15.18, p. 274). Myxomatosis limits rabbit density.

Population regulation implies some form of negative feedback between increasing density and factors such as predation, disease, food shortage, or territoriality. The effects of a regulating factor must be density dependent, as defined in Figure 16.2. But the problem is that not all density-dependent processes will achieve population regulation; they may not be quantitatively large enough. A predator that eats one lizard out of a total population of 1000 and three lizards out of 2000 is inflicting mortality that is density dependent, but it is also quantitatively trivial for population regulation in this species. Population regulation can be inferred only from a comprehensive model that includes all the factors affecting a population.

Compensation can complicate inferences about population regulation. Compensation occurs when a change in one factor produces the opposite change of identical magnitude in another factor, such that their combined effects on the population remain unchanged. One factor can essentially take the place of another factor. The opposite of compensation is *additivity*. Compensation is most easily seen experimentally by comparing, say, mortality rates with and without a particular factor: For example, measure overwinter mortality rates in two populations:

Population A: disease and food shortage

Population B: disease and no food shortage (food supplemented)

If the overwinter mortality rates are identical, food shortage and disease are completely compensatory. If processes are compensatory, population regulation is *either-or*, rather than *both-and*.

only a very few factors, including insect parasitism, were truly facultative.

Furthermore, Howard and Fiske said, a large proportion of the controlling factors, such as destruction by storms, high temperatures, and other climatic conditions, should be classed as *catastrophic*, because their actions are wholly independent of whether the insect is rare or abundant. For example, a storm that kills 10 out of 50 caterpillars on a tree would undoubtedly have destroyed 20 if 100 had been there, or 100 if 500 had been there. Thus the average percentage of destruction (the per capita death rate) remains the same no matter what the abundance of the insect.

Finally, Howard and Fiske noted[3] that other agencies, such as birds and other predators, work in a radically different manner: they maintain constant populations from year to year and destroy a constant number of prey. Consequently, when the prey species increases, the predators will destroy a smaller and smaller percentage of the prey (that is., they work in a manner that is the opposite of facultative agents). Howard and Fiske did not give factors of this type a distinct name.

---

[3]Incorrectly, as we now know from our historical vantage point: see Figure 13.19 (page 225).

They concluded that a natural balance can be maintained only through the operation of facultative agencies that destroy a greater proportion of individuals as the insect population in question increases in abundance. Facultative agencies thus cause the per capita death rate to increase with prey density, as in Figure 16.2a and 16.2c. Howard and Fiske believed that *insect parasitoids* were the most effective of the facultative agencies; that *disease* operated only rarely, when densities got very high; and that *starvation* was the ultimate facultative agency, but it almost never operated.

Howard and Fiske were the initial proponents of the *biotic school* of population regulation, which proposed that biotic agents, principally predators and parasitoids, were the main agents of natural regulation.

Meanwhile, another school of thought, the *climate school*, was in the process of formation. F. S. Bodenheimer (1928) was one of the first to hold that the population density of insects is regulated primarily by the effects of weather on both development and survival. Bodenheimer was impressed by all the work done in the 1920s on the environmental physiology of insects, which showed, for example, how low temperatures affect the rate of egg laying and the speed of development. He was also impressed by the fact that weather was responsible for the largest part of the mortality of insects; often 85–90% of the insects in their early stages were killed by weather factors.

B. P. Uvarov (1931) published a large paper, "Insects and Climate," in which he reviewed the effects of climatic factors on growth, fertility, and mortality of insects. He emphasized the correlation between population fluctuations of insects and the weather, and he regarded these weather factors as the prime agents that control populations. Uvarov questioned the idea that all populations are in a stable equilibrium in nature and emphasized the instability of field populations.

Three important ideas were central to the early climate school: (1) Insect population parameters are strongly affected by the weather, (2) insect outbreaks could be correlated with the weather, and (3) insect population fluctuations, not stability, was the rule.

It is important to realize here that all this controversy was over *insect* populations and their regulation; work on vertebrate populations had hardly begun by 1930, and no work had been done on the populations of other invertebrates or plants.

In 1933 the *Journal of Animal Ecology* published a supplement titled "The Balance of Animal Popula-

tions," written by A. J. Nicholson, an Australian economic entomologist. This paper is one of the great classics in ecology, and with it Nicholson ignited a controversy that continues to the present day. Nicholson was interested in the parasite-host system of insects, and he teamed up with mathematician V. Bailey to construct a model of this system. Nicholson was critical of the predator-prey models of Lotka and Volterra because they did not allow for time lags in the system, because they ignored age groups (and assumed instead that all individuals are equivalent), and because Lotka and Volterra used calculus rather than finite methods of mathematical analysis. Nicholson expanded his ideas on the parasite-host system to cover all interactions between animals.

According to Nicholson, the controlling factor was always *competition*—competition for food, competition for a place to live, or the competition of predators or parasites. Nicholson's theory was pre-

A. J. Nicholson *(1895–1969) CSIRO Division of Entomology, Canberra*

dominantly a biotic one, and his work is usually considered the cornerstone of the biotic school.

Nicholson's main ideas were essentially the same as those of Howard and Fiske. To these he added a mathematical model and the notion of competition as the controlling factor. Nicholson's ideas were given much stronger emphasis by H. S. Smith (1935), who considered the problem of population regulation in some detail. Smith pointed out first of all that populations are characterized by both stability and continual change. Population densities are continually changing, but their values tend to vary about some characteristic density that itself may vary. Smith compared a population to the surface of the sea, which paradoxically is a universal standard for altitude measurements that is continuously being changed by tides and waves. Thus Smith reaffirmed Nicholson's ideas on balance.

Different species of animals tend to have different average densities, and a given species will have different average densities in different environments. The variations about the average density are stable because there is always a tendency to return to the average density (that is, populations seldom become extinct or increase to infinity). This is what is loosely termed the "balance of nature."

That the equilibrium position, or average density, may itself change with time is what caused the economic entomologists so much trouble. The equilibrium position of an introduced pest may be so high that constant damage occurs to crop plants. So Smith set out to analyze the factors that determine the equilibrium position or average density. He pointed out that the number of injurious insects is very small relative to the total number of insects, and that we must study both *common* and *rare* species if we hope to understand the reasons for the abundance of species.

Smith recognized the distinction Howard and Fiske made between *facultative* and *catastrophic* agencies, and he renamed them *density-dependent* mortality factors and *density-independent* mortality factors. The average density of a population, Smith concluded, can never be determined by density-independent factors. Only if a plot of the death rate has a slope (that is, is a density-dependent component) can the population reach equilibrium. Thus only density-dependent mortality factors can determine the equilibrium density of a population.

Smith went on to point out that the density-dependent factors—disease, competition, and predation—are mainly *biotic*, and that the density-independent factors (in particular, climate) are mainly physical, or *abiotic*, factors. But Smith also pointed out that we should not conclude from this that the average population densities of species are *never* determined by climate. Climate, he stated, may act as a density-dependent factor under some circumstances. As an example, he suggested the case of *protective refuges:* If only so many refuges are available and all the unprotected individuals are killed by climate, this climatic mortality would be density dependent.

To summarize: Smith restated Nicholson's main points, adding the terms *density-independent factors* and *density-dependent factors*, and he stated, in contrast to Nicholson, that climate might act as a density-dependent factor in some cases. By this time, then, the main tenets of the biotic school—the idea of balance in nature, that this balance was produced by density-dependent factors, and that these factors were usually biotic agents such as predators and diseases—had been crystallized.

In 1954 H. G. Andrewartha and L. C. Birch, two Australian zoologists who completely disagreed with Nicholson's ideas, published an important book, *The Distribution and Abundance of Animals*, which attacked the ideas of the biotic school. They revived in their book a highly modified version of the climate school's ideas. Andrewartha and Birch concentrated on the individual organism and based their whole approach on this question: *What are the factors that influence an animal's chance to survive and multiply?* Given this question, they proceeded to classify environmental factors.

First, they rejected the distinction Howard and Fiske and others made between physical (abiotic) and biotic factors. Because, for example, food and shelter may sometimes be biotic, sometimes abiotic, this distinction does not help us much to classify the environment.

Second, they rejected the classification of the environment based on density-dependent factors and density-independent factors, because they believed that no component of the environment is likely to exert an influence that is independent of the density of the population (that is, all factors are density dependent). Here they are attacking a key principle of the biotic school. Consider, they said, the action of frost. Between a large population and a small population there may be genetic differences in cold hardiness, and the places where the insects live may differ with respect to the degree of protection from frost. Thus because large populations may be forced to occupy marginal habitats, they may suffer more from frost. Andrewartha and Birch concluded that density-independent factors do not exist, so there is no need to attach any special importance to density-dependent factors in classifying the effects of the environment on a population.

How, then, can one classify environmental factors? Andrewartha and Birch suggested that the environment may be divided into four components: weather, food, other animals and pathogens, and a place in which to live. (Andrewartha and Birch were concerned only with animal abundance, so plants appear principally as food in this classification.) These

FIGURE 16.4

The separation of population regulation processes into (a) extrinsic factors, and (b) combined extrinsic and intrinsic factors. Extrinsic processes (for example, disease) interact with the properties of individuals that make up the population (intrinsic processes), so that population regulation results from an interplay between these two kinds of processes.

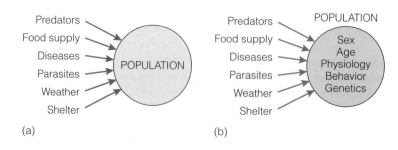

components of the environment are nonoverlapping, and they may be subdivided if necessary, for example, into other animals of the same species and other animals of different species. Together these four components and the interactions among them completely describe the environment of any animal.

Consequently, said Andrewartha and Birch, for any given species we must ask which of the four components of environment affect the animal's chances to survive and multiply. Once we can answer this question, we will be able to determine the reasons for the animal's abundance and distribution in nature. Andrewartha and Birch presented a general theory of the abundance of animals in natural populations. First, they stated that one cannot use expressions like "balance," "steady states," "equilibrium densities," or "ultimate limits" because there is no empirical way of giving meaning to these words. Second, they noted that one must take into account the fact that animals are always distributed patchily in nature, never uniformly. These "patches," or *local populations*, constituted the basic component of interest for these investigators.

According to Andrewartha and Birch, the numbers of animals in a natural population may be limited in three ways: (1) by a limited supply of material resources, such as food, places in which to nest, and so on; (2) by inaccessibility of these material resources relative to the animal's capacities for dispersal and searching; and (3) by shortage of time when the rate of increase (*r*) is positive. Of these three ways, they contended that the last is probably the most important in nature, and that the first is probably the least important. Regarding the third case, fluctuations in the rate of increase may be caused by weather, predators, or any component of the environment.

Andrewartha and Birch were principally concerned with insect populations, and their field experience was with insects in the severe desert and semi-arid

areas of Australia. Their main contribution to ecology has been to reemphasize the importance of getting *empirical data* on the problems of population regulation. They continually raised the ever-bothersome question, how can a particular idea be *tested* in real populations?

The two main schools of population regulation—the biotic school and the climate school—have concentrated on the role of the *extrinsic* factors in population control: food supply, natural enemies, weather, parasites and diseases, and shelter (Figure 16.4) . Many of these theories tend to assume that the individuals that make up the population are all identical, like atoms or marbles. This neglect of the importance of individual differences in population regulation has been challenged by a group of workers from diverse disciplines who have proposed theories of self-regulation. Their rallying point has been a search for *intrinsic* changes in populations, that might be important in natural control. Dennis and Helen Chitty, working at the Bureau of Animal Population in Oxford, were two of the pioneers in proposing ideas of self-regulation via intrinsic changes in population.

Two basic types of changes can occur in individuals, *phenotypic* and *genotypic* changes, and the proponents of self-regulatory mechanisms differ in the importance they attach to each of these basic types. Of course, no matter what the mechanism operating, it must have evolved in the species concerned, and consequently these theories of self-regulation all become concerned with evolutionary arguments.

Dennis Chitty (1955) presented the fundamental premise underlying all ideas on self-regulatory mechanisms. Suppose, Chitty argued, that we observe a population twice, at times 1 and 2, and that at time 2 the death rate (D_2) is higher than the death rate at time 1 (D_1). This death rate is the result of the interaction of the organisms (O) with their mortality factors (M).

Our problem now is to determine why D_2 is greater than D_1. The first hypothesis to be explored is that we are dealing with organisms whose biological properties are identical at times 1 and 2. In this case, we must look for a difference between the mortality factors at the two times. In other words, we might expect to find at time 2 that there are more predators or parasites, or that the weather is less favorable. Some population changes can certainly be explained in this manner, but in other cases this method has failed. We must look at the matter from another angle.

Consider, Chitty continued, the possibility that the environmental conditions are much the same at all times, that there is no real difference between the mortality factors at times 1 and 2. In this case, any change in the death rate must be due to a change in

Dennis Chitty (1912–)
Helen Chitty (1910–1987) *Bureau of Animal Population, Oxford and University of British Columbia*

the nature of the organisms, a change such that they become less resistant to their normal mortality factors. For example, the animals might die in cold weather at time 2, weather they might have survived at time 1.

These ideas can be summarized as follows:

	First Hypothesis		Second Hypothesis	
Time	1	2	1	2
Death rates	$D_1 < D_2$		$D_1 < D_2$	
Organisms	$O_1 = O_2$		$O_1 \neq O_2$	
Mortality Factors	$M_1 \neq M_2$		$M_1 = M_2$	

Changes in the individual organisms in the population may be physiological or behavioral, and they may be phenotypic changes or genotypic changes.

The first hypothesis describes the classic approach to population regulation used by both the

biotic school and the climate school (see Figure 16.4a). The second hypothesis describes an ideal self-regulatory approach to population regulation. It is unlikely in nature that this second situation would occur in such a pure form, more likely some mixture of these two situations would be found in self-regulatory populations (see Figure 16.4 b). Note that the concept of density dependence becomes ambiguous under the second hypothesis, because the idea that the environment can be subdivided into density-dependent and density-independent factors has meaning only insofar as the properties of the individuals in the population are constant. Self-regulatory systems have added an additional degree of freedom to the system—the individual with variable properties.

Variation among the individuals in a population may be either genetically based or environmentally induced. The British geneticist E. B. Ford (1931) was one of the first to point out the possible importance of genetic changes in population regulation. He suggested that natural selection is relaxed during population increases, with the result that variability increases within the population and many inferior genotypes survive. When conditions return to normal, these inferior individuals are eliminated through increased natural selection, simultaneously causing the population to decline and reducing variability within it. Thus, Ford argued, population increase inevitably paves the way for population decline.

From a study of population fluctuations in small rodents, Dennis Chitty (1960) set up the general hypothesis that *all species are capable of regulating their own population densities without destroying the renewable resources of their environment or requiring their enemies or bad weather to keep them from doing so.* Not all populations of a given species will necessarily be self-regulated, and the mechanisms evolved will be adapted to only a restricted range of environments. The species may well live in poor habitats where this mechanism seldom if ever comes into effect.

The actual mechanisms by which self-regulation can be achieved in natural populations involve some form of mutual interference between individuals or intraspecific hostility in general. Mechanisms of self-regulation do not require genetic changes and may be entirely phenotypic. This hypothesis can be applied only to species that show mutual interference or spacing behavior. The most important environmental factor for such populations is *other organisms of the same species.*

The problem of self-regulation has been approached from another angle by V. C. Wynne-Edwards, a British ecologist whose major work has been on birds. Wynne-Edwards (1962) began his analysis with the observation that most animals have highly effective mechanisms of movement. If we look in nature, we will usually find that organisms concentrate at places of abundant resources and avoid unfavorable areas—*that animals are dispersed in close relation to their essential resources.*

The essential resource most critical to animals is clearly *food*, Wynne-Edwards observed. Of course, many other requirements must be met before a species can survive in an area, but food is almost always the critical factor that ultimately limits population density in a given habitat. We must then study the food resource as the key to understanding population control.

Wynne-Edwards suggested that some artificial and harmless type of competition has evolved in many species as a buffer mechanism to halt population growth at a level below that imposed by exhaustion of the food supply. The best example of this kind of buffer mechanism is the territorial system of birds. The territories that birds defend so fiercely are just a parcel of ground, but the possession of a territory eliminates competition for food, since the owner and its dependents enjoy undisputed feeding rights on that area. Provided that the size of the territory varies with the productivity of the habitat, we get a perfect illustration of this model: Population density is controlled by territoriality, which ensures that the food supply will not be exhausted.

The margins of a species's range will probably not show this self-regulation, Wynne-Edwards stated. Physical factors will predominate in these harsh environments, and hence we should concentrate our attention on the more typical parts of the range, where self-regulation is the usual situation. Also, a few species will ultimately fail to be limited by food, and these will not fit into Wynne-Edwards's scheme.

A Modern Synthesis of Population Regulation

There has been a great deal of controversy in ecology over the concepts of population regulation (Sinclair 1989), and we need to highlight the areas of agreement and disagreement.

The definition of terms has always plagued discussions about population regulation. Let us start with a clear definition of two confusing terms:

1. *Limiting factor:* a factor is defined as limiting if a change in the factor produces a change in average or equilibrium density. For example, a disease may be a limiting factor for a deer population if deer abundance is higher when the disease is absent.

2. *Regulating factor:* a factor is defined as potentially regulating if the percentage mortality caused by the factor increases with population density.[4] For example, a disease may be a potential regulating factor only if it causes a higher fraction of losses as deer density increases.

The distinction between a potential regulating factor and an actual regulating factor is quantitative. Unless the change in mortality is large enough, a regulating factor will not stop population growth. Regulation is much more difficult to study than limitation. Most experimental manipulations of populations involve studies of limitation, and most practical problems in population ecology are problems of limitation not regulation.

The simple model of population regulation shown in Figure 16.2 is critically focused on the concept of equilibrium, and we must begin by asking whether natural populations can be equilibrium systems. Recent work on ecological stability has given us a more comprehensive view of the factors that affect stability (Figure 16.5). There is no reason to expect all populations to show stable equilibria (Wiens 1984a). Strong environmental fluctuations in weather can produce instability, but biotic interactions may also promote instability. We have seen examples in Chapter 13 of predator-prey interactions that are unstable. Time lags can also affect population stability. We should expect real-world populations to fall along the continuum from stable, equilibrium dynamics to unstable, nonequilibrium dynamics. The simple model shown in Figure 16.2 will be difficult to detect in a real population that shows unstable dynamics.

---

[4]Or alternatively, if the reproductive rate declines as population density rises.

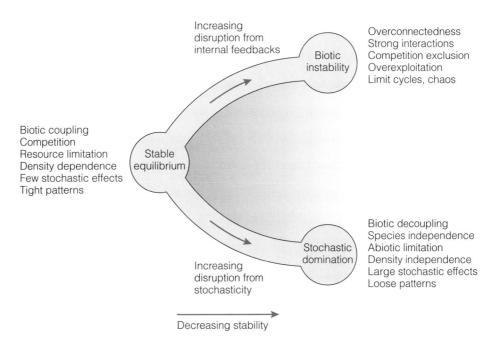

FIGURE 16.5
Schematic representation of ecological systems along a continuum from stable to unstable. Both biotic instability, caused by internal feedbacks, and stochastic domination, caused by strong environmental fluctuations, can result in instability. (From DeAngelis and Waterhouse 1987.)

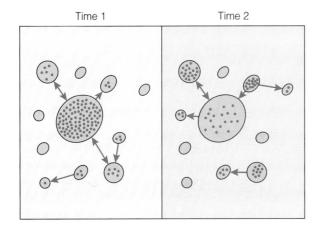

FIGURE 16.6
Hypothetical metapopulation dynamics. Closed circles represent habitat patches; dots represent individual plants or animals. Arrows indicate dispersal between patches. Over time the regional metapopulation changes less than each local subpopulation.

The spatial scale can also be important in considerations of stability. If we study a very small population on a small area, it may fluctuate widely and even go extinct; at the same time, a large population on a large study area may be stable in density. The important concept here is that local populations are linked together through dispersal into *metapopulations* (Figure 16.6). To study population regulation, we must know if a population is subdivided and, if so, how the patches are linked (Kareiva 1990). Ensembles of randomly fluctuating subpopulations, loosely linked by dispersal, will persist if irruptions at some sites occur at the same time as extinctions at other sites. The result can be that at a regional level the population appears stable while the individual subpopulations fluctuate greatly.

Butterflies on islands are a good example of metapopulations. To show that a set of local populations is a metapopulation, we must show that some

ESSAY 16.2

WHY IS POPULATION REGULATION SO CONTROVERSIAL?

The controversies over population regulation are legendary in the history of ecology. During the 1950s and 1960s highly charged exchanges in the literature and strong public verbal attacks at scientific meetings were the order of the day. While most of the vituperative attacks have stopped as time has passed, exchanges still occur in scientific journals (Turchin 1999). It is interesting to ask why this subject has been so controversial.

There are two aspects to any such controversy, one scientific and one personal. The scientific issue behind the population regulation controversy has been focused on the identification of density-dependent regulating factors as biotic agents—predators, diseases, parasites, food supplies—and density-independent nonregulating factors such as weather and other physical factors. The side issue was always that density-dependent factors are important and density-independent factors are not important, which we now know is not correct (see Figure 16.3). The difficulty of identifying density-dependent effects in real-world data has greatly prolonged the arguments (Wolda and Dennis 1993). The conclusion after all the controversy was that regulation is an empirical question for each population, and that

one cannot *a priori* assign factors like predators or weather to one category or another. The critical thing is to measure what effect a particular factor is having on a particular population, preferably in an experimental setting with proper controls. The realization that intrinsic processes could impinge on regulation, and that mortality could be compensatory rather than additive, also made the original 1950s controversy obsolete.

The personal element to scientific controversy is fascinating because many leading scientists are forceful personalities with large egos. This element is not so easily captured in the written word, but it is apparent at scientific meetings in which proponents of differing paradigms come face to face. Controversy galvanizes people, and population ecologists are indeed human. Population ecology has had an array of fascinating scientists that historians are now beginning to evaluate (Kingsland 1995). The important message is that not every scientist, no matter how distinguished, is right about everything, and in science we should appeal not to authority or personality but to experiment and observation, to empirical tests of ideas, not dogmatic assertions, no matter how articulate the speaker.

metapopulations go extinct in ecological time, and that these can be recolonized by dispersing individuals from nearby populations. Hanski et al. (1996) studied 1502 small populations of the Glanville fritillary butterfly (*Melitaea cinxia*) on islands in the Åland Archipelago between Finland and Sweden. This butterfly is an endangered species that has recently become extinct on mainland Finland and now exists only on islands in the Åland Archipelago. Larval caterpillars feed on two host plants and spin a web, which is easy to detect in field surveys. These butterfly populations ranged in size from 1 to 65 larval groups per meadow, but most populations are small,

averaging four larval groups per patch (corresponding to about 5–50 butterflies). From 1991 to 1993 an average of 45% of these local populations went extinct; smaller patches supported smaller populations and had a greater chance of going extinct (Figure 16.7). Small populations went extinct more often for two reasons. First, male and female butterflies tend to leave small patches, in which they presumably perceive a reduced chance of mating (Kuussaari et al. 1998). Figure 16.8 shows the residence time for female butterflies in populations of different sizes, and the fraction of mated females. Small butterfly populations suffered reduced population growth rates, an effect called the

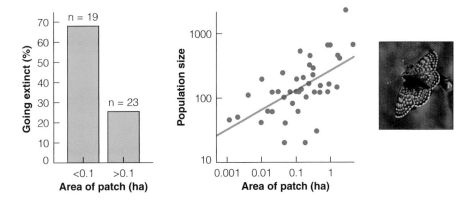

FIGURE 16.7
Probability of extinction over three years in relation to the patch area for metapopulations of the Glanville fritillary butterfly in the Åland Archipelago, Finland. Small patches are much more likely to go extinct, and small patches tend to have smaller populations of this endangered butterfly. (Data from Hanski et al. 1994, 1995.)

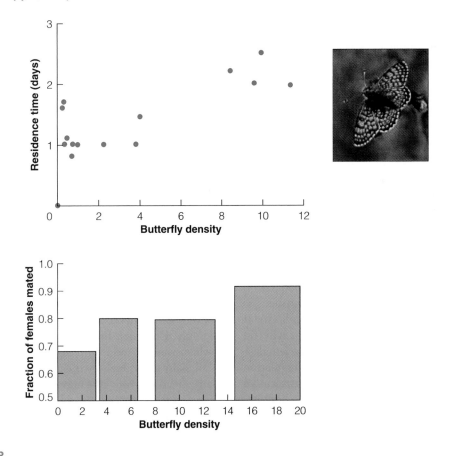

FIGURE 16.8
Effects of population density on the residence time of females and on the fraction of females mated in Glanville fritillary butterlflies in the Åland Archipelago, Finland. The net result of these processes is that small populations suffer decreased population growth rates. (Data from Kuussaari et al. 1998.)

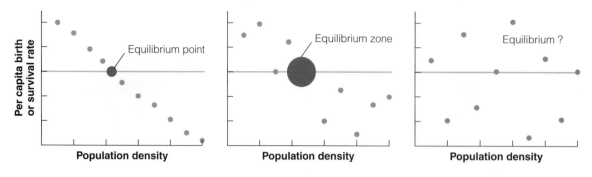

FIGURE 16.9

Types of density-dependent relationships for survival rates likely to be observed in real-world populations. For simplicity, a density-independent birth rate is shown (blue line). Increasingly, scattered points make it difficult to determine if there is a stable equilibrium point, or an equilibrium zone, or any equilibrium at all. (Modified after Strong 1984.)

Allee effect (Dennis 1989), the exact opposite of what is predicted by the simple density-dependent model (shown in Figure 16.2.).

An additional complication for the analysis of population regulation is that real-world populations rarely show smooth curves like those in Figure 16.2. A more usual observation is of a cloud of points, such that density dependence is either "vague" or absent (Strong 1984). Figure 16.9 illustrates the type of density-dependent relationships that might be observed in the real world. It may be very difficult to find density-dependent relationships in natural populations (Gaston and Lawton 1987).

If a population does not continue to increase, it is axiomatic that births, deaths, or movements must change at high density. The first step is to ask which of these parameters changes with increasing population density (Sinclair 1989). Does reproductive rate decline at high density, or does mortality increase (or both)? If mortality increases, does this fall more heavily on younger or on older animals, on males or on females? The first step to understanding population regulation in animals, then, is to see whether these patterns of changing reproduction and mortality with changing population density occur in a variety of populations.

The second step is to determine the reason for the changes in reproduction or mortality. Determining the cause of death of plants or animals in natural populations is not always simple. If a fox or a bat has rabies, a fatal disease, the cause of death is clear; a caterpillar with a tachinid parasite is certain to die from this infection. But as we examine more complex cases, decisions about causes of death are not clear. If a moose has inadequate winter food and the snow is deep, it may be killed by wolves (Peterson 1992). Is predation the cause of death? Yes, but only in the immediate sense. Malnutrition and deep snow have increased the probability of the moose's death. Because many components of the environment can affect one another and not be independent, mortality can be *compensatory*, as distinguished from *additive*. Because the concepts of compensatory and additive mortality are crucial to our understanding of population regulation, we must distinguish between them.

Additive mortality is applicable to the agriculture model of population arithmetic. If a farmer keeps sheep and a coyote kills one of them, the farmer's flock is smaller by one. In this model, deaths are additive, and to measure their total effect on a population, we simply add them up. But in natural populations, in which several causes of death operate, the arithmetic is not so simple. Consider, for analogy, a sheep population in which winter food is limiting such that starvation will kill many individuals by the end of winter. In this case, any sheep a coyote kills may have been doomed to die anyway from starvation, and the number of sheep left at the end of winter will be the same, whether predation occurs or not (in this hypothetical scenario). In this case, predation mortality is not additive but is compensatory, and simple arithmetic does not work.

Figure 16.10 illustrates how additive and compensatory effects can be recognized. Consider, for example, what happens if wolf predation increases elk calf mortality from 10% per year to 20% per year. If

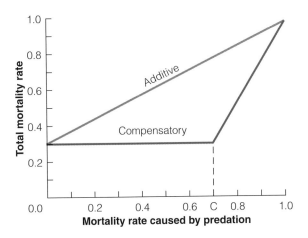

FIGURE 16.10
Schematic illustration of additive and compensatory mortality for losses due to predation. The additive hypothesis (blue) predicts that for any increase in predation mortality, total mortality increases by a constant amount. The compensatory hypothesis (red) predicts that, below a threshold mortality rate C, any change in predation losses has no effect on total mortality. This model can be applied to any mortality agent— predation, disease, starvation, or hunting. (Modified after Nichols et al. 1984.)

this mortality is additive, total elk calf mortality will increase from 45% to 55% per year (in this hypothetical example). If this mortality is compensatory, total elk calf mortality will remain unchanged at 45% per year. Clearly if mortality from predation is very high, compensation is not possible, as shown on the right side of Figure 16.10.

Compensatory mortality is the reason behind many ecological anomalies that puzzle the average person. If we kill pests, they will not necessarily become less abundant. Chapter 18 discusses this question of pest control. If we shoot grouse or catch fish, their numbers may not necessarily fall (Chapter 17). Compensatory mortality has practical consequences when it occurs.

In natural populations, mortality agents will rarely be completely additive or completely compensatory. We can determine if a particular cause of mortality is compensatory only by doing an experiment in which total losses are measured with and without the particular cause of death. Few of these experiments have been conducted, and there is an unfortunate tendency to assume the agricultural model of population arithmetic—that all losses are additive—for natural populations.

If birth rates change with population density, it is important to identify the factors that cause reproduction to change. Food supply is usually the first hypothesis to be tested for animals; nutrient availability is the first to be tested for plants. But other factors may cause birth rates to change as well. Social interactions can inhibit reproduction in vertebrates (Ishibashi et al. 1998), and risk of predation can change the behavior of animals such that they can gather less energy and thus produce fewer offspring (Lima 1998). These factors can most easily be identified experimentally by manipulations of field populations, or by careful descriptive studies of processes in unmanipulated populations.

The bottom line is that inferences about population limitation and population regulation are both important and difficult to come by. Given these problems, how might one develop a systematic approach to answer these key questions of population dynamics?

Two Approaches to Studying Population Dynamics

There are two competing paradigms about how best to study population dynamics to uncover the causes of population change. *Key factor analysis* is a method of analyzing populations through the preparation of life tables and a retrospective analysis of year-to-year changes in mortality and reproduction. *Experimental analysis* forms a second method of analyzing population changes that approach questions of limitation and regulation directly. Let us consider the advantages and disadvantages each of these two approaches.

Key Factor Analysis

Morris (1957) developed key factor analysis as a technique for determining the cause of population irruptions in the spruce budworm, which periodically defoliates large areas of balsam fir forests in eastern Canada. Varley and Gradwell (1960) improved Morris's method, and their approach is now used.

Key factor analysis begins with a series of life tables of the type shown in Table 16.1. The life table data are most easily obtained for organisms with one discrete generation per year. The life cycle is broken down into a series of stages (eggs, larvae, pupae, adults)

TABLE 16.1 **Life Table for the Winter Moth in Wytham Woods, near Oxford, England, 1955–1956.**

	Percentage of previous stage killed	No. killed (per m²)	No. alive (per m²)	Log no. alive (per m²)	k Value
Adult Stage					
Females climbing trees, 1955			**4.39**		
Egg Stage					
Females × 150			658.0	2.82	
Larval Stage					$0.84 = k_1$
Full-grown larvae	86.9	551.6	**96.4**	1.98	$0.03 = k_2$
Attacked by *Cyzenis*	6.7	**6.2**	90.2	1.95	$0.01 = k_3$
Attacked by other parasites	2.3	**2.6**	87.6	1.94	$0.02 = k_4$
Infected by microsporidian	4.5	**4.6**	83.0	1.92	
Pupal Stage					$0.47 = k_5$
Killed by predators	66.1	54.6	28.4	1.45	$0.27 = k_6$
Killed by *Cratichneumon*	46.3	**13.4**	15.0	1.18	
Adult Stage					
Females climbing trees, 1956			**7.5**		

Note: The figures in bold are those actually measured. The rest of the life table is derived from these.
Source: After Varley et al. (1973).

on which a sequence of mortality factors operate. We define for each drop in numbers in the life table:

$$k = \log_e(N_s) - \log_e(N_e) \qquad (16.1)$$

where k = Instantaneous mortality coefficient[5]

N_s = Number of individuals starting the stage

N_e = Number of individuals ending the stage

For example, from Table 16.1 we see that 83.0 winter moth larvae entered the pupal stage in 1955, and of these, 54.6 were killed by pupal predators (shrews, mice, beetles) during late summer, which reduced the population to 28.4 per m². Thus

$$k_5 = \begin{pmatrix} \text{instantaneous mortality} \\ \text{coefficient for pupal} \\ \text{predation} \end{pmatrix}$$

$$= \log_e(83.0) - \log_e(28.4) = 0.47$$

---

[5]Note that these k values are the same as the instantaneous mortality rate defined in Appendix III, without the minus sign.

We do these calculations in logarithms to preserve the additivities of the mortality factors (see Appendix III). Thus we can define *generation mortality K* as

$$K = k_1 + k_2 + k_3 + k_4 + k_5 + \cdots \qquad (16.2)$$

Key factor analysis assumes that all mortality factors are additive and ignores compensatory mortality, and this is an important limitation to this method. For our sample data in Table 16.1,

$$K = \log_e(658) - \log_e(15) = 1.64$$
$$\text{(no. eggs)} \quad \text{(no. adults of both sexes)}$$

which is identical to:

$$K = 0.84 + 0.03 + 0.01 + 0.02_4 + 0.47 + 0.27$$

$$= 1.64$$

Note that since the k values are instantaneous rates, they may take on any value between zero and infinity. Larger k values represent higher mortality rates. Varley et al. (1973) give a detailed description of these calculations.

Given a series of life tables like Table 16.1 over several years, we can proceed to the second step of key factor analysis, in which we ask an important question: *What causes the population to change in density from year to year?* Simple visual inspection of Figure 16.11

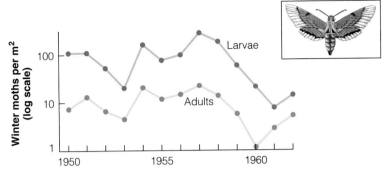

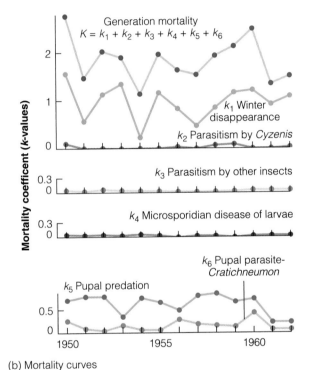

(b) Mortality curves

FIGURE 16.11
Key factor analysis of the winter moth in Wytham Woods near Oxford, 1950–1962. (a) Winter moth population fluctuations for larvae and adults. (b) Changes in mortality, expressed as k values, for the six mortality factors listed in Table 16.1. The biggest contribution to change in the generation mortality K comes from changes in k_1, winter disappearance, which is the key factor for this population. (After Varley et al. 1973.)

shows that k_1 (winter disappearance) is the *key factor* causing population fluctuations. A *key factor* is defined as the component of the life table that causes the major fluctuations in population size. An implication of this definition is that key factors can be used to predict population trends (Morris 1963).

Finally, we can use the *k* values to answer a second important question: *Which mortality factors are density dependent and thus might halt population increase?* By plotting the *k* values against the population density of the life cycle stage on which they operate, we can estimate density dependence. Figure 16.12 shows these data for the winter moth, and Figure 16.13 shows the idealized types of curves that can arise from this type

of key factor analysis. Note that the key factor need not be density dependent and need not be involved in population regulation. In this example for the winter moth, winter disappearance is the key factor, but pupal predation is the major density-dependent factor.

Key factor analysis has been widely applied to insect populations (Varley et al. 1973), but it has some important limitations. It cannot be applied to organisms with overlapping generations, including birds and mammals. Mortality factors may be difficult to separate into discrete effects that operate in a linear sequence, do not overlap, and are completely additive (Åström et al. 1996). In addition, key factor analysis is sensitive to the number of stages in the life cycle that are lumped

FIGURE 16.12

Relationship of winter moth mortality coefficients to population density. The k *values for the different mortalities are plotted against the population densities of the stage on which they acted.* k_1 *and* k_6 *are density independent and quite variable,* k_2 *and* k_4 *are density independent but constant,* k_3 *is inversely density dependent, and* k_5 *is strongly density dependent. Compare these data with the idealized curves in Figure 16.13. (After Varley et al. 1973.)*

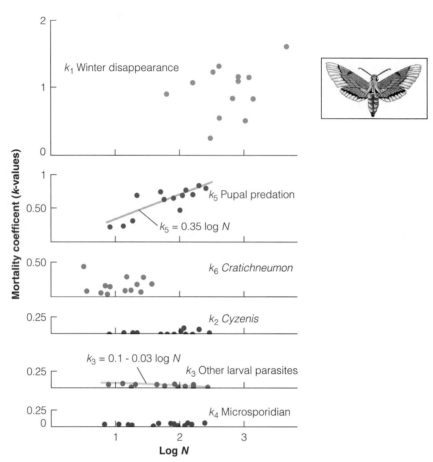

together into one k value. For example, in the winter moth data, winter disappearance (k_1) includes many distinct mortality processes that are grouped together in this stage of the life cycle (Royama 1996). Such groupings may blur the interpretation of key factors.

Finally, density dependence may be difficult to detect if the equilibrium density (see Figure 16.3) varies greatly from year to year (Moss et al. 1982). Nevertheless, key factor analysis has provided for some populations a reliable quantitative framework within which the problems of natural regulation can be discussed.

Experimental Analysis

An alternative approach to population regulation is to ask the empirical question: *What factors limit population density during a particular study?* This approach does not utilize the density-dependent paradigm because density dependence is often impossible to demonstrate with field data. Instead we try to identify *limiting factors* and

study them with manipulative experiments. A population may be held down by one or more limiting factors, and these factors can be recognized empirically by a manipulation—by adding to or reducing the relevant factor. If we suspect that food is limiting a population, we can increase the food supply and see if population size increases accordingly (Watson and Moss 1970). Alternatively, we can observe changes in population density and the supposed limiting factors over several years and see if they vary together. This is another way of testing hypotheses about limitation, but gathering the relevant data may take a long time. Observations of this type, however, always provide weaker evidence than manipulations involving experimental and control populations.

The experimental approach uses the most direct and empirical techniques for answering the two central questions of regulation—what determines average abundance, and what stops population growth? If we think that parasites reduce the average abundance of pheasants, we can increase or reduce parasite loads

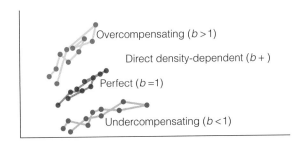

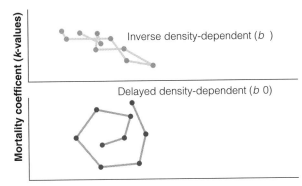

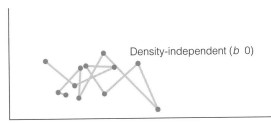

Log population density of stage on which factor acts

FIGURE 16.13
Idealized forms of the possible relationships between k
values determined from key factor analysis and
population density. The points are connected in a time
sequence, and b *is the slope of the regression line.*
Compare with Figure 16.12. (After Southwood 1978.)

and observe the changes in pheasant numbers. If we
think that food shortage halts population growth in
cabbage aphids, we can manipulate the food resources
and measure aphid population growth. It is important
to realize that more than one factor may be involved
in population limitation. Perhaps both parasite levels
and food supplies affect average abundance, so if we
have shown one factor to be significant, we cannot
assume that only one factor is involved.

Often it is not possible to manipulate a suspected factor, either because it is physically impossible (for example, the weather) or because it is not possible biologically
or politically. Experimental analysis can be carried out

without manipulations if one sets up a hypothesis and
makes a prediction about what can be observed. For
example, if we postulate that cold spring weather halts
population growth in spruce budworm (Morris 1963), we
can measure spring weather and population trends in several sites in several years and test this prediction.

Experimental analysis is forward looking
(prospective) and oriented toward testing hypotheses
about regulation mechanisms. Key factor analysis is
retrospective looking and is confined to a descriptive
analysis of a population. Theoretically, both methods
should converge to provide an understanding about
population regulation.

Plant Population Regulation

Because most plants are modular organisms, population regulation in plants must be discussed as the regulation of biomass rather than of numbers. Plant
ecologists have not usually addressed the problem of
population regulation in the same way as have animal
ecologists (Crawley 1990, 1997), but the same principles can be applied. As a plant population increases in
numbers and biomass, either reproduction or survival
will be reduced by a shortage of nutrients, water, or
light; by herbivore damage; by parasites and diseases;
or by a shortage of space. Because plants are typically
fixed in one location, competition for light or nutrients is often implicated in population regulation. This
competition has been described by the -3/2 power rule
(also called *Yoda's law* or the *self-thinning rule*).

The self-thinning rule describes the relationship
between individual plant size and density in even-aged
populations of a single species. Mortality, or "thinning," from competition within the population is postulated to fit a theoretical line with a slope of -3/2:

$$\log(\overline{m}) = -\frac{3}{2}(\log N) + K \qquad (16.3)$$

where $\overline{m}$ = average plant weight (grams)

N = plant density (individuals/m^2)

K = a constant

This line has been suggested as an ecological law
(Westoby 1984, Hutchings 1983) that applies both
within any given plant species and among different

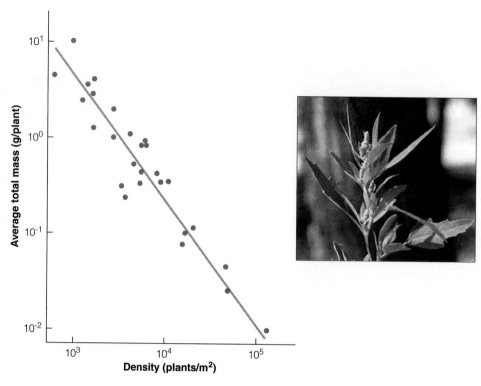

FIGURE 16.14

The self-thinning line for the herb Chenopodium album. *The slope of this line is − 1.33, close to the theoretical − 3/2 of the self-thinning rule. Populations started at densities to either side of this line would be expected to move to the line and then reach equilibrium along the line. (Data from Yoda et al. 1963.)*

plant species. Figure 16.14 illustrates the -3/2 power rule. The self-thinning rule highlights the tradeoffs that can occur in organisms having plastic growth, such that the size of an individual can become smaller as population density increases.

Recent evaluations of the self-thinning rule have found many exceptions to it (Weller 1987, 1991). However, the principle of a tradeoff between average plant size and total plant population density is supported by all plant studies. The self-thinning rule has been replaced by a more general "−3/2 boundary rule," which postulates the self-thinning line as an upper limit for the relationship between plant size and population density in monocultures (Hamilton et al. 1995). The self-thinning rule expresses competition between plants for essential resources. If this competition is largely for light, the self-thinning rule should predict that leaf area should remain constant during thinning. This type of formulation would have practical consequences for stand densities of forest trees in plantations (Cao 1994). Verwijst (1989) found that in the mountain

birch tree (*Betula pubescens*) in Sweden, tree mass decreased with density, but the slope was –1.37, close to but not exactly –3/2 (Figure 16.15). Mountain birch trees allocate most of their resources to radial growth rather than to increasing crown depth, resulting in a slightly lower slope than predicted by the self-thinning rule. These results argue for viewing the self-thinning rule as a boundary rule rather than an absolute thinning law for all plants. The slope of the thinning line is variable but gives us further insight into species differences under strong competition for light and nutrients.

The self-thinning rule has been applied to animals as well. Animals with plastic growth rates, such as fish, can respond to changes in population density by changing growth rates and body size (Dunham and Vinyard 1997). Animals of larger body size use more energy, and when populations are food limited or space limited, a tradeoff can occur between average size and population density. Salmon and trout fingerlings living in streams are a good example, and for these kinds of animals with plastic growth, the self-

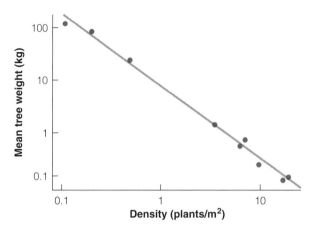

FIGURE 16.15
Self-thinning line for mountain birch trees (Betula pubescens) *in central Sweden. All sites were dense pure stands of birch. The slope is −1.39, slightly less than the expected slope of −1.5 if the self-thinning rule applied. The densest stands were about ten years old, and the least dense stands were about 60 years old. (Data from Verwijst 1989.)*

thinning rule is a useful empirical description of these tradeoffs between body size and population density.

Source and Sink Populations

Local populations can be classified as *source* populations, in which there is a net excess of reproduction over mortality, and *sink* populations, in which there is a net excess of mortality over reproduction. Left to themselves, source populations would grow to infinity, and sink populations would shrink to extinction. But in practice, source populations do not increase forever, but are regulated. This regulation may involve a net export of animals via dispersal, such that in a source population emigration exceeds immigration. Sink populations may indeed go to extinction, so we would not necessarily know about them, but more typically they are in negative balance in that immigration exceeds emigration. Sink populations continue to exist only if they attract immigrants from nearby source populations.

Sources and sinks have become more important in human-impacted landscapes, such as formerly large continuous areas of forest or grassland that have been dissected by modern agriculture into a series of small fragments (Pulliam 1988). Source and sink dynamics are thus often part and parcel of habitat fragmentation. Forests in agricultural landscapes have been particularly fragmented, and there is much concern that fragmentation can turn source populations into sink populations.

To identify source and sink populations we need to measure reproduction, mortality, and movements among a whole set of local populations. Much of the

concern about source and sink populations has concerned migratory birds in North America (Robbins et al. 1989). For the simplest model of population change for birds, we can estimate the finite rate of increase from three parameters:

$$\lambda = P_A + P_J \beta \qquad (16.4)$$

where λ = finite rate of population growth ($\lambda = 1$ for stable populations)

P_A = adult survival rate during the year

P_J = juvenile survival rate during the year

β = number of juveniles produced per adult by the end of the breeding season

and assuming an equal sex ratio of males to females.

If we can estimate these three parameters for any population, we can determine if that population is a source ($\lambda > 1$) or a sink ($\lambda < 1$). For example, a population of house sparrows on an island off the coast of Norway produced 6.33 fledglings per female in 1993 (or 3.165 fledglings per adult bird), and the nonbreeding-period finite survival rate was 0.579 for juveniles and 0.758 for adults. Assuming a 1:1 sex ratio, from Equation 16.4 we obtain for this population

$$\lambda = P_A + P_J \beta$$

$$= 0.758 + (0.579)\left(\frac{6.333}{2}\right)$$

$$= 2.59$$

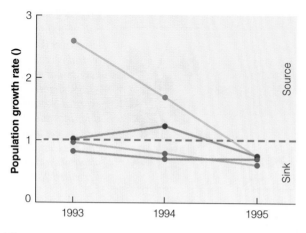

FIGURE 16.16
Population growth rates of house sparrows (Passer domesticus) *on four islands off northern Norway from 1993 to 1995. All populations were sinks in 1995, and one island (Indre Kvarøy, green) was particularly productive on average. Two islands were always sink populations because of poor juvenile survival. (From Sæther et al. 1999.)*

Given these demographic rates, this population will more than double each year and must be a source population.

Source-sink dynamics may be characteristic of particular metapopulations, or they may be a product of variation in weather from year to year. Sæther et al. (1999) studied house sparrows on four islands off Norway to measure variation in population growth rate among the islands and over time. Figure 16.16 shows that some islands on average were much more productive than others, but that all islands are sink populations in particularly severe years. Populations on each of the four islands remained nearly constant from 1993 to 1996, with immigration boosting the sinks and emigration evening the source populations. The dynamics of source-sink populations are graphic illustrations of how immigration and emigration can be just as important as reproduction and mortality as agents of population change.

Evolutionary Implications of Population Regulation

How are systems of population regulation affected by evolutionary changes? We have already discussed some of the problems involved in coevolution of predator-prey systems (see Chapter 13) and herbi-vore-plant systems (see Chapter 14). In many of these interactions, evolutionary changes operate very slowly and are difficult to detect. But recent work in ecological genetics (Endler 1986) has shown that evolutionary changes may occur very rapidly, such that the evolutionary time scale approaches the ecological time scale. Natural selection may thus impinge upon population regulation in some organisms.

Many changes in average abundance can be attributed to changes in extrinsic factors such as weather, disease, or predation. But some changes in abundance are the result of changes in the genetic properties of the organisms in a population. Such evolutionary changes can be produced by natural selection. Pimentel (1961) catalogs some spectacular examples of genetic changes playing a role in population limitation. For example, the population of the herbivorous Hessian fly was reduced drastically in Kansas after 1942 when genetically altered, fly-resistant varieties of wheat were introduced. Another example is the myxomatosis-rabbit interaction in Australia (discussed in Chapter 15) in which evolutionary changes occurred in both the virus and the rabbit.

Genetic changes in populations can affect the interspecific interactions that limit abundance. The coevolution of interacting populations of predator and prey, disease and host, and food plant and herbivore may have implications for population dynamics. The important point is that we should not assume that the ecological traits of species are constant and unchanging in ecological time. In particular, an evolutionary perspective on questions of population limitation and regulation serves as a warning with respect to the continual introduction of new species into ecological communities of distant areas.

Self-regulatory populations present yet another problem in evolutionary ecology. Under what conditions should we expect a population to be regulated by intrinsic processes involving spacing behavior in the broad sense, including territoriality, dispersal, and reproductive inhibition? We might expect that vertebrates, with their relatively complex behavior, would be the most obvious species to show intrinsic regulation. Wolff (1997) has suggested a conceptual model that predicts which vertebrates have the potential for intrinsic regulation. He discusses mammals in particular, but similar arguments could be made for birds and other vertebrates. The key to Wolff's model is that territoriality in female mammals has evolved as a counterstrategy to infanticide committed by strange females. Infanticide is a mechanism of competition by which intruders usurp the breeding space of residents and increase their fitness by killing the offspring of resident females.

Female mammals should evolve territorial behavior to defend their young from infanticide only if young are not mobile at birth. Females with precocial young, which have their eyes open and can move very soon after birth, will not be susceptible to infanticide and will not defend territories. These predictions from Wolff's model are consistent with most of what is known about mammalian social systems. For example, hares have precocial young while rabbits have altricial [1] young. Infanticide is unknown in hares but is known to occur in rabbits. Many carnivores (for example, lions) have altricial young, are subject to infanticide, and are territorial. By contrast, kangaroos have altricial young but carry them around in a pouch so that they are not vulnerable to infanticide. None of the kangaroo species are territorial.

Another feature of self-regulation in mammals is reproductive suppression of juveniles (Wolff 1997). If juveniles do not disperse from their natal area, they risk the possibility of breeding with close relatives. Selection against inbreeding has molded the dispersal pattern of mammals such that male juveniles will emigrate while female offspring remain near their natal site (Dobson 1982, Greenwood 1980). But high density may make dispersal costly due to aggressive encounters

---

[6]Altricial young are typically blind, naked, and cannot move around at birth.

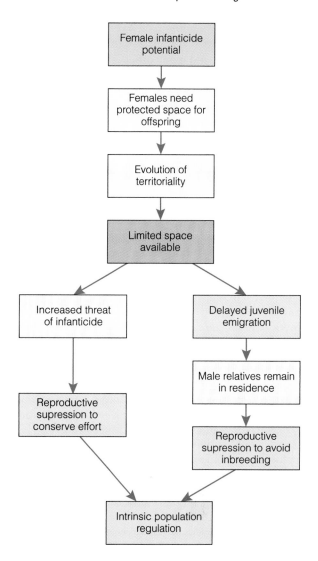

FIGURE 16.17

Wolff's hypothesis for the evolution of intrinsic population regulation in mammals. Spacing behavior is the key mechanism to evolve in species that compete for space free from infanticidal individuals. The demographic attributes that contribute to population regulation are shown in blue boxes. (From Wolff 1997.)

such that all juveniles stay near the birth place. At high density, adults may suppress sexual maturation of their offspring through pheromones in order to prevent inbreeding, especially if space for breeding is limited. The result can be that a large fraction of the population is not breeding, as has been observed in many rodents, primates, and wolves. This reproductive suppression of juveniles at high density acts as a density-dependent factor to potentially regulate the population.

Figure 16.17 summarizes Wolff's model for the evolution of intrinsic regulation in mammals. Many mammal species and many other vertebrates will not be subject to potential infanticide, and these species would be expected to be subject to extrinsic regulation by predators, food shortage, disease, or weather. Note that intrinsic regulation is not in itself an evolved strategy. What evolves are behavioral strategies such as territoriality, dispersal, and reproductive inhibition, and these individual strategies can result in population regulation at the level of the population. Evolution works at the level of the individual, in most cases, and not at the population level.

Summary

Populations of plants and animals do not increase without limits but show more or less restricted fluctuations. Two general questions may be raised for all populations: (1) What stops population growth? (2) What determines average abundance?

Three general theories answer these two questions by focusing on the interactions between the population and the environmental factors of weather, food, shelter, and enemies (predators, parasites, and diseases). The biotic school suggests that density-dependent factors are critical in preventing population increase and in determining average abundance. Natural enemies are postulated to be the main density-dependent factors in many populations. The climate school emphasizes the role of weather affecting population size and suggests that weather may act as a density-dependent control. By contrast, the self-regulation school focuses on events within a population, on individual differences in behavior and physiology. The general premise of this school is that abundance may change because the quality of individuals changes. Population increase may be stopped by deterioration in the quality of individuals as density rises, rather than by a change in environmental factors. Average abundance may be altered by genetic changes in populations. Quality and quantity are both important aspects of populations.

Population regulation theory has focused on equilibrium conditions, and many ecologists now emphasize non-equilibrium concepts and ask what factors reduce stability for populations. The spatial scale of a study affects conclusions about stability, and if a population is subdivided into local populations, stability may be increased for the entire population. *Metapopulations*, or clusters of local populations, are critical foci for conservation as habitats are broken up into small, isolated blocks.

The mortality agents affecting a particular population may cause *additive* or *compensatory* losses. Additive losses allow simple population arithmetic, and additive mortality factors may limit or regulate population density. Compensatory mortality occurs when the causes of death are replaceable, so that if one factor does not cause mortality, some other factor will. Compensatory mortality may be irrelevant to both limitation and regulation.

The theories of population regulation are not mutually exclusive but overlap, and a synthesis of several approaches may be most useful in attempting to answer practical questions. The limitation and regulation of populations are critical areas of theoretical ecology because they are central to many questions of community ecology and because they have enormous practical consequences, which we will explore in the next three chapters.

Key Concepts

1. Two questions are central to population dynamics: What stops population growth? and What determines average abundance?

2. To stop population growth, natality, mortality, or movement rates must depend on population density. Population regulation requires density dependence.

3. Biotic agents such as predators and diseases can limit or regulate populations, as can climatic and physical factors such as temperature, water, and nutrients.

4. Individual differences in physiology, genetics, or behavior can limit or regulate populations through intraspecific competition or self-regulation.

5. Populations may be subdivided into local populations or metapopulations that may go extinct and be recolonized by dispersing individuals. Local populations may be unstable while the entire metapopulation is stable.

6. Mortality agents may be additive or compensatory. Compensatory mortality is replaceable and thus may not limit or regulate population density.

Selected References

Brawn, J. D and S. K. Robinson, 1996. Source-sink population dynamics may complicate the interpretation of long-term census data. *Ecology* 77:3–12.

Hixon, M. A. 1998. Population dynamics of coral-reef fishes: Controversial concepts and hypotheses. *Australian Journal of Ecology* 23:192–201.

Matthiopoulos, J., R. Moss, and X. Lambin. 1998. Models of red grouse cycles. A family affair? *Oikos* 82:574–590.

Murdoch, W. W. 1994. Population regulation in theory and practice. *Ecology* 75:271–287.

Peterson, R. O., N. J. Thomas, J. M. Thurber, J. A. Vucetich, and T. Waite. 1998. Population limitation and the wolves of Isle Royale. *Journal of Mammology* 79:828–841.

Pierson, E. A. and R. M. Turner. 1998. An 85-year study of saguaro (*Carnegiea gigantea*) demography. *Ecology* 79:2676–2693.

Porter, W. F. and H. B. Underwood. 1999. Of elephants and blind men: Deer management in the U.S. National Parks. *Ecological Applications* 9:3–9.

Pulliam, H. R. 1988. Sources, sinks, and population regulation. *American Naturalist* 132:652–661.

Sinclair, A. R. E. and R. P. Pech 1996. Density dependence, stochasticity, compensation and predator regulation. *Oikos* 75:164–173.

Singer, F. J. 1998. Thunder on the Yellowstone revisited: An assessment of management of native ungulates by natural regulation. *Wildlife Society Bulletin* 26:375–390.

Turchin, P. 1999. Population regulation: A synthetic view. *Oikos* 84:160–163.

Watson, A., and R. Moss. 1970. Dominance, spacing behaviour and aggression in relation to population limitation in vertebrates. In *Animal Populations in Relation to Their Food Resources*, ed. A. Watson, pp. 167–218. Blackwell, Oxford, England.

Wolff, J. O. 1997. Population regulation in mammals: An evolutionary perspective. *Journal of Animal Ecology* 66:1–13.

Questions and Problems

16.1 Morris (1957, p. 49), in discussing the interpretation of mortality data in population studies, states:

> We tend to overlook the fact that these mortality estimates do not represent an ultimate objective in population work. Long columns of percentages, which are sometimes presented only with the conclusion that high percentages indicate important mortality factors and low percentages indicate unimportant ones, contribute little to our understanding of population dynamics.

Discuss this claim.

16.2 Density-dependent relationships can be looked for by studying different local populations living in different patches (spatial density dependence) or by following one local population over several years (temporal density dependence). Discuss the interpretation of these two types of data with regard to the problem of regulation.

16.3 Singer et al. (1997) reported on the population dynamics of elk in Yellowstone National Park, with the following data from 1975–1991. Calf recruitment is the number of calves per adult female in autumn, survival rates are finite annual rates. Population estimates are for autumn of each year, and calf data are from the following summer and winter. There is a gap in the data between 1978 and 1982.

Year	Summer calf recruitment rate	Summer calf survival rate	Winter calf survival rate	Population size
1975-6	0.18	0.22	0.36	15,797
1976-7	0.30	0.38	1.00	13,305
1977-8	0.27	0.32	0.59	15,350
1982-3	?	?	0.76	19,523
1983-4	?	?	0.52	20,837
1984-5	0.52	0.80	0.28	21,115
1985-6	0.37	0.61	0.54	22,115
1986-7	0.44	0.69	0.38	19,825
1987-8	0.26	0.40	0.32	21,706
1988-9	0.20	0.31	0.17	20,619
1989-90	0.30	0.37	0.35	17,843
1990-91	0.78	?	1.00	17,950

Are any of these three measures of recruitment or mortality density dependent? What can you conclude about population regulation in Yellowstone elk? Compare your conclusions with those of Singer et al. (1997).

16.4 Darwin (1859) wrote in *The Origin of Species* (Chapter 2): "Rarity is the attribute of a vast number of species of all classes, in all countries." Discuss the possible effects of rarity and abundance on population-regulation mechanisms.

16.5 Mourning doves (*Zenaida macroura*) are hunted in eastern North America. McGowan and Otis (1998) reported the following data for two populations of doves in South Carolina:

Area	Recruitment per adult bird	Adult survival rate (annual)	Juvenile survival rate (annual)
Bennettsville	3.400	0.359	0.118
Eutawville	2.325	0.359	0.118

Calculate the finite rate of population growth for these two dove populations from Equation (16.4), and discuss what management action these results might indicate.

16.6 Murray (1979, p. 66) states: "The fact that a particular population exhibits sigmoid growth does not constitute evidence that density-dependent factors are acting." Is this correct? Why or why not?

16.7 The saguaro is a prominent columnar cactus of the Sonoran Desert of Arizona and northern Mexico. Saguaro are long-lived perennials, and individuals may reach 150–200 years of age. Pierson and Turner (1998) reported the following data from a long-term study of four populations in an ungrazed desert preserve:

		Census		
Plot	1964	1970	1987	1993
North	284	265	232	221
South	1308	1316	—	1087
East	1367	1394	—	1277
West	603	586	—	459

What would you conclude about the population dynamics of these cacti from these data? What additional data would you like to have to predict future population trends?

16.8 Can a population persist without regulation? How could you determine if a population was persisting without regulation? Read Strong (1984) and Reddingius and den Boer (1970) and discuss.

16.9 Sinclair (1989) tabulates the causes for density dependence under six categories: space, food, predators, parasites, disease, and social interactions. Compare and contrast these categories with those proposed by Andrewartha and Birch (1954).

16.10 Read Wynne-Edwards's (1986) ideas on population regulation and group selection and the review of his book by Bell (1987). Are there any circumstances under which group selection could produce mechanisms of population regulation?

16.11 Is the dispersal rate in mammals density dependent? Read Wolff's (1997) arguments and discuss under what conditions emigration might regulate population density in mammals.

Overview Question

Local populations can be classified as source populations ($\lambda > 1$) or sink populations ($\lambda < 1$). How would you determine for a metapopulation of plants which local populations were sources and which were sinks? Discuss the application of population regulation theories to a metapopulation of plants containing sources and sinks.

CHAPTER 17

Applied Problems I:
Harvesting Populations

To MANAGE A POPULATION EFFECTIVELY, we must have some understanding of its dynamics. The list of populations destroyed by inadequate management throughout human history should serve as both a warning and as a stimulus for us to achieve a better understanding of harvesting principles. The central problem of economically oriented fields such as forestry, agriculture, fisheries, and wildlife management is how to produce the largest crop without endangering the resource being harvested. The problem may be illustrated with a simple example from forestry. If you were managing a forest woodlot that was growing to maturity, you obviously would not cut the trees when they were saplings because this would yield little wood production and less profit. At the other extreme, you would not let the trees grow too old and begin to rot because you would get little timber to sell. Somewhere between these two extremes is some optimum point to harvest the trees, and the problem is how to identify it.

Next to forestry and agriculture, the greatest amount of work on the problem of optimum harvesting has been done in fishery biology, especially because of the tremendous economic importance of marine fisheries in particular. Many marine fisheries have dwindled in size since the 1920s because of overfishing, and this has stimulated a great deal of research on "the overfishing problem."

For any harvested population, the important unit of measure is the crop or *yield*. The yield may be expressed in *numbers* or in *weight* of organisms, and it always involves some unit of time (often a year). We are interested in obtaining the optimum yield from any harvested population. We will begin by defining *optimum yield* very specifically, and at the end of the chapter we will reconsider other ways of defining *optimum*. The concept of *maximum sustained yield* has been the basis of scientific resource management since the 1930s (Larkin 1977). Let us consider first the simple situation in which maximum yield in biomass is defined as the optimum yield. Implicit in this concept is the idea of a sustained yield over a long time period.

Russell (1931) was one of the first to deal in detail with the harvesting problem in fisheries. In any exploited fish population, there is usually a portion of the population that cannot be caught by the type of gear used or is purposely not harvested. The harvestable sector of the population is called the *stock*. For a fishery, interest normally centers on yield in weight, so instead of individuals we will deal in biomass units. Russell pointed out that two factors decrease the weight of the stock during a year: natural mortality and fishing mortality. Similarly, two factors increase the weight of the stock: growth and recruitment.[1]

---

[1]Recruitment in fisheries is usually measured when the fish reach a certain size or age. Recruitment thus includes natality and early life history survival and growth.

Consequently, one can write the following simple equation to describe this relationship:

$$S_2 = S_1 + R + G - M - F \qquad (17.1)$$

where S_2 = weight of the stock at the end of the year

   S_1 = weight of the stock at the start of the year

   R = weight of new recruits

   G = growth in weight of fish remaining alive

   M = weight of fish removed by natural deaths

   G = yield to fishery

If we wish to balance the fish population, $S_1 = S_2$, and hence

$$R + G = M + F \qquad (17.2)$$

This means that in an unexploited stage ($F = 0$), in which the stock biomass remains approximately constant from one year to the next, all growth and recruitment is on the average balanced by natural mortality. When exploitation begins, the size of the exploited population is usually reduced, and the loss to the fishery is made up by compensatory changes such as (1) greater recruitment rate, (2) greater growth rate, or (3) reduced natural mortality. In some populations, none of these three occurs, and the population is exploited to extinction because the right side of Equation (17.2) always exceeds the left side.

Note that stability at *any* level of population density is described by the equation:

Recruitment + growth = natural losses + fishing yield

Thus a crucial question arises: What level of population stabilization provides the greatest weight of catch to the fishery? One of the first attempts to solve this problem was made by Graham (1935), who proposed the *sigmoid-curve theory*.

Start by considering a very small stock of fish in an empty area of the sea, said Graham. At what rate will such a stock increase in size? Graham suggested that the growth of this population would follow a sigmoid curve like the one described by the logistic equation (Figure 17.1) . Initially, the population grows more slowly in absolute size, reaches a maxi-

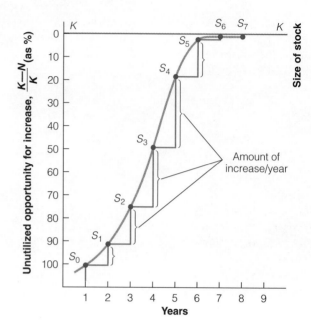

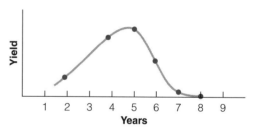

FIGURE 17.1

Use of sigmoid curve theory to describe the growth of a population that could be exploited. The amount of increase per year is the yield that could be taken by the fishery. Maximum yield is obtained by keeping the population at one-half of carrying capacity. (After Graham 1939.)

mum rate of increase near the middle of the curve, and grows slowly again as it approaches the asymptote of maximal density. We can use the terminology of the logistic equation to show that two factors interact to determine the amount of increase per year. For symplicity, let $K = 200$ units and $r = 1.0$:

Point on curve	Population size	$\dfrac{K-N}{K}$	rN	Amount of increase per year
S_1	20	0.90	20	18
S_2	50	0.75	50	38
S_3	100	0.50	100	50
S_4	150	0.25	150	38
S_5	180	0.10	180	18

According to the logistic equation, the amount of population increase depends on the carrying capacity (K), the intrinsic rate of increase (r), and the current population size (N):

$$\frac{dN}{dt} = r N \left(\frac{K - N}{K} \right) \qquad (17.3)$$

and this is maximal at the midpoint of the curve (S_3).

If we wish to maintain the maximal yield from such a population, Graham pointed out, we should keep the stock around point S_3 of the curve. The important point here is that the highest production from such a population is not near the top of the curve, where the fish population is relatively dense, but at a lower density. This can be expressed as the first rule of exploitation: *Maximum yield is obtained from populations at less than maximum density.*

All the vital statistics of an exploited population—recruitment, growth, and natural mortality—may be a function of population density and also of age composition. Because in most fisheries we do not know how these vital statistics relate to density or age, we make some simplifying assumptions. Two alternative approaches have been developed for determining optimum yield: *logistic models* and *dynamic pool models* (Schaefer 1968). We next discuss each of them in turn.

Logistic Models

In logistic models[2], we do not distinguish among growth, recruitment, and natural mortality but instead combine them into a single measure, *rate of population increase*, which is a function of population size. Graham's *sigmoid-curve theory* is a classic example of this type of model. The general case can be written as:

Rate of population increase = f (population size) − amount of fishing losses

If we specify that the function of population size in this equation is a simple linear function,

$$f \text{(population size)} = r \left(\frac{K - N}{K} \right) = r - \left(\frac{rN}{K} \right) \quad (17.4)$$

we obtain the logistic equation modified for fishing losses:

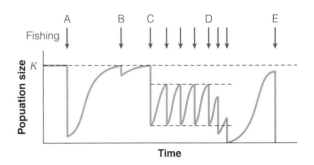

FIGURE 17.2

Schematic diagram of the assumed response of a fish population to exploitation according to logistic-type models. Periodic fishing of different intensity and frequency is indicated at the top of the diagram. At point A fishing is intensive, but the time interval is long enough that the fish population recovers to carrying capacity (K). At C a moderate intensity fishery is operating, and at D this fishing intensity is applied more frequently, causing the stock to collapse. At E excessive fishing drives the stock to extinction. Note that during every recovery phase the fish population increases logistically.

$$\frac{dN}{dt} = r N \left(\frac{K - N}{K} \right) - qXN \qquad (17.5)$$

where N = population size

t = time

r = per capita rate of population growth

K = asymptotic density (in absence of fishing)

q = catchability (a constant)

X = amount of fishing effort (so qX = fishing mortality rate)

The ecological assumptions of logistic models are that no time lags operate in the system, that age structure has no effect on the rate of population increase, and that catchability remains constant at all densities of fish. Figure 17.2 illustrates how an exploited population is postulated to respond to a series of fishing episodes in this model. This model, although crude, may be useful for populations that are in approximately steady states in the absence of fishing and that

---

[2]Also called surplus *yield models, stock production models,* or *Schaefer models.*

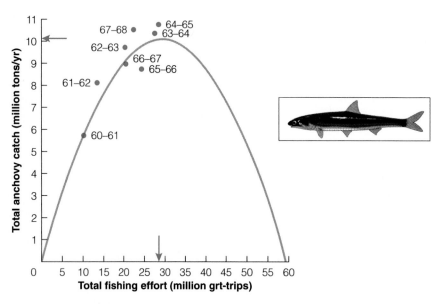

FIGURE 17.3
Relation between total fishing effort and total catch for the Peruvian anchovy fishery, 1960–1968. The effects of humans and seabirds are combined in these data. The parabola represents the logistic model fitted to these data, as in Figure 17.1. Arrow indicates maximum sustained yield and appropriate fishing effort. (After Boerema and Gulland 1973.)

do not change greatly from year to year. Because of their simplicity, logistic models can be used on fisheries with relatively few data available. The following example illustrates how this can be done.

The Peruvian anchovy (*Engraulis ringens*) is restricted in distribution to areas of upwelling of cool, nutrient-rich water along the coasts of Peru and northern Chile. The upwelling causes very high productivity in the coastal zone. The Peruvian anchovy is a short-lived fish, spawning first at about one year of age and rarely living beyond three years. It is a small fish, about 12 cm in length at one year and seldom reaching 20 cm in length. Young anchovies enter the fishery at only five months of age (8–10 cm). Anchovies occur in schools and are caught near the surface.

The Peruvian anchovy fishery was the largest fishery in the world until 1972, when it collapsed. From 1955, when the major fishery first began, the anchovy catch doubled every year until 1961. In 1970, 12.3 million metric tons were harvested, and this single-species fishery constituted 18% of total global harvest of fish. Figure 17.3 shows the total catch and the total fishing effort. These two parameters were used to fit a logistic model to the fishery (Boerema and Gulland 1973). The logistic model predicts a parabolic relationship

between fishing effort and total catch, with an optimal catch at half the carrying capacity (K). Anchovy are taken both by fishermen and by large colonies of seabirds, and these two were combined to measure the total "catch." Figure 17.3 indicates a maximum sustained yield between 10 million and 11 million metric tons, which, after subtraction of the bird share, left about 9 to 10 million tons for the fishery. From 1964 to 1971 the catch was close to the supposed maximum indicated in Figure 17.3. Note that the estimate of maximum sustainable yield in Figure 17.3 refers to average conditions over a number of years.

In 1972 average conditions disappeared, and the Peruvian anchovy fishery collapsed. Early in 1972 the upwelling system off the coast of Peru weakened, and warm tropical water moved into the area. This phenomenon—known as "El Niño" (The Child) because it often happens around Christmas—occurs about every five years and greatly changes the regional ecosystem (Mysak 1986). The productivity of the sea drops, seabirds starve, and anchovies move south to cooler waters and may congregate. In early 1972 very few young fish were found; the spawning of 1971 had been very poor, only one-seventh of normal. Adult fish were highly concentrated in cooler waters in early

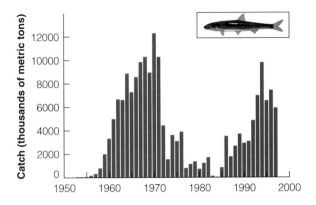

FIGURE 17.4
Total catch for the Peruvian anchovy fishery, 1950–1997.
This fishery was the largest in the world until it collapsed in
1972 during an El Niño event. In spite of reduced fishing,
it took 25 years for the fishery to recover. (Data from FAO
Yearbooks of Fishery Statistics.)

1972, and these concentrations produced large catches
for the fishermen. By June 1972 the anchovy stocks
had fallen to a low level, catches had declined drastically, and no young fish were entering the population.

The fishery was suspended to allow the stocks to
recover, but from 1972 to 1985 there was little sign of
a return of the anchovy to its former abundance.
Catches fell to low levels and began to recover only
during the 1990s, after 20 years of low catches (Figure
17.4). The economic consequences of the fishery collapse of 1972 were very great, and some of them might
have been avoided if the fishery had been closed a few
months earlier or if the fishing intensity had been
slightly less than the maximum shown in Figure 17.3.

The Peruvian anchovy has become a model case
of overfishing and has raised the important question
of how to manage fisheries in a sustainable manner.
The important message the collapse of the Peruvian
anchovy fishery conveys is the fragility of the assumptions that fish populations are in a state of equilibrium
and that average conditions never change.

Dynamic Pool Models

Dynamic pool models of harvested populations are
more biologically explicit because they include estimates of growth, recruitment, and mortality for the
population being harvested. These models originated
in a classic fisheries book by Beverton and Holt in
1957 and represented a biologically realistic approach

to fisheries management that appealed strongly to
fishery scientists. Ray Beverton and Sydney Holt revolutionized fishery science in the ten years following
World War II by applying mathematics to the problem of defining the optimum yield from a fishery. In
these models, various simplifying assumptions are
made. Natural mortality rate is assumed to be constant, independent of density, and the same for all
ages. Growth rates are assumed to be age specific but
unrelated to population density. Fishing mortality
(effort) is assumed to act just like natural mortality—
to be independent of density and constant for all ages
of fish. These assumptions are unrealistic, but they are
useful as a starting point, and they can be relaxed later
in the analysis. The object is to determine what yield
a given level of fishing mortality will produce. In this

Raymond J. H. Beverton *(1922–1995) Fisheries Laboratory,*
Lowestoft, England

simple model, the population size of R recruits after t
years in the fished population is given by the formula
for geometric decrease:

$$N_t = \mathrm{Re}^{-(F + M)t} \qquad (17.6)$$

where N_t = number of recruits alive at t years
 after entering fishery

 t = time in years since recruits
 entered fishery

 R = number of original recruits

 F = instantaneous fishing mortality rate

 M = instantaneous natural mortality rate

This is the familiar curve of geometric increase (or
decrease). If $R = 1$, this formula gives the fraction of

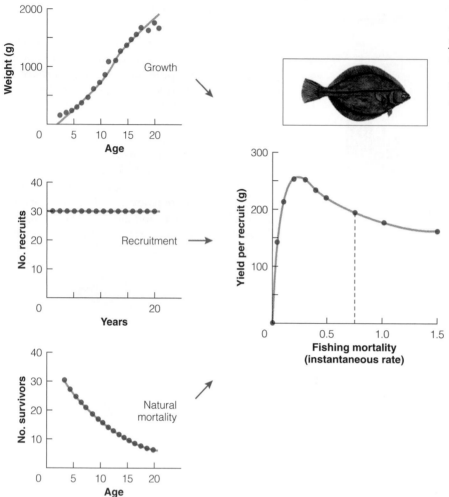

FIGURE 17.5
Simple model of equilibrium yield for plaice in the North Sea. The fishing intensity before World War II is indicated by the dashed line. (After Beverton and Holt 1957.)

recruits alive at any time since entering the fishery. The yield to the fishery in this simple model is defined as

Yield = (number in age class) x (average weight)
        x (fishing mortality rate)

summed over all age classes caught in the fishery. This can be written

$$Y = \sum_{t=t_c}^{\infty} FN_t W_t \qquad (17.7)$$

where Y = yield in weight for a year

        F = instantaneous fishing mortality rate
               per year

N_t = population size of age t fish

W_t = average weight of age t fish

t_c = age at which fish enter the fishery

Let us illustrate this simple dynamic pool model with an example from the plaice (*Pleuronectes platessa*) fishery in the North Sea. The plaice is a shallow-water flatfish that is an important commercial species in the North Sea. Plaice spawn in midwinter when females are five to seven years old and males are four to six years old. Females can lay up to 350,000 fertile eggs, an enormous reproductive potential that is balanced by an equally high mortality. On the average, all but

TABLE 17.1 **Calculation of equilibrium yield per recruit for North Sea plaice for a fishing mortality of 0.5.**

Fishing year	Age at midpoint (yr)[a]	(1) = W_t average weight (g)	(2) = N_t fraction of recruits surviving to this age, $e^{-(F+M)t}$	(3) Yield to fishery, F	Product (1) × (2) × (3)
0–1	4.2	158	0.741	0.5	58.54
1–2	5.2	237	0.407	0.5	48.23
2–3	6.2	331	0.223	0.5	36.91
3–4	7.2	435	0.122	0.5	26.54
4–5	8.2	546	0.067	0.5	18.29
5–6	9.2	664	0.037	0.5	12.28
6–7	10.2	784	0.020	0.5	7.84
7–8	11.2	904	0.011	0.5	4.97
8–9	12.2	1024	0.006	0.5	3.07
9–10	13.2	1143	0.003	0.5	1.71

Total yield per recruit 218.38 g

[a]The age at recruitment is 3.7 years.

Note: The average weight is obtained from the growth curve shown in Figure 17.5.
The fraction of recruits is calculated by applying a constant loss per year of $(F + M)$, which in this example is $(0.5 + 0.1)$.
The total yield per recruit is obtained from the formula

$$Y = \sum_{t_c}^{\infty} F N_t W_t$$

ten fish out of every million eggs laid must die before reaching maturity, and the actual range observed by Beverton (1962) during 26 years was between 999,970 and 999,995 dying for every 1 million eggs laid. Much of this loss occurs during the pelagic phase, when the eggs float as plankton until hatching, and the larval plaice are carried about by water currents in the North Sea. After about two months the larval plaice settle out on nursery areas off the sandy coasts of the Netherlands, Denmark, and Germany. There the young plaice remain until between two and three years of age, when they begin to move off the coast and toward the middle of the North Sea. They enter the commercial fishery between three and five years of age, at a length of 20–30 cm.

The plaice population has remained fairly stable, with the exception of the periods during the world wars, when fishing was reduced and stocks increased. We can illustrate a dynamic pool model most easily in this type of near-equilibrium condition. First, we must determine growth rate with respect to age in the plaice, and we can do that with samples from the fishery (Figure 17.5). We assume in this simple model that growth does not depend on population density. Second, we need to specify recruitment, and we assume a constant number of recruits each year. For

the plaice, this is not an unreasonable first approximation (see Figure 17.5). Third, we must determine the natural mortality rate. We can do this by mark-and-recapture techniques (see Chapter 9) or by indirect means. We assume that in the simple case natural mortality is constant at all ages and at all population densities. For the plaice, Beverton and Holt (1957) estimated M = 0.10, and Figure 17.5 shows how a cohort of recruits would decline according to predictions of natural mortality *only*.

Because recruitment is assumed constant, we can express the yield as yield per recruit, and by combining the three factors we obtain the yield curve shown in Figure 17.5. An example of how this yield for plaice was calculated for a fishing mortality of 0.5 is given in Table 17.1 . The yield per recruit was then calculated for several values of fishing mortality to obtain the curve in Figure 17.5. This is only an approximate calculation because we should use calculus instead of finite summation to find the yield (details in Beverton and Holt 1957). Figure 17.5 also shows the pre-World War II fishing intensity (F = 0.73), which was clearly not at the point of optimum yield.

We have treated fishing mortality in the same way that we have treated natural mortality, using humans as just another "predator" in the system. This fishing

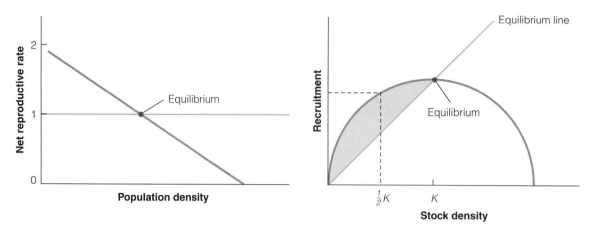

FIGURE 17.6

Recruitment curve for a hypothetical population growing according to the logistic model. The graph at left is the same as Figure 11.2, and the graph at right is the same data plotted in a slightly different way (recruitment as population density multiplied by net reproductive rate). The hatched area shows the potential surplus that could be fished. Maximum yield is obtained at 0.5 K in the logistic model, as shown by the dashed line. An equilibrium occurs at the point at which recruitment into the population just balances losses from the population.

mortality rate must be converted into fishing effort before the results of a yield analysis, such as that in Figure 17.5, can be applied to an operating fishery. This application is a complex problem that revolves around the types of equipment used, the equipment's efficiency, the interactions between different units of equipment, the spatial and seasonal patterns of exploitation, and the area occupied by the stock (Winters and Wheeler 1985). Gulland (1988) discusses these problems in some detail, and the analysis depends on the type of fishery operation.

Once we have built a dynamic pool model of a fishery, we can test it by regulating the fishery accordingly. Thus for the North Sea plaice we would predict from Figure 17.5 that an increased yield would result from lowering fishing mortality to one-third or one-half the prewar level of 0.73. This is the critical test of any model: Does it predict accurately? The North Sea plaice is one of the success stories of fisheries management. After World War II, fishing effort was reduced to the level of maximum yield predicted by Beverton and Holt (see Figure 17.5), and the landings from the fishery nearly tripled over the next 30 years (Rijnsdorp and Millner 1996). By the 1990s fishing effort increased again, the stock declined, and yield fell 35%, as the model illustrated in Figure 17.5 would predict.

This approach can identify the annual equilibrium yield of the fishery, but within it are several potential

problems. For one thing, it assumes that a constant number of recruits enter the usable stock every year. But does any fishery in fact have a constant recruitment? A constant recruitment implies that the number of recruits does not depend on population size; to put it another way, it assumes that two adult fish could produce the same number of progeny as 10,000 adults. This is on the face of it quite impossible, and thus we are led to inquire into the relationship between population size (stock) and recruitment, which is just another way of discussing the problem of population regulation. Fish populations, even when exploited, are still subject to population regulation. Recruitment in exploited populations is always measured as a rate, such as the number of young fish entering a fishery per year.

Some component of the vital statistics—births, deaths, or dispersal—must be related to population density in order to prevent unlimited population growth. As we saw in Chapter 11, population growth cannot be curtailed unless the net reproduction curve is depressed below 1.0 at high population densities (see Figure 11.2, p. 158). This can occur if adult mortality increases with density, but fishery ecologists think that natural mortality of adult fish is independent of density. Fecundity does decline at high population density in some fishes (Bagenal 1973), but most of the regulation in fish populations is believed to occur in the early life-cycle stages. One of the axioms of modern fisheries ecology is that the important density-dependent processes in fish occur during the first few weeks or months of life (Cushing and Harris 1973).

Some of the most important early research on the relationship between stock and recruitment was done in the 1950s by Bill Ricker working for the Fisheries Research Board of Canada at the Pacific Biological Station in British Columbia. Figure 17.6 shows a stock-recruitment graph for a population subject to logistic population growth (compare it with Figure 12.2). Two points on this curve are fixed. Where there is no stock, there is no recruitment. The point at which stock equals recruitment is an equilibrium point. The shape of recruitment curves is important for fisheries management. Two general shapes

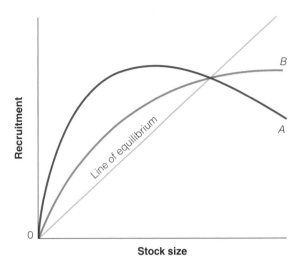

FIGURE 17.7

Two possible relations between stock and recruitment for an exploited population. The straight line represents all the conditions in which recruitment balances stock losses, and the point at which the recruitment curves cross this diagonal line represents an equilibrium point. Curve A, with a descending upper section, is the Ricker model, which resembles in shape the logistic model shown in Figure 17.6; curve B, which tends to plateau, is the Beverton and Holt (1957) model.

William E. Ricker *(1908–) Fisheries Scientist, Pacific Biological Station, Nanaimo, B.C.*

may occur (Figure 17.7). Beverton and Holt (1957) suggested a curve that rises to an asymptote at very high stock densities. Maximum recruitment in the Beverton-Holt model always occurs at maximum stock size. Ricker first suggested in 1958 that the recruitment curve may peak below equilibrium density, so that there would be a maximum in recruitment at intermediate stock sizes (Ricker 1975, Hilborn and Walters 1992).

The Beverton-Holt recruitment curve (see Figure 17.7) is essentially a logistic population model that leads to a smooth asymptotic population growth curve (as in Figure 11.4). The Ricker recruitment curve is

more closely related to the discrete generation analogs of the logistic equation in which population growth may show large oscillations about the carrying capacity (see Figure 11.3; Hilborn and Walters 1992). Species that are short-lived are more likely to show Ricker-type recruitment curves, whereas long-lived species will show Beverton-Holt–type recruitment.

Recruitment in fish populations is highly variable from one year to the next, and most stock-recruitment relations show great scatter. Figure 17.8 illustrates this for North Sea plaice, and Figure 17.9 for sockeye salmon in British Columbia. This variation is presumed to be caused by oceanographic effects on the survival of young fish. The variation seems to obscure any density effects and makes it difficult to fit the recruitment curves of Figure 17.7 to field data (Walters 1986, Walters and Korman 1999). The variability in recruitment may be quite different in different species. For example, in the North Sea haddock, the abundance of recruits has varied 500-fold in 30 years of study, whereas the variation in recruitment of the North Sea plaice has been

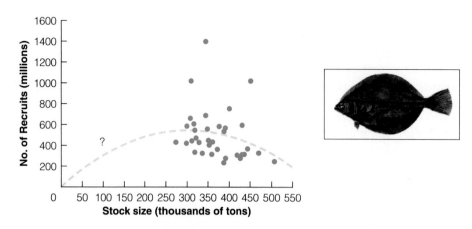

FIGURE 17.8

Stock and recruitment relationship in plaice in the North Sea, 1957–1991. The number of progeny surviving to recruitment bears little relation to the biomass of adult fish. The stock-recruitment relationship can only be guessed at because of a lack of data from the lower part of the line and the great scatter of data points. (Data from International Council for the Exploration of the Sea, North Sea Flatfish Working Group, 1998.)

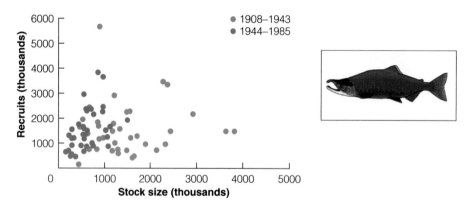

FIGURE 17.9

Stock and recruitment relationship of sockeye salmon of the Skeena River, British Columbia, from 1908–1985. Recruits are measured as the number of salmon returning four to five years after spawning, and stock size is number of spawning salmon. There is little relationship between stock size and recruitment, and no clear indication that allowing more spawners past the fishery (above a minimum stock size around 500,000) will produce a larger return. (Data from Department of Fisheries and Oceans, Canada.)

only sixfold in 26 years of study (Beverton 1962). If the amount of recruitment is highly variable, a population being exploited may be susceptible to overfishing. The Peruvian anchovy is a good example of this problem.

Why do some year-classes fail? This is one of the most important and difficult problems being addressed by fisheries ecologists. The critical period or match/mismatch hypothesis (Hjort 1914) postulates that early in the life of most fishes there is a short time period of maximum sensitivity to environmental factors. It is commonly assumed that oceanographic effects (particularly current patterns, winds, and water temperature) on food availability for newly hatched fry are critical in determining year-class size in fishes and that either temporal or spatial mismatching can produce recruitment failure.

One way to search for explanations of year-class failures in fish is to look for correlations between environmental factors such as water temperature and the relative success of recruitment of young fish. Fisheries

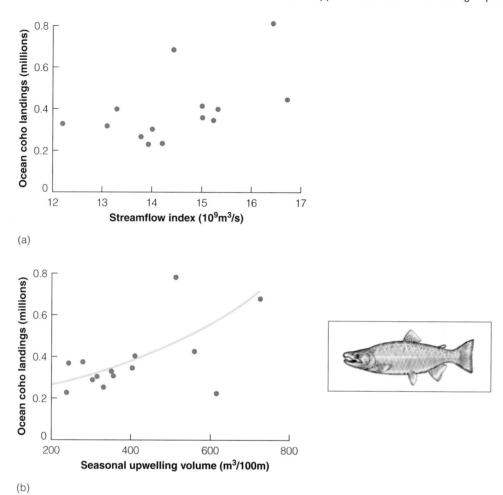

FIGURE 17.10
Correlations between coho salmon (Oncorhynchus kisutch) *production in Oregon and two environmental variables: (a) streamflow rates during freshwater residence of young salmon; and (b) coastal upwelling when young salmon move out into the ocean. The correlation with coastal upwelling is better than that with streamflow. (After Nickelson and Lichatowich 1983.)*

ecologists have found a great number of correlations between environmental factors and recruitment success (Shepherd et al. 1984). Figure 17.10 illustrates one example of the kinds of correlations observed. Coho salmon in Oregon spend their first year of life in freshwater, move into the ocean to grow to adult size, and then return to freshwater to spawn at age 4. Survival during the early life-history phases of salmon appear to be critical in determining population sizes of adult fish, and there is considerable controversy about exactly when in the larval-juvenile stage of life limitation occurs. For Oregon coho salmon Figure 17.10 would suggest that the early marine stages are the most criti-

cal. However, all the correlations between fish abundance and environmental factors show a wide scatter of points that makes it difficult to predict accurately. But more importantly, these correlations do not specify the mechanisms by which populations are limited, and if we are to understand fish population dynamics we must uncover the mechanisms that cause variations in recruitment (Rothschild 1986, Myers 1998).

The details of how environmental factors affect recruitment are known for relatively few fishes. The English sole (*Paraphrys vetulus*) is an important commercial bottom fish off the coast of Oregon. Kruse and Tyler (1989) constructed a model of recruitment

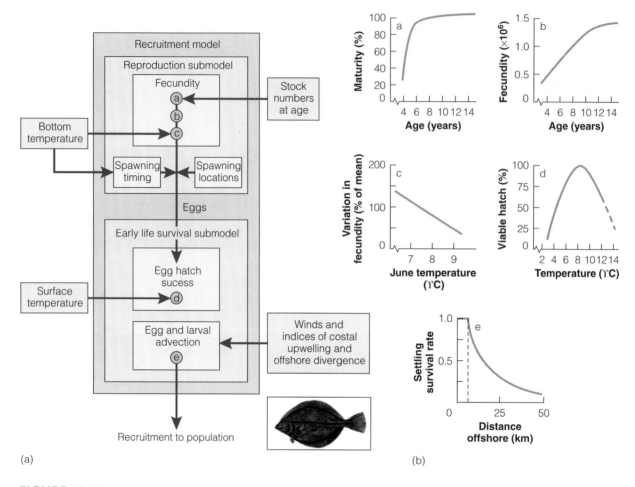

(a)

(b)

FIGURE 17.11

Recruitment model for the English sole off the coast of Oregon. (a) Possible components and processes associated with reproduction and early survival of a year-class. Functional relationships (a) to (e) are shown in (b). Temperature in the coastal zone waters and the upwelling currents along the coast are the two major determinants of recruitment. If currents carry larval fish too far offshore, they cannot settle to the bottom and survive. (From Kruse and Tyler 1989.)

in English sole by describing the components that together determine the success or failure of young sole to survive to age 4 when they enter the fishery. Figure 17.11 illustrates their recruitment model and the relationships it incorporates. Recruitment in the English sole is limited by several factors acting on spawning adults, eggs, larval, and juvenile fish, and can be understood and predicted only by studying the complex ecological relationships that determine reproduction and survival at each stage.

Laboratory Studies on Harvesting Theory

Animal populations vary greatly in their ability to withstand sustained losses. This ability is related to their size and to their capacity for increase. Laboratory studies of insects and small fishes have been particularly useful in analyzing the basic principles of harvesting theory.

A particularly clear demonstration of the relationships among rate of exploitation, population size,

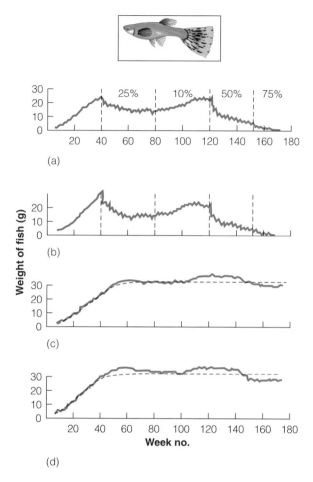

FIGURE 17.12

Population biomass changes in guppies maintained in the laboratory. (a) and (b): Experimental populations subjected to harvesting after week 40 at the indicated rates. (c) and (d): Control populations that were not exploited.(Silliman and Gutsell 1958.)

and yield is shown in the experimental work of Silliman and Gutsell (1958) on guppies (*Lebistes reticulatus*) in laboratory aquariums. They maintained two populations as unmanipulated controls and two populations as experimental fisheries subjected to a sequence of four rates of fishing (Figure 17.12). Populations were counted once each week and cropped every third week, so that (for example) a 25% cropping rate would mean that every fourth fish was removed from this population during the census at weeks 3, 6, 9, and so on.

Control guppy populations reached a stationary plateau by week 60 and remained there until the end of the experiment in week 174. Cropping at 25% once every three weeks reduced the experimental populations to about 15 grams of biomass, compared with 32 grams for the controls. Reduction of the cropping to 10% increased both experimental populations to about 23 grams of biomass, and the imposition of 50% cropping in week 121 caused a decline in population size to about 7 grams. A cropping intensity of 75% every third week was too great for these fish to withstand, and both experimental populations were driven extinct by "overfishing."

We can use these data to construct a yield curve directly. We weigh the fish removed at each cropping and obtain the following results:

Exploitation Rate (%)	Weeks of the experiment (Fig. 17.12)	Experimental Population (avg. wt. In g/cropping)	
		A	B
25	61-76	3.35	3.58
10	100-118	2.20	2.51
50	136-148	3.88	3.58
75	163-172	0.82	0.40

Only data for the last half of each exploitation period are used to approximate an equilibrium fishery condition. If we plot fishing mortality against yield, we obtain the yield curve for guppies shown in Figure 17.13 . A maximum yield is obtained at exploitation rates between 30% and 40%, with a population biomass of 8–12 grams, compared with 32 grams in unexploited controls (Figure 17.12).

The experiments on guppies by Silliman and Gutsell (1958) clearly illustrate four principles of exploitation:

1. Exploitation of a population reduces its abundance, and the greater the exploitation, the smaller the population becomes.

2. Below a certain level of exploitation, populations are resilient and compensate for removals by surviving or growing at increased rates.

3. Exploitation rates may be raised to a point at which they cause extinction of the resource.

4. Somewhere between no exploitation and excessive exploitation is a level of maximum sustained yield.

Before we see how well these principles apply to natural populations, let us consider in more detail the concept of maximum sustained yield.

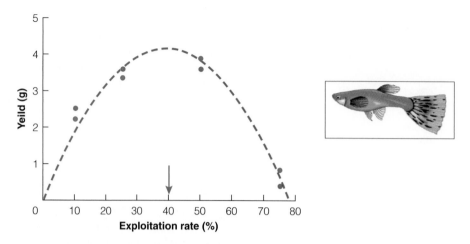

FIGURE 17.13
Equilibrium yield in relation to fishing intensity for two laboratory populations of guppies studied by Silliman and Gutsell (1958). Maximum yield could be obtained at a fishing mortality of approximately F = 0.5 (an exploitation rate of 40% removal triweekly), indicated by the red arrow. Population A is represented by a red dot, population B by a blue dot.(After Silliman and Gutsell 1958.)

The Concept of Optimum Yield

The concept of maximum sustained yield has been considered the "optimum" or "best" yield and has dominated fisheries management since the 1930s (Larkin 1977). In many situations, maximum yield is not a desirable goal. In sport fisheries, for example, the object is to maximize recreation, and the desirable fish are often the large ones. Hunters of large mammals may place more emphasis on the trophy status of the animals they harvest, and the harvesting of wildlife populations is often done without the goal of maximum sustained yield.

In any fishery that harvests several species at the same time, it is impossible to harvest at maximum sustained yield for all species. One species may be overharvested while another caught in the same nets may be underharvested. Even within a single species, there are often subpopulations, or *stocks*, that have different resilience to harvesting. Harvesting of Pacific salmon operates on mixtures of stocks from different river systems and different spawning areas within one system. The result is that less productive salmon stocks are overfished, and even driven extinct, while more productive stocks are not fully utilized (Walters 1986).

Additionally, any specification of optimum yield must include economic factors. The real yield from fisheries is not fish but dollars, and economists have

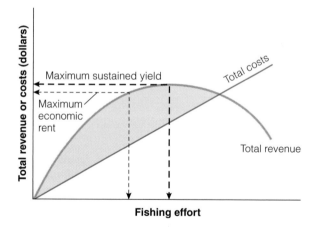

FIGURE 17.14
A simple economic model of a fishery in which costs are directly related to fishing effort, and revenue is directly related to yield. The shaded zone is the area in which revenue exceeds costs. Maximum economic rent is achieved when the difference between revenue and cost is maximal, and this is below maximum sustained yield in this simple model.

long recognized that it is poor business to operate a fishery at maximum yield. H. Scott Gordon (1954) was one of the first to show that there is a level of harvesting associated with maximum sustained economic revenue, and that this usually occurs at a lower fishing intensity than the maximum sustained yield.

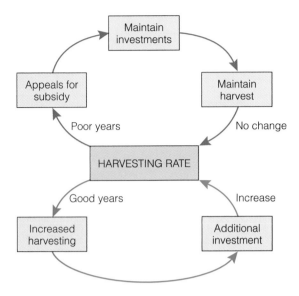

FIGURE 17.15
Ludwig's ratchet: For a fluctuating resource, continuing economic investment and ecological optimism fuel positive feedback that ratchets up the harvest rate to unsustainable levels and causes the eventual collapse of the fishery. (Modified from Ludwig et al. 1993.)

What is optimal to an economist is not necessarily optimal to a biologist.

Figure 17.14 shows a simple economic model for a fishery. Total costs are assumed to be proportional to fishing effort. The revenue or benefit from fishing is assumed to be directly proportional to the yield. Thus the yield curve of Figure 17.13 is identical to the revenue curve of Figure 17.14 in this simple model. But the important point is that *maximum sustained yield*, the peak of the curve, is not at the same point as *maximum economic rent* (total revenue – total cost). The maximum economic profit will always occur at a lower fishing intensity than maximum yield. If this simple model prevailed, the economic management of fisheries would always be a safe biological management strategy. Alas, it is not always such. Scott Gordon (1954) showed that in an unmanaged fishery the only social equilibrium that will be reached occurs at the point where total costs equal total revenue, which is beyond the point of maximum sustained yield.

Clark (1990) has shown in an elegant analysis that under some situations it will pay fishermen to deplete the fishery to extinction, as might have happened to whales had the International Whaling Commission not intervened. The key economic idea

in these cases is that of discounting future returns. If fishermen are given the choice of making $1000 today by overfishing, or $1500 in ten years by delaying the harvest, most fishermen will take the money now and not wait. This type of exploitation makes perfect economic sense under our current economic theories, but it leads to overexploited populations and ecological disaster. Sustained yields can rarely be achieved without strong social or political controls on the allowable harvest.

Most of the world's fisheries are overexploited beyond the limits of sustainability, and a historical perspective on fisheries management leads to pessimistic conclusions about the future (Ludwig et al. 1993). The problem is that the concept of optimum yield is an equilibrium concept, that works well when a harvestable population is stable over time. But if the harvestable resource fluctuates, a ratchet effect begins to operate (Figure 17.15). Estimates of sustainable harvest rates are nearly always too high, and if profit margins are good, additional investment in equipment is made, and the harvesting industry becomes increasingly susceptible to a sequence of poor years. Because of job losses, government will typically step in to subsidize the harvesting during the poor years, and this encourages even more overharvesting. The long-term result is a heavily subsidized industry that overharvests the resource until it completely collapses. The following three examples illustrate the difficulties of applying these harvesting theories and the concept of optimum yield to real-world fisheries.

Case Study: The King Crab Fishery

The harvesting of king crabs in the North Pacific Ocean began commercially early in the twentieth century, when the Japanese began canning and exporting crabs to the United States. The Japanese gradually moved the crab fishery east and began harvesting in the eastern Bering Sea around 1930 using tangle nets dragged across the bottom. The king crab fishery was interrupted during World War II, and then both the Soviet Union and Japan took king crabs until the early 1970s, when the United States took over the king crab fishery. King crabs are now taken only in pot traps by U.S. fishermen.

The history of the king crab fishery in the eastern Bering Sea is shown in Figure 17.16. From initially low catches the crab fishery reached a peak

FIGURE 17.16

Catch of a red king crab (Paralithodes camtschatica) *from the Bristol Bay area of the Bering Sea, 1950–1998. The catch was taken mostly by Japanese and Soviet boats before 1969. The fishery collapsed in 1981, was closed in 1983 and again in 1994 and 1995, and has shown no signs of recovery. (Data from Alaska Department of Fish and Game, Commercial Fisheries, 1999.)*

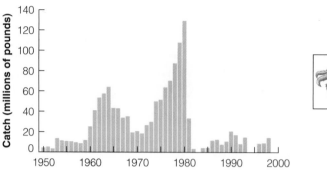

around 60 million pounds (9 million crabs) in 1964, and the catch was reduced over the next seven years while Japanese and Soviet boats were gradually eliminated from this fishery. The catch by U.S. fishermen grew rapidly during the 1970s to a peak catch of 130 million pounds (21 million crabs) in 1980. The king crab fishery collapsed in 1981–1982 and has recovered only very slightly since then. In 1994 and 1995 the major king crab fishery in Alaska was closed to all fishing, and fishermen have switched to snow crabs (*Chionoecetes opilio*) and other crab species. The snow crab fishery peaked in 1999 and collapsed, and the quota for 2000 has been cut 85%.

The early life cycle of the king crab is complex. Eggs are brooded by females for 11 months, and individual females lay from tens of thousands to hundreds of thousands of eggs (Larkin et al. 1990). Larval king crabs go through four stages while swimming in coastal waters, and then transform into the adult form at a length of 2 mm. Mortality is very high in the larval stages. Growth occurs slowly (Figure 17.17), and individuals mature at about age 5 years and at a carapace length of 90 mm. Young crabs aggregate in large pods of hundreds to thousands of individuals that move as a wave across the ocean floor. King crabs are predators and scavengers on a wide variety of invertebrates and fish. Their growth rate is affected by water temperature and can change with oceanographic conditions.

Why did the king crab fishery collapse in 1981? One assumption underlying the management of this fishery, which is unusual in several respects, has been the belief that by harvesting only males, the productivity of the population could be preserved. This assumption is true only if one mature male is able to service a large number of females. For behavioral reasons this assumption is now known to be faulty for king crabs. Fertilization occurs only after female have molted, and a male who has clasped a female may have to wait a few days for her to molt. After copulation the

male may retain his clasp on the female. Females prefer to mate with larger males, and if a female cannot copulate within nine days of molting, the entire brood for that year is lost. Behavioral complications resulting from a male-only harvest may thus compromise the reproductive potential of the population, since the larger males are removed by the fishery.

King crab growth and survival rates in the juvenile stages are also highly variable, perhaps due to changes in water temperature. During the 1970s the king crab fishery was relying on several very large cohorts from the 1960s, and these abundant recruitments provided a false sense of optimism in fishery managers and encouraged overinvestment in the fishery (see Figure 17.15).

Another factor affecting king crab abundance is the loss of immature crabs in pots (traps). For every legal-sized male crab (over 136 mm carapace length) taken, about seven immature males are caught and subsequently released. These immature crabs may be stressed or damaged during capture and may subsequently die. Other groundfish fisheries in the Bering Sea may capture king crabs incidentally or damage smaller crabs that are not retained in the nets. Finally, about 10% of the crab pots are lost each year due to storms and faulty ropes, and these lost pots continue to catch crabs, producing a hidden mortality on immature and mature crabs of both sexes.

In retrospect, the collapse of the king crab fishery was unavoidable because it built up too rapidly on the basis of a series of good year-classes, such that the maximum sustained yield was overestimated. The momentum associated with large capital investments in fishing boats makes it difficult to reduce rapidly the catch quotas when environmental conditions produce a series of poor year-classes. The compromise reached between socioeconomic realities and biological conservation measures almost always causes a further decline in stock abundance, and consequently an even longer time is required for recovery.

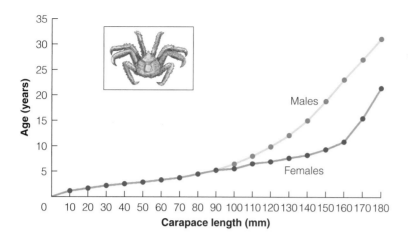

FIGURE 17.17

Average growth curves for male and female red king crabs (Paralithodes camtschatica). *Sexual maturity is reached at age 5 years and 90 mm carapace in length. Only males over 136 mm in length are retained. (Data from Larkin et al., 1990.)*

The "tragedy of the commons" was the term coined by Hardin (1968) to describe this type of exploitation. Whenever a resource is held in common by all the people, the best policy for each individual is to harvest as much of the resource as possible. There can be no incentive for individuals to stop harvesting at some optimum point because they can always make more money by overharvesting, and if you do not overharvest, your neighbor will. This tragedy of the overexploitation of common-property resources can be averted only by some form of management that restricts harvest, or by converting a common property resource to a private resource. Social control of harvesting is required for all large-scale fisheries, and for this reason good resource management is a creative mix of ecology, economics, and sociology. Alas this mix does not always produce good fisheries management, as the next example shows.

Case Study: The Northern Cod Fishery

The Atlantic cod (*Gadus morhua*) is a marine fish that occurs in cool northern waters, living on nearshore areas and out on the continental shelf to a depth of 600m. Cod played a major role in the early colonization of North America by Europeans. When John Cabot came to Newfoundland from England in 1497 he found the sea "swarming with fish—which can be taken not only with nets but in baskets let down with a stone." Basque fishermen from northern Spain had preceded Cabot to the Grand Banks off Newfoundland to catch cod, salt them, and carry them back to Europe. Salted cod was a delicacy in Europe during the fif-

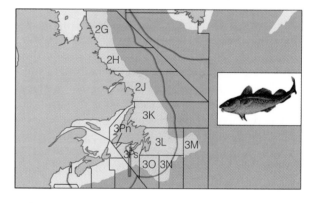

FIGURE 17.18

Boundaries of the North Atlantic Fisheries Organization regions for the cod fishery off eastern Canada. The northern cod stock largely occupies the Grand Banks of Newfoundland—regions 2J, 3K, and 3L. The edge between the pink and blue areas marks the edge of the continental shelf. The red line marks the 200-mile limit over which Canada has fisheries jurisdiction. (Map from Fisheries and Oceans Canada 1999.)

teenth and sixteenth centuries, and the need to be on land to dry cod, and salt it drove the settlement of the northeastern part of North America. Since Cabot's time the Atlantic cod has been the dominant commercial species of the northwest Atlantic. Now it borders on extinction, a victim of overfishing, and the collapse of the cod fishery has been a social, economic, and ecological disaster for the people of Newfoundland.

Cod populations of the northwest Atlantic can be subdivided into stocks, and fishery managers have tried to place boundaries on the stocks so that they could be managed separately (Figure 17.18). We will concentrate here on the northern cod stock occupying areas 2J, 3K, and 3L on this map. This cod stock

FIGURE 17.19
Total harvest of northern cod from the 2J, 3K, and 3L regions off Newfoundland, 1959–1998. The fishery completely collapsed between 1989 and 1992, when it was closed. (Data from Fisheries and Oceans Canada 1999.)

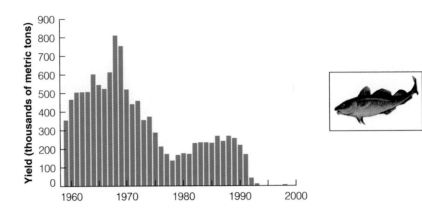

undergoes extensive spawning migrations up to 800 km from their winter spawning grounds, moving north to the coast of Labrador to spawn and then south to the inshore areas of Newfoundland in summer. Cod are predators that will eat nearly anything. Young cod feed on small crustaceans in the plankton, and as they grow they feed on shrimp, amphipods, and other juvenile fishes. Older cod feed on herring, capelin, sand lance, and a variety of smaller fishes, as well as on shrimp and crabs. Northern cod reach maturity at six to seven years of age in females, and an 11-year-old female will lay about 2 million eggs. Fecundity increases geometrically with size, and a 16-year-old female cod lays about 11 million eggs. Fertilized eggs rise to the ocean surface, where they are fed upon by a variety of predators. Only about one egg in a million will survive to complete the life cycle, and mortality of larval cod is also extremely high.

The cod fishery operated sustainably for nearly 500 years. During the 1600s the annual catch of cod was about 100,000 metric tons per year, and the annual harvest rose to as high as 200,000 tons in the 1700s due to the demand in Europe for salted cod. During the 1800s the catch ranged from 150,000 tons to 400,000 tons annually. After 1900 fishing boats became larger and had more efficient nets, and the efficiency of the fishing fleet continued to increase due to technological enhancements. Until 1900 all the cod were salted and dried, but then freezing replaced the older methods of preservation, and cod fishing intensity continued to grow. During the 1950s about 900,000 tons of cod were harvested in the northwest Atlantic, and this increased to 2 million tons in the 1960s. Of this total harvest the northern cod stock

contributed about 40%. Figure 17.19 shows the harvest of northern cod from 1959 to 1998. The very high catches of the 1960s followed a series of good years and in retrospect were clearly unsustainable. In 1977 Canada extended control over coastal fishing from 12 nautical miles to 200, and most of the cod fishery off Newfoundland came under Canadian management, which has been a disaster (Myers et al. 1997, Walters and Maguire 1996). Spawning biomass of cod in 1991 was 4% of what it was in 1962. With the collapse of the cod fishery, the dominant commercial fishery of the northwest Atlantic was closed in 1992, throwing 35,000 Newfoundlanders out of work. What went wrong with the management of this fishery?

Two major scientific errors occurred in the management of northern cod stocks during the 1980s and early 1990s. First, the estimates of the size of the cod stocks were far too high. Cod stocks are estimated by measuring the catch-per-unit-of-effort in a series of fishing surveys off the coast. There appeared to be a change in the behavior of cod as their numbers collapsed. Cod concentrated in high-density aggregations as their numbers fell (Hutchings 1996), and fishing in these aggregations, once they were located, produced large catches. Fishermen do not fish at random, and a change in the behavior of the cod could result in the concentration of fishing boats where the cod congregated, such that the catch-per-unit-effort would not be a good index of the size of the fish population. Second, the mortality rate of cod from fishing was grossly underestimated as a result of two sources of mortality that are typically not measured by fishery scientists. Catches of young cod that are too small to sell legally are usually discarded at sea, causing mortality that is due to fishing

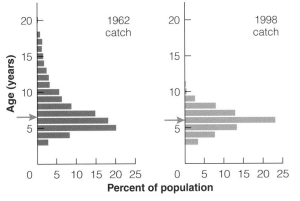

FIGURE 17.20
Age structure of northern cod in 1962 and after the stock collapsed in 1998. Note the absence of fish older than 10 years old in the 1998 sample. The average age of the catch has decreased from about seven years in 1962 to five years in 1998 (red arrows). (Data from Fisheries and Oceans Canada 1999.)

but does not contribute to the yield to the fishery. As the abundance of older fish was reduced (Figure 17.20), the population of the catch that is undersized would increase. The size of this discarded catch can be enormous; some fishing trawlers reported having to catch 500,000 cod and discard 300,000 undersized fish to get 200,000 legal-sized cod (Hutchings 1996). This increase in mortality rates of young fish would directly effect how many cod reached adult age of six to seven years and begin to reproduce.

There has been enormous controversy over the causes of the northern cod collapse that devastated the economy of Newfoundland and cost the Canadian taxpayers at least $ 4 billion. There are two major hypotheses:

1. *Changing ocean environment hypothesis:* For almost every fishery crisis a potential explanation is that the crisis resulted from a change in environmental conditions. For marine fisheries, in particular, the favorite culprits are postulated to be changes in water temperature and salinity. In this case, the particular hypothesis attributes the collapse of the northern cod stock to changes in oceanographic conditions during the late 1980s and early 1990s. If this idea were correct, it would remove all the blame for the collapse from fishery managers and the fishermen. But this hypothesis does not appear to be correct for the northern cod. Long-term data on ocean temperatures and salinity off Newfoundland, ice data from Labrador, and air-temperature data for the past century do not suggest that conditions during these recent years were extraordinary (Hutchings 1996).

A variant of this hypothesis attributes the cod decline to predation by seals. Harp seals (*Phoca groen-landica*), which eat primarily one and two-year-old cod, increased in abundance in the north Atlantic during the 1980s. But surveys of these small cod (too small for the commercial fishery to catch) have shown no change in survival rates during the 1980s and early 1990s (Hutchings 1996). There is no evidence that predation by harp seals had anything to do with the cod collapse. Many politicians have called for culling the harp seal population to "save" the cod, but there is no scientific support for such a policy (Myers et al. 1997).

2. *Overfishing hypothesis:* All the scientific evidence points to overfishing as the cause for the collapse of the northern cod. The decline occurred throughout the 1980s, at the same time that fishing effort was increasing and fishing gear underwent continual technological improvement. The mortality rates exerted on cod by fishing were excessive such that few fish reached reproductive size (age 6–7), and few large cod were left to lay large numbers of eggs. The commercial cod fishery was highly efficient, and able to find the last concentrations of cod as its populations declined. Finally, many undersized cod were discarded as the fishery went into decline, and this unreported mortality added to the collapse of the stock.

Even though the recovery of the northern cod will take decades, there is continuing political pressure to reopen the fishery. Only limited evidence of stock recovery has been observed during the last seven years, and there is concern that the cod stocks may never recover. If we define recovery as a spawner biomass of 1.5 million tons and a recruitment of 1 billion three-year-old-cod each year, then recovery will take approximately 35 years (Walters 1996).

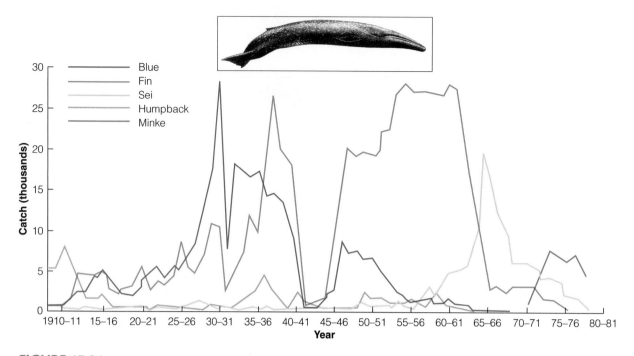

FIGURE 17.21

Catches of baleen whales in the Southern Hemisphere, 1910–1977. The usual lengths of whales in the commercial catches were: blue, 21–30 m; fin, 17–26 m; sei, 14–16 m; humpback, 11–15 m; and minke, 7–10 m. The blue whale is illustrated. (After Allen 1980.)

Case Study: Antarctic Whaling

The exploitation of whale populations was the subject of vigorous and heated debate during the 1970s and 1980s. At the present time almost all commercial whaling has been stopped and most whales are protected. The large whales comprise ten species divided into two unequal groups. The sperm whale was the only toothed whale hunted commercially; the other nine species were all baleen whales, which have bony plates (baleen) in the roof of the mouth. Baleen whales are filter feeders whose principal food in the Antarctic is krill (shrimplike crustaceans) and other plankton.

The history of whaling is characterized by a progression from more valuable species to less attractive species as stocks of the original targets were reduced. Modern whaling dates from 1868, when a Norwegian, Svend Foyn, invented the harpoon gun and the explosive harpoon. In about 1905 whalers pushed south into the Antarctic and discovered large populations of blue whales and fin whales. Blue whales dominated the catches through the 1930s, but by 1955 few were being taken (Figure 17.21). Attention was turned to the fin whale, originally the most abundant whale in the southern oceans. Fin whale numbers collapsed in the early 1960s. Sei whales, which were ignored as long as the bigger species were available, were not harvested until 1958. Sei whale catches were restricted after 1972 by the International Whaling Commission to prevent the collapse of these populations.

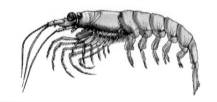

Krill (Euphausia superba)

Harvesting models for whales have been developed extensively since 1961 (Allen 1980). Logistic-type models have proved inadequate (Figure 17.22). Maximum sustained yield seems to occur at a density about 80% of equilibrium density, rather higher than

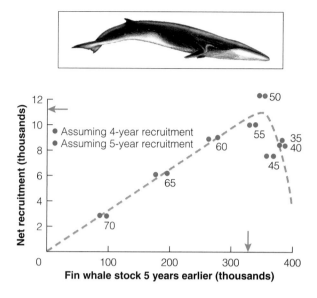

FIGURE 17.22
Logistic-type model for Antarctic fin whales. Estimated stock size and net recruitment at maximum sustained yield are indicated by arrows. The logistic model would predict maximum yield at 0.5 K (200,000), but these data suggest maximum yield around 0.8 K (330,000). (After Chapman 1981.)

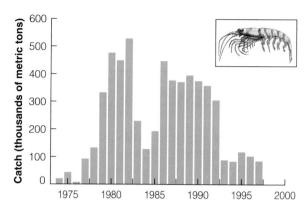

FIGURE 17.23
Krill commercial production in the Antarctic 1974–1997. The krill fishery could become the largest fishery in the world, with an estimated potential harvest of 150 million tons a year. The biomass of krill in the Antarctic may be the largest of any animal species on the planet. (Data from FAO, Rome 1999.)

the 50% predicted in the simple logistic model (Chapman 1981). Complications with these simple models are not difficult to find. Figure 17.22 assumes that all fin whales in the southern ocean belong to one population; it is now known that several subpopulations occur (Chapman 1981). Whales may interact, and most whale models are single-species models that do not recognize that many different species of whales and seals feed on krill in the Antarctic.

The current management of whales is directed to measuring the recovery rate of the depleted whale populations. Paradoxically, most of the data we have on whales came from whaling operations, and now that commercial whaling has stopped, additional research must be mounted to monitor how whale populations respond. Whale populations change slowly, and even ten years is a short time to estimate accurately a population's response to protection from exploitation.

The principal food of the baleen whales is krill, a group of 85 species of shrimplike crustaceans that are on average about 6 cm long and weigh about 1–2 grams. Krill are so abundant in Antarctic waters that they were considered for potential harvest for many years before commercial harvesting began in the

1970s (Figure 17.23). Estimates of the sustainable harvest for krill are extremely large, nearly equal to the total production of all other fisheries on the planet (Ross and Quetin 1986). Commercial harvesting has been hampered by the remote location of the Antarctic and by processing problems once the krill are captured. Krill have powerful digestive enzymes that tend to spoil the catch by breaking down the edible tissues immediately after death. Krill also contain high amounts of fluoride, which must be removed before they can be used for human food. One of the emerging conservation problems of the southern oceans is to estimate the effect of krill harvesting on the recovery of whale populations in the Antarctic (Rosenberg et al. 1986, Nichol and de la Mare 1993).

Risk-Aversive Management Strategies

Because of the many failures of resource management in the past, resource managers have begun to search for management strategies that are designed to minimize risk. Two general approaches have been suggested. The first is to redirect management toward a harvest strategy that does not simply seek to achieve the maximum sustained yield (Hilborn and Walters 1992, Lande et al. 1997). In this approach we try to find a harvesting strategy that maximizes average yield

ESSAY 17.1

PRINCIPLES OF EFFECTIVE RESOURCE MANAGEMENT

Even though renewable resource management has historically failed, as many examples of fish harvesting will testify, we are currently committed to the general principle of sustainable use of resources. How can we do a better job in the future? According to Ludwig, Hilborn, and Walters (1993), five principles should underlie good resource management:

1. *Include humans as part of the system.* Human motivation, shortsightedness, and greed can underlie many of the problems of resource management. Instead of thinking of humans managing resources, we should think of resources managing human behavior, often with a short time frame.

2. *Act before scientific consensus is achieved.* For many management problems we do not need additional research to decide on management policies. Examples would include pollution impacts in the Great Lakes, tree harvesting on slopes subject to erosion, and harvesting of undersized fish. Calls for additional research on many topics are often just delaying tactics.

3. *Rely on scientists to recognize problems, but not to remedy them.* Good science is important for resource management, but it is not enough. The management of *human* activities is what is essential and

this is a sociological, psychological, and political problem.

4. *Distrust claims of sustainable resource use.* Because we have failed in the past to harvest sustainably, any new plan that represents itself as sustainable should be suspect and subject to detailed scrutiny. The linkage between basic research on fish populations and sustainable fisheries policies is a loose one, and good basic research does not automatically lead to better management.

5. *Confront uncertainty.* We often operate under the illusion that if we do enough research with enough funding we will be able to identify a solution to harvesting problems. But the large levels of natural variation found in most populations preclude any exact predictions about future dynamics. We need to favor management actions that are strong in the face of uncertainty and that are reversible if found to be damaging.

"Sustainable development" is the buzzword of the moment, and we must not pretend that scientific or technological advances alone will be sufficient to solve resource management problems. These are human problems that we have created many times in the past and under many types of political systems, and they will not necessarily be solved by more scientific data.

over a long time period during which the population fluctuates naturally. At the same time that we try to maximize long-term yields, we need to minimize the risk of resource collapse or extinction. Two popular strategies for risk-aversive harvesting are (1) to impose a constant percentage harvest on a population, or (2) to harvest all individuals above a threshold population size, and no individuals below that threshold. In both cases there is a threshold or "escapement"[3] level below which harvesting stops (see Box 17.1). The problem with many fisheries is that this thresh-

old is set too low to sustain the resource, and often the management authorities do not know very accurately where the population is with respect to the threshold. The problem with complete threshold harvesting is that in years when the population is below the threshold there will be no harvest, with the attendant economic dislocations. For this reason Hilborn and Walters (1992) prefer a constant harvest-rate strategy.

The second general approach is to impose protected areas, or "no-take" zones, on the resource. This is a bet-hedging strategy in which we incur the cost of reducing the catch for the benefit of a reduced risk of catastrophic collapse of the fishery. This strategy has been discussed particularly for

---

[3]Escapement is the unharvested portion of the population and may be defined as a percentage or as an absolute number.

B O X   1 7 . 1

WHAT ARE THE HARVEST STRATEGIES FOR A FISHERY?

The management of any harvested resource requires a clearly specified harvest strategy that is understood by biologists, managers, and harvesters alike. For a fishery, a harvest strategy is a plan stating how the catch will be adjusted from year to year depending on the size of the stock, and the economic condition of the fishery, the condition of other stocks, and the uncertainty regarding our biological knowledge of the stock. Fisheries harvest strategies should be quantitative and explicit, not vague statements and wishes, and they need to be developed with the active participation of the fisherman and the industry that will be affected.

Three basic harvest strategies can be applied to a fishery (see the following graphs)

1 Constant quota: This strategy specifies a fixed catch or constant quota that does not change with stock size (blue area). If the stock is large, the catch will be a small fraction of the population; if the stock is small,, the entire stock may be taken.

2 Constant exploitation rate: For this strategy the catch does not rise one-to-one with the stock, but instead at some lower rate (b). For example, the catch might rise 0.5 times the stock rise. The escapement thus increases as the stock grows larger. A variant of this strategy uses a threshold or lower limit point so that the fishery is closed at low stock abundance.

3 Constant escapement This strategy implies that the catch will rise one-to-one with the stock size and that the stock size or escapement will not vary. This strategy has an implicit threshold so harvest begins only when the size of the stock exceeds this threshold (red arrow).

Different fisheries are managed with different harvest strategies. From a theoretical perspective, the optimal harvest strategy is usually a constant escapement strategy (graph c) (Hilborn and Walters 1992). But there is a trade-off, because if we wish to maximize average yield to the fishery, we also maximize the variation in yields from year to year. Fishermen often prefer less variance in catch, so an optimal strategy from a human perspective might be to have a less-than-maximum yield with reduced variability from year to year

Other types of harvesting strategies can be applied. In some cases it is more efficient to use a periodic harvest in which the fishery is not operated every year. This can be a useful strategy when it is more economically advantageous to take a large catch every few years than a smaller catch every year, or when older animals are much more valuable than younger animals. Some clam and abalone fisheries operate this way, as do most aquaculture industries. The most important message is the need to tailor the harvesting strategy to the particular resource being exploited by means of full consultation between industry workers and management biologists.

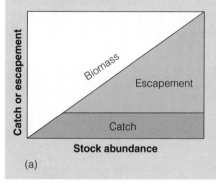

(a)

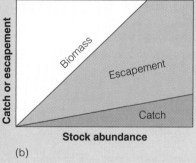

(b)

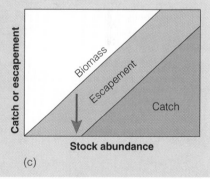

(c)

marine fishes (Hall 1998, Lauck et al. 1998). The idea of a protected area in the aquatic realm is equivalent to the idea of national parks on land and is most useful for demersal fish that inhabit large areas of the ocean floor and are nonmigratory. The idea is simple: Set aside a large enough "no-take" zone to ensure that the stock will remain at greater than 60% of carrying capacity over a given time horizon (for

example, 20 years). Fishermen could harvest at a specified rate outside the "no-take" zone but would not be permitted to fish inside this protected area. The details of how to achieve these simple goals must be worked out for each resource, and the detailed trade-off of costs and benefits must be identified if this strategy is to obtain practical support among fishermen.

Summary

To harvest a population in an optimal way, we must understand the factors that regulate the abundance of that population. That humans so frequently mismanage exploited populations like the northern cod is partly a measure of our ignorance of population dynamics. When humans harvest a population, its abundance must decline, and the losses caused by harvesting are compensated for by increased growth, increased reproduction, or a reduced natural mortality. Harvested populations lose the older and larger individuals and often respond by a reduction in the age at sexual maturity. Species vary greatly in the amount of harvesting they can withstand.

Maximum sustained yield is often the goal of resource managers. Simple and complex models alike have been developed to estimate the maximum sustained yield for fisheries and forestry. Most models contain the hidden assumption that the environment remains constant, and for this reason they often fail in practice to prevent overexploitation and collapse. Economics and politics add further difficulties to achieving maximum sustained yield for valuable populations. Harvesting is subject to a ratchet effect in which the exploitation rate is pushed by ecological optimism and economic and social pressures toward overexploitation and collapse.

Management of forestry, fishery, and wildlife resources is at present based more on political and economic pressures than on scientific knowledge and forecasting. Because of the inherent ecological uncertainty in anticipating future changes in populations, resource management must adopt more risk-aversive strategies. The imposition of a protected-area strategy or "no-take" zones is one approach that may work for some fisheries. One of the great challenges of modern ecology is to help place resource management on a sustainable basis. We can all be very good at managing yesterday's populations. When will we be equally adept at managing tomorrow's?

Key Concepts

1. Any harvested population must decline in abundance, and the losses due to harvesting must be compensated for by increased growth, increased reproduction, or decreased natural mortality.

2. Simple population growth models like the logistic equation can be used to estimate the maximum sustainable yield of a harvested population.

3. All simple yield models assume equilibrium conditions and fail when there are changes in ocean conditions, weather, predators, or diseases.

4. Uncertainty in our knowledge of population dynamics and the variable effects of weather argue for more conservative harvesting goals than maximum yield.

5. Incessant political pressure for increased harvests coupled with uncertainty concerning biological information has produced a global crisis in overfishing and overharvesting of renewable natural resources.

Selected References

Caddy, J. F. 1999. Fisheries management in the twenty-first century: Will new paradigms apply? *Reviews in Fish Biology and Fisheries* 9:1–43

Clark, C. W. 1996. Marine reserves and the precautionary management of fisheries. *Ecological Applications* 6: 369-370.

Hall, S. J. 1998. Closed areas for fisheries management—The case consolidates. *Trends in Ecology and Evolution* 13:297–298.

Hilborn, R. and C. J. Walters. 1992. *Quantitative Fisheries Stock Assessment: Choice, Dynamics, and Uncertainty*. Chapman and Hall, New York. 570 pp.

Keiner, C. 1998. W. K. Brooks and the oyster question: Science, politics, and resource management in Maryland, 1880–1930. *Journal of the History of Biology* 31:383–424.

Lande, R., B. E. Sæther, and S. Engen. 1997. Threshold harvesting for sustainability of fluctuating resources. *Ecology* 78:1341–1350.

Ludwig, D., R. Hilborn, and C. Walters. 1993. Uncertainty, resource exploitation, and conservation: Lessons from history. *Science* 260:17, 36.

Myers, R. A., J. A. Hutchings, And N. J. Barrowman. 1996. Hypotheses for the decline of cod in the North Atlantic. *Marine Ecology Progress Series* 138:293–308.

Pascual, M. A., and M. D. Adkison. 1994. The decline of the Steller sea lion in the northeast Pacific: Demography, harvest, or environment? *Ecological Applications* 4:393–403 .

Pauly, D., V. Christensen, J. Dalsgaard, R. Froese, and F. Torres Jr. 1998. Fishing down marine food webs. *Science* 279:860–863.

Pitcher, T. J., P. Hart, and D. Pauly, eds. 1998. *Reinventing Fisheries Management*. Kluwer Academic Publishers, Boston. 435 pp.

Roberts, C. M. 1997. Ecological advice for the global fisheries crisis. *Trends in Ecology & Evolution* 12:35–38.

Silliman, R. P. and J.S. Gutsell. 1958. Experimental exploitation of fish populations. Fishery Bulletin (U.S.) 58(133):215–252.

Walters, C. and J. J. Maguire. 1996. Lessons for stock assessment from the northern cod collapse. *Reviews in Fish Biology and Fisheries* 6:125–137.

Questions and Problems

17.1 Murphy (1967, p. 733) gives the following estimates[4] for the vital statistics of the Pacific sardine living under optimal conditions:

Age, X (yr)	l_x	m_x
2	1.0000	0.5147
3	0.6703	1.3618
4	0.4493	1.6819
5	0.3012	1.8816
6	0.2019	2.0257
7	0.1353	2.1358
8	0.0907	2.2357
9	0.0608	2.2686
10	0.0408	2.2686
11	0.0273	2.2686
12	0.0183	2.2686
13	0.0123	2.2686

Calculate the net reproductive rate (R_0) and the capacity for increase (r) of this sardine population. Use this value of r to calculate a logistic curve for an unexploited sardine population (assume that $K = 2,400,000$ tons and $a = 7.785$). How many years would it take for this unexploited sardine population to recover from a starting point of 10,000 tons to a level of 2,000,000 tons if it followed this logistic curve?

17.2 Forest resources are another major natural resource subject to harvesting regimes. Are forest resources in your area being harvested in a sustainable manner? How could you determine if this were true or not? Which of the admonitions given in this chapter for fisheries would apply equally well to forest harvesting?

17.3 Suppose that in fact no relationship exists between stock and recruitment in the sockeye salmon (see Figure 17.9). Discuss the implications of this with respect to the various theories of population control (see Chapter 16).

17.4 Construct an equilibrium yield curve showing the relationship between yield (biomass) and hunting mortality rate for a hypothetical deer population with constant recruitment of 1000 fawns per year, and an instantaneous natural mortality rate of 0.7 per year for age 0 to 1 and of 0.4 per year for older deer. Assume for simplicity that all growth, recruitment, and losses occur at a single point each year and that hunting operates only on deer age 2 and over. The growth curve data are as follows:

Age (yr)	Weight (lb)	No. of Survivors (natural mortality only)
0	10	1000
1	50	497
2	80	333
3	90	223
4	100	150
5	110	100
6	120	67
7	130	45
8	140	30
9	150	20
10	160	14
11	170	9

Calculate the equilibrium yield in numbers for this population at a harvest rate of 20% per year.

---

[4]This is a relative life table and fertility table (starting at age 2), and it can be converted to the usual format by dividing the l_x values by 71,000 and multiplying the m_x values by this amount. It is given in this way to avoid cumbersome numbers.

17.5 Plot the following hypothetical reproduction curves (like in Figure 17.7) for a species that reproduces only once (numbers are arbitrary "egg units"):

Population A		Population B	
Adult egg production	Progeny egg production	Adult egg production	Progeny egg production
2	4	1	5
4	6	2	8
6	7.5	4	11
8	8	5	12
10	7.5	6	11
12	6	8	8
14	3	10	5
		12	2
		14	1

Use these plots to determine graphically the population changes for 20 generations from starting adult stocks of 14, 6, and 2. Introduce random environmental variations to this simple method by flipping a coin each generation and then multiplying progeny egg production by 0.5 for heads and 1.5 for tails. What effect does this random factor have on the population curves?

17.6 Examine the catch statistics for a fishery in your area or in an area of interest to you. Sources of data on the Web might be the *Fisheries Statistics of the United States*, *Fisheries Statistics of Canada*, or the *Food and Agricultural Organization's Yearbook of Fishery Statistics* published by the United Nations. If the fishery you choose has been managed, is there any evidence of overfishing?

17.7 Black duck populations have been declining in North America since the mid-1950s. One hypothesis for this decline is that it is due to overhunting. Review the evidence for and against this hypothesis. Nichols (1991) provides references and an overview of this management problem.

17.8 One of the assumptions of maximum sustained yield models is that birth, death, and growth responses to population density are repeatable, such that a given population density will always be characterized by the same vital statistics. What mechanisms may make this assumption false?

17.9 Ricker (1982) showed that since 1950, sockeye salmon caught in British Columbia have decreased in size by 140–180 grams on average, about 5% of their weight. Discuss mechanisms by which a fishery might select in favor of smaller salmon.

17.10 Ludwig and Walters (1985) showed in a computer simulation that the management of a hypothetical fishery could be done better using simple yield models like the logistic equation than by using more realistic, detailed models like dynamic pool models. Discuss why this might be correct for a real fishery.

Overview Question

Suppose that you are in charge of a newly established fishery. Discuss the criteria you might use to detect when the population is being overexploited, and outline the relative merits of the different criteria.

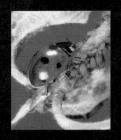

CHAPTER 18

Applied Problems II:
Pest Control

SOME SPECIES INTERFERE with human activities, in which case they are assigned the label "pests." The most damaging pests we have are introduced species. The first response to pests is to control them, which in this context means to control damage and not necessarily to regulate the pest population around some equilibrium density. One of the obvious ways of controlling damage is to reduce the average abundance of the pest species, but there are other ways of reducing damage by pests without affecting abundance (such as using insect repellents).

A population is defined as being controlled when it is not causing excessive economic damage, and as being uncontrolled when it is. The boundary between these two states will depend on the particular pest. An insect that destroys 4-5 percent of an apple crop may be insignificant biologically but may destroy the grower's profit margin. Conversely, forest insect pests may defoliate whole areas of forest without bankrupting the lumbering industry. To all questions of pest control, we must apply the concept of economic thresholds, including the cost of the damage caused by the pest, the costs of control measures, the profit to be gained with the crop, and interactions with other pests and their associated costs.

Pest control in most agricultural systems is achieved by the use of toxic chemicals, or *pesticides*. An estimated 2.5 billion kilograms (nearly 5 billion pounds) of toxic chemicals are being used annually world wide to control plant and animal pests (Pimentel et al. 1992). Despite the use of these pesticides, about 48% of the world's crops are lost to pests

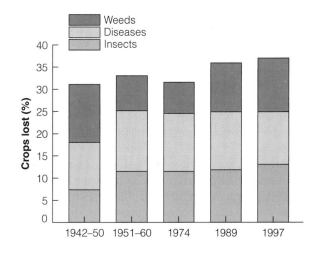

FIGURE 18.1
Percent of crops estimated to have been lost before harvest in the United States because of insect pests, plant diseases, and weeds, 1942-1997. During this time period pesticide use increased 33-fold. After harvest about 20% more is lost to other pests. (Data from Pimentel et al. 1991, and Pimentel 1997.)

before and after harvesting, and despite increasing pesticide use in the last 60 years, crop losses have not been reduced (Figure 18.1).

Pesticides are only a short-term solution to the problem of pest control for several reasons. First, toxic chemicals have strong effects on many species other than pests. Rachel Carson gained fame as the first naturalist to point out to the public at large the

331

FIGURE 18.2
Increases in California red scale
(Aonidiella aurantii) *infestation on
lemon trees caused by monthly
applications of DDT spray. Nearby
untreated lemon trees suffer no
damage because in the absence of
DDT a variety of insect parasites
and predators keep the red scale
under biological control in southern
California. (After DeBach 1974.)*

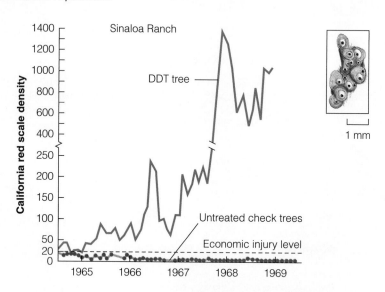

ecological consequences of toxic chemicals. The well-known effects of DDT on bird populations, which Carson highlighted in *Silent Spring* (1962), is a good example of how pesticides can degrade environmental quality. Second, many pest species are becoming genetically resistant to toxic chemicals that formerly killed them. Insects that attack cotton have evolved resistance to so many pesticides that it is no longer possible to grow cotton in parts of Central America, Mexico, and southern Texas. Third, and perhaps most surprisingly, the use of toxic chemicals in some situations can actually produce a pest problem where none previously existed. Figure 18.2 illustrates how lemon trees can become infested by massive outbreaks of a scale insect when sprayed with DDT. Toxic chemicals such as DDT destroy many insect parasites and predators that cause mortality in the pest species, and after treatment the few pest individuals that survive can multiply without limitation.

How can we achieve pest control without these problems? There are four primary strategies for dealing with pests:

1. *Natural control:* Pest populations are exposed to naturally occurring predators, parasites, diseases, and competitors.

2. *Pesticide suppression:* Pest populations are treated with herbicides, fungicides, insecticides, or other chemical poisons to reduce their abundance.

3. *Cultural control:* Pests are reduced by agricultural manipulations involving crop rotation, strip cropping, burning of crop residues, staggered plantings, or other agricultural practices.

Rachel Carson *(1907–1964) Author of Silent Spring (1962)
(©Magnum Photos)*

4. *Biological control:* pests are reduced by introductions of predators, parasites, or diseases; by genetic manipulations of crops or pests; by sterilization of pests; or by mating disruption using pheromones.

Integrated control, or integrated pest management (IPM), is the use of all four of these strategies; the goal is to reduce pest damage by minimizing pesticide use and maximizing natural control.

In this chapter we discuss the principles used in biological control and cultural control and relate them to ecological theory. Biological and cultural controls

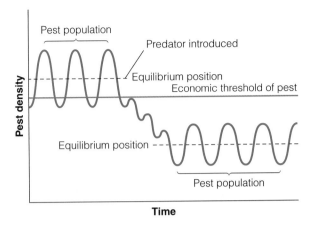

FIGURE 18.3
Classical biological control in which the average abundance of an insect pest is reduced after the introduction of a predator. The economic threshold is determined by humans' activities, and its position is not changed by biological control programs. (After van den Bosch et al. 1982.)

aim to reduce the average density of a pest population (Figure 18.3), and may be viewed as a practical application of the problem of what determines average abundance (see Chapter 16). The aim of pest management is not to eradicate the pest, which is usually impossible, but to reduce its effects to an acceptable level.

Examples of Biological Control

In this section we consider three examples of biological control. The general situation in which biological control is applied unfolds as follows: A pest, often an introduced species, is causing heavy damage. Efforts are then made to find predators and parasites from the pest's native habitat that can be introduced into its new habitat. If these introductions are successful, the pest population is reduced to a level at which no economic damage occurs. The first example of biological control we will consider involves the cottony-cushion scale.

Cottony-Cushion Scale (*Icerya purchasi*)

One of the most striking successes of biological control concerned the cottony-cushion scale, a small coccid insect that sucks sap from leaves and twigs of citrus trees. This scale insect was first discovered in California in 1872, and by 1887 the whole citrus industry of

southern California was threatened with destruction. Because of the size of the infested area, chemical control by cyanide and other sprays was a failure. In 1888 Albert Koebele of the Division of Entomology was sent to Australia by the U.S. government to represent the State Department at an international exposition in Melbourne. (All foreign travel for the Division of Entomology had been restricted to save money, this was the only subterfuge by which an entomologist could travel to Australia to search for parasites of the cottony-cushion scale, a native of Australia.) Koebele sent back to California two species of insects, a small dipteran parasite, *Cryptochaetum iceryae*, and a predaceous ladybird beetle called the vedalia (*Rodolia cardinalis*). The dipteran parasite was thought to be a potentially important agent for control, but Koebele sent the ladybird beetles along as well, apparently without thinking that they could be very useful.

Cottony-cushion scale (Icerya purchasi) (©J.H. Robinson/Photo Researchers, Inc.)

In late 1888 the first ladybird beetles were received in California, and by January 1889 a total of 129 individuals had been released near Los Angeles under an infested orange tree covered by a large tent. By April 1889 all the cottony-cushion scales on this tree had been destroyed; the tent was then opened. By June 1889 over 10,000 beetles had been sent to other citrus orchards from this first release point. By October 1889, scarcely one year since *Rodolia* was found in Australia by Koebele, the cottony-cushion scale was virtually eliminated from large areas of citrus orchards in southern California. Within two years it was difficult to find a single individual of the scale *Icerya*, and this control continued so that the pest was effectively eliminated. The cost: about $1500; the saving: millions of dollars every year. The California legislature was impressed, and California became a center of activism in promoting the value of biological control (Caltagirone and Doutt 1989).

The cottony-cushion scale reappeared with the advent of DDT. Infestations of the scale that had not been seen in over 50 years were found after DDT had eliminated the vedalia beetle from some local areas. Under these circumstances, the beetle must be continually reintroduced.

Some of the host plants of the cottony-cushion scale are not suitable for the vedalia. For example, the scale infests scotch broom (*Cytisus scoparius*) and maples in central California, but the vedalia will not become established on these plants, for unknown reasons (Clausen 1956). Such host plants serve as a refuge or reservoir for the scale, which can then recolonize citrus trees.

The great success in controlling the cottony-cushion scale ushered in an era in which biological control was viewed as a panacea for all insect-pest problems. Large numbers of insects were collected from all over the world and released in North America without any testing or quarantine procedures. Fortunately, only time and money were wasted with all these importations, and this dangerous policy was eventually stopped - not, however, because of its dangers, but because of a sequence of repeated failures at control (Howarth 1991).

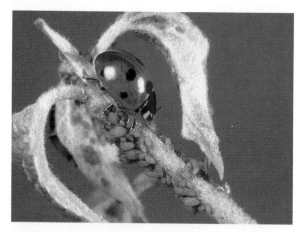

Vedalia beetle (Rodolia cardinalis) (©N.Cattlin, Holt Studios International/Photo Researchers, Inc.)

Prickly Pear (*Opuntia* spp)

Prickly pear is a cactus native to North and South America. Of the several hundred species of prickly pear, about 26 have been introduced into Australia as garden plants. One species, *Opuntia stricta*, has become a serious weed in Australia.

In 1839 *O. stricta* was brought from the southern United States to Australia as a potted plant and it was planted as a hedge plant in eastern Australia. It gradually got out of control and was recognized as a pest by 1880. By 1900 it occupied some 40,000 km (15,600 mi), and thereafter it spread rapidly in Queensland and New South Wales (Figure 18.4):

Area Infested with Opuntia

	km^2	mi^2
1900	40,000	15,600
1920	235,000	90,600
1925	243,000	93,700

About half this area consisted of dense growth completely covering the ground (Figure 18.5), Sometimes rising 1 - 2 meters in height and often too dense for anyone to walk through

Prickly pear is propagated by seeds and by segments. The cactus pads, when detached from the parent plant by wind or people, can root and begin a new plant. Seeds are viable for at least 15 years. The problem of eradicating this weed was largely one of cost. The grazing land it occupied in eastern Australia was worth only a few dollars an acre, and poisoning the cactus cost about $25 to $100 an acre. Consequently, homesteads had to be abandoned to this invasion.

In 1912 two entomologists were sent from Australia to visit the native habitats of *Opuntia* and to learn of possible biological control agents that could be introduced. They sent back from Sri Lanka a mealybug, *Dactylopius indicus*, which was released and, in a few years, had destroyed a minor pest cactus, *Opuntia vulgaris*. But the major pest, *O. stricta*, continued to spread, and after World War I it was subjected to a more intensive effort of biological control. Investigations in the United States, Mexico, and Argentina resulted in 50 insect species being shipped to Australia for evaluation as possible control agents. Of these, only 12 species were released; three were of some help in controlling *O. stricta*, but only one, the moth *Cactoblastis cactorum*, was capable of eradicating it.

Cactoblastis cactorum is a moth native to northern Argentina. Two generations occur each year. Females lay about 100 eggs on average, and the adults live about two weeks. The larvae damage the cacti by burrowing into and feeding within the pads and by introducing bacterial and fungal infections with these activities. Two introductions of *Cactoblastis* were made. The first introduction, in 1914, failed (Osmond and Monro 1981). For

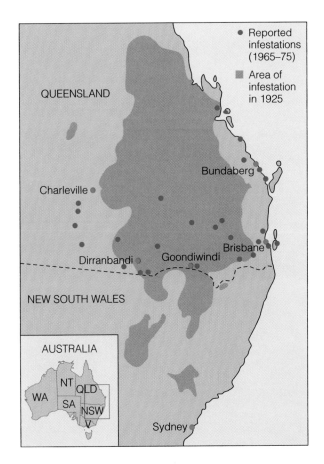

the second introduction approximately 2750 eggs were shipped from Argentina in 1925, and two generations were raised in cages until March 1926, when 2 million eggs were set out at 19 localities in eastern Australia. The moth was immediately successful, and further efforts were expended from 1927 to 1930 in spreading eggs and pupae from one field area to another.

By 1928 it was obvious that *Cactoblastis* would control *O. stricta*, so further introductions were curtailed. *Cactoblastis* multiplied rapidly up to 1930, and between 1930 and 1931 the *Opuntia* stands were ravaged by an enormous *Cactoblastis* population. This collapse of the prickly pear population caused the moth population to fall steeply during 1932 and 1933, and the cactus then began to recover in some areas. Between 1935 and 1940 *Cactoblastis* recovered and completely controlled the cactus. After 1940, prickly pear survived only as a scattered plant in the community (Dodd 1940, 1959).

The present picture is that *Opuntia* exists in a stable metapopulation at low density maintained by *Cactoblastis* grazing. The eggs of *Cactoblastis* are not laid at random but are clumped on some plants; other plants escape infestation entirely (Myers et al. 1981). Plants heavily loaded with larvae are subsequently completely destroyed, and many *Cactoblastis* larvae thus starve and die. Larvae cannot move from one plant to another if cacti are 2 meters or more apart. The clumped distribution of the eggs of *Cactoblastis* thus both destroys *Opuntia* plants and ensures that not all plants are killed; as a result, the metapopulation does not go to extinction, although local populations do disappear.

Most of the areas where prickly pear is now periodically considered a pest are outside the original area of dense cactus infestation, and plants in these areas seem to be partly resistant to *Cactoblastis* attack. Without *Cactoblastis*, prickly pear would make a rapid recovery.

Why was *Opuntia* such a successful plant in eastern Australia? Three important physiological properties of *Opuntia* determine its success (Osmond and

FIGURE 18.5
(a) A dense stand of prickly pear, October 1926, Chinchilla, Queensland, Australia. (b) The same stand, is shown three years later, after attack by the moth (Cacto-blastis), *(After Dodd 1940; photographs courtesy of A. P. Dodd and Commonwealth Prickly Pear Board.)*

Monro 1981). First, the tissues of this cactus are almost entirely photosynthetic. There is minimal investment in structural tissues, and the root system is shallow and small. Second, *Opuntia* is capable of crassulacean acid metabolism (CAM), a process in which CO_2 fixation largely occurs at night, when minimal water vapor is lost to the atmosphere. Thus photosynthesis can be done with minimal water loss. Third, CAM plants retain photosynthetically competent tissues throughout periods of stress. When the rains come, CAM plants can immediately begin to photosynthesize and grow. Because of this combination of characteristics, *Opuntia* proved a near perfect opportunist with superior competitive ability over the native plants that lacked CAM metabolism.

Floating Fern (*Salvinia molesta*)

The floating fern *Salvinia*, a plant native to South America, was introduced to Sri Lanka in 1939 through the Botany Department at the University of Colombo, and over the next 50 years it spread to Africa, India, Southeast Asia, and Australia. It is an important aquatic weed, forming mats up to 1 meter. thick and covering lakes, rivers, canals, and irrigation channels. Because *Salvinia* clogged the waterways, all water transport and fishing was disrupted.

Salvinia had been incorrectly identified until 1972, when it was recognized to be a new species (Room 1990). Because of this taxonomic uncertainty, ecologists were not able to look for specialized herbivores of this plant in its native habitat until the plant was found in southeastern Brazil in 1978. *Salvinia molesta* is unusual in being sterile, and all ramets appear to be genetically identical no matter where they occur in its geographical range. Plants of *Salvinia* are colonies of ramets held together by horizontal, branching rhizomes. The rate of growth of *Salvinia* on the water's surface is limited by temperature and the nitrogen concentration of the water.

Three insect species (a weevil *Cyrtobagous singularis*, a moth, and a grasshopper) found attacking *Salvinia auriculata* in Trinidad in the 1960s were introduced into Sri Lanka, India, Africa, and Fiji in the 1970s. None of these introductions had any effect on the weed. *S. molesta* was located in Brazil, what was thought to be the same insect species were collected and the weevil was released in north Queensland, Australia, in 1981. The weevil increased dramatically and destroyed the *Salvinia* within one year (Figure 18.6). The weevil was then discovered to be a new species, *Cyrtobagous salviniae*, and it proved highly successful at control in Australia, India, Sri Lanka, Botswana, and Namibia (Room 1990).

The success of the weevil in controlling *Salvinia* is partly explained by its tolerance of high population densities before it shows interference competition and then emigrates. The weevil reaches densities of 1000 adults per square meter, and by feeding on the buds as adults and on roots and rhizomes as larvae, the weevil either kills the plants or greatly reduces their size.

Theory of Biological Control

Most biological control has operated empirically with a few rules-of-thumb, and this approach has achieved some spectacular successes. But if we are to avoid a case-by-case approach, we need to develop some general theory of biological control that could guide empirical work (Murdoch and Briggs 1996). Most of the theory that has developed comes from the Nicholson-Bailey model of predator-prey interactions, similar to the discrete predator-prey model discussed in Chapter 13 (see page 207). The premise of this approach is that successful biological control resulted from the predator imposing on the prey a low, stable equilibrium (see Figure 18.3). Theoretical evaluation

of these predator-prey models by Beddington et al. (1976), May (1978), and May and Hassell (1981) suggests an array of properties of successful biocontrol agents: (1) They are host specific; (2) they are synchronous with the pest; (3) they have a high intrinsic rate of increase (r); (4) they are able to survive when few prey are available; and (5) they have great searching ability. These properties are more typical of insect parasitoids than of predators in general. Most predators are considered by this theory to be poor candidates for biological control because they are generalists (not host-specific), they are rarely synchronized with the pest, they have relatively low r values, and they may feed on other, beneficial prey species.

Four issues are crucial for setting up a theory of biological control:

1. Parasitoid and predator aggregation. The classical theory of biological control is an equilibrium theory in which the main issue is how to produce stability in parasitoid-host interactions. Nicholson and Bailey (1935) produced a model of parasitoid-host interaction that was a discrete-generation model with one generation of host and parasitoid each year. The result of this simple model was instability, much as the discrete population model discussed in Chapter 11 shows unstable dynamics over a wide range of parameters. Part of this instability arises from the assumption in the original Nicholson-Bailey model that encounters between parasitoids and hosts are random. If we introduce nonrandom search and allow the parasitoids to spatially aggregate to high densities of hosts, the models produce a stable interaction (Hassell and May 1974). But this stability is achieved in the model only if parasitoids are not allowed to move from patch to patch within a generation. Parasitoids in the model are always revisiting patches of the host that have already been heavily attacked. If we relax this assumption and allow parasitoids to move among patches within a generation, these models do not induce stability in population dynamics (Murdoch and Stewart-Oaten 1989). The trade-off with stability is effectiveness: Parasitoids that concentrate their attacks on high-density patches of prey are more likely to be effective biological control agents because they suppress the pest, but they do not induce stability in population dynamics. We need to look for stabilizing effects elsewhere.

2. Metapopulations of hosts. If a pest population is a metapopulation composed of many local

FIGURE 18.6
Biological control of the floating fern
Salvinia molesta. *(a) The water*
above Wappa Dam, Nambour,
Queensland, completely covered by
Salvinia, *October 1982. (b) The*
same scene in September 1983 after
the population explosion of the weevil
Cyrtobagous salviniae *(c) and the*
subsequent crash of both Salvinia
and the weevil. (Photos courtesy of
Dr. P. A. Room, CSIRO Division of
Entomology.)

populations, then even if great instability of predator-prey interactions occurred at the local level, great stability could occur at the regional or metapopulation level. This suggestion is analogous to the observations of Huffaker's mite predator-prey system in the laboratory (see Figures 13.8 and 13.9). The question is whether or not parasitoid-host systems in biological control are in fact distributed as metapopulations. At present few data are available to decide this issue, and this remains an important theoretical idea to be tested.

3. Refuges for the host populations. If the host population has a refuge habitat in which the parasitoid or predator cannot reach them, there is a potential for a stable interaction of parasite and host similar to that Gause (1934) observed with *Didinium* and *Paramecium* (see Figure 13.7 on p. 213). The question is whether or not successful biological control agents operate best in systems containing prey refuges. Only one case has been tested, and this explanation for stability was rejected (Murdoch et al. 1995).

4. Density dependence in the parasitoid attack rate. If parasitoids attack a pest species in a density-dependent manner such that the parasitoids interfere with one another when they are at high density, the host population will be stabilized. This mechanism is an attractive one for maintaining a stable predator-prey interaction, but it may achieve stability only at high pest densities, rather than at the low pest densities we desire for pest control. Consequently, we must first determine if density dependence occurs in parasitoid attack rates, and then determine how this can act to suppress the pest population to low density.

Successful control agents cause density-dependent losses in the host population, and this density dependence may be spatial or temporal or both. Spatial density dependence occurs when predators or parasitoids cause a higher fraction of losses in dense host patches than in sparse host patches (Murdoch et al. 1995). If predators can aggregate in patches of high host density, then, according to this theory, biological control of the pest is much more likely.

The classical theory of biological control based on the Nicholson-Bailey model is an equilibrium theory, and it is preoccupied with stability of the predator-prey interaction. It has been challenged recently by an alternate view based on a nonequilibrium model of predator-prey interactions (Murdoch et al. 1985). Bill Murdoch, working at the University of California at

Santa Barbara, has been one of the leaders in testing theories of biological control with field data and then devising new theories that better fit the data.

William W. Murdoch *(1938–) Professor of Ecology, University of California, Santa Barbara*

The nonequilibrium model begins with the assumption that a stable equilibrium of predator and prey is not necessary for satisfactory biological control. Pest populations may fluctuate wildly without pest densities exceeding the economic threshold. Some local populations of pests may be driven extinct in this model. The nonequilibrium model is a metapopulation model, and as such it emphasizes that populations in different patches may fluctuate independently, and that movements between patches may play an important role in metapopulation stability.

We can test these two alternative views by comparing their predictions with observations on case histories of successful biological control efforts. Murdoch et al. (1985) compiled data on several highly successful control efforts (Table 18.1), and even though in most cases biological control was successful, the mechanisms of success were not those predicted by the classical theory. Let us look in detail at one example, the California red scale.

The California red scale (*Aonidiella aurantii*) and its wasp parasitoid *Aphytis melinus* interact in a classic example of successful biological control. Once a serious pest of citrus crops in southern California before *Aphytis* was released in 1957, the red scale now exists at low and constant densities. Reeve and Murdoch (1985, 1986) analyzed three possible causes for this successful control program. First, mortality caused by the wasp *Aphytis* was not density dependent in time (Figure 18.7), as

TABLE 18.1   **Predictions of two alternative models of successful biological control with observations from five case studies.**

	Stable equilibrium of pest population at low density	Parasitoids produce density-dependent mortality in host	Parasitoids spatially aggregate to host density	Parasitoids host-specific	Parasitoids synchronized with host
Predictions of:					
Classical theory	Yes	Yes	Yes	Yes	Yes
Nonequilibrium theory	No	Not necessary	Not necessary	Not necessary	No
Successful cases of biological control					
Winter moth in Nova Scotia	No	?	?	No	?
Olive scale in California	No	No	No	Yes	No
Larch sawfly in Manitoba	No	Yes	?	Yes	Yes
Red scale in California	Yes	No	No	Yes	Yes
Mosquitoes in ponds and irrigation ditches and mosquitofish (*Gambusia* spp.)	No	Yes?	?	No	No

Note: The nonequilibrium theory fits these data better than does the classical model.
Source: Data from Murdoch et al. (1985).

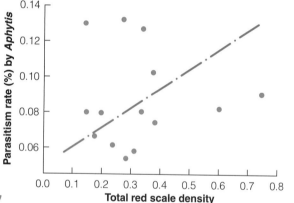

FIGURE 18.7

A test for temporal density-dependent predation by the wasp Aphytis melinus *on California red scale over three years. The blue line shows the expected line if density dependence were occurring. The data do not fit the line, so there is no indication of temporal density-dependence in this predator-prey system (r= 0), despite its success as a biological control program. The photo shows an* Aphytis *female laying an egg on a red scale. (From Reeve and Murdoch 1986.)*

classical theory would predict. Second, even though stability of host numbers can also be achieved by spatial density dependence (Hassell 1978), for the California red scale there was no sign of density-dependent mortality in space; *Aphytis* did not aggregate on high-density clumps of red scale again as classical theory would predict (Figure 18.8). The third possible mechanism for stability is the existence of a refuge for the prey. Reeve and Murdoch (1986) found that the red scale has a refuge in the interior branches of citrus trees (Figure 18.9), and in this refuge the density of red scales is 100-fold higher than it is on exterior, exposed branches of the grapefruit trees. They postulated that in this refuge the red scale is protected from parasitoids by the Argentine ant. The ants do not actively protect the red scales but seem to disturb *Aphytis* and prevent them from attacking some scales. To test this they removed ants from some trees and compared those trees with trees open to ant colonization. Even though removing ants opened up the refuge areas to higher parasitism rates, a large difference between the refuge and the exterior bark still existed (Figure 18.10) . A second refuge exists

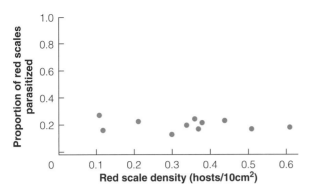

(a) Whole trees

Aphytis wasp

0.5 mm

FIGURE 18.8
A test for spatial density dependence in predation by the wasp Aphytis melinus *on California red scale. Three levels of analysis were done, from the whole tree as a spatial unit (a) down to the individual branchlet (c). There is no indication of density dependence in space, despite the fact that this is a highly successful biological control program. (From Reeve and Murdoch 1985.)*

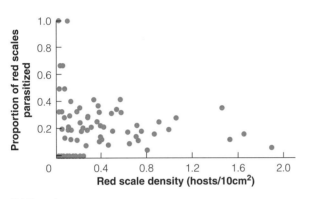

(b) Branches

Red scale

1 mm

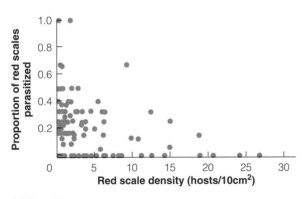

(c) Branchlets

because the red scale cannot be attacked by *Aphytis* once the scale has become adult (Murdoch et al. 1987). Because of these refuges, red scales do not decline to extinction on citrus trees. If the refuge areas could be eliminated from trees, the red scale would have no spatial refuge, and Reeve and Murdoch (1986) predicted that under these circumstances the red scale would be completely eliminated by parasitoids.

The results of analyzing successful biological control programs seem to favor the nonequilibrium view of control and suggest that density-dependent interactions between predators and their prey are not necessary for successful biocontrol. Two strategies of predators could promote local pest extinction (Murdoch et al. 1985). A "lying-in-wait" strategy requires the continuous presence of the predator in local areas and a high attack rate on the pest when it reinvades the area. Predators that adopt this strategy typically are not host specific but feed on a variety of prey. A second strategy is the "search-and-destroy" strategy, in which a predator finds and attacks the pest in a hide-and-seek pattern. This strategy involves host-

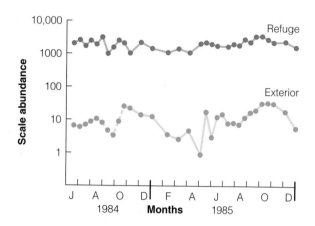

FIGURE 18.9
The density of the California red scale per average twig in the exterior and in refuge sections of grapefruit trees in southern California. Abundance is plotted on a log scale. The refuge area under the bark harbors populations of red scale about 100 times the density of those inhabiting exposed areas of the trees. (From Murdoch et al. 1995.)

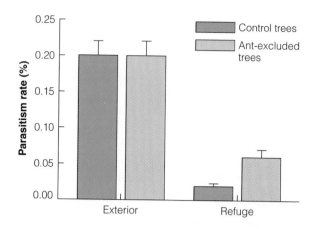

FIGURE 18.10
The parasitism rate by Aphytis melinus *on California red scale in the exterior branches of grapefruit trees in Southern California and in the refuge areas under the bark in the interior part of the trees. Argentine ant removal from trees removes some of the protection of the refuge but not all of it, so that the refuge still harbored a higher density of scale than the exterior. The ant removal produced a large drop (about 50%) in scale density in both the exterior and the refuge areas. (Data from Murdoch et al. 1995.)*

specific predators with high search rates and effective dispersal powers. The system avoids extinction because of spatial patchiness or the presence of refuges in the natural habitats of the prey species. The important message is that there may be several different ways to achieve success in biological control.

Genetic Controls of Pests

Another alternative method of pest control is *genetic control*, a type of biological control that uses two strategies to reduce pest problems. Either crop plants can be genetically manipulated to increase their resistance to pests, or pests may be changed genetically so that they become sterile or less vigorous and thus decline in numbers.

The use of crop varieties resistant to attack by pests is one of the oldest and most useful techniques of pest control. In 1861 the grape *phylloxera*, an aphid that feeds on the roots of grape plants, was accidentally introduced into Europe from North America. The European grape (*Vitis vinifera*) was extremely susceptible to the *phylloxera*, and the wine-making industry of France was on the brink of collapse by 1880. The American grape (*Vitis labrusca*) is resistant to phylloxera attack, so European grape vines were

grafted onto American rootstocks to produce an artificial hybrid grape plant that was resistant to phylloxera attack.

Resistant varieties of many crop plants have been developed by selective breeding (Maxwell and Jennings 1980). The method used is, in principle, very simple. Individual plants that are not being damaged are sought in an area where the pest species is common, and these plants are removed to the greenhouse for selective breeding. If resistance is inherited in the greenhouse lines, the new selected variety may be used for commercial production.

Selective breeding can be a two-edged sword, however, and must be used with care. For example, all species of cotton produce a plant chemical called *gossypol* (a sesquiterpene), which occurs in the green parts and seed of the cotton plant and is toxic when fed to chickens and pigs. To increase the value of cottonseed as an animal feed, plant breeders bred strains of cotton with low gossypol content and were able to reduce the concentration of gossypol to only one-fourth that of normal cotton. But breeding gossypol out of cotton deprived the plant of much of its resistance to insect pests and also made the cotton plant susceptible to a whole set of new pests (Klun 1974).

Resistant plants do not necessarily have chemical defenses. Morphological defenses can be highly effective (Myers and Bazely 1991). Soybeans are a major crop in the midwestern United States despite the presence of a serious potential pest, the potato leafhopper (*Empoasca fabae*). The potato leafhopper will not attack soybean varieties that have leaves covered with short hairs, whereas they attack and nearly destroy soybeans that have hairless, smooth leaves. The hairs deter insect movement and are highly effective as a defense mechanism.

Breeding resistant varieties of plants has been an important factor in limiting pest damage in many crops, but the rapid adaptability of plant pathogens has compromised much effort (Lupton 1977). For example, potato blight, a disease caused by a fungus, *Phytophthora infestans*, first appeared in Europe in the 1840s, and it spread rapidly, causing famine in Ireland. Plant geneticists have attempted to induce a high level of resistance to blight in cultivated potato plants by introducing into them single genes derived from closely related wild species. Four genes have been used in this way, but after the commercial introduction of each new gene for resistance, new races of the fungus appeared that could attack the "resistant" potatoes. Sexual recombination or asexual mutation of fungal pathogens results in rapid evolutionary changes in field populations, so that crop resistance breaks down over time.

One promising area of intense development currently is the production of resistant crop plants by means of genetic engineering. Genes that produce resistance in one species can be inserted into a crop plant to make the crop genetically resistant to specific pests. Alternatively, bacteria may be used as vehicles to carry biopesticide genes. *Bacillus thuringiensis* (Bt) is the main focus at present for developing insect-resistant crops (Lambert and Peferoen 1992). This bacteria normally lives in the soil and carries a gene for a protein that is toxic to the larvae of butterflies and moths. By splicing this gene into bacteria that normally live on crop plants, genetic engineers can produce insect-resistant crops. Insect pests ingest the bacteria while feeding on the plant and are thereby poisoned. Alternatively, the Bt genes that code for the toxins can be transferred directly into the plant's genome, so the plant would protect itself.

Transgenic plants with Bt genes are considered by some to be the future in crop protection by providing plants that are genetically resistant to insect pests (Hilder and Boulter 1999). One anticipated problem with this technology is that pest insects will become resistant to the biopesticide, just as they became resistant to chemical pesticides. More than a dozen species of insects are already resistant to the toxins produced by Bt (Tabashnik et al. 1998). Diamondback moth larvae have evolved resistance in field crops (Tabashnik et al. 1990). To avoid the development of resistance, farmers have been advised to plant 20% of cotton or corn crops in nontransgenic plants, so that the majority of the pests develop in this unmanipulated crop. But this strategy results in economic losses for the farmer. One approach to get around this problem is to "piggyback" two toxins into transgenic plants so that if one toxin does not kill the pest, the second one will (Roush 1998). This strategy will work if the two toxins act independently in the pest insects. The development of resistance to both natural and artificial pesticides will continue to be a major problem in pest management in agriculture (Pimentel 1991).

In addition to changing the genetic makeup of the plants, we can attempt to alter the genome of the pest species. The simplest genetic manipulation that can be carried out on a pest species is sterilization. Sterility can be produced in several ways, but the usual procedure is to sterilize large numbers of pest individuals by irradiation or by treatment with chemicals and then to release them into the wild, where they can mate with normal individuals. Because of these matings, the number of progeny produced in the next generation is greatly reduced, and control can be achieved. The sterile-insect method cannot be used on all pest populations because it requires the rearing and sterilizing of large numbers of individuals and a situation in which immigration of fertile individuals is not a major factor.

One example of the successful use of the sterile-insect method was the suppression of the mosquito *Culex pipiens quinquefasciatus* on a small island off Florida (Patterson et al. 1970). Between 8400 and 18,000 sterile males were released each day over a ten-week period during midsummer on a 0.3 km island, and by the end of the experiment 95% of the eggs sampled on the island were sterile (Table 18.2). Thus the experiment was a success, but because the island was only 3 km from the mainland, recolonization by dispersing females occurred quickly once the experiment ended.

The sterile-mating technique of control may be rendered less effective if pest populations are genetically subdivided. One example of this difficulty is the screwworm control program in the southern United States (Richardson et al. 1982). Screwworms are larvae of several species of blowflies that lay their eggs on

TABLE 18.2   **Sterile-male release experiment with the mosquito *Culex pipiens quinquefasciatus* on Seahorse Key, a small island off Florida.**

Generation	Ratio of sterile to normal males	Eggs expected to be sterile (%)	Eggs actually sterile (%)	Reduction in eggs laid (%)
1	All normal	0	0	0
2 (begin releases)	3:1	75	62	36
3	4:1	80	85	34
4	12:1	92	82	79
5	100:1	99	84	96
6 (end)	100:1	99	95	96

Note: Each generation of mosquitoes took about two weeks during this summer period.
Source: Data from Patterson et al. (1970).

open wounds of warm-blooded animals. The larvae enter the wound and feed on the living tissue, possiby leading to the death of the host animal (often cattle, sheep, or deer) because of physical damage or secondary infections. A program to eradicate screwworms in the southern United States was begun in the 1950s. These programs were very effective until 1968, when a series of unexplained outbreaks began. Serious outbreaks occurred again in 1972, 1976, and 1978.

At least 11 chromosomal types of screwworms can be recognized (Richardson et al. 1982). Many of these types occur together geographically and could possibly be different species of screwworms. If there is a genetic mismatch between the sterile flies raised in the laboratory and the wild type causing an outbreak, then clearly the sterile-mating technique will not work because the flies will not mate. All sterile screwworm flies are now produced in a single factory in Chiapas, Mexico, and recent failures of sterile-insect releases to stop screwworm outbreaks have probably been due to genetic differences between the cultured screwworms and the screwworms in the outbreak area.

Immunocontraception

A new method of biological control for vertebrates, *immunocontraception*, has emerged with recent developments in biotechnology. While much of biological control has aimed at ways of increasing the mortality rate in a pest population, an alternative approach is to reduce the fertility of the pest. This is the approach that underlies immunocontraception. Fertility could be reduced by the use of immunocontraceptive vaccines delivered in a bait, or by a virus or other contagious agent that spreads naturally through the target pest population (Figure 18.11). There are many dif-

ferent sperm and egg surface proteins that could be used for immunocontraception. One of the most commonly used set of proteins are the zona pellucida glycoproteins (ZPG), which facilitate sperm penetration of the egg. Antigens against zona pellucida proteins prevent sperm from attaching to the surface of the egg, thereby preventing fertilization.

Immunocontraception works. Figure 18.12 illustrates the decline in fertility of white-tailed deer on Fire Island National Seashore in New York State (Kirkpatrick et al. 1997). White-tailed deer are overabundant in many areas of eastern North America where all their predators have been removed and agricultural areas provide superabundant food. Given that immunocontraception works, we next need to ask what effect it will have on the productivity of the pest species. We might expect that the higher the level of

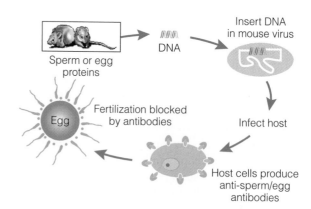

FIGURE 18.11

General procedure for immunocontraception in mammals, illustrated for the house mouse. The aim is to immunize the pest species against its own sperm or egg proteins, so that fertility is blocked. The same general procedure could be applied to any mammal or bird pest. (After Pech et al. 1997.)

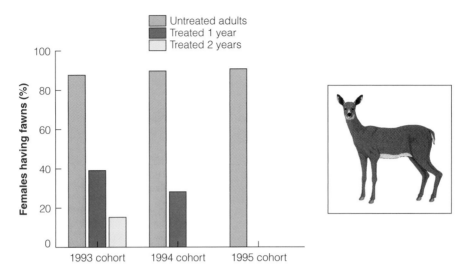

FIGURE 18.12

Efficiency of contraception against a free-ranging herd of white tailed deer (Odocoileus virginianus) *on Fire Island National Seashore, New York, 1993-95. Deer were inoculated with darts containing 65 μg of porcine zona pellucida protein. New adult recruits were treated each year, and the bars show the fertility in subsequent years of the study. The fertility rate was cut dramatically, and this could be one approach to control overabundant deer populations. (Data from Kirkpatrick et al. 1997.)*

sterilization, the fewer the litters produced, but Caughley et al. (1992) pointed out that this was an oversimplification for mammals which have a social system. Consider the case in which there is a linear dominance hierarchy among females, such as a pecking order. If only the most dominant female breeds, and sterilization removes her from the pecking order (allowing subordinate females to breed), then the effect of sterilization on productivity is much less pronounced (Figure 18.13). The main point is that imposing sterility on a population may not have a simple effect on the population.

The best illustration of the effect of sterilization of pests on population trends comes from studies on the European rabbit in Australia. Female rabbits in wild populations were live trapped and surgically sterilized at two sites, one in Western Australia and one in southeastern Australia (Twigg and Williams 1999). Populations with 0%, 40%, 60%, and 80% sterility were then followed for four years. The reproductive output of rabbits decreased with increasing levels of sterility imposed (Figure 18.14a, page 346). The juveniles produced, however, survived much better in populations given the sterility treatments (Figure 18.14b), compensating partly for the imposed sterility. The adult rabbits that were sterilized also survived much better than fertile females (Figure 18.14c), again compensating for the

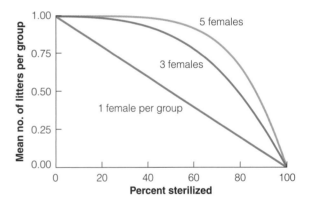

FIGURE 18.13

Expected number of litters produced per breeding season for a hypothetical mammal subject to various rates of sterilization. If there is one female in the group, we expect a linear decline. But if there is a dominance hierarchy and only the top female will breed, the results are not linear. This simple model assumes that sterilization of a female removes her from the dominance hierarchy. (Modified from Caughley et al. 1992.)

sterility imposed. The net result was that this sterilization program changed the population density very little until a level of 80% imposed sterility was reached. The practical message from these experiments is that immunocontraception will not be

FIGURE 18.14
Sterility experiment on the European rabbit in Australia. (a) The reproductive output under the four levels of imposed female sterility. The number of rabbit kittens emerging from burrows is close to that expected by the percentage of sterility (blue line). (b) Percentage of juvenile rabbits surviving to become adults in the four treatments. Juvenile survival compensated for the sterility by increasing above the expected (control) line shown in black. (c) Adult female survival during the experiment. Adult rabbits sterilized surgically in the first year of the study survived much better than females that bred, so there is a survival cost of reproduction in females that compensates for the imposed sterility. (Data from Twigg and Williams 1999.)

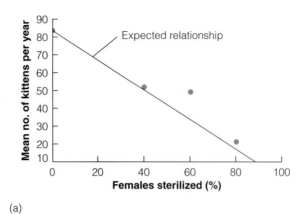

(a)

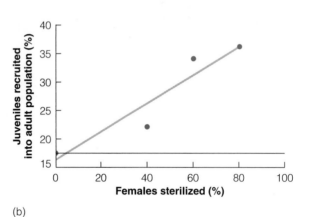

(b)

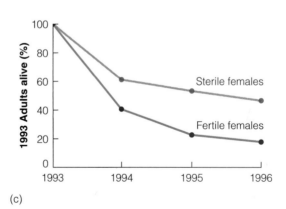

(c)

effective in reducing rabbit numbers in Australia unless it can reach about 80% of the rabbit population annually (Figure 18.15).

Immunocontraception is an innovative idea that can assist in a coordinated plan of pest control for a variety of wildlife species, and much work is now underway in the United States, Australia, and New Zealand to determine its potential (Chambers et al. 1997, Muller et al. 1997).

Integrated Control

Many important pests cannot be controlled by any one technique, so biologists concerned with pest management have been forced to take a wider view of pest problems. A unified approach, called integrated control or integrated pest management (IPM), uses biological, chemical, and cultural methods of control in an orderly sequence (Figure 18.16).

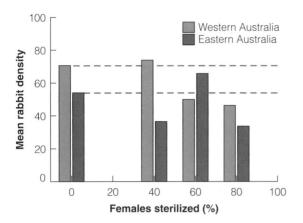

FIGURE 18.15
Density changes during the sterility experiment on the European rabbit in Australia from 1993 to 1996. The average number of rabbits over the entire experiment showed little change except at the most extreme level of 80% sterility, where rabbits declined in density 34-37%. Control densities are shown by the horizontal dashed lines. (Data from Twigg and Williams 1999.)

The objective of integrated pest management is to minimize economic, environmental, and health risks. Integrated control can be achieved only if the population ecology of the pest and its associated species, and the dynamics of the crop system, are known. Integrated control systems are ecologically sound because they rely on natural biological control as much as possible and resort to chemical treatments only when absolutely necessary. A considerable amount of information is needed to permit the effective use of an integrated control program. Density levels of the potential pest populations, stage of plant development, and weather data are often required to enable the pest manager to predict the future development of the crop and to judge the necessity for pesticide application.

An example of an integrated control program is the alfalfa pest management project developed in Indiana and in use in much of North America (Giese et al. 1975). Alfalfa is an important crop because it produces high-quality feed for cattle and also improves the soil by fixing nitrogen. Alfalfa is a perennial crop and is relatively long-lasting. Several hundred species of insects can be found in alfalfa fields, yet only a few are serious pests. The alfalfa weevil (*Hypera postica*) is the most important single alfalfa pest in the world, and as an illustration we can design an integrated control program for this weevil (Armbrust and Gyrisco 1982).

The life cycle of the alfalfa weevil in the eastern United States is shown in Figure 18.17 . Eggs are laid in the fall and winter, and they hatch in the spring. New adults that emerge in the spring feed for a short time and then move into wooded areas to estivate during the summer. In the fall, adults return to the alfalfa fields and become sexual. Many eggs

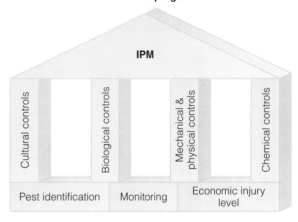

FIGURE 18.16
Foundations of an integrated pest management system. No one type of pest control will be sufficient for many agricultural and forestry pests, and the various alternative methods of control need to be integrated ecologically. Detailed taxonomic identification of the pest is an important foundation, as is careful monitoring of the pest population and damage. (Diagram courtesy of U.S. Department of Agriculture.)

are laid over the winter, and larvae hatch and start to feed just as alfalfa plants begin to grow in the spring. Damage can be severe, and spring weather is critical. Low temperatures retard larval growth more than plant growth such that little damage occurs; higher temperatures speed larval development and increase damage.

Weather conditions are critical for determining the timing of control procedures against the alfalfa

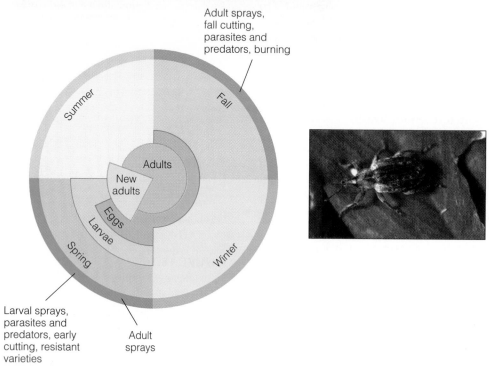

Adult sprays,
fall cutting,
parasites and
predators, burning

Larval sprays,
parasites and
predators, early
cutting, resistant
varieties

Adult
sprays

FIGURE 18.17
*Life cycle of the alfalfa weevil living on alfalfa in the eastern United States. Severe
defoliation shown in the photo at top can occur in the spring from larval feeding. In the
summer, adults move into woody habitats and then return to alfalfa fields in the fall.
Possible seasonal control methods are listed. (After Armbrust and Gyrisco 1982; photo
courtesy of University of California Statewide IPM Project.)*

weevil. Because temperature is the major variable for
both the plant and insect populations is so variable
seasonally, the best way to assess control requirements
is to count weevil larvae on the alfalfa stems. For
example, farmers can be given the following rules for
determining control action:

Collect 30 alfalfa stems at random, remove the
weevil larvae by vigorous shaking of the stems, and
then count the larvae:

- If weevil abundance is less than one larva per
  alfalfa stem, no action is needed.

- If weevil abundance is greater than three larvae per stem, immediate action is required..
- If weevil abundance is between one and three larvae per stem, continued monitoring is needed, and the cost of must be compared to the potential value of the alfalfa crop.

Timing becomes a critical element in all integrated control programs; hence a detailed understanding of the life cycle of the pest and its host is necessary. The proper timing for spraying an insecticide can reduce the alfalfa weevil population, allow the alfalfa to grow, and delay the onset of high weevil densities until later in the spring, when insect parasites can attack weevil larvae and maintain densities below the damage threshold. Emphasis in integrated control has shifted from trying to eradicate the pest to asking how much damage can be tolerated. Thus a key point in pest control is to determine the damage. We cannot eliminate pests and must learn how best to live with them, taking advantage of all techniques available to keep them below the threshold of economic damage.

Integrated control programs derive their validity from field studies and are thus empirical ecology in action. They have not been developed as theoretical strategies but as working programs, and they hold great promise for the future because they retain biological control as a core element of the integrated program (Benbrook 1996).

Generalizations About Biological Control

Why can we not control all pests by biological control? Biological control is something akin to a gambling system: It works sometimes. But how often? Table 18.3 summarizes data from a global appraisal of the success rates of classic biological control against insect and arachnid pests. About one-third of the parasites and predators introduced get established more or less permanently after introduction (Hall and Ehler 1979). If we define success in biological control according to economic benefits, only 16% of classic biological control attempts qualify as complete successes (Hall et al. 1980). Why is this? What makes some biological control agents such as the vedalia work so well while others completely fail? A number of empirical generalizations have been suggested.

Most successful biological control programs have operated quickly. Clausen (1951) suggests that three generations (or a maximum of three years) is the outside limit and that if definite control is not achieved in the vicinity of the colonization point within this time, the control agent will be a failure. This rule of thumb suggests that colonization projects should be discontinued after three years if no success is achieved and that prolonged efforts at establishment are wastes of money. Most examples of successful biological control examples to date support this rule, which suggests that major evolutionary changes in the host-parasite system seldom occur in introduced pests. If a parasite is not already adapted to control the host, it will not evolve quickly into a successful control agent.

The unfortunate truth is that we can evaluate a biological control agent only in retrospect, and biological control programs are a gamble. We release a predator or parasite and hope for the best. A vital historical lesson is the frequency with which a species such as the vedalia was released more on faith than on any evidence that it could control the pest. There currently is no evidence that biological control would not be just as successful if we were to release a random sample of a pest species enemies. But ecological theory can learn from the past successes of control programs and develop new insights into how to select biological control agents (Murdoch and Briggs 1996, Waage 1990).

Most successful biological control programs have resulted from a single species of parasite or predator, which raises a question: If one parasite species is good, are two species better? Turnbull and Chant (1961) argued that only one species should be released at a time for pest control, because two parasites might interfere with each other when the pest is reduced to low numbers. This argument follows from the observation that native insect pests have numerous predators and parasites. The spruce budworm, for example, which is a serious forest pest, has over 35 species of parasites and many predators. Is the spruce budworm a pest because it has many parasites? Or does it have many parasites because it is moderately abundant?

If competitive interactions occur between introduced parasitoids and predators, biological control should be more successful when fewer enemies are released. Figure 18.18 shows that the more species of enemies that were released per program, the fewer programs were successful in getting established (Ehler

TABLE 18.3 Summary of the success of biological control efforts against insect and arachnid pests throughout the world.

Category	No. of attempts	Established efforts		
		Established (%)	Partial or complete successes (%)	Complete successes (%)
Total	2295	34	58	16
By order of insects introduced				
Homoptera	819	43	80	30
Diptera	258	37	31	0
Hymenoptera	105	34	56	0
Lepidoptera	628	27	48	6
Coleoptera	364	23	36	4
By demographic origin of pest				
Exotic pests	2163	34	60	17
Native pests	132	25	29	6
By geographic isolation				
Islands	827	40	60	14
Continents	1468	30	56	17
By habitat stability				
Unstable habitats (vegetable and field crops)	640	28	43	3
Intermediate (orchards)	916	32	72	30
Stable habitats (forests, rangelands)	535	36	47	8

Source: Data compiled by Hall and Ehler (1979) and Hall et al. (1980).

and Hall 1982). Thus competitive exclusion of introduced natural enemies has probably occurred and has contributed to the low rate of success in biological control (see Chapter 12).

What can we conclude regarding the problem of natural regulation from these examples of biological control? This is a difficult question. Belief in the success of biological control is based on a thoroughly biased sample. Economic pressures run high in this field, because crops worth millions of dollars may be destroyed by a single pest. Consequently, some states, such as California, have full-time bureaus devoted solely to searching the world for insects to control current agricultural pests. Candidates for control are carefully screened before they are released to ensure that they will not destroy the native fauna instead of the pest. However, once these control agents are introduced, little further work is usually done. Either the agents work and the pest decreases, or they do not work and the entomologists go looking for other parasites or predators. Consequently, the literature is full of all sorts of spurious correlations that are seldom further investigated.

The contrast between the restricted fluctuations of natural ecosystems and the recurrent pest outbreaks in agricultural systems suggests another way of looking at pest control problems. Pest species thrive in our agricultural systems for three possible reasons. First, the agricultural systems are typically monocultures, often of genetically similar plant varieties, whereas natural ecosystems have many species and a great deal of spatial complexity. The hazards of dispersal and habitat selection are greatly reduced when the habitat becomes a monoculture. Root (1973) called this general effect the *resource concentration hypothesis* and suggested that monocultures permitted higher herbivore densities. Second, the plants, herbivores, and predators of agricultural systems do not always form a coevolved system, and thus the normal processes of evolutionary integration are not achieved in agricultural systems. Third, the much more numerous disturbances in agricultural systems compared to natural systems leads to a reduced diversity of species in agricultural systems and makes natural communities and agricultural systems fundamentally different (Murdoch 1975). The best analog of an agricultural system may be simplified laboratory systems that

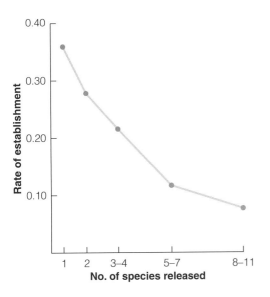

FIGURE 18.18
Rates of establishment of insect enemies in biological control for single-species releases and for simultaneous, multiple-species releases for 548 projects classified as failures. The same trend is seen if only the successful projects are analyzed. (After Ehler and Hall 1982.)

include only a few species. Thus spatial complexity may be important in crop systems, just as it is in laboratory populations.

Integrated pest management, through the use of biological control with other types of control tactics, is rapidly becoming one of the most important practical applications of ecological theory to modern problems of food production. We are gradually replacing an outmoded version of attempted pest eradicationusing toxic chemicals with a new view of crop management with minimal environmental disturbance. To achieve this goal, we need to know the population biology of both the crops and their associated pests. The challenge is great but the payoffs are vital.

Risks of Biological Control

Introducing a nonnative species into an ecological system to control a pest is not without some ecological risks. The danger is that the introduced species will attack nontarget organisms and cause more damage than it prevents. The clearest examples of this involve generalized predators released for biological control (Simberloff and Stiling 1996). The Indian mongoose (*Herpestes auropunctatus*), which was introduced into Hawaii and many islands in the West Indies to control rats in sugarcane fields, has become an important predator of native birds on all these islands and is suspected of causing the extinction of some reptiles in the West Indies (Lever 1985).

In other cases, the predatory snail *Euglandina rosea* has been introduced from Florida and Central America to many islands in the Pacific and Indian Oceans to control another introduced snail, the Giant African snail *Achatina fulica*, which can reach 15 cm in length and was originally introduced for food. It eats hundreds of plants and is considered an agricultural pest on many islands. Biological control has consisted of introductions of predatory snails that are not restricted in their feeding, so they have driven native snail species to low numbers or to extinction. Hawaii

Indian mongoose (Herpestes auropunctatus) (©Gregory G. Dimijian/Photo Researchers, Inc.)

Achatina fulica

once had 931 species of land snails, but about 600 species have disappeared since European colonization (mostly because of deforestation for agriculture). The introduced snail *Euglandina rosea* is now causing the extinction of many of the remaining native snails (Civeyrel and Simberloff 1996). The lesson learned from these mistakes is not to use generalist predators as biological control agents.

Three aspects of biological control programs have come under scrutiny because of these potential problems. First, pest problems must be quantified before a biological control program is initiated. How much damage is being caused, and is the problem more esthetic than economic? Some species that are thought of as pests do not actually cause economic or environmental damage (Thomas and Willis 1998). Second, nontarget species must be tested more widely before a potential biocontrol agent can be considered for release. A broad array of nontarget species, not just other agricultural crops, must be considered for potential harm, because the risk of introductions typically involves conservation problems with native species. The potential of the biocontrol agent to attack new hosts also needs to be considered, for example, by determining the genetic basis of host preference. Third, more research is required after a biocontrol agent has been released. Studies of the demographic processes by which biological control is achieved are typically severely limited to control costs, so after the release we know only that it is a success or a failure but never why. As more and more pest species are spread around the world, the need for strict guidelines for the release of nonnative species becomes stronger, so that a clear estimate of the costs as well as the benefits can be evaluated (Ehler 1998).

Summary

Pests are species that interfere with human activities and hence need to be controlled. Most pest control in agricultural systems is achieved in a temporary manner using pesticides, but these toxic chemicals affect other important species and become ineffective because pests develop genetic resistance to the chemicals. Biological control makes use of predators and diseases to reduce the average abundance of a pest species.

There are many cases of major reductions in numbers of introduced pests by predators or by insect parasitoids that are introduced specifically for the purposes of control. Many other introductions have failed, leaving the pest to be controlled by chemical means. We cannot adequately explain most of the successes; nor can we explain why failure is so common. The classical model of pest control through density-dependent predation does not seem to describe the situation of successful control programs in the field.

Genetic control of pests can be accomplished by producing resistant crop plants or by interfering with the fertility or longevity of the pest. Many techniques for the genetic control of pests have been proposed, but few have been used successfully in the field. Immunocontraception is a new method of pest control being developed for overabundant mammalian species. All forms of pest control raise ecological questions of how the pest may compensate for increased mortality or reduced fertility. Genetic engineering holds great promise for producing both new methods of reducing pest numbers and crops that are more resistant to insect pests.

Integrated control combines the best features of biological and chemical control methods in an effort to minimize the environmental degradation that has been typical of modern commercial agriculture. To achieve integrated control, we need to understand the population dynamics of the pest species; this is, at present, one of the greatest challenges in applied ecology.

Biological control programs entail the potential risk that the control agent will attack other native species in addition to the targeted pest. The cure must be better than the disease, and extensive testing must be carried out before and after any biological control program is activated.

Key Concepts

1. Pest control is applied population ecology that asks what limits the average density of a pest species, and what we can do to change average density. Most pest control utilizes chemical poisons.

2. Pest control is the reduction of damage caused by a particular pest below an economic threshold; consequently it involves information on the ecology of the pest and the economics of the damage.

3. Biological control involves the introduction of nonnative predators or diseases to reduce the population density of the pest. Even though in some cases it is spectacularly successful, in most cases it does not work.

4. Genetic control involves either changing the genetic makeup of the host species to make it more resistant to pest attack, or changing the pest to make it sterile.

5. Integrated control is the coordination of chemical control, biological control, and cultural control in an overall strategy to control a pest.

6. Predators introduced for biological control may themselves become pests of native species and careful evaluation must precede introductions of alien species

Selected References

Butler, V. 1998. Elephants: Trimming the herd. *BioScience* 48:76–81.

Chambers, L., G. Singleton, and G. Hood. 1997. Immunocontraception as a potential control method of wild rodent populations. *Belgian Journal of Zoology* 127: 145–156.

Flint, M. L. and R. Van Den Bosch. 1981. A history of pest control. *An Introduction to Integrated Pest Management*, pp. 51–58. Plenum Press, New York.

Hilder, V. A. and D. Boulter. 1999. Genetic engineering of crop plants for insect resistance - a critical review. *Crop Protection* 18:177–191.

Lafferty, K. D. and A. M. Kuris. 1996. Biological control of marine pests. *Ecology* 77:1989–2000.

Louda, S. M., D. Kendall, J. Connor, and D. Simberloff. 1997. Ecological effects of an insect introduced for the biological control of weeds. *Science* 277:1088–1090.

Murdoch, W. W., and C. J. Briggs. 1996. Theory for biological control: Recent developments. *Ecology* 77:2001–2013.

Myers, J. H., C. Higgins, and E. Kovacs. 1989. How many insect species are necessary for the biological control of insects? *Environmental Entomology* 18:541–547.

Pimentel, D. 1997. Pest management in agriculture. In D. Pimentel, ed. *Techniques for Reducing Pesticide Use: Economic and Environmental Benefits* pp. 1–11 Wiley, Chichester, England.

Pimentel, D., L. McLaughlin, A. Zepp, B. Lakitan, T. Kraus, P. Kleinman, F. Vancini, W. J. Roach, E. Graap, W. S. Keeton, and G. Selig. 1991. Environmental and economic effects of reducing pesticide use. *Bioscience* 41:402–409.

Simberloff, D., and P. Stiling. 1996. How risky is biological control? *Ecology* 77:1965–1974.

Singleton, G. R., L. Hinds, H. Leirs, and Z. Zhang. 1999. *Ecologically-based Management of Rodent Pests*. Australian Centre for International Agricultural Research, Canberra, Australia. 494 pp.

Twigg, L. E., and C. K. Williams. 1999. Fertility control of overabundant species: can it work for feral rabbits? *Ecology Letters* 2:281–285.

Van Lenteren, J. C. and W. A. Overholt. 1994. Ecology and integrated pest management. *Insect Science and Its Applications* 15:557–582.

Questions and Problems

18.1 Purple loosestrife (*Lythrum salicaria*) is a wetland plant introduced to North America from Europe in the early 1800s. It has been declared a severe environmental problem in Canada and the United States because it is believed to take over wetlands, displacing native vegetation and adversely affecting wildlife species. Discuss how you would test the hypothesis that purple loosestrife has detrimental effects on other species in wetlands. Compare your action plan with the data presented by Hager and McCoy (1998), who questioned whether purple loosestrife does have adverse effects.

18.2 Why does *Bacillus thuringiensis* produce proteins that are toxic to insects? Review the biology of this bacterium and its geographical distribution, and discuss the evolution of its protein toxins. Lambert and Peferoen (1992) provide references.

18.3 Review the evidence for and against the idea that biological control is much more successful on islands like Hawaii than on continental areas (see DeBach 1964, p. 136, and Hall et al. 1980).

18.4 Elton (1958) showed that introduced species often increase enormously and then subside to a more static, lower density level. How might this occur in a species that was not the subject of introductions for biological control? How could you distinguish this case from a decline that followed the introduction of some parasitoids for biological control?

18.5 It is customary to obtain insect parasitoids from the native country of an introduced pest and to use only these parasitoids for possible biological control. Pimentel (1963) suggests another strategy of introducing the parasitoids of other species closely related to the pest you want to control. What might be the rationale for this recommendation? See Room (1981) for a discussion of how these ideas apply to weed control, and Waage (1990) for a critique of this strategy.

18.6 Fire ants were introduced into the southern United States around 1918 and are a serious pest (Porter et al. 1997, Vinson 1997). Efforts to control fire ants have been very controversial and of limited success. Review the biology of fire ants, and discuss the reasons for the poor success of control policies. Lewis et al. (1992) provide an overview of the problem in California.

18.7 Contact your local municipal authorities and find out how Norway rats are controlled in your area. Discuss any ecological problems you can see with their methods and approach. Buckle and Smith (1994) provide background material on rat control methods.

18.8 In selection for crops resistant to insect attack, should an agricultural ecologist recommend maximal crop resistance as a goal, or could there be an advantage to having crops that are only partly resistant to pests? Discuss this issue before and after reading van Emden (1991).

18.9 For r-selected pests, Stenseth (1981) suggests that optimal control can be achieved by reducing reproduction rather than by increasing mortality. Discuss the ecological reasons behind this recommendation.

Overview Question
How do pests evolve resistance to chemicals used to control them? Will this problem arise with the most recent techniques that utilize genetic engineering and immunocontraception? How could you overcome the evolution of resistance in pest populations?

CHAPTER 19

Applied Problems III:
Conservation Biology

CONSERVATION BIOLOGY is the biology of population decline and scarcity and is a central focus of much public concern. Much of population ecology has been focused on abundant species and the factors preventing population growth. Many population biologists have pointed out that most species are rare, and rarity itself ought to be a focus for population research. Species that have become endangered or threatened are either rare or in sharp decline, and in this chapter we explore the causes of decline and rarity of species and what can we do to alleviate the problems of threatened populations.

Conservation biology divides rather cleanly into two separate approaches that operate with different ideas and different goals. Caughley (1994) called these two paradigms the *small-population paradigm* and the *declining-population paradigm*. We examine the small population paradigm first.

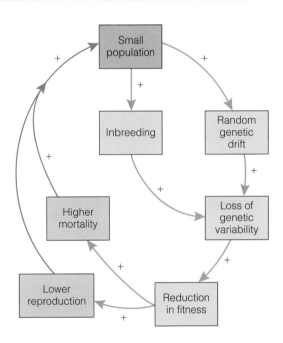

FIGURE 19.1
The extinction vortex of the small–population paradigm. Small populations, such as those on islands or in zoos, fall into a vortex of positive feedback loops in which small population size leads to inbreeding and genetic drift, resulting in loss of genetic variability. Because genetic variability is necessary for viability, fitness falls and the population size is reduced further.

Small-Population Paradigm

This paradigm focuses on the population consequences of rareness and the abilities of small populations to deal with smallness as such. The ideal is a small island population, or a small group of endangered species of animals or plants in a zoo or a botanical garden, and the questions arising from this paradigm deal largely with population genetics and demographic models of extinction in small popula-

tions. The essence of the small-population paradigm is encapsulated in the extinction vortex (Figure 19.1). Small populations risk positive feedback loops of inbreeding depression, genetic drift, and chance demographic events that lead inexorably to extinction. An essential feature of the small-population paradigm is a set of strong theoretical predictions that follow from population genetics theory. The key element is the maintenance of genetic variability, which is essential for future evolution and thus for long-term persistence. Michael Soulé of the University of California, Santa Cruz, has been a leader in promoting the conservation of endangered species and ecosystems, and is the father of conservation biology. Soulè was particularly concerned with the problems

Michael Soulé *(1935–) Conservation biologist, University of California, Santa Cruz*

faced by small populations of rare species, and this consideration has led to the concept of minimum-viable populations.

Minimum Viable Populations

Rare species will still be able to sustain their numbers at some population density. This idea has been formalized as the concept of the *minimum viable population* (MVP)—that population size that will ensure at some acceptable level of risk that the population will persist for a specified time (Gilpin and Soulé 1986). The analysis of minimum viable populations involves the analysis of extinction. What factors cause a species to go extinct? Some extinctions are due to chance, and Shaffer (1981) recognized three kinds of variation that can contribute to population loss.

1. *Demographic stochasticity.* This source of variation reflects random variation in birth and death rates that can lead by chance to extinction. If only a few individuals make up the population, the fate of each individual can be critical to population survival. If a female produces only male offspring and then dies, the population goes extinct. These are examples of demographic stochasticity. In general, demographic variability is critical to extinction only when populations are less than about 30-50 individuals (Caughley 1994).

2. *Genetic stochasticity.* Because evolution cannot occur without genetic variability, any loss of genetic variation can be a cause of extinction. Many genetic studies have shown that individuals with more heterozygous loci are fitter than individuals with less genetic variation (Allendorf and Leary 1986, Futuyma 1998). Genetic variability is lost by genetic drift, the nonrandom assortment of genes during reproduction, and by inbreeding. Both drift and inbreeding are minimized when populations become sufficiently large, so these phenomena are paradigms of the small-population problem.

3. *Environmental stochasticity and natural catastrophes.* These factors include variation in population growth rates imposed by changes in weather and biotic factors, as well as by fire, floods, hurricanes, and landslides, which can also be responsible for species declines. The key concept is how much variation the environment imposes on the rate of increase of the population. If the variance in the rate of population growth is greater than the population growth rate itself, environmental stochasticity can cause extinction (Lande 1993).

Small populations may become small because of habitat changes, but much of the small-population paradigm focuses on small populations of rare species. Rare species are particularly important in conservation biology, but they are not always identical to what a conservation biologist is concerned with in the small-population paradigm. When a naturalist says that a bird or a plant is rare, he or she may mean several different things (Harper 1981, Rabinowitz 1981). The concept of rarity can be described best according to three characteristics: geographic range, habitat specificity, and local population size. If we divide each characteristic into two levels, we can classify rare species as in Table 19.1. Of the eight classes, only one

TABLE 19.1 **A classification of rare species based on three characteristics: geographic range, habitat specificity, and local population size.**

Population size	Geographic Range			
	Large		Small	
	Habitat Specificity			
	Wide	Narrow	Wide	Narrow
Large, dominant somewhere	Locally abundant over a large range in several habitats	Locally abundant over a large range in a specific habitat	Locally abundant in several habitats but restricted geographically	Locally abundant in a specific habitat but restricted geographically
Small, nondominant	Constantly sparse over a large range and in several habitats	Constantly sparse in a specific habitat but over a large range	Constantly sparse and geographically restricted in several habitats	Constantly sparse and geographically restricted in a specific habitat

Source: Modified after Rabinowitz (1981).

describes "common" organisms, and thus "rare" may mean seven different things to an ecologist. If there are seven different classes of rare species, we must recognize different kinds of management to protect species that are threatened with extinction.

Classic rare species are often those with small geographic ranges and habitat specificity. Many plants of this type are restricted endemics, and are often endangered or threatened (Rabinowitz 1981). Other rare species have very large geographical ranges and occur widely in different habitats but are always at low density. These species are ecologically interesting but almost never appear on lists of endangered species. The important point is that not all rare species are concerns of conservation biology.

Detailed studies of small populations that are going extinct are rare, and the few known cases illustrate how a mix of chance events can doom a species to extinction. The prairie chicken (*Tympanuchus cupido*) was a common grouse from New England and Virginia, west to Kanasas, when Europeans first arrived in North America. The prairie chicken was distributed widely across the central plains of the United States, but it has been fragmented by agriculture into scattered populations in the central United States. In Illinois, prairie chickens numbered in the millions in the nineteenth century but declined to 25,000 birds by 1933 (Westemeier et al. 1998). By 1993 only 50 prairie chickens survived in Illinois, but large populations remained in Kansas, Minnesota, and Nebraska. Figure 19.2 shows the decline of one population in Jasper County, central Illinois, from 1970 to 1997. The population decline was mirrored in a decline in the hatching rate of eggs, which was thought to be due to low levels of genetic diversity. In 1992 a translocation program was begun to move prairie chickens from the large populations in Kansas and Nebraska into Illinois. Over the next five years a total of 271 prairie chickens were translocated to Illinois. Egg viability quickly improved (see Figure 19.2), and the population rebounded. Reduced genetic variability in the declining Illinois population was verified by analyzing microsatellite loci isolated from feather roots of museum specimens and recent collections (Bouzat et al. 1998). Table 19.2 shows that fewer alleles were found in the recent Illinois population compared with the number found either in other large populations or in Illinois before the population decline of the past 50 years. These genetic data confirm that the collapse of the Illinois prairie chicken population followed the extinction vortex until it was rescued from imminent extinction in 1992 by translocating new genetic stock into the Illinois population (Bouzat et al. 1998a).

One general principle is that the smaller the population, the greater the risk that chance events can lead to extinction. But how small is small, and how can we determine the minimum viable population for a particular species? The task is not easy (Soulé 1987). Genetics and demography provide

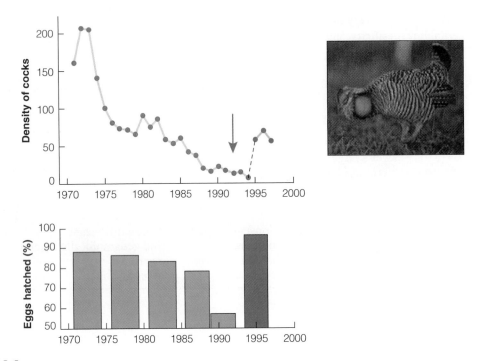

FIGURE 19.2
The decline of the prairie chicken (Tympanuchus cupido) *in central Illinois, 1970–1997. The population collapse was mirrored in a reduction in fertility. In 1992 prairie chickens from Minnesota, Kansas, and Nebraska were translocated to increase genetic variability (blue arrow). The population rebounded strongly after this introduction. (Modified after Westemeier et al. 1998)*

TABLE 19.2 Number of alleles per locus in the greater prairie chicken from a survey of the current Illinois population (1974–1993 birds) and the large populations from Kansas, Minnesota, and Nebraska.

A sample of museum specimens was used to obtain the pre-1950 Illinois data.
Six microsatellite loci were used from DNA isolated from feather roots.

	Illinois before 1950	Illinois after 1974	Kansas	Minnesota	Nebraska
Mean number of alleles	5.12	3.67	5.83	5.33	5.83
Standard error	0.87	0.56	0.75	0.84	1.05
Sample size	15	32	37	38	20

Source: From Bouzat et al. (1998b).

two very different approaches for estimating minimum viable populations. Population genetics has provided a very simple rule for minimum viable populations: the 50/500 rule. Franklin (1980) pointed out that inbreeding could be kept to a low level with a minimum population of about 50 animals, and that this rule worked well for animal breeders working in agriculture. He suggested that this level was high enough to prevent inbreeding depression, one fac-

tor in the extinction vortex (see Figure 19.1). To prevent genetic drift a larger population is needed, and Franklin (1980) suggested that a minimum of 500 animals or plants would be sufficient to allow evolution to proceed unimpeded. Note that population geneticists count individuals in units of *effective population size,* and that these units are not the same as the individuals population ecologists count (Box 19.1). Real populations would often

B O X   1 9 . 1

WHAT IS EFFECTIVE POPULATION SIZE?

Geneticists and ecologists talk about population sizes in two quite different ways. To an ecologist a population consists of immature and mature plants and animals, some of which are breeding and some of which are not. A geneticist, by contrast, is concerned about the genetic population—those individuals that contribute genes to the next generation. If an individual does not breed or breeds unsuccessfully, it does not exist in a geneticist's count, even though it uses resources and is potential food for predators.

We begin with a definition from Stewart Wright (1931), "The effective size of a population is the size of an ideal population that would undergo the same amount of genetic drift as the population under consideration." *Genetic drift* is change in the genetic composition of a population arising as a consequence of sampling of gametes in a finite population. Drift is a sampling problem, and sampling problems are always more difficult in small populations. Genetic drift produces random changes in population composition, and a gradual increase in homozygosity, and thus a loss of genetic variation. For a given size of population, the rate of genetic drift will depend on the mating structure and the variation in number of successful offspring. In the ideal world of genetics, we can imagine a world in which each individual contributes gametes equally to a pool for the next generation—this is the *effective population size* or N_e. The important point for conservation is the relationship between actual population size and effective population size (Franklin 1980).

1. *Variation in progeny number.* If the population consists of N individuals who have variance in progeny number σ^2, we obtain

$$N_e = \frac{4N}{2 + \sigma^2} \quad (19.1)$$

If the mean family size is 2 and the variance is 4, the effective population size is 2/3 observed numbers.

2. *Unequal breeding numbers in the two sexes.* If the breeding system is a lek or harem system, as in fur seals or zebras, and only a few males do most of the breeding, the effective population size is given by

$$N_e = \frac{1}{\left(\frac{1}{4N_m} + \frac{1}{4N_f}\right)} \quad (19.2)$$

where N_m = number of breeding males

N_f = number of breeding females

Thus if we have a population of 100 fur seals, with 95 breeding females and five harem bulls, you have an effective population size of 19 seals, not 100.

3. Fluctuating population size. If the population fluctuates in size from generation to generation, the effective population size is again reduced. Crow and Kimura (1970) show that for t generations:

$$\frac{1}{N_e} = \frac{1}{t}\left(\frac{1}{N_1} + \frac{1}{N_2} + \frac{1}{N_3} + \frac{1}{N_4} + \cdots + \frac{1}{N_t}\right) \quad (19.3)$$

The effective population size is strongly affected by the lowest observed population. For example, if we have a lion population that is stable for nine generations at 100 individuals but because of a severe drought drops to ten individuals for one generation, the effective population size for this time period is 53 lions, not 100.

These are all simple illustrations of the general notion that effective population sizes are almost always much less than observed population sizes, often by margins of three or four times or even more.

have to be three to ten times larger than their effective population size counterpart. The 50/500 rule was proposed as a rule of thumb and it should not be applied as a law for all species (Boyce 1992). Because it is based purely on genetic concepts, it cannot be applied to animals and plants that are subject to varying levels of environmental variation and different breeding systems. Soulè and Mills (1992) have pointed out that simple rules for minimum viable populations

need critical evaluation before we can use them in practical decisions about endangered species.

The small-population paradigm in conservation biology has a strong theoretical base in population genetics, and is useful to conservation biologists for exploring the problems that small populations face if they are to survive in the short term and in the long term. It does not often *solve* the problem of small populations, which is the focus of the next paradigm.

E S S A Y   1 9 . 1

DIAGNOSING A DECLINING POPULATION

Much of the practical conservation biology depends on the careful diagnosis of declining populations. Caughley (1994) laid out a series of logical steps to determine what is driving a species toward extinction.

1. *Confirm that the species is presently in decline, or that it was formerly more widely distributed or more abundant.* This will require some qualitative or quantitative assessment of population trends and distribution. A species in decline may be common or rare. Both types can become conservation problems if the decline continues.

2. *Study the species's natural history and collect all information on its ecology and status.* For many species a considerable amount of background knowledge, both formal and informal, exists. Information on related species may be useful here.

3. *List all the possible causes of the decline, if enough background information is available.* This is the method of multiple working hypotheses; cast a wide net to consider all possible causes. Remember that direct human actions may be an agent of decline, but do not restrict hypotheses to human causes.

4. *List the predictions of each hypothesis for the decline, and try to specify contrasting predictions from the different hypotheses.* Do not assume that the answer is already known by scientific or folk wisdom.

5. *Test the most likely hypothesis by experiment to confirm that this factor is indeed the cause of the decline.* Often factors are correlated with the decline but are not causing it. The best experiment involves removing the suspected agent of decline.

6. *Apply these findings to the management of the threatened species.* This will involve monitoring subsequent recovery until the problem of decline is resolved.

Applying this approach to an endangered species already in low abundance will be difficult, but there is no alternative. Several suspected agents of decline may have to be removed at once, and additional studies undertaken to identify exactly which one was most responsible. It is better to save the species than to achieve scientific purity.

The Declining-Population Paradigm

The declining-population paradigm focuses on the ways of detecting, diagnosing, and halting a population decline. The problem is viewed in demographic terms—as a population in trouble—and for this reason this paradigm is action oriented. Some external agent must be identified as the cause of the decline, and research efforts focus on what can be done about it. Because it is action oriented, there is almost no theory in this paradigm. Research efforts are concentrated on each specific case study, and at least in the short term no great theoretical advances in understanding the causes of extinction will occur. But if done properly, the research gets the job done. This paradigm does not consider the current size of the population to be as important—it is the downward trend that is the main concern.

Paul Ehrlich of Stanford University has been instrumental in putting declining populations and extinction on the worldwide agenda through his books and popular writings on the biodiversity crisis. He has highlighted the many ways in which humans are contributing to population losses. Ehrlich's main point is that many extinctions are not due to "chance" in the broad sense. Many population declines are completely determined by some inexorable change from which there is no escape without action. Shaffer (1981) called these *deterministic extinctions.* Deforestation is one such change; glaciation is another. If an area is deforested, all species that require trees are eliminated. Deterministic extinctions occur when some essential resource is removed or when something lethal is introduced into the environment. Loss of habitat leads to deterministic extinctions and is a major problem in almost every ecosystem on Earth (Ehrlich and Ehrlich 1981, Sinclair 1998).

Paul Ehrlich *(1932–) Conservation biologist, Stanford University*

Why do deterministic extinctions occur? The four causes of extinction, called the "evil quartet" by Diamond (1989), are:

Overkill
Habitat destruction and fragmentation
Introduced species
Chains of extinction

Next we consider each of these in turn.

Overkill

Overkill consists of fishing or hunting at a rate that exceeds a population's capacity to rebound. The species that are most susceptible to overkill are the large species with low intrinsic rates of natural increase (r) — elephants, whales, rhinoceros, and others that are considered valuable by humans. Species on islands are also vulnerable to extinction if the island is small. The great auk, a large flightless seabird, was hunted to extinction on islands in the Atlantic Ocean in the 1840s because of a demand for feathers, eggs, and meat (Montevecchi and Kirk 1996).

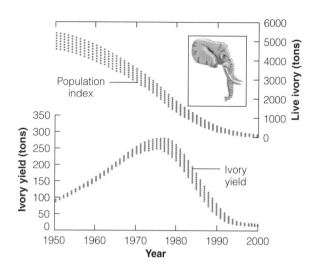

FIGURE 19.3
Estimates of the decline of the African elephant since 1950, deduced from the partially illegal ivory trade. The continuing decline in elephant numbers across Africa is not always reflected in changes in individual national parks, which may protect local populations. (From Caughley et al. 1990.)

Great Auk (Pinguinis impennis) *Extinct by 1844. (©The Academy of Natural Sciences of Philadelphia/Corbis)*

The decline of the African elephant is a classic example of the effect of hunting on a large mammal. The African elephant is the largest living terrestrial mammal, weighing up to 7500 kg. Sexual maturity is reached only after 10–11 years, and a single calf is born every three to nine years. The potential rate of increase was estimated by Sinclair (1997) to be about 6% per year, a low population growth rate. Figure 19.3 shows the decline of the African elephant population during the past 50 years. Only in South

Africa have elephant populations been stable or increasing.

Illegal hunting for ivory has been the major cause of the recent collapse of the elephant population (Milner-Gulland and Beddington 1993). The demand for elephant ivory has been highly variable, and when the price increased during the 1970s the amount of poaching for ivory grew dramatically. There has been much controversy over the need to legalize the ivory trade and control it in African countries such as Botswana that have protected their elephant population. The current ban on the ivory trade is having little effect in central and east Africa, where poaching is rampant.

Overkill, or excessive human exploitation, will remain a problem for all animals and plants that are valuable or large.

Habitat Destruction and Fragmentation

The second factor in the "evil quartet" that promotes extinctions is habitat loss. Habitats may simply be destroyed to make way for housing developments or agricultural fields. Cases of habitat destruction appear to provide the simplest examples of the declining-population paradigm. An example can illustrate how subtle the effects of habitat destruction can be.

The red-cockaded woodpecker is an endangered species endemic to the southeastern United States. It was once abundant from New Jersey to Texas and inland to Missouri. It is now nearly extinct in the northern and inland parts of its geographic range. The red-cockaded woodpecker is adapted to pine savannas, but most of this woodland has been destroyed for agriculture and timber production. These birds feed on insects under pine bark and nest in cavities in old pine trees. Because most old pines have been cut down, the availability of nesting holes has become limiting (Walters 1991).

Designing a recovery program for the red-cockaded woodpecker has been complicated by the social organization of this species. They live in groups of a breeding pair and up to four helpers, nearly all males. Helpers do not breed but assist in incubation and feeding. Young birds have a choice of dispersing or staying to help in a breeding group. If they stay, they become breeders by inheriting breeding status upon the death of older birds. Helpers may wait many years before they acquire breeding status. Figure 19.4 shows a schematic of the life history events of the red-cockaded woodpecker, along with the probabilities of moving between states.

From a conservation viewpoint, the problem is that red-cockaded woodpeckers compete for breeding vacancies in existing groups, instead of forming new groups that might occupy abandoned territories or start at a new site by excavating nesting cavities. The key problem is the excavation of new breeding cavities. Because of the time (typically several years) and energy needed to excavate new cavities, birds are better off competing for existing territories than establishing new ones. Habitat loss appeared to be the main factor causing population decline.

To test this idea, Walters (1991) and his colleagues artificially constructed cavities in pine trees at 20 sites in North Carolina. The results were dramatic—18 of 20 sites were colonized by red-cockaded woodpeckers, and new breeding groups were formed only on areas containing artificial cavities. This experiment showed clearly that much suitable habitat is not occupied by this woodpecker because of a shortage of cavities. Management of this endangered species should not be directed toward reducing mortality of these birds but instead should focus on the providing tree cavities suitable for nesting.

An additional complication of cavity-nesting species is competition for cavities. The endangered red-cockaded woodpecker population at the Savannah River Site in South Carolina was rescued from near extinction by a combination of adding artificial nest cavities and translocating birds from larger nearby populations (Franzreb 1997). To prevent competition for the artificial cavities, 2304 southern flying squirrels (*Glaucomys volans*) were removed between 1986 and 1995. The woodpecker population responded dramatically, increasing from four to 99 individuals in response to these management actions.

The rescue of the red-cockaded woodpecker is a good example of how successful conservation biology must depend on a detailed understanding of population dynamics and social organization, so that limiting factors can be identified and made more abundant. There are no general prescriptions for rescuing endangered species, and we must operate on a case-by-case approach. Detailed information on resource requirements, social organization, and dispersal powers are required before recovery plans can be specified for species suffering from habitat loss and fragmentation.

Humans have appropriated a large fraction of the land surface of the Earth for agriculture, and many plants and animals cannot survive in an agricultural landscape. Of the remaining areas, many have been fragmented, or broken up into small patches. Figure

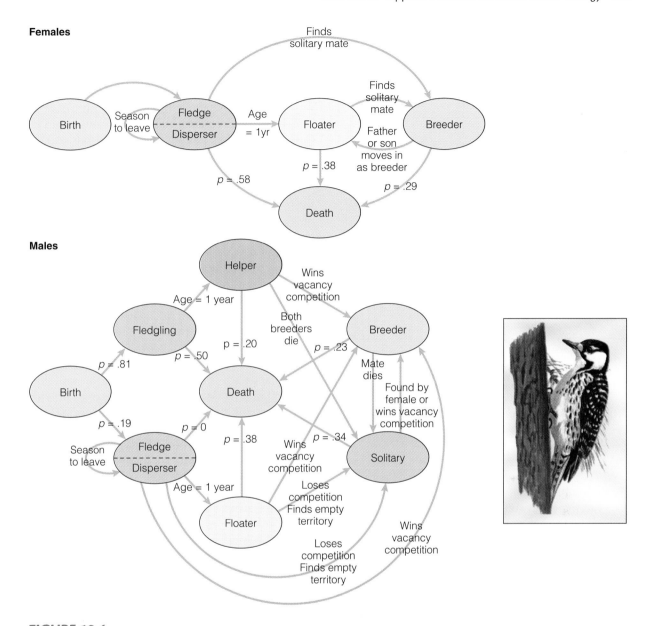

FIGURE 19.4
Annual transition probabilities of the red-cockaded woodpecker in the sandhills of North Carolina: females above, males below. Whereas females disperse to new territories, males must decide whether to remain on the same territory or disperse to another. Most males tend to remain as helpers on their natal territory. (From Letcher et al. 1998.)

19.5 illustrates how forest areas in southern Wisconsin have been fragmented since the 1830s. Forest fragmentation, which is occurring at a rapid rate in tropical forests, has been documented by aerial photos and more recently by satellite imagery. Figure 19.6 shows tropical rain forest losses in the Sierra de Los Tuxtlas, Veracruz, Mexico, between 1967 and 1986. Dirzo and Garcia (1992) showed that deforestation

has proceeded up from the lowlands, and by 1986 about 84% of the original forest had been lost. They predicted that by the year 2000 only 8% of the original forest would remain in the form of an archipelago of small forest islands. Deforestation in this area is caused mostly by clearing for cattle ranches. The human population of this region has more than doubled in the past 25 years.

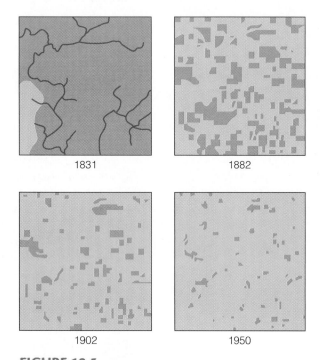

FIGURE 19.5
Reduction and fragmentation of the woodland in Cadiz Township, Wisconsin, 1831–1950. More than 95% of the original forest was lost, and the remaining 5% was cut up into small, isolated blocks. (After Curtis 1959.)

Habitat fragmentation has many components with varying effects on population dynamics (Table 19.3). The impact of fragmentation is species specific. A habitat is called *fine-grained* for a species if the patches are short distances apart and the species can move back and forth between patches with little cost. Conversely, a *coarse-grained* habitat for a species requires long-distance dispersal, and individuals in coarse-grained habitat typically live most or all their life in one patch. Species such as eagles that move over large areas may treat a fragmented habitat as continuous, whereas the exact same habitats may appear very coarse-grained to a plant with limited dispersal powers (Rolstad 1991). Scale is critical in fragmentation, and ecological scales are highly species specific.

Fragmentation of habitats can be analyzed by considering the dynamics of populations subdivided into small patches. At one extreme, when patches are too small the species cannot survive. We can see this

very clearly by looking at incidence functions, the occupancy rate of a species in habitats of differing size. The simplest incidence functions come from island studies. Figure 19.7 illustrates this concept with data on shrews in Finland. Shrews are tiny mammals with very high metabolic rates, and because of their high food requirements, small islands might not be suitable. Extinction rates of shrews on islands are correlated with body size, and smaller species such as *Sorex caecutiens* go extinct more easily in small habitat patches. Incidence functions are species specific, and the occupation of patches will depend on the ability of the species to move between patches and survive in those patches (Peltonen and Hanski 1991).

Small patches are also subject to chance extinction due to weather or disease more often than are large patches. In western Europe the European red squirrel (*Sciurus vulgaris*) occupies patches of forest interspersed in a mosaic of agricultural land (Verboom and van Apeldoorn 1990). Home ranges of this squirrel range from 1.5 ha to 13.4 ha. Figure 19.8 shows that red squirrels in the eastern part of the Netherlands are almost always present in woodlots larger than 3 ha. Woodlots with pine trees provide more food for squirrels, and this improvement in habitat quality is also important in maintaining populations. These small woodlots may be connected by fence rows or trees along roads, and what is crucial for all fragmented populations is how readily individuals can move between patches. The study of fragmented patches thus becomes a study of metapopulations (see page 289); when subpopulations in patches become extinct, the patches can be recolonized by dispersing individuals.

Recolonization may not always occur in isolated patches. The Bogor Botanical Garden was established in 1817 on 86 ha in west Java. Until 1936 the Botanical Garden was connected with other forest areas to the east, but for the past 60 years it has been an isolated patch of forest, with the nearest patch 5 km away (Diamond et al. 1987). Of the 62 bird species recorded as breeding in the Botanical Garden during 1932–1952, 20 species had disappeared by 1980–1985 and four more were close to extinction. The species that were lost were the less common species, and their low abundance combined with the lack of recolonization from surrounding areas has been the main cause of extinction (Diamond et al. 1987). The result is that much of the conservation

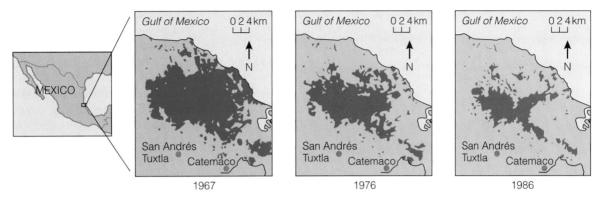

FIGURE 19.6
Deforestation in Veracruz, southeast Mexico, 1967–1986. Tropical rain forest (blue) has been removed at a rate of 4.2% per year. (From Dirzo and Garcia 1992.)

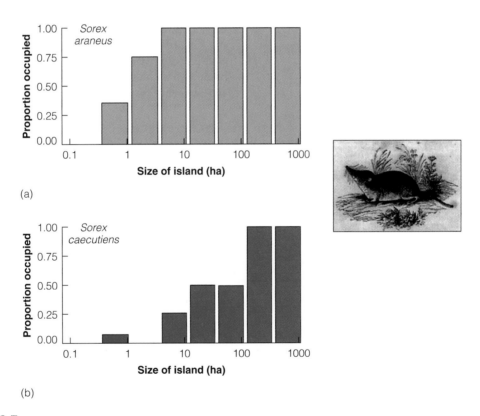

FIGURE 19.7
Incidence functions for two species of shrews on islands in Finnish lakes. A survey of 108 islands in three lakes provided an estimate of the proportion of islands of different sizes that were occupied by each species. (a) Sorex araneus *occurs on most islands above 3 ha in size. (b)* Sorex caecutiens *occurs on fewer small islands and is always present only on islands above 100 ha in area. (From Peltonen and Hanski 1991.)*

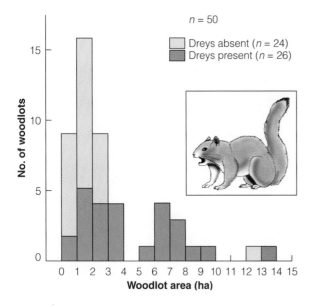

FIGURE 19.8
Distribution of woodlot sizes for 50 woodlots from the Twente area of the Netherlands. Red squirrel (Sciurus vulgaris) *presence was determined from dreys (nests) On few woodlots smaller than 3 ha are red squirrels present. (From Verboom and van Apeldoorn 1990.)*

value of the Botanical Garden for birds has been lost because it is too small by itself to support a secure population of many tropical forest birds.

In almost all cases habitat fragmentation leads to species loss. The prairies of North America are a good example. Prairie covered about 800,000 ha of southern Wisconsin when Europeans first arrived, and now prairie occupies less than 0.1% of its original area (Leach and Givnish 1996). Plant surveys of 54 Wisconsin prairie remnants studied between 1948 and 1954 were repeated in 1987–88. Between 8% and 60% of the plant species were lost during these four decades, at average rates between 0.5% and 1.0% per year. At this rate of extinction approximately half the plant species would disappear in 50 to 100 years. Losses were particularly high among the shorter plant species and the rare species. The control of fire in prairies seems to be the agent of decline for prairie plants, and controlled burns should be done to reverse these population declines (Leach and Givnish 1996).

One of the important consequences of fragmentation is that it increases the amount of edge in a habitat (see Table 19.3). If predators search habitat edges, higher predation rates might occur in smaller

TABLE 19.3   Changes associated with habitat fragmentation and their possible effects on population dynamics.

	Habitat change	*Consequences for population dynamics*
Population-level effects	Reduced connectivity, insularization, increased interfragment distance	Directly affects dispersal and reduces the immigration rate
	Reduced fragment size, reduced total area	Directly affects population size and increases the extinction rate
Landscape or community-level effects	Reduced interior-edge ratio	Indirectly affects mortality and production through increased pressure from predators, competitors, parasites, and disease
	Reduced habitat heterogeneity within fragments	
	Increased habitat heterogeneity in surrounding matrix	Indirectly affects population size through reduced carrying capacity within the fragment
	Loss of keystone species from the habitat	Indirectly affects mortality and production through increased carrying capacity of predators, competitors, etc. in the surrounding matrix
		Indirect effect through disruption of mutualistic guilds or food webs

Source: From Rolstad (1991).

E S S A Y   1 9 . 2

FRAGMENTATION OF HABITATS AND AREA-SENSITIVE SPECIES

When landscapes are broken up by agriculture or other human activities, an array of habitat fragments are typically left. Species vary dramatically in their sensitivity to this fragmentation, and the central question for each species of conservation concern becomes *How small a fragment will it occupy successfully?* The best data available are from birds. In southeastern Australia the eastern yellow robin is highly sensitive to the size of woodland remnant left after clearing for agriculture. In central New South Wales, Briggs, Seddon, and Doyle (1999) obtained the following data for 36 woodlots: (see below).

Eastern yellow robins do not occur in remnant woodlands smaller than 15 ha in area. Consequently, protecting many small woodlots will not help preserve this declining species. The cause of this area-dependent occupancy is not known, but excessive predation or food shortage are commonly suspected causes.

In the corn belt of central Illinois, only a few grasslands remain, and only nine grassland remnants larger than 40 ha still exist (Walk and Warner 1999). Greater prairie chickens, savannah sparrows, and upland sandpipers all nest only in grasslands larger than 40 ha. These birds do not recognize smaller grassland remnants as suitable nesting habitat, and this habitat selection has implications for the design of reserves to conserve grassland birds in agricultural areas.

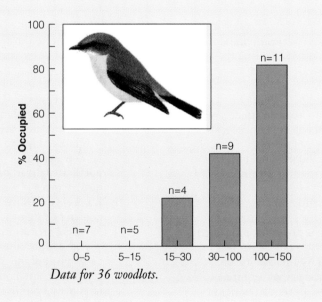

Data for 36 woodlots.

fragments because of the edge effect. Andrén and Angelstam (1988) tested this idea in central Sweden in a mosaic of conifer forest and farmland. Fifty artificial ground nests containing two brown chicken eggs each were placed in farmland and in forest patches each year for three years. Predation rates were much higher on nests in farmland and at the edge of forest patches (Figure 19.9). This predation

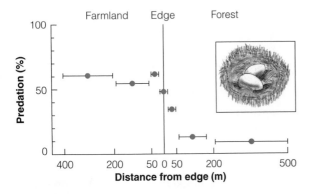

FIGURE 19.9
Predation rate on artifical bird nests containing two brown chicken eggs in relation to distance from the farmland forest edge in central Sweden, 1984–1986. Each point is based on 13 to 30 nests. Generalist predators such as red foxes do not seem to go more than 50 m into these coniferous forest patches (From Andrén and Angelstam 1988.)

effect extended about 50m into the forest, so that nests in small forest patches were affected much more than nests in large patches.

One of the most critical variables in the dynamics of populations in fragmented habitats is migration between patches. At present we have few data on movements of animals and plants between patches. Much discussion in conservation agencies has focused on providing corridors between refuges so that species can disperse from one patch to the next. Dan Simberloff of the University of Tennessee has been one of the key conservation ecologists in identifying the advantages and disadvantages of corridors. Corridors, if used, help to prevent inbreeding depression and allow recolonization (Simberloff and Cox 1987). But there are potential costs to corridors, because they may fascilitate disease transmission,, conduct fires, and expose individuals to increased predation risk (Table 19.4). The Florida panther (*Felis concolor*) has been

TABLE 19.4 Potential advantages and disadvantages of conservation corridors.

Potential advantages	*Potential disadvantages*

1. Increase immigration rate to a reserve, which could:
  a. Increase or maintain species richness and diversity (as predicted by island biogeography theory).
  b. Increase population sizes of particular species and decrease probability of extinction (provide a "rescue effect") or permit reestablishment of extinct local populations.
  c. Prevent inbreeding depression and maintain genetic variation within populations.
2. Provide increased foraging area for wide-ranging species.
3. Provide predator-escape cover for movements between patches.
4. Provide a mix of habitats and successional stages accessible to species that require a variety of habitats for different activities or stages of their life cycles.
5. Provide alternative refuges from large disturbances (a "fire escape").
6. Provide "greenbelts" to limit urban sprawl, abate pollution, provide recreational opportunities, and enhance scenery and land values.
	1. Increase immigration rate to a reserve, which could:
	  a. Facilitate the spread of epidemic diseases, insect pests, exotic species, weeds, and other undesirable species into reserves and across the landscape.
	  b. Decrease the level of genetic variation among population or subpopulations, or disrupt local adaptations and coadapted gene complexes ("outbreeding depression").
	2. Facilitate spread of fire and other abiotic disturbances ("contagious catastrophes").
	3. Increase exposure of wildlife to hunters, poachers, and other predators.
	4. Riparian strips, often recommended as corridor sites, might not enhance dispersal or survival of upland species.
	5. High cost, and conflicts with conventional land preservation strategy for preserving endangered species habitat (when inherent quality of corridor habitat is low).

Source: From Noss (1987).

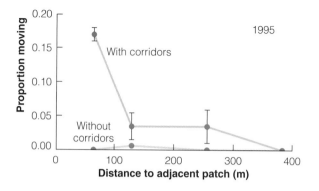

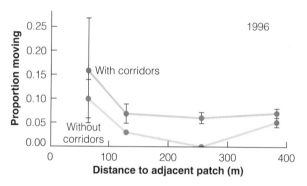

FIGURE 19.10

Proportion of marked butterflies (the variegated fritillary, Euptoieta claudia) *that moved various distances between adjacent habitat patches that were either connected by corridors or separated by unsuitable forest habitat. In both years of the study, corridors facilitated butterfly movements. (From Haddad 1999.)*

Daniel Simberloff *(1942–) Professor of Environmental Studies, University of Tennessee*

reduced from approximately 1400 individuals to about 30 animals isolated in undeveloped areas of south Florida. By providing a corridor system between wildlife refuges, managers hope to increase the effective population size of panthers (Simberloff and Cox 1987, Simberloff et al. 1997). But there are no data to determine how wide a corridor must be before large mammals like the panther will use them. Moreover, it may be difficult to stop poaching in corridors, which may be expensive to purchase and maintain.

Detailed studies of the movements of individuals between patches and along corridors are rare. A critical hypothesis of conservation biology is that corridors increase animal and plant movement between fragments of habitat. Haddad (1999) tested this idea with two butterfly species in pine plantations in South Carolina. Butterflies live in the open habitats between closed pine forests, and by harvesting the forest in patches Haddad could construct butterfly habitat in two kinds of 1.64-ha blocks: isolated square blocks and square blocks connected by a corridor of habitat. The distance between connected and unconnected patches varied from 64 m to 384 m. Figure 19.10 shows the results from two summers. Corridors facilitated movements of butterflies in both years, validating the presumed value of corridors for conservation. Corridors also increased butterfly density. The variegated fritillary (*Euptoieta claudia*) was more than twice

FIGURE 19.11
Frequency of two forest herbs found in hedgerows in relation to distance to the adjacent forest patch in central New York State. The decline in herb frequency with increasing distance implies that hedgereows could serve as movement corridors for herbs in fragmented forest landscapes. Left photo: Smilacina racemosa; *Right photo:* Arisaema triphyllum. *(From Corbit et al. 1999.)*

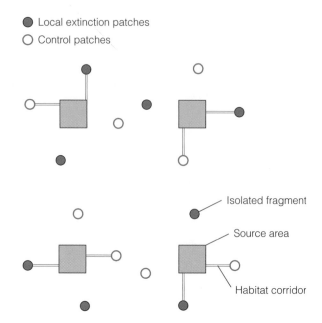

FIGURE 19.12
Experimental design for testing effectiveness of corridors for conservation. The squares represent source areas for recolonization. In local extinction patches the species being studied is experimentally removed, and the rate of colonization is then measured. (From Nicholls and Margules 1991.)

as abundant in connected patches (0.45 butterflies/ha) than it was in isolated patches (0.22/ha) (Haddad and Baum 1999).

The utility of corridors for plant dispersal has been questioned for plants that disperse by seeds. Corbit et al. (1999) investigated the utility of hedgerows as corridors for the movement of herbs between forest patches in a fragmented landscape. The frequency of many species of forest herbs declines with distance from the forest in hedgerows studied in central New York State (Figure 19.11), implying that hedgerows serve as a corridor of colonization for many forest herbs. Hedgerows may also serve as a reservoir for some species of forest herbs, which can then recolonize a new forest that is regenerating on old agricultural land.

Corridors are not necessarily useful for the conservation of all species in all situations, and thus conservation recommendations will not be the same for all species affected by fragmentation (Beier and Noss 1998). Experimental manipulations of local populations could be used to test the general hypothesis that

patches of remnant habitat connected to source areas by habitat corridors will be recolonized more readily than patches without corridors. More well-designed studies are needed to measure the effects of corridors on plants and animals (Nicholls and Margules 1991). Figure 19.12 illustrates the type of experimental design that is needed to measure corridor effectiveness for conservation. The species of interest can be experimentally driven extinct on patches that are either connected by corridors or lack corridors to see what effect corridors have for this particular species. Beier and Noss (1998) lament that many corridor studies are impossible to interpret because they have not incorporated a proper experimental design. Corridors can be an effective adjunct to conservation planning in fragmented landscapes, and it is prudent to retain landscape connectivity where possible.

When landscapes are fragmented, species on the smaller patches may begin to go extinct. If extinction occurs at random among all the species, the extinction pattern of the remaining species will be random. But

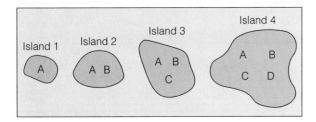

FIGURE 19.13
Hypothetical island faunas forming a series of nested subsets. Islands represent any insular habitats, whether true islands or habitat fragments. A through D represent species occurring on the islands. Because all species in smaller faunas also occur in all larger faunas, the smaller faunas are subsets of the larger faunas. (From Cutler 1991.)

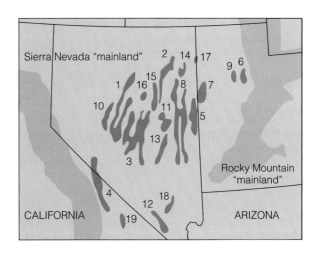

FIGURE 19.14
Distribution of montane forests above 2300 m (7500 ft.) in the Great Basin region studied by Brown (1971, 1978). Stippled areas indicate relic forests on isolated ranges: 1, Toiyabe–Shoshone; 2, Ruby; 3, Toquima–Monitor; 4, White–Inyo; 5, Snake; 6, Oquirrh; 7, Deep Creek; 8, Schell Creek–Egan; 9, Stanbury; 10, Desatoya; 11, White Pine; 12, Spring; 13, Grant–Quinn Canyon; 14, Spruce–South Pequop; 15, Diamond; 16, Robert Creek; 17, Pilot; 18, Sheep; 19, Panamint. Areas of lighter blue indicate "mainland" forests in the Sierra Nevada and Rocky Mountains. (Modified from Brown 1978.)

for many faunas extinction does not occur at random and the resulting patterns form *nested subsets* (Patterson 1987). Figure 19.13 illustrates the concept of a nested subset. A nested subset can be considered as a time series—species D, which occurs only on island 4, is the first to go extinct, followed by species C, which occurs on islands 3 and 4, and so on, in sequence (Cutler 1991). If this is correct, extinction is more predictable than random, and conservation biologists can focus on those species that need special protection.

The coniferous mountain forests in the Great Basin of the western United States (Figure 19.14) contain a good example of nested subsets. Brown (1978) tabulated the occurrence of 14 mammal species on these isolated areas, and Skaggs and Boecklen (1996) added some additional records (Table 19.5). These forest populations are relicts from the Ice Age when coniferous forests were more widespread; the forests now constitute patches or islands in a sea of relatively unsuitable desert habitat. If the mammal patterns were completely nested, there would be no holes or outliers in the table. For example, the chipmunk *Eutamias dorsalis* is "missing" from mountain range 2 (a hole), and the rabbit *Sylvilagus nutalli* is present in mountain range 19 (an outlier). The pattern shown in Table 19.5 is clearly nonrandom and is a good example of a nested subset.

Nested subsets may result from selective extinction or selective colonization. In the Great Basin mammals, selective extinction is usually given as the explanation of the nested structure, but data on potential colonization movements across the intervening barriers are lacking. Wright et al. (1998) found that about half of the 279 data sets like that shown in Table 19.5 were significantly nested, and they suggested that extinction is more often the process that leads to nested subsets. It is important to determine if nested subsets occur in fragmented habitats. Not all species are equally vulnerable to extinction, and it is important to direct conservation efforts toward the most vulnerable species (Cutler 1991).

Impacts of Introduced Species

Introduced animals are responsible for about 40% of historic extinctions. Most of these data involve mammals and birds, for which we have more detailed information, and these are no doubt biased (Caughley and Gunn 1996). But no one doubts the adverse effects of

TABLE 19.5 Species distribution matrix for boreal mammals of Great Basin mountain ranges.

Mountain range numbers refer to those shown in Figure 19.14.

Species	|1	2	3	4	5	6	7	8	9	10	11	12	13	14	15	16	17	18	19	No. of occurrences
Eutamias umbrinus	x	x	x	x	x	x	x	x	x	x	x	x	x	x	x	x		x		17
Neotoma cinerea	x	x	x	x	x	x	x	x	x	x	x	x		x	x	x	x	x	x	18
Eutamias dorsalis	x		x	x	x	x	x	x	x	x	x	x	x	x	x		x	x	x	17
Spermophilus lateralis	x	x	x	x	x		x	x		x	x	x	x	x	x	x	x			15
Microtus longicaudus	x	x	x	x	x	x	x	x	x	x	x		x		x					13
Sylvilagus nutalli	x	x	x	x	x		x	x		x	x	x		x	x	x			x	14
Marmota flaviventris	x	x	x	x	x	x	x	x	x	x	x			x	x					13
Sorex vagrans	x	x	x	x	x	x	x	x	x				x							10
Sorex palustris	x	x	x	x	x	x			x						x					8
Mustela erminea	x		x	x	x	x			x											6
Ochotona princeps	x	x	x	x						x										5
Zapus princeps	x	x			x				x						x					5
Spermophilus beldingi	x	x	x																	3
Lepus townsendi		x				x			x	x	x	x								6
No. of Species	13	12	11	11	10	10	9	8	10	9	8	7	5	4	6	8	3	3	3	

Source: Data from Skaggs and Boecklen (1996).

introduced species. The Nile perch, which was introduced into Lake Victoria in the early 1980s, caused the extinction of over 200 endemic species of cichlid fish between 1984 and 1997 (Seehausen et al. 1997).

Nearly 50% of the mammal extinctions of the past 200 years occurred in Australia. Figure 19.15 shows the weight distribution of the threatened and extinct mammals of Western Australia. Neither very small nor very large mammals have been affected in these recent losses. A critical weight range from 35 to 4200 g contains all the missing mammals (Burbidge and McKenzie 1989). Many causes can be suggested to explain these extinctions, from habitat clearing associated with agriculture, to changes in the fire regime, to introduced herbivores as competitors, to introduced predators. The main culprit seems to be introduced predators, particularly the red fox (Short and Smith 1994, Kinnear et al. 1998). The details of the loss of medium–sized marsupials in Australia is a mirror image of the spread of the red fox (Short 1998). If the red fox can be controlled, some of the threatened species, now confined to offshore fox–free islands, could be reintroduced to their former range.

*Eastern Hare Wallaby (*Lagorchestes leporides*) extinct in 1890*

There are many examples of introduced predators causing conservation problems. Box 19.2 gives one illustration for introduced cats.

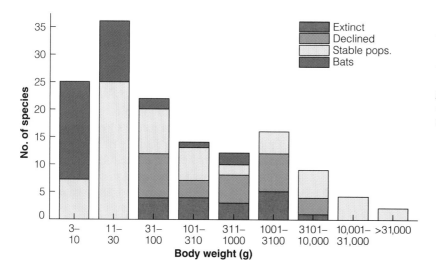

FIGURE 19.15
Frequency distribution of the weights of Western Australian mammals. Body weight is on a log scale. Recent extinctions (red) and declining populations (green) have been concentrated in the medium-sized mammals between 35 and 4200 g. (From Burbidge and MacKenzie.)

Introduced species are one of the most serious conservation problems today. As global trade has increased, many inadvertent or deliberate introductions are occurring, often with little regard for their conservation consequences (Ruesink et al. 1995).

Chains of Extinctions

The last of the "evil quartet" causing extinctions is a set of secondary extinctions that follow from a primary extinction. If other species depend on a lost species for survival, these other species must also go extinct. Chains of extinctions require obligate specialist relationships that are more typical of tropical areas than of temperate or polar zones. One obvious chain of extinctions involves the loss of parasite species when their host goes extinct. This matter has received scant attention to date.

The clearest examples of chains of extinctions involve large predators that disappeared when their prey went extinct. The extinct forest eagle of New Zealand (*Harpagornis moorei*), which weighed 10–13kg, and preyed on large ground birds, died out around A.D. 1400 when moas became extinct in New Zealand (Holdaway 1989). The decline of the black-footed ferret in North America was associated with the decline of its main food, prairie dogs, on the Great Plains (Caughley and Gunn 1996, p. 91). Currently the black-footed ferret is being reintroduced into areas where prairie dog colonies are safe (Biggins et al. 1998), but its future is not secure because it is highly susceptible to canine distemper, which is endemic in carnivores on the Great Plains.

Reserve Design and Reserve Selection

One way to conserve species in danger of extinction is to set up reserves or protected places. National parks in many countries have been viewed as protected areas for populations and communities. The selection and design of nature reserves is an important part of conservation biology, and much effort has gone into developing good methods of reserve selection and design. To begin we need to specify exactly what a reserve is intended to accomplish. Two quite divergent aims are often stated for reserves:

1. To conserve specific animal and plant communities that are subject to change because of fire, grazing, or predation. These reserves must be managed by intervention to set the permissible levels of fire, grazing and predation.

2. To allow the system to exist in its natural state and to change as governed by undisturbed ecological processes, so that no attempt will be made to influence the resulting changes in populations and communities.

Often reserves such as national parks have both these aims, creating a recipe for conflict over what kinds of

B O X   1 9 . 2

RECOVERY OF PETRELS AFTER ERADICATION OF FERAL CATS ON MARION ISLAND, INDIAN OCEAN

This box provides an example of the application of the declining population paradigm described by Caughley and Gunn (1996) to solve a particular conservation problem.

Problem

Marion Island (290 km^2) in the southern part of the Indian Ocean had breeding populations of 12 petrel species, which breed in burrows. Five house cats were introduced in 1948 to control introduced house mice on the island, and the cat population increased at 23% per year to reach 3045 cats in 1977. The cats preyed on the adults, chicks, and eggs of eight species of burrowing petrels. As the petrel populations shrank, the cats shifted their attention to house mice. The great-winged petrel, *Pterodroma macroptera*, was especially vulnerable to the cats because it breeds in winter, has a long breeding season, and used larger burrows than other petrels.

Diagnosis of the cause of the decline

In 1975 cats killed around 48,000 great-winged petrels, which became relatively rare compared to numbers on the neighboring (and cat-free) Prince Edward Island.

Prince Edward Island:	No cats	33% nests (*n* = 30) with chicks in 1979
Marion Island:	Cats present	1% nests (*n* = 109) with chicks in 1979

An introduced disease reduced cats, but then the survivors again increased in numbers. Outside a cat-proof exclosure, no petrel nests contained chicks, compared to 50% of nests inside the predator exclosure. On this evidence the factor driving the petrel decline was postulated to be cat predation.

Recovery treatment

1977	Feline panleukopenia (FLP) introduced
1982	Cats reduced to 620; FLP antibodies subsequently decreased in the cats
1986–1990	952 cats removed by shooting and trapping
1990	Petrel survival increased from 100% chick mortality in 1979-1984 to 0% chick mortality in 1990
1991	Cat eradication believed complete
1992	Reports of increases in house mice abundance

changes are acceptable and what kinds are unacceptable to the managers or to the general public (Caughley and Sinclair 1994).

Approximately 7% of the world's land area is now set aside as some form of a reserve, and the goal of many governments is to protect about 12% of terrestrial habitats. If we are given the job of selecting and locating reserves, how should we proceed? One way is to identify "hotspots" that are particularly rich in species, and to locate reserves in these areas (Reid 1998). One problem with this approach is that areas that are hotspots for birds are typically not hotspots for butterflies, so we cannot choose reserves on the basis of only one taxonomic group and expect that it will protect other groups as well. Nevertheless, some small areas are much richer in species than others, and we should use this kind of information to help select reserves. Caughley and Gunn (1996, p. 321) have given the following overview of how to proceed in reserve selection:

Step 1: Decide on the objective of the reserve system clearly and unambiguously.

Step 2: Identify which areas of land are available for designation as reserves within the terms of the objectives decided in step 1.

Step 3: Survey each patch that might become a reserve and obtain a list of species present, and if possible, an estimate of abundance of each.

Step 4: Formulate a starting rule for selecting the first reserve, and how subsequent patches will be chosen in sequence.

Box 19.3 illustrates one method of formulating objective rules for reserve selection (Nicholls and Margules 1993). It is important to realize that many different ways of selecting reserves are possible, depending on the objectives. The preference criterion may be to preserve rare species, or to preserve sites with many different species, or to preserve the largest number of taxonomic units such as genera or families. Most reserve selection algorithms use presence/absence as the relevant criterion rather than species abundance because it is easier to determine presence/absence than it is to estimate abundances for many species.

To create a reserve system that is useful for conservation, it is necessary to know the ecological requirements of the species of concern. A special problem exists for species that use temporary habitats.

BOX 19.3

AN ALGORITHM FOR CHOOSING RESERVES FOR A TAXONOMIC GROUP.

Many methods exist for selecting reserves for conservation. Once the objective of the reserves is decided, we must specify objective rules for evaluating which areas are best to select. Nicholls and Margules (1992) have suggested the following method for selecting reserves for conservation.

Step 1. State the objective as clearly and specifically as possible; for example, "To create a reserve system that captures 10 % of the range occupied within a region by each species in the genus *Eucalyptus*."

Step 2. This is an optional step that allows the inclusion of some sites before the selection process begins. Examples might be reserves already set aside, national parks, or protected sites with known rare and endangered species.

Step 3. Select all sites that have a species that occurs in no other site.

Step 4. Find the next rarest species and select the sites that, when added to those already selected, will represent that species plus the greatest number of additional species at or above the required proportion (10%) of their area of distribution.

Step 5. If there is a choice, select the site that is closest in proximity to a site already selected.

Step 6. If there is still a choice, select the site that also contributes the largest number of as yet inadequately represented species.

Step 7. If there is still a choice, select the site that achieves the required level of representation of the rarest species remaining underrepresented.

Step 8. If there is still a choice, select the site that contributes the most to achieving the required level of representation of the rarest group of species remaining underrepresented.

Step 9. If there is still a choice, select the site that either contains the smallest percentage area needed to achieve the required level of representation of the species under consideration or that contributes the largest percentage of that species's range if no one site achieves adequate representation.

Step 10. If there is still a choice, select the smallest site.

Step 11. If there is still a choice, select the first suitable site on the list.

Step 12. Go to Step 4.

This objective method of site selection assumes a list of available sites and the species that occur in them.

Source: Modified from Caughley and Gunn (1996).

Many butterflies use areas for egg laying and larval development that are temporary. If the protected area set aside in a reserve—for example, from a meadow to a forest—the butterfly loses its host plants (Warren 1994, Hanski et al. 1995). Butterflies are often distributed as metapopulations, and movement between suitable patches of habitat is critical to survival.

One of the most significant contributions of conservation biology has been to show that viable populations of some species are large so that it may be impossible to maintain the required number of animals in parks or sanctuaries (Soulé 1987). Figure 19.16 shows the situation for the grizzly bear in the area containing Yellowstone and Grand Teton National Park. If we draw a biotic boundary for a minimum viable population of 500 grizzly bears, the area needed to support this population is 122,330 sq. km, about 12 times the actual park area of 10,328 sq. km.

Our existing parks are far too small to maintain large mammals and birds on the scale we now expect (Newmark 1985; Soulé 1987). Areas of private land outside of parks must also contribute to the preservation of diversity, and the integration of land use for agriculture and forestry with conservation is an important area of focus.

About 6–7% of the Earth's land area is now protected (Green and Paine 1997). Most of the protected areas in the world are small (Figure 19.17), with 59% being smaller than 1000 ha in area and occupying only 0.2% of the total protected area. By contrast, the six largest protected areas (including Greenland National Park at 972,000 km^2) constitute nearly 20% of the total protected area. Protected areas are not always protected from poaching and hunting, and setting aside land for conservation is an important first step but not the end point of conservation.

FIGURE 19.16
The legal and biotic boundaries for the grizzly bear (Ursus arctos) in the Yellowstone-Grand Teton National Park assemblage. The biotic boundaries are defined by the entire watershed for the parks and the area necessary to support a minimum viable population (MVP) of 50 bears for short term survival, and 500 bears for long–term survival. (From Newmark 1985.)

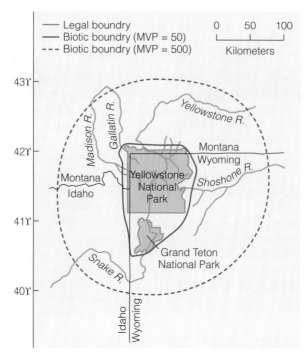

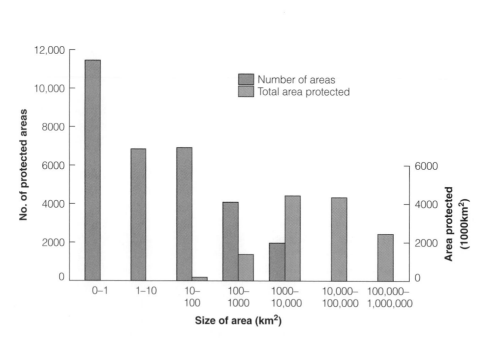

FIGURE 19.17
The number of protected areas in the world, and the area of land protected in various size classes in 1997. The International Union for the Conservation of Nature (IUCN) has defined six categories of protected areas. Most protected areas are small, and the largest protected area (in Greenland) may protect more ice and rock than biodiversity. A total of 30,350 areas are protected. (From Green and Paine 1997.)

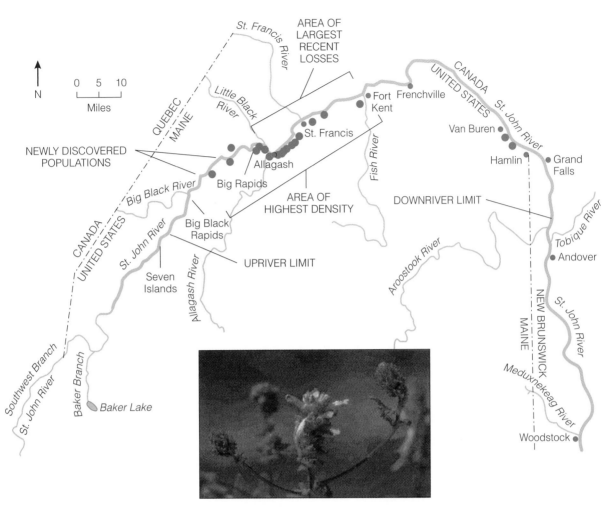

FIGURE 19.18
Geographic range of the endangered plant Pedicularis furbishiae *along the St. John River in northern Maine.*

Examples of Conservation Problems

Two examples of conservation problems will illustrate the practical realities of applying conservation principles to endangered plants and animals. Many more examples are given in Caughley and Gunn (1996).

Furbish's Lousewort

Furbish's lousewort (*Pedicularis furbishiae*) was once thought extinct, and when rediscovered in northern Maine it became one of the first plant species to be listed as endangered (Menges 1990). Furbish's louse-

wort, a herbaceous perennial that reproduces only by seed, lives along riverbanks and only in disturbed habitats. It is found only along the St. John River in northern Maine (Figure 19.18). Disturbance is frequent there because of ice jams and ice scour in the spring. Ice scour is a benefit for Furbish's lousewort because it opens new habitat, but it is also a cost because it kills many plants.

To estimate population viability, Menges (1990) studied 6000 individually marked and mapped plants in 15 local populations along the St. John River. He used a stage-based projection matrix (see Figure 11.18, p. 173) to estimate the finite rate of population change for each population. Figure 19.19 shows how

FIGURE 19.19

Finite rate of population growth for Furbish's lousewort for populations growing in several conditions: (a) percent vegetative cover, (b) presence or absence of woody vegetation, and (c) soils with different moisture level. Note the year–to–year changes in the finite rate of increase (λ). If $\lambda < 1$, the population will decline. If $\lambda > 1$, the population will increase. (From Menges 1990.)

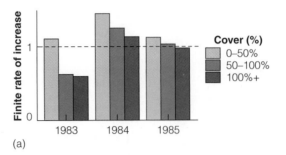

(a)

Cover (%)
0–50%
50–100%
100%+

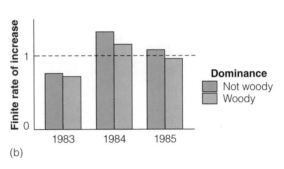

(b)

Dominance
Not woody
Woody

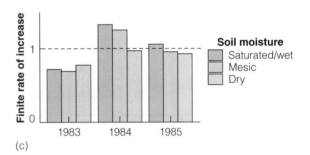

(c)

Soil moisture
Saturated/wet
Mesic
Dry

the rate of population change fluctuated in different habitats and over the three years of the study. Furbish's lousewort is a poor competitor, so the higher the vegetative cover, the less well the population does. Woody, shrub-dominated habitats also are poor for this lousewort, as are dry soils. The prediction from the study of these local populations is that open habitats with good soil moisture will support viable populations of Furbish's lousewort.

This prediction assumes that viable populations are not destroyed by catastrophic events like ice scour. Menges (1990) surveyed 32 local populations and found that 2–12% of these disappeared each year because of ice scour or riverbank collapse. New populations were established at a rate of 3% of empty sites per year. Population viability depends on the balance between extinction rates and establishment rates, and for the years of the study extinction rates seemed to be higher than establishment rates.

Species like Furbish's lousewort present a challenge for conservation biologists. We cannot simply protect the best local populations and ignore others because the disturbance regime of the river causes local extinctions by chance, and new sites for colonization must be available. A reserve system that would protect only the existing populations of this plant may not provide enough recolonization sites. By contrast, too little disturbance would also doom this species because woody vegetation would take over the riverbank habitats. Table 19.6 provides estimates of the likelihood of survival of individual populations of Furbish's lousewort. There is some optimum point of disturbance; too much or too little could be detrimental to the long-term survival of this endangered plant.

Genetic models of population viability are not relevant for Furbish's lousewort because this species does not seem to have any detectable genetic varia-

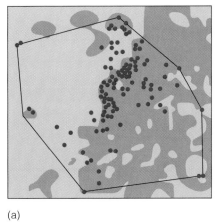

(a)

• Owl location
■ Old forest

FIGURE 19.20
Two examples of areas used by northern spotted owls in southwestern Oregon. (a) A lightly fragmented old–growth forest; (b) a heavily fragmented old–growth forest. The owls make very little use of young forest. (From Carey et al. 1992a.)

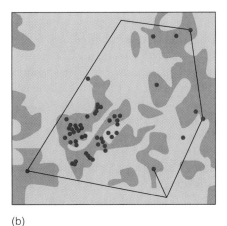

(b)

tion. Menges' (1990) analysis deals only with demographic and environmental stochasticity and natural catastrophes as potential causes of extinction for this plant species.

The Northern Spotted Owl

The northern spotted owl (*Strix occidentalis caurina*) has been the focus of intense debate and confrontations over how the remaining old-growth forests of the western United States should be managed. The northern spotted owl is a territorial owl that lives in old growth conifer forests. Each pair of owls utilizes about 250–1000 ha (1–4 sq. miles) of valuable old-growth forest, nesting in hollow trees and feeding on small mammals, birds, and insects. Heavy logging on private land in the past 40 years has destroyed most of the old–growth forest upon which these owls depend. Most of the remaining old growth is on lands man-

aged by the U.S. Forest Service and the National Park Service. The northern spotted owl is not now an endangered species, and the total population in the Pacific Northwest is roughly 2500 pairs (Lande 1988).

Old-growth forests are being rapidly reduced in the Pacific Northwest, as they are elsewhere on the globe. A large part of the controversy over the northern spotted owl concerns the questions of what type of habitat this owl requires, and how much its habitat can be fragmented by logging without causing a population decline. Northern spotted owls highly prefer old-growth forests for feeding and for roosting (Carey et al. 1992). In fragmented forests owls move more but still feed and roost only in old growth (Figure 19.20). The home range size of owls varies with the prey base. The most common prey in Washington and Oregon is the northern flying squirrel (*Glaucomys sabrinus*). In Washington State, owl home ranges include about 1700 ha of old-growth forests, but in Oregon ranges

TABLE 19.6 Probabilities of survival for local populations of Furbish's lousewort in northern Maine.

Conditions	Assumed annual probability of natural catastrophe				
	6%	2%	0.5%	0.1%	0%
1983–1986, all areas	0.00	0.09	0.57	0.92	1.00
1985–1986:					
Low cover	0.00	0.16	0.53	0.93	1.00
Intermediate cover	0.00	0.06	0.34	0.42	0.66
High cover	0.00	0.00	0.00	0.00	0.00

Source: After Menges (1990).

are less than half that size. These differences in home ranges are directly related to the prey base:

	Home range (ha)	Prey available (g/ha)
Washington	~ 1700	61
Oregon		
Douglas fir	813	244
Mixed conifer	454	338

Additional diet studies in northern California showed that the woodrat (*Neotoma fuscipes*) was a major prey item, and that owls preferred larger prey like woodrats (mean weight: 230 g) when they were available, with flying squirrels (110 g) a second choice (Ward et al. 1998).

Bart and Forsman (1992) surveyed 11,057 sq. km throughout the range of the northern spotted owl. They found no owls in forests that were only 50–80 years old and confirmed that owls occurred only where old–growth stands were present. Figure 19.21 shows that northern spotted owls were both more common and more successful reproductively in old-growth forests. Landscapes with less than 20% old-growth forest rarely supported an owl population. Spotted owls nest in trees that are much larger and older than the average tree in old-growth stands (LaHaye and Gutierrez 1999). In northern California more than 80% of their nest trees were older than 300 years old, and most were greater than 1.2 m in diameter.

One surprising result of studies on the northern spotted owl is that wilderness areas are not very suitable as habitat for the owls (Bart and Forsman 1992). Productivity within protected wilderness areas was only 30–50% as of that in old-growth forest outside these designated areas. Much of the wilderness areas and national parks in the Pacific Northwest are high-elevation areas that are less suitable for these owls. The surprising result is that currently protected stands of old-growth forest in parks and wilderness areas may be unable to sustain the northern spotted owl.

How much old-growth forest must be kept to preserve the northern spotted owl? The key parameters for making this estimate are the dispersal and colonization success of young owls, and the survival and reproductive rates of territorial owls living in landscapes with variable amounts of old forest. The projections of population growth rates of the northern spotted owl are most sensitive to the adult survival rate, which is estimated to be 0.942 per year (Lande 1988). This survival rate, if constant through adult life, could set a maximum age of 72 years for these owls. Lande (1988) estimated the annual finite rate of population growth (λ) for the northern spotted owl to be 0.96, very close to the equilibrium value of 1.0.

All analyses of the northern spotted owl concur in recognizing that a large part of the remaining old-growth forests in the Pacific Northwest must be preserved if we wish this species to persist. The problem thus passes from the conservation biologist to the general public as a matter of policy. The competing land use for these forests is logging and the associated jobs in the timber industry. The conflict over the northern spotted owl is a conflict over short-term needs and long–term goals. At the current rate of harvesting, most of the old-growth forests in the Pacific Northwest will be gone within 20 years, and at that time the problems of the timber industry will still be with us, but the northern spotted owl may not. The present conflict over land use in old-growth forests is but one example of a much broader question: How can human populations and the Earth's biota coexist without serious disruptions? This is the central issue for conservation biology in the twenty-first century.

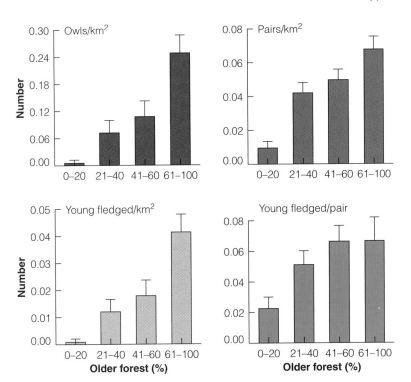

FIGURE 19.21
Density and reproductive success of northern spotted owls in relation to the amount of older forest on 145 forest areas in Washington, Oregon, and northern California. (From Bart and Forsman 1992)

Conclusion

In Part Three we have considered a complex set of ecological questions about the abundance of populations. We used population mathematics to illustrate how we can deal with populations in a precise, quantitative manner. Herein lies the strength and the weakness of population ecology, because to some degree we must abstract the population from the matrix of other species in the community in order to describe its dynamics.

For many populations, other species in the community are essential neighbors, and hence we need to broaden our frame of reference beyond the population level. Thus we are led to consider the whole biological community and, in particular, to ask how distribution and abundance interact to structure the biological communities that cover the globe. This is the subject of Part Four of the book.

Summary

Conservation biology is focused on the ecology of rare and declining species. Two threads of conservation biology are a focus on small populations and the consequences of being small (the small-population paradigm), and a focus on declining populations (the declining-population paradigm). Small populations are subject to an array of uncertainties, from chance demographic events (having all male offspring) to chance environmental events (a flood), to chance genetic events (genetic drift). Even though not all small populations are conservation problems, being small increases the chances of extinction for many populations and can lead a species into an extinction vortex powered by positive feedback of chance processes. An elegant body of theory has given us a good description of the hazards of being a small population.

The declining-population paradigm focuses on identifying the ecological causes of decline and designing alleviation measures to stop the decline. It contains almost no ecological theory but is focused on individual action plans. Only by understanding the population biology of an endangered plant or animal can we provide a rescue plan for a declining population. In some cases, such as the African elephant, the causes of population decline are clear. In other cases we do not have the ecological understanding to recommend action, and we need to develop insights for action plans.

Extinction is the ultimate conservation focus, and four causes are prominent: excessive hunting or harvesting, habitat destruction and fragmentation, introduced species, and chains of extinctions. The major causes of recent extinctions are habitat destruction and introduced species. Habitat destruction leads to population reductions that may trigger the extinction vortex, so protecting habitat is a major goal for all conservation efforts. At present about 6–7% of the world's land areas is protected, but most protected areas are small. Existing parks and reserves are seldom large enough to contain viable populations of larger vertebrates, and conservation efforts on private lands are essential to maintaining populations of flora and fauna.

Habitat fragmentation has been a side effect of agriculture and forestry and has many adverse effects on populations. Populations in isolated patches may go extinct, and unless recolonization occurs, a species may be lost. Corridors between reserves may assist dispersal between patches, but some potential problems, such as the spread of disease, can be aggravated by corridors. Maintaining connectedness of reserves has become an important goal of conservation biology.

The ecological challenge of conservation biology is to develop specific management plans for individual species, whereas the political challenge to the broader conservation movement is to protect large natural areas from destruction. Without parks and reserves there can be no conservation, but with them there is no guarantee of success unless conservation biology can solve the challenging ecological problems of endangered species.

Key Concepts

1. Conservation biology is the applied ecology of endangered species. It rests on two themes—the effects of small population size on fitness, and the causes for population decline and extinction.

2. Small populations are subject to chance events associated with demography, environmental accidents, and genetic drift. All these events can contribute to an extinction vortex of positive feedbacks that result in declining fitness and finally extinction.

3. Declining populations must be studied to diagnose the causes of the decline and to prescribe a remedy. Much detailed information on the endangered species is required to achieve successful conservation.

4. Extinctions are increasing worldwide, primarily as a result of habitat loss and fragmentation and the introduction of nonnative species. Providing corridors that connect habitat fragments can facilitate movements, but this approach does not work for every species at risk.

5. Reserves are part of an effective conservation strategy, but because none are large enough, they cannot conserve large vertebrates without explicit management.

6. The effects of increasing human populations and the continued loss of habitat for natural communities are the root causes behind the current conservation crisis.

Selected References

Balmford, A. 1998. On hotspots and the use of indicators for reserve selection. *Trends in Ecology and Evolution* 13:409.

Beier, P. and R. F. Noss. 1998. Do habitat corridors provide connectivity? *Conservation Biology* 12:1241–1252.

Caughley, G. 1994. Directions in conservation biology. *Journal of Animal Ecology* 63:215-244.

Caughley, G. and A. Gunn. 1996. *Conservation Biology in Theory and Practice.* Blackwell Science, Oxford. 459 pp.

Dobson, A. P., J. P. Rodriguez, W. M. Roberts, and D. S. Wilcove. 1997. Geographic distribution of endangered species in the United States. *Science* 275:550–553.

Ehrlich, P. R. and A. Ehrlich. 1981. *Extinction.* Random House, New York. 305 pp.

Frankham, R. 1998. Inbreeding and extinction: Island populations. *Conservation Biology* 12:665-675.

Laurance, W. F., L. V. Ferreira, J. M. Rankin-De Merona, and S. G. Laurance. 1998. Rain forest fragmentation and the dynamics of Amazonian tree communities. *Ecology* 79:2032-2040.

Newmark, W. D. 1995. Extinction of mammal populations in western North American national parks. *Conservation Biology* 9:512-526.

Reid, W. V. 1998. Biodiversity hotpots. *Trends in Ecology and Evolution* 13:275-280.

Simberloff, D. and J. Cox. 1987. Consequences and costs of conservation corridors. *Conservation Biology* 1:63–71.

Soulé, M. and L. S. Mills. 1992. Conservation genetics and conservation biology: A troubled marriage. Pages 55–69 in O. T. Sandlund, K. Hindar, and A. H. D. Brown, eds. *Conservation of Biodiversity for Sustainable Development*. Scandinavian University Press, Oslo, Norway.

Woodroffe, R., and J. R. Ginsberg. 1998. Edge effects and the extinction of populations inside protected areas. *Science* 280:2126–2128.

Young, A., T. Boyle, and T. Brown. 1996. The population genetic consequence of habitat fragmentation for plants. *Trends in Ecology and Evolution* 11:413–418.

Questions and Problems

19.1 Barro Colorado Island was formed 85 years ago in central Panama when Gatun Lake was created as part of the Panama Canal. Since that time 65 of 394 species of birds have disappeared from the island, 21 of them in the past 25 years. Discuss what mechanisms might cause extinctions of birds that can fly in an undisturbed area of tropical forest. Robinson (1999) discusses these changes.

19.2 Much of conservation biology focuses on rare species, yet Tilman et al. (1994) suggested that the species that are the best competitors and are the most abundant are at greatest risk from habitat loss. Discuss why more-abundant species might be at greater risk than rare ones when habitat area is reduced. McCarthy et al. (1997) discuss this problem.

19.3 When organisms of the same species are brought together to breed from divergent geographical areas, outbreeding depression may occur in which the fertility or viability of the offspring is impaired (Templeton 1986). Discuss the reasons outbreeding depression occurs and its implications for conservation biology.

19.4 The creation of an abrupt edge in a forest because of clearing land for agriculture seems to increase the number of species in temperate zone forests and to decrease the number of species in tropical Amazonian forests (Lovejoy et al. 1986). What factors change when an edge is created, and why might there be different effects in tropical vs. temperate forests?

19.5 Kirtland's warbler is an endangered species that breeds in northern Michigan jack-pine forests. Since 1951 the population of this species has been declining, and it now numbers about 200 individuals. The most important factor in the population decline seems to be increasing parasitism of nests by brown-headed cowbirds. Cowbirds were removed from the breeding area of Kirtland's warbler starting in 1971, but no change has occurred in warbler numbers (Ryel 1981). Read Walkinshaw (1983) and Ryel (1981) and discuss why this might be. What management plan you would now recommend for this endangered species?

19.6 Calculate the intrinsic rate of natural increase for a northern spotted owl population in which annual fecundity is 0.24 eggs/female, the age at first breeding is three years, the survival probability to fledging is 0.60, the probability of successful dispersal is 0.18, the subadult annual survival rate is 0.71, and the annual adult survival rate is 0.942. How much does r change if you truncate the life table at 10 years? At 20 years? Lande (1988) discusses the life table for this species.

19.7 Review the history of the successful rehabilitation of the endangered Lord Howe Island woodhen (*Tricholimnas sylvestris*) on Lord Howe Island in the Pacific (Caughley and Gunn 1996, pp. 75–81). Discuss the reasons for the success of this project and the general principles it illustrates for conservation problems.

19.8 Captive breeding programs are one technique used to rescue endangered species. Under what conditions should captive breeding be used? Discuss the limitations of captive breeding as a conservation strategy. Western and Pearl (1989) provide references.

19.9 Discuss the assumptions underlying the nested subset model of patch occupancy (see Figure 19.13). Explain what ecological processes could produce "holes" in the data matrix (see Table 19.5), and what processes could produce "outliers."

19.10 Amphibian populations have been declining in many parts of the world during the past ten years (Blaustein and Wake 1990). Discuss the hypotheses proposed to explain these declines and suggest a research plan to rescue these populations. Fisher and Shaffer (1996) discuss the amphibian declines in central California, and Wake (1998) gives an overview of the problem.

Overview Question

Debate the following proposal: Resolved, that conservation biology is a crisis–oriented discipline and consequently should not be subject to the normal procedures of science for creating hypotheses and testing them experimentally.

P A R T F O U R

Distribution and Abundance at the Community Level

CHAPTER 20 THE NATURE OF THE COMMUNITY

CHAPTER 21 COMMUNITY CHANGE

CHAPTER 22 COMMUNITY ORGANIZATION I: BIODIVERSITY

CHAPTER 23 COMMUNITY ORGANIZATION II: PREDATION AND COMPETITION
            IN EQUILIBRIAL COMMUNITIES

CHAPTER 24 COMMUNITY ORGANIZATION III: DISTURBANCE AND NONEQUILIBRIUM
            COMMUNITIES

CHAPTER 25 ECOSYSTEM METABOLISM I: PRIMARY PRODUCTION

CHAPTER 26 ECOSYSTEM METABOLISM II: SECONDARY PRODUCTION

CHAPTER 27 ECOSYSTEM METABOLISM III: NUTRIENT CYCLES

CHAPTER 28 ECOSYSTEM HEATH: HUMAN IMPACTS

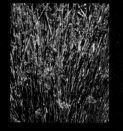

CHAPTER 20

The Nature of the Community

NEITHER ORGANISMS NOR SPECIES populations exist by themselves in nature; they are always part of an assemblage of populations living together in the same area. When we previously discussed the interactions of two or more of these populations in predation and competition for food, the focus was on individual populations. Now we focus on the assemblage of populations in an area, the *community*. Most generally, a community is *any assemblage of populations of living organisms in a prescribed area or habitat*. So we can speak of the community of animals in a rotting log or the community of plants in the beech-maple deciduous forest. A community may be of any size.

Why do we need to be concerned about communities? The key to answering this question lies in the relationships between species in a community.

Dynamic Relations Between Populations

If the community is a complex ecological unit, a kind of superorganism, the populations should be bound together in a network and organized by obligate interrelations. This is the basic idea of the "web of life." To evaluate how strong the network of interrelations is, we must go back to a discussion of the factors that limit the distribution and abundance of populations. This is the first and most important question we must

ask about any community: How strong are the connections between species?

Two models describe the answer to this question for a community: the rivet model and the redundancy model. Both of these models are illustrated in the schematic design below.

1. Rivet Model of Communities. Paul and Anne Ehrlich (1981) first described this model in discussing conservation problems. They suggested that species in a community were like the rivets in the wings of airplanes—not all the rivets were necessary to hold the wing together, but if someone started taking out rivets one by one, while we were flying in the airplane, we would become concerned. The rivet model is a metaphor, not a formal mathematical model of a community, and we can illustrate it as schemes A and C in the following schematic diagram:

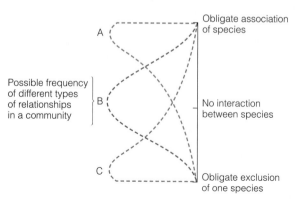

In scheme A, most of the species in the community are associated tightly with other species, such that the web of relationships is relatively tight. If one species in the community is reduced or increased, many others are affected. Scheme C is the competitive version of scheme A in which species interact negatively, so that one species excludes others. In both models the community web is relatively tight, so changes in one species reverberate into many others.

2. Redundancy Model of Communities. The alternative model of how species interact in a community is diagrammed as scheme B above. Brian Walker (1992) first suggested the redundancy model in relation to conservation goals. According to this model, most species in the community have little to do with one another, so that the "web of life" is very loose. An increase or decrease in one species has little effect on other species in the community. Species, and ecological processes, are redundant: If one predator disappears, another species in the community takes its place.

Note that it is still important to study ecological communities, even if the redundancy model is correct. Species in communities do interact, even if these interactions are not obligate. These two models are polar-opposite views of communities, and much of what we discuss in the next eight chapters can be viewed as attempts to answer the question of how tightly linked species are in particular communities. There is no reason to assume that all communities will fit into one or the other of these models. But there are important practical consequences that flow from these models if we are trying to manage a community (Walker 1992). If species are tightly linked in natural communities, as schemes A and C suggest, losses of species may have cascading effects on other species. In these cases, conservation biology must be concerned about community dynamics rather than single-species dynamics. Within a type A or type C community some species are "drivers" and others are "passengers," and the loss of some species is more critical than the loss of others. By contrast, if we are trying to conserve a species in a type B community, we can concentrate on the single-species approach discussed in Part Three of this text.

The rivet model and the redundancy model of community interactions are similar to the historical conflict in plant ecology over the organismic and individualistic views of plant communities. Plant ecologists put the question this way: *Is the community an organized system of recurrent species, or a haphazard collection of populations with minimal integration?* Two schools developed over this question during the early part of the twentieth century. At one extreme were the views of F. E. Clements and A. G. Tansley that the community is a superorganism or a quasi-organism. At the other extreme is the individualistic view of H. A. Gleason that the community is a collection of populations with the same environmental requirements.

Arthur Tansley and Frederic Clements were two of the founding fathers of plant ecology in the English-speaking world. Tansley was influential in establishing plant community ecology in Britain and can be considered the father of British plant community ecology (Sheail 1987). Clements, working in the United States, was equally influential in shaping the early course of American plant community ecology (Tansley 1947).

Sir Arthur G. Tansley *(1871–1955) Founder of British Plant Ecology, Cambridge University*

A major assumption of many community ecologists, including Clements and Tansley, is that some "fundamental unit" of natural communities really exists, and that this unit is natural in the sense that it is present in nature. This assumption led some ecologists to draw the analogy between the "species" concept and the "community" concept. If fundamental units exist in nature, we should be able to discover these units and

ESSAY 20.1

WHAT IS THE GAIA HYPOTHESIS?

The superorganismic view of communities that was advanced by F. E. Clements and A. G. Tansley over 70 years ago is similar to the Gaia Hypothesis proposed by James Lovelock during the past 20 years. (Gaia is the name the Greeks gave to their Goddess of the Earth.) The Gaia Hypothesis arose from Lovelock's observations of the atmospheres on Mars and other planets. His answer to the question, Why is the Earth so different and so suitable for life? is that the Earth's atmosphere and organisms are tightly coupled in a feedback loop, such that organisms control the makeup of the atmosphere and keep it at or near the chemical composition that favors life. Thus the living biota and the physical atmosphere act in a feedback system to control oxygen and carbon dioxide levels on Earth.

Two crucial aspects of the Gaia Hypothesis require investigation. First, we need to ascertain if in fact there are feedback loops between the biota and the Earth's atmosphere that act to control changes in oxygen and carbon dioxide levels. One suggested mechanism of climate control is cloud production due to dimethylsulphide (DMS) production by marine phytoplankton. Plankton produce DMS, which aids in cloud formation, which in turn affects global climate by stabilizing temperature (Lovelock 1988).

Second, if the mechanisms of the Gaia Hypoythesis do exist, we need to find out how such a system could have evolved. Most evolutionary ecologists are skeptical of the Gaia Hypothesis because natural selection to maximize fitness at the level of the individual not at the level of groups of species. Traits that operate for a good of the species or for the good of the whole biota are believed to have evolved by individual selection, not by group selection, which is typically very weak compared to individual selection. The Gaia Hypothesis postulates group selection to the extreme, such that evolution selects for systems that operate for the good of all living things. The challenge is to derive individual-based selective advantages for mechanisms that act to control climate (Wilkinson 1999).

A second criticism of the Gaia Hypothesis is that if such mechanisms do exist, they have not been very effective in the past in controlling climatic changes. Large changes in the concentrations of carbon dioxide over the past million years (Adams et al. 1990, Watson et al. 1998) do not suggest any effective stabilizing mechanisms that the Gaia Hypothesis postulates. But the control of the Earth's atmosphere postulated by the Gaia Hypothesis could allow large swings in gas composition within some limits that are never exceeded. From a human perspective, even if the Gaia Hypothesis is correct, we must not neglect the causes of climate change in the naive hope that Gaia will rescue us from our folly of increasing emissions of greenhouse gases.

classify them, perhaps in the way we classify species. This fundamental-unit view has been held by many plant ecologists in Europe and North America, and the famous plant ecologists J. Braun-Blanquet (France), Clements (United States), and Tansley (United Kingdom) all strongly supported this assumption, which is equivalent to schemes A and C above.

The fundamental-unit assumption of plant community ecology was attacked almost simultaneously by L. G. Ramensky in Russia, Henry Gleason in the United States, and F. Lenoble in France (Whittaker 1962). Henry Gleason in 1926 attacked the prevailing views of Clements in the United States that communities were similar to superorganisms. Gleason and the European plant ecologists emphasized the principles of vegetational continuity and of species individuality. Each species has its own specific range. A plant community can be regarded as an assemblage of wandering populations and is an arbitrary unit, unlike a species. This individualistic school argues that communities can be recognized and classified, but any classification is for the convenience of the human observer and is not a description of the obligate interactions that tie the community together.

Henry A. Gleason *(1882–1975) Plant ecologist and plant geographer, New York Botanical Garden*

For each community we can ask where on this scale most of the species in the community would fall. Another way of stating this question is to ask how frequently the distribution and abundance of one species is determined by interactions with other species. No one knows the answer to this at present, but we can make a few general statements.

Obligate associations may exist for certain parasites that have a single host species, or for certain animals that feed on only one species of plant. But few plants and only a small number of animals in temperate and polar regions seem to have life cycles so tightly coupled to one other species. Most species depend only partly on others. An insect may feed on one of several plant species, and predators may eat a variety of prey species. Partial dependence of this type seems most common in nature and grades off into a state of indifference in which species do not interact (Whittaker 1962). Studies of the composition of plant communities show this very clearly. Consider the following example.

Juncus effusus is an important weed in upland pasture in Wales, and Agnew (1961) studied 99 quadrats spread through all community types that contained this weed. After species that were found fewer than five times in the 99 one-square-meter quadrats were eliminated, 53 plant species remained. The results of Chi–square tests of association run on all species pairs can be presented graphically in a species constellation,

as in Figure 20.1. Three "groups" of species (shaded in Figure 20.1) might be recognized as communities. If plant communities are tightly integrated, as Clements suggested, species groups should be highly integrated and distinct from one another. If communities are loose aggregations of species, groups will not be highly integrated or distinct. Obviously there are many intermediate species that fit into none of the three communities, and the communities are not completely independent of one another. Figure 20.1 shows that a majority of the species pairs in *Juncus effusus* communities show no evidence of interaction, and thus these communities seem closer to scheme B than either A or C.

In tropical regions, particularly in species-rich tropical rain forests, the importance of obligate associations is much less clear. The taxonomy of many species groups is poorly known, and for the groups that have been properly classified, there are few ecological studies to determine the sensitivity of the system to species removals.

How much dynamic integration is present in a community is a critical question that arises repeatedly in community ecology and will be addressed in the next eight chapters. The current view of the nature of the community lies closer to scheme B or Gleason's individualistic view than to Clements's superorganismic interpretation or schemes A and C. Species are distributed individualistically according to their own genetic characteristics. Populations of most species tend to change gradually along environmental gradients. Most species are not in obligatory association with other species, which suggests that communities can be formed with many combinations of species and vary continuously in space and in time. To classify such communities into discrete units is a highly artificial undertaking, and attention has turned from classification of communities to analyzing community dynamics and functional organization.

The individualistic nature of the community is well supported by historical analyses of plant communities. Historical changes in vegetation can be interpreted in some detail using fossil pollen grains in lake sediments. The organismic view of plant communities would predict that the fossil record would show stable communities and then sudden shifts to a new type of community. The individualistic view would predict gradual changes in species composition over time, with no clear community types that remained as a unit. The

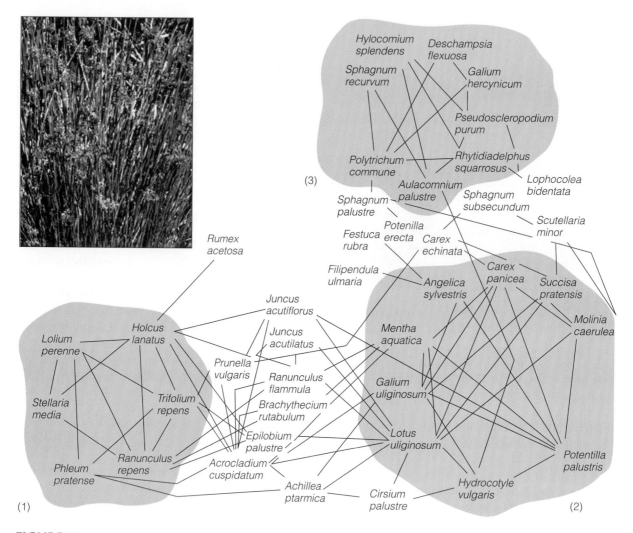

FIGURE 20.1

Species constellations between plant species found in 99 samples of communities in which the rush Juncus effusus *occurred in northern Wales. Lines represent significant positive associations between the species. If species in communities are highly integrated, they should show many associations with other species. Three separate communities (shaded) can be recognized, but they are not discontinuous or highly integrated, and there are many species that do not fit conveniently into one community or the other. Consequently these plant communities are best described by the individualistic model of Gleason. (After Agnew 1961.)*

data available for fossil plant communities support the individualistic view. If we reconstruct the forest history of an area such as Minnesota (Wright 1968), we find a continuous series of species coming and going. Some modern forest communities have no analog in the past, and conversely, some communities found in the past do not exist anywhere at the present time.

Fossil pollen studies have proved very successful at describing the sequence of plant communities over the past 30,000 years. Because of the Pleistocene ice sheets in the Northern Hemisphere, climatic and vegetational shifts occurred around the globe. Figure 20.2 shows the pollen from a bay in North Carolina about 500 kilometers south of the line of maximal glacial advance (Whitehead 1973). In this bay, about 5 meters of sediment has been deposited over the past 30,000 years. From 30,000 to 21,000 years before the present (BP), temperate forests of oaks, birches, and pines occupied the area. From 21,000 to 10,000 years BP, boreal forest containing spruce and northern pines was present, and the climate was colder and drier than at present. Deciduous forests, with a predominance of oak, replaced the boreal forests about 10,000 years ago. Swamp forests featuring blackgum, cedar, magnolia, and red maple began to develop about 7000 years BP. The swamp forests were essentially modern about 4000 years ago.

Pollen studies of the type shown in Figure 20.2 are important because they have established that the displacement of species by the ice sheet was individualistic—associations of species did not move north and south as a unit, but instead each species moved independently (Whitehead 1981). The displacement of the boreal forest was approximately 1300 kilometers from its present location. Modern plant communities in the temperate zone have existed in their present form for only a very few years.

How stable are communities? Fossil pollen studies clearly show changes in communities over geological time, but do these changes occur over ecological time? Modern community ecologists are divided over the issue of whether communities should be considered *equilibrium* or *nonequilibrium* systems (Schoener 1987b). Closely related to this issue is whether we study a community as an *open* system or a *closed* system. *Closed* systems consist of a single homogeneous patch of habitat, whereas *open* systems are collections of patches connected by *dispersal*. Equilibrium models are restricted to the behavior of a system near an equilibrium point, whereas *nonequilibrium* models consider the transient behavior of a community as it moves from one state to another (DeAngelis and Waterhouse 1987). Many of the models we discussed in the population chapters are closed, equilibrium models; the competition equations of Lotka

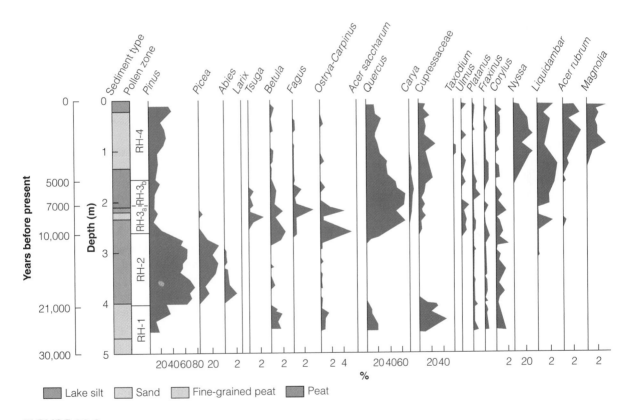

FIGURE 20.2
Pollen diagram for fossil tree pollen in Rockyhock Bay, northeastern North Carolina. The sediment type in the core varied from fine silt to coarse- and fine-grained peat deposits. The rate of sedimentation varies with the sediment type, and thus the time scale (at left) is not linear. The approximate ages are based on radiocarbon dating. The pollen zones refer to community types: RH-1 = oaks, birches, and pines; RH-2 = boreal spruce and pines; RH-3 = oak, beech, and birches; RH-4 = swamp forests of black gum, cedar, magnolia, and red maple. (After Whitehead 1981.)

and Volterra (see page 180) and the predation equations (see page 207) are two good examples. *Stability* is a dominant preoccupation of equilibrium models.

But one characteristic of natural communities is change. Communities are open systems, and a general property of open systems is that transient, nonequilibrium conditions may persist for long periods of time. The critical question thus becomes: How long does it take a community to reach equilibrium? If there are many linkages between the species in a community, equilibrium may never be achieved. Thus it may be more useful to view communities as open, nonequilibrium systems (Paine 1983, DeAngelis and Waterhouse 1987).

The rivet model and the redundancy model of communities are important because they focus our attention on the dynamic interrelations of the species that make up a community. What are the attributes of communities that we can measure to look for interrelations among species?

Community Characteristics

Like a population, a community has a series of attributes that do not reside in its individual species components and have meaning only with reference to the community level of integration. Four traditional characteristics of communities have been measured and studied:

Biodiversity: We can ask, What species of animals and plants live in a particular community? This species list is a simple measure of species richness or species diversity and leads us to the more critical question of what controls biodiversity.

Growth form and structure: We can describe the type of plant community by major categories of growth forms: trees, shrubs, herbs, and mosses. We can further sort the growth forms into categories such as broadleaf trees and needle-leaved trees. These different growth forms determine the stratification, or vertical layering, of a community.

Relative abundance: We can measure the relative proportions of different species in a community, and ask whether all species are equally common in communities.

Trophic structure: We can ask, Who eats whom? The feeding relations of the species in a community determine the flow of energy and materials from plants to herbivores to carnivores, and determine the biological organization of the community.

These attributes can all be studied either in communities that are in equilibrium or in communities that are changing. The changes may be temporal ones, in which case the changes are called *succession* and lead to a stable *climax community*. Or the changes may be spatial, along environmental gradients, and we may study, for example, how the characteristics of a community change along a moisture or temperature gradient.

We discuss the techniques of measuring the four characteristics of communities in subsequent chapters because these characteristics are difficult to quantify, even though they are intuitively clear. We next look at three features of communities.

Community Boundaries?

If communities are discrete, natural units, then contacts between stands of two different communities should be sharp and discontinuous. Three types of boundaries between communities might occur: sharp, diffuse, or mosaic. All three have been found in nature. The boundary between the prairie and the deciduous forest in the eastern United States was apparently very sharp. However, this sharp boundary can be interpreted in two ways. Clements would say that it shows that communities are discrete natural units, whereas Gleason would interpret it as an artificial boundary maintained by fire disturbance. One difficulty of interpreting boundaries is that the human observer fixes on a few dominant species (often trees), whereas a more analytical assessment of shrubs, herbs, or grasses might give a different view of how sharp the boundary in fact is.

We could try to eliminate from this discussion all boundaries caused by sharp environmental discontinuities and by disturbance, but this is difficult to do. Striking breaks in soil type will produce discontinuity, but all the competing schools of thought about the nature of plant communities agree about this type of sharp discontinuity. Other sharp boundaries may be due to environmental discontinuities but need more study. For example, the boreal forest in northern Canada gives way to the tundra over a broad zone of overlap, which constitutes a mosaic of sharp community boundaries. This sharp boundary may mark a sharp break between areas that have sufficient soil, drained-well and areas that have little soil that drain poorly (Ritchie 1992). Should we interpret forest-tundra boundaries as evidence for discrete associations á la Clements?

A number of plant ecologists almost simultaneously began to question the assumption that plant communities had sharp edges and instead emphasized that vegetation was a complex *continuum* of populations rather than a mosaic of discontinuous units. Robert Whittaker of Cornell University was one of the most influential members of this group, and his early work in the mountainous regions of the United States did much to establish the continuum view of vegetation. The continuum school of plant ecologists developed a series of techniques, called *gradient analysis*, to study the continuous variation of vegetation in relation to environmental factors (Whittaker 1967, Ter Braak and Prentice 1988).

The simplest application of gradient analysis is to take samples at intervals along an environmental gradient, such as elevation on a mountain slope. Figure 20.3 shows the relative abundance of three species of *Pinus* along an altitudinal gradient from 425 meters to 1430 meters (1400–4700 feet) on south-facing slopes in the Great Smoky Mountains of Tennessee. No discontinuity is evident in Figure 20.3, and Whittaker (1956) presents data from 24 other tree species to show a continual gradation from the Virginia pine forest at low elevations, to the pitch pine heath at middle elevations, to the table-mountain pine heath at higher elevations.

Robert H. Whittaker *(1926–1972) Professor of Plant Ecology, Cornell University*

Elevation is a complex environmental gradient because it includes gradients of temperature, rainfall, wind, and snow cover. Other gradients can be used as well to show the continuity of vegetation. For example, Whittaker (1960) grouped stands along a soil-moisture gradient at a fixed elevation in the Siskiyou Mountains of southern Oregon and northern California (Figure 20.4). Some species, such as Port Orford cedar (*Chamaecyparis lawsoniana*), are found only in moist sites; others, such as Pacific madrone (*Arbutus menziesii*), are most common on dry sites.

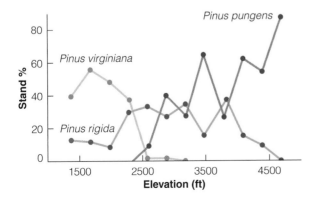

FIGURE 20.3
Transect of the elevation gradient along dry, south-facing slopes in the Great Smoky Mountains of Tennessee. Along this gradient, no boundaries separate the three community types an ecologist is likely to recognize: Pinus virginiana *forest at low elevations,* Pinus rigida *heath at middle elevations, and* Pinus pungens *heath at high elevations. The lack of sharp boundaries between these communities supports the continuum concept of communities. (Data from Whittaker 1956.)*

FIGURE 20.4

Distribution of trees along a soil–moisture gradient at low elevations in the central Siskiyou Mountains of Oregon and California. Fifty stands were sampled between an elevation of 610 and 915 meters (2000–3000 ft). Only three of 20 tree species are shown here to illustrate the types of responses observed. There is extensive overlap of tree distributions and no signs of sharp boundaries. (Data from Whittaker 1960.)

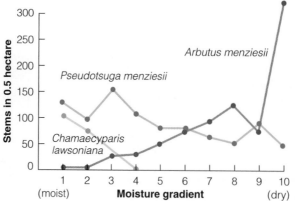

We can use information from gradient analysis to determine which model of the community is most appropriate for natural vegetation (Austin 1985). Figure 20.5 shows four alternative models for vegetation organization along a gradient. The first model is the classical community-unit model of Clements and Tansley in which communities are discrete and have sharp boundaries (Figure 20.5a). The other three models are variants of the vegetational continuum model of Gleason, Lenoble, and Whittaker. The individualistic continuum model (Figure 20.5b) has neither sharp boundaries nor well-defined groups of species. If there is strong competition for the gradient resource competition would be expected to lead to an even distribution of species along an elevational gradient (Figure 20.5c). If there are several strata in a community (such as trees, shrubs, and herbs), each stratum may operate independently of the others (Figure 20.5d) and produce a pattern similar to that shown by the individualistic continuum model. The data shown in Figures 20.3 and 20.4 are most similar to the models of a vegetation continuum shown in Figures 20.5b and 20.5d.

Plant ecologists have found strong support for the individualistic model of the plant community, in which most species are not tightly bound by obligate relationships, in keeping with scheme B articulated on page 386. Sharp boundaries between communities are rare unless there is an environmental edge, and species distributions fit the continuum model. But we should remember that the individualistic model of the community is not an absolute one, specifying that all species in a community are independent. Scheme B, the individualistic continuum model, in fact allows a fraction of species to be mutually dependent, so that the differences between these views are quantitative, not absolute.

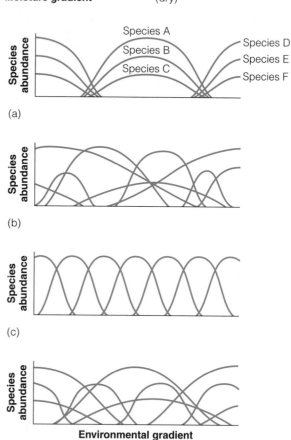

FIGURE 20.5

Alternative models for vegetation organization along an environmental gradient. Each curve represents a single hypothetical species. (a) The fundamental unit concept of a plant community with sharp boundaries, suggested by Clements and Tansley. (b) The individualistic continuum model suggested by Whittaker. (c) The resource-partitioned continuum model. (d) The resource-partitioned continuum model with several strata. (After Austin 1985.)

Distributional Relations of Species in Communities

If the separate stands that make up a community are similar, all or many of the species in the community must have similar geographic distributions. Plant species that make up a community should have distributional maps that closely coincide on a local level, and the geographic limits of the species should coincide with the continental limits of the community. These predictions from the organismic view of communities as integrated wholes have stimulated an examination of the geographical boundaries of species.

Floristic provinces can be recognized on a continental scale by major vegetational changes. Figure 20.6 illustrates a subdivision of North America into ten floristic provinces. The exact position of these boundaries is often debated, and some workers recognize more or fewer provinces, but no one questions the existence of large areas of similar vegetation in which the ranges of many species coincide. Figure 20.7 illustrates the coincidence of ranges for six tree species in the eastern deciduous forest province.

Boundaries between floristic provinces are called tension zones, and they coincide with the distributional limits of many species. Curtis (1959) has analyzed in detail a tension zone in Wisconsin between two parts of the deciduous forest, the prairie hardwoods province and the northern hardwoods province. Figure 20.8 shows the range limits for 182 plant species that abut at this boundary. The width of this tension zone in Wisconsin is variable, from as little as 16 km (10 miles) to as much as 48 km (30 miles).

A floristic province is composed of many different communities, and the critical analysis of the distributions of species cannot be at the continental scale of a floristic province, but must instead occur at the local level of a community. Suppose that we study a number of stands of a particular community in Wisconsin and another group of stands in Michigan. How can we compare these two samples? Several measures of community similarity are available (Krebs 1999, p. 375); here we shall discuss one simple measure based on species presence only. In two communities, one with x number of species, and the other with y number of species, and with z species occurring in both communities, we define

$$\text{Index of similarity} = \frac{2z}{x + y} \qquad (20.1)$$

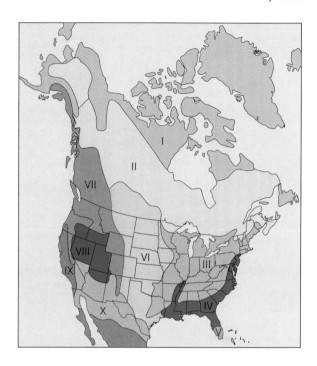

FIGURE 20.6

Floristic provinces of the continental United States and Canada. I, tundra province; II, northern conifer province; III, eastern deciduous forest province; IV, coastal plain province; V, West Indian province; VI, grassland province; VII, cordilleran forest province; VIII, Great Basin province; IX, California province; X, Sonoran province. For this map, the lines between the provinces have been drawn boldly to show the general outlines rather than details. The actual boundaries are generally not sharp; they overlap and interfinger extensively, and small enclaves of one province may be wholly surrounded by another. (After Gleason and Cronquist 1964.)

This index ranges from 0 (no similarity) to 1 (complete similarity). For example, the southern mesic forests of Wisconsin contain 26 tree species (dominated by sugar maple, basswood, beech, and red oak), and the northern mesic forests of Wisconsin contain 27 tree species (dominated by sugar maple, eastern hemlock, beech, yellow birch, and basswood). Seventeen species occur in both communities, so the index of similarity for tree species is calculated as

$$\text{Index of similarity} = \frac{(2)(17)}{26 + 27} = 0.64$$

We can also illustrate the use of the index of similarity for the animal communities of the Great Lakes

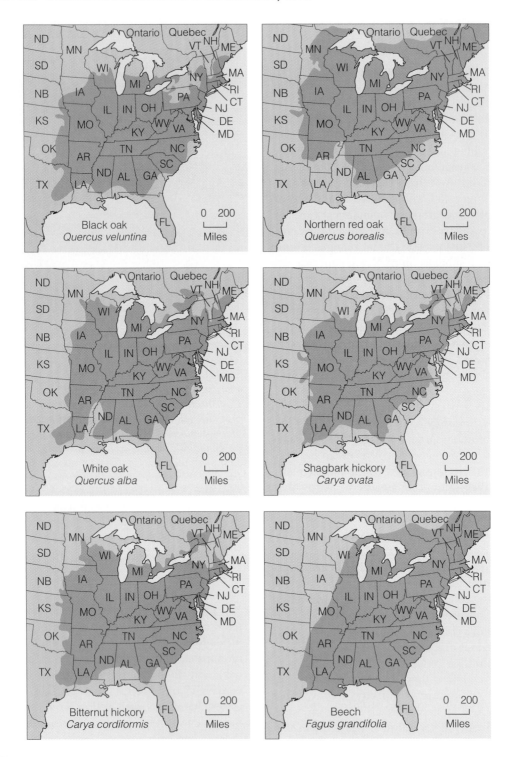

FIGURE 20.7
Ranges of six forest trees of the eastern deciduous forest province of North America.
Geographical ranges are largely coincident but differ slightly at the edges. (After Gleason
and Cronquist 1964.)

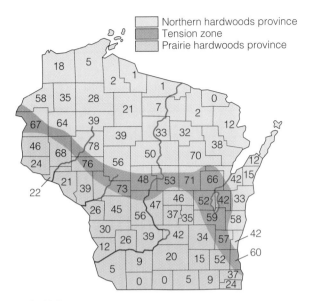

FIGURE 20.8
Tension zone (shaded area) between two floristic provinces in Wisconsin. Summary of range limits for 182 species of plants. The figures in each county indicate the number of species attaining a range boundary there. (After Curtis 1959.)

of North America, which form the largest inland water system in the world (Figure 20.9). Water flows from Lake Superior and Lake Michigan through Lake Huron into Lake Erie and Lake Ontario, from which the St. Lawrence River carries the flow into the North Atlantic. The Great Lakes have been studied in considerable detail by biologists in the United States and Canada because of pollution problems associated with the large cities that border the lakes.

Table 20.1 gives the records of crustacean zooplankton species for the Great Lakes. Most of the 25 species occur in all the Great Lakes, and hence all the indices of similarity for this group of species are high:

	Index of similarity
Lake Superior and Lake Michigan	0.81
Lake Michigan and Lake Huron	0.93
Lake Erie and Lake Ontario	0.90

High similarity does not occur in all of the species groups that constitute the freshwater community of the Great Lakes. Rotifers show more differences among the

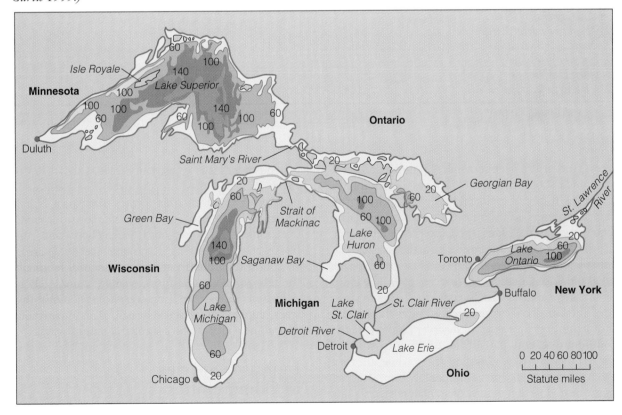

FIGURE 20.9
The Great Lakes of North America. Major depth contours are given in fathoms (1 fathom = 6 feet = 1.8 meters)

TABLE 20.1   Crustacean zooplankton species recorded from the Great Lakes of North America.

An asterisk indicates the presence of a species in a particular lake.

Species	Lake Superior	Lake Michigan	Lake Huron	Lake St. Clair	Lake Erie	Lake Ontario
Senecella calanoiddes Juday	*	*	*	*		
Limnocalanus macrurus Sars	*	*	*	*	*	*
Eurytemora affinis (Poppe)	*	*	*	*	*	*
Epischura lacustris Forbes	*	*	*	*	*	*
Diaptomus sicilis Forbes	*	*	*	*	*	*
D. ashlandi Marsh	*	*	*	*	*	*
D. minutus Lillj.	*	*	*	*	*	*
D. oregonensis Lillj.	*	*	*	*	*	*
D. siciloides Lillj.	*	*	*	*	*	*
D. pallidus Hennrick	*	*	*			
Diacyclops bicuspidatus thomasi Forbes	*	*	*	*	*	*
Acanthocyclops vernalis Fischer	*	*	*	*	*	*
Mesocyclops edax (Forbes)	*	*	*	*	*	
Tropocyclops prasinus mexicanus Keifer	*					
Osphranticum labronectum Forbes	*					
Alona spp.	*	*	*	*	*	
Bosmina longirostris O.F.M.	*	*	*	*	*	*
Ceriodaphnia lacustris Birge	*	*	*	*	*	
Chydorus sphaericus O.F.M.	*	*	*	*	*	
Daphnia ambigua Scour.	*					
D. galeata mendotae Birge	*	*	*	*	*	*
D. longiremis Sara	*	*				
D. parvula Fordyce	*	*				
D. pulex DeGeer	*					
D. retrocurva Forbes	*	*	*	*	*	*

Source: After Watson (1974).

lakes (Watson 1974). For example, 15 species of rotifers are found in both Lake Erie (25 species total) and Lake Ontario (30 species total), and the index of similarity based on rotifers is 0.55 for these two lakes. Similarity among some of the lakes is even lower for some groups of fish. Of the 11 species of whitefish in the Great Lakes, two species are common to both Lake Erie (three species total) and Lake Ontario (seven species total). The index of similarity based on whitefish is 0.4 for these two lakes.

The Great Lakes example illustrates the general problem of trying to define discrete assemblages as communities. One part of the community may be very similar in two adjoining lakes while another group of species may vary considerably from one lake to the next. The community of animals changes gradually in composition, with no sharp demarcations.

If we repeat this kind of analysis for many different associations, we find that some are more homogeneous over large areas than others. Curtis (1959), for example, pointed out the great floristic homogeneity of the deciduous forests of eastern North America. On mesic sites, sugar maple is dominant over most of this area, and beech, basswood, and buckeye are the leading dominants over large areas. By contrast, the conifer swamp community, originally thought to be very uniform, changes completely in floristic composition from west to east. Only two trees, tamarack and black spruce, remain constant; they alone appear to be a similar community type.

Indicator Species in Communities

Because communities contain so many different species in many taxonomic groups, there has been a continuing search for shortcuts in both defining communities and monitoring changes in them. The various shortcuts all involve selecting one or a few species that reflect the community. The oldest and most venerable method recognizes *indicator species*. The presence and changes in numbers of the indicator species are postulated to reflect changes in other members of the community (Landres et al. 1988, Noss 1990). The use of an indicator species makes the job of land managers or conservation biologists much simpler because they need not measure and catalog all the species in the community. By concentrating on a few important species, the practical side of community management is simplified.

Indicator species can be useful in studies of communities if they are selected by rigorous criteria. The first and most important step is to decide exactly what the objectives are, what exactly is to be indicated (Simberloff 1998). Are the indicator species to serve as a signpost for a particular community type, or are they to reflect the 'health' of the community? Box 20.1 lists some criteria that are required to assess a species as a possible indicator species.

Indicator species are a general category of species that could be used as monitors of the state of communities. Several kinds of indicator species have been used in defining communities for conservation purposes. *Umbrella species* are indicator species with large area requirements; these species can be used in conservation to bring many other species under protection. The grizzly bear is a good example; if we use it as

BOX 20.1

CRITERIA FOR INDICATOR SPECIES

Indicator species have a long history of use in plant ecology (Muller-Dombois and Ellenberg 1974), and plant ecologists have developed methods for defining indicator species in studies of communities.

Choosing a species as an indicator species requires that we go through a rigorous process to select the indicators objectively. We assume that we have first clearly identified the objective of using indicator species (for example, as a signpost of a community type or as an indicator of the health of a community). The following criteria should be assessed:

1. The species should be taxonomically well known and stable, so that individuals can be readily recognized.

2. The biology and natural history of the species should be well understood so that we know its tolerances and requirements. Indicator species should be permanent residents of the community.

3. The species should be able to be surveyed easily, so that even inexperienced or local people can be involved in surveys.

4. The species should be specialized to one community or habitat, or to one set of conditions it is supposed to indicate. Specialist species are always preferred to generalists because they are more precise indicators.

5. The species should be closely associated with a group of other taxa that it serves to indicate.

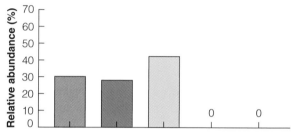

(a) Primary forest

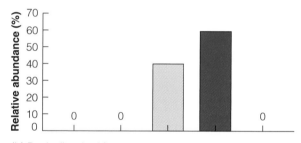

(b) Partly disturbed forest

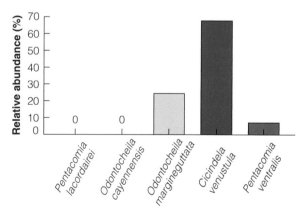

(c) Disturbed forest

FIGURE 20.10

The relative abundance of 5 species of tiger beetles along a gradient of disturbance in evergreen tropical forests in Venezuela. Two species occur only in undisturbed forest communities and two occur only in disturbed forest stands. (From Rodriguez et al. 1998.)

an umbrella species and design conservation plans around it, the presumption is that other plant and smaller animal species will also be protected. *Flagship species* are popular, charismatic species that serve as conservation symbols and rallying points for the protection of areas. The panda is the most well-known example of a flagship species. *Keystone species* are pivotal species in a community that maintain the structure of the community. Keystone species are the antithesis of redundant species, because if they are lost a large part of the existing community is lost with them. They can be very useful indicator species, if they are recognized easily. (Keystone species will be discussed more in Chapter 24.) In general these categories of indicators (especially umbrella species) are often used by land managers who must make decisions about land use in the absence of detailed information.

An attempt is being made within the conservation movement to select indicator species that are useful on a global scale for monitoring habitat changes. One example of this approach involves tiger beetles (*Coleoptera: Cicindelidae*), widespread insects with desirable attributes as indicators (Rodríguez et al. 1998). Each species of tiger beetle occurs in only one or a few habitat types, and these species are well known taxonomically. Because they are colorful it is relatively easy to train people in the field to recognize the different species of tiger beetles, and their natural history as predators of other insects is quite well known. Rodríguez et al. (1998) used tiger beetles to monitor habitat degradation in the tropical forests of Venezuela. They found that five species of tiger beetles living on the forest floor serve as good indicators of forest disturbance (Figure 20.10 on page 399).

Summary

A community is an assemblage of populations living in a prescribed area. The key question about communities is how much integration exists among the species in a community. Two broad models have been suggested to answer this question. The first model, the *rivet model*, suggests that communities are tightly integrated groups of interacting species that act almost like a superorganism. The web of life is tightly woven, so changing a single species has repercussions for the whole community. The second model, the *redundancy model*, suggests that communities are loose associations of species that happen to have similar ecological requirements. The web of life is loosely woven in this model, and many species can disappear with only minimal effect on the rest of the community. We can distinguish these models by determining the number of obligate species associations in communities. Most ecologists now subscribe to the redundancy model, also called the *individualistic continuum model* of the community. Species are distributed individualistically, that there are no sharp boundaries between communities, which vary continuously across landscapes. We classify and name communities for our own convenience, not because they are clear and distinct entities in nature.

Four community parameters can be estimated for a community: biodiversity, growth form and structure, relative abundance, and trophic structure. Community ecologists ana-

lyze the dynamics of the community, how it is organized, the role of dominant and keystone species, and how it changes in time. We need to know how much redundancy exists in a community if we are to predict the effects of species removals. By using our understanding of population ecology we can begin to build an understanding of how communities work.

Communities can be viewed as either *open* or *closed* systems and as either *equilibrium* or *nonequilibrium* systems. Closed, equilibrium models view community stability as the norm, whereas open, nonequilibrium models emphasize change as the norm. Paleoecological data show conclusively that communities change dramatically over time, and that community stability is never reached even on short time scales.

To avoid the complexity of dealing with all the species in a community, ecologists often use one of several types of indicator species to provide a rapid way of identifying a community or assessing its health. If chosen carefully, indicator species can be useful in classifying areas for conservation management or preservation.

Most of our understanding of plant and animal communities comes from studies in the temperate zones, and this leads to an inevitable bias in our conceptions. More research on tropical communities may change our vision of the nature of communities.

Key Concepts

1. A community is an assemblage of populations living in a prescribed area.

2. The key question is: How strong are the connections between the species in a community? The rivet model suggests a tight integration of interdependent species, whereas the redundancy model suggests a loose confederation of independent species.

3. Most community ecologists now support the redundancy model. Communities lack sharp boundaries in space and change over ecological time.

4. Communities change gradually with environmental conditions, and species are distributed as a continuum along these environmental gradients.

5. Indicator species can be used as a shortcut for identifying communities or recognizing community degradation.

6. Most community studies have been done in temperate areas, so we have a biased sample. We know much less about complex tropical communities

Selected References

Austin, M. P. and T. M. Smith. 1989. A new model for the continuum concept. *Vegetatio* 83:35–47.

Lenton, T. M. 1998. Gaia and natural selection. *Nature* 394:439–447.

McIntosh, R. P. 1998. The myth of community as organism. *Perspectives in Biology and Medicine* 41:426–438.

Niemi, G. J., J. M. Hanowski, A. R. Lima, T. Nicholls, and N. Weiland. 1997. A critical analysis on the use of indicator species in management. *Journal of Wildlife Management* 61:1240–1251.

Pielou, E. C. 1984. *The Interpretation of Ecological Data: A Primer on Classification and Ordination.* Wiley, New York. 263pp.

Roughgarden, J. and J. Diamond. 1986. Overview: The role of species interactions in community ecology. In *Community Ecology*, ed. J. Diamond and T. J. Case, pp. 333–343. Harper & Row, New York.

Simberloff, D. 1998. Flagships, umbrellas, and keystones: Is single-species management passé in the landscape era? *Biological Conservation* 83:247–257.

Wiens, J. A. 1984. On understanding a non-equilibrium world: Myth and reality in community patterns and processes. In *Ecological Communities*, ed. D. R. Strong Jr., D. Simberloff, L. G. Abele and A. B. Thistle, pp. 439–457. Princeton University Press, Princeton, New Jersey.

Questions and Problems

20.1 Caswell (1978, p. 133) gives the following example to illustrate how long it takes a system to reach equilibrium:

> A system consists of 100 light bulbs, each of which can be either on or off. For each bulb in each second the transition (on-off) occurs with probability 0.5. The reverse transition occurs with probability 0.5 only if at least one bulb in the on state is connected to the bulb in question. Otherwise no bulb will turn on. The equilibrium state of the system is all lights off. If the system begins with all lights on, guess how long it will take to reach equilibrium if:

> (a) there are no connections among the bulbs;

> (b) all the bulbs are connected to each other.

Compare your guesses with the values in Caswell, and discuss the possible relevance for community ecology.

20.2 Discuss which of the models of the nature of the community would be more appropriate if (a) all the controls of community structure came from the physical environment, or (b) all the controls of community structure came from biological interactions among the species in the community.

20.3 Discuss the significance of ecotypes to the problem of community description and classification, and suggest possible ways of alleviating any difficulties that you uncover in your analysis.

20.4 If communities are loose confederations of species distributed in a continuum across resource gradients, should we attempt to classify and name them? Discuss the theoretical and the practical aspects of classifying and naming communities from the perspective of conserving protected areas.

20.5 Morey (1936, p. 54) studied two stands of virgin forest in northwestern Pennsylvania and found 93 species of plants common to both stands, 91 species confined to Heart's Content stand, and 57 species confined to Cook Forest stand. Calculate the index of similarity from Equation (20.1) for these two virgin stands.

20.6   The controversy over the nature of the community was for the most part a controversy among plant ecologists. Discuss why this might be.

20.7   Discuss the relative advantages and disadvantages of field experiments and natural experiments for analyzing community dynamics. Natural experiments differ from field experiments in that the ecologist does not do a manipulation but instead selects sites where a manipulation is occurring or has already occurred (Diamond 1986). For example, natural experiments could examine the recovery of the vegetation after a volcanic explosion or a fire. Compare your analysis of field experiments and natural experiments with that of Diamond (1986).

20.8   An underlying assumption of many discussions of the response of plants to gradients (see, for example, Figures 20.3 and 20.4) is that the response curves are symmetric, bell-shaped curves. What mechanisms might make this assumption questionable? Austin (1990) discusses this question.

20.9   What factors might make it difficult to find good indicator species for bird communities?

20.10   Community ecologists often talk about the "health" of a community. Discuss the analogy between human health and community health. How might you measure the health of a biological community? Cole et al. (1998) and Ferretti (1997) discuss the application of health indicators to communities.

Overview Question

What is the relationship between the concept of the ecological niche and the concept of the community? Does the controversy over the nature of the community have any important consequences for niche theory? Explain.

CHAPTER 21

Community Change

ONE OF THE MOST IMPORTANT FEATURES of communities is change, and in this chapter we focus on the factors that cause communities to change in ecological time. If we sat in a prairie for ten years, or in a forest for 20 years, we would see the surrounding community change, whether slowly or dramatically. The consequences for land and water management and conservation can be severe. If we designate a prairie grassland as a protected area, for example, and keep grazing animals and fire off the site, the grassland will turn into shrubland and finally forest, and we will have lost the community you set out to protect.

Community change has important repercussions for conservation, and for all forms of water and land management. Two main types of changes occur in communities: directional changes and cyclic changes. Throughout this chapter we focus on two questions: (1) What factors cause community changes? and (2) How predictable are community changes?

When stripped of its vegetation by fire, flood, glaciation, or volcanic activity, the resulting area of bare ground does not long remain devoid of plants and animals. The area is rapidly colonized by a variety of species that subsequently modify one or more environmental factors. This modification of the environment may in turn allow additional species to become established. This development of the community by the action of vegetation on the environment leading to the establishment of new species is termed *succession*. Succession is the universal process of directional change in vegetation during ecological time. It can be recognized by a progressive change in the species composition of the community.

Most observed successions are called *secondary succession*, the recovery of disturbed sites. A few successions are called *primary succession* because they occur on a new sterile area, such as that uncovered by a retreating glacier or created by an erupting volcano. It is this latter situation that we explore next.

Primary Succession on Mount St. Helens

Mount St. Helens in southwest Washington state erupted catastrophically on May 18, 1980. About 400 m was blown off the cone of this volcano, and the blast from the eruption devastated a wide arc extending some 18 km north of the crater (Franklin et al. 1985). Three main areas were affected by the eruption: the blast zone, in which trees and vegetation were blown down but not eliminated; the pyroclastic flows (a hot mixture of volcanic gas, pumice, and ash) to the north of the crater; and the extensive mudflows and ash deposits away from the crater toward the south, east, and west. In addition, the eruption spewed tephra (ash) over thousands of square kilometers. The eruption produced a landscape with low nutrient availability, intense drought, and frequent surface movements and erosion, a great variety of conditions for vegetative recolonization (del Moral and Wood 1988).

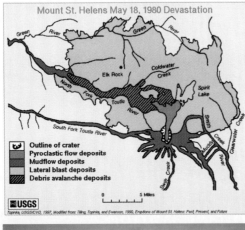

FIGURE 21.1

The eruption of Mount St. Helens, Washington. Photo of Mount St. Helens before the eruption. (Photo of Mount St. Helens before the eruption courtesy of Roger del Moral. Photo of the eruption courtesy of U.S. Geological Survey.)

Primary succession following volcanic eruptions has been studied less than other forms of succession, and Mount St. Helens provided a good opportunity to study the mechanisms determining the rate of primary succession. Permanent plots have been established at several sites above treeline around the crater, and early succession has been described by del Moral and Wood (1993), del Moral et al. (1995), and del Moral (1999).

Colonization of habitats above treeline on Mount St. Helens has been slow. Figure 21.2 shows the vegetation on a small mudflow southwest of the crater at Butte Camp, from 1981 to 1998. There were no surviving plants in this site in 1980, but 19 species had colonized the area by 1982 (Figure 21.3). Only a few additional species invaded after 1982, and plant cover has increased very slowly so that none of the species are common 19 years later.

Early primary succession on volcanic substrates rarely produces plant densities sufficient to inhibit the colonization of other species. Neither space nor light are limiting resources for plants in this environment. So-called nurse plants facilitate the establishment of other species. Lupines (*Lupinus lepidus*) have heavy seeds and are poorly dispersed, but they have become locally common on mudflows and pyroclastic surfaces (del Moral 1992). Before lupines get very common, other wind-dispersed plants such as *Aster ledophyllus* and *Epilobium angustifolium* become established in lupine clumps and survive better in the shelter of these nurse plants. Lupines die after four or five years, and because they fix nitrogen they contribute on a local scale to increased soil-nitrogen levels.

Chance events strongly affected primary succession on Mount St. Helens. Biological mechanisms are initially very weak in the severe environments produced by volcanic flows. The ability to become established in these severe environments is directly related to seed size (Wood and del Moral 1987), but dispersal ability is inversely related to seed size. Consequently, subalpine areas on Mount St. Helens received many

(a)   1981

(b)   1985

(c)   1989

(d)   1992

(e)   1998

FIGURE 21.2
Primary succession on mudflow deposits on the southwest side of Mount St. Helens following the eruption of May 1980. In 1981 plants had already colonized this devastated area, and plant cover slowly increased as the site underwent primary succession. Figure 21.3 shows data from these plots. (Photos courtesy of R. del Moral, 1999.)

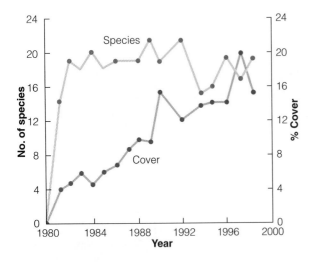

FIGURE 21.3
Number of species occurring in 1-m² quadrats in the Butte Camp area on the south cone of Mount St. Helens. The number of species reached a plateau around 20 relatively quickly, but the total plant cover on this area has increased only slowly because of the harsh conditions on these volcanic deposits. Photos of these plots are shown in Figure 21.2. (Data courtesy of R. del Moral, 1999.)

wind-blown seeds, but almost none of these small seeds germinated and achieved colonization under the stressful conditions. When plants with large seeds colonize by chance, they become a focus of further community development. If by chance a single individual plant survives in the devastated landscape, it quickly becomes a locus for seed dispersal to adjacent areas, so positive feedback occurs during the early years of succession. Primary succession on Mount St. Helens has been very slow because of erosion, low-nutrient soils, chronic drought stress, and limited dispersal of larger seeds to areas distant from undisturbed vegetation.

Mount St. Helens provides a graphic example of plant succession after extreme disturbance. Measuring the speed of change on the mudflow areas allows us to estimate that it will require more than 100 years for the landscape on Mount St. Helens to return to a stable plant community. Understanding succession also requires understanding the mechanisms that drive changes in communities, and one focus has been on the effects that early successional species have on later successional species. Early species can help, hinder, or not affect the establishment of later species. Competition between individual plants for resources such as

water, light, or nitrogen may drive succession. On Mount St. Helens we can see these processes occurring slowly, which enables us to understand them more easily. Understanding how naturally disturbed landscapes renew themselves can help us to understand how human-disturbed landscapes might respond. We now turn to these broader issues, and to the theory of succession and the mechanisms involved.

Concepts of Succession

The concept of succession was largely developed by the botanists J. E. B. Warming (1896) in Denmark and Henry C. Cowles (1901) in the United States. Warming's book greatly influenced Cowles, who studied the stages of sand-dune development at the southern edge of Lake Michigan. Henry Cowles was one of the most

Henry C. Cowles *(1869–1939) Professor of Plant Ecology, University of Chicago*

influential plant ecologists in the United States in the early years of the twentieth century, and the students he taught at the University of Chicago became a who's who of American plant ecology.

Successional studies pioneered by Warming and Cowles have led to four major hypotheses of succession. The first is the classical theory of succession, which was called *relay floristics* by Egler (1954) because it postulates an orderly hierarchical system of change in the community (Figure 21.4a). The classical ideas of succession were elaborated in great detail by F. E. Clements (1916, 1936), who devel-

	Clements (1916)	Egler (1954)	Connell and Slatyer (1977)	Lawton (1987)
(a₁) A →⁺ B → C → D (D replaces itself)	Primary succession	Relay floristics	Facilitation model	
(a₂) A_BCD →⁺ B_CD → C_D → D	Secondary succession			
(b) A_BCD →⁻ B_CD; D ←⁻ C_D		Initial floristic composition	Inhibition model	
(c) A_BCD →⁰ B_CD; D ←⁰ C_D	Subclimax		Tolerance model	
(d) A_BCD ⇄⁰ B_CDA; D_ABC ⇄⁰ C_DAB				Random colonization model

FIGURE 21.4
Four models of succession (a–d) proposed by different authors. The letters A–D represent hypothetical vegetation types or dominant species; subscript letters indicate species that are present as minor components or as propagules. Light arrows represent species or vegetation sequences in time; bold arrows represent alternative starting points for succession after disturbance. Curved arrows indicate that the species replaces itself. + = facilitation; − = inhibition, 0 = no effect. The relay floristics model has two patterns (a₁, a₂), depending on whether primary or secondary succession is occurring. (Modified after Noble 1981.)

oped a complete theory of plant succession and community development called the *monoclimax hypothesis*. The biotic community, according to Clements, is a highly integrated superorganism that develops through a process of succession to a single end point in any given area—the *climatic climax*. The development of the community is gradual and progressive, from simple pioneer communities to the ultimate or climax stage. This succession is due to biotic reactions only; the plants and animals of the pioneer stages alter the environment such that it favors a new set of species, and this cycle recurs until the climax is reached. Development through succession in a community is therefore analogous to development in an individual organism, according to Clements's view (Clements 1936, Phillips 1934–1935). Thus reverse succession (retrogression) is not possible unless some disturbance such as

Frederic E. Clements *(1874–1945) Professor of Botany, University of Nebraska*

fire, grazing, or erosion intervenes. Secondary succession differs from primary succession in having a seed bank from plants that occur later in succession, so that late succession species are already present in the early stages of secondary succession (Figure 21.4 a_1, and 21.4a_2).

The key assumption of the classical theory of succession is that species replace one another because at each stage they modify the environment to make it less suitable for them and more suitable for others. Thus species replacement is orderly and predictable and provides directionality for succession. These characteristics led Connell and Slatyer (1977) to call this the *facilitation* model of succession. The early species in succession facilitate the arrival of the later species.

The climax community in any region is determined by climate, in Clements' view. Other communities may result from particular soil types, fire, or grazing, but these *subclimaxes* are understandable only with reference to the end point of the climatic climax. Therefore, the natural classification of communities must be based on the climatic climax, which represents the state of equilibrium for the area.

A second major hypothesis of succession was proposed by Egler (1954), who called it *initial floristic composition*. In this view, succession is very heterogeneous because the development at any one site depends on who gets there first. Species replacement is not necessarily orderly because each species excludes or suppresses any new colonists (Figure 21.4b, c). Thus succession becomes more individualistic and less predictable because communities are not always converging toward the climatic climax. Egler's hypothesis of initial floristics actually contained two ideas. Part of his hypothesis was called the *inhibition* model of succession by Connell and Slatyer (1977): The species present early in succession inhibit the establishment of the later species (see Figure 21.4b). No species in this model is competitively superior to another; whoever colonizes the site first holds it against all comers until it dies. Succession in this model proceeds from short-lived species to long-lived species and is not an orderly replacement because "who gets there first" is a matter of chance. Wilson et al. (1992) called this model the *preemptive initial floristics model* to emphasize that the first species at a site preempt the course of succession.

Egler's initial floristics model can also describe the third major model of succession proposed by Connell and Slatyer (1977), who called it the *tolerance* model. In the tolerance model, the presence of early successional species is not essential—any species can start the succession (see Figure 21.4 c). Some species are competitively superior, however, and they eventually come to predominate in the climax community. Species are replaced by other species that are more tolerant of limiting resources. Succession proceeds either by the invasion of later species or by a thinning out of the initial colonists, depending on the starting conditions. The tolerance model includes Egler's emphasis on the initial floristics as a major influence on how succession proceeds.

A fourth model of succession was proposed by Lawton (1987) to provide a null model with no ecological interactions. The *random colonization* model suggests that succession involves only the chance survival of different species and the random colonization by new species. There is no facilitation and no interspecific competition (Figure 21.4d) and succession can move in any direction.

The first three hypotheses of succession agree in predicting that many of the pioneer species in a succession will appear first because these species have evolved colonizing characteristics, such as rapid growth, abundant seed production, and high dispersal powers (Table 21.1). The critical feature of the life-history traits listed in Table 21.1 is that there is a trade-off, or inverse correlation, between traits that promote success in early succession and traits that are advantageous in late succession (Huston and Smith 1987).

The critical distinction among the four hypotheses of succession is in the mechanisms that determine subsequent establishment. In the classical facilitation model, species replacement is *facilitated* by the previous stages. In the inhibition model, species replacement is *inhibited* by the present residents until they are damaged or killed. In the tolerance and random colonization models, species replacement is *not affected* by the present residents.

The utility of these four models of succession is that they immediately suggest experimental manipulations to test them. Removing or excluding early colonizers, transplanting seeds or seedlings of late succession species into earlier stages, and other experiments can shed light on the mechanisms involved in succession.

To explain a successional sequence we must add to these idealized models of succession some additional

TABLE 21.1 Physiological and life history characteristics of early- and late-successional plants.

Characteristic	Early succession	Late succession
Photosynthesis		
Light-saturation intensity	high	low
Light-compensation point	high	low
Efficiency at low light	low	high
Photosynthetic rate	high	low
Respiration rate	high	low
Water-use efficiency		
Transpiration rate	high	low
Mesophyll resistance	low	high
Seeds		
Number	many	few
Size	small	large
Dispersal distance	large	small
Dispersal mechanism	wind, birds, bats	gravity, mammals
Viability	long	short
Induced dormancy	common	uncommon?
Resource-acquisition rate	high	low?
Recovery from nutrient stress	fast	slow
Root-to-shoot ratio	low	high
Mature size	small	large
Structural strength	low	high
Growth rate	rapid	slow
Maximum life span	short	long

Source: From Huston and Smith (1987).

information on seed availability, insect and mammal herbivory, mycorrhizal fungi, and plant pathogens (Walker and Chapin 1987). Figure 21.5 summarizes the relative importance of a variety of factors in succession. The primary processes underlying successional changes could be competition or mutualistic interactions between plants, but these plant-plant interactions are affected by animal grazing and diseases, and by seed dispersal and storage. The resulting successional sequences are thus complex and do not proceed in a single direction to a fixed end point (Burrows 1990).

A Simple Mathematical Model of Succession

We can construct a simple mathematical model of succession by assuming that succession is a replacement process (Horn 1981). For each plant, we ask a simple question: What is the probability that this plant will be replaced in a given time by another plant of the same species or of another species? A matrix of replacement probabilities can be constructed in forests by counting the number of saplings of various species growing under the canopy of the mature trees. For example, of a total of 837 saplings growing beneath large gray birch trees, Horn (1975a) found that (to mention but a few species) zero were gray birch saplings, 142 were red maple saplings, and 25 were beech saplings. Thus, the probability that a gray birch will be replaced by another gray birch (self-replacement) is 0/837 = 0, by a red maple is 142/837 = 0.17, and by a beech is 25/837 = 0.03. These probabilities are entered as percentages in Table 21.2.

We can use these replacement probabilities to calculate what will happen to this forest community

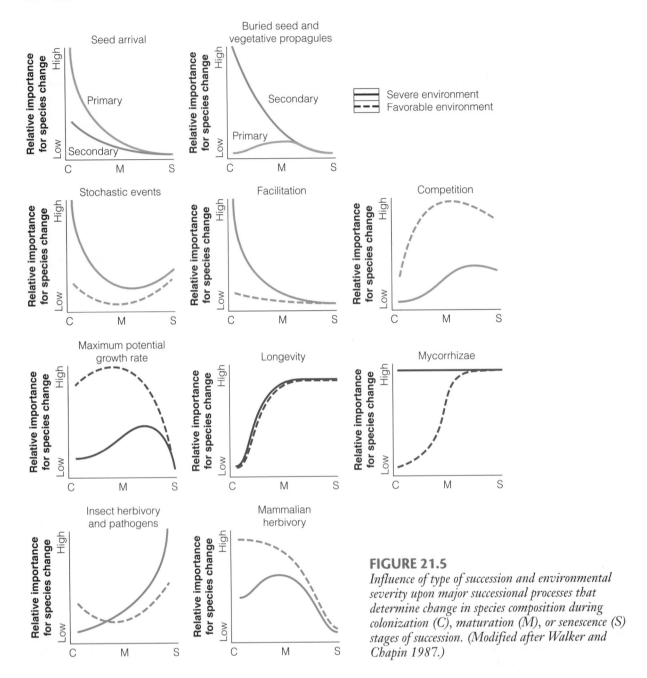

FIGURE 21.5
Influence of type of succession and environmental severity upon major successional processes that determine change in species composition during colonization (C), maturation (M), or senescence (S) stages of succession. (Modified after Walker and Chapin 1987.)

in the future. For example, working down the columns of the table, we can calculate:

Proportion of the next generation that will be white oak = 0.02 gray birch + 0.03 sassafras + 0.01 blackgum + 0.10 white oak + 0.06 red oak + 0.03 hickory + 0.02 red maple + 0.01 beech

where all the species on the right side of this equation refer to the current abundance of that species. We multiply the current abundances of each species by the replacement probabilities in Table 21.2 and add them together to get the predicted abundance of each canopy species in the next generation. We can cycle through these calculations, which are more tedious than difficult, and predict the abundances of all

TABLE 21.2 Transition matrix for saplings growing beneath various species of trees at Institute Woods in Princeton, New Jersey.

Values are expressed as percentages of the total number of saplings (final column) found growing beneath the canopy species listed. Entries are interpreted as the percentages of canopy species that will be replaced one generation hence by the sapling species listed across the top; the percentages of "self-replacements" are in boldface.

Canopy species	Sapling species											
	BTA	GB	SF	BG	SG	WO	RO	HI	TT	RM	BE	Total No.
Big-toothed aspen (BTA)	**3**	5	9	6	6	—	2	4	2	60	3	104
Gray birch (GB)	—	—	47	12	8	2	8	—	3	17	3	837
Sassafras (SF)	3	1	**10**	3	6	3	10	12	—	37	15	68
Blackgum (BG)	1	1	3	**20**	9	1	7	6	10	25	17	80
Sweetgum (SG)	—	—	16	—	**31**	—	7	7	5	27	7	662
White oak (WO)	—	—	6	7	4	**10**	7	3	14	32	17	71
Red oak (RO)	—	—	2	11	7	6	**8**	8	8	33	17	266
Hickories (HI)	—	—	1	3	1	3	13	**4**	9	49	17	223
Tulip tree (TT)	—	—	2	4	4	—	11	7	**9**	29	34	81
Red maple (RM)	—	—	13	10	9	2	8	19	3	**13**	23	489
Beech (BE)	—	—	—	2	1	1	1	1	8	6	**80**	405

Source: After Horn (1975b).

species any number of generations into the future (Table 21.3). After several generations, the abundances of all species settle down to a stationary distribution that will not change over time. In this example, the stationary distribution predicted will contain 50% beech, 16% red maple, 7% tulip tree, and so on.

To use this simple model of succession, we must make a number of assumptions. The most critical assumption is that the table of replacement probabilities does not change over time. Under this assumption, the community will approach a steady state that is independent of both the community's initial composition and of the type of disturbance that starts the succession going. In this form, the model is a statement of any type of succession (facilitation, inhibition, or tolerance) that predicts a regular, repeatable change culminating in a stable climax.

We must also assume that we can calculate replacement probabilities in a realistic way. For forests, this involves assuming that abundance in the sapling stage is a sufficient predictor of the chances that trees will reach the canopy. There is also a problem of overlapping generations in different species. In this example, some trees live longer than others, and one must correct these pre-

dictions for variable life spans (see Table 21.3). This correction is tedious but not difficult (Horn 1975b).

Can we alter the replacement model to describe the inhibition model of succession that does not culminate in a stable, fixed climax? If the table of replacement probabilities depends on the present composition of the forest, predictions from the model change dramatically. Assume, for example, that the recruitment of young plants depends on the density of trees of their own species. The transition probabilities for any one tree are not constant under this assumption but change depending on who neighbors are. Succession in this model does not converge to one point; instead, alternative communities could be produced, depending on the accidents of history, as suggested by the inhibition model.

The replacement model of succession is useful because it focuses our attention on the local regeneration of species, and how species replace themselves and other species as disturbances open up new areas for succession. It is not, however, a mechanistic model of succession, and to understand why succession is occurring we need to focus on the mechanisms controlling succession. One possible mechanism driving succession is competition for limiting resources (Tilman 1985).

TABLE 21.3 Theoretical approach of the Institute Woods in Princeton, New Jersey, to a stationary distribution ("climax forest").

The starting point is an observed 25-year-old forest stand dominated by gray birch. The theoretical predictions are obtained by multiplying the starting tree composition (generation 0) by the transition matrix of probabilities in Table 21.2.

	BTA	GB	SF	BG	SG	WO	RO	HI	TT	RM	BE
					Species[a] (%)						
Theoretical generation											
0	—	49	2	7	18	—	3	—	—	20	1
1	—	—	29	10	12	2	8	6	4	19	10
2	1	1	8	6	9	2	8	10	5	27	23
3	—	—	7	6	8	2	7	9	6	22	33
4	—	—	6	6	7	2	6	8	6	20	39
5	—	—	5	5	6	2	6	7	7	19	43
...	...	...	...	...	...	...	...	...	...	...	...
...	...	...	...	...	...	...	...	...	...	...	...
...	...	...	...	...	...	...	...	...	...	...	...
Stationary distribution (%)	—	—	4	5	5	2	5	6	7	16	50
Longevity (yr)	80	50	100	150	200	300	200	250	200	150	300
Age-corrected stationary distribution (%)	—	—	2	3	4	2	4	6	6	10	63

[a] Abbreviations refer to the names of tree species listed in Table 21.2.
Source: After Horn (1975b).

One way to model succession mechanistically is to use an individual-based plant model that explicitly incorporates light as the limiting resource (Huston and Smith 1987). Each individual plant is given species-specific traits of maximum size and age, maximum growth rate, and tolerance to shading. Most of these models have been used for trees to model forest succession, but they could be applied to other kinds of plants as well. The key variable in these models is light availability, and each individual plant is analyzed to see how much shading is produced by its neighbors. If light is limited, growth rates and survival rates are reduced accordingly. This type of simple mechanistic model can produce successional sequences of tree species that resemble natural succession (Shugart 1984). Figure 21.6 illustrates three scenarios with species of different life-history traits. In every case the species that is most shade tolerant and grows to the largest size wins out in succession. The seral stages vary greatly depending on which trees are present.

The additional effects of competition for soil nitrogen can be added to these simple models, such that both light and nitrogen become limiting resources (Tilman 1985, Shugart 1984). Mechanisms of nutrient limitation make these models more realistic, but also more complex. The most successful models of succession are for forests. Because of their economic importance a wealth of details are known about the life-history traits of individual tree species.

How well do natural communities fit the four hypotheses of succession? Does succession in a region converge to a single end point, or are there multiple stable states? For Mount St. Helens we have seen that the facilitation model of succession is a good description of the early stages of succession. As vegetation cover on the mudflows becomes more complete, inhibition may begin to operate. In the next section we look at some additional examples of succession to evaluate these models.

Case Studies of Succession

Although numerous studies of succession have been conducted, in few of them can succession be related to a time scale. Here we examine three investigations of succession in which the time scale is known: glacial

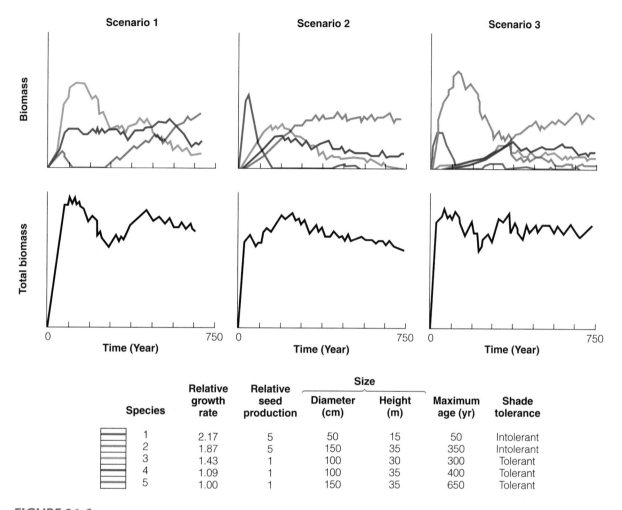

FIGURE 21.6

Species biomass dynamics and community biomass for hypothetical successional sequences with three, four, and five species. In scenario 1, all three species have late-successional characteristics but differ sufficiently in relative competitive abilities to produce a "typical" successional replacement. In scenario 2, an early-successional species is added to the three species in scenario 1. In scenario 3, a "super species" (species 2) with a high growth rate and large size is added to the four species in scenario 2. (From Huston and Smith 1987.)

moraines in Alaska, sand dunes near Lake Michigan, and abandoned farmland in North Carolina.

Glacial Moraine Succession in Southeastern Alaska

In general, glaciers in the Northern Hemisphere have been retreating during the past 200 years. As the glaciers retreat, they leave moraines, whose age can be determined by the age of the new trees growing on

them or, in the past 80 years, by direct observation. The most intensive work on moraine succession has been done at Glacier Bay in southeastern Alaska. Since about 1750 the glaciers there have retreated about 98 kilometers, an extraordinary rate of retreat (Figure 21.7).

The pattern of primary succession in this area proceeds as follows (Cooper 1939, Lawrence 1958). The exposed glacial till is colonized first by mosses, fireweed, *Dryas*, willows, and cottonwood. The wil-

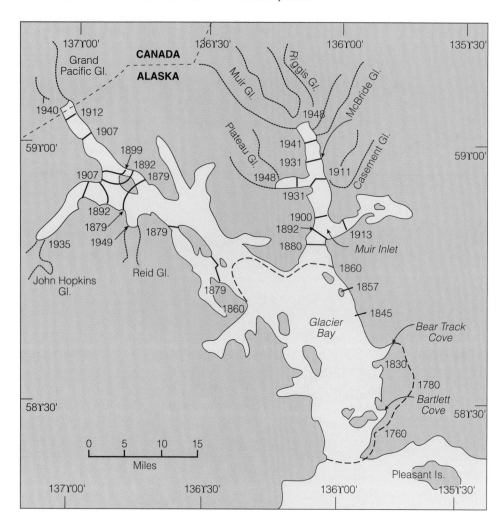

FIGURE 21.7
Map of Glacier Bay fjord complex of southeastern Alaska, showing the rate of ice recession since 1760. As the ice retreats, it leaves moraines along the edge of the bay on which primary succession occurs. The dashed lines show the approximate edge of the ice in 1760 and 1860 as gleaned from historical descriptions. (After Crocker and Major 1955.)

lows begin as prostrate plants but later grow into erect shrubs. Very quickly the area is invaded by alder (*Alnus*), which eventually (in about 50 years) forms dense pure thickets up to 9 meters tall. These alder stands are invaded by Sitka spruce, which, after another 120 years, forms a dense forest. Western hemlock and mountain hemlock invade the spruce stands, and after another 80 years the situation has stabilized with a climax spruce-hemlock forest. This forest, however, persists on well-drained slopes only.

Sitka spruce needles (Picea sitchensis) *(©Adam Hart-Davis/Photo Researchers, Inc.)*

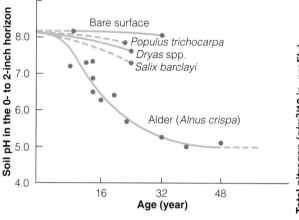

FIGURE 21.8
Soil pH change at Glacier Bay, Alaska, under different types of pioneer vegetation. The soil becomes acid very rapidly under alder. (After Crocker and Major 1955.)

Dryas integrifolia, *pioneering nitrogen fixer*

In areas of poor drainage, the forest floor of this spruce-hemlock forest is invaded by *Sphagnum* mosses, which hold large amounts of water and acidify the soil greatly. With the spread of conditions associated with *Sphagnum*, the trees die out because the soil is waterlogged and too oxygen-deficient for tree roots, and the area becomes a *Sphagnum bog*, or *muskeg*. The climax vegetation, then, seems to be

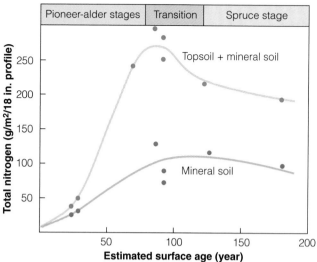

FIGURE 21.9
Total nitrogen content of soils recently uncovered by glacial retreat at Glacier Bay, Alaska. Plant succession is shown along the top. (After Crocker and Major 1955.)

muskeg on the poorly drained areas and spruce-hemlock forest on the well-drained areas.

The bare soil exposed as the glacier retreats is quite basic, with a pH of 8.0 to 8.4 because of the carbonates contained in the parent rocks. Soil pH falls rapidly with the advent of vegetation, and the rate of change depends on the vegetation type (Figure 21.8). Almost no change in the pH occurs due to leaching in bare soil. The most striking change is caused by alder, which reduces the pH from 8.0 to 5.0 in 30–50 years. The leaves of alder are slightly acid, and as they decompose they become more acid. As the spruce begins to take over from the alder, soil pH stabilizes at about 5.0, and it does not change in the next 150 years.

The organic carbon and total nitrogen concentrations in the soil also show marked changes with time. Figure 21.9 shows the changes in nitrogen levels. One of the characteristic features of the bare soil is its low nitrogen content. Almost all the pioneer species begin succession with very poor growth and yellow leaves due to inadequate nitrogen supply. The exceptions to this are *Dryas* and, particularly, alder; these species can fix atmospheric nitrogen (Lawrence et al. 1967). The rapid increase in soil nitrogen in the alder stage is caused by the presence of nodules on the alder roots that contain microorganisms that actively fix nitrogen from the air. Spruce trees have no such adaptations; consequently, the soil nitrogen level falls when alders are eliminated. The spruce for-

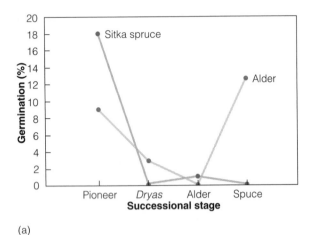

(a)

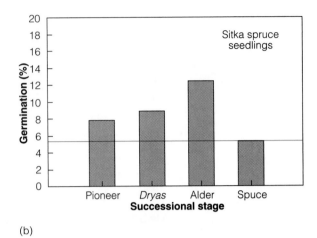

(b)

FIGURE 21.10
Primary succession at Glacier Bay, Alaska. (a) Germination of alder and Sitka spruce seeds sewn on undisturbed soil in the four stages of succession. Germination of spruce and alder was inhibited at all stages of succession (except for alder in the spruce stage), compared with the pioneer stage with less than 5% vegetative cover. (b) Biomass of Sitka spruce seedlings transplanted for three years in the four stages of succession. The red line marks the expected biomass if there is no effect of the pioneer species. Both Dryas *and alder facilitate the growth of spruce seedlings. (After Chapin et al. 1994.)*

est develops by using the capital of nitrogen accumulated by the alder.

One way to test experimentally for the effects of species on each other is to seed or plant the later successional species into the earlier stages of succession. Chapin et al. (1994) introduced seeds of alder and sitka spruce into all the early successional stages at Glacier Bay, and planted sitka spruce seedlings as well. Inhibition of germination and of seedling survival was the dominant process in the early successional stages (Fig-

Alder leaves, *pioneering tree that fixes nitrogen (©John Frett/University of Delaware)*

ure 21.10), but once established spruce seedlings were facilitated during early succession and grew better than they did in the spruce stage of the succession.

The important point here is the reciprocal interrelations of the vegetation and the soil. The pioneer plants alter soil properties, which in turn permit new species to grow, and these species in turn alter the environment in different ways, bringing about succession. Succession on the moraines at Glacier Bay contains elements of both the classical facilitation theory (see Figure 21.4a$_1$) and the inhibition model (see Figure 21.4b). Life-history traits affect succession because all the early pioneers have very light, wind-dispersed seeds. *Dryas* seeds weigh about 10 milligrams, spruce seeds about 270 milligrams. No single model of succession accounts for primary succession at Glacier Bay, and both facilitation and inhibition are involved.

Lake Michigan Sand-Dune Succession

Henry Cowles (1899) worked on the sand-dune vegetation of Lake Michigan and made a classic contribution to our understanding of plant succession. The sand dunes around the southern edge of Lake Michigan

E S S A Y   2 1 . 1

WHY IS SPHAGNUM MOSS SO COMMON?

More carbon may be locked up in *Sphagnum* bogs than in any other plant genus. Peat is largely poorly decomposed *Sphagnum*, and the surface of many bogs is dominated by living *Sphagnum* moss. Peat bogs are a prominent feature of many north temperate and south temperate areas, and the dominance of *Sphagnum* in these bogs makes them an interesting example of plant succession to a simple climax community. In cool climates peat bogs have been an important source of fuel for humans, and both Ireland and Scotland ran on peat fuel before the era of oil and gas.

Bogs are permanently wet areas that are fed by rainwater, and thus there is a shortage of essential nutrients for plant growth. In particular, all vascular plants growing in bogs must be able to tolerate very low levels of nitrogen and phosphorus. *Sphagnum* bogs have a 10—40 cm surface layer where ground water fluctuates depending on rainfall. Because *Sphagnum* lacks roots and has no internal water conducting tissues, it is very susceptible to desiccation.

Sphagnum is successful in bogs because it creates an environment in which few other plants can live. "Nothing eats *Sphagnum*" is a general rule of bog ecology. Fresh *Sphagnum* has no lignin, which is the main constituent of cell walls in other plants. *Sphagnum* consists mostly of polysaccharides made up of glucose and galacturonic acids, which make it highly acidic. *Sphagnum* is also rich in phenols, defensive chemicals against herbivores. *Sphagnum* decomposes so slowly that anatomical details can be seen even after 70,000 years (Van Breemen 1995). Almost every other plant decomposes more rapidly than *Sphagnum* so that the proportion of bog material that is *Sphagnum* increases with depth in a bog. *Sphagnum* peat is famous for its excellent preservation of human and animal bodies and of other organic artifacts. This preservation is produced by 5-keto-*D*-mannuronic acid, which suppresses microbial activity and aids in a tanning-like process as long as the peat is waterlogged.

Sphagnum is possibly the climax vegetation for many areas of pine and spruce forest on relatively flat areas in the north temperate zone. Stunted trees and shrubs are common on bogs, and poor growth is caused by adverse conditions below ground—low nutrients, high acidity, low oxygen, and excess water. *Sphagnum* also conducts heat very poorly, so that soil temperatures remain low, giving a short growing season to vascular plants. *Sphagnum* has a negative effect on vascular plants but a positive effect on its own growth. It has very low nutrient requirements, and it competes successfully for light, not by growing tall but by attacking its possible competitors at the root level. And because *Sphagnum* bogs contain about 20–30% of the world's carbon locked up in soils, the fate the bog plant community in the future can impact on atmospheric CO_2 levels and ultimately climate change in the next centuries.

have been a model system for examining the theories of succession in a dynamic landscape. Cowles did not have access to radiocarbon dating methods and Olson (1958) reexamined the successional stages in this area in relation to an absolute time scale determined by radiocarbon dates.

During and after the retreat of the glaciers from the Great Lakes area, the resulting fall in lake level left several distinct "raised beaches" and their associated dune systems. These systems, which run roughly parallel to the present shoreline of Lake Michigan, are about 7, 12, and 17 meters above the present lake level (Figure 21.11). Olson dated the older dunes by radiocarbon techniques and the younger areas by tree-ring counts and recorded historical changes since 1893.

The dunes, like glacial moraines, offer a near-ideal system for studying plant succession because many of the complicating variables are absent. The initial substrate for all the area is dune sand, the climate for the whole area is similar, the relief is similar, and the available flora and fauna are the same. So the differences between the dunes should be due only to *time*, *to biological processes of succession*, and to *chance events* associated with dispersal and colonization.

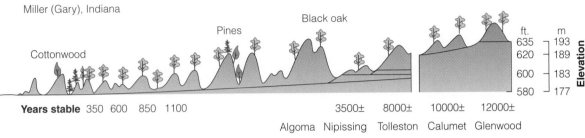

FIGURE 21.11
Diagrammatic profiles across sand dunes at the southern end of Lake Michigan. Successively older dune systems originated along earlier and higher beaches. (After Olson 1958.) One of Henry Cowles's botany field classes from the University of Chicago on the Indiana Dunes about 1902; Lake Michigan is to the right. (Photo courtesy the University of Chicago.)

Two processes produce bare sand surfaces ready for colonization. One is the slow process of a fall in lake level; the other is a rapid process, the *blowout* of an established dune that results from the strong winds that come off the lake. This wind erosion creates a moving dune that is gradually stabilized (only by vegetation) after migrating inland.

The bare sand surface is colonized first by dune-building grasses, of which the most important is marram grass (*Ammophila breviligulata*). Marram grass usually propagates by rhizome migration, only rarely by seed. It spreads very quickly and can stabilize a bare area in six years. After the sand is stabilized, marram grass declines in vigor and dies out. The reason for this is not known, but the result is that this grass is not found in stable dune areas after about 20 years.

Two other grasses are important in dune formation and stabilization: sand reed grass (*Calamovilfa longifolia*) and the little bluestem (*Andropogon scoparius*). The sand cherry (*Prunus pumila*) and willows (*Salix* spp.) also play a role in dune stabilization. The first tree to appear in the young dune is usually the cottonwood (*Populus deltoides*), which may also help to stabilize the sand.

Once the dune is stabilized, it may be invaded very quickly by jack pine and white pine if seed is available; normally pines are found after 50 to 100 years of development. Under normal conditions, black oak replaces the pines, entering the succession at about 100 to 150 years. A whole group of shrubs that require considerable light invade the early pine and oak stands; they are replaced by more shade-tolerant shrubs as the forest of black oak becomes denser.

Cowles believed that this succession to black oak might be part of the succession sequence that would proceed to a white oak–red oak–hickory forest and finally to the "climatic climax," a beech-maple forest. But Olson questioned whether this could ever occur. The oldest dunes Olson studied (12,000 years old) still had black oak associations, and he could see no tendency for any further succession. Moreover, the black oak community was very heterogeneous, and Olson recognized four different types of understory communities that could occur under black oak. Figure 21.12 summarizes the successional patterns on the dunes.

Olson also studied the changes in the soil in relation to this time sequence. The pH of the soil decreased with dune age, from high values of 7.6 at the start of succession to 4.0 after 10,000 years. The initial drop in pH is caused by carbonates being leached from the soil very quickly. Soil nitrogen increases rapidly in the first 1000 years of development, from very low values initially to approximately 0.1%, and then remains unchanged in older dunes. Organic carbon in the soil develops similarly.

Thus, most of the soil improvements of the original barren dune sand occur within about 1000 years of stabilization. As a result of these trends in the soil, Olson pointed out, the nutritional conditions for succession toward beech and maple probably become *less* favorable with time. (These trees require more calcium, near neutral pH, and larger amounts of water.) It seems improbable that this succession will move beyond the black oak stage, contrary to what Cowles had suggested. Beech and maple associations in this area are found only in favorable situations such as moist lowlands, where the soil characteristics differ from those of the dry dunes. The low fertility of the dune soils favors vegetation, such as the black oak, that has limited nutrient and water requirements. But this sort of vegetation is ineffective in returning nutrients to the dune surface in its litter, which continues the cycle of low fertility.

As far as the dunes are concerned, probably the most striking vegetational changes occur in the first 100 years, and the system seems to become stabilized with the black oak association by about 100 years. Olson pointed out that it was a mistake to distort the different dune successions into a single linear sequence leading from pine to black oak to oak-hickory to beech-maple. Successions in the dunes proceed in different directions with different destinations, depending on the various soil, water, and biotic factors involved at a particular site. Instead of convergence to a single climax, Olson suggested, we may get *divergence* of different communities on different sites. Human use of the beach areas along the dunes tends to disturb the sand and slow down the whole successional sequence (Poulson and McClung 1999).

Thus, dune succession begins as the classical model suggests but changes to the inhibition model at the black oak stage and culminates in several different stable communities (Poulson 1999).

Abandoned Farmland in North Carolina

When upland farm fields are abandoned in the Piedmont area of North Carolina, the following sequence of plant species colonizes the area:

Years after last cultivation	Dominant plant	Other common species
0 (autumn)	Crabgrass ⇓	
1	Horseweed ⇓	Ragweed
2	Aster ⇓	Ragweed
3	Broomsedge ⇓	
5-15	Shortleaf pine ⇓	Loblolly pine
50-150	Hardwoods (oaks)	Hickories

This striking sequence initially features rapid replacements of herbaceous species, and Keever (1950) attempted to find out why the initial species die out and why entry into the succession is delayed for the later colonizers.

The sequence of this succession is dictated by the life history of the dominant plants. Horseweed (*Erigeron canadensis*) seeds will mature as early as August and germinate immediately. It overwinters as a drought-resistant rosette plant, and it grows, blooms,

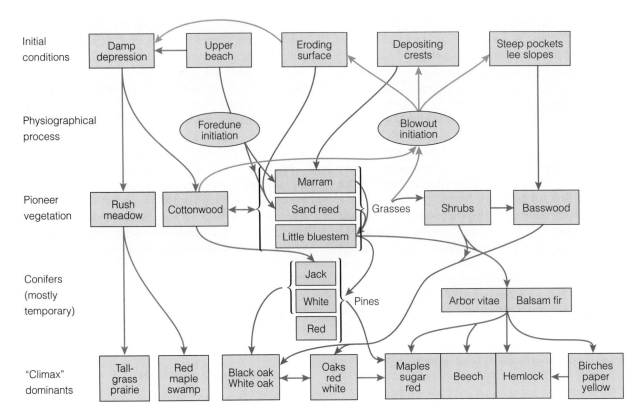

FIGURE 21.12

Alternative dune successions in Lake Michigan sand dunes. Beaches, foredunes, and blowout dunes provide diverse sites that undergo different successions. The center of the diagram gives an oversimplified outline of "normal" succession, from dune builders to jack pine or white pine to black oak—white oak with several understory types. Blue lines indicate the most common successional pathways, red lines the succession on wetter sites. (After Olson 1958.)

and dies the following summer (it is a biennial). The second generation of horseweed plants is stunted and does not grow well in the second year of the succession. Decaying horseweed roots inhibit the growth of horseweed seedlings. The density of horseweed individuals is much greater in second-year fields, but these small individuals do not do well with increased competition from aster. The result is a great reduction of horseweed dominance in second-year fields.

Aster (*Aster pilosus*) seeds mature in the fall, too late to germinate in the year that plowing ceases. Seeds germinate the following spring, and seedlings grow slowly during their first year to 5–8 cm (2–3 inches) by autumn. This slow growth is partly caused by shading by horseweed, and decaying horseweed roots stunt aster growth. So horseweeds do not pave the way for asters; if anything, they have a detrimental effect on them. Asters enter the succession in spite of horseweeds, not because of them.

Aster (a perennial) blooms in its second year, after horseweed declines, but it is not drought resistant. Seedlings of aster are present in large numbers in third-year fields but succumb to competition for moisture with the drought-resistant broomsedge (*Andropogon virginicus*). In fields with more available water, aster is able to last into the third year, but eventually broomsedge overwhelms it.

Broomsedge seeds will not germinate without a period of cold dormancy. A few broomsedge plants are found in one-year fields, but they do not drop seed until the fall of the second year. Broomsedge is a very drought-resistant perennial and competes very well for soil moisture. Few broomsedge plants grow in one- and two-year fields because few seeds are present. Once some plants begin seeding, broomsedge rapidly increases in numbers (in the third year). Broomsedge grows better in soil with organic matter, especially in soil with aster roots. It grows very poorly in the shade.

Horseweed (Erigeron canadensis) *(©Angelia Lax/Photo Researchers, Inc.)*

Thus, early succession in Piedmont old fields is governed more by competition than by cooperation between plants. The early pioneers do *not* make the environment more suitable for later species, and the later species achieve dominance in spite of the changes caused by the early species, rather than because of them. If seeds were available, broomsedge could colonize an abandoned field immediately rather than following horseweed and aster. Old-field succession is not described well by the facilitation or tolerance models but seems to be an excellent illustration of the inhibition model.

After this succession by herbs and grasses, abandoned farmland of the Piedmont of North Carolina is invaded in great numbers by shortleaf pine (*Pinus echinata*). Pine seeds are light and are blown by the wind, so they effectively disperse into abandoned fields if large pines grow anywhere nearby. Pine seeds can germinate only on mineral soil and are able to become established in bare sites among the herbs and grasses (Christensen and Peet 1981). Pine seedlings are very effective competitors for soil water, and as the pine seedlings grow, they shade the herbs and grasses in the understory. The density of pines is very high but falls rapidly as they lose their dominance to hardwoods (Figure 21.13). After approximately 50 years several species of oaks become important trees

Broomsedge (Andropogon virginicus) *(©Michael P. Gadowski/Photo Researchers, Inc.)*

in the understory, and the hardwoods gradually fill in the community. Reproduction of shortleaf pine is almost completely lacking after about 20 years (Figure 21.14) because there is no bare soil for seed germination, and pine seedlings cannot live in shade. The networks of pine roots in the soil become closed very quickly, and the accumulation of litter under the pines causes the old-field herbs to die out (Figure 21.15). Oak seedlings first appear after 20 years, when enough litter has accumulated to protect the acorns from desiccation and the soil is able to retain

FIGURE 21.13
Decline in the abundance of shortleaf pine, and increase in the density of hardwood tree seedlings during succession on abandoned farmland in the Piedmont area of North Carolina. (After Billings 1938.)

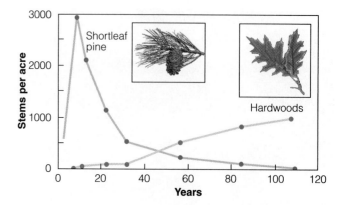

more moisture. Hardwood seedlings persist in the understory because they develop a root system deep enough to exploit soil water, because they are shade tolerant, and because they compete more effectively than pines for water and nutrients (see p. 97).

Soil properties change dramatically along with this plant succession on Piedmont soils. Organic matter accumulates in the surface layers of the soil and increases in the deeper soil layers. Because the moisture-holding capacity of the soil increases with organic content, the soil becomes more able to hold colloidal water for use by the plants.

Thus, shortleaf pine is independent of the early succession in that it requires only bare soil for germination. The elimination of all herbaceous species from the early succession would not affect coloniza-

tion by pines, which fit the tolerance model depicted Figure 21.4c. Pines can invade old fields as soon as seeds become available. Oaks and other hardwoods, by contrast, depend on the soil changes caused by pine litter, so oak seedlings could not become established without the environmental changes produced by pines. Thus, the latter part of this succession seems to fit the classical facilitation model.

The examples we have just discussed do not always fit any single model of succession, as some replacements are facilitated whereas others are inhibited. Drury and Nisbet (1973) reviewed succession in forested regions and concluded that most forest succession does not conform to the classical model. If the classical model is not always correct, we must reconsider the nature of the climax state, the "end point" of succession.

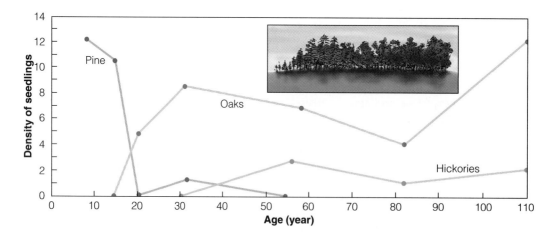

FIGURE 21.14
Density of pine, oak, and hickory seedlings during the shortleaf pine succession in the Piedmont area of North Carolina. (After Billings 1938.)

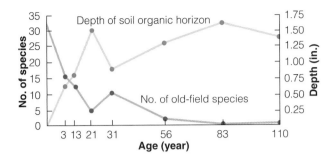

FIGURE 21.15
Decline in the number of old-field herbaceous species and increase in soil organic matter during the succession from old-field to shortleaf pine in the Piedmont area of North Carolina. (After Billings 1938.)

The Climax State

In the examples of succession described in the preceding section, the vegetation has developed to a certain stage of equilibrium. This final stage of succession is called the *climax*, defined as *the final or stable community in a successional series. It is self-perpetuating and in equilibrium with the physical and biotic environment.* There are three schools of thought about the climax state: the monoclimax school, the polyclimax school, and the climax-pattern view.

The *monoclimax* theory was an invention of the American F. E. Clements (1916, 1936). According to the monoclimax theory, every region has only one climax community toward which all communities are developing. The fundamental assumption of Clements—that given time and freedom from interference, a climax vegetation of the same general type will be produced and stabilized, irrespective of earlier site conditions. Climate, Clements believed, was the determining factor for vegetation, and the climax of any area was solely a function of its climate.

However, it was clear in the field that certain areas had communities that were not climax communities but appeared to be stable. For example, tongues of tallgrass prairie extended into Indiana from the west, and isolated stands of hemlock occurred in what is supposed to be deciduous forest. In other words, we observe communities in nature that are nonclimax according to Clements but apparently in equilibrium nevertheless. These communities are determined by topographic, edaphic (soil), or biotic factors.

The *polyclimax* theory arose as the obvious reaction to Clements's monolithic system. Tansley (1939) was one of the early proponents of the polyclimax idea—that many different climax communities may be recognized in a given area, including climaxes controlled by soil moisture, soil nutrients, activity of animals, and other factors. Daubenmire (1966) also suggested that several stable communities may be found in a given area such that no single climax existed for a region.

The real difference between these two schools of thought lies in the time frame for measuring relative stability. Given enough time, say the monoclimax proponents, a single climax community would develop, eventually overcoming the edaphic climaxes. The question is, Should we consider time on a geological scale, or on an ecological scale? If we view the problem on a geological time scale, we would classify communities such as the coniferous forest as a seral stage to the establishment of deciduous forest. The important point is that climate fluctuates and is never constant. We see this vividly in the Pleistocene glaciations, and more recently in the advances and retreats of mountain glaciers in the past 1000 years. As a result, *equilibrium can never be reached because the vegetation is subject not to a constant climate but to a variable one.* Climate varies on an ecological time scale as well. In a sense, then, succession is continuous because we have a variable vegetation interacting with a variable climate.

Whittaker (1953) proposed a variation of the polyclimax idea, the *climax-pattern hypothesis.* He emphasized that a natural community is adapted to the whole pattern of environmental factors in which it exists—climate, soil, fire, biotic factors, and wind. Whereas the monoclimax theory allows for only one climatic climax in a region and the polyclimax theory allows for several climaxes, the climax-pattern hypothesis allows for a continuum of climax types that varies gradually along environmental gradients and is not neatly divisible into discrete climax types. Thus, the climax-pattern hypothesis is an extension of the continuum idea and the approach of gradient analysis to vegetation (Whittaker 1953). The climax is recognized as a steady-state community in which its constituent populations are in dynamic balance with environmental gradients. We do not speak of a climatic climax, but of prevailing climaxes that are the end result of climate, soil, topography, and biotic factors, as well as fire, wind, salt spray, and other influences, including chance. The utility of the climax as an operational concept is that similar sites in a region should produce similar climax stands. This stand-to-stand regularity should allow prediction for new sites of known environment, and we can say that in 100 years a particular site should develop, for example, to a stand of sugar maple and beech of specified density.

A good example of how biotic factors may affect plant succession is found in the uplands of northwest Scotland (Miles 1987). Sheep grazing selects against trees and favors grassland. Several vegetation types occur in these Scottish uplands. Under low grazing pressure and no fire (Figure 21.16a) Scot's pine and birch woodlands tend to be favored. Under high grazing pressure and frequent fires (Figure 21.16b) only grassland is self-sustaining, and when grazing is moderate and fires are occasional (Figure 21.16c), both heather moor (*Calluna vulgaris*) and bracken fern (*Pteridium aquilinum*) communities are favored. There is no one climatic climax on these Scottish uplands, and the plant communities are a mosaic, changing as sheep grazing and fire frequency change. These communities are best described by Whittaker's climax-pattern hypothesis.

How, then, can we recognize climax communities? The operational criterion is the attainment of a steady state over time. Because the time scale involved is very long, observations are lacking for most presumed successional sequences. We assume, for example, that we can determine the time course of succession for a spatial study of younger and older dune systems around Lake Michigan (see Figure 21.11), but this translation of space and time may not be valid. In forests, we can use the understory of young trees to look for changes in species composition, because the large trees must reproduce themselves on a one-for-one basis if steady state has been achieved. Forest changes may be very slow. Lertzman (1992) studied a subalpine forest stand at 1100 meters elevation near Vancouver, Canada. This site was undisturbed by fire for almost 2000 years and still was not in equilibrium: the dominant hemlocks

were not replacing themselves while amabilis fir seedlings were invading the understory in large numbers. Climax vegetation on this site was not reached after two millennia.

Some communities may appear to be stable in time and yet may not be in equilibrium with climatic and soil factors. A striking example of this occurred after the outbreak of the disease myxomatosis in the European rabbit in Britain. Before 1954, rabbits were common in many grassland areas. Myxomatosis devastated the rabbit population in 1954, and the consequent release of grazing pressure caused dramatic changes in grassland communities (Thomas 1960, 1963). The most obvious change was an increase in the abundance of flowers. Species that had not been seen for many years suddenly appeared in large numbers. There was also an increase in woody plants, including tree seedlings that were commonly grazed by rabbits. No one anticipated these effects following the removal of the rabbits.

We conclude from this discussion that climax vegetation is an abstract ideal that is, in fact, seldom reached, owing to the continuous fluctuations of climate. The climate of an area has clear overall control of the vegetation, but within each of the broad climatic zones are many modifications caused by soil, topography, and animals that lead to many climax situations. The rate of change in a community is rapid in early succession but becomes very slow as it nears the potential climax community. But the climax community, like Nirvana, may never quite be attained.

Patch Dynamics

We have seen that communities are dynamic and changing continually. In 1947 A. S. Watt, a British plant ecologist, first called attention to cyclic events that occur repeatedly in communities occupying spatially small patches. A plant community over a region may be moving slowly toward a climax state, while on a local scale the internal dynamics of the community are producing more-rapid cyclic changes in patches or gaps in the community. The study of gap dynamics has shed interesting light on the overall processes of community change. Here we discuss three examples of patch dynamics.

Watt (1947) studied several examples of cyclic changes in British vegetation. One of these was the

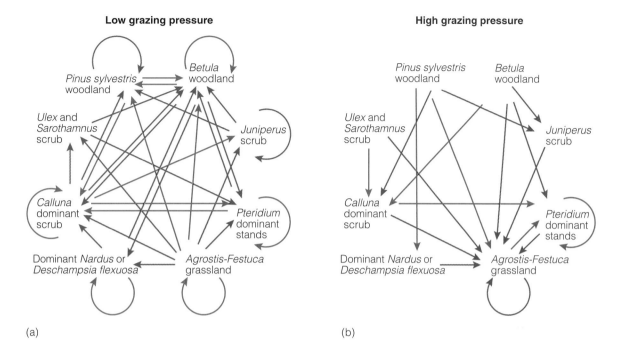

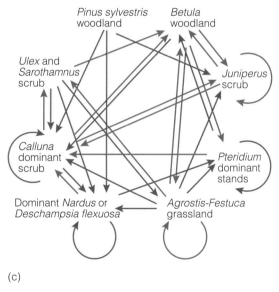

FIGURE 21.16
Successional transitions in the British uplands (particularly northwest Scotland) between eight vegetation types given (a) low grazing pressures and no burning; (b) high grazing pressures and frequent burning; and (c) intermediate levels of grazing and occasional burning. Blue arrows represent common transitions, red arrows less frequent transitions, and curved arrows self-replacement. The vegetation types are arranged so that types tending to acidify soils are on the left, and types with contrasting soil effects are on the right. (Modified after Miles 1985.)

Calluna heath that covers large areas in Scotland and has made heather almost synonymous with Scotland. The dominant shrub in this community is heather (*Calluna*), which loses its vigor as it ages and is invaded by lichens of the genus *Cladonia*. The lichen mat in time dies back, leaving bare ground. This bare area is invaded by bearberry (*Arctostaphylos*), which in turn is invaded by *Calluna*.

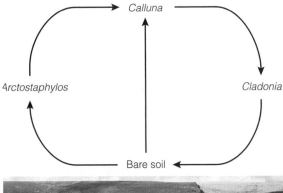

Calluna vulgaris (©Porter Field/Chickering/Photo Researchers, Inc.)

Heather (*Calluna*) is the dominant plant, and *Arctostaphylos* and *Cladonia* are allowed to occupy the area that is temporarily vacated by *Calluna*.

The cycle of change can be divided into four phases (Figure 21.17):

- *Pioneer:* establishment and early growth in *Calluna;* open patches, with many plant species (years 6 to 10).
- *Building:* maximum cover of *Calluna* with vigorous flowering; few associated plants (years 7 to 15).
- *Mature:* gap begins in *Calluna* canopy and more species invade the area (years 14 to 25).
- *Degenerate:* central branches of *Calluna* die; lichens and bryophytes become very common (years 20 to 30).

Barclay-Estrup and Gimingham (1969) describe this sequence in detail from maps of permanent quadrats in Scotland. The life history of the dominant plant *Calluna* controls the sequence.

A second example of patch dynamics that Watt studied involved cyclic changes associated with microtopography in a grassland in England: the hummock-and-hollow cycle. The vegetation of the grassland Watt studied was very patchy, and he could recognize four stages (Figure 21.18). The whole scheme centers around the grass *Festuca ovina*. Once the seedlings of this grass become established in the bare soil of the hollow stage, the plant builds a "tussock" by trapping windborne soil particles and by its own growth. The vigor of this grass declines with age; it begins to degenerate in the mature phase and is invaded by lichens in the early degenerate phase. These lichens

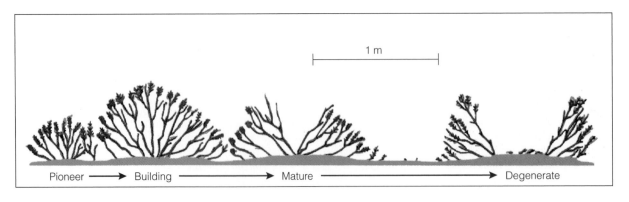

FIGURE 21.17
Profile of the four phases in the Calluna *cycle in Britain. Like many perennial plants, heather loses vigor with age. (After Watt 1955.)*

use up the organic matter and in turn die, and the hummock is eroded down to base level, only to begin the process again. At any given time, all four stages can be found in a *Festuca* grassland. Seedlings cannot usually get established except in the hollow and building phases. Lichens dominate the degenerating phase, when they can use the organic matter that has accumulated. Bryophytes seem to suffer competition from fescue and cannot get established except in the degenerate or hollow phase.

Watt also studied Bracken fern (*Pteridium aquilinum*), a cosmopolitan species that lives in a great variety of soil types and climatic conditions. It forms rhizomes in the soil, and once established it spreads vegetatively. Bracken is fire resistant because of its underground rhizomes. Watt (1947a) studied bracken in the process of invading grassland and found a vigorous "front" of invading bracken but reduced vigor in older fronds (Figure 21.19). The obvious explanation for this marginal effect is that some soil nutrient is depleted by the advancing fern and is in short supply in the older stands. However, Watt (1940) could find no soil change to account for the reduced vigor, and the addition of fertilizers to sample plots in the older stands produced no effect. The significant variable seems to be rhizome age: Younger rhizomes produce more vigorous fronds. The explanation for this is not known.

Watt divided all these cycles of change into an *upgrade* series and a *downgrade* series and pointed out that the total productivity of the series increases to the mature phase of the cycle and then decreases. What initiates the downgrade phase? A possible explanation there seems to be a general relationship between age and performance (vigor) in most perennial plants, and consequently between age and competitive ability (Kershaw and Looney 1985). Several studies on the relation of leaf diameter to age also support this idea.

For this reason, a stable community will be in a constant state of phasic fluctuation, one species becoming locally more abundant as another species reaches its degenerate phase. All these dynamic interrelationships in natural communities tend to operate on small spatial scales and may not be apparent without detailed measurement.

Patch dynamics will result in monocultures if one species is able to replace itself and all other species as well. As early as 1905, French foresters suggested that in virgin forests, individual trees tended to be succeeded in time by those of another species (Fox 1977). This phenomenon, called *reciprocal replacement*, occurs at the individual tree level, and could be one explanation of why old-growth forests have a mixture of tree species rather than a single dominant. If, for example, seedlings of species A were found predominately under large species B trees, and seedlings of species B were found predominately under large species A trees, we would have reciprocal replacement at the individual tree level.

American beech and sugar maple are codominant trees in old growth forests in southern Michigan, and reciprocal replacement of individual trees has been suggested as one possible mechanism of codominance. Poulson and Platt (1996) found that reciprocal replace-

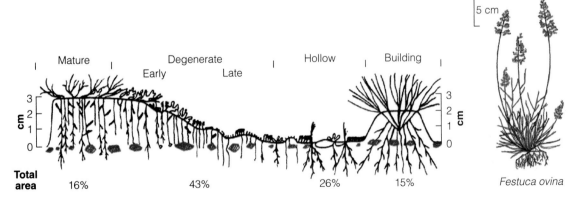

FIGURE 21.18
Phases of the hummock-and-hollow cycle, showing change in flora and habitat and indicating the "fossil" shoot bases and detached roots of Festuca ovina *in the soil. The whole cycle centers around the grass* Festuca ovina. *(After Watt 1947b.)*

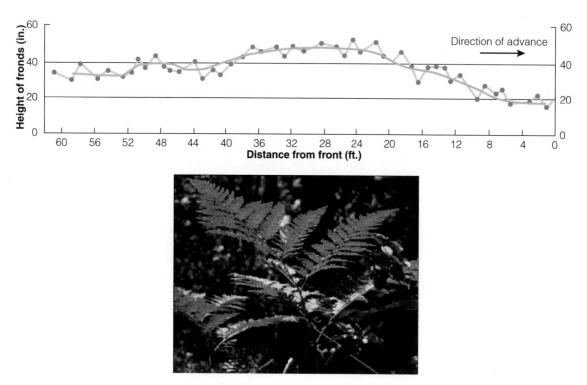

FIGURE 21.19
Average height above the soil surface of bracken fern fronds behind an advancing front as bracken invades a grassland in England. (After Watt 1940.)

ment did not occur at the individual tree level, but that codominance was caused by differing light intensities in regenerating gaps once older trees die. Sugar maples grow quickly upward in gaps in the forest, whereas beech seedlings spread laterally and capture light in flecks in the understory. Thus beech does better in the understory of these forests and will decrease in relative abundance when tree-fall gaps become more frequent. Figure 21.20 illustrates schematically the mechanisms of coexistence of these two trees. Stable coexistence depends on a mixture of gaps of different sizes being created over periods of hundreds of years.

Tropical rain forests are another example of a community that may be a floristic mosaic of tree-fall-created patches undergoing cyclic changes (Richards 1952). Forest regeneration occurs by the production of a gap, then colonization and growth in the gap, and finally by a mature tree phase during which eventual treefalls renew the cycle. Rain and wind cause treefalls.

On Barro Colorado Island in Panama, gaps averaged 86 sq. meters, and about 3% of the forest area was gaps (Brokaw 1985). Gaps of this size were created at a rate of about one per hectare per year on Barro Colorado. Trees in tropical forests are loosely classified in two groups. Primary, or shade-tolerant, species live as suppressed small trees in the understory until a canopy gap opens above them; pioneer, or shade-intolerant, species germinate only in gaps and grow rapidly. Brokaw (1987) tested the idea that each pioneer species occupies a separate regeneration niche on Barro Colorado. He studied three species that were most abundant in 30 natural gaps. Table 21.4 gives the details of his results, and Figure 21.21 shows the growth rates of the three species. *Trema micrantha* recruits to gaps only in the first year and grows most rapidly; *Miconia argentea* grows more slowly but recruits over several years to the gaps. But there is overlap in these pioneer species, and the regeneration

25 years before treefall

No gaps nearby

Local gaps every 100–200 years

< 1% full sun

10 x 10 meters

Huge beech canopies cast deep shade; rare sapling beech are suppressed

25 years after treefall

Multiple sugar maple winners in a huge beech treefall

Nearby gaps every 0–20 years

Local gaps every 0–40 years

1–30% full sun

More sugar maple than beech in understory; sugar maple rarely completely suppressed

Gaps grow with time; sugar maple replaces most beech and some sugar maple; beech replaces some sugar maple

FIGURE 21.20
Conceptual model of how the American beech-sugar maple forests of Michigan maintain codominance of two "climax" tree species. Left column is before treefall, right column is after treefall. Infrequent and small treefall gaps (top panel) favor beech (open tree silhouettes), where large and frequent gaps (bottom panel) favor sugar maple (green tree silhouettes). Xs mark the former positions of crowns of fallen trees. Maples from the understory were suppressed on average 20 years before being released to grow in a gap, whereas beech from the understory were suppressed an average of 121 years. (From Poulson and Platt 1996.)

niches of these trees are not discrete. The floristic mosaic of lowland tropical forests is partly caused by the cyclic changes in tree-fall gaps driven by variation in shade tolerance among different species.

Gap dynamics play an important role in the regeneration of forest and other communities in which space is a key resource. Space in forests (and in many other plant communities) is equivalent to light, and competition for space is competition for light. But light is not the only resource for which plants compete, and soil nitrogen levels, other soil nutrients, and water may modify competition among plants. Thus communities are dynamic and change because of the interactions between the life-history traits of the dominant species and patterns of physical disturbances. A stable community at a regional level may be a mosaic of patches undergoing changes at a local level. We consider next how biological communities are organized, and what mechanisms control their structure.

TABLE 21.4   Relative behavior of three tree species that colonize treefall gaps in tropical rain forest, Barro Colorado Island, Panama.

Species	Family	Type of fruit	Seed dispersal agents
Trema micrantha	Ulmaceae	Drupe	Many bird species
Cecropia insignis	Moraceae	Seeds compressed in peduncles	Birds, bats
Miconia argentea	Melastomataceae	Berry	Monkeys, birds

	Recruitment		
	1st year after gap formation	*2nd year*	*Survivors to years 8, 9*
Trema	Nearly all in first year	None after year 2	Only those established in first year
Cecropia	Most in first year	Many in year 2, none after year 3	From each of these years
Miconia	Begin in first year	Highest in years 2–5 and continue to years 7–9	From each of these years

	Sizes of gap occupied	
	Years 1–3 (seedlings)	*Years 8, 9 (saplings, poles)*
Trema	50 m² and larger	largest gaps only (> 376 m²)
Cecropia	20 m² and larger	in largest and mid-sized gaps (> 215 m²)
Miconia	20 m² and larger	all sizes of gaps down to 102 m²

Source: Data from Brokaw (1987).

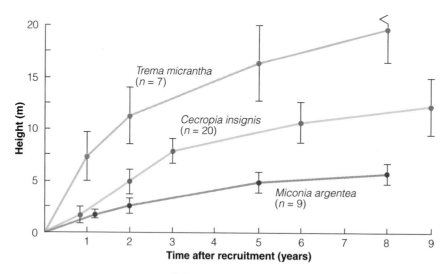

FIGURE 21.21

Mean height (± 1 standard deviation) of three pioneer tree species at intervals after recruitment in treefall gaps on Barro Colorado Island, Panama. Sample sizes are given in parentheses. (From Brokaw 1987.)

Summary

Communities change over time, and the study of succession has been an important focus of community ecologists for a century. Succession is the process of directional change in communities. Most of the work on succession has been done on plants, but the same principles apply to animal communities.

Four conceptual models of succession have been proposed to explain directional vegetation changes. All models agree that pioneer species in a succession are usually fugitive or opportunistic species with high dispersal rates and rapid growth. How are these pioneer species replaced? The classical model states that species replacements in later stages of succession are *facilitated* by organisms present in earlier stages. The inhibition model, at the other extreme, suggests that species replacements are *inhibited* by earlier colonizers and that successional sequences are controlled by "who gets there first." The tolerance model suggests that species replacements are not affected by earlier colonizers, and that later species in succession are those able to *tolerate* lower levels of resources than earlier species. The *random colonization* model is a null model that suggests that species replacements occur completely randomly, with no interspecific interactions. No single model explains an entire successional sequence. We now view succession as a dynamic process resulting from a balance between the colonizing ability of some species and the competitive ability of others. Succession does not always involve progressive changes from simple to complex communities.

Succession proceeds through a series of seral stages from the pioneer stage to the climax stage. The *monoclimax hypothesis* suggested that a single predictable end point existed for whole regions and that given time, all communities would converge to the climatic climax. This hypothesis has been superseded by the *climax pattern hypothesis*, which suggests a continuity of different climaxes varying along environmental gradients controlled by soil moisture, nutrients, herbivores, fires, or other factors.

A stable community contains patches repeatedly undergoing cyclic changes that are part of the internal dynamics of the community. The life cycle of the dominant organisms dictates the cyclic changes, many of which are caused by the decline in vigor of perennial plants with age. In many forests, tree-fall gaps create a mosaic of patches undergoing cyclic changes within a relatively stable climax community.

Communities are not stable for long periods in nature because of disturbances—short-term changes in climate, fires, windstorms, diseases, or other environmental factors. For most communities we can observe changes over time, but we need to determine the mechanisms that cause the changes. Unless we understand the mechanisms behind succession, we will be unable to suggest manipulations to alleviate undesirable trends caused by human activities.

Key Concepts

1. Succession is the process of directional change in communities.

2. Succession proceeds through a series of seral stages toward a climax community that remains stable. There are many different climax communities for a region, depending on soil type, water, grazing, and other environmental gradients.

3. The key question is how species interact during succession. Existing species may facilitate, inhibit, or not influence invading species.

4. Communities contain a mosaic of patches undergoing local dynamics, often in cycles. Patch or gap dynamics are controlled by the life history-traits of the species.

5. Plant succession is largely controlled by competition among plants for light, and shade tolerance is a key parameter for succession models.

Selected References

Burrows, C. J. 1990. *Processes of Vegetation Change*. Chapters 12 and 13, pp. 420–489. Unwin Hyman, London.

Chapin, F. S., III, L. R. Walker, C. L. Fastie, and L. C. Sharman. 1994. Mechanisms of primary succession following deglaciation at Glacier Bay, Alaska. *Ecological Monographs* 64:149–176.

Clements, F. E. 1949. *Dynamics of Vegetation*. Macmillan (Hafner Press), New York. 296 pp.

Connell, J. H. and R. O. Slatyer. 1977. Mechanisms of succession in natural communities and their role in community stability and organization. *American Naturalist* 111:1119–1144.

del Moral, R. 1999. Plant succession on pumice at Mount St. Helens, Washington. *American Midland Naturalist* 141:101–114.

Finegan, B. 1996. Pattern and process in neotropical secondary rain forests: The first 100 years of succession. *Trends in Ecology and Evolution* 11:119–124.

Henry, H. and L. Aarssen. 1997. On the relationship between shade tolerance and shade avoidance strategies in woodland plants. *Oikos* 80:575–582.

Horn, H. S. 1981. Succession. In *Theoretical Ecology*, ed. R. M. May, pp. 253–271. Blackwell, Oxford, England.

Huston, M. and T. Smith. 1987. Plant succession: Life history and competition. *American Naturalist* 130:168–198.

McCook, L. J. 1994. Understanding ecological community succession: Causal models and theories, a review. *Vegetatio* 110:115–147.

Tanner, J. E., T. P. Hughes, and J. H. Connell. 1996. The role of history in community dynamics: A modelling approach. *Ecology* 77:108–117.

van der Maarel, E. 1996. Pattern and process in the plant community: Fifty years after A. S. Watt. *Journal of Vegetation Science* 7:19–28.

Questions and Problems

21.1 Discuss the inhibition, facilitation, and tolerance models of succession with respect to the following simple experiment. In this hypothetical succession, species A normally precedes species B. Two treatments are applied: (1) All of species A are removed from a series of replicate plots, and (2) a portion of species B equal to the biomass of A is removed from another set of plots. Growth in biomass is then measured over several years. Interpret all the possible outcomes of this experiment. Compare your analysis with that of Botkin (1981).

21.2 Culver (1981) observed the following transition probabilities for a forest in Great Smoky Mountains National Park in North Carolina and Tennessee:

	Species in Saplings		
Species in canopy	*Yellow birch*	*Red spruce*	*Fraser fir*
Yellow Birch	0.01	0.45	0.54
Red spruce	0.13	0.20	0.67
Fraser fir	0.12	0.28	0.61

Calculate the changes over five generations in the composition of a forest containing these three species, starting from equal numbers of each species. What is the climax forest in this area?

21.3 Abundance within size classes in an undisturbed hemlock-beech association at Heart's Content, Pennsylvania., was measured by Lutz (1930, p. 27):

	Abundance (%) per size class				
Tree Species	*0–0.9 ft[a]*	*1.0 ft[a] to 0.9 in. DBH[b]*	*1–3.5 in. DBH[b]*	*3.6–9.5 in. DBH[b]*	*≥ 10 in. DBH[b]*
Hemlock	23.4	44.0	19.2	13.5	36.1
White pine	0.2	0.1	—	—	11.1
Beech	4.9	22.7	59.9	50.3	24.0
Red maple	66.5	17.0	6.6	9.2	10.6
Chestnut	0.2	1.5	0.4	9.2	8.8
White oak	—	—	—	—	1.6
Red oak	0.1	0.2	—	0.7	1.8
Black birch	0.8	1.4	4.8	10.4	2.7
Black cherry	0.3	1.0	1.7	0.9	0.8
Yellow birch	0.5	0.1	0.6	1.7	0.1
Sugar maple	0.4	1.2	1.1	0.7	0.3
White ash	0.1	—	—	0.1	—
Misc.	2.6	10.8	5.7	3.3	2.1
Total	100.0	100.0	100.0	100.0	100.0

[a]Height of tree.
[b]Diameter at breast height (1.4 m above ground).

Lutz stated that "the hemlock-beech association is believed to represent a stage in forest succession somewhat less advanced than the climatic climax of the region. Probably in the climax forest the amount of white pine entering into the stand in considerably smaller." Do these data support this conclusion? If so, how? If not, what additional data could support it?

21.4 Langford and Buell (1969, p. 130) state:

> Whereas biotic influences play an outstanding role in determining the nature of climax vegetation in moist temperate areas, abiotic factors are outstandingly preeminent in controlling vegetation in arid or very cold regions. In such regions succession, which is essentially due to modification of the environment by organisms, with its direction of course somewhat variable according to the availability of various propogules, may be almost absent.

Search for data on succession and the climax for either desert or arctic plant communities, and discuss them with reference to this statement.

21.5 Discuss the application of the concept of succession to marine communities.

21.6 Relate the adaptive strategies of species in early and late stages of succession to the ideas of r and K selection discussed in Chapter 12, pages 199–200.

21.7 In discussing forest succession as a plant-by-plant replacement process, Horn (1975b, p. 210) states: "Copious self-replacement does not guarantee a species' abundance or even its persistence in late stages of succession." How can this be true?

21.8 Williamson and Black (1981) suggest that succession from pines to oaks (see Figure 21.14) can be interrupted by fire, which kills the oaks but not the pines. Natural selection has produced pine foliage and pine litter that facilitates fires and thus inhibits oaks from taking over a succession. If this is correct, suggest how this type of adaptation could evolve.

21.9 In the primeval forest landscape, where were the plants that are abundant today in old fields (see p. 419)? Discuss the evolution of colonizing ability in plants that evolved in temporary forest openings, and those that evolved in persistent open, marginal habitats. Compare your conclusions with those of Marks (1983).

21.10 Discuss how much the current state of a plant community such as a forest depends on history. Does the simple model of succession presented on page 409 have any history in it?

21.11 How can species that facilitate other species in a successional sequence evolve? For example, why should species that fix nitrogen from the air leak this nutrient into the soil to assist their competitors who will replace them in the successional sequence? Is this an example of altruistic behavior?

Overview Question

Is it possible to construct a theory of succession in plant communities solely on the mechanism of competition between species? What would be missing from such a theory?

CHAPTER 22

Community Organization I:
Biodiversity

ECOLOGICAL COMMUNITIES do not all contain the same number of species, and one of the currently active areas of research in community ecology is the study of species richness or *biodiversity*. A. W. Wallace (1878) recognized that animal life was on the whole more abundant and varied in the tropics than in other parts of the globe, and that the same applies to plants. Other patterns of variation have long been known on islands; small or remote islands have fewer species than large islands or those nearer continents (MacArthur and Wilson 1967). The regularity of these patterns for many taxonomic groups suggests that they have been produced in conformity with a set of basic principles rather than as accidents of history. How can we explain these trends in species diversity?

Biodiversity measurement is an important part of conservation biology, because we need an inventory of what is to be protected. Whereas conservation biologists often concern themselves with particular species, community ecologists tend to lump the species and condense information into counts of species. Often this is done within specific groups, such as the bird species or the tree species of an area. This community-based approach looks for large patterns in groups of species and tries to understand what has caused them. To do this we first need to know how to identify species of plants and animals, and then how to measure biodiversity.

Measurement of Biodiversity

The simplest measure of biodiversity is the *number of species*. In such a count we include only resident species, not accidental or temporary immigrants. It may not always be easy to decide which species are accidentals: Is a bottomland tree species growing on a ridge top an accidental species or a resident one? The number of species is the first and oldest concept of species diversity and is called *species richness*.

A second concept of species diversity is that of *heterogeneity*. One problem with counting the number of species as a measure of diversity is that it treats rare species and common species equally. A community with two species might be divided in two extreme ways:

	Community 1	Community 2
Species A	99	50
Species B	1	50

The first community is very nearly a monoculture, and the second community intuitively seems to be more diverse than the first. We can combine the concepts of number of species and relative abundance into a single concept of heterogeneity: Heterogeneity is higher in a community when there are more species and when the species are more nearly equally abundant.

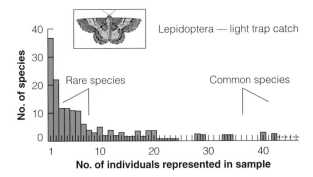

FIGURE 22.1
Relative abundance of Lepidoptera (butterflies and moths) captured in a light trap in Rothamsted, England, in 1935. A total of 6814 individuals of 197 species were caught (some of the abundant species are not shown). Thirty-seven species were represented in the catch by only a single specimen, and six common species constituted 50% of the catch. One very common species was represented by 1799 individuals in the catch. (Modified from Williams 1964.)

A difficult problem arises in trying to determine the number of species in a biological community: *Species counts depend on sample size.* Adequate sampling can usually get around this difficulty, particularly with vertebrate species, but not always with insects and other arthropods, in which species counts cannot be complete.

Ecologists have adopted two different strategies to deal with these problems. First, a variety of statistical distributions can be fitted to data on the relative abundances of species. One very characteristic feature of communities is that they contain comparatively few species that are common, and comparatively many species that are rare. Because it is relatively easy to sample any given area and count both the *number of species* on the area and the *number of individuals* in each of these species, a great deal of information of this type has accumulated (Williams 1964). The first attempt to analyze these data was made by Fisher, Corbet, and Williams (1943).

In many faunal samples, the number of species represented by a single specimen is very large, species represented by two specimens are less numerous, and so on, such that only a few species are represented by many specimens. When Fisher, Corbet, and Williams (1943) plotted the data, the result was a "hollow curve"

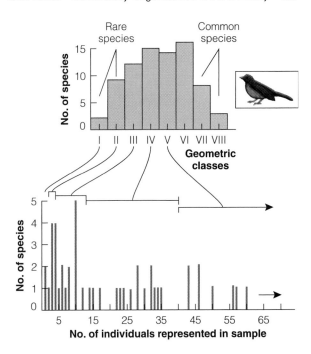

FIGURE 22.2
Relative abundance of nesting bird species in Quaker Run Valley, New York. The lower portion of the figure shows the distribution on an arithmetic scale, and the upper portion shows the same data on a geometric scale (class I = 1; class II = 2–4, class III = 5–13, class IV = 14–40, class V = 41–121, etc.). Most species have intermediate abundances, and fewer are very common or very rare. (After Williams 1964.)

(Figure 22.1) that could be described mathematically by a logarithmic series. The most significant ecological observation is that the largest number of species in a community fall into the "very rare" category.

The logarithmic series arises in communities with relatively few species in which a single environmental factor is dominant (May 1975). It describes an extreme type of "niche preemption" in which the most successful species preempts a fraction k of the total resources, the next species a fraction k of the remaining resources, and so on. This type of niche-preemption hypothesis predicts a logarithmic series distribution or "hollow-curve" as a description of species abundances in natural communities.

Even though the logarithmic series implies that the greatest number of species have minimal abundance—that the number of species represented by a

ESSAY 22.1

BIODIVERSITY: A BRIEF HISTORY

Between 5 million and 30 million species of animals and plants live on the earth. About 1.4 million of these are described by taxonomists, perhaps 10% of all life. This situation is a scandal that few nonbiologists seem to recognize. If only 10% of the companies being traded on Wall Street were known, or if the catalog of the Louve Museum included only 10% of its paintings, right-thinking people would be outraged. Not so with biodiversity.

Edward O. Wilson *(1922–) Entomologist, Ecologist, Conservation biologist, Harvard University*

Taxonomists are the heroes of biodiversity, and without them working quietly in the background we would not know even the 10% we do, and our appreciation of community organization and dynamics would be much reduced. Fortunately a few taxonomists have risen to public recognition, including Edward O. Wilson of Harvard University. Wilson is an ant taxonomist by training and a naturalist by nature. While working on ant distributions on islands, he met and joined forces in 1961 with Robert MacArthur from the University of Pennsylvania to produce one of the most famous books on community ecology, *The Theory of Island Biogeography* (1967). Wilson has become a champion of biodiversity through his books on ants, and more recently through a series of popular books on biodiversity and its conservation. He is one of the few ecologists to have written an autobiography (*Naturalist*, Island Press, Washington D.C., 1994).

Many other ecologists cooperated to bring biodiversity into the public eye at the close of the twentieth century. Some of them, including Paul Ehrlich, we have already met in previous chapters. But many who are less well known work hard to bring biodiversity to the fore in biological research agendas, and in the realm of political and social action. Given our ignorance of biodiversity, the exploding human population and its expanding effects on the globe is producing extinctions of species we will never have named or even described, a loss that we should not bequeath to our children and grandchildren.

single specimen is always maximal—this is not the case in all communities. Figure 22.2 shows the relative abundance of nesting birds in Quaker Run Valley, New York. The greatest number of bird species are represented by ten breeding pairs, and the relative abundance pattern does not fit the hollow-curve pattern of Figure 22.1. Preston (1948) suggested expressing the x axis (number of individuals represented in the sample) on a geometric (logarithmic) scale rather than an arithmetic scale. When this conversion of

scale is done and the species are combined into classes whose ranges of species abundances increase geometrically, relative abundance data take the form of a bell-shaped, normal distribution, and because the x axis is expressed on a geometric or logarithmic scale, this distribution is called *log-normal*. The essential point is that populations tend to increase geometrically rather than arithmetically (see Chapter 11, p. 157), so the natural way to analyze abundances is as the *logarithm* of population density.

The log-normal distribution fits a variety of data from surprisingly diverse communities. Figure 22.3 gives two more examples of relative abundance patterns in different communities. The log-normal distribution arises in all communities in which the total number of species is large, and the relative abundances of these species is determined by many factors operating independently. The log-normal distribution is thus the expected statistical distribution for many biological communities (May 1975). There is something very compelling about the log-normal distribution. The fact that moths in England, freshwater algae in Spain, snakes in Panama, and birds in New York all have a similar type of species abundance curve suggests regularity in community structure. The log-normal distribution can describe all the data that fit the logarithmic series and is a more general model of species abundance patterns in natural communities.

The log-normal distribution was recognized as an empirical regularity before it was given a theoretical justification (Preston 1962). Sugihara (1980) has provided one explicit biological mechanism that leads to log-normal distribution. Assume that a community has a set of total niche requirements, which enables us to define a communal niche space. This niche space can be likened to a unit mass that is sequentially split up by the various component species such that each fragment denotes relative species abundance. Consider Sugihara's model for the simple case of a three-species community. First, the total niche space is broken randomly to produce two fragments (Figure 22.4). The larger of the two fragments must range in size from 0.5 to 1.0 (of the original unit mass) and statistically will average 0.75 units. Next,

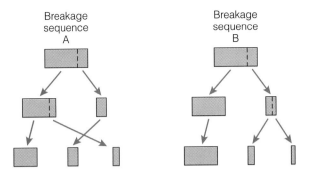

FIGURE 22.4
A hypothetical illustration of the sequential breakage hypothesis of Sugihara (1980). Two possible breakage sequences are illustrated for a hypothetical three-species community. If niches are subdivided in this manner in natural communities, the resulting abundance patterns for the species in the community will be log-normal. (From Sugihara 1980.)

one of these two fragments is now chosen at random and broken to yield a third fragment. If the larger fragment is broken in the second step (breakage sequence A in Figure 22.4), we end up with three "species" with average relative abundances of 0.57, 0.28, and 0.15 (Sugihara 1980). If the smaller of the original two fragments is broken (breakage sequence B in Figure 22.4), we end up with three "species" with average relative abundances of 0.75, 0.19, and 0.06. (These relative abundance estimates are averages that would apply to a large sample of three-species communities; any given community will vary because the breakage occurs at random.) Multispecies communities are more difficult to do these calculations for, but the principles remain the same. The

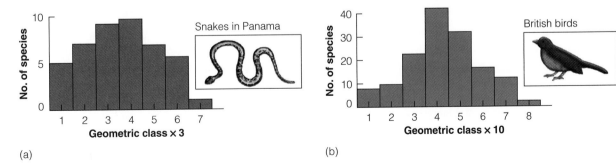

(a)   (b)

FIGURE 22.3
Log-normal distribution of relative abundances in two diverse communities: (a) snake species in Panama and (b) British birds. Most species are intermediate in abundance in both these communities. (Data from Williams 1964.)

important point is that the subdividing is done sequentially and not instantaneously, which corresponds with the biological assumption that the niche structure for communities is hierarchical. The niche space of a community has many dimensions and must not be thought of as a single resource axis.

The sequential breakage hypothesis predicts relative abundance patterns that are log-normal. Data from various communities fit this hypothesis very well, and the empirical findings of Preston (1962) can thus be biologically interpreted as a consequence of sequential niche subdivision.

A second approach to species diversity involves measures of the heterogeneity of a community. Several measures of heterogeneity are in use (Magurran 1988, Krebs 1999), and the most popular has been borrowed from information theory. The main objective of information theory is to measure the amount of order (or disorder) contained in a system. We ask the question, How difficult would it be to predict correctly the species of the next individual collected? This is the same problem faced by communication engineers interested in predicting correctly the next letter in a message. This uncertainty can be quantified by a measure of information content, the Shannon-Wiener function; the details of how to calculate this measure of heterogeneity are provided in Appendix IV (p. 618).

To summarize, to measure biodiversity we need a combination of two types of data: (1) the number of species in the community, and (2) the relative abundance of the species making up the community.

Some Examples of Diversity Gradients

Tropical habitats support large numbers of species of plants and animals, and this diversity of life in the tropics contrasts starkly with the relatively impoverished faunas of temperate and polar areas. A few examples will illustrate this global gradient. A 50-ha plot of tropical rain forest in Malaysia contained 830 species of trees, and a 6.6 ha area in Sarawak contained 711 tree species (Whitmore 1998). A deciduous forest in Michigan contains 10 to 15 species on a plot of 2 hectares, and the whole of Europe north of the Alps has 50 tree species.

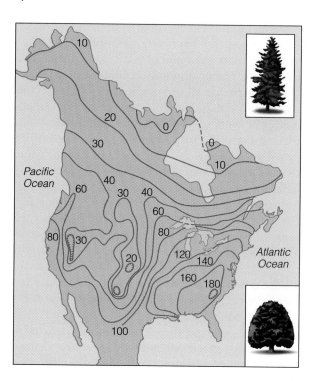

FIGURE 22.5
Number of tree species in Canada and the United States. Contours connect points with the same number of species. (From Currie and Paquin 1987.)

The 620 native tree species in North America north of Mexico are arrayed along a gradient that roughly follows latitude (Figure 22.5). More species in the United States occur in southeastern forests than in western forests, and minima occur in the rain shadows just east of the Rocky Mountains and the Sierra Nevada (Currie and Paquin 1987).

Ants are much more diverse in the tropics than in the high latitudes (Fischer 1960):

	No. of ant species
Brazil	222
Trinidad	134
Cuba	101
Utah	63
Iowa	73
Alaska	7
Arctic Alaska	3

There are 293 species of snakes in Mexico, 126 in the United States, and 22 in Canada. Figure 22.6 shows

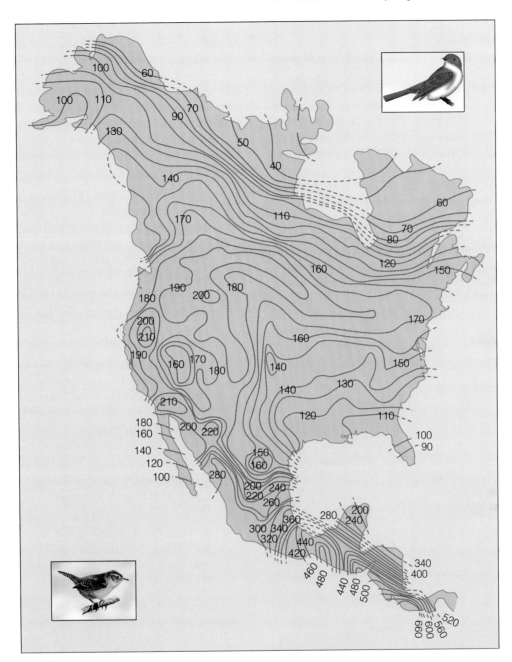

FIGURE 22.6
Geographic pattern of biodiversity in the land birds of North and Central America.
Numbers for contour lines indicate the numbers of species present. (From Cook 1969.)

the number of breeding land-bird species in different parts of North America.

Freshwater fishes are much more diverse in tropical rivers and lakes. Lakes Victoria, Tanganyika, and Malawi in east Africa each contain about 1450 species of freshwater fish. Over 1000 species of fishes have been found in the Amazon River in South America, and exploration is still incomplete in this region. By contrast, Central America has 456 fish species, and the Great Lakes of North America have 173 species (Rodhe 1998). Lake Baikal in Asia has 39 fish species; Great Bear Lake in northwestern Canada has 14 species of fish.

Marine invertebrates also have high diversity in the tropics. Figure 22.7 shows that calanoid copepods (planktonic crustaceans) are most diverse in the tropical Pacific Ocean and least diverse in the Bering Sea and Arctic Ocean.

Not all floras and faunas show a smooth trend of biodiversity with changing latitude. Figure 22.8 shows the diversity of Australian carnivorous marsupials. For small carnivorous marsupials, species diversity is highest in the arid zone of central Australia. Tropical Australia is species-poor for both small and large carnivorous marsupials (Dickman 1989).

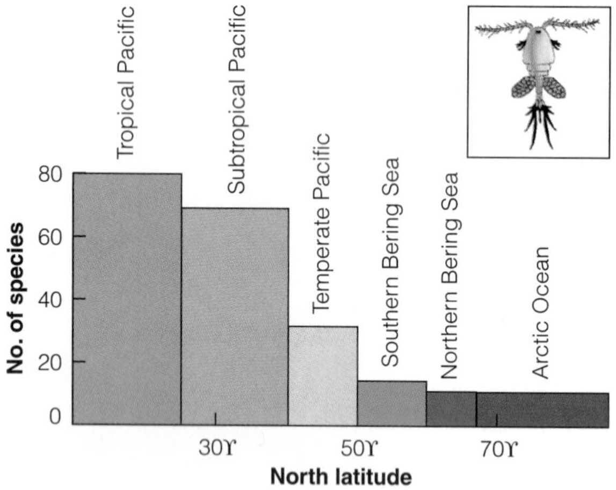

Species diversity patterns of North American mammals, analyzed in detail by Simpson (1964), are a good example of a complex gradient. Figure 22.9 shows that the number of land-mammal species increases from 15 in northern Canada to over 150 in Central America. Simpson recognized five notable features of this pattern:

- *North-south gradient:* The north-south gradient is not smooth. Some mammal groups— pocket gophers, shrews, and ungulates—are most diverse in the temperate zone and become less diverse toward the tropics. Bats contribute most of the high species richness for mammals in the tropics (Wilson 1974).

- *Topographic relief:* Areas like the Rocky Mountains or the Appalachians support a higher-than-average number of mammal species.

- *East-west trends:* Superimposed on the topographic variation is a general trend toward more species in the west than in the east. The topographically uniform Great Plains contain as many mammal species as the topographically diverse Appalachian Mountains.

- *Fronts of abrupt change:* Areas of rapid change in species diversity are often (but not always) associated with mountain ranges.

- *Peninsular "lows":* On peninsular areas such as Florida, Baja California, the Alaska Peninsula, and Nova Scotia, the number of mammal species is smaller than on adjacent continental areas.

The species diversity gradient seen in North America is apparent in South America as well (Figure 22.10), even though the mammals of these two continents have distinct evolutionary histories (Kaufman and Willig 1998). Box 22.1 illustrates a simple model that can be used to predict the diversity gradient with respect to latitude for any group of organisms.

This brief look at some details of species-diversity gradients can be useful in the next section, in which we examine some factors proposed to affect latitudinal gradients in species diversity.

FIGURE 22.7
Tropical-to-polar gradient in species diversity for the calanoid copepods of the upper 50 meters of a transect from the tropical Pacific Ocean to the Arctic Ocean. (After Fischer 1960.)

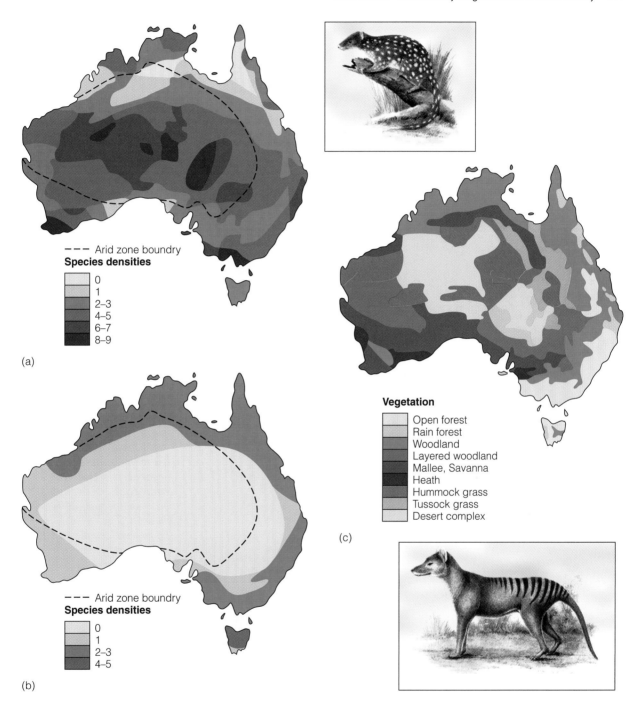

FIGURE 22.8

Species diversity of Australian carnivorous marsupials. (a) Small species (< 250 g). (b) Large species (> 500 g). (c) Generalized vegetation map. The arid center of Australia is shown by the dashed line. There is no clear latitudinal gradient in diversity in these species. (From Dickman 1989.)

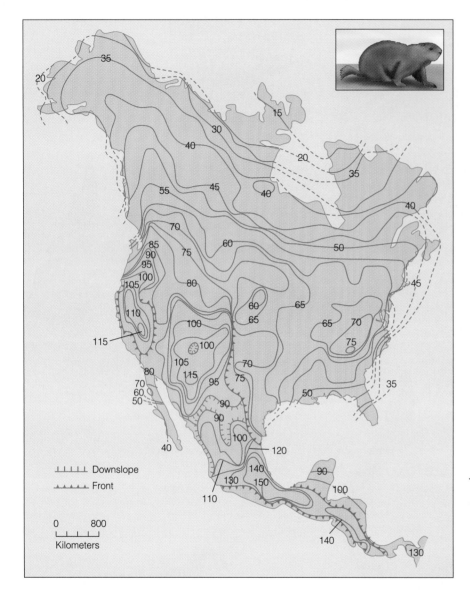

FIGURE 22.9
Species-density contours for existing mammals of continental North America. The contour lines are isograms for numbers of continental (nonmarine and noninsular) species in 150 mi^2 (240 km^2) quadrats. The "fronts" are lines of exceptionally rapid change that are multiples of the contour level for the given region. (After Simpson 1964.)

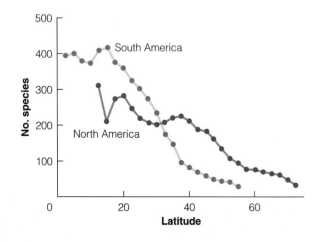

FIGURE 22.10
Species richness of mammals in North and South America in relation to latitude. The number of mammal species declines from the equator to the poles at the same rate in both continents, despite the fact that the mammal faunas of the two continents have quite separate evolutionary histories. Species richness was measured in 2.5° latitudinal bands across each continent. (After Kaufman and Willig 1998.)

B O X 2 2 . 1

A SIMPLE MODEL OF LATITUDINAL GRADIENTS IN BIODIVERSITY.

One strategy ecologists have used in trying to model complex ecological problems is to construct a *null model*, a model that is completely random and makes no biological assumptions about competition or predation or evolutionary history. Willig and Lyons (1998) did this for the latitudinal gradient in species diversity. Their model predicts the number of species at each latitude in a completely probabilistic manner; it contains no environmental gradients or any of the eight factors affecting diversity that are discussed in this chapter.

We begin by assuming that we know the northern and southern limits of latitude for our community. The southern limit of latitude is coded as 0, and the northern limit as 1. Any latitude in between these limits can be expressed as a fraction p of these ranges. For example, bats range in the New World from 66°N to 55°S , so we would set 55°S to be 0 and 66°N to be 1. Hence an intermediate latitude in the Southern Hemisphere (for example, 23°S) would be given as a fraction:

$$p = \frac{\text{southern latitude} - \text{selected limit}}{\text{northern limit} + \text{southern limit}} \quad (22.1)$$

$$= \frac{55 - 23}{66 + 55} = 0.264$$

If the selected limit is in the Northern Hemisphere (for example, 44°N for bats), we use the reciprocal relationship:

$$p = 1 - \left(\frac{\text{northern latitude} - \text{selected limit}}{\text{northern limit} + \text{southern limit}}\right) \quad (22.2)$$

$$= 1 - \left(\frac{66 - 44}{66 + 55}\right) = 0.818$$

We can obtain the probability that any point of latitude L includes the range of a species from the binomial distribution

$$\text{Pr}(L) = 2pq \quad (22.3)$$

where L = latitude selected

p = proportion of ranges to the south (as defined in the equations above)

$q = 1 - p =$ proportion of ranges to the north

For the bats at 44°N, we obtain:

$$\text{Pr}(L) = 2pq = 2(0.818)(1 - 0.818) \quad (22.3)$$

$$= 0.2978$$

Thus, under the null model assumptions of complete randomness in range limits, about 30% of all bats would be expected to extend their geographic ranges to 44°N. If we wish to express this fraction as a species (S) count, we multiply by the number of species. For example, for bats (255 species total in the New World) at 44°N we predict:

No. of species at latitude $L = [\text{Pr}(L)]S = 2pqS$

$$= 2(0.818)(1 - 0.818)(255)$$

$$= 0.2978(255)$$

$$= 75.94 \text{ species}$$

The results of this simple model can be illustrated schematically as a diagram from Willis and Lyons (1998): (the bats' range is shown by line a)

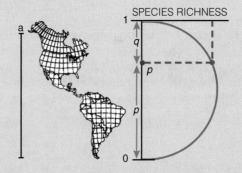

The predicted result of this model is a parabola that reaches its maximum at the halfway point between the northern and southern extremes of the latitudinal range. Deviations from this random model can be examined in relation to the factors limiting geographic ranges and the factors controlling biodiversity in communities.

Factors That Might Cause Diversity Gradients

Tropical-to-polar gradients in species richness may be produced by up to eight interrelated causal factors (Table 22.1.) These gradients represent a complex community property in which we cannot look for a single explanation involving only one causal factor. Many causes have interacted over evolutionary and ecological time to produce the assemblages we see today. For any particular diversity gradient we can ask which of these eight factors are involved and which are most important.

TABLE 22.1   Factors hypothesized to influence biodiversity.

Factor	Rationale
1. History	More time permits more complete colonization and the evolution of new species
2. Habitat heterogeneity	Physically or biologically complex habitats provide more niches
3. Competition	Competition affects niche partitioning
4. Predation	Predation retards competitive exclusion
5. Climate	Fewer species can tolerate climatically unfavorable conditions
6. Climatic variability	Fewer species are adapted to tolerate variable environments
7. Productivity	Richness is limited by the partitioning of production or energy among species
8. Disturbance	Moderate disturbance retards competitive exclusion

Source: Modified after Currie (1991), and Fraser and Currie (1996).

History Factor

The idea that history affects diversity, proposed chiefly by zoogeographers and paleontologists, has two main components. First, biotas in the warm, humid tropics are likely to evolve and diversify more rapidly than those in the temperate and polar regions (Figure 22.11) because of a constant, favorable environment and a relative freedom from climatic disasters like glaciation. Second, biotic diversity is a product of evolution and therefore is dependent on the length of time through which the biota has developed in an uninterrupted fashion (Fischer 1960). Tropical biotas are examples of mature biotic evolution, whereas temperate and polar biotas are immature communities. In short, all communities diversify over time, and older communities consequently have more species than younger ones.

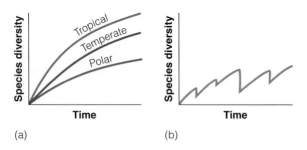

(a)   (b)

FIGURE 22.11

History as a factor in biodiversity: (a) Hypothetical increase in species diversity with decreasing latitude in the absence of interruptions; (b) actual pattern of change in species diversity of a temperate or polar habitat subjected to glaciation and climatic variations. (After Fischer 1960.)

History may operate on ecological or evolutionary time scales. The ecological time scale is a shorter time scale, operating over a few generations or over a few tens of generations. Ecological time involves situations in which a given species could occupy an environment but has not had time to disperse there. The evolutionary time scale is a longer time scale, operating over hundreds and thousands of generations. Evolutionary time applies to cases in which a position in the community exists but is not occupied because insufficient time has elapsed for speciation and evolution to have occurred.

Lake Baikal in the former Soviet Union is a particularly striking illustration of the role of time in generating species diversity. Baikal situated in the temperate zone, is one of the oldest lakes in the world, and contains a very diverse fauna (Kozhov 1963). For example, there are 580 species of benthic invertebrates in the deep waters of Lake Baikal. A lake of comparable area in glaciated northern Canada, Great Slave Lake, contains only four species in this same zone (Sanders 1968).

Some palaeontological data support the assumption that species diversity increases over geological time. The number of species of terrestrial plants, as reflected in the fossil record, appears to have increased in two waves during the past 450 million years (Figure 22.12). No plateau in biodiversity has yet been reached for terrestrial plants (Knoll 1986).

Note that the species diversity of a community is a function not only of the rate of addition of species through evolution but also of the rate of loss of species through extinction or emigration. Compared with polar communities, the tropics could have both a more rapid rate of evolution and a lower rate of

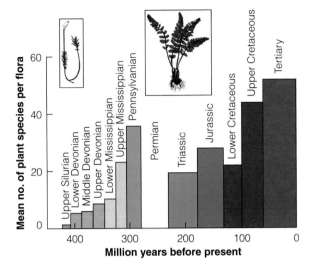

FIGURE 22.12
Historical pattern of increase in the number of terrestrial plant species over evolutionary time; data are derived from fossils. (Data from Nicklas et al. 1980.)

extinction, and these two rates act together to determine species diversity. If we accept this analysis, we are faced with a second problem: Why is the rate of evolution greater in the tropics, or the rate of extinction lower? History can work only through one or more of the other seven ecological factors that affect species diversity. Because it often involves geological time scales and is not amenable to direct experimentation, history is the most difficult of the eight causal factors to assess.

The historical factor can also be used to try to explain differences in diversity on an ecological time scale. One way to do this is to study recently colonized areas and see if species richness increases over time. For example, sugarcane, which is native to New Guinea, has been introduced into at least 75 regions around the world over the past 3000 years. Strong et al. (1977) could find no tendency for the number of arthropod pests of sugarcane to increase with time. Pest species must accumulate in this system very quickly, and the history hypothesis does not explain why sugarcane has more pests in some areas and fewer pests in others. The best predictor of the number of pests on sugarcane is the area of cane under cultivation.

The history hypothesis suggests that species richness never reaches a limit but continues rising over time. We do not know if this is a correct interpreta-

tion of the fossil record (Gould 1981). Because history is a difficult factor to test, we should use it only after we have exhausted simpler hypotheses to explain diversity trends. Factors must operate in ecological time to maintain biodiversity in communities, and we need to analyze these factors to understand the best ways of protecting biodiversity in the future.

Spatial Heterogeneity

A general increase in environmental complexity may occur as one proceeds toward the tropics. The more heterogeneous and complex the physical environment, the more complex the plant and animal communities and the higher the species diversity. This factor can be considered on both a large and a small scale.

Topographic relief, one aspect of spatial heterogeneity, may have a strong effect on species diversity in some groups of organisms. Simpson (1964) has shown that the highest diversities of mammals in the United States occur in mountainous areas (Figure 22.9). The explanation for this seems quite simple: Areas of high topographic relief contain many different habitats and hence more species. Also, mountainous areas produce more geographic isolation of populations and may thus promote speciation. But this conclusion does not fit all taxonomic groups. Neither the trees (see Figure 22.5) nor the land birds (see Figure 22.6) of North America show diversity patterns related to topographic relief.

MacArthur (1965) suggested that we should recognize two components in trying to analyze latitudinal gradients in species diversity: *within-habitat* diversity (also called α diversity), and *between-habitat* diversity (called β diversity). We can illustrate this distinction by two simple schemes to explain tropical diversity:

Hypothetical scheme A	Temperate location	Tropical location
No. of species per habitat	10	10
No. of different habitats	10	50

Hypothetical scheme B	Temperate location	Tropical location
No. of species per habitat	10	50
No. of different habitats	10	10

In scheme A, between-habitat diversity or β diversity accounts for all the increase in diversity for tropical

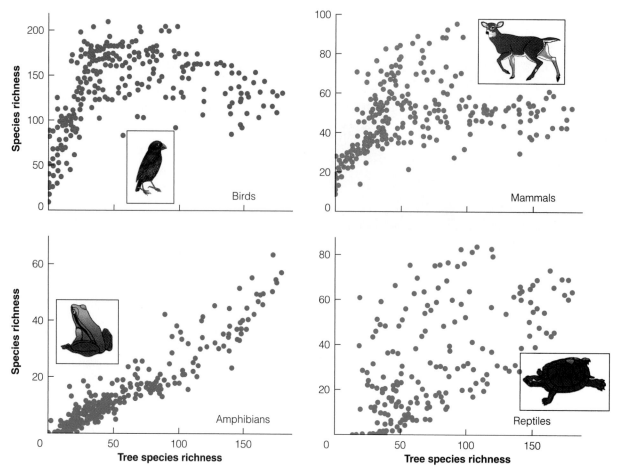

FIGURE 22.13
Species richness (number of species) of North American vertebrates in relation to tree species richness. (The number of tree species is a good index to the total number of plant species in a habitat.) Only for amphibians is there any correlation with tree species richness. (From Currie 1991.)

species; in scheme B, all the increase in tropical diversity is due to within-habitat diversity or α diversity.

Can topographic relief provide some explanation for latitudinal variation in species diversity? There is some evidence that this is part of the reason that the tropics are so rich in species. MacArthur (1969) showed that, for land birds, 2 hectares of Panama forest supported $2\frac{1}{2}$ times as many bird species as 2 hectares of Vermont forest. But larger areas in the tropics support proportionately even more species. Ecuador has seven times the number of bird species as New England, even though the areas are both approximately 260,000 km². For land birds, there are both more species per habitat in Ecuador and also more habitats per unit of area, so that scheme A and scheme B are both correct for birds.

If spatial heterogeneity means more habitats, we ought to be able to measure this by the number of plant species in a large region. One of the simplest models for animal biodiversity is that it is determined by plant biodiversity. Unfortunately this model does not seem to be correct. For North America, neither mammal, bird, nor reptile diversity is related to plant biodiversity (Figure 22.13). Only amphibian diversity is closely correlated with plant diversity, but this correlation is probably due to some third factor related to climate that correlates well with both amphibian diversity and plant diversity (Currie 1991).

Tropical-to-polar gradients in the oceans seem unlikely to be explained by spatial heterogeneity. The oceans are not uniform water masses, yet they provide fewer opportunities for habitat specialization. Benthic marine invertebrates become more diverse as one moves from shallow waters on the continental shelf to deeper waters at the edge of the shelf (Sanders 1968). There is no obvious change in bottom sediments to explain this increase in biodiversity.

We conclude that spatial heterogeneity does not explain very many of the observed tropical-to-polar diversity gradients (Rohde 1992). It cannot be a general explanation because many aquatic habitats such as shallow saltwater mudflats show these gradients in the absence of any change in spatial heterogeneity. In cases in which spatial heterogeneity can be used to explain latitudinal gradients in species diversity, we must still identify the ecological "machinery" behind this explanation.

Competition

Many naturalists have argued that natural selection in the temperate and polar zones is controlled mainly by physical factors of the environment, whereas biological competition becomes a more important part of evolution in the tropics. For this reason, the argument goes, animals and plants are more restricted in their habitat requirements in the tropics, and this increases between-habitat (β) diversity. Animals may also have more restricted diets in each habitat, increasing within-habitat (α) diversity. Competition is keener in the tropics, and niches are smaller. Tropical species are more highly evolved and possess finer adaptations than do temperate species. Consequently, more species can occupy a given habitat in the tropics (Dobzhansky 1950).

Competition theory (see Chapter 12) can be expanded in an attempt to explain species diversity in an equilibrium world (Chesson and Case 1986). The key prediction that emerges is that at least n limiting resources are needed for the coexistence of n species in a community. For plants there are at most four or five limiting resources (Tilman 1986), and thus competition by itself cannot explain the large number of plant species in natural communities. The conclusion is that this hypothesis is not a possible explanation for within-habitat diversity of plants.

For animal species, many more limiting resources potentially exist. The effect of competition on species richness can be made apparent by looking at the niche relations of the species in a community. Consider the simple case of one resource, such as soil water for plants or food-item size for animals. Two niche measurements are critical: *niche breadth* and *niche overlap* (Figure 22.14a). We can recognize two extreme cases.

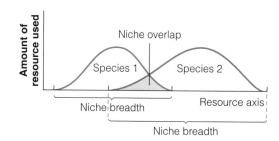

(a)

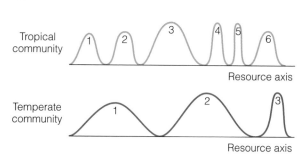

(b) No niche overlap example

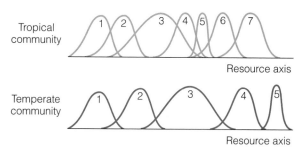

(c) Constant niche breadth example

FIGURE 22.14

Diagram to illustrate two extreme hypothetical cases of how niche parameters may differ in tropical and temperate communities. Both niche breadth and niche overlap are determined by competition within the communities. If there is no niche overlap (b), the number of species that can be in a community is determined by niche breadth. If there is constant niche breadth (c), the number of species is determined by the amount of niche overlap.

If there is no niche overlap between species, then the wider the average niche breadth, the fewer the number of species in the community (Figure 22.14b). At the other extreme, if niche breadth is constant, then the smaller the niche overlap, the fewer the species in the community (Figure 22.14c). In this hypothetical analysis, tropical animal communities might have more species because tropical species have smaller

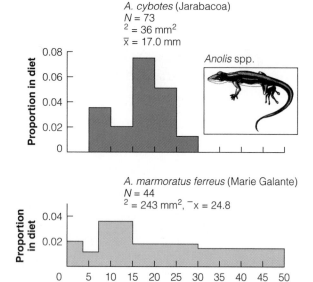

FIGURE 22.15
Niche breadth of two Anolis *lizard species on islands in the Caribbean.* Anolis cybotes *coexists with five other* Anolis *species on Jarabacoa and has a narrow niche.* Anolis marmoratus *is the only species on Marie Galante. (After Roughgarden 1974.)*

niche breadths or greater niche overlaps. Both these arguments assume that Gause's hypothesis (see p. 190) is true for natural communities.

To evaluate the competition factor, we must measure these niche parameters in a variety of tropical and temperate animal communities. The problems of measuring niche overlap and niche breadth are discussed in detail by Magurran (1988) and Krebs (1999). The basic problem is to decide which resource axes are relevant to any particular group of species; if the resource axes can be linearly ordered and measured, these niche parameters can be measured as indicated in Figure 22.14.

In relatively few cases have detailed measurements been made to test the schematic model of Figure 22.14. The best example is the work on lizards summarized by Roughgarden (1986). Lizards of the genus *Anolis* are small, diurnal, insect-eating iguanid lizards that are a dominant component of the vertebrate community on islands in the Caribbean. Most species perch on tree trunks or bushes. They are sit-and-wait predators, and food size is a critical niche dimension. Roughgarden (1974) tested the prediction that niche breadth would decrease as more species occurred together on an island. Figure 22.15 shows the results

for two *Anolis* species. The results are consistent with the predictions from competition theory and support the suggestion that niche breadth is reduced in species-rich communities. Pacala and Roughgarden (1985) showed in enclosure experiments that *Anolis* species show strong effects of competition when their diets are similar; competition for food is a major factor determining the species diversity of these lizards.

Competition between species is an important process in population dynamics (see Chapter 12), and in Chapter 23 we will discuss the role of competition in community organization. It is clear that competition does not play a large role in the maintenance of plant biodiversity (Austin 1990), but it is less clear how much it affects animal biodiversity in modern communities. The evolution of niche parameters due to competition could be a cause of greater tropical diversity, or it could be an effect of higher species numbers (Rohde 1992).

Predation

Paine (1966) argued that predators and parasites are more abundant in the tropics than elsewhere, and they hold their prey populations to such low levels that competition among prey organisms is reduced. This reduced competition allows the addition of more prey species, which in turn support new predators. Thus, in contrast to the competition proposal, *less* competition should exist among prey animals in the tropics. Providing we can measure "intensity of competition," we can distinguish quite clearly between these two ideas.

Paine (1966) supported his ideas with some experimental manipulations of rocky intertidal invertebrates of the Washington coast. The food web of these intertidal areas on the Pacific coast is remarkably constant: Paine removed the starfish *Pisaster* from a section of the shore and observed a *decrease* in diversity from a 15-species system to an 8-species system. A bivalve,

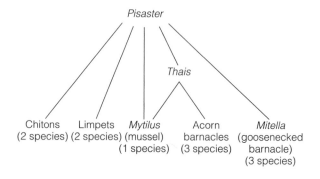

Mytilus, tended to dominate the area, crowding out the other species. Four of the species that disappeared were not eaten by *Pisaster* but were affected by the increase in *Mytilus*. "Succession" in this instance is toward a simpler community. By continual predation, the starfish prevent the barnacles and bivalves from monopolizing space. Thus local species diversity in intertidal rocky zones appears to be directly related to predation intensity. Paine called the starfish a *keystone species* in this community.

The prediction from Paine's work that increased predation will lead to greater diversity of prey species depends on the ability of one prey species to be competitively dominant. For the predation hypothesis to operate on a broad scale, the predators involved must be very efficient at regulating the abundance of their prey species. In terrestrial food webs, predators are usually specialized and in some cases do not seem to regulate prey abundance (see Chapter 13). Note that the predation hypothesis cannot be a sufficient explanation for tropical species diversity unless it can be applied to all trophic levels. If the species diversity of herbivores is determined by their predators, we are left with explaining the diversity of the primary producers. Keystone species, such as the starfish *Pisaster*, should be more common in tropical communities, but currently there is no evidence that this is correct (Power et al. 1996).

The effect of predation can be extended to the primary-producer level. Tropical lowland forests contain many species of trees and corresponding low densities of adult trees of each species. Most adult trees of a given species are also spread out in a regular pattern in the tropical forest, leading Janzen (1970) and Connell (1971) to suggest that these characteristics of tropical trees can be explained by the predation hypothesis, with the species that eat seeds or seedlings

Joseph H. Connell *(1923–) Research Professor of Biology, University of California, Santa Barbara*

filling the role of the predators. The Janzen-Connell model for the maintenance of tropical tree species diversity, shown schematically in Figure 22.16, predicts that tree seedlings will do poorly if they are close to a large tree of the same species. To test this, Steve Hubbell and Robin Foster in 1981 established a 50-ha forest plot on Barro Colorado Island in Panama. This plot was censused in 1981–1983, 1985, and 1990–1991, and 244,000 stems of 303 species were measured and geographically mapped in 1990–91 (Condit 1995). Many of the species in this rain-forest plot show intraspecific effects of density on recruitment (Wills et al. 1997). Most of the negative effects occur within a species, so interspecific competition for resources does not seem important in maintaining

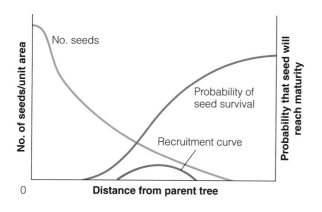

FIGURE 22.16

The Janzen-Connell model to account for high tropical-forest diversity. The amount of seed dispersed declines rapidly with distance from the parent tree, and the activity of host-specific seed and seedling herbivores and diseases is most evident near the parent tree. The product of these two factors determines a recruitment curve that peaks at the distance from the parent tree at which a new adult tree is likely to appear. (Modified after Janzen 1970.)

FIGURE 22.17
Mortality rate of 843 Ocotea whitei saplings on Barro Colorado Island in Panama in relation to distance of each sapling from the nearest adult tree of the same species. This mortality is caused primarily by a stem canker disease, and was measured over the interval from 1982 to 1991. The Janzen-Connell model predicts higher mortality nearer to conspecific adults, exactly as shown here. (Data from Gilbert et al. 1994.)

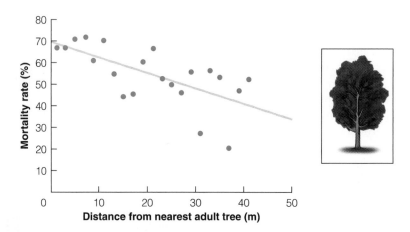

diversity. Pests and pathogens may be particularly significant. Figure 22.17 shows that the effect of stem canker disease in a laurel tree (*Ocotea whitei*) sapling is more severe near an adult tree, as the Janzen-Connell model would predict. Thus, each tree casts a "seed shadow," in which survival of its own kind is reduced. As one moves from the lowland tropical forests to temperate forests, the seed and seedling herbivores are hypothesized to be less efficient at preventing establishment of seedlings close to the parent tree. Data from tropical rain forests have supported the Janzen-Connell model in some but not all cases (Wills et al. 1997).

The effects of predation and competition on species diversity may be complementary (Lubchenco 1986). Competition may be more important in maintaining high diversity among parasites and predators, whereas the process of predation and disease may be more important among herbivores and plants, respectively. Superimposed on these effects is another pattern: In complex communities with many species, predation may be the dominant interaction affecting diversity, whereas competition may be the dominant interaction in simple communities.

Climate and Climatic Variability

The more stable the climatic parameters and the more favorable the climate, the more species will be present. According to this idea, regions with stable climates allow the evolution of finer specialization and adaptations than do areas with erratic climates, resulting in smaller niches and more species occupying a unit space of habitat. Species should be more flexible in temperate and polar areas and should be more specialized in the tropics.

The factor of climatic stability may combine with history, with which it has much in common. The *stability-time* hypothesis (Sanders 1968) emphasizes the role of all environmental parameters—temperature, moisture, salinity, oxygen, pH—and so on—in permitting diversity. Low-diversity habitats may be *severe, unpredictable*, or both. Some severe environments, such as hot springs or the Great Salt Lake in Utah, can be very predictable (physical conditions are constant from day to day) but have low species diversity. A desert environment with irregular rainfall would be an example of an unpredictable and severe environment. Sanders (1968) applied the stability-time hypothesis to the marine fauna of deep-sea muddy bottoms. The deep sea represents a stable environment of long standing, and the diversity of bivalve and polychaete species in bottom samples in the deep sea was almost equal to that in tropical shallow-water areas. Abele and Walters (1979) criticized Sanders's (1968) analysis and concluded that marine invertebrate biodiversity was better explained by the spatial heterogeneity of their habitats.

One of the simplest explanations of the polar-to-tropical gradient in terrestrial biodiversity is the *species richness-energy model* (Brown 1981), which suggests that species richness is limited by the available energy. Recent data support this hypothesis for trees (Currie and Paquin 1987), British birds (Turner et al. 1988), and vertebrates from North America (Currie 1991). Figure 22.18 illustrates the relationship between biodiversity and available energy for trees and for vertebrates. Available energy can be measured by annual evapotranspiration, which measures the energy bal-

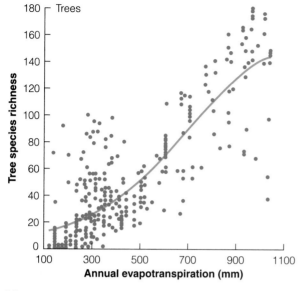

(a)

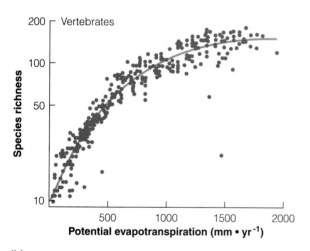

(b)

FIGURE 22.18
The species richness–energy hypothesis for biodiversity. Species richness of (a) trees and (b) vertebrates from North America are related to annual available energy at each site, as measured by evapotranspiration (which combines solar radiation and temperature). (From Currie 1991.)

ance at a site and is a function of solar radiation and temperature. Vertebrate biodiversity in North America is not correlated with plant productivity or with climatic variability and is only weakly correlated with spatial heterogeneity (Currie 1991). The species richness–energy hypothesis can be tested readily because it makes specific predictions about seasonal bird migrants (Turner et al. 1988). In temperate areas, energy levels in summer should control the diversity of summer birds, whereas energy levels in winter should control the numbers of winter residents. In 75 localities in Britain, the biodiversity of summer birds is correlated with summer temperature, and the diver-

sity of winter birds is correlated with winter temperature, exactly as this energy hypothesis predicts.

Coral reefs are species rich in tropical waters, and the number of taxa rapidly declines as one moves from warm tropical seas to cooler temperate waters (Figure 22.19). High species diversity in corals has usually been attributed to the high rates of speciation in the Indo-Pacific region associated with the historical factor. Fraser and Currie (1996) found that the best predictors of coral species diversity were ocean temperature and coral biomass, so energy-rich areas had more coral genera and species. Historical factors are responsible for the major differences between the

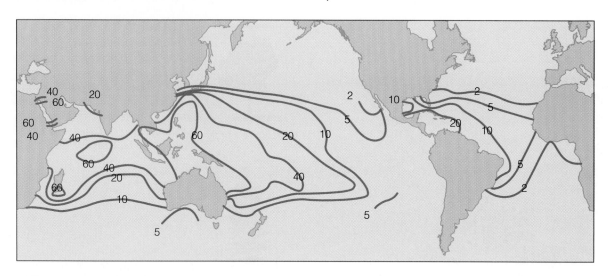

FIGURE 22.19
Abundance of corals in all of the tropical and subtropical areas of the world. Contour lines connect areas with the same number of genera. The Indo-Pacific region is much richer in corals than is either the eastern Pacific or the Atlantic regions. (From Fraser and Currie 1996.)

number of coral species in the Atlantic and the Indo-Pacific regions, so that both history and available energy were important overall.

Favorable climates on a broad geographic scale thus support high biodiversity. This idea explains a large fraction of the global tropical-to-polar diversity gradient, but it cannot explain local, habitat-level variations in species diversity. No single factor can explain all biodiversity gradients biodiversity from the local to the regional scale.

Productivity

The productivity hypothesis in its pure form states that greater production results in greater diversity, everything else being equal. The data available do not support this idea. For example, Tilman (1986) describes several examples in which plant biodiversity is maximal in resource-poor habitats of low productivity. Two of the world's most diverse plant communities, the fynbos of South Africa and the heath scrublands of southeastern Australia, both occur on nutrient-poor soils, and in both cases adjacent areas with better soils and more productive vegetation have fewer species. Productivity in plant communities seems to lead to reduced biodiversity on a local scale (Tilman 1986).

Productivity has been considered more important for animal communities on a global scale, but again

the available data do not agree with this conclusion. Currie (1991) could find no relationship between productivity and vertebrate biodiversity in North America. Productivity by itself does not seem to be the key to understanding diversity gradients on a global scale.

A common modification of the productivity factor is the idea of increased temporal partitioning in the tropics. The main argument is that the longer growing season of tropical areas allows the component species to partition the environment temporally as well as spatially, thereby permitting the coexistence of more species. This idea combines the stability hypothesis with the productivity hypothesis and suggests that the stability of primary production is a major determinant of the species diversity in a community (Connell and Orias 1964). One way of testing this idea is by looking at primary production in different communities over an annual cycle; another way is to look at the way organisms thrive in different communities.

Disturbance

If natural communities exist at equilibrium and the world is spatially uniform, then competitive exclusion ought to be the rule, and each community should come to be dominated by a few species—the best competitors (Crawley 1986). But if communities exist in a nonequilibrium state, competitive equilibrium is prevented. A whole range of factors can prevent equilib-

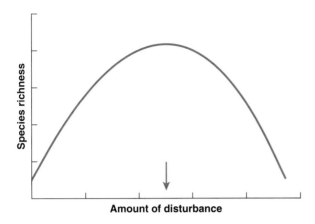

FIGURE 22.20
The intermediate disturbance hypothesis of species diversity. This model predicts that biodiversity is maximized at some intermediate level of disturbance. (red arrow). (Modified from Huston 1979.)

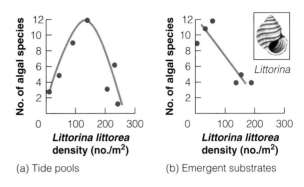

(a) Tide pools   (b) Emergent substrates

FIGURE 22.21
The effect of periwinkle snail grazing on the diversity of algae in (a) high-tide pools and (b) on emergent rocks in the low intertidal zone in Massachusetts and Maine. The intermediate disturbance hypothesis applies only in the tide pools. (From Lubchenco 1978.)

rium, including predation, herbivory, fluctuations in physical factors, and catastrophes such as fires, and we lump these together as "disturbance." When disturbances occur too often, species go extinct if they have low rates of increase. When disturbances are rare, the system goes to competitive equilibrium and species of low competitive ability are lost. The idea that in between is a level of disturbance that maximizes biodiversity (Figure 22.20) is called the *intermediate disturbance hypothesis* (Grime 1973, Horn 1975, Connell 1978). If population growth rates are low for all members of a community, the competitive equilibrium is approached so slowly that it is never reached. Thus, species diversity is maintained by periodic disturbance or by environmental fluctuations. If this model is in fact correct, the worst thing we can do to a community is to prevent disturbances such as fire.

Disturbance can also operate on a local scale to produce patches that undergo succession. Within each patch on a local area the species composition may be changing, but on a larger spatial scale the species composition may be constant and include both pioneer species and climax species (Connell 1987).

Disturbance does not always produce maximum diversity at intermediate levels of disturbance, as predicted by the intermediate disturbance hypothesis (Wootton 1998); some data are at variance with this prediction. Three examples illustrate some of the patterns. On rocky shores in Massachusetts the periwinkle snail *Littorina littorea* is the most common herbivore (Lubchenco 1978). In tide pools, moderate grazing by *Littorina* on the algae that are competitively dominant permits many competitively inferior algae to survive (Figure 22.21a), as predicted by the intermediate disturbance hypothesis. But on emergent rocks, the snails do not eat the perennial brown and red algae that are competitively superior, but instead feed on the competitively inferior algae. Consequently, on emergent rocks *Littorina* grazing reduces algal diversity (Figure 22.21b). The critical factors are the food preferences of the grazer and the competitive abilities of the plants.

Streams may have variable water flow and changeable water temperatures. Death and Winterbourn (1995) found that aquatic invertebrates did best when there was minimal disturbance, contrary to the predictions of the intermediate disturbance hypothesis (Figure 22.22a). Similarly, in prairie grassland fire is a major disturbance, and plant diversity declines with more fires (Figure 22.22b). The intermediate disturbance hypothesis does not apply to some grazing systems (Collins et al. 1995, Stohlgren et al. 1999).

The intermediate disturbance hypothesis is an attractive hypothesis for the maintenance of high species diversity in communities, but it does not apply to all communities, and further work is needed to delimit its range of application. In particular, land managers should not assume the validity of the intermediate disturbance hypothesis in making management plans for national parks or other protected areas.

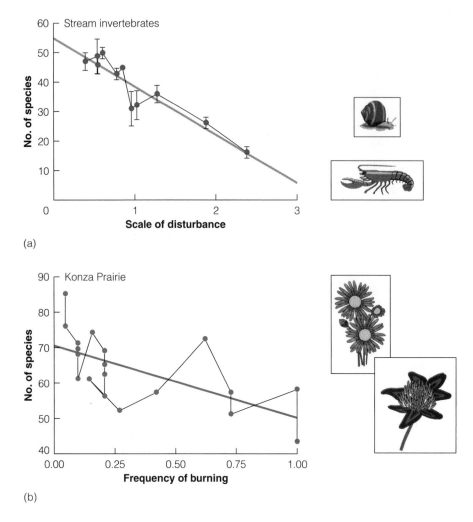

FIGURE 22.22
The effect of disturbance on species richness in two contrasting communities. (a) Aquatic invertebrates in streams on the South Island of New Zealand. The scale of disturbance is a composite measure of variation in temperature, stream flow, and bottom stability. (b) Plant species richness in tallgrass prairie in Kansas. The frequency of burning is the probability of being burned each year between 1972 and 1990. Because species richness declines with the amount of disturbance, the intermediate disturbance hypothesis does not apply to either of these communities. ((a) From Death and Winterbourn 1995. (b) From Collins et al. 1995.)

Local and Regional Diversity

Biodiversity in local habitats could be limited by either evolutionary or ecological causes. The mixing of evolutionary processes on a long time scale and ecological processes on a short time scale has made it difficult to untangle the reasons for the latitudinal change in species diversity, as we have just seen. One way to separate out evolutionary and ecological causes is to see if each community is saturated with species by plotting local species diversity against regional species diversity.

To do this we need to define what is local and what is regional (Srivastava 1999). Local diversity is measured on a scale in which all the species in the community could interact with each other in some unit of ecological time, typically a generation. For example, fish species in a lake or stream, herb species in a meadow, or bird species in woodlots are all examples of local diversity. Regional diversity, by contrast, refers to a larger spatial scale, typically 100 or more times that of the local scale. Within the region, species could disperse to and colonize a local patch through dispersal over tens of generations. Examples of regional diversity would be the fish species of the Great Barrier Reef, the grass species of Britain, or the bird species of the boreal forest of Canada and Alaska. Regional species richness can be specified only if the flora and fauna of a region are well known. For this reason, studies of local and regional diversity have concentrated on the better known groups such as birds, butterflies, and trees.

Communities would be saturated if there were intense competition among the existing species such that no new species could fit into the suite of species. Figure 22.23 illustrates the idea of testing for local community saturation; the key is that we expect a linear relationship if communities are unsaturated, and a curvilinear relationship if communities are saturated. The best data for comparisons are from a single defined habitat sampled in several geographically distinct regions. Figure 22.24 shows a comparison of local and regional diversity at the global scale for a whole range of taxa (Caley and Schluter 1997). In a broad, global sense, there is no evidence of local community saturation, which implies that biodiversity at the local level is not constrained by intensive competition, and that communities are not closed to new invaders.

On a smaller spatial scale, Pearson (1977) studied the bird communities of six undisturbed lowland tropical forest sites in Amazonia, Borneo, New Guinea, and West Africa. He censused local plots of about 15 ha, spending 200–700 hours on each plot to census the birds. He found that local and regional richness were linearly related, suggesting that bird communities in these tropical forests were not saturated with species (Figure 22.25).

The majority of studies to date have shown communities to be unsaturated. Srivastava (1999) summarized 36 studies, of which about two-thirds reported

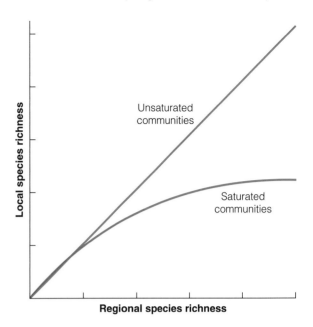

Regional species richness

FIGURE 22.23
Local and regional biodiversity plots. If local communities are unsaturated, community diversity will continue increasing with regional diversity in a linear manner (blue line); richer regions will have richer local communities. By contrast, local communities that are saturated with species will reach an asymptote or maximum species richness (red line) set by competition for resources and niche overlap. Testing for saturation requires examining several communities in several different regions.

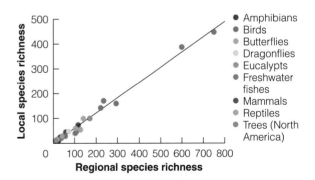

FIGURE 22.24
Local and regional species richness across continents for nine different taxonomic groups. No indication of saturation is apparent in this relationship, and remarkably all the different groups on different continents appear to follow the same linear regression. (Data from Caley and Schluter 1997.)

communities to be unsaturated. But several pitfalls exist in the analyses of local-regional diversity plots. Sample sizes must be large, or species will be missed (Caley and Schluter 1997). Determining the number of species in the regional pool must be done carefully, and the size of the regions should be equal to avoid bias. At present the assumption that ecological communities are saturated with species does not appear to be correct for many species groups (Ricklefs and Schluter 1993). We explore the consequences of this hypothesis in the next two chapters.

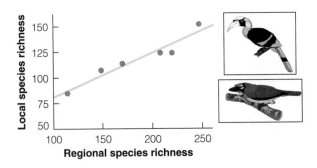

FIGURE 22.25
The relation of local and regional species richness of lowland tropical forest birds in Central America. The linear relationship suggests that bird communities are unsaturated. (Data from Pearson 1977 and Srivastava 1999.)

Summary

Biodiversity can be measured by simply counting all the species in a collection or by weighting each species by its relative abundance. Most communities consist of a few common species and many rare species.

Tropical environments support more species in almost all taxonomic groups than do temperate and polar areas. The latitudinal gradient in biodiversity from the tropics to the polar regions is one of the great patterns found in community ecology. The trees, birds, and mammals of North America and Australia illustrate the complexity of species diversity gradients, which are not always gradual trends from the equator to the poles.

Combinations of eight factors—history, spatial heterogeneity, competition, predation, climate, climatic variability, productivity and disturbance—postulated to control biodiversity operate in many natural communities. Three factors seem most important: (1) *History*, which summarizes the evolutionary history of a region, is the most difficult factor to evaluate and is potentially important on a regional scale. (2) *Climate*, particularly solar energy, operates on a regional scale and is a good predictor of diversity for both terrestrial and aquatic communities. (3) *Disturbance*, can operate on a local scale to maximize diversity by preventing competitive exclusion.

There is no simple, general answer to the question, *What controls biodiversity?* As we have seen in most ecological systems, the answer depends on both the taxonomic group and the scale of analysis. Evolutionary factors are needed to explain the origin of high-diversity communities, whereas ecological factors help explain how this diversity is maintained without being lost (Givnish 1999). Some factors (*history*) may provide insight on a global scale, whereas others (*predation*) help to explain biodiversity in local habitats. Communities do not appear to be saturated with species, and we are led to consider the broader question of what controls community organization.

To protect and preserve biodiversity we need to know what controls community organization, the central question of the next two chapters.

Key Concepts

1. Biodiversity can be measured most simply by counting the number of species in an area (species richness).

2. There is a strong gradient in species diversity from the tropical regions toward the poles in most groups of plants and animals. This is one of the most striking patterns in community ecology.

3. Eight factors act jointly to enhance species richness in communities. Some factors (history and climatic stability) help to explain the origin of high diversity in tropical areas; others (predation, competition, and spatial heterogeneity) can help to maintain high diversity once it is present.

4. Climatic favorability, measured by incoming solar energy, is the single best predictor of species richness in both aquatic and terrestrial communities.

5. Local species richness tends to increase linearly with regional species richness, suggesting that local communities are never saturated with species.

6. The answer to the general question, *What controls biodiversity?* depends on the species group and the scale of study. What is important at the local level will not necessarily be critical at the global level.

Selected References

Angel, M. V. 1993. Biodiversity of the pelagic ocean. *Conservation Biology* 7:760–772.

Blackburn, T. M. and K. J. Gaston. 1996. Spatial patterns in the species richness of birds in the New World. *Ecography* 19:369–376.

Condit, R. 1995. Research in large, long-term tropical forest plots. *Trends in Ecology and Evolution* 10:18–22.

Currie, D. J. 1991. Energy and large-scale patterns of animal–and plant–species richness. *American Naturalist* 137:27–49.

Danell, K., P. Lundberg, and P. Niemela. 1996. Species richness in mammalian herbivores: Patterns in the boreal zone. *Ecography* 19:404–409.

Ehrlich, P. R. and E. O. Wilson. 1991. Biodiversity studies: Science and policy. *Science* 253:758–762.

Givnish, T. J. 1999. On the causes of gradients in tropical tree diversity. *Journal of Ecology* 87:193–210.

Grubb, P. J. 1987. Global trends in species-richness in terrestrial vegetation: A view from the northern hemisphere. In *Organization of Communities Past and Present*, ed. J. H. R. Gee and P. S. Giller, pp. 99–118. Blackwell Scientific Publications, Oxford.

Magurran, A. E. 1988. *Ecological Diversity and its Measurement*. Croon Helm, London. 179 pp.

Rohde, K. 1992. Latitudinal gradients in species diversity: the search for the primary cause. *Oikos* 65:514–527.

Rosenzweig, M. L. 1995. *Species Diversity in Space and Time*. Cambridge University Press, Cambridge, England. 436 pp.

Srivastava, D. S. 1999. Using local-regional richness plots to test for species saturation: Pitfalls and potentials. *Journal of Animal Ecology* 68:1–16.

Wills, C., R. Condit, R. B. Foster, and S. P. Hubbell. 1997. Strong density–and diversity-related effects help to maintain tree species diversity in a neotropical forest. *Proceedings of the National Academy of Sciences of the USA* 94:1252–1257.

Wootton, J. T. 1998. Effects of disturbance on species diversity: a multitrophic perspective. *American Naturalist* 152:803–825.

Questions and Problems

22.1 The tree flora of Europe is less diverse compared with that of eastern North America or eastern Asia (Grubb 1987). Why should this be? Compare your explanations with those of Grubb (1987) and of Currie and Paquin (1987).

22.2 The number of taxa recorded in the fossil record tends to increase irregularly but steadily with geological time (Gould 1981, Knoll 1986). Strong et al. (1977, p. 173) state, "Where the fossil records exist there is no evidence that species number increases inexorably over long periods in the history of communities." Discuss these different views of whether the history hypothesis fits the fossil record.

22.3 Marine algae along the west coast of North America do not increase in species richness toward the tropics but peak at about 70 species per 100 km of coastline around $40°$ N latitude (Gaines and Lubchenco 1982). Along the east coast of North America, species richness gradually increases as you move toward the tropics. Discuss why these patterns might hold.

22.4 In analyzing the role of fire as a disturbance in tallgrass prairie, Collins et al. (1995) found that the intermediate disturbance hypothesis was supported if, instead of plotting fire frequency as in Figure 22.22b, they plotted time since the last fire on the X-axis. Why should they get different results for these two plots of the same data?

22.5 Would you expect to have latitudinal gradients in the species richness of macroparasites of mammals and birds? What factors might control species richness in macro- and microparasites?

22.6 Whittaker (1972) argues that among terrestrial plants and insects, species diversity can increase without any upper limit because the evolution of diversity is a self-augmenting process. Evaluate Whittaker's argument, and discuss its implications with respect to the analysis of diversity gradients.

22.7 The longest experiment in ecology is the Park Grass Experiment begun in 1856 at Rothamsted, England. A mowed pasture was divided into 20 plots, and a series of plots were fertilized annually with a variety of nutrients, including nitrogen. Discuss the predictions you would make regarding biodiversity on fertilized and unfertilized plots for this experiment, using the eight factors discussed in this chapter. Tilman (1986, pp. p. 62–63) shows the observed results.

22.8 In Antarctica, species richness in soft-bottom invertebrates (sponges, bryozoans, polychaetes, and amphipods) is higher than that of almost all other tropical- and temperate-zone soft-bottom communities (Clarke 1990). What observations or experiments would you perform to find out why this high biodiversity occurs in Antarctica?

22.9 Figure 22.18 shows that on a global scale, species richness increases smoothly with solar energy and temperature. Why should this occur? Why is the available energy not monopolized by a few superspecies? Compare your ideas with those of Currie (1991, p. 46).

22.10 Grazing by cattle is a disturbance that has long caused conflict with conservation biologists. Is there any evidence that grazing increases plant species diversity, at least up to intermediate levels of grazing? Stohlgren et al. (1999) reports data from grazing exclosures in place for up to 60 years and discusses this issue.

22.11 Does the Janzen-Connell hypothesis for diversity maintenance in tropical rain forests imply that mortality in small trees should be density dependent? Describe an experimental design (including the time frame required) that would allow you to test this hypothesis.

22.12 Use the null model described in Box 22.1 to predict the geographic ranges for New World marsupials (70 species in North and South America), which range from 47°N to 55°S. Compare your predictions with the actual data for marsupials given in Willig and Lyons (1998, p. 97).

Overview Question

To *preserve* biodiversity, how much do we need to understand about the factors that *control* biodiversity? Sketch the outlines of a management plan for preserving biodiversity in a large national park in a tropical rain forest.

CHAPTER 23

Community Organization II:
Predation and Competition in Equilibrial Communities

COMMUNITIES CAN BE ORGANIZED by four biological processes—competition, predation, herbivory, and mutualism. Competition among plants, herbivores, and carnivores might control the diversity and abundance of species in a community. Predation and herbivory might organize the community according to "who-eats-whom," such that the framework of community organization is set by the animals. Mutualism, an important process that links species, might serve to increase community organization by linking species to the benefit of all. Physical processes set limits to these four biological processes, and variation in temperature, salinity, and other physical factors have potential implications for the species in a community. To study community organization, we need to look at the component species and the processes that interconnect them.

To speak of community organization implies that there is some regularity in the biomass or the numbers of species that make up the community. Naturalists looking for particular birds, butterflies, or flowers have an implicit model of community organization in mind. Conservation biologists have an implied model of community organization when they discuss the preservation of the Florida Everglades or other natural landscapes. Natural communities could be very loosely organized, or be very tightly organized. How can we determine this for any particular community?

Communities contain so many different species that we cannot study each species separately. One way to reduce the complexity of communities is to measure the biodiversity of the community, as we saw in Chapter 22. If we measure the species richness of a community, we implicitly assume that each species is equal to every other species in the community. A second way to simplify the analysis is to define feeding roles in the community and to group species according to their roles. We can group species into *trophic levels* (such as herbivores), or at a finer level into *feeding guilds*. A third way is to look at particular types of species and to ask; *Are all species of equal importance in a community?* This question is purposely vague in that we must define *importance*, and we can do this in several ways (see Box 23.1 for measuring community importance). We could consider a species important if, when we remove it, the diversity or abundance of other species in the community changes. Such species are *keystone species*. Alternatively, we could determine which species are most common in the community—the *dominant species*. Dominant species could be major players in defining the organization of the community. In this chapter we discuss each of these approaches to understanding community organization.

We begin this analysis with the classical assumption that communities are in equilibrium. Communities are in equilibrium when species abundances remain constant over time, when nature is in a "state of balance." In most cases equilibrium means *stable equilibrium* (Figure 23.1a). In different habitats the

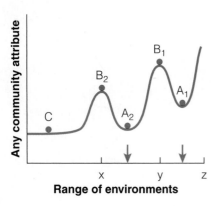

(a)

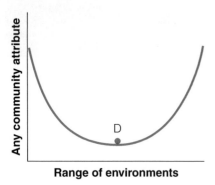

(b)

FIGURE 23.1
Schematic view of types of equilibrium points: (a) Two locally stable equilibrium points A_1 and A_2 are shown. B_1 and B_2 are unstable equilibrium points, and C is a neutrally stable equilibrium point. The community at A_1 is locally stable between the environmental range from y to z, and A_2 is locally stable between x and y. (b) A globally stable community at D will come to the same point no matter what the environmental change. Most real-world communities are only locally stable; only a few are globally stable. (Modified from DeAngelis and Waterhouse 1987.)

equilibrium point may differ, such that there is spatial variation in species numbers, but the key point is that at each spatial location the community is in equilibrium and remains constant (Chesson and Case 1986). This equilibrium will usually be *locally stable* within a specified environmental range. In some cases the equilibrium can be *globally stable*, such that over all environmental conditions the system will return to the equilibrium point following any disturbance (Figure 23.1b).

The classical equilibrium assumption of community ecology is an abstraction and will not be found in its pure state in natural communities. Real communities will be spread along a continuum from equilibrium to non-equilibrium (Figure 23.2). Equilibrium communities purportedly show stability, and stability can be measured in several different ways (Pimm 1991). The mathematician's idea of *local stability* (points A_1 and A_2 in Figure 23.1a) is the simplest meaning. Stability can be measured by the *time* it takes for a community to recover from disturbance; accordingly, stable communities recover quickly from disturbances. Stability can also be measured as the *variability* of a community over time, so that if the populations that make up the community fluctuate in size dramatically from year to year, the community would be considered unstable. (This is the most common meaning ecologists attach to the word *stability*.) Stability can also be measured as the *persistence* of a community over time.

An ideal equilibrium community would score high on all these measures of stability. Such a community would have many biotic interactions involving competition and predation, and these processes would operate in a density-dependent manner to regulate population size (see Chapter 16). Equilibrium communities would also be saturated with species, such that species invasions would be rare. Weather catastrophes would rarely occur, and the community would form a tightly coupled biotic unit, an interlocking web of life.

By contrast, ideal nonequilibrium communities would score low on all these measures of stability. Species would operate individualistically, and density-dependent population regulation would be difficult to find. Climatic catastrophes would occur frequently, and species would come and go regularly, such that the composition of the community would be highly variable. We discuss nonequilibrium community dynamics and attempt to ascertain where most natural communities fall on this continuum in Chapter 24.

The three major equilibrium theories of community organization are the classical competition theory, the competition-predation theory, and the competition-spatial patchiness theory:

1. *Classical competition theory:* Hutchinson (1959) argued that competition is the major biological process controlling community structure. The subsequent development of this theory is reviewed by Armstrong and McGehee (1980). The essential assumptions of this theory are that:

B O X   2 3 . 1

MEASURING COMMUNITY IMPORTANCE

Even if all species in a community are not equally important, we need to develop a measure of community importance. Mills et al. (1993) were the first to define community importance values for particular species:

$$CI_x = \begin{bmatrix} \text{Percentage of species lost} \\ \text{from a community upon} \\ \text{removal of species } x \end{bmatrix} \quad (23.1)$$

Thus, if you remove the starfish *Pisaster* from a rocky intertidal area, and nine of 23 invertebrate species are lost from that area, the community importance value of *Pisaster* would be 39%. By contrast, if a redundant species were removed, nothing would happen and the community importance value for that species would be 0%.

Power et al. (1996) recognized that not all of the effects of a keystone species would show up as species losses, so they devised the following metric of community importance to make it more general:

$$CI_x = \frac{(t_N - t_D)/t_N}{p_x} \quad (23.2)$$

where CI_x = community importance of species x

t_N = quantiative measure of community trait in intact community

t_D = quantiative measure of community trait after species x is removed

p_x = proportional abundance of species x before removal

Any community trait can be used—species richness, productivity, or the abundance of indicator species. For example, Fagan

and Hurd (1994) studied the effects of praying mantids on the numbers of other arthropods and found that the arthropod community without mantids had 316 individuals in 4 m2, whereas plots with mantids had 194 individuals. Mantids averaged 14 individuals, or 7.2% of the arthropods. Consequently, for these mantids:

$$CI_x = \frac{(t_N - t_D)/t_N}{p_x} = \frac{(194 - 316)/194}{0.072} = -8.73$$

Negative values indicate that species x reduces the community measure when it is present.

Community importance measurements are similar to Paine's measurement of interaction strength in communities (Paine 1992):

$$I_x = \frac{(t_N - t_D)/t_D}{n_x} \quad (23.3)$$

where terms are as previously defined and n_x = number of individuals of species x in unmanipulated plots. For the mantid data just given, the interaction strength is

$$I_x = \frac{(t_N - t_D)/t_D}{n_x} = \frac{(194 - 316)/316}{14} = -0.028$$

The interaction strength is a per capita estimate of effects that measures how much a single individual of species x changes the community. Negative values of interaction strength indicate that species x reduces the abundance or other trait of the community being analyzed, in this case by about 3% per mantid individual. Community importance values and interaction strengths are two similar ways of measuring effects of species, and because they are highly correlated measures, either one may be used to quantify effects of species removals on community structure.

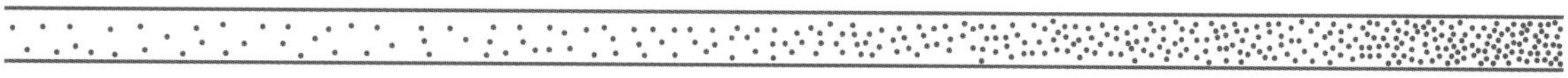

NONEQUILIBRIUM	EQUILIBRIUM
Biotic decoupling	Biotic coupling
Species independence	Competition
Unsaturated	Saturated
Abiotic limitation	Resource limitation
Density independence	Density dependence
Opportunism	Optimality
Large stochastic effects	Few stochastic effects
Loose patterns	Tight patterns

FIGURE 23.2

Natural communities may be arrayed along a continuum of states from nonequilibrium to equilibrium. At either extreme, several attributes of community organization and dynamics can be anticipated. In this chapter we discuss equilibrium models; we discuss nonequilibrium models of community organization in Chapter 24. (From Wiens 1984b.)

FIGURE 23.3
A simplified version of the Antarctic marine food chain. Blue arrows indicate the major trophic interactions. (Modified after Knox 1970.)

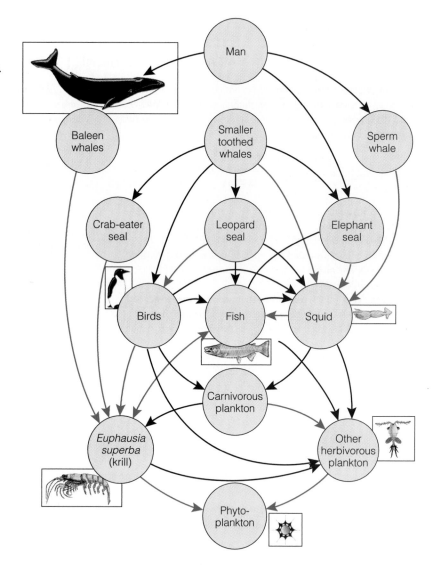

a. Population growth rates can be described with deterministic equations, and environmental fluctuations can be ignored

b. The environment is spatially homogeneous, and migration is unimportant

c. Competition between species is the only significant biological interaction.

d. The coexistence of competing species requires a stable equilibrium point for population densities (see, for example, Figure 12.3, case 3, p. 183).

This theory predicts that *n* limiting resources are required for the coexistence of *n* species. More-

over, there will be a limiting similarity of species such that species differ in their use of the available resources (Chesson and Case 1986).

2. *Competition-predation theory:* The classical theory was clearly deficient in allowing only competition to operate. Adding predation to assumption (c) of the classical theory produces a new equilibrium model that will allow *n* species to coexist on fewer than *n* resources (Levin 1970).

3. *Competition-spatial patchiness theory:* Another equilibrium model that was modified from the classical theory allowed the environment to be subdivided into patches, such that different species would be favored in different patches (Levin 1974).

Each patch has its own distinct stable equilibrium, and the resulting model is similar to a metapopulation model (see Figure 16.6, p. 289).

In real communities, competition for nutrients or food and competition for space both occur (Yodzis 1986). By adding predation and spatial patchiness we can construct more realistic models of community organization, and the equilibrium model of community organization includes competition, predation, and spatial patchiness. Despite the fact that mutualism between species could also help to structure ecological communities, almost no attention has been paid to it as a factor in community organization.

Given this background, let us consider the ways in which we can classify species in communities; then we can see how much of this structure we can explain by classical equilibrium theories involving competition, predation, and spatial patchiness.

Food Chains and Trophic Levels

One component of community organization is "who-eats-whom." The transfer of food energy from its source in plants through herbivores to carnivores is referred to as the *food chain*. Elton (1927), one of the first to apply this idea to ecology and to analyze its consequences, pointed out the great importance of food to organisms, and he recognized that the length of food chains was limited to four or five links. Thus, we may have a pine tree–aphids–spiders–warblers–hawks food chain. Elton recognized as well that these food chains were not isolated units but were linked into *food webs*. Let us look at a few examples of food chains.

The Antarctic pelagic food chain is a good example of a food chain found in seasonally productive oceans. Phytoplankton are fed on by the dominant herbivores, euphausids (krill), and copepods. These zooplankton species are fed on by an array of carnivores, including fish, penguins, seals, and baleen whales (Figure 23.3). Squid, which are carnivores that feed on fish as well as zooplankton, are another important component of this food chain because seals and the toothed whales feed on them in turn. During the whaling years, humans became the top predator of this food chain. Having reduced the whales to low numbers, humans are now harvesting krill (see p. 325).

In the boreal forests of northern Canada and Alaska, snowshoe hares are the dominant herbivores, along with red squirrels (Krebs et al. 1995, Boutin et al. 1995). An array of mammalian and avian predators feed on the herbivores in the forest, and the herbivores are in turn supported by an array of trees, shrubs, and herbs that provide food (Figure 23.4). This food web has a few unusual aspects. Red squirrels and arctic ground squirrels are herbivores, but they also kill and eat baby snowshoe hares when they are available (O'Donoghue 1994). Also, some predators will kill other predators when they are hungry and have the chance. Coyotes will kill lynx, lynx will kill foxes, and great-horned owls will kill a variety of other birds of prey (O'Donoghue et al. 1995). The important point is that not all the links in food webs are linear from predators to herbivores to plants.

In many cases ecologists simplify food webs, typically by taking two approaches. First, some taxonomic groups are lumped together. Often all the vertebrate species are identified individually, but plants or insects are lumped together. The boreal forest food web in Figure 23.4 shows this approach. Second, only a part of the whole food web is isolated for analysis to keep things relatively simple. Figure 23.5 shows a partial food web for a portion of the Chesapeake Bay estuary. The central core of this food web is fish larvae (bay anchovy) in the estuary, and the predation on them by sea nettles (jellyfish) and by juvenile striped bass. Because all these species interact near the bottom of the estuary, where oxygen concentrations fall during summer, Breitburg et al. (1997) studied the effects of changing oxygen levels in the water on this food web. They found that as oxygen levels fell, sea nettle predation rates increased on fish larvae but decreased on fish eggs; they concluded that the structure of the food web was potentially affected by oxygen-depleting pollution in the upper Chesapeake Bay.

Within food webs we can recognize several different *trophic levels*:

Producers	=	Green plants	=	First trophic level
Primary consumers	=	Herbivores	=	Second trophic level
Secondary consumers	=	Carnivores, insect parasitoids	=	Third trophic level
Teritary consumers	=	Higher carnivores, insect hyperparasites	=	Fourth trophic level

For long food chains, fifth and even higher trophic levels are possible.

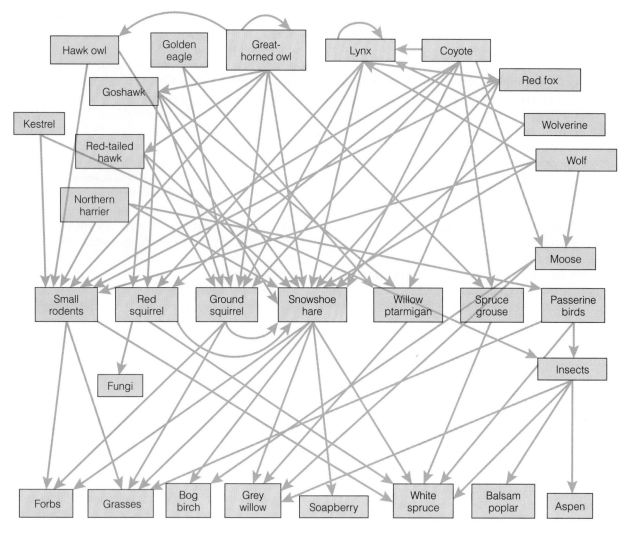

FIGURE 23.4
Feeding relationships of the snowshoe hare-dominated food web in the boreal forests of northwestern Canada. The dominant species in this community are shown in yellow. (After Krebs et al. 2001.)

The classification of organisms by trophic levels is one of *function* and not of species as such, because a given species may occupy more than one trophic level. For example, male horseflies feed on nectar and plant juices, whereas the females are blood-sucking ectoparasites. Sea nettles in Chesapeake Bay are secondary consumers of zooplankton and tertiary consumers of fish larvae, which themselves are secondary consumers of zooplankton.

The size of organisms has a great effect on the organization of food chains, as Elton (1927) recognized. Animals of successive trophic levels in a food chain tend to be larger (except for parasites). Of course, definite upper and lower limits exist for the size of food a carnivorous animal can eat. The size and structure of an animal puts some limit on the size of food it can ingest. Except in a few cases, large carnivores cannot live on very small food items because they cannot catch enough of them in a given time to meet their metabolic needs. The one obvious exception to this general size rule is the omnivore *Homo sapiens*, and part of the reason for our biological success is that we can prey upon almost any level of the food chain and can eat any size of prey.

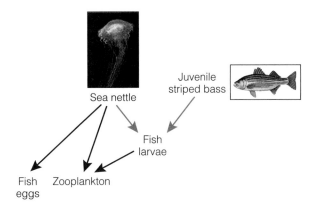

FIGURE 23.5
*Partial food web for the bottom waters of the Chesapeake Bay estuary on the Atlantic Coast of the United States. The sea nettle (*Chrysaora quinquecirrha*) and juvenile striped bass (*Morone saxatilis*) are the main predators of fish larvae (bay anchovy and several other species). (After Breitburg et al. 1997.)*

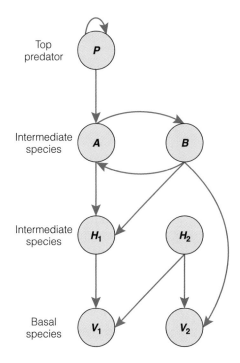

FIGURE 23.6
Hypothetical food web to illustrate definitions of food web properties defined in Table 23.1. Species A and B illustrate a cycle in a food web. Top predator P illustrates cannibalism. There are ten interactions in this hypothetical web of seven species, including cannibalism. There are 7^2 or 49 possible interactions, so connectance is 10/49 or 0.20. The linkage density is 10/7 or 1.43. Species A and B are both omnivores.

Food webs can form a useful starting point for the theoretical analysis of community organization (Pimm et al. 1991). Table 23.1 provides definitions for terms used in food web theory, and Figure 23.6. illustrates these properties schematically. Over 200 food webs have now been elucidated, and attempts have been made to draw generalizations about food web structure (Pimm et al. 1991).

TABLE 23.1 Definitions of food web terminology.

Figure 23.6 illustrates these definitions.

Top predators: species eaten by nothing else in the food web
Basal species: Species that feed on nothing within the web (usually plants)
Intermediate species: species that have both predators and prey within the web
Trophic species: groups of organisms that have identical sets of predators and prey
Cycles within a food web: species *A* eats species *B* and species *B* eats *A*
Cannibalism: a cycle in which a species feeds upon itself
Interactions: any feeding relationship (line with an arrow in a food web diagram)
Possible interactions: among *s* species in a food web, there can be s^2 possible interactions, including cannibalism
Connectance: number of actual interactions in a food web divided by the number of possible interactions
Linkage density: average number of links or interactions per species in the web
Omnivores: species that feed on more than one trophic level
Compartments: groups of species with strong linkages among group members but weak linkages to other groups of species

Source: Modified from Cohen (1978) and Pimm (1982).

ESSAY 23.1

USE OF STABLE ISOTOPES TO ANALYZE FOOD CHAINS

Stable isotopes are alternate forms of elements. For example, nitrogen in the air contains two isotopes, ^{14}N and ^{15}N, only a small fraction (0.4%) of which is ^{14}N. Carbon contains two isotopes, ^{13}C and ^{12}C, and only about 1% of carbon in nature is ^{13}C. The ratio of the isotopes in any material is expressed as δ values, which are parts-per-thousand differences from a standard substance. For nitrogen, we have

$$\delta\,^{15}N = \left[\left(\frac{^{15}N\big/^{14}N_{sample}}{^{15}N\big/^{14}N_{air}}\right) - 1\right] 1000 \qquad (23.4)$$

Nitrogen in the air is the standard for nitrogen analysis, and carbon in a particular limestone is the standard for carbon isotopes. The $\delta^{15}N$ values are in parts per thousand, and positive values indicate that the sample material is richer in the heavy isotope than the standard is. So far, this is interesting chemistry, but what does it have to do with ecology?

Different organisms take up nitrogen and carbon in ways that discriminate among these isotopes, so that the isotopic composition of plants and animals varies. On average, $\delta^{15}N$ increases 3.4‰ in animals relative to their diet. The result is that animals in different trophic levels have different isotopic signatures. For example, in Lake Ontario, the pelagic food chain shows the following pattern (Cabana and Rasmussen 1994):

The crustacean *Mysis relicta* eats zooplankton and has a $\delta^{15}N$ ratio of about 9‰, alewife and smelt feed on *Mysis* and have a $\delta^{15}N$ of 13–14‰, and lake trout that feed on alewife and smelt have a $\delta^{15}N$ of 16‰. The key point is that if we have a species of unknown position in the food chain, we can determine its trophic level by measuring its $\delta^{15}N$ ratio.

Marine and terrestrial plants differ in their $\delta^{15}N$ values, and this difference can be used to see if coastal animals utilize marine base foods more than terrestrial foods. Anderson and Polis (1998) found that coastal spiders on islands in the Gulf of California had $\delta^{15}N$ values of 20‰, compared with inland spiders with values of 12‰, showing that marine-based resources were sustaining these coastal spider populations.

One of the most important advantages of the stable isotope method of diet analysis is that it integrates the diet of the animals over a long time period, as opposed to measuring diet from stomach samples, which identifies the contents of the last meal only. In communities that process detritus, the use of stable isotopes can be helpful in separating the sources of the organic material in the system. For wide-ranging seabirds, stable isotopes can distinguish birds that feed inshore from those that feed offshore (Hobson et al. 1994). Stable isotopes provide another tool in the toolbox of ecologists trying to decipher food webs.

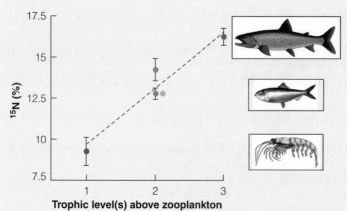

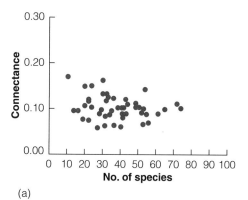

(a)

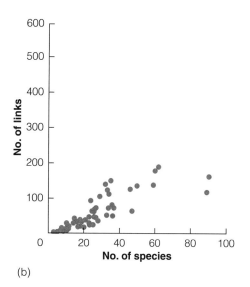

(b)

FIGURE 23.7
Relationship between connectance and the size of the food web for the pelagic zone of 50 lakes in New York State. (a) Connectance (actual interactions/ potential interactions) is relatively constant around 0.10 over food webs of different species richness. (b) Linkages increase as the food web size increases. (Data from Havens 1992.)

One question we can ask about food webs is whether there are limits to their complexity. As more and more species are involved in a food web, the number of possible linkages increases, and the empirical question is whether or not the number of *actual* linkages also goes up. Figure 23.7 shows that the number of linkages does indeed increase as food webs with greater diversity are considered, such that the connectance in food webs remains relatively constant for a range of species richness in aquatic food webs (Warren 1994). The result of this constant connectance is that

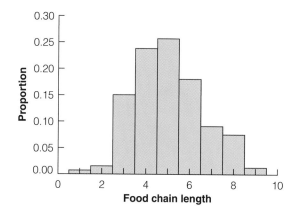

FIGURE 23.8
Distribution of food chain lengths in the Ythan Estuary of northeast Scotland. A total of 95 species in this community have been studied in detail to construct a complex food web from which 5518 food chain lengths were counted. The most common chain length was five. (From Hall and Raffaelli 1997.)

each species is on average connected to more and more species as the species richness of the food web increases.

A second generalization about food webs is that food chains are short (Elton 1927). (The length of a food chain is defined as the number of links running from a top predator to a basal species.) If we count for each food web all the possible routes from a basal species to a top predator, we can get a set of chain lengths and identify the maximum chain length for that food web. Hall and Raffaelli (1997) have done this for the Ythan Estuary in northeast Scotland, where they have detailed studies of 95 species in this community.[1] Figure 23.8 shows the distribution of all the 5518 possible chains in this community. The most common food chain length is five links, and the range was one to nine links. There is some suggestion that the length of food chains increases as food webs contain more species. But there is a limit on food chain length; few chains exceed eight or nine links, and the mean chain length rarely exceeds five links.

There are several hypotheses for why food chains should be relatively short like this. (Hall and Raffaelli 1997). The *energetic hypothesis*, the most popular explanation for food chain length, suggests that the length of food chains is limited by the inefficiency of energy transfer along the chain (Pimm 1991). This

---

[1]The food web for this community is too complex to illustrate.

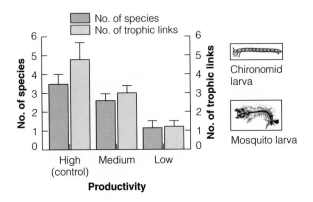

FIGURE 23.9

Experimental test of the energetic hypothesis for the restriction on food chain length. Tree-hole communities in Queensland were simulated using litter input at three levels: high litter input = natural (control) rate of litter fall, medium = 1/10 natural rate, and low = 1/100 natural rate. Reducing energy input reduced food chain length, in agreement with the hypothesis. The tree-hole community consists of microbes that break down leaf litter, mosquito larvae that feed on these microbes, predatory midges (chironomids), and other insects that feed directly on leaf litter. (From Jenkins et al. 1992.)

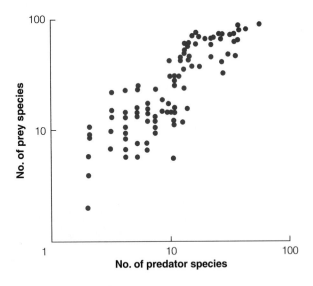

FIGURE 23.10

The relationship between the number of predator species and the number of prey species in 92 freshwater invertebrate food webs. The prey-predator ratio varied between 2:1 and 3.5:1 in these communities. (Data from Jeffries and Lawton 1985.)

classical hypothesis for food chain length was articulated by Charles Elton in 1924. If this idea is correct, food chains should be longer in habitats of higher productivity, a clear prediction that can be tested.

The *dynamic stability hypothesis* explains short food chains by the fact that because longer food chains are not stable, fluctuations at lower levels are magnified at higher levels and top predators go extinct. Moreover, in a variable environment top predators must be able to recover from catastrophes, and the longer the food chain, the slower the recovery rate from catastrophes for top predators. If catastrophes occur too often, the top predators will again go extinct. This hypothesis predicts shorter food chains in unpredictable environments, a prediction that again can be tested when detailed data are available.

These two hypotheses to explain food chain length are difficult to test in large communities. Jenkins et al. (1991) used organisms that inhabit natural, water-filled tree holes in the subtropical rainforest of Queensland to test these ideas in a small system. They simulated tree holes using 1-liter plastic containers to which they added leaf litter in various amounts. By reducing leaf litter input to 1/10 and 1/100 the natural rate over one year, Jenkins et al. (1991) found that both the number of species supported and the number of trophic links were reduced as leaf-litter input was reduced, results that support the prediction of the energetics hypothesis (Figure 23.9) that reduced energy input will result in reduced food chain lengths.

A third generalization about food webs is that the proportions of species that are top predators, intermediate species, and basal species are nearly constant, regardless of the size of the food web. Figure 23.10 shows this relationship for predator-prey ratios. There is an approximately constant ratio of two to three prey species for every predator species in food webs, regardless of the total number of species in the web (Martinez 1991). In the Ythan Estuary food web analyzed in Figure 23.8, the ratio of intermediate to top species is 2.4 (Hall and Raffaelli 1997).

A fourth generalization is that omnivory—feeding on more than one trophic level—seems to be rare in food webs. This generalization has been challenged by new, more-detailed data (Polis 1991, Hall and Raffaelli 1993) suggesting that omnivory is very common in some food webs. Aquatic communities often have fishes that eat their way up the food chain as they grow in size. Also, the detritus that sustains some organisms originates in several trophic levels (Pimm et al. 1991). Omnivory is difficult to estimate accurately unless the taxonomic breakdown of species in the food web is comprehensive. Lumping species together can increase the estimate of links that are classified as omnivorous. For example, if we group all insects as together in a food web, the "insect" group will appear to be omnivorous because some species of insects are herbivores, some are predators, and some are secondary predators. A careful separation of species is necessary for food web definition (and for a careful analysis of diets as well). The general conclusion from more detailed food web studies is that omnivory may be more common than originally thought, but that more precise data are needed to decide this issue (Polis and Winemiller 1996).

Many of the early generalizations about food webs have now been revised as more-detailed data on food web structure becomes available (Hall and Raffaelli 1997). It is clear that real food webs are often very complex, and that their properties are not simple. The importance of understanding food webs was emphasized by Pimm (1991) because the structure of food webs has implications for community persistence. Some food webs can support additional species without suffering any losses, whereas other food webs are unstable and thus subject to species losses. If we can better understand the structure of food webs, we can design better management strategies for conservation.

Functional Roles and Guilds

Trophic levels provide a good description of a community, but by themselves they are not sufficient for defining community organization. A better approach is to use food webs to subdivide each trophic level into *guilds*, which are groups of species exploiting a common resource base in a similar fashion (Root 1967). For example, hummingbirds and other tropical nectar-feeding birds form a guild exploiting a set of flowering plants (Feinsinger 1976). We expect competitive interactions to be potentially strong between the members of a guild. By grouping species into guilds, we may also identify the basic functional roles played in the community.

There are four advantages to using guilds in the study of community organization (Root 1967):

- Guilds focus attention on all competing species living in the same community, regardless of their taxonomic relationship.
- The use of the term *guild* clarifies the concept of niche: Groups of species having similar ecological roles can be members of the same guild but cannot be occupants of the same niche.
- Guilds allow us to compare communities by concentrating on specific functional groups. We need not study the entire community but can concentrate on a manageable unit.
- Guilds might represent the basic building blocks of communities and thus aid our analysis of community organization.

A community can be viewed as a complex assembly of component guilds, each containing one or more species. Guilds may interact with one another within the community and thus provide the organization we observe. No one has yet been able to analyze all the guilds in a community, and at present we can deal only with a few guilds that make up part of an entire community. Two examples of the organization of guilds illustrate how this concept can be applied to communities.

Root (1973) grew collards (*Brassica oleracea* var. *acephala*) in two experimental habitats: in pure stands and in single rows bounded on each side by meadow vegetation. Three herbivore guilds were associated with collard stands (Figure 23.11). Pit feeders are insects that rasp small pits into leaf surfaces; they comprise 18 species, of which two chrysomelid beetles were abundant. Strip feeders, insects that chew holes in the leaves, included 17 species, of which only one was abundant. Sap feeders suck the juices of the collard plants and included 59 species, many of them aphids. The pit feeders usually formed the most important herbivore guild, particularly in the pure collard stands.

The species composition of the three herbivore guilds changed from year to year, and these changes were most striking among the sap feeders. The cabbage aphid (*Brevicoryne brassicae*) was the most abundant aphid in 1966 and 1968 but was absent entirely in 1967, when other aphids increased in abundance. The implication is that within some guilds, species can replace one another and perform the same functional role.

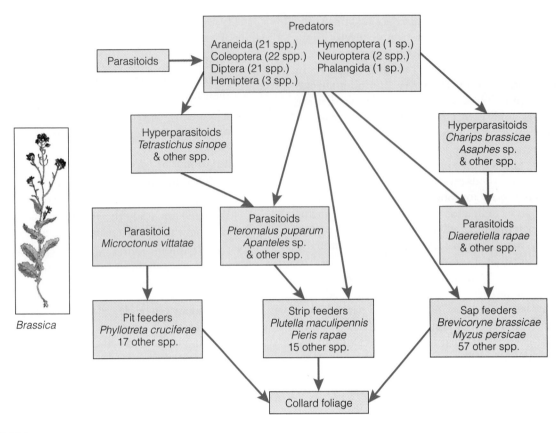

FIGURE 23.11
*The food web formed by the major arthropod species associated with collard stands at Ithaca,
New York. The herbivores are divided into guilds. (After Root 1973.)*

The nectar-eating birds of successional montane forests in Costa Rica form a guild clearly organized around competition for food (Feinsinger 1976). This guild of hummingbirds is organized around the dominant species, *Amazilia saucerottei*, the blue-vented hummingbird. *Amazilia* specializes on plants that produce large quantities of nectar and sets up individual feeding territories that each bird defends against other hummingbirds. *Amazilia* is aggressively dominant over most other hummingbird species. A second common species, *Chlorostilbon canivetii*, is excluded from this rich flower resource by aggressive *Amazilia* individuals and exhibits "trap-line" feeding, following a regular route between scattered flowers. *Chlorostilbon* spends much time in flight and only rarely defends any flowers. Two other hummingbirds complete the core group of this guild. *Philodice bryantae* sets up feeding territories in flower-rich areas but defends these territories only against other *Philodice*. This species seldom elicits attack behav-

ior from the dominant *Amazilia*, partly because *Philodice* looks more like a bee than a bird. The final member of the core species in this guild is *Colibri thalassinus*, which is highly migratory and moves in to exploit seasonal flowering. Ten other hummingbird species forage in the study area, and most of these species were important in adjacent communities. The foraging of all these species is affected by the territorial behavior of the dominant *Amazilia*, and the high diversity of this bird guild is related to the highly migratory strategy of many of these hummingbird species.

Most of the hummingbird species in the guild that Feinsinger (1976) analyzed were general nectar feeders, and hence functional equivalents. If one of these species were removed from the community, we would predict that the other hummingbird species would take its place, and the community would be little changed.

The guild or role concept of community organization is not yet fully developed (Simberloff and

Dayan 1991). We can recognize four hypotheses that require testing in natural communities:

- Many species form interchangeable members of a guild from the point of view of the rest of the community. These species are functional equivalents.

- The number of functional roles within a community is small in relation to the number of species and might be constant among different communities.

- There may be a limit on the number of species that can simultaneously fill a given functional role. A community always has a set of roles, but the guilds may be packed with different numbers of species.

- Species within guilds fluctuate in abundance in such a way that the total biomass or density of the guild remains stable.

At present we can define roles or guilds only crudely via the analysis of food webs. There is a need to define objectively the criteria used to assign species to guilds in natural communities (Simberloff and Dayan 1991). The utility of the guild concept is that by reducing the number of components in a community, it should help us to study how communities are organized. It also emphasizes that ecological units are not taxonomic units. Ants, rodents, and birds can all eat seeds in desert habitats, and thus they form a single guild of great taxonomic diversity (Brown and Davidson 1979).

Keystone Species

A role may be occupied by a single species, and the presence of that role may be critical to the community. Such important species are called *keystone species* because their activities determine community structure. Bob Paine was the first ecologist to recognize keystone species in his research on the rocky intertidal zone (Paine 1969). Keystone species typically are not the most common species in a community, and their effects are is much larger than would be predicted from their relative abundance. One way to recognize keystone species is through removal experiments.

The starfish *Pisaster ochraceous* is a keystone species in rocky intertidal communities of western North America (Paine 1974). When *Pisaster* was removed

Robert T. Paine *(1933–) Professor of Zoology, University of Washington*

manually from intertidal areas, the mussel *Mytilus californianus* was able to monopolize space and exclude other invertebrates and algae from attachment sites (see p. 448). *M. californianus* is an abundant species that is able to compete for space effectively in the intertidal zone. Predation by *Pisaster* removes this competitive edge and allows other species to use the space vacated by *Mytilus*. *Pisaster* is not able to eliminate mussels because *Mytilus* can grow too large to be eaten by starfish. Size-limited predation provides a refuge for the prey species, and these large mussels are able to produce large numbers of fertilized eggs (Paine 1974).

Sea otters are a keystone predator in the North Pacific. Once extremely abundant, they were reduced by the fur trade to near extinction by 1900. Once they were protected by international treaty, sea otters began to increase and by 1970 had recovered in most areas to near maximum densities (Estes and Duggins 1995). Sea otters feed on sea urchins, which in turn feed largely on macroalgae (kelp). Early natural history observations showed that in areas where sea otters were abundant, sea urchins were rare and kelp forests were well developed. Similarly, where sea otters were rare, sea urchins were common and kelp was nonexistent. Sea otters are thus a good example of a keystone species in a marine subtidal community. Since about 1990 sea otters have declined precipitously in large areas of western Alaska (Figure 23.12), often at rates of 25% per

FIGURE 23.12

Sea otters as keystone predators in the North Pacific. (a) Changes in sea otter abundance over time at several Aleutian islands and concurrent changes in (b) sea urchin biomass, (c) grazing intensity, and (d) kelp density measured from kelp forests at Adak Island. Error bars in (b) and (c) indicate one standard error. The proposed mechanisms of change are portrayed in the marginal drawings: The one on the left shows how the kelp forest ecosystem was organized before the sea otter's decline, and the one on the right shows how this ecosystem changed with the addition of killer whales as a top predator. Heavy arrows represent strong trophic interactions; light arrows represent weak interactions. (From Estes et al. 1998.)

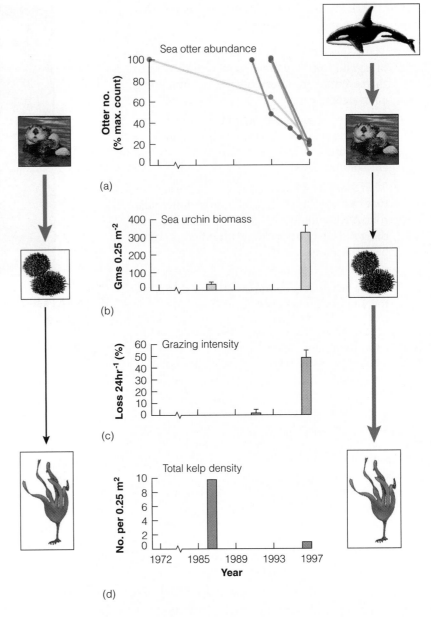

year. The loss of this keystone species has allowed sea urchins to increase and has resulted in the destruction of kelp forests. Killer whales are the suspected cause of the sea otter decline (Estes et al. 1998). Killer whales have begun to attack sea otters in the last 10–15 years because their prey base (seals, sea lions) has declined along with the fishes that constitute the seals' prey base. Fish have probably declined from human overharvesting in the North Pacific, illustrating that the interactions in food webs can propagate from top predators to basal species in unexpected ways.

A third example of a keystone species is the African elephant (Laws 1970), a relatively unspecialized herbivore that relies on a diet of browse supplemented by grass. By their feeding activities, elephants (Figure 23.13) destroy shrubs and small trees and push woodland habitats toward open grassland. Elephants feeding on the bark can destroy even large mature trees. As more grasses invade the woodland habitats, the frequency of fires increases, which accelerates the conversion of woods to grassland. This change works to the elephants' disadvantage, how-

FIGURE 23.13
African elephants browsing in open woodland in the Serengeti area of East Africa. (Photo courtesy of A. R. E. Sinclair.)

ever, because grass alone is not a sufficient diet for elephants, and they begin to starve as woody species are eliminated. Other ungulates that graze the grasses are favored by the elephants' activities.

The critical effect of keystone predators is that they can reverse the outcome of competitive interactions. The impact of keystone predators is clearly evident in aquatic communities. Amphibians are a major component of temporary ponds. In the coastal plain of North Carolina a single pond can support five species of salamanders and 16 species of frogs and toads (Fauth and Resetarits 1991). Salamanders are the major predators in these temporary ponds, and the broken-striped newt *Notophthalmus viridescens* acts as a keystone predator. By selectively preying on the dominant competitors *Rana utricularia* and *Bufo americanus*, it allows less competitive frogs such as the cricket frog (*Hyla crucifer*) to survive.

Kangaroo rats in the Chihuahuan Desert form a keystone guild (Brown and Heske 1990). Kangaroo rats (*Dipodomys* spp.) prefer to eat large seeds. When they were excluded by experimental fences from areas

of desert shrubland, large-seeded winter annuals increased greatly in abundance, raising the vegetative cover and reducing the ability of ground-feeding birds to feed on seeds. After 12 years grasses increased, and the area became desert grassland. Grassland species of rodents, such as the cotton rat, which were previously absent, began colonizing the habitat, from which kangaroo rats were excluded. Fifteen species of rodents live in the Chihuahuan Desert, but only the three species of kangaroo rats seem to play this keystone role.

Keystone species may be rare in natural communities, or they may be common but unrecognized. At present, few terrestrial communities are believed to be organized by keystone species, but in aquatic communities keystone species may be more common. There seems to be no simple way of recognizing keystone species in food webs without doing detailed studies (Power et al. 1996). The important message is that some species of low abundance can have strong effects on community structure, so land managers and conservationists must be concerned with both common and uncommon species in communities. We

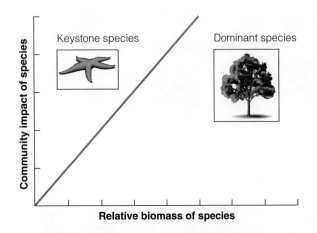

FIGURE 23.14
Schematic illustration of the difference between dominant species and keystone species. Species whose total impact is exactly proportional to their abundance would fall on the red line. Both the dominant and keystone species in a community are assumed to have a high community impact, but keystone species have low biomass. Trees, giant kelp, and corals are examples of dominant species in their communities. (Modified after Power et al. 1996.)

cannot determine which species might be keystone species unless we gain a detailed understanding of food web structure by analyzing it experimentally.

Dominant Species

Dominant species in a community may exert powerful control over the occurrence of other species, and the concept of dominance has long been engrained in community ecology. Dominant species are recognized by their numerical abundance or biomass and are usually defined separately for each trophic level. For example, the sugar maple is the dominant plant species in part of the climax forest in eastern North America; its abundance determines in part the physical conditions of the forest community. Dominance usually means numerical superiority, and keystone species are not usually the dominant species in a community (Figure 23.14).

Dominant species are usually assumed to achieve their dominance by competitive exclusion. Buss (1980) identified several possible configurations of competing species (Figure 23.15). The simplest case, *transitive* competition, occurs when a linear hierarchy exists (species *A* outcompetes *B* and *B* outcompetes *C*). In this case, competitive exclusion can occur, as we saw in Chapter 12. A more complex case of *intransitive* com-

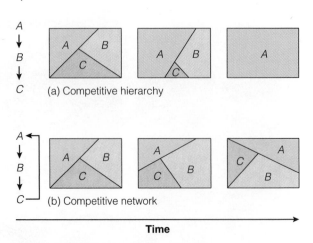

FIGURE 23.15
Schematic illustration of competitive relationships among three species competing for space. Each rectangle represents a plot of ground or a rock surface in the intertidal zone. (a) The conventional competitive hierarchy showing transitive competition, in which species A outcompetes species B, which in turn outcompetes species C. As time progresses, competitive exclusion will occur. (b) A competitive network showing intransitive competition, in which species A outcompetes B and B outcompetes C, but C is able to outcompete A. This system changes in time but does not move toward competitive exclusion. (After Buss and Jackson 1979.)

petition occurs when a circular network exists in which no one species can be called dominant. In circular networks of spatial competition, species *A* outcompetes species *B*, *B* outcompetes *C*, but *C* in turn is able to outcompete *A*. This type of competitive interaction is called *intransitive* because it has no end point. Competitive exclusion does not occur in circular networks, and if intransitive competition is the rule in natural communities, species diversity need not decline because of competitive exclusion (Buss and Jackson 1979).

Competitive dominance is not the only explanation for a species becoming dominant. Australian mangrove forests show a complex zonation across the intertidal zone (Smith 1987). This zonation has usually been explained either by mechanisms of physiological tolerance to seawater inundation or by tidal sorting of seeds by size. Smith (1987) found that seed predation by small grapsid crabs was very high, and dominance in four of five mangrove species was correlated with the amount of seed predation (Figure 23.16). Predation may override competition in some communities.

Communities that develop under similar ecological conditions in a given geographic region are expected

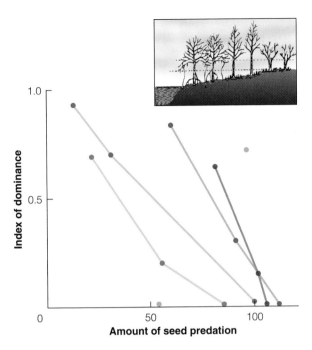

FIGURE 23.16

The amount of predation on seeds of five species of Australian mangrove trees in relation to community dominance. Each point represents one species on one site. Four of the five mangroves can become dominant only in areas of low seed predation by crabs. Only Ceriops tagal *is not affected by seed predation.* Avicennia marina; *purple.* Bruguiera exaristata; *blue.* B. gymnorrhiza; *red.* Ceriops tagal; *orange.* Rhizophora stylosa; *green. (From Smith 1987.)*

to be dominated by the same species. For example, a deciduous forest in Ohio is expected to be dominated by beech and sugar maple, and botanists would be surprised if a rare species such as black walnut or white ash became dominant. Ecologists have long wondered if the same pattern occurs in aquatic communities, particularly those in the open ocean. The central gyre of the North Pacific Ocean has a rich diversity of phytoplankton, zooplankton, and fish species (McGowan and Walker 1979). Because the ocean mixes and the gyre is so large, the dominant species might be expected to vary from place to place within this large area, but this does not seem to occur. McGowan and Walker (1985) found that about 30 species of copepods were abundant, and that the dominance structure of this community remained the same in samples collected up to 16 years apart. The same constancy of community structure was evident in the phytoplankton and fish communities of the central gyre. These oceanic communities appear to

be as constant in their dominant species as are temperate-zone forests.

The human-induced removal of a dominant species in a community has occurred frequently, but unfortunately few of these removals have been studied in detail. The American chestnut was a dominant tree in the eastern deciduous forests of North America before 1910, making up more than 40% of overstory trees. This species has now been eliminated as a canopy tree by chestnut blight. The effects of this removal have been negligible, as far as anyone can tell, and various oaks, hickories, beech, and red maple have replaced the chestnut (Keever 1953). Of the 56 species of Lepidoptera that fed on the American chestnut, seven species went extinct, but the other 49 species apparently did not rely only on the chestnut for food and still survive (Pimm 1991).

Dominance has been studied in freshwater communities in considerable detail. The zooplankton community of many temperate-zone lakes is dominated by large-sized species when fish are absent and by small-sized species when fish are present. Brooks and Dodson (1965) observed this change in Crystal Lake, Connecticut, after the introduction of a herring-like fish, the alewife (*Alosa pseudoharengus*) (Figure 23.17). They proposed the *size-efficiency hypothesis* as a wide-ranging explanation of the observed shift in dominance in the zooplankton community. The size-efficiency hypothesis is based on two assumptions: (1) that planktonic herbivores (zooplankton) all compete for small algal cells (1–15 μm) in the open water, and (2) that larger zooplankton feed more efficiently on small algae than do smaller zooplankton, and that large animals are able to eat larger algal particles that small zooplankton cannot eat.

Given these two assumptions, Brooks and Dodson (1965) made three predictions:

- When predation on zooplankton is of low intensity or absent, the small zooplankton herbivores will be completely eliminated by large forms (dominance of large cladocera and calanoid copepods).

- When predation is of high intensity, predators will eliminate the large zooplankton and allow the small zooplankton (rotifers, small cladocera, and small copepods) to become dominant.

- When predation is of moderate intensity, predators will reduce the abundance of the large zooplankton such that the small zooplankton species are not eliminated by competition.

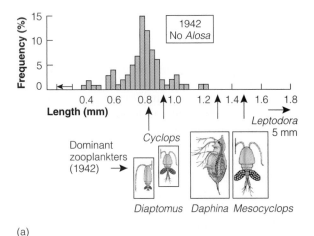

(a)

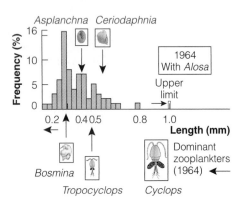

(b)

FIGURE 23.17

*The composition of the crustacean zooplankton of Crystal Lake, Connecticut, (a) before and (b) after the introduction of the alewife (*Alosa*), a plankton-feeding fish.* Daphnia *is a large cladoceran;* Ceriodaphnia *and* Bosmina *are small cladocerans.* Mesocyclops *is a large copepod;* Tropocyclops *is a small copepod. Some of the larger zooplankton species are not represented in the 1942 histogram because they were not abundant. (After Brooks and Dodson 1965.)*

Thus competition forces communities toward larger-bodied zooplankton, where as fish predation forces them toward smaller-bodied species. These three predictions of the size-efficiency hypothesis are consistent with the keystone-species idea discussed in the preceding section.

The second and third predictions of the size-efficiency hypothesis have been tested in several lakes, and the predictions seem to describe adequately the zooplankton distributions in many lakes (Kerfoot 1987). Fish predation does seem to fall more heavily on the larger zooplankton species, but invertebrate predators in the plankton seem to prey more heavily on the smaller zooplankton species. Large zooplankton may predominate in lakes with no fish either because they are superior competitors (as the size-efficiency hypothesis predicts) or because small zooplankton are selectively removed by invertebrate predators.

The second assumption of the size-efficiency hypothesis is that large-sized zooplankton are superior competitors for food resources. In a laboratory microcosm, Gliwicz (1990) fed eight species of *Daphnia* on constant food levels and measured the concentration of algae they needed to maintain body weight. Figure 23.18 shows that large copepods can subsist on much lower algal concentrations than small copepods, as assumed by the size-efficiency hypothesis.

The importance of fish predation in structuring zooplankton communities is now well established (Kerfoot 1987), but the competitive nature of feeding relationships among the zooplankton may not always favor large species because of fluctuating food conditions in lakes and ponds. In some ponds and lakes small zooplankton predominate in the absence of fish predators, and further work on the feeding strategies of zooplankton of different sizes is required (DeMott and Kerfoot 1982).

Dominance is an important, although poorly understood component of community organization. Dominant species may be the focal point of interactions that structure much of the species makeup of a community. Moreover, the characteristics of dominant species may affect the stability of the community as well.

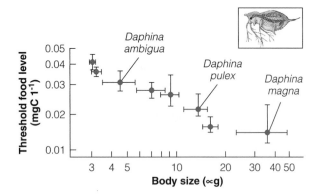

FIGURE 23.18

A test of the central assumption of the size-efficiency hypothesis that large zooplankton are competitively superior to small zooplankton. Eight Daphnia *species of different sizes were raised in the laboratory on a constant food source. Larger* Daphnia *can maintain weight on a lower food concentration and are thus competitively dominant. (From Gliwicz 1990.)*

ESSAY 23.2

FISHING DOWN FOOD WEBS

Overfishing in both marine and freshwater fisheries is having a predictably devastating effect on aquatic food webs. Fisheries do not operate at random within the food web. The valuable fishes humans prefer to eat, such as tuna and cod, tend to be top predators in marine food webs. Pauly et al. (1998) asked whether there were any systematic global trends in overfishing. To do this they had to assign each species in the commercial catch a trophic value from 2.0 (for herbivorous fishes like anchovies) to 5.0 (for top predators like killer whales). Secondary predators would be assigned trophic value 3.0, and tertiary predators like cod would be assigned 4.0. To make these assignments they had to know at least approximately the diet of each of the main species in the fishery. In general, the higher the trophic level, the larger the size of the fish.

Many of the world's fisheries are overexploited, and overall catches have been declining since 1989. To their surprise, Pauly et al. also found a global pattern of collapse in average trophic level for both marine and freshwater fisheries:

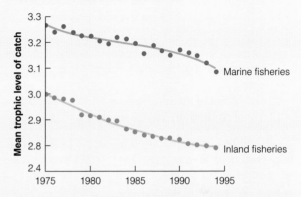

This change in trophic status reflects a gradual global shift from catching long-lived predatory fish to catching short-lived plankton-feeding fish and invertebrates. Clearly this pattern of decline in the trophic level of the catch is not sustainable, in the future unless we wish to dine on zooplankton.

Community Stability

Stability is a dynamic concept that refers to the ability of a system to recover from disturbances (see p. 460). If a brick is raised slightly from the floor and then released, it will fall back to its original position. This is the physicists' concept of *neighborhood stability* or local stability, in which the system responds to small, temporary disturbances by returning to its original position. Thus, for example, a rabbit population may show neighborhood stability to moderate hunting pressure if it returns to its normal density after hunting is prohibited.

Physicists discuss stability in terms of small perturbations, but ecological systems are subject to large disturbances. To deal with these, we must consider a second type of stability, *global stability*. A system that has local stability shows global stability only if the system returns to the same point after large disturbances. That brick, for example, shows both local and global stability because if we raise it either 10 mm or 10 meters from the floor and release it, it will fall back to the floor. Ecological communities are not passive

objects like bricks, and global stability is probably rare. One of the problems of ecology is to identify the limits of stability for various communities (Lewontin 1969). If in Figure 23.1a we schematically move a community at point A_2 beyond point B_2, it will move to a new state somewhere near point C. In this hypothetical case community A_2 is locally stable, but not globally stable, if forced beyond point B_2. All equilibrium theories of community organization assume that the equilibrium is stable, and the usual assumption is that the equilibrium is globally stable. Note that the shape of stability "basins" need not be circular in cross section. There may be great stability to disturbances in one direction but little stability to disturbances in other directions.

If equilibrium theories of community organization are correct, there are four important consequences for our understanding of community dynamics (Chesson and Case 1986):

- *Community conservation:* An equilibrium community will show no tendency to lose species over time. Global stability implies that in the absence of external perturbations no losses of species will ever occur.

- *Community recovery:* An equilibrium community can recover from events that drive any of its constituent species to low density.

- *Community composition:* An equilibrium community can be built up by immigration of species from outside the system. Combinations of species that can coexist will increase to their equilibrium values (see Figure 12.3).

- *Independence of history:* Because of global stability, past events have no effect on community structure, and within a broad range the order of arrival of member species is irrelevant to the final community composition.

Equilibrium communities are stable in the sense of persistence, but they may not be stable in the sense of being resilient to disturbances, particularly disturbances caused by humans.

One of the classic tenets of community ecology and a hallowed tenet of conservation biology has been that biodiversity *promotes* stability. Elton (1958) suggested several lines of circumstantial evidence that support this conclusion:

- Mathematical models of simple systems show how difficult it is to achieve numerical stability (see Chapters 12–14).

- Gause's laboratory experiments on protozoa confirm the difficulty of achieving numerical stability in simple systems.

- Small islands are much more vulnerable to invading species than are continents.

- Outbreaks of pests are most often found in simple communities on cultivated land or on land disturbed by humans.

- Tropical rain forests do not have insect irruptions like those common to temperate forests.

- Pesticides have caused irruptions by eliminating predators and parasites from the insect component of crop plant communities.

Many of these statements are only partly correct, and the simple, intuitive, and appealing notion that biodiversity leads to stability has been questioned as a general conclusion (Pimm 1984).

The intuitive argument that increasing community complexity in the food web automatically leads to increased stability was attacked by May (1973), who showed that increasing complexity *reduces* stability in

general mathematical models. In hypothetical communities in which the trophic links are assembled at random, the more diverse communities are more unstable than the simple communities. Thus, May cautioned community ecologists that if species diversity does indeed result in stability in the real world, it is not an automatic mathematical consequence of species interactions. Natural communities are products of evolution, and evolution may have produced nonrandom assemblages of interacting species in which diversity and stability are related.

Even though theoretical ecologists produced mathematical models showing that greater species diversity does not lead to greater stability, field ecologists followed Elton in believing that complex communities were indeed more stable than simple ones. As more data on the composition and complexity of food webs became available during the 1990s, empirical ecologists became more and more convinced that complexity does indeed favor stability. McCann et al. (1998) have constructed a set of theoretical models that predict that complex food webs will be more stable. The key to this stability is the interaction strength between species in the food web. If there are many weak links between species in a food web, complex communities are more stable than simple ones. The conclusions reached by McCann et al.'s model are exactly the reverse of those reached by May's (1973) model. The reasons for these differences are that McCann et al. used nonlinear models for interactions between species and assumed realistic optimal foraging constraints for feeding interactions. The result is that we now have both theoretical and empirical agreement that greater diversity contributes to greater stability in ecological communities (Polis 1998).

Are there any data from field experiments that quantify the diversity-stability relationship? The few detailed studies conducted on the relationship between stability and diversity suggest that in many systems greater diversity does lead to greater stability (Johnson et al. 1996). All the studies to date are on plant communities.

In 1982 David Tilman began a long-term study of four Minnesota grasslands to identify the relationships between species diversity in plants and community functions. By measuring the variability in plant biomass on 207 plots at the peak of 11 growing seasons, Tilman (1996) directly assessed the diversity-stability hypothesis. Figure 23.19 shows the data from 54 plots in one field. Plots with greater plant diversity showed less fluctuation in yield than plots with low diversity, in keeping with the expectations of the diversity-stability hypothesis of Elton.

David Tilman *Professor of Plant Ecology, University of Minnesota*

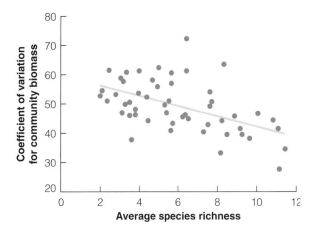

FIGURE 23.19
*The variability of total plant biomass in relation to species richness on 54 plots of Minnesota grasslands, averaged over 11 years. Each point represents one 4m x 4m plot. Aboveground living biomass of all plant species was clipped at the height of the growing season each summer and dried to estimate the community biomass. A significant negative relationship (*r = −0.39*) exists between variability in biomass and species diversity, as predicted by the diversity-stability hypothesis. (After Tilman 1996.)*

Stability is usually measured as variability in numbers or biomass, but it can also be measured as *resistance* to change. More-stable communities will change less when external stress is imposed on them. For temperate plant communities, drought is a major stress, and plant ecologists have used droughts to test the diversity-stability hypothesis. In Yellowstone National Park a severe drought in 1988 allowed Frank and McNaughton (1991) to study the effects of diversity on stability in grassland communities. Figure 23.20 shows that species-rich communities

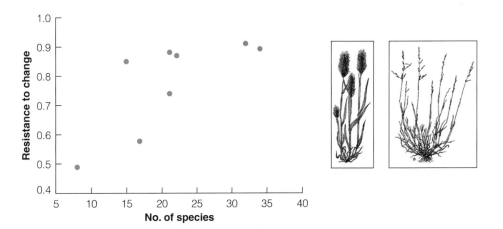

FIGURE 23.20
The relationship between stability and diversity for grasslands in Yellowstone National Park. Eight plots of 0.5 m² were measured in 1988, a year of severe summer drought, and 1989, a year of normal rainfall. Resistance is a measure of change in species abundances (high resistance = little change, low resistance = much change). Biomass was measured on grazed plots at the peak of the summer growth season. (Data from Frank and McNaughton 1991.)

had higher resistance to change, as predicted by the diversity-stability hypothesis.

The conjecture of Charles Elton in 1958 that species diversity indeed imparts stability to ecological communities has been supported for many, but not all, ecological systems. The practical application of this idea to our human-degraded landscapes is the focus of the applied area of restoration ecology. What progress have we made in restoring damaged ecological systems?

Restoration Ecology

Communities recover from disturbances through a whole series of biological restoration mechanisms. Succession is a major pathway in restoration ecology, which aims to harness natural processes to restore systems adversely affected by humans. The key starting principle in restoration ecology is that the spatial scale of the impact and the recovery time are related, such that the larger the scale of the disturbance, the longer the time frame for restoration (Figure 23.21). There appears to be no difference in this relationship between man-made and natural disturbances (Dob-

son et al. 1997a). If we can identify the processes that limit the speed of recovery, we can alter this curve to reduce the effects of human disturbances.

The first principle of restoration ecology is that environmental damage is not irreversible. This optimistic principle must be tempered by the second principle of restoration ecology—that communities are not infinitely resilient to damage. Figure 23.1(a) illustrates these ideas schematically—there is only a finite range of environments over which a community is locally stable. To illustrate the first principle of restoration ecology, let us consider one example of lake restoration that has been successful. Aquatic communities have been disturbed by pollution of human origin for many years, and the stability of aquatic systems under pollution stress is a critical focus of restoration ecology today. Several large-scale uncontrolled experiments involving the diversion of sewage into large lakes near cities have been performed. Lake Washington is one such instance.

Lake Washington is a large, formerly unproductive lake in Seattle, Washington. In the early phases of city's development, Lake Washington was used for raw sewage disposal, but this practice was stopped between 1926 and 1936 (Figure 23.22). However, with additional population pressure, a number of sewage-treatment plants built between 1941 and 1959 began discharging treated sewage into the lake in increasing amounts. By 1955 it was clear that the sewage was destroying the clear-water lake, and a plan to divert sewage from the lake was voted into action. More and more sewage was diverted to the ocean from 1963 through 1968, and almost all was diverted

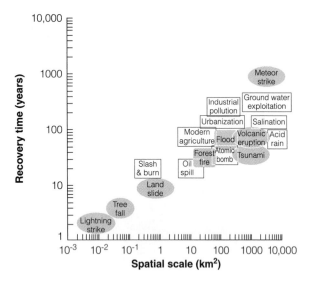

FIGURE 23.21
The relationship between the spatial scale of natural and artificial disasters and the approximate expected time to recovery. Natural disasters are depicted as red ellipses, and human-caused changes are represented by black rectangles. The aim of restoration ecology is to reduce the recovery time by manipulating ecological factors restricting the time sequence of recovery. (From Dobson et al. 1997.)

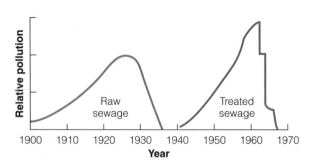

FIGURE 23.22
Sewage history of Seattle's Lake Washington. Raw sewage was diverted from the lake gradually over the period 1926–1936; then treated sewage was added at an increasing rate until a second diversion was made in 1963–1967. (After Edmondson 1969.)

1.   **EITHER**

    a)   Describe the mechanisms that are thought to underlie activity-dependent modifications of synaptic strength; include descriptions of the experimental evidence on which these hypotheses are based. Evaluate the extent to which these modifications of synaptic strength can account for the multiple forms and durations of learning and memory.

OR

    b)   Describe the role that the patient known as HM played in establishing the importance of the hippocampus in learning and memory. Describe the experiments that you consider are the most convincing of the role of the hippocampus in learning and memory.

2. Describe the evidence for the involvement of neuroimmune mechanisms in demyelinating and neurodegenerative diseases.

3. Describe the structure and function of ionotropic and metabotropic kainate receptors in the hippocampus, and their role in modifying neuronal excitability. What are the underlying mechanisms and signalling cascades?

END OF PAPER

UNIVERSITY OF SOUTHAMPTON   BIOL 3020WI

SEMESTER 2 EXAMINATION 2004-5

SYSTEMS NEUROSCIENCE

Duration: 120 mins

Answer **TWO** questions

Write each answer in a separate book

Number of
                                                 Pages 2

from March 1967 onward. Thus the recent history of Lake Washington consists of two pulses of nutrient additions followed by complete diversion.

What happened to the organisms in Lake Washington during this time? Some information can be obtained by looking at the sediments in the bottom of the lake. After sewage had been added to the lake, the sedimentation rate rose to about 3 mm per year. The organic content of this sediment (Figure 23.23, top) progressively increased since the early 1900s, which suggests an accelerated rate of primary production. The recent lake sediments also contain a greater amount of phosphorus (Figure 23.23, middle). Because phosphorus is one of the two main nutrients added by sewage, this change parallels that of the organic matter. The composition of the diatom community in Lake Washington has also changed (Figure 23.23, bottom). (The "shells" of diatoms are made of silica and are preserved well in sediments.) Stockner and Benson (1967) showed that a group of diatom species (the Araphidinae) varied in abundance in association with the sewage history and consequently can be used as *indicator species* of pollution in this lake.

Edmondson (1991) has recorded the changes in Lake Washington in detail since the diversion of sewage began in 1963. Figure 23.24 shows the rapid drop in

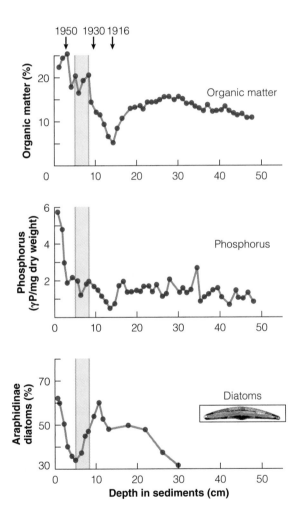

FIGURE 23.23
Historical changes in Lake Washington as revealed by core samples of lake-bottom sediments taken in 1958. The shaded area represents the approximate position in the core of the time period 1930–1940, when nutrient pollution from sewage was temporarily halted. (After Edmondson 1969.)

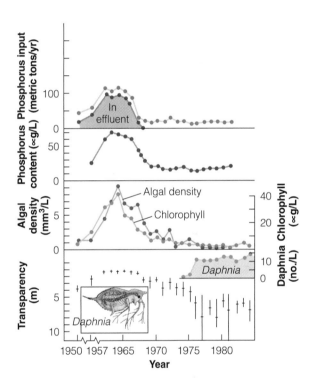

FIGURE 23.24
Recovery of Lake Washington, 1950–1984. Treated sewage flowed into the lake in increasing amounts during the 1950s. Sewage was diverted from the lake gradually from 1963 to 1968. The phosphorus content of the lake decreased rapidly after sewage was diverted. Algal density dropped in parallel with phosphorus levels because phosphate is the nutrient that limits algal growth in this freshwater lake. In 1976 the small crustacean Daphnia increased greatly in abundance, and because it eats small algae, algal abundance fell even more, making the lake water clearer. Transparency is the depth to which a standard white plate can be seen in the water during midsummer. (Data courtesy of W. T. Edmondson.)

phosphorus in the surface waters and the closely associated drop in the standing crop of phytoplankton. Nitrogen content of the water has dropped very little, which suggests that phosphorus is a limiting nutrient to phytoplankton growth. The water of the lake has become noticeably clearer since the sewage diversion. The phosphorus tied up in the lake sediments is apparently released back into the water column rather slowly.

In 1976 Lake Washington suddenly became much clearer than had been recorded previously. This clearing of the water coincided with the colonization of Lake Washington by *Daphnia* sp., filter feeders that sweep small green algae out of the water and in the process made the lake clearer. *Daphnia* have always been present in lakes around Lake Washington, so their sudden appearance in 1976 was not a simple matter of colonization and dispersal. Two factors interacted to hold *Daphnia* numbers low before 1976. The blue-green alga *Oscillatoria*, which was common in Lake Washington during its polluted phase, gradually declined after 1968, when sewage was diverted. *Oscillatoria* clogs the filter-feeding apparatus of *Daphnia* because it forms long filaments, so *Daphnia* could not feed properly until *Oscillatoria* became scarce. *Daphnia* are also a major prey item for the shrimp *Neomysis mer-*

cedis. *Neomysis* decreased during the 1960s because of fish predation by increased populations of longfin smelt in the lake. Smelt abundance had increased during the 1960s, when their spawning grounds in the Cedar River were protected from dredging and construction in the river. Thus, the aquatic community of Lake Washington contains a complex web of predator-prey interactions that have profound effects on community structure (Edmondson 1991).

The Lake Washington experiment is of considerable interest because it suggests that detrimental changes in lakes may be *stopped and reversed* if the input of nutrients is halted. The restoration of Lake Washington shows that this aquatic community displays a considerable degree of global stability and is a good example of an equilibrium community.

Restoration ecology can depend on the natural time scale of succession, as occurred in Lake Washington, or it can speed processes of recovery by adding nutrients when they are deficient, by seeding areas that have a shortage of available colonists, and by adding microbes to break down organic compounds such as oil. This area of applied ecology will assume more importance in the years to come as we try to speed the recovery of degraded landscapes.

Summary

Communities can be organized by competition, predation, and mutualism working within a framework set by the physical components of the environment. Of these three processes, the greatest emphasis has been placed on the roles of competition and predation in organizing communities, and the role of mutualism is largely unstudied.

Two broad views of communities are postulated as explanations of community organization. The classical model is that communities are in equilibrium, and their species composition and relative abundances are controlled by biotic interactions. According to this equilibrium model, interspecific competition, predation, and spatial heterogeneity are the major processes controlling organization. The nonequilibrium model does not assume stable equilibria but holds instead that communities are always recovering from disturbances.

Species in a community can be organized into food webs based on "who-eats-whom." Trophic levels may be recognized in all communities from the level of producers (green plants) to the higher carnivores and hyperparasites. Within a trophic level, we can recognize *guilds* of species exploiting a common resource base. Guilds may serve to pinpoint the functional roles species play in a

community, and species within guilds may be interchangeable in some communities.

Food webs can be analyzed theoretically and empirically. Patterns of food chain length and complexity can be described, but all of them may vary with the complexity of the food web. The analysis of complex tropical food webs containing more than 100 species is a difficult task and is only just beginning.

Keystone species single-handedly determine community structure and can be recognized by removal experiments. Dominant species are the species of highest abundance or biomass in a community. Dominance is often achieved by competitive superiority, and some dominant species can be removed from the community and be replaced by subdominants with little effect on community organization. In aquatic communities, dominance in zooplankton herbivores may be determined by competition when fish predators are absent, and by predation when fish are present.

The characteristics of dominant species may affect community stability, with respect to the system's ability to return to its original configuration after disturbance. The ecological generalization that *diversity promotes stability* is supported by field data and by theoretical analyses, but we do not know if it is true for all communities. The attributes of individual species and compartments in food webs may be significant in determining community stability.

Restoration ecology strives to apply ecological knowledge of community dynamics to restore damaged landscapes. Succession can heal damaged landscapes but may take more time that we might like. We can try to speed recovery by knowing how succession operates in a community and what limits its rate of progress. Communities can recover from disasters, but there is a limit to their resilience.

Key Concepts

1. Communities can be organized by competition, predation, or mutualism. Almost all discussion centers on competition and predation, and mutualism has been largely ignored in studies of community integration.

2. Two general models of community organization exist. The *equilibrium model* focuses on community stability and biotic coupling, the *nonequilibrium model* on stochastic effects and species independence.

3. Food webs describe who eats whom in a community. Within food webs, food chains are short, predator-prey ratios relatively constant, and connectance approximately constant, regardless of the number of species involved.

4. Species may be grouped into *guilds*, competitive webs of similar species that exploit a common resource base.

5. Not all species in a food web are of equal importance. *Keystone species* are low in abundance, but their removal causes high community impact. *Dominant species* are high in abundance and help determine community structure.

6. Stability of community composition and dynamics is produced by species diversity in many communities. The hypothesis that *diversity promotes stability* is a key argument in conservation programs.

7. The restoration of degraded communities can occur through the natural processes of succession, or we can speed recovery through our understanding of what limits community composition and species abundances.

Selected References

Brown, J. H. and E. J. Heske. 1990. Control of a desert-grassland transition by a keystone rodent guild. *Science* 250:1705–1707.

Connell, J. H. and W. P. Sousa. 1983. On the evidence needed to judge ecological stability or persistence. *American Naturalist* 121:789–824.

Connell, J. H., T. P. Hughes, and C. C. Wallace. 1997. A 30-year study of coral abundance, recruitment, and disturbance at several scales in space and time. *Ecological Monographs* 67: 461–488.

Doak, D. F., D. Bigger, E. K. Harding, M. A. Marvier, R. E. O'Malley, and D. Thomson. 1998. The statistical inevitability of stability-diversity relationships in community ecology. *American Naturalist* 151:264–276.

Estes, J. A., M. T. Tinker, T. M. Williams, and D. F. Doak. 1998. Killer whale predation on sea otters linking oceanic and nearshore ecosystems. *Science* 282:473–476.

France, R., K. Westcott, P. Del Giorgio, G. Klein, And J. Kalff. 1996. Vertical foodweb structure of freshwater zooplankton assemblages estimated by stable nitrogen isotopes. *Researches in Population Ecology* 38:283–287.

Hall, S. J. and D. G. Raffaelli. 1997. Food-web patterns: What do we really know? In A. C. Gange and V. K. Brown, eds. *Multitrophic Interactions in Terrestrial Systems.* Blackwell Science, London.

Johnson, K. H., K. A. Vogt, H. J. Clark, O. J. Schmitz, and D. J. Vogt. 1996. Biodiversity and the productivity and stability of ecosystems. *Trends in Ecology and Evolution* 11:372–377.

McGowan, J. A. and P. W. Walker. 1985. Dominance and diversity maintenance in an oceanic ecosystem. *Ecological Monographs* 55:103–118.

McPeek, M. A. 1998. The consequences of changing the top predator in a food web: A comparative experimental approach. *Ecological Monographs* 68: 1-23.

Polis, G. A. and D. R. Strong. 1996. Food web complexity and community dynamics. *American Naturalist* 147:813–846.

Power, M. E., D. Tilman, J. A. Estes, B. A. Menge, W. J. Bond, L. S. Mills, G. Daily, J. C. Castilla, J. Lubchenco, and R. T. Paine. 1996. Challenges in the quest for keystones. *BioScience* 46:609–620.

Tilman, D. 1996. Biodiversity: Population versus ecosystem stability. *Ecology* 77:350–363.

Questions and Problems

23.1 Elton (1958, p. 147) claims that natural habitats on small islands are much more vulnerable to invading species than natural habitats on continents. Find evidence that is relevant to this assertion, and evaluate its importance for the question of community stability.

23.2 Doak et al. (1998) argued that there must be a positive relationship between community stability and species richness because of statistical averaging of the fluctuations in species abundances. Construct a simple model to convince yourself of the truth of this criticism. What assumptions does it make about community dynamics? Read the rebuttal by Tilman et al. (1998) for additional information on this problem.

23.3 In discussing community organization, Hairston (1964, p. 238) states:

> It would be possible, presumably, to build a picture of community organization by separate complete studies of each species present, but such an approach would be comparable to describing an organism cell by cell.

Is this a proper analogy?

23.4 Compare and contrast the following statements of an evolutionist and an ecologist about species diversity and the stability of biological communities:

a. Simpson (1969, p. 175) states: "If indeed the earth's ecosystems are tending toward long-range stabilization or static equilibrium, three billion years has been too short a time to reach that condition."

b. Recher (1969, p. 79) states: "The avifaunas of forest and scrub habitats in the temperate zone of Australia and North America have reached equilibrium and are probably saturated."

23.5 How is it possible for stable isotope ratios to change between trophic levels? List several possible physiological mechanisms that might cause such changes. Are there any population mechanisms for achieving these changes? Would you expect differences in isotope ratios if you measured different parts of an animal or plant?

23.6 How long a time does a community need to be studied before all components of its food web are identified? Discuss the implications of constructing a time-specific food web versus a cumulative food web over a long period. Compare your analysis with that of Schoenly and Cohen (1991).

23.7 Compare the definitions of a *trophic species* (see p. 465) and a *guild* (see p. 469). How does the aggregation of species into trophic species affect the analysis of a food web? Conversely, how would poor taxonomic resolution within species groups affect estimates of connectance in food webs—for example, if species are grouped into categories such as "insects"? Polis (1991) and Hall and Raffaelli (1993) discuss these problems of aggregation and taxonomic resolution in food web analysis.

23.8 In some tropical rain forests, 50–100% of the canopy trees are one species (Connell and Lowman 1989). List all the possible mechanisms by which a single species can maintain its dominance in a species-rich forest system. Which of these mechanisms could not operate in species-poor temperate forests? Compare your list with that of Connell and Lowman (1989, p. 97).

23.9 Bracken fern (*Pteridium aquilinum*) often invades lowland heaths in Britain and develops a dense, uniform stand with a very depauperate flora and fauna. To reverse this habitat deterioration, various chemical and physical control methods were carried out over 18 years, but the objective of this restoration scheme (to restore heather heathland) was not achieved. Read Marrs et al. (1998) and discuss the reasons for the failure of this restoration program.

23.10 Adult lake trout from 24 lakes in eastern Canada had $\delta^{15}N$ values ranging from 7.5‰ to 17.5‰. Compare this variation with that shown in the graph in Essay 23.1 (see p. 466), and discuss whether this variation invalidates the use of $\delta^{15}N$ values as trophic level indicators. Cabana and Rasmussen (1994) discuss this problem.

Overview Question

You are asked to determine the food web for a lake that is proposed to be part of a national park and to assess the web's stability. Discuss how you would construct this food web and measure its stability, and what operational decisions you would have to make about what to include in the web.

CHAPTER 24

Community Organization III:
Disturbance and Nonequilibrium Communities

THE EQUILIBRIUM MODEL OF COMMUNITY organization has been the classical model of community organization for the past 50 years. Like many classical models in ecology, more and more exceptions have been found to its predictions. The equilibrium model is focused on stability, and in some communities it is a good description of community organization. But in many other communities of a few hectares in area, *change* seems to be the rule rather than stability, causing many ecologists to search for a broader model for communities. To replace the classical equilibrium model, ecologists are now forging a new model of community organization called the *nonequilibrium model*, which focuses on the small spatial scale and emphasizes two central ideas—the concepts of *patches* and *disturbance* (DeAngelis and Waterhouse 1987). In this chapter we explore some of the new concepts the nonequilibrium model has stimulated.

Patches and Disturbance

Patchiness, a term that refers to the spatial scale of a system, has been recognized as a factor in how a system is described. The patchiness of different communities varies, and conclusions that apply to one spatial scale will not necessarily apply to others. We can recognize five spatial scales at which ecologists work (Wiens et al. 1986):

- Space occupied by one plant or sessile animal, or the home range of an individual animal

- Local patch, occupied by many individual plants or animals

- Region, occupied by many local patches or by local populations linked by dispersal

- Closed system (if it exists), or a region large enough to be closed to immigration or emigration

- Biogeographical scale, including zones of different climate and different communities

At sufficiently small spatial scales, all ecological systems are short-lived and can never be at equilibrium. Understanding community dynamics at small spatial scales of a few hectares and aggregating the resulting dynamics into a regional scale is an important approach to predicting large-scale community dynamics (Chesson and Case 1986).

Most field studies of communities and virtually all experimental manipulations of communities are conducted at the local patch scale. No single general definition of a "patch" can cover all ecological communities, but in general a patch will cover a few square meters to a few hectares. Note that patches need not be completely spatially discrete, nor do they need to be completely homogeneous (Pickett and White 1985).

A *disturbance* is any discrete event that disrupts community structure and changes available resources, substrate availability, or the physical environment. Note that disturbances can be destructive events like fires or an environmental fluctuation like a severe frost. The notion of "normal" is excluded from ecologists'

485

TABLE 24.1 Definitions of measures of disturbance.

Measure	Definition
Distribution	Spatial distribution, including relationship to geographic, topographic, environmental, and community gradients
Frequency	Mean number of events per time period
Return interval, or turnover time	The inverse of frequency: mean time between disturbances
Rotation period	Mean time needed to disturb an area equivalent to the study area (the study area must be defined)
Predictability	An inverse function of variance in the return interval
Area or size	Area disturbed. This can be expressed as area per event, area per time period, area per event per time period, or total area per disturbance type per time period
Magnitude:	
Intensity	Physical force of the event per area per time (e.g., windspeed for hurricanes)
Severity	Effect on the community (e.g., basal area removed)
Synergism	Effects on the occurrence of other disturbances (e.g., drought increases fire intensity, or insect damage increases susceptibility to windstorm)

Source: Modified from Pickett and White (1985).

view of disturbance in communities, an important change of view that has implications for conservation and management. We cannot assume under the non-equilibrium model that communities in the "good old days" were "normal," and that the job of conservationists or land managers is to get back to what the community was like in the "good old days." For some communities disturbances are frequent, but in others disturbances are rare.

Disturbances can be measured in a variety of ways (Table 24.1), most of which provide either a spatial or temporal perspective. Disturbances may also be classified as *exogenous* (arising from outside the community, like fire) or *endogenous* (resulting from biological interactions, like predation). These two classes are the extremes of a continuum of types of disturbances, and many communities are affected by a combination of endogenous and exogenous disturbances. The challenge to ecologists is to incorporate the various measures of disturbance in the field in their efforts to understand how disturbances affect particular communities.

The Role of Disturbance in Communities

In this section we examine the effects of disturbance in two communities: coral reef communities and rocky intertidal communities.

Coral Reef Communities

Coral reefs have existed in tropical oceans for at least 60 million years, and this long history is postulated to have produced the great diversity of organisms that are present on reefs today. Even though coral reefs have long been viewed as the classical equilibrium community living in tropical waters, our view has been changed by long-term studies of coral reefs. We consider two aspects of coral reefs: the corals and the reef fishes.

Coral reefs are subject to a variety of physical disturbances associated with tropical storms. At the Heron Island reef at the southern edge of the Great Barrier Reef off Queensland, Connell et al. (1997) followed changes in coral cover over a 30-year period using permanently marked quadrats. They measured the amount of coral cover to estimate abundance because corals are modular organisms and colonies vary greatly in size, and they measured larval recruitment of corals by sequential photographs of the permanent quadrats.

Violent storms were the main source of disturbance to Heron Island reefs, and the amount of damage caused by cyclones was strongly affected by the position of the coral colonies on the reef (Figure 24.1). Five cyclones passed near Heron Island during the 30 years of study from 1963 to 1992. Of the four study areas shown in Figure 24.1, only the protected area of the inner flat was relatively unaffected by cyclones. Virtually every cyclone caused a reduction in coral cover in the exposed pools. The 1972 cyclone

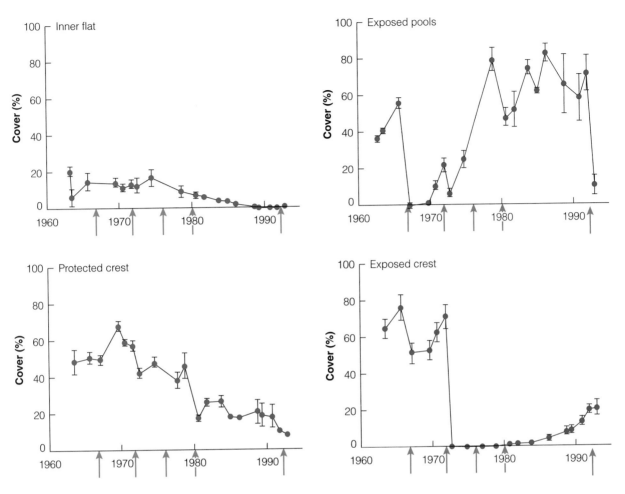

FIGURE 24.1
Percentage cover of corals in four areas of the coral reefs surrounding Heron Island at the southern edge of the Great Barrier Reef in Australia. Years with tropical cyclones are indicated by red arrows. Cover was measured on permanent quadrats in these shallow-water sites from 1963 to 1992. Damage from cyclones was highly variable depending on how much the site was protected by the island. (From Connell et al. 1997.)

completely removed coral cover on the exposed crest, the most severe disturbance observed. Recovery on the exposed crest was slow for the next 25 years. Gradual declines in coral cover on the protected sites was caused by increasing exposure to air as the corals grew upward over the 30 years of study.

Recruitment rates of corals were highly variable, and this is typical of many marine invertebrates whose larvae drift before settling. Figure 24.2 illustrates the differences in recruitment rates among years and among sites on the Heron Island reef. There were no particularly good or bad years for coral recruitment

in the sense that the whole reef varied in unison. The variability in recruitment was partly associated with how much free space was available in different areas. Coral larvae cannot attach to other living coral or macroalgae and need free space to settle.

The picture that emerges from this work on the Great Barrier Reef is of a coral community that changes continually because of exogenous disturbance caused by tropical cyclones and the internal processes of growth and recruitment. The coral community is not in equilibrium at the spatial scale of the reef because the frequency of disturbance exceeds the rate

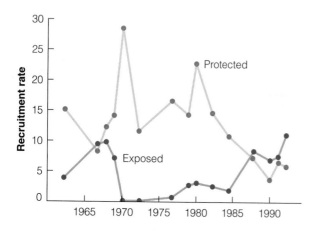

FIGURE 24.2
Coral recruitment rate, measured as number of recruits per m² per year, on permanent plots on protected and exposed crests of the Heron Island reef off Queensland, 1963–1992. Because recruitment is highly variable among sites and among years and does not correlate among the different sites, there were no generally "good" or "bad" recruitment years. (After Connell et al. 1997.)

of recovery. Coral reefs are a good example of a nonequilibrium community.

Coral reef fishes are one of the primary drawing cards for ecotourism to coral reef areas. The diversity of coral reef fishes is astonishing. For example, at One Tree Reef, a small island at the southern edge of the Great Barrier Reef, Talbot and colleagues (1978) recorded nearly 800 species of fish. At the northern edge of the Great Barrier Reef, over 1500 species of fish have been recorded. What determines the community structure of a coral reef fish community? Are these fish communities stable in time and space?

There are two extreme schools of thought about what controls the organization of coral reef fish communities. The first view suggests that coral reef fish communities are equilibrium systems in which fish populations are controlled by density-dependent processes. Within this equilibrium viewpoint are two hypotheses concerning the mechanism by which equilibrium is achieved. The *niche-diversification hypothesis* suggests that coral reef fish communities are equilibrium competitive systems in which each species has evolved a very specific niche with respect to food and microhabitat. According to this view, current competition among species is strong and maintains niche differences, species segregation, and high diversity (Anderson et al. 1981a); alternatively, density-dependent predation on adults could control populations of reef fish and maintain equilibrium communities. The second school of thought champions the *variable recruitment* or *lottery hypothesis*, which suggests that coral reefs are nonequilibrium systems in which larval recruitment is as unpredictable as a lottery. Competition among species is present, but the winner in competition cannot be predicted, and the local community present on a reef is a random sample from a common larval pool (Sale 1977). Mortality after larval settlement is density independent, and consequently local populations fluctuate under the control of unpredictable recruitment. How can we evaluate these equilibrium and nonequilibrium hypotheses of community organization?

The first question we may ask is, How specialized are reef fishes? Reef fish exhibit both food and habitat specialization, but often several species are found within one restricted niche, and many generalist feeders are present (Sale 1977). Table 24.2 gives a sample of data on the feeding specializations of butterfly fishes. Herbivorous fishes are more generalized feeders than predatory fishes. But even among the specialist feeders,

TABLE 24.2 **Feeding specializations among 20 species of butterfly fishes near Lizard Island, Great Barrier Reef, Australia.**

Hard coral	Soft coral and some hard coral	Noncoralline invertebrates	Generalists[a]
Chaetodon aureofasciatus	*Chaetodon kleinii*[b]	*Chaetodon auriga*	*Chaetodon citrinellus*
C. baronessa	*C. lineolatus*	*C. auriga*	*C. ephippium*
C. ornatissimus	*C. melannotus*	*Chelmon rostratus*	*C. ulietensis*
C. plebeius	*C. unimaculatus*	*Forcipiger* spp.	*C. vagabundus*
C. rainfordi	*C. speculum*	*C. trifascialis*	*C. trifasciatus*

[a]Includes plankton in the diet.
[b]The most common food items for this species were polychaetes and crustaceans.
Source: After Anderson et al. (1981).

ESSAY 24.1

WHY ARE CORALS BLEACHING?

Corals are animals that contain within their cells symbiotic algae that both contribute to coral growth (by conducting photosynthesis) and impart color to the corals. When corals lose their symbiotic algae, they lose their color—a process called bleaching—and often the corals die.

Coral reef bleaching has increased dramatically in many tropical areas around the globe over the past 20 years. Widespread bleaching can cause the death of whole coral reefs. The primary cause of coral bleaching is elevated sea surface temperatures. Many reef-building corals live very close to their upper lethal temperatures, and small increases of 0.5–1.5°C over a few weeks, or large increases of 3–4°C over several days, can kill corals (Huppert and Stone 1998).

What might cause increased surface temperatures? Water temperatures in the equatorial Pacific Ocean follow a roughly cyclical pattern of warm and cool phases; broad-scale warming occurs every three to seven years, a phenomenon called El Niño. It has been suggested that if we know that El Niño conditions are due, perhaps we can predict episodes of coral bleaching. But part of the difficulty in accepting a simple temperature model of coral reef bleaching has been the observation that not all corals bleach in a given area, and that areas outside the Pacific (such as the Caribbean) are also affected. One possible explanation is that these large-scale oceanographic events have worldwide climatic repercussions and thus are not confined to the traditional El Niño regions of the Pacific and Indian Oceans.

If the temperature explanation for bleaching is correct, why should bleaching have increased dramatically in the past 20 years? Three factors may be involved. Global warming is a key suspect. In addition, degradation of coral reefs from pollution may have reduced the general resilience of reefs to damage, including bleaching. Finally, increased ultraviolet radiation could be combining with increased temperatures to induce bleaching. El Niño years are often associated with clear skies and calm seas, which increase the penetration of UV radiation to greater ocean depths.

The ecological consequences of the destruction of richly diversified coral reefs are large, and this problem deserves global attention.

Bleaching in corals on the Maldive Islands. Within five months this reef went from healthy corals to bleached corals to corals overgrown by brown algae.

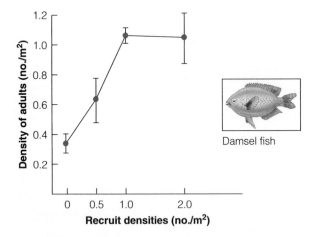

FIGURE 24.3

Test of the variable recruitment hypothesis for coral reef fish. Four different recruitment levels were experimentally provided for the damselfish Pomacentrus amboinensis *for three years. Adult densities increased directly with recruitment, as predicted by the variable recruitment hypothesis, up to 1 recruit/m², but above 1 recruit/m² density was limited. (From Jones 1990.)*

it is common to find two or three species with identical specializations, as seen in Table 24.2. Thus, feeding niches are not organized as tightly as the niche-diversification hypothesis would predict, and competition for food does not appear to be an important process of importance in these fish communities.

Habitat specialization could be another way that reef fishes have evolved niche differences. Adult reef fishes are very sedentary and thus could have very narrow habitat requirements, but this does not seem to be the case. Habitat partitioning does occur to the extent that few species range over all regions of the reef. Species tend to occur in broadly defined habitats, such as "reef flat" or "surge channel," but when microhabitats are assigned more carefully, extensive overlap of species is observed. Thus, we do not find a high degree of habitat specialization among coral reef fishes (Sale 1977).

Natural history information thus does not tend to support the niche-diversification hypothesis. How can we test these hypotheses experimentally? If the stochastic recruitment hypothesis is correct, then reef fish communities ought to be unstable in species composition and highly variable from reef to reef. Also, the species structure at a given site should not recover

following artificial removals or additions of species, and we should be able to predict population size on a reef from the number of recruits that arrive.

To test the first prediction, Talbot et al. (1978) put out standard cement building blocks to create artificial reefs of constant size and shape. Forty-two fish species colonized these artificial reefs, averaging 17 species per reef. The similarity among replicate reefs was only 32% which means that even though these reefs were set out in the same lagoon at the same time within a few meters of one another, only about 32% of the fishes colonizing them were of the same species. A survey of natural coral isolates of about the same size as the artificial reefs (0.6 m³) showed only a 37% similarity (Talbot et al. 1978). Moreover, as one followed the artificial reefs over time, very high turnover occurred from month to month. Of the species on a reef one month, 20–40% would have disappeared by the next month and been replaced by a species that was not previously present. Clearly, reef fish communities are very unstable and highly variable from one small reef to the next.

The variable recruitment hypothesis assumes the absence of any resource limitations on populations and the lack of competitive effects; population size is set by how many recruits arrive at a reef. One prediction from this hypothesis is that if recruits are experimentally added to a coral reef population, adult numbers will rise. Jones (1990) transplanted juveniles of the damselfish *Pomacentrus amboinensis* for three years to natural patches of reefs of approximately 8 m² at the southern edge of the Great Barrier Reef in Australia. Figure 24.3 shows that adult densities increased as more recruits were added, but only to a ceiling. At high recruitment levels, density-dependent interactions between adults and potential recruits put a ceiling on numbers, contrary to the predictions of the variable recruitment hypothesis. The important point illustrated by Figure 24.3 is the existence of both a range of recruitment that fits the variable recruitment hypothesis, and a threshold above which this hypothesis does not hold.

Another way to test the variable recruitment hypothesis is to survey a variety of reefs and measure recruitment rates. Doherty and Fowler (1994) did this for nine years on reefs at seven scattered islands in the southern part of the Great Barrier Reef. They found that average recruitment of the damselfish *Pomacentrus moluccensis* over the nine years was highly correlated with the average population size in each lagoon (Figure 24.4). These results agree with the predictions of the variable recruitment hypothesis. The mortality rate of damselfish after recruitment appeared to be constant rather than density dependent, suggesting a nonequilibrium model.

Discussions among coral reef ecologists about the importance of prerecruitment and postrecruitment processes in limiting adult fish densities have remained controversial because some results favor the variable recruitment hypothesis and others do not (Hixon 1998). Both recruitment and postrecruitment processes affect the abundance of coral reef fishes. Moreover, the results from these studies may depend on the spatial scale at which the fish populations were observed. Most studies have been done on patch reefs that were small enough to be censused by one or two divers. Ault and Johnson (1998) suggested that the variable recruitment model may apply to many small, isolated patch reefs, but that it cannot necessarily be extrapolated to fish communities in large sections of continuous coral reefs.

Coral reef fishes are similar to many marine organisms in having a life history that includes a planktonic larval phase that is transported by ocean currents. This type of life cycle implies that local reproduction is not linked to local recruitment, in complete contrast to the life cycle of birds and mammals. For these marine organisms the population or community can never be a closed system, and the physical factors controlling recruitment may control the system. This has been called "supply-side ecology" by Roughgarden et al. (1987). The ideas of supply-side ecology have a long history (Underwood and Denley 1984). The structure of an ecological community driven by supply-side ecology cannot be understood solely as a result of competition, predation, and disturbance, but only by identifying what controls variable recruitment, which keeps populations under the carrying capacity (Underwood and Fairweather 1989).

How is the coexistence of so many species permitted if the variable recruitment hypothesis is correct? Sale (1982) calls the reef fish community a *lottery competition*. Individuals compete for access to units of resources (for these fish, space) without which they cannot join the breeding population. A lottery competition is a type of interference competition in which an individual's chances of winning or losing are determined by who gets access to the resource first. Lottery competitive systems are very unstable but can persist if there is high environmental variability in birth rates (Chesson 1986). Because recruitment of

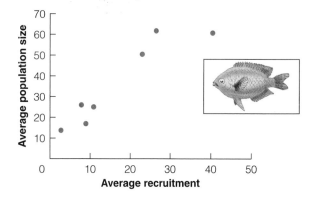

FIGURE 24.4
Survey of the recruitment of the damselfish Pomacentrus moluccensis *on the southern Great Barrier Reef. Ten patch reefs in seven lagoons were surveyed for nine years. Population size units are number of fish per 100 m². Measuring recruitment allows predictions of subsequent population densities, as suggested by the variable recruitment hypothesis. (From Doherty and Fowler 1994.)*

reef fishes depends on larvae settling from the plankton, high variability is the rule, and vacated space is allocated at random to the first recruit to arrive from the larval pool (Sale 1982). The lottery competition model has three important requirements if it is to explain the coexistence of many species: (1) Environmental variation must be such that it permits each species to have high recruitment rates at low population densities, (2) generations must overlap, and (3) adult death rates should be unaffected by competition (Chesson and Warner 1981).

Thus, the high diversity of coral reef fish communities is not achieved by precise niche diversification in an equilibrium community, but rather by highly variable larval recruitment resulting in a competitive lottery for vacant living spaces in which the first to arrive wins. Reef fish communities are thus not in equilibrium, but instead continually fluctuate in local species composition while retaining high regional diversity.

Rocky Intertidal Communities

The rocky intertidal zone is a tension zone between land and sea featuring disturbances in the form of waves and storms. Space is the key limiting resource in the intertidal zone, and competition for space has been a key component of many studies of this community (Dayton 1971, Sousa 1985). The key concepts of keystone species and the intermediate disturbance hypothesis have their origins in the rocky intertidal zone. Two examples illustrate the ways in which rocky intertidal communities can be organized.

Jane Lubchenco conducted classic work on communities of seaweeds and invertebrates on the rocky coasts of New England to measure the effects of predation and physical disturbance on seaweed abundance. Seaweeds in New England are of two groups: Ephemeral seaweeds live for weeks or months, grow rapidly, and are eaten rapidly by herbivores such as limpets; perennial seaweeds can live for many years, grow more slowly, and except in their juvenile stages are relatively inedible (Lubchenco 1986). Seaweeds compete for space and for light, but the primary resource in the rocky intertidal zone is space. Using wire-mesh cages to exclude herbivores, Lubchenco (1986) found no simple answer to the question of what controlled seaweed abundance (Figure 24.5). On protected areas in summer, limpet grazing reduced the abundance of ephemeral algae soo much that competition for space among the algae was eliminated. On wave-exposed sites, limpets cannot easily live and algae are washed away by wave action, so the amount of competition among seaweeds is reduced. In this system herbivores set the stage for competitive interactions, and the exact dynamics of a small patch of rock depends on the

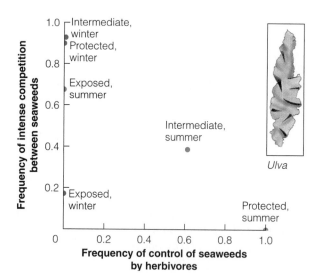

FIGURE 24.5
Effects of herbivores on the frequency of competition among ephemeral seaweeds (including Ulva *sp.) in the rocky intertidal zone of New England. Season and wave exposure are indicated for each point. The effects of herbivores depend on the season and the physical environment. (From Lubchenco 1986.)*

Jane Lubchenco *(1947–) Professor of Marine Biology, Oregon State University*

physical environment (wave action) and the number of herbivores present.

Coralline algae are encrusting algae in the rocky intertidal zone that compete with each other via overgrowth. In the absence of herbivores like chitons and limpets on a smooth surface, Paine (1984) found a clear dominance of competitive interactions (Figure

24.6)—that is, competition for space was *transitive*. But in natural communities with and without grazers, three of the coralline algae showed intransitive competition, winning some encounters and losing others. Grazers thus act to slow down the rate of succession of coralline algae by inducing competitive uncertainty, and this acts to promote species diversity. In the absence of disturbance (grazing), a single competitive dominant would monopolize space in the rocky intertidal (Paine 1984). This algal community is not an equilibrium assemblage under natural conditions because grazing changes the system from a transitive competitive network with a definite stable outcome to an intransitive network with no stable equilibrium.

Theoretical Nonequilibrium Models

These two examples of community dynamics illustrate some of the ideas that have been central to the development of nonequilibrium models of communities. Communities can be positioned along a gradient from stable, biotically interactive communities that are equilibrium centered, to unstable, interactive communities in which biotic interactions do not lead to a stable equilibrium, to weakly interactive commu-

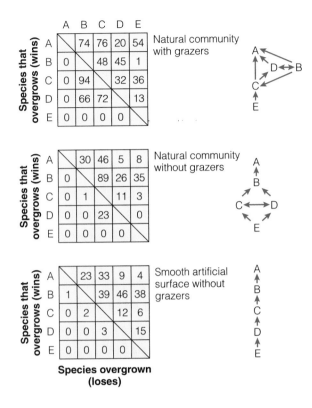

FIGURE 24.6

Interspecific relationships within a guild of five coralline algal species under three different conditions: (top) with grazers, on a natural surface; (center) without grazers, on a natural surface; (bottom) without grazers, on a smooth artificial surface. Letters refer to species: A, Pseudolithophyllum lichenare; B, Lithothamnium phymatodeum; C, Pseudolithophyllum whidbeyense; D, Lithophyllum impressum; E, Bossiella sp. Numbers in the array are observed overgrowths when two guild members come into contact. The diagrammed competitive interactions indicate the change of position induced under the various conditions. Arrows point from toward competitive winners. Two-headed arrows indicate that no significant bias exists in the interaction's direction. (From Paine 1984.)

nities in which physical factors such as temperature, salinity, or fire prevent any stable equilibrium. Models have been developed all along this gradient to describe the ecological complexities of these systems. Chesson and Case (1986) recognize four types of non-equilibrium models of communities:

1. *Fluctuating environment models:* The simplest deviation from the classical equilibrium model of a community is a model with temporal variability. Competition is viewed in these models as the major biological interaction, but the environment

changes seasonally or irregularly and the competitive rankings of species also fluctuates such that no one species can win out. These models may include movements between patches, each of which may have a different environment. Fluctuating environment models are similar to equilibrium models in that they produce stable communities, but they differ in emphasizing the dynamics of dispersal among patches and variable life history traits.

2. *Density-independent models:* These models assume that population densities change, but the classical models, fluctuations, are often density independent. Density-vague dynamics (see Figure 16.9, p. 292) predominates in these theoretical communities, and populations are typically at levels at which competition for resources is rare (Strong 1986). Spatial patchiness is added to some of these models as another feature promoting fluctuations in the community.

3. *Directional changing environment models:* Variable environment models usually consider environments to fluctuate about some mean value that remains constant over time. When the mean itself changes, the result heavily depends on the amount of fluctuation and the speed of the community in reacting to change. The current concern over global climate change (see Chapter 28) makes these models very significant for the future. Unlike many community models, these models cannot ignore history, and the response of a community to, for example, changing climate, depends on its past history. Modern communities, these theories argue, cannot be understood only by looking at present environmental conditions. Life history characteristics and dispersal abilities strongly affect the ability of a species to respond to environmental changes.

4. *Slow competitive displacement models:* If competitive abilities of species are nearly equal, the process of competitive exclusion will be very slow, and random variation in success will obscure any obvious displacements. Hubbell and Foster (1986) argue that the tropical rain forest has many species that are ecologically identical, and that community composition is the net consequence of a slow random-walk of tree species densities. Competition in these models occurs all the time, but because all species are identical in competitive abilities, there is no time trend or succession. Under these

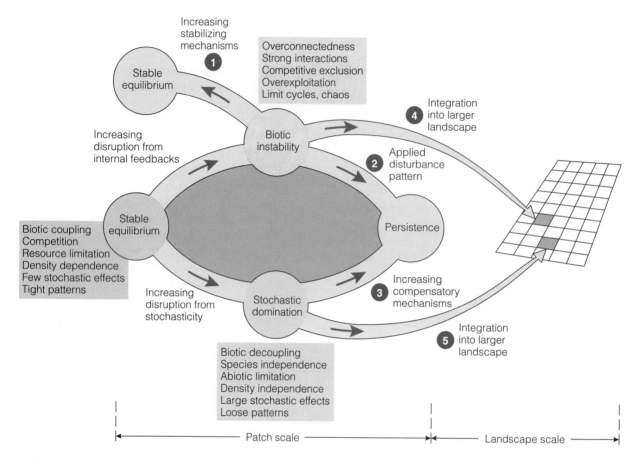

FIGURE 24.7

A schematic diagram showing five general mechanisms that may explain why communities tend to persist despite the prevalence of biotic instabilities and environmental fluctuations. (From DeAngelis and Waterhouse 1987.)

models, community structure is strongly affected by chance and by history, and changes occur only on a geological time scale.

The purpose behind all of these models is to understand what enables a community to persist over time. These four models utilize five general types of mechanisms to explain why communities tend to persist (Figure 24.7). The first mechanism can be characterized as mechanisms that stabilize interactions and that are usually additions to classical equilibrium models of community organization. If predator functional responses are sigmoid (see Figure 13.16, p. 222), or if consumers are self-regulated, the stability of the community could be increased (DeAngelis and Waterhouse 1987). The

second mechanism that promotes community persistence is (paradoxically) disturbances that create nonequilibrium landscapes and prevent competitive exclusion. The third mechanism involves compensatory changes in reproduction, survival, or movements when populations reach low densities. Such changes could favor rare species over common ones. The fourth and fifth mechanisms that promote community persistence both involve spatial patchiness. If local populations are connected into metapopulations at the landscape level (see Figure 24.7), the fact that each local patch is unstable may not matter, because species can recolonize by dispersal between patches. Stability in metapopulations is seen at the landscape level, not at the local population level. Species may go extinct in local patches, but as long as local patches are out of phase with one another, the species will persist in the landscape.

To translate these ideas on nonequilibrium community dynamics into the real world, ecologists have developed a series of conceptual models that can be tested in field studies.

Conceptual Models of Community Organization

A series of models have been proposed by field ecologists to try to capture the interrelations between physical factors and biological interactions in organizing communities. All of these models recognize that many different kinds of ecological communities exist and that the important processes will not be the same in all ecological systems (Schoener 1986a).

The most comprehensive model of community organization was proposed by Menge and Sutherland (1987). They recognized three ecological processes as the main determinants of community organization—physical disturbance, predation, and competition—and they included variable recruitment as a part of the model (Figure 24.8). A central assumption of this model is that food web complexity decreases with increasing environmental stress. This model makes three predictions for communities that have high recruitment. First, in stressful environments herbivores have little effect because they are rare or absent, and plants are regulated directly by environmental stress. Neither predation nor competition is significant. An example of such a community is the arctic tundra or a desert. Second, in moderately stressful environments, consumers are ineffective at controlling plants, and plants attain high densities. Competition among plants is the dominant biological interaction in these communities. Third, in benign environments, consumers control plant numbers, and plant competition is rare. Predation is the dominant biological interaction under these benign conditions.

The Menge-Sutherland model is a general model of community organization that incorporates some ideas from a model proposed by Hairston, Smith, and Slobodkin (1960). Hairston et al. predicted that for terrestrial communities in benign environments, predators must limit herbivores, which are then unable to limit plants. Consequently, plants compete

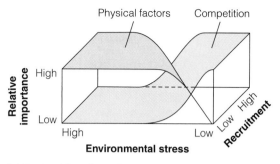

(a) Top level (carnivores)

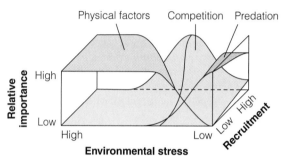

(b) Intermediate level (herbivores)

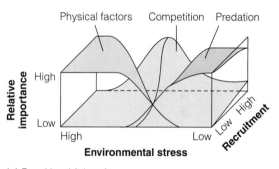

(c) Basal level (plants)

FIGURE 24.8
The Menge-Sutherland model, in which three factors—interspecies competition, predation, and physical factors—drive community organization. The relative importance of these factors changes with trophic level, harshness of the environment, and level of recruitment. (From Menge and Sutherland 1987.)

for nutrients and light, but herbivores do not compete. (They restricted their model to herbivores that feed on green plants and excluded seed and fruit eaters.) This model predicts that whereas herbivore removals will have little effect on plants, predator removals will strongly affect herbivore numbers.

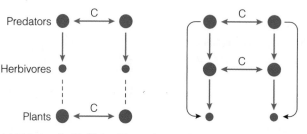

Predators, Herbivores, Plants

(a) Hairston-Smith-Slobodkin model   (b) Menge-Sutherland model

FIGURE 24.9

Schematic comparison of the (a) Hairston-Smith-Slobodkin (HSS) model and the (b) Menge-Sutherland (MS) model of community organization for benign environments. Size of the circles indicates the relative abundance of the trophic levels. Vertical arrows indicate feeding relationships, whereas horizontal (C) arrows indicate competition. Dashed lines indicate weak interactions. In a world driven by the HSS model, herbivores should be rare and plants abundant. In a world driven by the MS model, herbivores should be abundant and plants relatively rare. (Modified after Pimm 1991.))

The Hairston-Smith-Slobodkin (HSS) model makes the same predictions as the Menge-Sutherland (MS) model for predator removal experiments, but not for herbivore removals. Figure 24.9 compares these two models and shows that the Menge-Sutherland model assumes omnivory to be a common feature of the food web (Pimm 1991). We can test these models by measuring the frequency of competitive effects in different communities in benign environments. Hairston et al. (1960) predict intense competition among plants, whereas Menge and Sutherland predict little competition among plants. Similarly, herbivores in benign environments are expected to compete in the MS model but rarely compete in the HSS model (see Figure 24.9).

Predator removal experiments provide one way to test these two models. Herbivore removals should to have no effects on plants, if the HSS model is correct (see Figure 24.9). Sih et al. (1985) surveyed removal experiments; the results are given in Table 24.3. The vast majority of herbivore removal experiments both in marine and terrestrial systems had large effects on the plants. This evidence favors the MS model and is contrary to the HS model.

Freshwater ecologists have proposed several models for community organization in freshwater ecosystems that parallel the MS and the HSS models. The key to these conceptual models is to consider the interactions between any two adjacent trophic levels.

For example, for vegetation (V) and herbivores (H), three possible relationships are possible:

$$V \longrightarrow H \quad V \longleftarrow H \quad V \longleftrightarrow H$$

$\longrightarrow$ means that an increase in vegetation will cause an increase in the numbers or biomass of herbivores, but not vice versa. Similarly, $\longleftarrow$ means that an increase in herbivore numbers will cause an effect on vegetation (a decrease), but not vice versa. $A \longleftrightarrow$ means that an increase in vegetation will cause an increase in herbivore numbers and an increase in herbivore numbers will decrease the vegetation (a reciprocal interaction).

Given these simple interactions, we can define two polar views of community organization: the *bottom-up* model and the *top-down* model. The bottom-up model postulates $V \longrightarrow H$ linkages, which means that nutrients control community organization because nutrients control plant numbers, which in turn control herbivore numbers, which in turn control predator numbers. The simplified bottom-up model is thus $N \longrightarrow V \longrightarrow H \longrightarrow P$. By contrast, the top-down model postulates that predation controls community organization, because predators control herbivores, which in turn control plants, which in turn control nutrient levels. The simplified top-down model is thus $N \longleftarrow V \longleftarrow H \longleftarrow P$ The top-down model has been called the *trophic cascade model* by Carpenter et al. (1985). The cascade model, which is sim-

TABLE 24.3   The percentage of field experiments on predation showing large significant effects as a function of trophic level and system.

"Large effects" is defined as twofold or greater changes. Numbers in parentheses are sample sizes (= number of studies).

System	*Species removed*		
	Herbivore	*Primary carnivore*	*Secondary carnivore*
Intertidal	84 (120)	70 (67)	—
Other marine	95 (82)	68 (57)	—
Lakes	—	73 (22)	75 (95)
Rivers and streams	—	61 (106)	55 (29)
Terrestrial	74 (112)	61 (36)	—

Source: From Sih et al. (1985).

ilar to the HSS model (see Figure 24.9a), predicts for strong interactions among species a series of + / − effects across all the trophic levels. Thus predators will strongly depress herbivore numbers, and depressed herbivore numbers will have only a minor effect on plant abundance, so the abundant plants can strongly depress nutrients. The trophic cascade model predicts for freshwater systems with four trophic levels that removing the top (secondary) carnivores[1] will increase the abundance of primary carnivores, decrease herbivores, and increase phytoplankton. The effects of any manipulation will thus be propagated down or up the trophic structure as a series of positive or negative effects.

The top-down and the bottom-up models are clearly not the only models that can be postulated for food chains. Sinclair et al. (2000) derived 27 different models from various combinations of $\longrightarrow$, $\longleftarrow$ and $\longleftrightarrow$ arrows. For example, a *pure reciprocal model* would postulate two-way effect at all trophic levels: $N \longleftrightarrow V \longleftrightarrow H \longleftrightarrow P$. The important point is that we can start with two simple models, but we must realize that many intermediate types of models are in fact possible for communities, and that it is unlikely that all communities will fit only one model.

These models immediately suggest experimental manipulations of communities to search for trophic-level effects. Several experiments have attempted to test these models in freshwater lakes. An extensive

winterkill of fish in Lake St. George, Ontario, allowed McQueen et al. (1989) to test these models. Figure 24.10 shows the changes in community structure that occurred in the five years after the winterkill. The top predators (bass, pike, and yellow perch) recovered in five years to their former levels of abundance. Planktivorous fishes such as bluegill (primary carnivores) increased after the winterkill, and herbivorous zooplankton declined. Phytoplankton changes (measured by chlorophyll) occurred but were not correlated with zooplankton numbers; phosphorus levels seemed to determine phytoplankton numbers. The results shown in Figure 24.10 suggest that trophic cascades damp out as they move down the food chain, and at the level of phytoplankton and zooplankton both nutrient limitation and predation could be controlling (McQueen et al. 1986).

Trophic cascades can also occur in terrestrial communities. The lesser snow goose breeds in colonies in marshes along the west coast of Hudson Bay; it is the only significant herbivore in these marshes (Jefferies 1988, Kerbes et al. 1990, Hik et al. 1992). Geese numbers have tripled since 1968 and are increasing at 8% per year because of improved winter food supplies in southern areas. In 1985 there were 218,000 pairs of breeding snow geese on the west coast of Hudson Bay. Increased goose populations are having a strong effect on vegetative growth and composition on all the breeding grounds in northern Canada. In spring the geese grub for roots and rhizomes of graminoid plants, and a single goose can strip 1 m^2 of sedge in one hour. Because arctic sedges and grasses are perennials and few reproduce by seed,

---

[1]In aquatic systems, secondary carnivores are piscivorous (fish-eating) fish, and primary carnivores are planktivores (zooplankton-eating fish).

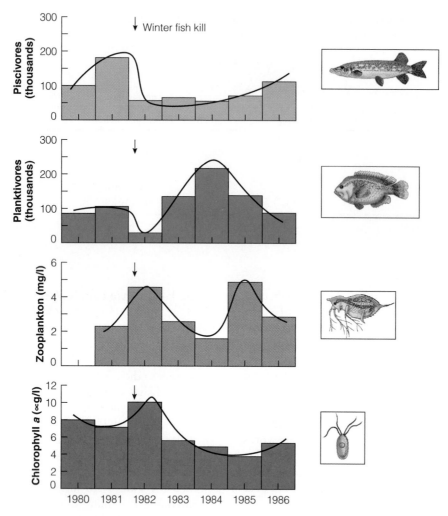

FIGURE 24.10
Changes in the densities of fish, zooplankton, and phytoplankton in Lake St. George, Ontario, before and after a winterkill of fish in 1982 (arrow). A trophic cascade propagated down the food chain over the next four years. (From McQueen et al. 1989.)

disturbed sites can be recolonized by clonal growth but only slowly, and when large areas are disturbed the plants are not able to reclaim the area. Erosion can make the problem worse, and large areas are now completely bare of grasses and sedges (Figure 24.11). This community appears to reach no equilibrium, and snow geese move to new areas as the vegetation is destroyed (Kerbes et al. 1990). Whether the plant community can recover in the absence of grazing and how long this might take are unknown, but the time

scale must be in decades, if not centuries. A positive feedback loop between goose grazing and salt marsh degradation exists, resulting in a runaway trophic cascade and habitat destruction (Srivastava and Jefferies 1996). The snow goose acts as a keystone herbivore in this tundra community. This situation appears to be an unusual case of a runaway trophic cascade in a simple arctic community, and Strong (1992) argues that this kind of destructive interaction is rare in diverse ecosystems.

E S S A Y   2 4 . 2

BIOMANIPULATION OF LAKES

Many freshwater lakes have been degraded by pollution, and one of the major thrusts of applied aquatic ecology has been to devise methods to aid lake recovery from pollution. The trophic cascade model of lake communities immediately suggests ways of improving water quality. In lakes with four trophic levels, adding top predators should improve water quality by reducing algal populations. In lakes with three trophic levels, removing fish should improve water quality. This tool for lake restoration, called *biomanipulation*, can be illustrated simply as follows:

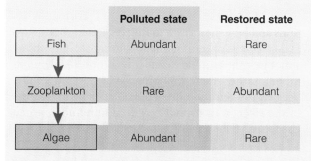

	Polluted state	Restored state
Fish	Abundant	Rare
Zooplankton	Rare	Abundant
Algae	Abundant	Rare

Many attempts at lake restoration using biomanipulation have been made, but mixed results have been obtained, possibly related to variation in the depth and size of the lake (McQueen 1998).

Horppilla et al. (1998) described one of the largest food web manipulation trials yet carried out. Lake Vesijärvi in southern Finland is a large lake (110 km²) with a mean depth of only 6 m. It was heavily polluted with city sewage and industrial wastewater until 1976; once these inputs were stopped, its water quality began to recover. But by 1986 massive blooms of blue-green algae began to appear, and these algal blooms coincided with a very dense population of roach, a planktivorous cyprinid fish that had built up during the years of nutrient input. To reverse these changes, from 1989 to 1993 some 1018 tons of fish were removed from Lake Vesijärvi, reducing roach to about 20% of their former abundance. At the same time the lake was stocked with pikeperch, a predatory fish that feeds on roach, adding a fourth trophic level to the lake.

Biomanipulation was a success in Lake Vesijärvi; the water became clear and blue-green algal blooms stopped in 1989. The lake continues to remain clear seven years later, even though fish removal stopped in 1993. But the mechanism was not as suggested in the previous diagram because zooplankton density in the lake did not change and the same zooplankton species were present. The reduction of algal blooms was achieved because nutrient-rich excretion by roach was greatly reduced, and it was this source of nutrients that was stimulating the excessive algal growth in the lake. An additional nutrient pathway from fish excretion directly to the phytoplankton could be an important additional mechanism for achieving of lake restoration. Lake Vesijärvi may be an example of a lake with two alternate stable states defined by nutrient transfer from fish to algae.

Another model of community organization has been suggested by Fretwell (1977) and elaborated by Oksanen et al. (1981). They recognized that different communities would vary in structure because of the number of trophic levels. In habitats of low productivity, herbivores will be scarce and have little effect on plants. As productivity rises, herbivores will be supported and will suppress plants. In more productive systems, three trophic levels will be present (as discussed by Hairston et al. (1960), see Figure 24.9a), and predators will regulate herbivores. When four trophic levels are present, herbivores will be released to suppress the plants. In this model, predictions regarding the effects of species removals depend on the number of trophic levels in the community. The Fretwell-Oksanen model is very similar to the trophic cascade model of Carpenter et al. (1987); it differs from the MS model by using *productivity* as a major variable instead of *environmental stress*.

FIGURE 24.11
A snow goose exclosure at La Perouse Bay, Hudson Bay, Canada. Over 200,000 snow geese have denuded the coastal salt marsh area shown here of grasses and sedges, and only a salt-tolerant inedible plant (Salicornia borealis) *remains on the saline soils outside the exclosure. Dead willow brushes in the background are a result of high soil salinity after grass removal by the geese. (Photo courtesy of R. Rockwell.)*

Riverine food webs provide a good model system for testing some of the predictions of community organization models, and Mary Power, working at the University of California, Berkeley, conducted a series of classic experiments to test these models. Figure 24.12 shows the food web of boulder-bedrock parts of the Eel River in northern California. Four trophic levels occur in this community, and by constructing cages in the river, Power (1990) was able to measure the effects of removing fish on community dynamics.

Table 24.4 summarizes the results of these manipulations and how they relate to the predictions of the two major models of community organization. The observations fit the trophic cascade (or Fretwell-Oksanen) model rather than the Menge-Sutherland model because the chironomids in the boulder-bedrock areas of the river were strongly reduced in numbers when predatory fishes were excluded. Fish removals in

Mary E. Power *(1949–) Professor of Integrative Biology, University of California, Berkeley*

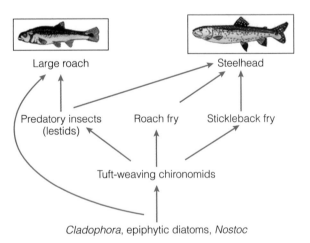

FIGURE 24.12
Food web of the South Fork of the Eel River in northern California during the summer period of low water flow. Four trophic levels are present. By removing the top predators it is possible to test the trophic cascade model of community organization. (From Power 1990.)

TABLE 24.4 Changes in the abundances of the lower three trophic levels when top predators are removed from the community.

All changes are related to the intact, four-trophic-level food web illustrated in Figure 24.12.

	Predicted changes in abundance		
Trophic level	Menge-Sutherland model	Trophic-cascade model	Observations from Eel River, California
Plants (algae)	Increase	Increase	Increased about 3-fold for *Cladophora* and 120-fold for *Nostoc*
Herbivores (chironomids)	Increase	Decrease	Decreased about 80%
Primary carnivores (insects, fish fry)	Increase	Increase	Increased about 10-fold

Source: Data from Power (1990).

rivers can have major effects on all trophic levels via a trophic cascade.

The strong effect of fish predation in the Eel River was limited to boulder-bedrock substrates (Power 1992). In gravel beds of the river, fish predation had very little effect on algae or invertebrate abundance. Fish predation is relatively inefficient in gravel bars because invertebrate prey are relatively scarce and can hide in the spaces within the gravel.

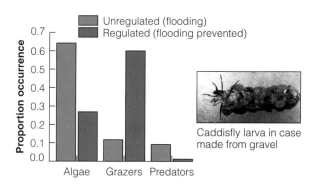

Caddisfly larva in case made from gravel

FIGURE 24.13
Food web structure of northern California rivers, illustrating a trophic cascade driven by flooding. In regulated (dammed) rivers in which no flooding occurs, algae are relatively scarce, grazers (caddisflies) are abundant, and predators are scarce during the summer. Disturbance by flooding (scouring) reduces the caddisflies (which allows other insect grazers to increase) and releases the algae from grazing. Flooding produces effects similar to those observed in a grazer-removal experiment. (Data from Wootton et al. 1996.)

The results of species removals are complex because the interactions among species are complex and may be habitat specific, as we have just seen. Food webs can also be affected directly by disturbances. Riverine food webs in northern California are of two types. In rivers regulated by dams, large caddisfly larvae become abundant because they have gravel cases that make them invulnerable to fish predators, and they graze algae to low levels. In unregulated rivers, floods move rocks, killing many caddisflies and, by reducing their numbers, allowing algae to increase (Figure 24.13). The food webs of dammed rivers thus change dramatically from their original composition, with implications for fish predators such as juvenile salmon (Wootton et al. 1996). The effects of disturbances in changing the dynamics of food webs can parallel those observed in species removal experiments. With frequent disturbances the community structure will differ dramatically from that expected under disturbance-free equilibrium conditions.

The Special Case of Island Species

Islands can be viewed as special kinds of traps that catch species that are able to disperse there and colonize successfully. Since Darwin's visit to the Galápagos Islands, biologists have been using islands as microcosms for studying evolutionary and ecological

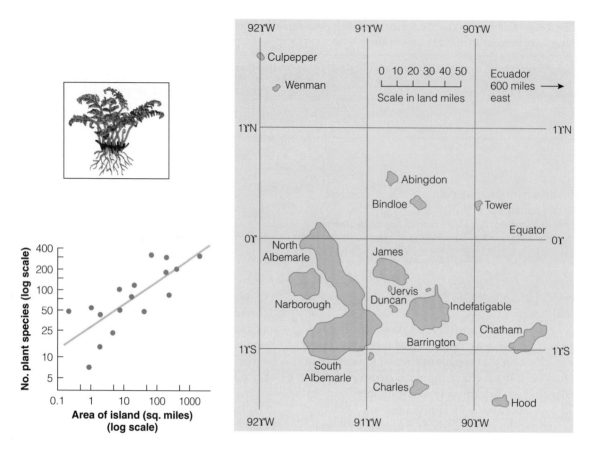

FIGURE 24.14
Number of land-plant species on the Galapagos Islands in relation to the area of the island. The islands range in area from 0.2 to 2249 mi² (0.5–5850 km²) and contain from seven to 325 plant species. (After Preston 1962.)

problems. Because they are bounded, islands are useful for analyzing community structure, and for determining the role of disturbances in affecting communities.

One of the oldest generalizations of ecology is that the number of species on an island is related to the size of the island. Alexander von Humboldt wrote in 1807 that larger areas harbor more species than smaller ones. This phenomenon can be seen most easily in a group of islands like the Galápagos (Figure 24.14). The relationship between species and area can be described by a simple equation called the *species-area curve*:

$$S = c A^z \qquad (24.1)$$

or, taking logarithms,

$$\log S = (\log c) = z(\log A) \qquad (24.2)$$

where S = Number of species

c = A constant measuring the number of species per unit area

A = Area of island (in square units)

z = A constant measuring the slope[2] of the line relating to S and A

[2]Many species-area curves are reported in English units rather than metric units. Because the scales are logarithmic, the z value (slope) is independent of scale and thus does not depend on whether English or metric units are used. The c value, however, is completely dependent on the units used to measure area.

For the Galápagos land plants shown in Figure 24.14 we obtain

$$S = 28.6\,A^{0.32}$$

For these plants, the slope (z) of the species-area curve is 0.32, and the number of plant species on 1 km^2 of island (c) is predicted to be 28.6 on average.

Michael Rosenzweig has championed the species-area relationship as a fundamental ecological law, claiming that the species-area curve is a useful descriptive model for both plants and animals (Rosenzweig 1995). Figure 24.15 illustrates this basic relationship between species and area for the amphibian and reptile fauna of the West Indies, where the relationship is

$$S = 3.3A^{0.30}$$

Preston (1962) noted that the slope of the species-area curve (z) tended to be about 0.3 for a variety of island situations, from beetles in the West Indies and ants in Melanesia to vertebrates on islands in Lake Michigan and land plants on the Galápagos. This raises an interesting question: What is the species-area curve for continental areas? Is its slope the same as that for islands—and is z some sort of ecological constant?

The number of species increases with area on continental areas as well as on islands. Figure 24.16 shows the species-area curve for the breeding birds of North America. Note that the species-area curve is not a single straight line. On very small areas, the slope is greater, and the same is true for at very large areas, such that the whole relationship resembles a logistic curve (He and Legendre 1996). But for areas that range from approximately 10 acres to approximately a billion acres, the curve is a straight line of the form:

$$S = 40A^{0.17}$$

Michael L. Rosenzweig *(1941–) Professor of Ecology and Evolutionary Biology, University of Arizona*

for North American breeding birds. Preston (1962) noted that species-area curves for continental areas, or for *parts* of large islands, had slopes that range from 0.15 to 0.24, a range below the z values found in island studies. This means that as we sample larger and larger areas, we add fewer new species if we are sampling continental areas than if we are sampling a series of islands. The explanation for this is that islands are *isolates* with reduced immigration and emigration, whereas continental areas experience a continual flux of immigrants and emigrants. Thus, each sample area on a continent will probably contain some transient species from adjacent habitats, which acts to lower the slope of the species-area curve.

The number of species living on any plot, whether an island or an area on the mainland, is a balance

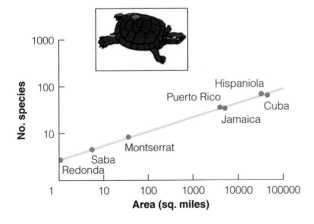

FIGURE 24.15
Species-area curve for the amphibians and reptiles of the West Indies. The slope of the species-area curve (z) is 0.30. (After MacArthur and Wilson 1967.)

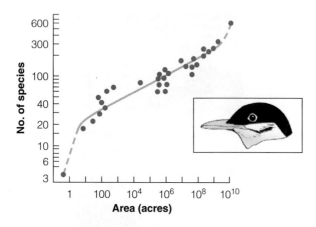

FIGURE 24.16
Species-area curve for North American breeding birds. The areas represented by points range in this graph from a 0.5-acre plot with three species in Pennsylvania, to all of the United States and Canada (4.6 billion acres) with 625 species. For the middle range of these data, the slope of the species-area curve is 0.17. (After Preston 1960.)

between immigration and extinction. If the immigration of new species exceeds the extinction of old species already present, the plot or island will gain species over time. Thus, we can treat the problem of species diversity on islands by an extension of the approach used in population dynamics (see Chapter 11), in which changes in population size were produced by the balance between immigration and births on the one hand, and

emigration and deaths on the other hand. Figure 24.17 shows the simplest model. MacArthur and Wilson (1967) discussed this approach in detail.

The immigration rate, expressed as the number of new species per unit time, falls continuously because as more species become established on the island, most of the immigrants will be from species already present there. The upper limit of the immigration curve is the total fauna for the region. The extinction rate (the number of species disappearing per unit time) rises because the chances of extinction depend on the number of species already present. The point at which the immigration curve crosses the extinction curve is by definition the equilibrium point for the number of species on the island.

The shapes of the immigration and extinction curves are critical for making any predictions about island situations (see Box 24.1). Assume for the moment that distance will affect the immigration curve only; near islands will receive more dispersing organisms per unit time than will distant islands. Assume also that extinction rates will differ between small islands and large islands such that the chances of becoming extinct are greater on small islands. Figure 24.18 illustrates these assumptions and shows why more-distant islands should have fewer species than nearer islands (if island size is constant), and why

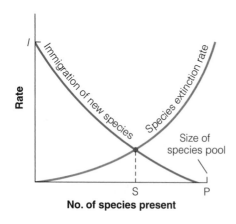

FIGURE 24.17
Equilibrium model of a biota of a single island. The equilibrial number of species (S) is reached at the point at which the curve of immigration of new species to the island intersects the curve of extinction of species on the island. (After MacArthur and Wilson 1967.)

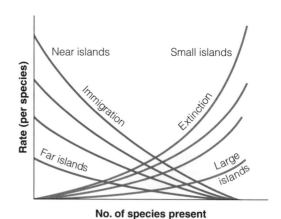

FIGURE 24.18
Equilibrium model of biotas of several islands of various sizes and distances from the principal source area. An increase in distance (near to far) lowers the immigration curve; an increase in island area (small to large) lowers the extinction curve. An equilibrium of species richness occurs at each intersection point of the immigration and extinction curves. (After MacArthur and Wilson 1967.)

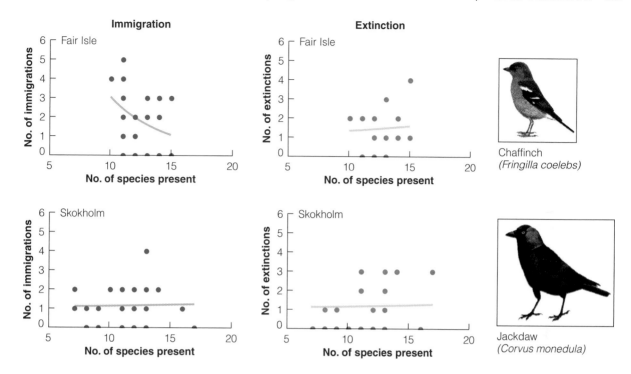

FIGURE 24.19

Island biogeography theory applied to British birds on two islands. Immigration curves are on the left, extinction curves on the right. Consecutive annual censuses established gains and losses for each island. The MacArthur-Wilson model assumes concave curves for these relationships (see Figure 24.17). Extinction curves rise with species richness as predicted, but immigration curves do not always fall as predicted; note the considerable noise in the data. (Data from Manne et al. 1998.)

small islands should have fewer species than large islands (if distance from the source area is constant).

We need long term studies to obtain observational data to test the MacArthur-Wilson model, and so far most of the data available come from bird studies. Britain is surrounded by many islands of various sizes for which bird census data are available for 50 years or more (Russell et al. 1995). Figure 24.19 shows the observed immigration and extinction curves for two islands. A central assumption of the MacArthur-Wilson model is that the immigration and extinction functions are concave, and while this seems to be correct for many islands, there are exceptions to the rule, and much variation occurs from year to year.

If the equilibrium theory of MacArthur and Wilson is correct, then the slopes of species-area curves for terrestrial mammals should increase progressively from the mainland to oceanic islands (Figure 24.20a). If, however, nonequilibrum conditions apply, the slope of the

species-area curve should be lower on oceanic islands than on landbridge islands (Figure 24.20b). (*Landbridge islands* are islands close to continents that were connected to the mainland at the end of the Ice Age; oceanic islands were never connected to the mainland.) Oceanic islands are colonized less frequently than landbridge islands because of isolation by distance. Figure 24.20a illustrates the expected differences in species-area curves under the equilibrium hypothesis depicted in Figure 24.18. On oceanic islands, colonization rates are lower than and extinction rates are the same as those on landbridge islands of the same size, resulting in a lower number of species at equilibrium. If, however, species numbers on oceanic islands are not in equilibrium because they are limited by colonization, then the species-area curve will be flat, as illustrated in Figure 24.20b. Under this hypothesis, oceanic islands never achieve the species richness predicted for them on the basis of their area (Lawlor 1986).

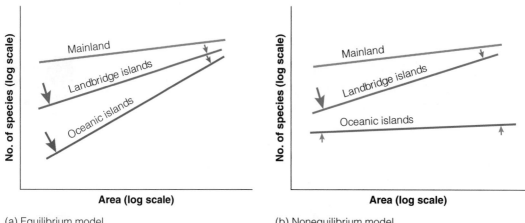

(a) Equilibrium model

(b) Nonequilibrium model

FIGURE 24.20
Species-area curves predicted for terrestrial mammals on islands. On all islands both immigration and extinction are occurring. Blue arrows indicate the relatively greater effect of extinctions; red arrows indicate the relatively greater effect of immigration. Smaller islands are much more affected than larger islands. (a) Equilibrium model of MacArthur and Wilson (the same model illustrated in Figure 24.18, plotted as a species-area curve). (b) Nonequilibrium model of Lawlor (1986). For oceanic islands the immigration rate limits species numbers, which are little affected by island size. The slope of the blue species-area curve for oceanic islands is critical for distinguishing between these two models. (From Lawlor 1986.)

FIGURE 24.21
A test of the nonequilibrium theory for island faunas of nonflying mammals. (a) Range of slope values (z) for species-area curves for oceanic (six island sets) and landbridge islands (12 island sets). Oceanic islands have lower slopes, as predicted by the nonequilibrium theory (see Figure 24.20b). (b) One example of species-area curves for the nonflying mammals of islands off Baja California. (From Lawlor 1986.)

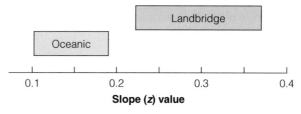

(a)

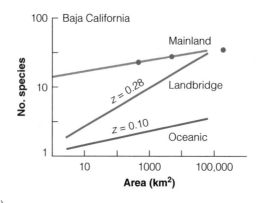

(b)

BOX 24.1

MEASURING IMMIGRATION AND EXTINCTION RATES

Immigration and extinction rates are key to all of the predictions of the MacArthur-Wilson theory of island biogeography, but these rates are not easy to measure. The most common approach is to count all the species on an island each year during the breeding season. If a species was present last year and is absent this year, it is counted as an extinction, and conversely if a species that is present this year was absent last year, it is counted as an immigration. But this simple arithmetic belies some ecological complexity. For example, a species that colonized the island after the last census and then died out before the current census would not be counted. More problems arise if the census cannot be done every year, because species can come and go, or even come and go and come back again, between the censuses. In measuring these rates there can be no substitute for detailed field data.

If we conduct an annual census, we can estimate immigration and extinction rates as follows. Assume for simplicity that for one species we have records of presence (P) and absence (A) in a series of 30 years, as follows:

PPPPPPAAPPPAPPAAAAAPPPAPAPPPAA

We define:

k = number of transitions from absence to presence = 5
l = number of transitions from absence to absence = 6
m = number of transitions from presence to absence = 6
n = number of transitions from presence to presence = 12

Given these raw data, there are two rates that follow directly from the observed transitions. The first, the immigration rate λ, is the probability that a species not present in the community will enter it in a given time interval (usually a year). Immigration rate is estimated as

$$\lambda = \frac{k}{k+l} \quad (24.3)$$

The second rate, the probability that a species becomes absent after being present the previous year (δ), is estimated as

$$\delta = \frac{m}{m+n} \quad (24.4)$$

From these rates we can estimate that for a very long series of observations, the population will be absent from the island for the fraction $\delta/(\lambda+\delta)$ of the total number of censuses.

To estimate the extinction rate we note that a species may go extinct and recolonize in between the census times, such that it appears from the records that nothing has changed. To take this into account we note that

$$\begin{pmatrix} \text{Probability} \\ \text{of} \\ \text{becoming absent} \end{pmatrix} = \begin{pmatrix} \text{probability} \\ \text{of} \\ \text{extinction} \end{pmatrix}\begin{pmatrix} \text{probability} \\ \text{of not} \\ \text{recolonizing} \end{pmatrix} \quad (24.5)$$

$$\delta = \mu(1-\lambda) \text{ or}$$

$$\mu = \frac{\delta}{1-\lambda}$$

where μ is the extinction rate. For these hypothetical data, λ=5/11 or 0.45, and δ=6/18 or 0.33, so the extinction rate μ is 0.33/0.55 or 0.60. For this hypothetical species on this island, clearly the extinction rate is on average greater than the immigration rate. We expect that in the long run the species will be absent from the island for $\delta/(\lambda+\delta)$ or 42% of the censuses.

If we assume there is a common immigration rate and extinction rate for all species, these transitions can simply be added to obtain estimates for the island (see Clark and Rosenzweig 1994). These annual rates are all probabilities and thus range from 0 to 1.0. We can use these rates to answer questions about the likelihood of extinction of individual species or of species groups on islands, or questions about whether the immigration rate varies with island size.

If a census is not conducted each year, so that only sporadic records of presence and absence are available, it is more difficult to estimate rates of immigration and extinction because much can happen between census periods. Clark and Rosenzweig (1994) discuss this problem of estimation as well.

One critical test of these two models involves a comparison of the slopes of the species-area curves for oceanic islands. Lawlor (1986) separated mammals into those that can fly (bats) and those that cannot (all others). Figure 24.21a shows that the slope values (z) for the species-area curves of nonflying terrestrial mammals on oceanic islands are lower than those of nonflying mammals on landbridge islands. This relationship is illustrated in Figure 24.21b for 14 oceanic islands and 20 landbridge islands off Baja California. By contrast, bats can fly to distant islands, and their species-area curves fit the equilibrium model. The mammal fauna of oceanic islands is incomplete and is not an equilibrium assemblage.

The MacArthur-Wilson theory has stimulated much work on island faunas. By concentrating on predictions of the *number* of species, it has ignored the more difficult questions of *which* particular species will occur where, which is often the question conservationists ask. Habitat heterogeneity is a major determinant of the species-area curve, and detailed studies of habitats are needed to further our understanding of islands (Rosenzweig 1995, Williamson 1989). Individual species differ greatly in their abilities to occupy islands, and by understanding the population and community dynamics of individual species we can improve our understanding of island faunas and floras.

Multiple Stable States in Communities

If communities are not equilibrium assemblages of species, their composition will change over time and we will not observe a single stable configuration. But if natural communities can exist in multiple stable states, changes in community composition that appear to be nonequilibrial may instead be the result of two or more alternative states for the same community (Sutherland 1990). What evidence is required to demonstrate that natural communities have multiple stable states? Connell and Sousa (1983) defined the following criteria that must be met if at a given time a community exhibits multiple stable states:

- The community must show an equilibrium point at which it remains, or to which it returns if perturbed by a disturbance.

- If perturbed sufficiently, the community will move to a second equilibrium point, *at which it will remain after the disturbance has disappeared.*

- When multiple stable states are believed to exist, the abiotic environments of the community must be similar for the various sites.

- The community on both sites that are postulated to be alternate stable states must persist for more than one generation of the dominant species.

A community may show multiple stable states on a single site over time or simultaneously at two or more sites. By applying these criteria, Connell and Sousa (1983) questioned many of the examples of communities purported to be multiple stable states. Many cases can be rejected because the physical environment differs on the two sites. In other cases the alternate state persists only when artificial inputs are maintained. For example, Lake Washington (see p. 480) is not an example of an aquatic community with multiple stable states because the enriched lake community could be maintained only by continually adding sewage nutrients. Connell and Sousa (1983) concluded that they could find no studies of natural communities that showed conclusive evidence for multiple stable states. They issued a challenge to ecologists to search for solid evidence of multiple stable states in natural communities.

Woodlands and grasslands of east Africa may represent multiple stable states of a grazing system (Dublin et al. 1990). Woodlands in parks and reserves over much of the savanna areas of east Africa have declined in the past 30 years. Three hypotheses have been proposed to explain the reduction in woodlands (Figure 24.22):

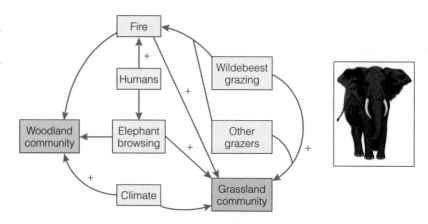

FIGURE 24.22
Factors affecting the establishment of woodland or grassland communities in the Serengeti-Mara region of East Africa. This community exists in two alternate stable states, woodland or grassland. (Based on Dublin et al. 1990.)

- Human-induced fires have eliminated woodland. There is one stable state; if fires were reduced woodlands would return.

- Elephants eliminate woodland, and the resulting grassland is maintained by fire. There are two stable states; if fires were eliminated, woodlands would return to their former abundance, even if elephants remained.

- Fire eliminated woodlands, and elephants hold tree regeneration in check by eating small trees, so woodlands can never return. Eliminating fire will not cause woodlands to return, and there are two stable states.

The available evidence supports the third hypothesis (Dublin et al. 1990). During the 1960s fire burned an average of 62% of the Serengeti each year (Sinclair 1975), and even in the absence of elephants or other browsing or grazing mammals, tree recruitment was too low to sustain woodlands. Elephant and wildebeest numbers increased in the parks and reserves by the 1980s. Wildebeest (see Figure 15.8, p. 267) grazed much of the grass each dry season, so the fuel load was reduced in the 1980s and fires burned only 5% of the area each year. Elephant browsing in the 1980s was severe and by itself was capable of preventing woodlands from reestablishing (thereby rejecting the first and second hypotheses listed above). If elephants and wildebeest are reduced by poaching in the future, fires will increase and woodlands will not return. The Serengeti-Mara ecosystem seems to be locked into a grassland state unless both fires and elephant browsing are reduced.

White-tailed deer may be creating alternate stable states of woody plant communities in the eastern United States (Stromayer and Warren 1997). Between 1890 and 1920 much of the hardwood forests in Pennsylvania were clear-cut. These stands contained valuable trees such as white ash, sugar maple, red maple, and black cherry. Deer populations increased rapidly in the regenerating stands, which produced much browse. At the same time, predators such as wolves were removed from the system, so deer numbers skyrocketed such that deer are now considered "overabundant." Deer browsing has been shown to reduce hardwood regeneration, particularly of valuable timber species. With sufficient browsing the seed bank of these hardwoods becomes exhausted within three or four years, and no regeneration is possible. Ferns and grasses invade the forest floor and complete suppress regeneration of desirable hardwoods (Tilghman 1989). The result is a community of trees dominated by black cherry and containing other trees less preferred by foresters and less browsed by deer, an alternate tree community that may be stable on the time frame of 300 years or more.

Multiple stable states may occur in other communities affected by humans and may be confused with nonequilibrium communities. In some of these cases the community may revert to its original configuration once human disturbance is removed, but in others the community may become locked into a changed configuration even after the disturbance is stopped. For the purposes conservation and land management, it is important to determine which model of community organization applies to natural communities. We cannot assume that all communities subjected to human disturbance will return to their original configuration once the disturbance is ameliorated.

Summary

Many community ecologists question the existence of a single equilibrium state for biological communities. Patchiness is an inherent property of natural communities, and the disturbances that lead to patchiness have been the main focus of nonequilibrium theories of community organization. Nonequilibrium communities exist when disturbance intervals are shorter than recovery times, so the community never reaches equilibrium. Disturbances may include fires or weather events, as well as grazing, predation, disease, or other biotic events.

Coral reefs were once thought to be classic examples of stable, equilibrium communities, but careful long-term studies have shown that reefs vary dramatically due to disturbances caused by cyclones and oceanographic changes resulting from weather fluctuations. Coral reef fish communities may be driven by the lottery of variable recruitment, which can cause irregular population fluctuations.

The two extreme conceptual models of community organization are the *top-down model* in which changes in food webs are driven from above by predators, and the *bottom-up model* in which nutrients and plants control the food web. Many intermediate models between these two extremes can be used to describe particular communities.

Studies in the rocky intertidal zone and in aquatic systems have stimulated several models of community organization. The Menge-Sutherland model includes the roles of environmental harshness and variable recruitment in its prediction of when community interactions will be dominated by competition, predation, or physical factors. The trophic cascade or Fretwell-Oksanen model emphasizes the alternation of positive and negative effects in food webs. When top predators are removed, the effects cascade as alternating positive and negative effects down the trophic ladder. Trophic cascades are common in aquatic systems but also occur in some terrestrial communities. But not all systems follow trophic cascades. Some communities are driven bottom-up by nutrients, and the lower trophic levels may be affected from below by plant dynamics and remain unaffected by changes in predator abundance.

Island communities are a special case in which species makeup is driven by the interaction between colonization and extinction. The species-area curve, which describes how biodiversity increases with island size, is one of the grand generalizations of community ecology. Not all islands are equilibrium systems in which immigration and extinction are balanced, however, and historical effects are an important component of many island faunas and floras.

Some communities may exist in multiple stable states, and these communities may be confused with nonequilibrium assemblages. If a community is perturbed sufficiently, it may change to a new configuration at which it will remain even when the disturbance is stopped. Considerable controversy existing concerning how common multiple stable states are in natural ecosystems, and the question is important for conservation and land management.

Key Concepts

1. Communities are not in equilibrium if their recovery times exceed the frequency of disturbance.

2. Patchiness in communities can result from disturbances caused by physical or biotic factors.

3. Two extreme alternative models of community organization are the *top-down model* in which predators drive all the lower trophic levels, and the *bottom-up model* in which nutrients drive all the higher trophic levels.

4. Many intermediate models are possible between these two extremes. Reciprocal interactions between trophic levels can complicate predictions based on food web manipulations.

5. The general rule that larger areas contain more species is one of the oldest generalizations in community ecology and is codified in the *species-area curve*.

6. Island communities can be equilibrium systems in which immigration balances extinction. Some islands never reach their expected species equilibrium because they are limited by colonization.

7. Some communities may exist in multiple stable states in which disturbances move them from one state to another. Whether a community has several stable states or is a nonequilibrium system is a critical distinction for conservation and management strategies.

Selected References

Brett, M. T. and C. R. Goldman. 1996. A meta-analysis of the freshwater trophic cascade. *Proceedings of the National Academy of Sciences USA* 93:7723–7726.

Carpenter, S. R., J. F. Kitchell, and J.R. Hodgson. 1985. Cascading trophic interactions and lake productivity. *Bioscience* 35:634–639.

Locke, A. 1996. Applications of the Menge-Sutherland Model to acid-stressed lake communities. *Ecological Applications* 6:797–805.

Menge, B. A. 1995. Joint "bottom-up" and "top-down" regulation of rocky intertidal algal beds in South Africa. *Trends in Ecology and Evolution* 10:431–432.

Moritz, M. A. 1997. Analyzing extreme disturbance events: Fire in Los Padres National Forest. *Ecological Applications* 7:1252–1262.

Petraitis, P. S. and S. R. Dudgeon. 1999. Experimental evidence for the origin of alternative communities on rocky intertidal shores. *Oikos* 84:239–245.

Polis, G. A. and K. O. Winemiller (eds.) 1996. *Food Webs: Integration of Patterns and Dynamics*. Chapman and Hall, New York. 472 pp.

Power, M. E. 1992. Top-down and bottom-up forces in food webs: Do plants have primacy? *Ecology* 73:733–746.

Reice, S. R. 1994. Nonequilibrium determinants of biological community structure. *American Scientist* 82:424–435.

Terborgh, J., R. B. Foster, and P. V. Nunez. 1996. Tropical tree communities: A test of the nonequilibrium hypothesis. *Ecology* 77:561–567.

Wootton, J. T., M. S. Parker, and M. E. Power. 1996. Effects of disturbance on river food webs. *Science* 273:1558–1561.

Questions and Problems

24.1 The species-area curve rises continually as area is increased, implying that there is no limit to the number of species in any community. Is this a correct interpretation? What hypotheses can you suggest to explain why the number of species rises as area increases?

24.2 The number of vascular-plant species and geographic parameters for several islands off California were given by Johnson, Mason, and Raven (1968, p. 300), as follows:

Island	Area (mi²)	Maximum. elevation (ft)	Latitude (°N)	Dist. from mainland (mi)	No. plant species
Credros	134.0	3950	28.2	14.0	205
Guadalupe	98.0	4600	29.0	165.0	163
Santa Cruz	96.0	2470	34.0	20.0	420
Santa Rosa	84.0	1560	34.0	27.0	340
Santa Cataline	75.0	2125	33.3	20.0	392
San Clemente	56.0	1965	32.9	49.0	235
San Nicolas	22.0	910	33.2	61.0	120
San Miguel	14.0	830	34.0	26.0	190
Natividad	2.8	490	27.9	5.0	42
Santa Barbara	1.0	635	33.4	38.0	40
San Martin	0.9	470	30.5	3.5	62
San Geronimo	0.2	130	29.8	6.0	4
South Farallon	0.1	360	37.7	27.0	12
Año Nuevo	0.02	60	37.1	0.25	40

Make four plots of the number of plants versus (1) area, (2) elevation, (3) latitude, and (4) distance from the mainland. Plot these first on arithmetic scales and then on logarithmic scales (log-log plot). What variable is most closely related to species numbers? Estimate graphically the slope of the species-area curve for this ensemble of islands, and compare it with those given in the text.

24.3 Mammals on mountain tops in the Great Basin of western North America have been cited as a model case at variance with the MacArthur-Wilson theory of island biogeography. Brown (1971) postulated that mammals in the Great Basin were remnant populations subject only to extinction (no immigration is possible). Lawlor (1998) rejects this explanation. Review Lawlor's data and the changing interpretations of these mountain top communities in relation to equilibrium and nonequilibrium theories of community dynamics.

24.4 Can nonequilibrium models of community organization be stable? Read Chesson and Case (1986) and DeAngelis and Waterhouse (1987) and discuss the relationship between stability and equilibrium/nonequilibrium concepts.

24.5 Are the data in Table 24.3 (see p. 497) suitable for testing the Menge-Sutherland model of community organization? Why or why not?

24.6 In discussing the MacArthur-Wilson theory of island biogeography, Brown (1986, p. 233) states:

> I suspect that the model has ultimately proven more valuable when its predictions have been refuted than when they have been supported.

Is this statement valid for other models of community organization?

24.7 Giller and Gee (1987, p. 540) state:

> A state of equilibrium can be attained at the landscape scale even when biotic instability and stochastic domination are unrestrained at the level of the local patch.

Under what conditions is this true? Discuss the effect of spatial scale on the concept of equilibrium.

24.8 Lawton (1984), in summarizing information on bracken fern and its insect herbivores, concluded that these insect communities were not saturated with species, and that at any particular site numerous vacant niches existed. Discuss this interpretation of community organization with respect to the nonequilibrium models of community organization discussed on pages 492–493.

24.9 Review the argument between Hairston (1991) and Sih (1991) over the interpretation of field data for testing the predictions of the Hairston et al. (1960) model of community organization. Discuss the problem of testing hypotheses about community organization within the framework of Figure 1.3 (see p. 13).

24.10 Analyze the elephant/fire multiple stable state model of Dublin et al. (1990) using the criteria for the existence of multiple stable states given by Connell and Sousa (1983). Would this example be acceptable to Connell and Sousa? Would the white-tailed deer example satisfy their criteria?

24.11 The immigration and extinction curves in the Mac-Arthur-Wilson theory of island biogeography are concave upwards (see Figure 24.17). What difference would it make if these curves were straight lines?

24.12 What role does history play in community organization as defined by the equilibrium model and the nonequilibrium model? Do we need to know anything about the history of a community to predict its future course? Tanner et al. (1996) discuss this issue for coral reefs.

Overview Question

List some of the possible manipulative experiments that could be applied to a community to test the Menge-Sutherland and the trophic cascade models of community organization. List the predictions each model would make for each possible manipulation. Are some manipulations more instructive than others?

CHAPTER 25

Ecosystem Metabolism I:
Primary Production

W E CAN TAKE TWO BROAD APPROACHES to the study of communities and ecosystems of plants and animals. First, we can treat species as biological entities, with all the specific adaptations and interrelationships they show; this has been the approach of the past four chapters and can be considered a population-ecological approach to community and ecosystem dynamics. Second, we can move beyond the details of particular species and concentrate on the physics of ecosystems as energy machines and nutrient processors. Exactly how plants and animals process energy and materials has important implications for humans. This second approach to ecosystem dynamics is the subject of the next three chapters.

The metabolism of ecosystems is most easily understood as the sum of the metabolism of individual animals and plants. Individual organisms require a continual input of new energy to balance losses resulting from metabolism, growth, and reproduction. Individuals can be viewed as complex machines that process energy and materials. Organisms pick up energy and materials in two main ways. *Autotrophs* pick up energy from the sun and materials from nonliving sources. Green plants are autotrophs. *Heterotrophs* pick up energy and materials by eating living matter. Herbivores are heterotrophs that live by eating plants, and carnivores are heterotrophs that live by eating other heterotrophs. Communities are mixtures of autotrophs and heterotrophs. Energy and materials enter a biological community, are used by the individuals, and are transformed into biological structure only to be ultimately released again into the environment. The *ecosys-*

tem level of integration includes both the organisms and their abiotic environment and is a comprehensive level at which to consider the movement of energy and materials. (We could also discuss the flow of matter and energy at the individual level or at the population level.) The basic unit of metabolism is always the individual organism, even when individuals are assembled into communities and ecosystems.

The first step in the study of ecosystem metabolism is to identify the food web of the community (see Chapter 23). Once we know the food web, we must decide how we can judge the significance of the different species to community metabolism. Even though some 5000 species of animals live on the 5 km^2 of Wytham Woods in Britain (Elton 1966), we feel intuitively that many of these 5000 species are not significant, and that many or most of them could be removed without affecting the metabolism of the woodland.

Three measurements might be used to define relative importance in an ecosystem:

1. *Biomass.* We could use the weight or standing crop of each species as a measure of importance. This is useful in some circumstances, such as the timber industry, but it cannot be used as a general measure. In a dynamic situation in which *yield* is important, we need to know how rapidly a community produces new biomass. When metabolic rates and reproductive rates are high, production may be very rapid, even from a small standing crop. Figure 25.1 illustrates the idea that yield need not be related to biomass.

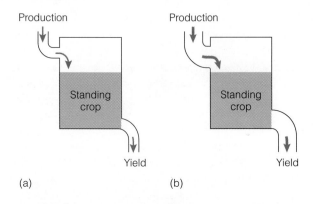

(a)   (b)

FIGURE 25.1
Hypothetical illustration of two equilibrium communities in which input equals output: (a) low input, low output, slow turnover; (b) high input, high output, rapid turnover. Standing crop is not related to production or yield because turnover time is not a constant for all systems. Production (input) must equal yield (output) in equilibrium communities, but many communities are not always in equilibrium, so standing crop may rise and fall.

2. *Flow of chemical materials.* We can view an ecosystem as a superorganism taking in food materials, using them, and passing them out. Note that all chemical materials can be recycled many times through the community. A molecule of phosphorus may be taken up by a plant root, used in a leaf, eaten by a grasshopper that dies, and released by bacterial decomposition to reenter the soil.

3. *Flow of energy.* We can view the ecosystem as an energy transformer that takes solar energy, fixes some of it in photosynthesis, and transfers this energy from green plants through herbivores to carnivores. Note that most energy flows through an ecosystem only once and is not recycled; instead it is transformed to heat and ultimately lost from the system. Only the continual input of new solar energy keeps the ecosystem operating. Again we may draw the analogy between an ecosystem and an organism that processes food energy.

To study the dynamics of ecosystem metabolism, we must decide what to use as the *base variable*. Most ecologists have decided to use either carbon or energy. Elements other than carbon are often tied up in biological peculiarities of organisms. For example, vertebrates and mollusks contain much more calcium than do most freshwater invertebrates because of the

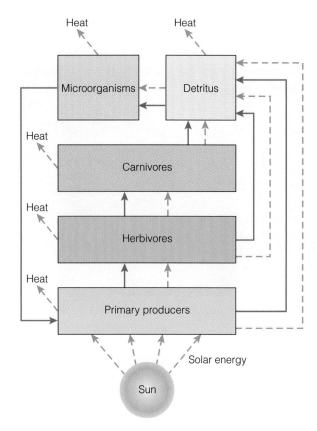

FIGURE 25.2
General representation of energy flows (dashed red lines) and material cycles (solid blue lines) in the biosphere. Energy flows included are solar radiation, chemical energy transfers (in the ecological food web), and radiation of heat into space. Materials flow through the trophic levels to detritus and eventually back to the primary producers. (From DeAngelis 1992.)

presence of bone or shell. Some marine invertebrates concentrate certain chemical elements. Even within an individual there are variations. Calcium in the teeth and bones of a mammal may be stable for long periods, whereas calcium in the blood serum may turn over rapidly because of ingestion and excretion. As a result, the description of the calcium flow through an ecosystem is very difficult. Because most energy is not recycled, it is easier to measure than are chemical materials. Figure 25.2 illustrates the flows of energy and materials through the food chain.

Energy or carbon content is just another way of describing an individual, a population, or a community, and the convenience and precision of these measures should not blind us to their limitations as a way of describing organisms. The great strength (and weakness) of the energetics approach is that it can allow us to add together different species in a community. It reduces the fundamental diversity of a biological community to a single unit—either the joule (for energy) or the gram (for carbon).

Primary Production

The process of photosynthesis is the cornerstone of all life and the starting point for studies of community metabolism. The bulk of the Earth's living mantle is green plants (99.9% by weight); only a small fraction of life consists of animals (Whittaker 1975).

Photosynthesis, the process of transforming solar energy into chemical energy, can be simplified as

$$12H_2O + 6CO_2 + \text{solar energy} \xrightarrow{\text{chlorophyll + enzymes}}$$
(from air)

$$C_6H_{12}O_6 + 6O_2 + 6H_2O$$
(carbohydrate)   (to air)

If photosynthesis were the only process occurring in plants, we could measure production by the rate of accumulation of carbohydrate, but unfortunately plants also respire, using energy for maintenance activities. In an overall view, respiration is the opposite of photosynthesis:

$$C_6H_{12}O_6 + O_2 \xrightarrow{\text{metabolic enzymes}} CO_2 +$$
(carbohydrate) (from air)   (to air)

$$H_2O + \text{energy for work and maintenance}$$

At metabolic equilibrium, photosynthesis equals respiration, and this is called the compensation point. Measures of photosynthesis and respiration are rates; they are always expressed as amount of material or energy per unit of time. If plants always existed at the compensation point, they would neither grow nor reproduce. We define two terms:

Gross primary production = energy (or carbon) fixed via photosynthesis per unit time

Net primary production = energy (or carbon) fixed in photosynthesis − energy (or carbon) lost via respiration per unit time

Production is always measured as a rate per unit of time. How can we measure these two aspects of primary production in natural systems?

For terrestrial plants, the direct way is to measure the change in CO_2 or O_2 concentrations in the air around plants. Most studies measure CO_2 uptake by an enclosed branch or a whole plant. During daylight conditions, CO_2 uptake rate is a measure of net production because both photosynthesis and respiration are operating simultaneously. At night only respiration occurs, and the rate at which CO_2 is released can be used to estimate the respiration component. Photosynthesis and respiration are both affected by temperature; photosynthesis is also affected by light intensity. The daily changes in leaf temperature and light intensity determine the daily net production for an individual plant.

We can determine the energetic equivalents of photosynthesis measurements from the chemical thermodynamics of the reaction

$$12H_2O + 6CO_2 + 2966 \text{ kJ} \xrightarrow{\text{solar energy}}$$

$$C_6H_{12}O_6 + 6O_2 + 6H_2O$$

Thus the absorption of 6 moles (134.4 liters at standard temperature and pressure) of CO_2 indicates that 2966 kJ has been absorbed. [1]

The measure of gas exchange around plants in the field has been used relatively little as an estimate of photosynthetic rates because it requires sophisticated and expensive instrumentation. A slightly different approach to measuring CO_2 uptake is to introduce radioactive ^{14}C – labeled CO_2 into the air surrounding a plant (covered by a transparent chamber) and after a time to harvest the whole plant and measure the quantity of radioactive ^{14}C taken up by photosynthesis.

The simplest method of measuring primary production is the *harvest method*. The amount of plant material produced in a unit of time can be determined from the difference between the amount present at the two times:

$$\Delta B = B_2 - B_1 \tag{25.1}$$

where ΔB = biomass change in the comunity between time *1* (t_1) and time *2* (t_2)
B_1 = biomass at t_1
B_2 = biomass at t_2

[1]Energy units have been reported in many forms in the literature, often in calories, and can be standardized to *joules* with the following conversion factors: 1 joule (J) = 0.2390 gram calorie (cal) = 0.000239 kilocalorie (kcal); conversely, 1 gram calorie = 4.184 joule and 1 kilocalorie = 4184 joules or 4.184 kilojoules (kJ).

Two possible losses must be recognized:

L = biomass losses by death of plants or plant parts
G = biomass losses to consumer organisms

If we know these values, we can determine net primary production:

$$\text{Net primary production} = NPP = \Delta B + L + G \qquad (25.2)$$

This may apply to the whole plant, or it may be specified as *aerial* production or *root* production.

The net primary production in biomass may then be converted to energy by measuring the caloric equivalent of the material in a bomb calorimeter. This measurement should be done for each particular species studied as well as for each season of the year. Golley (1961) showed that different parts of plants have different energy contents:

	Mean of 57 plant species	
	(cal/g dry wt)	(J/g dry wt)
Leaves	4,229	17,694
Roots	4,720	19,748
Seeds	5,065	21,192

Vegetation collected in different seasons also varied in energy content.

The harvesting technique of estimating production is used in a variety of situations. Foresters have used a modified version of it for timber estimation, and agricultural research workers use it to determine crop yield. The application of harvesting techniques to natural vegetation involves some specialized techniques that we will not describe here; Moore and Chapman (1986) and Pieper (1988) give details of techniques.

In aquatic systems, primary production can be measured in the same general way as in terrestrial systems. Gas-exchange techniques can be applied to water volumes, and usually oxygen release instead of carbon dioxide uptake is measured. This procedure is usually repeated with a dark bottle (respiration only) and a light bottle (photosynthesis and respiration), so that both gross and net production can be measured. Vollenweider (1974) discusses details of techniques for measuring production in aquatic habitats.

How does primary production vary over the different types of vegetation on the Earth? This is the first general question we can ask about community metabolism. Table 25.1 gives some average values for

TABLE 25.1 Net primary production for land and ocean estimated from satellite data, as illustrated in Figure 25.3.

All values of net primary production are in petagrams of carbon (1 petagram = 10^{15} grams = 10^9 metric tons).

Vegetation type	Annual net primary production
Ocean	48.5
Land	
Tropical rainforests	17.8
Broadleaf deciduous forests	1.5
Broadleaf and needleleaf forests	3.1
Needleleaf evergreen forests	3.1
Needleleaf deciduous forests	1.4
Savannas	16.8
Perennial grasslands	2.4
Broadleaf shrubs	1.0
Tundra	0.8
Desert	0.5
Cultivated areas	8.0
Total for land vegetation	56.4
Total for globe	104.9

Source: Field et al. (1998).

global net primary production in biomass for different vegetation types, and Figure 25.3 illustrates the yearly production for ocean and land areas of the globe. In general, primary production is highest in the tropical rain forest and decreases progressively toward the poles. Productivity of the open ocean is very low, approximately the same as that of the arctic tundra. But because oceans occupy about 71% of the total surface of the Earth, total oceanic primary production adds up to about 46% of the overall production of the globe. Grassland and tundra areas are less productive than forests of equivalent latitude.

How efficient is the vegetation of different communities as an energy converter? We can determine the efficiency of utilization of sunlight by the following ratio:

$$\text{Efficiency of gross primary production} = \frac{\text{energy fixed by gross primary production}}{\text{energy in incident sunlight}}$$

For example, Kozlovsky (1968) calculated the efficiency of the aquatic community of Lake Mendota, Wisconsin, as follows:

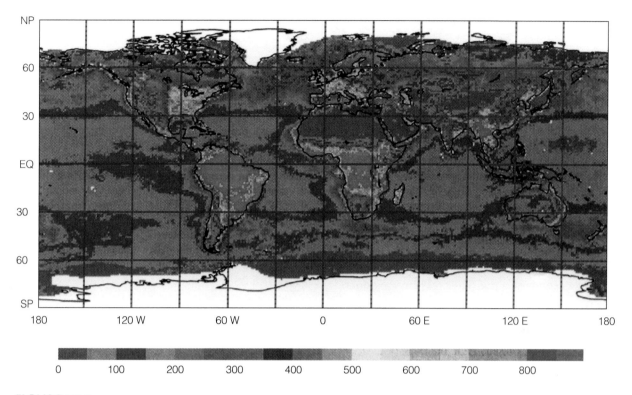

FIGURE 25.3
Annual net primary production (grams of carbon per m² per year) for the globe, as calculated from satellite imagery. Ocean data are averages from 1978 to 1983; land averages are from 1982 to 1990. Of total global primary production, the ocean contributes 46% and the land 54%. (After Field et al. 1998.)

$$\text{Efficiency of gross primary production} = \frac{\overset{\displaystyle 20{,}991 \text{ kJ/m}^2\text{/yr}}{\text{gross primary production}}}{4{,}973{,}604 \text{ kJ incident sunligh}}$$

$$= 0.42\%$$

Phytoplankton communities have very low efficiencies of primary production, usually less than 0.5%, although rooted aquatic plants and algae in shallow waters can have higher efficiencies. The efficiency of gross primary production is higher in forests (2.0–3.5%) than in herbaceous communities (1.0–2.0%) or in crops (less than 1.5%) (Kira 1975). Forest communities are relatively efficient at capturing solar energy.

How much of the energy plants fix by photosynthesis is subsequently lost via respiration? A great deal of energy is lost in converting solar radiation to gross primary production. Net primary production, which is what interests us, must therefore be even less efficient. In forests 50–75% of the gross primary

production is lost to respiration, such that net production may be only one-fourth that of gross production (Kira 1975). Forests have larger amounts of stems, branches, and roots to support than do herbs, and thus less energy is lost to respiration in herbaceous and crop communities (45–50%). The result of these losses is that for a broad range of terrestrial communities, about 1% of the sun's energy is converted into net primary production during the growing season.

Factors That Limit Primary Productivity

One important question about primary production is, *What controls the rate of primary production in natural communities?* Put another way, what factors could we change to increase the rate of primary production for a given community? Note that this question could be broken down into many questions of the same type for each population of plants. The control of primary production has been studied in greater detail for aquatic systems than for terrestrial systems, so we first look at some details of production in aquatic communities.

Aquatic Communities

The depth to which *light* penetrates in a lake or ocean is critical in defining the zone of primary production in aquatic communities. Water absorbs solar radiation very readily. More than half of the solar radiation is absorbed in the first meter of water, including almost all the infrared energy. Even in "clear" water, only 5–10% of the radiation may reach a depth of 20 meters. This decrease can be described reasonably well by a geometric curve of decrease in radiation:

$$\frac{dl}{dt} = kl \qquad (25.3)$$

where l = amount of solar radiation (joules per m^2 per unit of time)

t = depth

k = extinction coefficient (a constant)

This relationship is illustrated in Figure 25.4 for two values of k, the extinction coefficient. Large k values indicate less transparent waters. Figure 25.5 illustrates the decrease in photosynthesis with depth in three California lakes. Clear Lake is a *eutrophic* lake with high production and little light penetration. Castle Lake is a lake of intermediate productivity, in which the zone of photosynthesis extends below a depth of 20 meters. Lake Tahoe is an alpine *oligotrophic* lake of remarkably clear water in which the zone of photosynthesis extends to a depth of 100 meters, although there is little photosynthesis at any depth (Goldman 1988).

Very high light levels can also inhibit photosynthesis of green plants, and this inhibition can be found in tropical and subtropical surface waters throughout the year. When surface radiation is excessive, the maximum in primary production will occur several meters beneath the water surface of the sea.

Marine Communities

Light is a primary factor in limiting primary production in the ocean (Platt et al. 1992). If one knows the rate at which light decreases with depth (the extinction coefficient), the amount of solar radiation, and the amount of

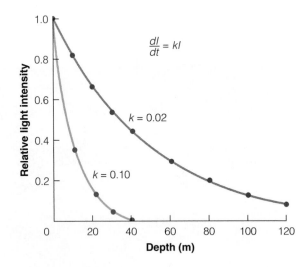

FIGURE 25.4

Theoretical attenuation of solar radiation (I) *with depth in a water column. Light intensity falls geometrically with depth, and the larger the extinction coefficient* (k), *the faster is the loss of light. An extinction coefficient of 0.02 occurs in pure water; one of 0.10 occurs in oceanic seawater. Coastal seawater has even higher extinction coefficients (approximately 0.30).*

chlorophyll in the aquatic plants, one can calculate the net production of phytoplankton. For any particular region of the ocean, the equations for estimating primary production are complex because of the need to correct for absorption and backscattering of light and for cloud cover (Platt and Sathyendranath 1988).

Considerable research on ocean production has occurred in the North Pacific Central Gyre (Figure 25.6). Because the Central Gyre is large and relatively stable, biological production can be measured in the absence of external inputs. Figure 25.7 shows the vertical distribution of temperature, nutrients, chlorophyll, and primary production in the Central Gyre. Virtually all the primary production occurs in the *euphotic zone*—the lower limit of which is the 1% light level—above 90 m depth. Primary production, measured by ^{14}C, is relatively low in the surface waters (due to excessive light) and is highest in the warm waters near the surface (10–30 m depth).

If light is the primary variable limiting primary production in the ocean, we would expect a gradient of productivity from the poles toward the equator. Figure 25.3, which shows the global distribution of primary production in the oceans, indicates that no such latitudinal gradient of production exists. Large parts of the tropics and subtropics, such as the

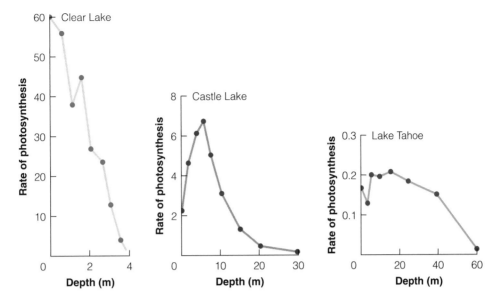

FIGURE 25.5
Change in photosynthesis with depth in three California lakes during the summer. Note changes in scale of depth and rate of photosynthesis. Rate of photosynthesis is measured in grams of carbon fixed per m² per day. (Data from Goldman 1968.)

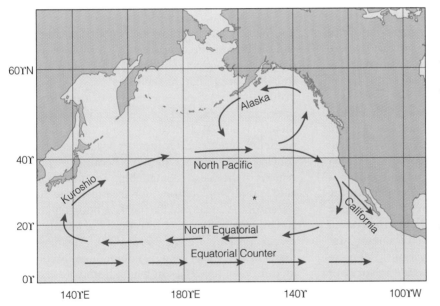

FIGURE 25.6
The major upper ocean currents in the North Pacific Ocean. The North Pacific Central Gyre is the large clockwise-flowing circulation system. Advection into the Central Gyre from outside is slight in comparison to other regions, such as coastal boundary current systems. The biological structure of the upper layers of the central North Pacific, most intensively studied at the location indicated by the asterisk, appears to be regulated primarily by in situ processes. (From Hayward 1991.)

Sargasso Sea, the Indian Ocean, and the Central Gyre of the North Pacific, are very unproductive. In contrast, the North Atlantic, the Gulf of Alaska, and the Southern Ocean off New Zealand are quite productive. The most productive areas are coastal areas off the western side of Africa and North and South America (Falkowski et al. 1998).

Why are tropical oceans unproductive when the light regime is good all year? *Nutrients* appear to be the primary limitation on primary production in the ocean through their effects on the biomass of chlorophyll in the phytoplankton. Two elements, nitrogen and phosphorus, often limit primary production in the oceans. One of the striking generalities of many

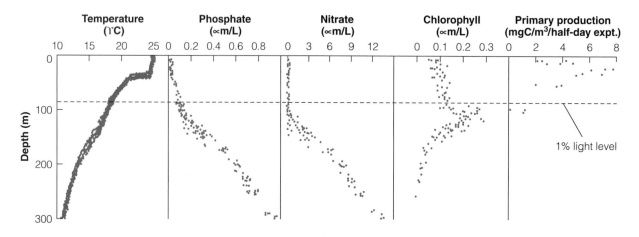

FIGURE 25.7

Typical vertical distribution of temperature, nutrients, and production in the upper layer of the North Pacific Central Gyre during summer. The curves are composites of several vertical profiles made over a two-day period at a single location (28°N, 155°W, indicated by the asterisk in Figure 25.6). The dashed line illustrates the depth to which only 1% of surface light penetrates (the traditional definition of the lower limit of the euphotic zone). Note that nitrate has been depleted to undetectable levels above the 1% light level, and that most of the primary production (measured from ^{14}C uptake) takes place above the depth where nitrate can be detected with conventional techniques. (From Hayward 1991.)

parts of the oceans is the very low concentrations of nitrogen and phosphorus in the surface layers where phytoplankton live (see Figure 25.7), whereas the deep water contains much higher concentrations of nutrients.

Nitrogen may be a limiting factor for phytoplankton in many parts of the ocean (Ryther and Dunstan 1971, Platt et al. 1992). Pollution (nutrient runoff) from duck farms along the bays of Long Island, New York, adds both nitrogen and phosphorus to the coastal water, but unlike phosphorus, the nitrogen added is immediately taken up by algae, and no trace of nitrogen can be measured in the coastal waters (Figure 25.8b) . That nitrogen was limiting was confirmed by nutrient-addition experiments (Figure 25.8c). The addition of nitrogen (in the form of ammonium) caused a heavy algal growth in bay water, but the addition of phosphate did not induce algal growth. This work has some obvious practical conclusions: If nitrogen is the factor currently limiting phytoplankton production, the elimination of phosphates from sewage entering the ocean will not help the problem of coastal pollution.

The discovery that nitrogen limits primary production in many parts of the ocean was completely

unexpected because nitrogen is abundant in the air and can be converted into a usable form by nitrogen-fixing cyanobacteria (Falkowski 1997). The expectation had been that phosphorus must be limiting productivity in the ocean because phosphorus does not occur in the air. But this has turned out to be completely wrong, and nitrogen seems to be the major nutrient limiting oceanic primary production. But this conclusion raises other questions because several large regions of the oceans contain high amounts of nitrate and few phytoplankton. For example, the surface waters of the equatorial Pacific have both high nitrate and high phosphate concentrations but low algal biomass (Behrenfeld et al. 1996). One explanation of these regions is that they are communities dominated by top-down processes in which herbivores control plant biomass. Alternatively, these could be bottom-up communities limited by some nutrient other than nitrogen or phosphorus.

The Sargasso Sea is an area of very low productivity in the subtropical part of the Atlantic Ocean. The seawater there is among the most transparent in the world, and the surface waters are very low in nutrients. Nitrogen and phosphorus, however, do not seem to be limiting primary production; instead, iron

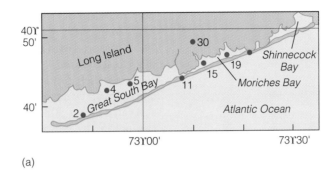

(a)

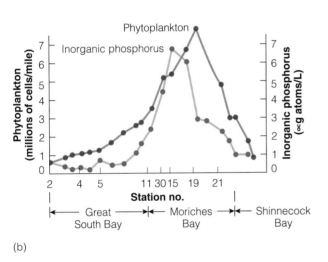

(b)

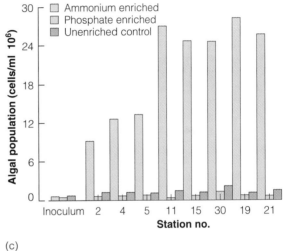

(c)

FIGURE 25.8

Effects of nutrient limitations on phytoplankton production in coastal waters of Long Island. (a) Coast of Long Island, New York; (b) abundance of phytoplankton and distribution of phosphorus arising from duck farms; (c) nutrient-enrichment experiments with alga Nannochloris atomus *in water from the bays. Phosphorus is superabundant, and nitrogen seems to limit algal growth. (After Ryther and Dunstan 1971.)*

seems to be critical (Menzel and Ryther 1961b), as was shown by a series of nutrient-enrichment experiments in which surface water from the Sargasso Sea was placed in bottles and enriched with various nutrients for three days. The results were as follows:

Nutrients added to experimental culture	Relative uptake of ^{14}C by cultures
None (controls)	1.00
N + P only	1.10
N + P + metals (excluding iron)	1.08
N + P + metals (including iron)	12.90
N + P + iron	12.00

This experimental demonstration of iron limitation in Sargasso Sea water stimulated the hypothesis that iron limitation could be responsible for the low productivity of the equatorial Pacific. Two large-scale open ocean iron-enrichment experiments were conducted in the equatorial Pacific in 1993 and 1995. Low concentrations of dissolved iron were spread over 72 km^2 of ocean over seven days, and the resulting changes in phytoplankton and zooplankton were measured. A massive bloom of phytoplankton developed in the iron-enriched area. Chlorophyll *a* levels increased to 27 times the starting value. At the same time, nitrate uptake increased 14-fold such that levels of nitrate in

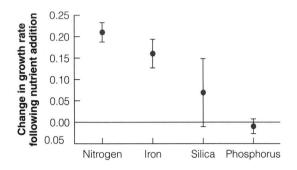

FIGURE 25.9

Effects of nutrient addition on marine phytoplankton growth rates in 303 experiments. Excess nutrients were added to a large water sample and followed for two to seven days. Doubling times for nitrogen addition were 3.3 days, for iron 4.1 days, and for silica 10 days. Silica limitation occurs when diatoms are the dominant species in the phytoplankton. The black line marks the line of zero effect. (After Downing et al. 1999.)

the seawater decreased by about 35% within one week (Coale et al. 1996).

Iron comes to the oceans largely as wind-blown dust from the land, and dust is particularly scarce in the Pacific Ocean and in the Southern Ocean. Iron is an essential component of the photosynthetic machinery of the cyanobacteria that fix nitrogen in the oceans. The effect of iron on primary production operates mainly through its role in nitrogen fixation, resulting in the following sequence of potential limitations in iron-poor parts of the ocean (Falkowski et al. 1998):

$$iron \xrightarrow{+} cyanobacteria \xrightarrow{+} \underset{fixation}{nitrogen} \xrightarrow{+} phytoplankton$$

In most of the open oceans, light is always available for photosynthesis, but nitrogen is not.

To quantify the relative effects of different limiting nutrients in the oceans, Downing et al. (1999) analyzed 303 controlled nutrient-addition experiments carried out over the past 30 years. They found that nitrogen addition stimulated phytoplankton growth most strongly, followed closely by iron addition (Figure 25.9) . Doubling times for phytoplankton populations averaged 3.3 days for nitrogen addition and 4.1 days for iron addition. In general these results are consistent with the conclusion that nitrogen and iron are key limiting resources in the oceans, and that silica

may limit diatom production when diatoms are a dominant component of the phytoplankton.

Compared with the land, the ocean is very unproductive; the reason seems to be that fewer nutrients are available. Rich, fertile soil contains 5% organic matter and up to 0.5% nitrogen. One square meter of surface soil can support 50 kg dry weight of plant matter. In the ocean, by contrast, the richest water contains 0.00005% nitrogen, four orders of magnitude less than that of fertile farmland soil. A column of rich seawater one square meter in cross section could support no more than 5 grams dry weight of phytoplankton (Ryther 1963). In terms of standing crops, the sea is a desert compared with the land. And although the maximal rate of primary production in the sea may be the same as that on land, these high rates in the sea can be maintained for a few days only, unless upwelling enriches the nutrient content of the water.

Areas of upwelling in the ocean are exceptions to the general rule of nutrient limitation. The largest area of upwelling occurs in the Antarctic Ocean (see Figure 25.3), where cold, nutrient-rich, deep water comes to the surface along a broad zone near the Antarctic continent. Other areas of upwelling occur off the coasts of Peru and California, as well as in many coastal areas where a combination of wind and currents moves the surface water away and allows the cold, deep water to move up to the surface. In these areas of upwelling, fishing is especially good, and in general a superabundance of nitrogen and phosphorus is available to the phytoplankton.

One of the most exciting recent developments in marine ecology is the ability to estimate primary production from satellite remote sensing data (Longhurst et al. 1995, Falkowski et al. 1998). Chlorophyll concentration in the surface water can be estimated by spectral reflectance using blue/green ratios. Figure 25.10 illustrates the detection of a plankton bloom using chlorophyll levels in coastal waters off British Columbia. Remote sensing allows marine ecologists to analyze large-scale production changes in the ocean that are impossible to detect by making a few measurements off a ship. These techniques promise to expand our understanding of how primary production in the oceans varies in space and time.

Total primary production in the ocean is thus rarely limited by light but instead by the shortage of nutrients, particularly nitrogen and iron, which are critical for plant growth. Limitation of primary production by phosphorus is rare in oceanic ecosystems.

FIGURE 25.10

Two satellite images covering 612 km² centered on Vancouver Island on the west coast of Canada. Coastlines (white) have been added to the satellite image. (a) A thermal image taken on June 23, 1992, showing a meandering stream of cold water moving south down the edge of the continental shelf and winding into an eddy. Warmer water is dark blue, colder water is green to yellow. Land is warmer than the water (black). The lower right of this image shows a thin band of cold upwelling water along the coast of Washington. Images like this are useful for understanding patterns of coastal currents and upwelling. (b) A processed image from May 22, 1995, showing a phytoplankton bloom in the center of the image, off the west coast of Vancouver Island. Clearwater is blue, with "bright" water from the algal bloom colored green to red. This plankton bloom spread and moved over a two-week period under the influence of coastal currents and then dissipated. The plume of muddy water from the Fraser River appears as a second white spot at the right edge of this image. (Images courtesy of the Institute of Ocean Sciences, Department of Fisheries and Oceans, Canada.)

Freshwater Communities

In freshwater communities the same conclusions do not seem to hold. Solar radiation limits primary production on a day-to-day basis in lakes, and within a given lake one can predict the daily primary productivity from the solar radiation (Horne and Goldman 1994). Temperature is closely linked with light intensity in aquatic systems and is difficult to evaluate as a separate factor. Nutrient limitations operate in freshwater lakes, and the great variety of lakes are associated with a great variety of potential limiting nutrients.

For growth, plants require nitrogen, calcium, phosphorus, potassium, sulfur, chlorine, sodium, magnesium, iron, manganese, copper, iodine, cobalt, zinc, boron, vanadium, and molybdenum. These nutrients do not all act independently, which has made the identification of causal influences very difficult (Wetzel 1983). The conclusion of early work—that nitrogen and phosphorus were the major limiting factors in freshwater lakes—was a practical one based on the fertilization of small farm ponds to increase fish production.

During the 1970s the problem of what controls primary production in freshwater lakes became acute because of increasing pollution. Nutrients added to lakes directly in sewage or indirectly as runoff had increased algal concentrations and had shifted many lakes from phytoplankton communities dominated by diatoms or green algae to those dominated by blue-green algae. This process is called *eutrophication*.

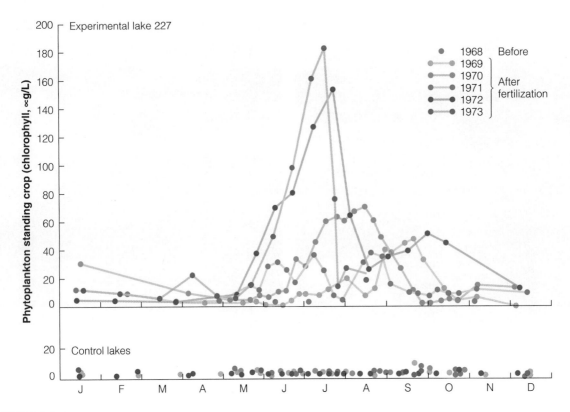

FIGURE 25.11

A comparison of phytoplankton standing crops (as chlorophyll a) in six natural unfertilized lakes (lower) and Lake 227, fertilized with 0.48 g of phosphorus and 6.29 g of nitrogen per m² annually (upper). All lakes are less than 13 meters deep. Note that Lake 227 had a standing crop similar to that of other lakes prior to fertilization. Large inputs of phosphorus and nitrogen will thus cause severe eutrophication problems regardless of how low carbon concentrations in the lake water are. The necessary carbon is drawn from the atmosphere. (After Schindler and Fee 1974.)

Before we can control eutrophication in lakes, we must decide which nutrients need to be controlled. Three major nutrients were considered: nitrogen, phosphorus, and carbon. Phosphorus is now believed to be the limiting nutrient for phytoplankton production in the majority of lakes (Edmondson 1991).

A series of elegant whole-lake nutrient-addition experiments conducted in the Experimental Lakes area of northwestern Ontario by David Schindler and his Winnipeg-based research group pinpointed the role of phosphorus in temperate-lake eutrophication (Schindler and Fee 1974). In one experiment, Lake 227 was fertilized for 5 years with phosphate and nitrate, and phytoplankton levels increased 50–100 times over those of control lakes (Figure 25.11) . To separate the effects of phosphate and nitrate, Lake 226 was split in half with a curtain and fertilized with

carbon and nitrogen in one half and with phosphorus, carbon, and nitrogen in the other. Within two months a highly visible algal bloom had developed in the basin to which phosphorus was added (Figure 25.12) . All this experimental evidence is consistent with the hypothesis that phosphorus is the master limiting nutrient for phytoplankton in freshwater lakes.

When phosphorus is added to a lake, algae may show signs of nitrogen or carbon limitation, but long-term processes cause these deficiencies to be corrected (Schindler 1977, 1990). Physical factors such as water turbulence and gas exchange seem to regulate CO_2 availability, so it rarely becomes limiting for algae. Nitrogen can be fixed by blue-green algae. These species, which are favored when nitrogen is in short supply, increase the availability of nitrogen to algae, and the lake eventually returns to a state of

FIGURE 25.12
Lake 226 in the Experimental Lakes area of northwestern Ontario, showing the role of phosphorus in eutrophication. The far basin, fertilized with phosphorus, nitrogen, and carbon, is covered with an algal bloom of the blue-green alga Anabaena spiroides. *The near basin, fertilized with nitrogen and carbon, showed no changes in algal abundance. Photo taken September 4, 1973. (Photo courtesy of D. W. Schindler.)*

phosphorus limitation. The net result is that the standing crop of phytoplankton in lakes is highly correlated with the total amount of phosphorus in the water (Figure 25.13) .

The practical significance of these and other experiments is the advisability of controlling phosphorus input to lakes and rivers as a simple means of checking eutrophication (Likens 1972, Schindler 1977). The amount of phosphorus that a lake can withstand can be calculated so that planners can determine the effects of human developments on a lake (Dillon and Rigler 1975).

Part of the difficulty of studying nutrient limitations of phytoplankton production is that nutrients may occur in several chemical states in aquatic systems. In some conditions, nutrients are present but not available to the organisms because they are bound up in organic complexes in the water or mud (Wetzel and Allen 1971). This has been shown dramatically in acid-bog lakes, which contain large amounts of phosphorus in forms not available to the phytoplankton. Waters (1957) showed that fertilizing acid-bog lakes in Michigan with lime ($CaCO_3$) increased the pH, allowed phosphorus to be released from sediments, and greatly increased phytoplankton abundance.

One of the changes that often accompany eutrophication in lakes is that blue-green algae tend to replace green algae (see Figure 25.12). Blue-green algae are "nuisance algae" because they become extremely abundant when nutrients are plentiful and form floating scums on highly eutrophic lakes. Blue-green algae become dominant in the phytoplankton

ESSAY 25.1

NUTRIENT RATIOS AND PHYTOPLANKTON

Chemistry is an important aspect of ecosystem science, and one example of its utility involves the nutrient ratios of primary producers. In 1958 the oceanographer A. C. Redfield discovered that samples of organisms from the open ocean consistently exhibited the atomic ratio $C_{106}N_{16}P_1$, which is now referred to as the Redfield ratio in his honor. In contrast to the constant Redfield ratio observed in the open ocean, the composition of phytoplankton from freshwater lakes is highly variable. The ratio of C:N:P in phytoplankton varies with the ratio present in the water and with the pH of the water (Sterner and Hessen 1994). In a series of 51 lakes surveyed, the C:N ratio varied from 4 to 20 (see the following graph), and the C:P ratio from 100 to 550, so that Redfield proportions are the exception rather than the rule in freshwater lakes.

In general, much more carbon is present in freshwater phytoplankton relative to nitrogen and phosphorus. Why should this variation matter? Algae with high C:P ratios are poor quality food for herbivores

such as zooplankton but are good-quality food for microbes. The C:N:P ratio could affect the structure of the food web. Different zooplankton species have different C:N:P ratios, and consequently survive better feeding on different algal species. In general, there is much variation in C:N:P ratios is plants, less variation in bacteria, even less in zooplankton, and still less in fish species (Sterner et al. 1998).

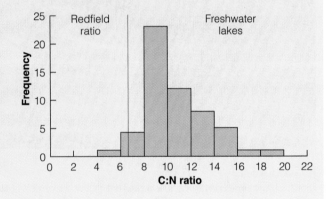

for several reasons. They are not heavily grazed by zooplankton or fish, which prefer other algae. Zooplankton often cannot manipulate the large colonies and filaments of blue-green algae. Some species of blue-green algae also produce secondary chemicals that are toxic to zooplankton (DeMott and Moxter 1991). Blue-green algae are also poorly digested by many herbivores, so they are low-quality food for them. Finally, many blue-green algae can fix atmospheric nitrogen, putting them at an advantage when nitrogen is relatively scarce. For example, in Lake Washington (see p. 477), blue-green algae were dominant when the nitrogen-to-phosphorus ratio was less than 23 but disappeared when this ratio exceeded 25 (Tilman et al. 1982). In eutrophication, more and more phosphorus is continually loaded into a lake, so the nitrogen-to-phosphorus ratio falls and nitrogen can become a limiting factor (Figure 25.14) (Smith 1982). The phytoplankton community in many temperate

freshwater lakes therefore may have two broad configurations at which it can exist, one with low nutrient levels (organized by predation and dominated by green algae) and one with high nutrient levels (organized by competition and dominated by blue-green algae).

Estuaries are often heavily polluted with nutrients from sewage and industrial wastes. Because they form an interface between saltwater, in which nitrogen is often limiting to phytoplankton, and freshwater, in which phosphorus is typically limiting, estuaries are complex gradients of nutrient limitation in which added phosphorus and nitrogen from pollution can strongly affect primary production throughout the estuary (Doering et al. 1995).

To summarize, the major controlling factors for primary production in freshwater communities are light (and temperature), phosphorus, and silicon (for diatoms), and occasional controlling factors include nitrogen and iron.

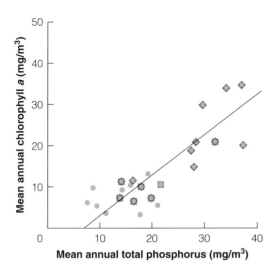

FIGURE 25.13
The relationship between total phosphorus concentration and phytoplankton standing crop (measured by chlorophyll a) in lakes of the Experimental Lakes area in northwestern Ontario. Both fertilized and unfertilized lakes are included. Squares indicate fertilized lakes deficient in nitrogen; diamonds indicate fertilized lakes deficient in carbon. (After Schindler 1977.)

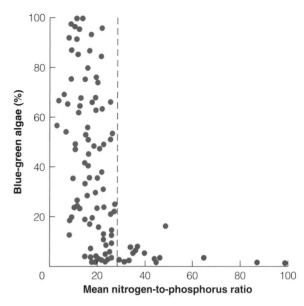

FIGURE 25.14
Relation between proportion of blue-green algae in the phytoplankton and nitrogen-to-phosphorus ratio in 17 lakes around the world. Each point represents data from one growing season. Blue-green algae are dominant when the nitrogen-to-phosphorus ratio is less than 25–30 (dashed line). (After Smith 1983.)

Terrestrial Communities

In terrestrial habitats, temperature ranges are much greater than in aquatic habitats, and the great variation in temperature from coastal to alpine or continental areas makes it possible to uncouple the solar radiation-temperature variable, which is so closely linked in aquatic systems. The large seasonal changes in radiation and temperature are reflected in the global patterns of primary production. Using satellite imagery, we can identify continental and global patterns of terrestrial productivity. Satellites, such as the NOAA[2] meteorological satellite operated by the United States, have onboard sensors that record spectral reflectance in the visible and infrared portions of the electromagnetic spectrum.

As green plants photosynthesize, they display a unique spectral reflectance pattern in the visible (0.4–0.7 μm) and the near-infrared (0.725–1.1 μm) wavelengths (Goward et al. 1985). Vegetation indices

that discriminate living vegetation from the surrounding rock, soil, or water have been developed by combining these spectral bands (see Box 25.1). One of the most common spectral vegetation indices, the

David W. Schindler *(1940–) Killam Professor of Zoology, University of Alberta*

---

[2]National Oceanic and Atmospheric Administration of the United States.

BOX 25.1

ESTIMATING PRIMARY PRODUCTION FROM SATELLITE DATA

Estimating net primary production is relatively easy on small areas or in small bodies of water, but estimating primary production on a global basis requires satellite imagery. The general approach has been to measure the amount of solar radiation and to correct it for the efficiency of light use by plants. The solar radiation that can potentially be used for photosynthesis is the radiation that is absorbed in the 400–700 nm wavelengths. Net primary production is calculated from the simple equation:

$$NPP = (APAR)(\varepsilon) \qquad (25.4)$$

where NPP = net primary productivity (g carbon per unit area per year)

$APAR$ = absorbed photosynthetically active radiation (joules per unit area per year)

ε = average light utilization efficiency (in g carbon per joule)

Satellites can measure radiation (both visible and invisible), and from measures of radiation we can derive estimates of $APAR$. For the oceans, $APAR$ can be correlated to measurements of surface chlorophyll. For land areas, satellites can measure greenness and obtain from it a vegetation index

$(NDVI)$ that corresponds to the amount of chlorophyll in land plants:

$$NDVI = \frac{NIR - RED}{NIR + RED} \qquad (25.5)$$

where $NDVI$ = normalized difference vegetation index

NIR = near-infrared reflectance (0.725-1.1 μm)

RED = red reflectance (0.6-0.7 μm)

The $NDVI$ is closely related to $APAR$ and can be used to estimate $APAR$.

The difficult parameter to be estimated from Equation (25.4) is ε, light use efficiency, and this estimate cannot be made from satellites but must be calculated from field measurements. This is relatively tedious, particularly on land, and much of the uncertainty in estimating net primary production comes from uncertainly of the exact value of ε. For the ocean, ε is estimated as a function of sea-surface temperature. For terrestrial systems, ε depends on ecosystem type (forest, grassland, tundra) and stresses from unfavorable levels of temperature, water, and nutrients (Field et al. 1998). There is now broad agreement among different models that estimate net primary productivity for terrestrial systems (Cramer et al. 1999).

---

normalized difference vegetation index ($NDVI$), is a ratio of near-infrared and visible red spectral bands:

$$NDVI = \frac{NIR - RED}{NIR + RED}$$

where $NDVI$ = normalized difference vegetation index

NIR = near-infrared reflectance (0.725-1.1 μm)

RED = red reflectance (0.6-0.7 μm)

This index is closely correlated with primary productivity (Graetz et al. 1992). The AVHRR (advanced very high resolution radiometer) sensor on the NOAA satellite and the SeaWiFS (Sea Viewing Wide Field-of-view Sensor) on the Sea Star spacecraft are especially useful for monitoring global vegetation (Figure 25.15) because they have global coverage at a

resolution of 1.1 km at least once per day in daylight hours (Signorini et al. 1999).

What limits primary production in terrestrial communities and produces the patterns shown in Figure 25.15? Rosenzweig (1968) showed that actual evapotranspiration could predict the aboveground production of terrestrial communities with good accuracy (Figure 25.16) . Actual evapotranspiration—the amount of water pumped into the atmosphere by evaporation from the ground and via transpiration from vegetation—is a measure of solar radiation, temperature, and rainfall. Rosenzweig used only climax vegetation in this analysis.

Primary production data for different types of forests have been summarized by Kira (1975). Figure 25.17 shows the range of primary production for five growth forms of forests in Japan. Evergreen broadleaf forests of the warm-temperate zone are among the most productive stands, but there is considerable variation

(a) June 1998 - August 1998

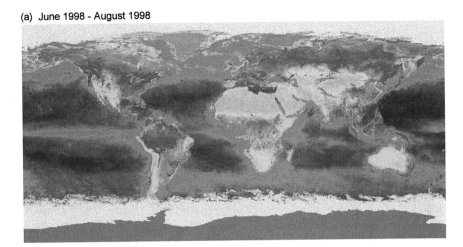

(b) December 1997 – February 1998

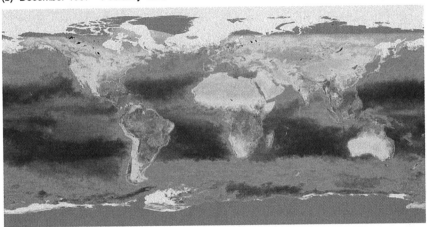

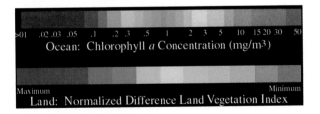

FIGURE 25.15
The world biosphere showing productivity of the land (NDVI) and oceans (chlorophyll concentration) for different seasons of the year. These images were obtained with the SeaWiFS (Sea-viewing Wide Field-of-view Sensor) on board the SeaStar spacecraft, which circles 705 km above the Earth in a sun-synchronous orbit. Spatial resolution is 1.1 km. (Images provided by the SeaWiFS Project, NASA/Goddard Space Flight Center, and ORBIMAGE.)

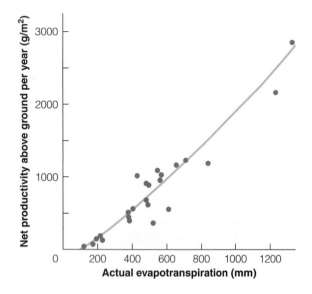

FIGURE 25.16
Prediction of net primary production of terrestrial communities from climatological data on solar radiation, temperature, and moisture (measured by actual evapotranspiration). (Data from Rosenzweig 1968.)

within forest types. Coniferous forests are on the average more productive than deciduous forests growing under the same climatic conditions. The differences in productivity among forests are due to variation in the length of the growing season and to differences in leaf-area index. Coniferous trees have a greater leaf surface area than do deciduous trees, and by retaining their leaves are able to achieve a longer growing season. The ratio of root biomass to shoot biomass is the same for both forest types. Figure 25.18 shows that the gross primary production of both broadleaf and needle-leaved forests is accurately predicted by the combined effects of leaf area-index and the length of the growing season. Temperature determines the length of the growing season.

Primary production in grasslands is strongly affected by the relative amounts of C_3 and C_4 grasses (see p. 99). In the Great Plains of the United States, C_3 grass production is correlated most closely with temperature, and the higher the temperature the lower the C_3 productivity (Epstein et al. 1997). C_4 grass production is most closely correlated with rainfall, and the higher the rainfall the higher the production. C_4 grasses dominate about 74% of the Great Plains, and C_3 grasses are dominant only in the northern, cooler parts of this area (Figure 25.19) . Within

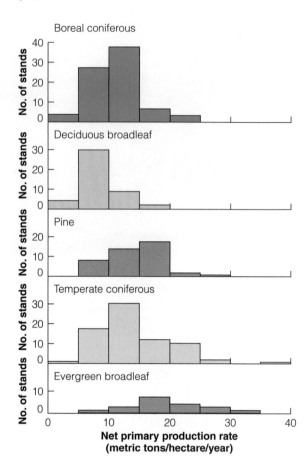

FIGURE 25.17
Frequency distributions of annual net primary production in 258 forest stands in Japan. Evergreen broadleaf forests of the warm-temperate zone are the most productive; in the cool-temperate zone, coniferous stands are more productive than deciduous forests. Only aboveground production is included. (After Kira 1975.)

the primary control exerted by temperature and moisture, secondary limitations are imposed by the type of soil and its water holding capacity, and by nutrient availability (Sala et al. 1988).

Nutrient-addition experiments on local sites can be used to determine how much primary production can be limited by nutrients. Cargill and Jefferies (1984) added nitrate and phosphate to salt-marsh sedges and grasses in the subarctic zone to test for nutrient limitation. Figure 25.20 shows that in the absence of grazing, the addition of nitrate doubled primary production of the sedges and grasses, and the joint addition of phosphate and nitrate quadrupled

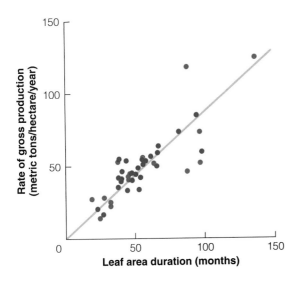

FIGURE 25.18
Relationship between the rate of gross primary production and leaf area duration (leaf-area index times length of growing season in months) in Japanese forest stands. Needle-leaved forests are represented by a purple dot, broadleaf forests by a blue dot. (After Kira 1975.)

FIGURE 25.19
Geographical areas of the Great Plains of the United States in which C_3 grasses (green) and C_4 grasses (yellow) dominate based on relative primary production. C_4 grasses dominate on 74% of the Great Plains. (After Epstein et al. 1997.)

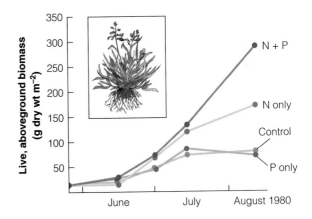

FIGURE 25.20
Effects of fertilization with nitrogen and phosphorus on primary production in a salt marsh dominated by Carex subspathacea, *southern Hudson Bay, Canada. Each treatment included four replicates. (From Cargill and Jefferies 1984.)*

production. In this marsh, as in many terrestrial communities, nitrogen is the major nutrient limiting productivity, and when nitrogen is suitably increased, phosphorus becomes limiting.

In unexploited virgin grassland or forest, all nutrients that the plants take up from the soil and hold in various plant parts are ultimately returned to the soil as litter that decomposes. The net flow of nutrients must be stabilized (such that input equals output), or the site would deteriorate over time. But in a harvested community, the situation is fundamentally different because nutrients are being continuously removed from the site. This makes it necessary to study the nutrient demands of crops so that the nutrient capital of soils will not be progressively exhausted. Coniferous forests use soil nutrients more efficiently than deciduous forests (Waring and Schlesinger 1985). Nutrients removed by harvesting trees are typically recovered from annual inputs in precipitation and in rock weathering (see p. 568).

Terrestrial communities, especially forests, have large nutrient stores tied up in the standing crop of plants. In this way they differ from marine and freshwater communities. This concentration of nutrients in the standing vegetation has important implications for nutrient cycles in forest communities. If the community is stable, the input of nutrients should equal the output, and a considerable amount of research effort is now being directed at studying nutrient cycles in terrestrial communities. We will discuss this work in Chapter 27.

E S S A Y   2 5 . 2

WHY DOES PRIMARY PRODUCTION DECLINE WITH AGE IN TREES?

Forest managers have long known that tree growth and wood production decline with tree age. Net primary production in forests reaches a peak early in succession and then gradually declines by as much as 76% from that peak (Gower et al. 1996). For example, in Russia primary production of 140-year-old Norway spruce trees declines 58% from the peak reached at about 70 years of age (see graph below).

Why should this occur? Three hypotheses have been advanced to explain age-related forest decline. The classical explanation is that a change occurs in the balance of photosynthesis and respiration. As trees grow larger with age, they have more tissues that respire and lose energy and proportionally less leaf area to photosynthesize. But this explanation is not supported by recent measurements showing that respiratory losses do not increase very much with tree age, because most of the sapwood uses little energy. The second hypothesis is nutrient limitation by nitrogen as the forest ages. Nitrogen is commonly found to be the limiting factor to tree growth, and as forests age, more woody litter accumulates on the soil surface. Woody litter decomposes very slowly compared to fine litter from leaves, so nitrogen becomes locked up in woody debris on the forest floor. The third and

newest explanation is that as trees grow larger, water transport to the leaves becomes limited because of increased hydraulic resistance associated with the greater distance between the roots and the stomata of the leaves. Trees reduce stomatal conductance to conserve water in their tissues, and because photosynthesis is tightly coupled with stomatal conductance, production declines. This hypothesis is consistent with the observation that leaf stomata close earlier in the day in older trees compared with young trees.

Current forest growth models suggest that nutrient limitation (hypothesis 2) and water flow limitations (hypothesis 3) are of nearly equal importance in reducing net primary production as trees age. An increase in respiration seems to contribute little to decreasing production (Gower et al. 1996). Knowing what limits primary production in forests is critical for understanding the effects of climate change on ecosystems.

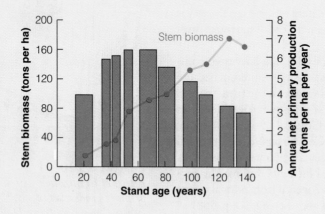

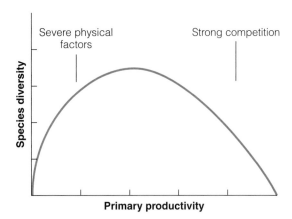

FIGURE 25.21
The relationship between primary productivity and biodiversity in plant communities. In areas of low productivity, we would expect productivity to rise with diversity, but in moderate to highly productive communities, productivity declines.

Plant Diversity and Productivity

One reason to conserve biodiversity might be that more-diverse sites are more productive. The hypothesis that diversity enhances productivity is an old one, and Charles Darwin suggested that it was true for terrestrial plants (Jolliffe 1997). The idea behind it is that different plant species have different resource needs such that the niches of several species would be complementary (Tilman et al. 1997). Under this hypothesis, resources would be more completely utilized when species richness is greater. But more species means more competition among plants, and it is possible that excessive competition could reduce productivity rather than permit it to increase. The net result of these opposing forces suggests a possible parabolic relationship between productivity and species diversity (Figure 25.21).

Over about half of the range of productivity we might expect a negative relationship between productivity and species richness. This has been observed many times in both plant and animal communities (Huston 1994). One example illustrates the point. The longest running experiment on productivity and diversity is the Park Grass Experiments at Rothamsted Experimental Station in England (Silvertown 1980). These experiments measured the effects of liming and fertilizing pastures on hay production.

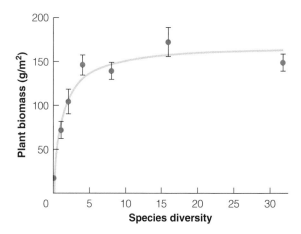

FIGURE 25.22
Relationship of species diversity to plant productivity in Minnesota grasslands. Varying numbers of species were seeded into 13m × 13m plots, and plants were harvested after completing growth over one summer. Only aboveground biomass was measured. Some plant biomass occurred even in plots with no species seeded because weeds invaded the plots before the final biomass clipping. (After Tilman et al. 1996.)

From the earliest measurements in 1856 to 1978, a negative relationship between pasture productivity and biodiversity has always existed. On one pasture fertilized with nitrogen each year since 1856, species numbers declined over time, as follows:

Year	Number of plant species
1856	49
1862	28
1872	16
1903	10
1919	8
1949	3

In spite of the loss in biodiversity, productivity in the fertilized plots remained high. The Park Grass Experiments illustrate the right side of the parabolic curve shown in Figure 25.21.

In plant communities in low nutrient soils or in severe physical environments, we expect a positive relationship between productivity and diversity. Tilman et al. (1997) have shown this in Minnesota grasslands on low-nitrogen soils. They seeded 289 plots with 1, 2, 4, 8, 16, or 32 perennial species and observed a positive relationship between plant productivity and species diversity (Figure 25.22) . It is clear from these experimental studies that in this prairie community, productivity has reached a plateau after about 10 species. The highest productivity in these experiments came from plots that had the most functional groups of plants—C_3 grasses, C_4 grasses, legumes, forbs, and woody plants. Productivity is driven by a few species of dominants that make up a large fraction of the plant biomass (Grime 1997). The key to ecosystem productivity is in the biological characteristics of the dominant plants in a community.

From a global perspective, primary production is largely driven by the physical environment in the form of light, temperature, rainfall, and nutrient availability. Plants have adapted to these environmental constraints to produce through photosynthesis the materials that drive all the subsequent biota, including ourselves.

Summary

A community can be viewed as a complex machine that processes energy and materials. To study community metabolism, we must identify the food web of the community and then trace the flows of chemical materials or energy through the food web. Many ecologists prefer to measure energy use in studying community metabolism because energy is not recycled within the community.

Primary production can be measured by the amount of energy or carbon fixed via photosynthesis by green plants per unit time. CO_2 uptake can be measured directly using radioactive carbon-14 or indirectly by harvesting new growth.

Only about 1% of solar energy is captured by green plants and converted into primary production. Forests are relatively efficient, and aquatic communities are relatively inefficient, at capturing solar energy.

Primary production varies greatly over the globe; it is highest in the tropical rain forests and lowest in arctic, alpine, and desert habitats. Global primary production is distributed nearly equally between the oceans and the land.

The sea is less productive than the land per unit of area (except for coastal areas and upwelling zones) because of limitations imposed by nitrogen and iron. In freshwater lakes and streams, light, temperature, and nutrients restrict primary production, and phosphorus is the limiting nutrient in many lakes.

Terrestrial primary productivity can be predicted from the length of the growing season, temperature, and rainfall. Nutrient limitations further restrict productivity levels set by these climatic factors, and the stimulation of plant growth achieved by fertilizing forests and crops indicates the importance of studying nutrient cycling in biological communities.

Remote sensing with satellites has provided new methods for measuring the spatial and temporal variability of primary production in the oceans and on land on large spatial scales. Identifying the factors that limit primary production is important for understanding how climatic changes will affect both natural and agricultural communities.

Key Concepts

1. Communities process solar energy through green plants, and the resulting energy fixed via photosynthesis sustains all the trophic levels in the food web.

2. Only 1% or less of solar energy reaching the Earth is captured by plants.

3. Primary production varies globally, and total production is nearly equally distributed between the land and the oceans, even though the oceans occupy more than twice as much surface area.

4. Primary production in aquatic environments is limited by nutrients—primarily by nitrogen and iron in the open ocean, and by phosphorus in freshwater lakes.

5. On land, temperature, moisture, and nutrients limit primary production. Nitrogen and phosphorus are often limiting, but trace metals can also be critical in some soils.

6. Satellite imagery now makes it possible to study large-scale changes in primary production on land and in the ocean.

Selected References

Arrigo, K. R., D. L. Worthen, M. P. Lizotte, P. Dixon, and G. Diekmann. 1997. Primary production in Antarctic sea ice. *Science* 276:394–397.

Cramer, W., D. W. Kicklighter, A. Bondeau, B. I. Moore, G. Churkina, B. Nemry, A. Ruimy, A. L. Schloss. 1999. Comparing global models of terrestrial net primary productivity (NPP): Overview and key results. *Global Change Biology* 5:1–15.

Epstein, H. E., W. K. Lauenroth, I. C. Burke, and D. P. Coffin. 1997. Productivity patterns of C_3 and C_4 functional types in the U.S. Great Plains. *Ecology* 78:722–731.

Falkowski, P., R. T. Barber, and V. Smetacek. 1998. Biogeochemical controls and feedbacks on ocean primary production. *Science* 281:200–220.

Field, C. B., M. J. Behrenfeld, J. T. Randerson, and P. Falkowski. 1998. Primary production of the biosphere: integrating terrestrial and oceanic components. *Science* 281:237–240.

Ganter, B., F. Cooke, and P. Mineau. 1996. Long-term vegetation changes in a snow goose nesting habitat. *Canadian Journal of Zoology* 74:965–969.

Hector, A., et al. 1999. Plant diversity and productivity experiments in European grasslands. *Science* 286:1123–1127.

Runyon, J., R. H. Waring, S. N. Goward, and J. M. Welles. 1994. Environmental limits on net primary production and light-use efficiency across the Oregon transect. *Ecological Applications* 4:226–237.

Sand-Jensen, K. and D. Krause-Jensen. 1997. Broad-scale comparison of photosynthesis in terrestrial and aquatic plant communities. *Oikos* 80:203–208.

Shaver, G. R. and F. S. Chapin, III. 1995. Long-term responses to factorial, NPK fertilizer treatment by Alaskan wet and moist tundra sedge species. *Ecography* 18:259–275.

Tanner, E. V. J., P. M. Vitousek, and E. Cuevas. 1998. Experimental investigation of nutrient limitation of forest growth on wet tropical mountains. *Ecology* 79:10–22.

Tilman, D., C. L. Lehman, and K. T. Thomson. 1997. Plant diversity and ecosystem productivity: Theoretical considerations. *Proceedings of the National Academy of Sciences of USA* 94:1857–1861.

Williams, P. J. I. B. 1998. The balance of plankton respiration and photosynthesis in the open oceans. *Nature* 394:55–57.

Questions and Problems

25.1 "Red tides" are spectacular dinoflagellate blooms that occur in the sea and often lead to mass mortality of marine fishes and invertebrates. Human deaths from eating shellfish poisoned with red tide algae are a worldwide problem. Review the evidence available about the origin of red tides, and discuss the implications for general ideas about what controls primary production in the sea. Anderson (1994) and Hallegraeff (1993) discuss this problem.

25.2 Calculate the attenuation of solar radiation through a water column (and the depth at which radiation falls to 1% of its surface value) for the following extinction coefficients:

	Extinction Coefficient (per meter)
Pure water	0.03
Coastal water, minimum	0.20
Coastal water, maximum	0.40
Eutrophic lake	1.50

25.3 In the Great Plains grasslands of the United States, Epstein et al. (1997) showed that primary production of C_3 grasses could be predicted from mean annual temperature, with minimal contribution from mean annual precipitation. Discuss why precipitation and soil nutrients do not appear to be relevant variables for C_3 grass production in this ecosystem.

25.4 In discussing the effect of light on primary productivity in the ocean, Nielsen and Jensen (1957, p. 108) state:

> It is thus quite likely that a permanent reduction of the light intensity at the surface (to, e.g., 50 percent of its normal value without the other factors being affected—a rather improbable condition in Nature) in the long run would have very little influence on the organic productivity as measured per surface area.

How could this possibly be true?

25.5 Even though the concentration of inorganic phosphate in the water of the North Atlantic Ocean is only about 50% of that found in the other oceans, the North Atlantic is more productive than most of the other oceans. How can one reconcile these observations if nutrients limit primary productivity in the oceans?

25.6 Review the "limiting-nutrient controversy," the argument about the relative importance of carbon and phosphorus in regulating algal growth in lakes (Likens 1972, Edmondson 1991). In particular, discuss the practical problem of what actions might be taken to reduce lake eutrophication.

25.7 In discussing the properties of ecosystems, Reichle et al. (1975, pp. 31–32) state:

> Unless populations in the ecosystem contribute to specific vital system functions such as photosynthesis or cycling of nutrients, strong negative selective pressure is exerted upon the individual populations by the system. Whatever individual population responses are established, the populations establish homeostatic feedback mechanisms so that the ecosystem survives and grows to a maximum persistent biomass.

Discuss the evolutionary implications of these statements.

25.8 Tilman et al. (1982, p. 367) state:

> We suggest that the spatial and temporal heterogeneity of pelagic environments will prevent us from meaningfully addressing questions on short time scales or small spatial scales.

Discuss the general issue of whether there are some questions in community ecology that we cannot answer because of scale.

25.9 Crop productivity has improved greatly during the past 60 years. Some of this improvement in production is due to genetic changes in the crops, and some is due to increased nutrients or water. How could one evaluate the relative contribution of these two components to increasing primary productivity in a particular crop? Boyer (1982) discusses this problem and provides some data for major U.S. crops. Calderomo and Slafer (1998) discuss the historical trends in wheat yields.

25.10 Many studies of nutrient limitation in freshwater lakes and in the ocean use small water bottles as experimental units to which nutrients of various types are added. Other aquatic ecologists use mesocosms of plastic that hold several cubic meters of water for their experiments. Discuss why these small-scale experiments might give less reliable results than whole-lake manipulations. Schindler (1998) discusses the problem for freshwater lakes.

25.11 Photosynthetic organisms produce about 300×10^{15} g of oxygen per year (Holland 1995). If this oxygen accumulated, the oxygen content of the atmosphere would double every 2000 years. Why does this not happen? Is the global system regulated? If so, how is this regulation accomplished?

Overview Question

What limits primary production in agricultural systems? List the differences in the controls of primary production in natural plant communities and agricultural crops, and discuss the implications for sustainable agriculture.

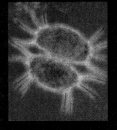

CHAPTER 26

Ecosystem Metabolism II:
Secondary Production

W̲E HAVE SEEN THAT PRIMARY PRODUCTION by green plants drives the entire biosphere. Primary production in agriculture is a particularly important part of global ecosystem metabolism, and we now ask how the energy fixed by green plants is dissipated by all the other parts of the food chain in both natural and agricultural ecosystems.

Measurement of Secondary Production

The biomass of plants that accumulates in an ecosystem as a result of photosynthesis can eventually move to one of two fates: consumption by herbivores or degradation by detritus feeders. The fate of the energy and materials captured in primary productivity can be illustrated most simply by looking at the metabolism of an individual herbivore.

The partitioning of food materials and energy for an individual animal can be seen as a series of dichotomies. With respect to energy, we have the energetic approach, championed by Eugene Odum, working at the University of Georgia, and his brother Howard Odum, that has formed the basis of major insights in how ecosystems work in both natural and agricultural systems:

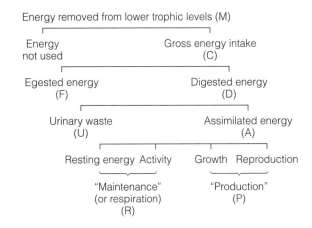

This scheme could be presented for carbon intake or any essential nutrient, and again we have the choice of using chemical materials or energy to study the system.

Let us look at this scheme in detail. Every animal will remove some energy or material from a lower trophic level for its food. Some of this energy will not be used, as for example, when a beaver fells a whole tree and eats only some of the bark. In this case, most of the energy removed from the plant trophic level is not used by the beaver but is left to decompose. Of the material consumed, some energy passes through the digestive tract (is egested) and is lost in the feces. Of the remaining digested energy, some is lost as urinary output, and the rest is available for assimilation or metabolic

Eugene P. Odum *(1913–) Professor of Ecology, University of Georgia*

energy. Assimilated energy can be subdivided into two general pathways, maintenance and production. All animals must expend energy in the process of respiration just to subsist. Production occurs by using assimilated energy for growth and for reproduction.

How can we measure the components of secondary productivity in an animal community? Several techniques are available (Petrusewicz and MacFadyen 1970), but the general procedure is as follows. Each species of animal is considered separately. To determine the gross energy intake of the population, we must know the feeding rate of individuals. This can be measured by confining a herbivorous animal to a feeding plot and measuring herbage biomass before and after feeding. In some predators, such as birds of prey, the number of food items being consumed can be counted by direct observation. Indirect techniques such as weight of stomach contents can also be used but require knowledge of the rates of digestion and feeding.

Assimilated or metabolizable energy can be measured very simply in the laboratory where gross intake can be regulated and feces and urine can be collected, but in the field it is extremely difficult to estimate assimilation directly. The usual approach is to measure it indirectly by use of the relation

$$\text{Assimilation rate} = \text{Respiration rate} + \text{net productivity} \quad (26.1)$$

If we can measure the rates of respiration and production, we can get assimilation rate by addition. Note that all these are rates (per unit time), and that assimilation rate is also called *gross productivity*.

Respiration can be measured very easily in laboratory situations by confining an animal to a small cage and measuring oxygen consumption, CO_2 output, and heat production directly. There is a minimum rate of metabolism, the basal metabolic rate, which increases with body size in warm-blooded animals. The relationship between basal metabolic rate and body size is not constant in all taxonomic groups, contrary to what is often reported (Hayssen and Lacy 1985). For example, within the mammals, bats of a given body size have lower basal metabolic rates than rodents. Basal metabolism is measured under conditions in which the animal is at rest and has no food in its stomach, at a temperature at which the animal is not required to expend energy for extra heat production or cooling. Measured in such an abstract way, basal metabolism is not closely related to respiration losses in field situations, in which activity is necessary, temperature varies, and digestion is occurring.

Energy metabolism in the field can be measured directly by means of doubly labeled water (Nagy 1987). This method involves injecting wild animals with water in the form $^3H_2O^{18}$ and then measuring the loss rates of the hydrogen (tritium) and oxygen isotopes. The hydrogen isotope relates to water loss, whereas the oxygen isotope is lost in both CO_2 and water. The difference between these isotope loss rates represents CO_2 loss alone, and is a field measure of metabolic rate (Box 26.1). Figure 26.1 shows how field metabolic rates vary with body size in mammals and birds. No single regression line describes this relationship for all groups of vertebrates, although for medium-sized endotherms (250–550 g) the field metabolic rates are similar for birds and mammals. For example, the regression for herbivorous mammals is

$$\log (\text{FMR}) = 0.774 + 0.727(\log \text{ body mass}) \quad (26.2)$$

where *FMR* = field metabolic rate (kJ/day/individual), and body mass is in grams.

The energetic costs of living for birds and mammals is very high compared with that for reptiles (see Figure 26.1d). A 250-g mammal or bird will use about 320 kJ of energy daily, whereas a 250-g iguanid lizard will use about 19 kJ per day, a 17-fold difference in energy use. Respiration in cold-blooded organisms is highly dependent on temperature, as well as body size.

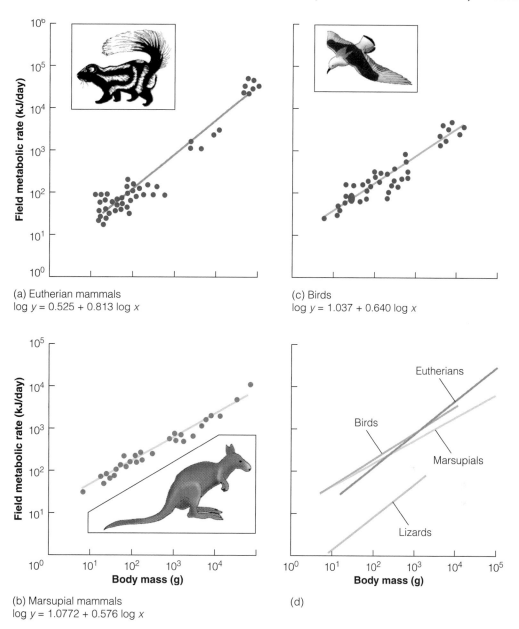

(a) Eutherian mammals
log y = 0.525 + 0.813 log x

(c) Birds
log y = 1.037 + 0.640 log x

(b) Marsupial mammals
log y = 1.0772 + 0.576 log x

(d)

FIGURE 26.1
Field metabolic rate in relation to body size for (a) eutherian mammals, (b) marsupial mammals, and (c) birds. In (d) the regression lines for these three groups are superimposed and contrasted with that for lizards. (From Nagy 1987.)

Net production in a population can be measured by the growth of individuals and the reproduction of new animals. We have discussed techniques for measuring changes in numbers in Chapter 9, and growth can be measured by weighing individuals at successive times. The only admonition we must make here is that sampling must be frequent enough so that individuals are not born and then die in the interval between samples. Net production is usually measured as biomass and converted to energy measures by determination of the caloric value of a unit of weight of the species.

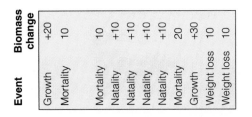

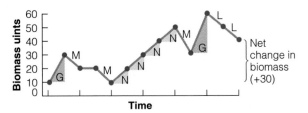

BOX 26.1

ESTIMATING ENERGY EXPENDITURE WITH DOUBLY LABELED WATER

Estimating energy expenditure in free-living animals has always been a Holy Grail for ecologists interested in bioenergetics. In 1966 N. Lifson and R. McClintock suggested that turnover rates of body water labeled with radioactive isotopes could be used to measure energy balance in animals. They showed that one could do this with doubly labeled water, $^2H_2\ ^{18}O$ or $^3H_2\ ^{18}O$. The method operates schematically in an individual animal, as follows:

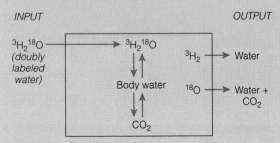

Tritium (^3H) leaves the body only in water, whereas ^{18}O leaves the body both in water and in carbon dioxide in respiration. We can get the amount of energy utilized by the difference:

$$^{18}O \text{ elimination} - {^3H_2} \text{ elimination} = CO_2 \text{ production}$$

In addition to measuring energy utilization, the doubly labeled water method can measure water turnover rates.

In practice this method is applied by injecting a known amount of the two isotopes, waiting 12–24 hours to allow the isotopes to mix throughout the body, and then taking a blood sample to define the start of the experiment. The animal is released and then recaptured after a specified time period, and a second blood sample is taken to define the end of the experiment. The isotope concentrations in the blood samples are measured on a liquid scintillation counter for tritium and proton activation analysis, or by mass spectrometry for ^{18}O.

The length of time over which a doubly labeled water experiment can be conducted depends on the turnover of the two isotopes. For studies of mammals and birds, the optimal metabolic interval is 7–21 days, but for reptiles and amphibians with lower metabolic rates, 1–2 months can be used for the sampling interval. By reinjecting individuals, field metabolic rates can be determined for long time periods (Nagy 1989).

FIGURE 26.2

Changes in biomass of a hypothetical population to illustrate the factors that contribute to secondary production. The net change in biomass over a given time period is the outcome of gains from net growth and reproduction and losses from death and emigration.

Figure 26.2 gives a schematic representation of how production can be determined from information on population changes. In this hypothetical population, production is the sum of growth and natality additions:

$$\text{Production} = \text{growth} + \text{natality} \qquad (26.3)$$

$$= 20 + 10 + 10 + 10 + 30 - 10 - 10$$

$$= 70 \text{ units of biomass}$$

Note that we can calculate production in a second way:

$$\text{Production} = \frac{\text{net change}}{\text{in biomass}} + \frac{\text{losses by}}{\text{mortality}} \qquad (26.4)$$

$$= 30 + 40 = 70 \text{ units of biomass}$$

Losses caused by mortality (including harvesting) or emigration are a part of production and should not be ignored. We can see this very clearly by looking at a population that is stable over time (net change in biomass zero): A ranch that has the same biomass of steers this year as last year does not necessarily have zero production over the year.

Now let's us look at an actual example of the calculations of the secondary productivity in an African elephant population. Petrides and Swank (1966) estimated the energy relations of the elephant herds in Queen Elizabeth National Park (now Ruwenzori

TABLE 26.1 **Elephant life table and production data, Queen Elizabeth (now Ruwenzori) National Park, Uganda, November 1956 to June 1957.**

Age, x	No. alive at beginning of age x	Mortality rate	Median no. alive	Weight average (kg)	Weight increment (kg)	Population weight increment[a] (kg)
1	1,000	0.03	850.0	91	91	77,350
2	700	0.02	630.0	205	114	71,820
3	560	0.10	532.0	318	114	60,648
4	504	0.10	478.5	455	136	60,076
5	453	0.10	430.5	614	159	68,450
6	408	0.10	387.5	795	182	70,525
7	367	0.10	348.5	1000	205	71,443
8	330	0.10	313.5	1205	205	64,268
9	297	0.10	282.0	1409	205	57,810
10	267	0.05	260.5	1614	205	53,403
11	254	0.05	247.5	1818	205	50,738
12	241	0.05	235.0	2023	205	48,175
13	229	0.05	223.5	2205	182	40,677
14	218	0.05	212.5	2386	182	38,675
15	207	0.02	205.0	2591	205	42,025
16	203	0.02	201.0	2795	205	41,205
17	199	0.02	197.0	3000	205	40,385
18	195	0.02	193.0	3182	182	35,126
19	191	0.02	189.0	3386	205	38,745
20	187	0.02	185.0	3591	205	37,925
21	183	0.02	181.0	3750	159	28,779
22	179	0.02	177.0	3864	114	20,178
23	175	0.02	173.0	3955	91	15,743
24	171	0.02	169.5	4045	91	15,425
25	168	0.02	166.5	4091	45	7,493
26–67	3,107[b]	0.02–0.20	3,026.5[b]	4091	0	0
Total	**10,933**		**10,487.5**	**2,291[c]**	**4,091**	**1,162,084**

[a]Median no. alive times weight increment.
[b]Sum of the numbers of animals in each year-class for ages 26 to 67.
[c]Average body weight.
Source: After Petrides and Swank (1966).

National Park) in Uganda. To do these calculations, we must assume a stable population of elephants with a stationary age distribution. For convenience we will also assume that no births or deaths occur during the study interval. The population was counted and the age structure estimated to construct a life table (see p. 133). The maximum age was estimated at 67 years, and the survivorship schedule is given in Table 26.1. Weight growth was estimated from some records of zoo animals and a limited amount of field data on weights. The average weight at each age and the average increase in weight from one year to the next are also given in Table 26.1. If the age structure is stationary, then

$$\begin{matrix} \text{Growth in} \\ \text{biomass} \end{matrix} = \sum_{\text{all ages}} \begin{pmatrix} \text{median no. alive} \\ \text{during age} \\ x \text{ to } x + 1 \end{pmatrix} \times \begin{pmatrix} \text{average weight} \\ \text{growth for ages} \\ x \text{ to } x + 1 \end{pmatrix} \quad (26.5)$$

This is calculated in the last column of Table 26.1. The caloric value of elephants is 6.276 kJ/g of live weight. We determine growth as follows:

From the bottom line in Table 26.1, 1000 elephants lived 10,487.5 elephant-years and produced 1,162,084 kg of growth.

$$\text{Average growth (in weight)/elephant/yr} = \frac{1,162,084}{10,487.5} = 110.8 \text{ kg}$$

$$\text{Average growth (in energy)/elephant/yr} = 110,800 \text{ g} \times 6.276 \text{ kJ/g}$$

$$= 695,381 \text{ kJ}$$

The population density of elephants was 2.077 elephants/km^2 or 0.000002077 elephant/m^2. Thus

$$\text{Growth} = 695,381 \text{ kJ} \times 0.000002077 = 1.44 \text{ kJ/m}^2/\text{yr}$$

A large amount of the food consumed by elephants passes through as feces. From studies on captive elephants, an average 2273-kg (5000-lb.) elephant would consume 23.59 kg dry weight of forage per day and produce from this 13.25 kg dry weight of feces. The food plants are worth approximately 16.736 kJ/g dry weight, so we can calculate

$$\text{Average consumption} = 23,590 \times 16.736$$

$$= 394,802 \text{ kJ}$$

$$\text{Average fecal production} = 13,250 \times 16.736$$

$$= 221,752 \text{ kJ}$$

Counting these for a whole year and multiplying by the number of elephants per square meter, we obtain:

$$\text{Food consumed} = 394,802 \text{ kJ/day/elephant} \times 365 \text{ days} \times 0.000002077 \text{ elephant/m}^2$$

$$= 299 \text{ kJ/m}^2/\text{yr}$$

$$\text{Feces produced} = 221,752 \text{ kJ/day/elephant} \times 365 \text{ days} \times 0.000002077 \text{ elephant/m}^2$$

$$= 168.1 \text{ kJ/m}^2/\text{yr}$$

We know that

$$\text{Food energy consumed} = \text{feces} + \text{growth} + \text{maintenance}$$

$$299 = 168.1 + 1.44 + \text{maintenance}$$

Maintenance must be about 130 kJ/m^2/yr if we ignore the losses due to urine production and the production due to newborn animals.

We can also estimate maintenance from equation (26.2) for the field metabolic rate of a standard 2273-kg (5000-lb.) elephant:

$$\text{log (field metabolic rate)} = 0.774 + 0.727 \text{ log (weight)}$$

$$= 0.774 + 0.727 \text{ log (2,273,000 g)}$$

$$= 5.395248$$

or

$$\text{Field metabolic rate} = 248,455 \text{ kJ/day/elephant}$$

$$\text{Estimated maintenance} = 248,455 \text{ kJ/day} \times 365 \text{ days} \times 0.000002077 \text{ elephants/m}^2$$

$$= 188 \text{ kJ/m}^2/\text{yr}$$

This is slightly greater than the maintenance estimate of 130 kJ/m^2/yr previously obtained.

Finally, we can determine the standing crop of elephants in energetic terms:

$$\text{Standing crop} = 0.000002077 \text{ elephant/m}^2 \times 2,291,000 \text{ g} \times 6.3 \text{ kJ}$$

$$= 30 \text{ kJ/m}^2$$

A rough estimate of primary productivity by the harvest method produced an estimate of net primary productivity of 3125 kJ/m^2/yr for the foraging area of the elephants.

We can summarize these estimates for energy dynamics of the African elephant population of Queen Elizabeth Park, as follows:

	Energy (kJ/m^2/yr)
Net primary production	3125[a]
Secondary production	
Food consumed	299
Fecal energy lost	168
Maintenance metabolism	130
Growth	1.44
Standing crop of elephants	30

[a] Probably a low estimate; compare Table 25.1, page 516.

Clearly, the greatest part of the energy intake of these elephants is used in maintenance or lost in fecal production.

The details of estimating secondary production obviously vary from species to species, and the number of assumptions that must be made depend on how well the species is studied. The procedure is to repeat these calculations for all dominant species in a community and by addition to obtain the secondary

ESSAY 26.1

THERMODYNAMICS AND ECOLOGY

The energetic approach to ecosystem science has a strong foundation in thermodynamics and in physics in general. Whatever happens in an ecosystem can be described as a transfer of energy from one place to another, or as a transformation of energy between different forms such as chemical energy to heat. In 1964 Eugene Odum dubbed energetics the "new ecology," and his brother Howard T. Odum has been a champion of the idea that energy is the central object of study in ecosystem ecology (Odum 1983, Patten 1993). The essential features of Howard Odum's approach to ecosystem analysis can be stated as five conjectures (Månsson and McGlade 1993):

Conjecture 1: All significant aspects of ecosystems can be captured by the single concept, energy.

Conjecture 2: The formalism of energy circuit language is sufficient for a holistic approach to ecosystem analysis. Odum invented energy cir-

cuit language as a substitute for linear mathematical models. A few examples of this symbolic language are shown below:

Conjecture 3: Ecosystems evolve such that power is maximized. This idea, the *maximum power principle*, is a new principle for the evolution of ecosystems.

Conjecture 4: Hierarchical structures can be deduced from the flow of energy in ecosystems.

Conjecture 5: Ecological succession is due to the maximum power principle and culminates in stable systems with maximum biomass and maximum gross production.

Even though these conjectures represented a bold attempt to bring thermodynamic theory into ecology, they have been superceded by new ecological concepts, and few ecologists now believe that energy can play such a central role in ecosystem analysis. Energy circuit language does not easily permit nonlinear systems analysis, which is now a common approach in defining system dynamics. There does not seem to be any evolutionary theory to support the maximum power principle as a guiding principle in ecosystem evolution, and ecological succession does not lead to maximum biomass and stable equilibrial systems. The complexities of interactions between species cannot be reduced to the single currency of energy.

The scope of energetics in the analysis of ecosystems has now become more limited than Howard Odum first proposed. Energetics has made an important contribution to ecological understanding, particularly with respect to the role of humans in ecosystems. Progress in science often involves the rejection of bold conjectures, and we gain deeper insights into how the world works via the process of conjecture and refutation.

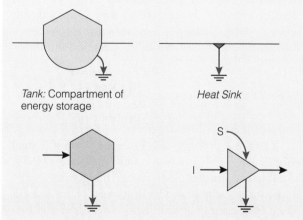

Tank: Compartment of energy storage

Heat Sink

Consumer: Transforms energy and stores it

Constant-gain amplifer: Delivers an output depending on input/and energy source *S*

production of the community. This procedure is more tedious than conceptually difficult, and we can now consider the results of this kind of analysis.

Problems in Estimating Secondary Production

In principle, ecologists can apply the kind of techniques just described for elephants to all the major consumer species in a community. In practice these calculations have led ecologists into three conceptual problems (Cousins 1987).

The first difficulty is that the individuals of a particular species do not fit clearly into discrete trophic levels. Plants are usually easily assigned to the producer level, trophic level 1. But the next trophic level includes animals that eat other animals as well as plants (Figure 26.3). House mice (*Mus musculus*) are thought of as herbivores, yet they consume substantial amounts of insects at some seasons of the year. The red fox (*Vulpes fulva*) eats herbivores such as rabbits, but they can also eat plant material, detritus, and other carnivores. In general, the higher up the food chain, the less clear it is how to categorize a species.

A second problem in estimating secondary production is what to do with detritus. The normal procedure has been to include detritus and dung in the first trophic level—to treat detritus as plant material—but this is not correct. Detritus from herbivores should be kept separate from plant detritus. Moreover, typically a complex food web exists within detritus itself, such that it is difficult to assign to the producer-herbivore-carnivore type of trophic organization.

The third major difficulty in estimating secondary production is the practical one of sampling a complex community adequately and in allowing for nonequilibrium conditions in natural ecosystems. Aquatic ecologists have been particularly aware of these problems (Kokkinn and Davis 1986), and the rapid changes that occur in plankton populations and benthic invertebrates may render accurate estimates of production impossible on a finite financial budget.

Given these problems, ecologists have moved away from trying to estimate secondary production for whole trophic levels, and have begun to analyze parts of food webs taxonomically (Cousins 1987). We can analyze single species like the elephant in detail to estimate their effects on the community. We now consider how efficient different species are in their use of energy.

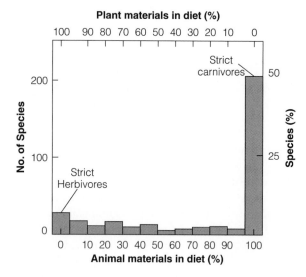

FIGURE 26.3
The composition of the diets of 430 species of North American birds. While 50% of birds are strictly carnivores and 8% are strict herbivores, over 40 percent are omnivores that feed on more than one trophic level. (Modified from Peters 1977.)

Ecological Efficiencies

If we view animals as energy transformers, we can ask questions about their relative efficiencies. A large number of ecological efficiencies can be defined (Kozlovsky 1968), and here we are first concerned with efficiency within a species on a particular trophic level. One useful measure of efficiency is defined as follows:

$$\text{Production efficiency} = \frac{\text{net productivity of species } n}{\text{assimilation of species } n} \quad (26.6)$$

Data on production efficiency for individual species are readily obtained, and Humphreys (1979, 1984) has summarized 235 energy budgets measured in natural populations. The first question we can ask about these energy budgets is whether different taxonomic groups have different efficiencies. Based on efficiency, homeotherms separate into four groups: insectivores, birds, small mammals, and other mammals; poikilotherms separate into three groups: fish and social insects, nonsocial insects, and other invertebrates. Within each of these groups, as respiration goes up 1 kJ, production likewise goes up by 1 kJ. This means that production efficiencies $P/(R + P)$ are the same for all sizes of animals within the seven

TABLE 26.2 Average production efficiencies (Equation 26.6) in order of increasing efficiency.

Group	Production efficiency (%)	No. of studies
Insectivores	0.86	6
Birds	1.29	9
Small mammals	1.51	8
Other mammals	3.14	56
Fish and social insects	9.77	22
Other invertebrates		
(excluding insects)	25.0	73
Herbivores	20.8	15
Carnivores	27.6	11
Detritivores	36.2	23
Nonsocial insects	40.7	61
Herbivores	38.8	49
Detritivores	47.0	6
Carnivores	55.6	5

Note: Data are from 235 natural populations. A breakdown into trophic groups is presented for two of the groups for which adequate data are available.
Source: After Humphreys (1979).

groups. Table 26.2 lists average production efficiencies and illustrates an important generalization in secondary production. For mammals and birds in general, respiration seems to utilize 97–99% of the energy assimilated, and consequently only 1–3% of the energy goes to net production in these groups. For insects the loss is less; approximately 59–90% of the energy assimilated is used for respiration. This difference between insects and mammals is a reflection of the cost of homeothermy. There appears to be no variation in production efficiency between animals in different habitats. Aquatic and terrestrial poikilotherms seem to have equal production efficiencies (Humphreys 1979).

Another measure of efficiency that we can use to describe ecosystems involves transfers between trophic levels and is called *trophic efficiency:*

$$\text{Trophic efficiency} = \frac{\text{net production at trophic level } i + 1}{\text{net production at trophic level } i} \quad (26.7)$$

This measure of the efficiency of energy transfer gives the fraction of production passing from one trophic level to the next (Pauly and Christensen 1995). The energy not transferred is lost in respiration or to detritus. For aquatic ecosystems, trophic efficiencies vary from 2% to 24% and average 10.1% (Figure 26.4). If we assume a trophic efficiency of 10%, we can calculate how much primary production is required to support a particular fishery. Consider the case of tuna caught in the open oceans. Tuna are top predators operating at trophic level 4, and in 1990 2,975,000 tons of tuna were taken, or 0.1 g carbon per m² of open ocean per year. To support this yield of tuna to the fishery, and assuming equilibrium conditions and trophic efficiency of 10%, we can calculate the

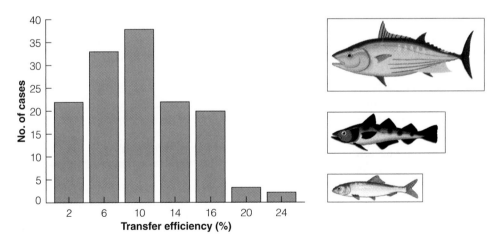

FIGURE 26.4
Frequency distribution of energy transfer efficiencies (Equation 26.7) for 48 trophic models of freshwater and marine aquatic ecosystems. The 140 estimates express for herbivores to carnivores the fraction of production passing from one trophic level to the next. The mean transfer efficiency is 10.1%. (From Pauly and Christensen 1995.)

production values of the other trophic levels in grams of carbon as follows:

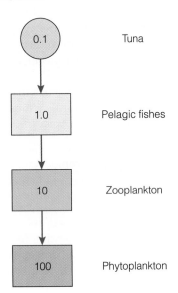

Note that these values are not standing crops but are production or yield values. For example, to provide 0.1 g C of tuna per m² we need to have 1 g C per m² of pelagic fishes to be eaten by the tuna, and 10 g C per m² of zooplankton to be eaten by the pelagic fishes, and finally 100 g C per m² of phytoplankton. Note that we do not know the standing crop in each box of this trophic chain, but instead only the production that comes out and moves up the chain. But if we know the net primary production of the plants, we can calculate what fraction of this production the tuna fishery is taking.

Using this approach, Pauly and Christensen (1995) aggregated all the data for the fisheries of the world and showed that on average 8% of global aquatic primary production was being used to produce the global fisheries catch. But this average masked high variation among different fisheries (Table 26.3). In continental shelf and upwelling ecosystems, fisheries harvest one-fourth to one-third of the net primary production, a very high fraction that leaves little margin for maintaining ecosystem integrity and a sustainable fishery (Pauly et al. 1998).

This analysis for aquatic ecosystems raises the question of whether terrestrial and aquatic ecosystems operate in the same way. Many terrestrial systems are dominated by decomposers, and most of the energy in the system flows through the decomposer link in the food web. Figure 26.5 illustrates this for a temperate deciduous forest. For a typical deciduous forest about 96% of the net primary production moves directly into dead organic matter and thence to the decomposers. This loss is greatly reduced at higher trophic levels, such that most of the production of herbivores is taken by carnivores and only 10% flows directly into the decomposer food chain (Hairston and Hairston 1993).

The amount of herbivory varies in different ecosystems. Herbivores in aquatic ecosystems consume a higher fraction of the primary production than they do in terrestrial ecosystems (Figure 26.6). Zooplankton in aquatic food webs consume an average of 79% of the net primary production of phytoplankton, whereas only 18% of the terrestrial primary production is eaten (Cyr and Pace 1993). Thus we can distinguish ecosystems dominated by grazers from those dominated by decomposers. Whittaker (1975) gives the following average values:

	Net primary production going to animal consumption (%)
Tropical rain forest	7
Temperate deciduous forest	5
Grassland	10
Open ocean	40
Oceanic upwelling zones	35

Thus in forest ecosystems, almost all of the primary production goes into the decomposer food chain. How do these differing consumption rates affect the standing crop of plants in different ecosystems? Lodge et al. (1998) summarized the reduction in standing crop of vascular plants in terrestrial, marine, and freshwater communities, citing the following averages:

	Reduction in standing crop of vascular plants from herbivores
Terrestrial ecosystems	26%
Marine ecosystems	65%
Freshwater ecosystems	31%

Herbivores reduce standing crops of plants in all systems by a significant amount. Thus, by excluding herbivores one would expect between a twofold and threefold larger effect in marine ecosystems compared with terrestrial ones.

Although much of the work on secondary production has centered on energy flow, an increasing amount of research on nutrient cycles is being done,

TABLE 26.3 **Global estimates of net primary production and the total catch to world fisheries (including discarded catches), and the calculated percent of primary production required to support the observed fishery catches.**

Data from 1988–1991 were used in these estimates.

Ecosystem type	Area ($10^6 km^2$)	Net primary production ($g\,C\,m^{-2}yr^{-1}$)	Fishery catch[a] ($g\,C\,m^{-2}yr^{-1}$)	Primary production required (%)
Open ocean	332.0	103	0.012	1.8
Upwellings	0.8	973	25.560	25.1
Tropical shelves	8.6	310	2.871	24.2
Temperate shelves	18.4	310	2.306	35.3
Coastal/reef systems	2.0	890	10.510	8.3
Rivers and lakes	2.0	290	4.300	23.6
Weighted means		126	0.330	8.0

[a]Includes an estimated 25% discards that are not counted in official fishery catch statistics.
Source: From Pauly and Christensen (1995).

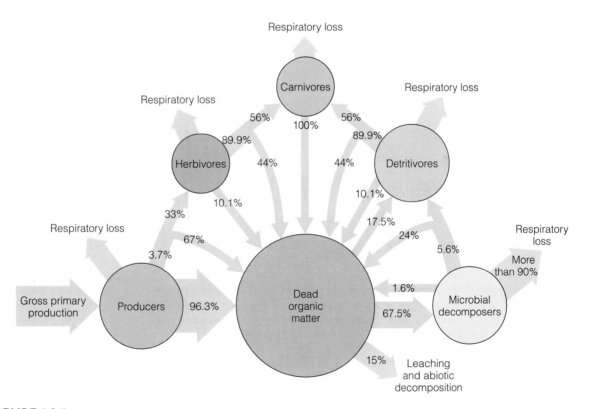

FIGURE 26.5
Estimates of energy flow through a temperate deciduous forest ecosystem. The fraction of energy going to the next trophic level (gross energy intake) is subdivided into two arrows representing egested energy and assimilated energy. The vast majority of the net primary production in temperate deciduous forests goes directly into detritus and then to decomposers, and about two-thirds of the organic matter is in the detritus pool. (From Hairston and Hairston 1993.)

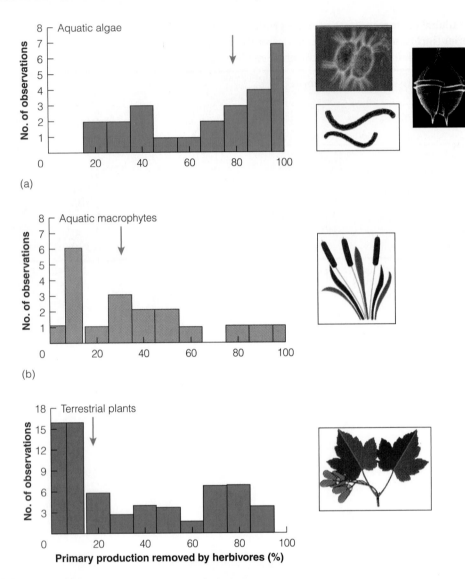

FIGURE 26.6

Percentage of net primary production removed by herbivores in ecosystems dominated by (a) algae (phytoplankton), (b) rooted aquatic plants, and (c) terrestrial plants. Red arrows indicate average values. Herbivores have a significantly greater effect on phytoplankton than on aquatic plants or terrestrial plants. (From Cyr and Pace 1993.)

FIGURE 26.7

Pyramids of numbers of forest floor invertebrates in a tropical forest in Panama. Very few invertebrates in this ecosystem are over 5 mm in length. Each tick mark on the x-axis represents 500 individuals. (After Cousins 1985.)

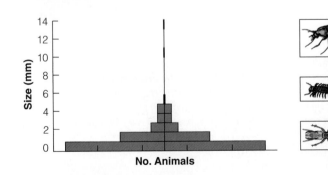

because work on individual populations has suggested that nutrients, not energy, may be limiting animal populations (Cousins 1987). We consider nutrient cycling in Chapter 27.

One consequence of low ecological efficiencies is that organisms at the base of the food web are much more abundant than those at higher trophic levels. Charles Elton recognized this in 1927, and when he put this observation together with the observation that predators are usually larger than the prey they consume, the result was a pyramid of numbers or of biomass that has been called an *Eltonian pyramid* in his honor. Figure 26.7 illustrates a pyramid of numbers of invertebrate individuals on a tropical forest floor in Panama. Note that pyramids can be constructed on the basis of numbers, biomass, or energy of standing crop. They illustrate graphically the rapid loss of numbers and biomass as one moves from smaller to larger animals in an ecosystem, a biological illustration of the second law of thermodynamics and the constraints of foraging (Cousins 1985).

What Limits Secondary Production?

This is one of the critical questions we need to answer. As a first approximation, we could state that secondary production is limited by primary production and by the second law of thermodynamics, which states that no process of energy conversion is 100% efficient. Figure 26.8 illustrates these broad patterns for a large array of 69 terrestrial communities from tundra to tropical forests. Herbivore biomass and consumption rise rapidly with increasing primary productivity, and secondary production increases with primary production in a 1:1 ratio (McNaughton et al. 1989). There is considerable scatter in the relationships shown in Figure 26.8, and, as one would predict from Figure 26.6, forest communities tend to fall below the regression line, whereas grassland communities fall above the line. To understand what limits secondary production in individual communities, we need to study the details of energy and nutrient flows. From 1965 to 1980 the International Biological Program (IBP) conducted out a series of studies of production in specific types of plant communities. Let us look at one example of these studies—grassland ecosystems—and then consider game ranching in Africa.

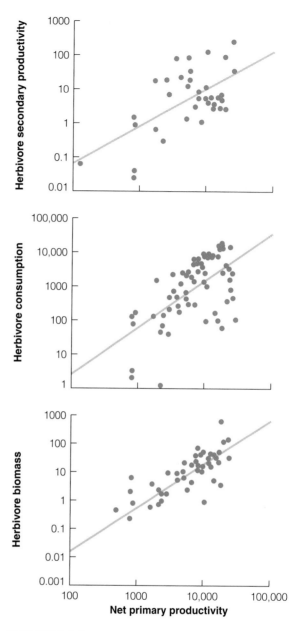

FIGURE 26.8
Relationships between net aboveground primary production and herbivore biomass, consumption, and net secondary productivity. Secondary production increases with primary production in a 1:1 ratio. Biomass measured as kJ/m², all others as kJ/m²/yr. Data from 69 studies from arctic tundra to tropical forests. (From McNaughton et al. 1989.)

Grassland Ecosystems

Grassland is the potential natural vegetation on 25% of the Earth's land surface. Grasslands occur in a great

FIGURE 26.9
Distribution of North American grasslands and the intensive study sites of the International Biological Program. (From French 1979.)

Bunchgrass
Annual
Desert
Shortgrass
Mixed
Tallgrass

diversity of climates and are defined by having a period of the year when soil water availability falls below the requirement for forest. In 1968 the IBP began a series of studies on grassland sites throughout the world (Coupland 1979). The grassland biome study of the North American International Biological Program was a major attempt to analyze how natural grasslands work (French 1979). Twelve sites were studied intensively in North America (Figure 26.9).

The primary productivity of grassland increases with precipitation, but the relevant variable is periodic drought, which arises as a combination of temperature and moisture. Grasslands range from arid desert grasslands with nearly continuous drought to tropical humid grasslands with nearly no drought. The average primary production of these six climatic types ranges from 100 g to 600 g dry weight/m^2/yr (Figure 26.10). Within these ranges, North American grasslands tend to fall at the lower end (Lauenroth 1979):

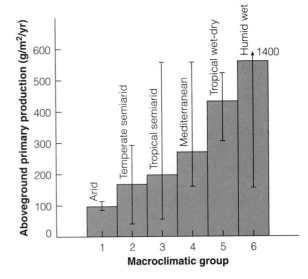

FIGURE 26.10
Net primary aboveground production for grasslands from six different climatic types. Vertical lines indicate the range of values within each group. (After Lauenroth 1979.)

	Aboveground net primary production (g dry wt/m^2/yr)
Tallgrass prairie	500
Annual grassland	400
Mixed grassland	300
Shortgrass prairie	200
Bunchgrass and desert grassland	100

Eltonian pyramids were calculated to determine the distribution of biomass among producers, consumers, and decomposers. Figure 26.11 illustrates the trophic pyramid for a tallgrass prairie site. Little variation

between years occurred in these pyramids, and the most striking finding was that below ground biomass in these grasslands greatly exceeds that above ground.

Herbivores and carnivores were grouped into taxonomic groups in order to summarize consumption and production figures for these ecosystems. Figure 26.12 shows the energy flow through tallgrass and shortgrass prairie sites. Several results stand out in these data. The most conspicuous animals are the least important energetically. Birds and mammals contribute

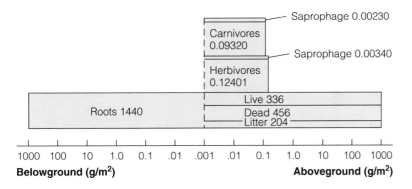

FIGURE 26.11
Eltonian pyramid as biomass for a tallgrass prairie site for the height of the growing season in mid-July. The base of the pyramid represents biomass (g dry wt/m²) of producers; the middle level, herbivores; and the top level, carnivores. Aboveground (right) and belowground (left) biomass are separated by the vertical dashed line. In this grassland, herbivores are only 0.04% of the aboveground biomass of living plants. Much of the plant biomass is below ground. (After French et al. 1979.)

E S S A Y   2 6 . 2

WHY IS THE WORLD GREEN?

The world is green, according to the *green world hypothesis*, because herbivores are held in check by their predators, parasites, and diseases such that they cannot consume all the plant biomass. On land about 83×10^{10} metric tons of carbon is tied up in plant biomass, and about 5×10^{10} metric tons of plant matter are produced each year (Polis 1999). Only 7–18% of this production on land is consumed by herbivores, so it is quite correct to say that grazing by herbivores plays a comparatively minor role for land plants on a global scale. But this conclusion must be tempered by the observation that because on occasion herbivores such as the gypsy moth do indeed destroy their plant resources, it is possible for the world not to be green. How can we reconcile these observations?

There are at least six reasons why the world is green:

1. *Plants are not passive agents waiting to be eaten.* Plants contain much woody lignin as well as many secondary compounds that inhibit herbivores. Not all that is green is edible.

2. *Nutrients limit herbivores, not energy.* Nutrients such as nitrogen are critical for animals and are often in short supply in plant materials (White 1993). Even in a world full of green energy, many herbivores cannot obtain sufficient nutrients to grow and reproduce.

3. *Abiotic factors limit herbivores.* Seasonal changes in temperature, precipitation, and other climatic factors depress herbivore numbers.

4. *Spatial and temporal heterogeneity reduce the availability of plants.* The world is not uniform; herbivores must search for food plants and cannot always locate them efficiently.

5. *Herbivores limit their own numbers.* Self-regulation through intraspecific competition—territoriality, cannibalism, or other forms of interference competition—can limit the numbers of some herbivores.

6. *Enemies limit herbivore numbers.* This limitation is the primary one suggested by the *green world hypothesis*. Enemies are effective in some communities in which predators, parasites, and diseases limit herbivores and prevent them from consuming all green plants. But not all herbivores are so limited, and the previous five mechanisms also act on predators to limit their numbers. On a global scale enemies may not be the most important limitation on herbivores.

All six of these mechanisms act in varying ways to limit herbivores, and the key issue for any particular plant community is the relative importance of each of these factors. No one mechanism by itself explains why the world is green.

FIGURE 26.12
Primary and secondary production and consumption for a tallgrass prairie and a shortgrass prairie in the United States. All values are expressed as energy (kcal/m²/yr). (a) Key to the components of energy flow measured. (b) Tallgrass data. (c) Shortgrass prairie. Much of the energy flow in these systems occurs under ground. (After Scott et al. 1979.)

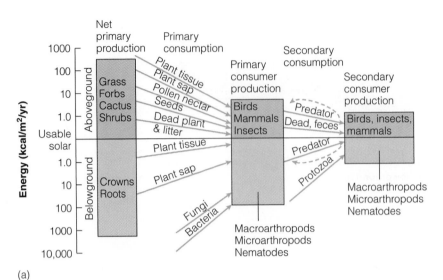

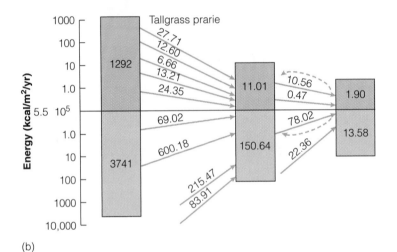

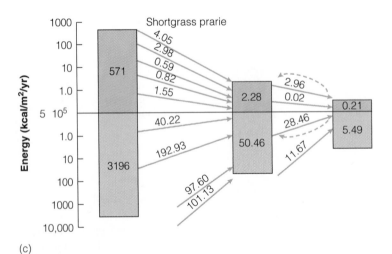

TABLE 26.4 Plant production eaten and wasted by herbivores, and herbivore production eaten and wasted by carnivores, for four grassland sites in the western United States.

	Desert grassland (%)	Shortgrass prairie (%)	Mixed (%)	Tallgrass prairie (%)
Plant Production				
Eaten by herbivores				
Aboveground	4.3	1.7	3.6	6.5
Belowground	—	7.3	26.4	17.9
Eaten and wasted				
Aboveground	6.3	3.4	8.4	10.2
Belowground	—	13.0	41.1	28.9
Herbivore Production				
Eaten by carnivores				
Aboveground	110.9[a]	119.7[a]	51.1	85.4
Belowground	—	50.9	76.1	47.5
Eaten and wasted				
Aboveground	120.5[a]	124.5[a]	60.5	97.2
Belowground	—	61.0	91.3	57.0

[a]Because consumption cannot exceed 100%, these values are too high, possibly because consumption was slightly overestimated or production underestimated.
Source: Scott et al. (1979).

almost nothing to production and very little to consumption (Scott et al. 1979). Birds consume 0.05% of the aboveground primary production in tallgrass prairie sites, and mammals consume 2.5%. The aboveground insects consume a greater amount, but the most important consumers are the soil animals—especially nematodes, which consume more than half of all plant tissue consumed. Nematodes are a major factor controlling total grassland primary production.

Only a small fraction of the primary production in grasslands is consumed by animals. Table 26.4 shows that, aboveground, only 2–7% of primary production is eaten by herbivores, but below ground, 7–26% is eaten. In contrast, virtually all of the secondary production by herbivores seems to be eaten by carnivores in grasslands. Predators may control consumer populations, at least above ground. Both production and consumption percentages increase in moving from shortgrass to tallgrass prairie.

The hypothesis that emerges from this analysis of grassland ecosystems is that grassland plants may be limited by nematode consumption of roots, by soil water, and by competition for nutrients and light, whereas consumer populations are controlled by predators (Scott et al. 1979).

This hypothesis was tested on a shortgrass prairie by experimentally providing water and nitrogen on 1-ha plots for six years (Dodd and Lauenroth 1979). Primary production increased dramatically in both treatments involving irrigation, especially in the water-plus-nitrogen plots (Figure 26.13). Nematode numbers increased about fourfold on the irrigated and water-plus-nitrogen treated areas. Small mammals (ground squirrels, voles, and mice) responded dramatically to the

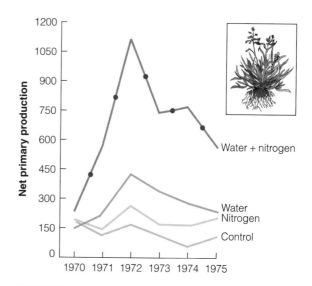

FIGURE 26.13
Net primary production over six years on plots of shortgrass prairie in north-central Colorado subjected to nitrogen fertilization, irrigation, and a combination of irrigation and fertilization. (After Dodd and Lauenroth 1979.)

water-plus-nitrogen plots because of the increased herbage cover. These experimental results tend to confirm the hypothesis that both water and nitrogen limit production in grassland ecosystems.

Game Ranching in Africa

The savanna and plains areas of Africa support a great diversity of ungulates, and the conservation of these ecosystems has been a primary concern of conservationists during the past 25 years. The history of human settlement on most continents has repeatedly involved the replacement of wild animals by domestic cattle and sheep, and during the 1950s Africa seemed next on the list. Against this background, Fraser Darling proposed in 1960 that in many areas of Africa, game animals were more productive than domestic cattle and sheep, and that a sustained yield of game would be more profitable than a sustained yield of cattle or sheep. Game-cropping or game-ranching schemes tried in many parts of Africa during the past 20 years provide a practical example of the problem of what limits secondary production in ecosystems.

Game ranching has been attractive to conservationists because it seems to rest firmly on an interlocking set of theoretical postulates (Caughley 1976b). The argument can be summarized as follows: African wildlife has evolved within its ecosystems for millions of years, and thus is uniquely adapted to the African environment. The numerous herbivores in Africa therefore use the vegetation more efficiently and are more productive than a cattle or sheep monoculture would be. Thus, African wildlife should attain a higher biomass than cattle or sheep on native African ranges. Furthermore, natural selection must have ensured that the different game species partition their food resources such that competition is reduced (Figure 26.14). So the diverse complex of wild species should cause less overgrazing than cattle or sheep and also be more resistant to disease. In brief, African wildlife should provide a higher sustained yield and higher net revenue to the African people.

Caughley (1976b) makes two comments on this theory. First, it is theoretically sound and eminently reasonable; second, attempts to demonstrate its validity since the 1960s have been unsuccessful. In no case has a sustained yield of game animals been shown to be more valuable economically than a sustained yield of livestock in a comparable area. How can we resolve the paradox within Caughley's comments?

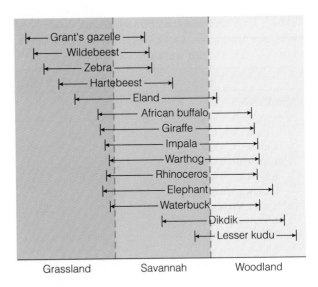

FIGURE 26.14
Habitat partitioning and overlap among 14 species of ungulates in the Serengeti region, Tanzania. Widespread niche overlap in habitat use occurs among the Serengeti ungulates. (From MacNab 1991.)

The initial enthusiasm for game ranching in Africa was derived from estimates of the carrying capacity of wild herbivores in African habitats (Petrides and Swank 1966). *Carrying capacity* is defined as the number or weight of animals of a single or mixed population that can be supported permanently on a given area (Sharkey 1970). Unfortunately, carrying capacity means different things to ecologists and economists. *Ecological carrying capacity* is the maximum density of animals that can be sustained in the absence of harvesting without inducing negative effects on vegetation, whereas *economic carrying capacity* is the density of animals that enables maximal sustained harvesting and is always lower than the ecological carrying capacity (Caughley 1976b). The confusion over these two types of carrying capacity has nullified some of the comparisons made between natural and domestic ecosystems in Africa.

Unfortunately, the early comparisons showing a higher ecological carrying capacity for wildlife than for livestock have been found to be in error, and, contrary to what Petrides and Swank (1966) reported, the carrying capacity of domestic ecosystems is not necessarily different from that of wildlife ecosystems (Walker 1976). Table 26.5 lists the standing crop of some wildlife and pastoral areas in the African savanna. In all these ecosystems, the standing crop of

TABLE 26.5 Standing crop, energy expenditure, and production of large herbivores from 30 savanna areas of east and southern Africa.

Locality	Large herbivore biomass (kg/km²)	Energy expenditure (kJ/km²/hr)	Production (live weight) (kJ/km²/acre)	Annual precipitation (mm)
Wildlife areas				
1. Rwindi Plain, Albert National Park, Zaire	17,448	62,479	1,936	863
2. Ruwenzori National Park, Uganda	19,928	76,253	2,554	1,010
3. Bunyoro North, Uganda	13,261	43,979	1,145	1,150
4. Manyara National Park, Tanzania	19,189	65,905	2,405	915
5. Ngorongoro Crater, Tanzania	7,561	56,105	1,503	893
6. Lake Nakuru National Park, Kenya	6,688	35,498	1,409	878
7. Amboseli Game Reserve, Kenya	4,848	22,970	934	350
8. Lochinvar Ranch, Zambia	7,568	40,591	1,983	813
9. Lake Rudolf (East) Kenya	405	2,436	87	165
10. Samburu-Isiolo, Kenya	2,018	9,745	402	375
11. Nairobi National Park, Kenya	4,824	24,728	1,008	844
12. Tsavo National Park (East, north of Voi R., Kenya)	4,033	12,968	351	553
13. Tsavo National Park (East, south of Voi R., Kenya)	4,388	15,239	438	553
14. Mkomasi Game Reserve, Tanzania	1,731	6,154	147	425
15. Loliondo Controlled Area, Tanzania	5,423	26,962	1,134	784
16. Serengeti National Park, Tanzania	8,352	43,063	1,743	803
17. Ruaha National Park, Tanzania	3,909	12,474	364	625
18. Akagera National Park, Rwanda	3,980	19,251	871	785
19. Sengwa Wildlife Research Area, Zimbabwe	4,315	18,299	722	597
20. Henderson's Ranch, Zimbabwe	2,869	15,148	684	406
21. Kruger National Park, northern section, South Africa	984	4,414	162	312
22. Kruger National Park, southern section, South Africa	3,783	20,552	884	650
23. William Pretorius Nature Reserve, South Africa	3,344	18,427	765	520
24. Mfolosi Game Reserve, South Africa	4,385	17,017	605	650
Pastoral Areas				
25. Mandera district, Kenya	1,901	—	—	228
26. Wajir district, Kenya	1,151	—	—	218
27. Garissa district, Kenya	3,818	—	—	398
28. Turkana district, Kenya	2,406	—	—	330
29. Samburu district, Kenya	6,514	—	—	500
30. Kaputei district, Kenya	7,884	—	—	710

Source: After Coe et al. (1976).

large herbivores, secondary production, and primary production are related to rainfall. Coe et al. (1976) found that the standing crop of herbivores on pastoral areas is always greater than that predicted for wildlife areas. Figure 26.15 shows that the same occurs in South America; for a given rate of plant production,

about ten times more secondary production occurs in areas stocked with cattle (Oesterheld et al. 1992).

Attempts have been made to domesticate wild ungulates as a superior alternative to cattle. One example is the eland (*Taurotragus oryx*), a large antelope that is easy to tame and requires less water than

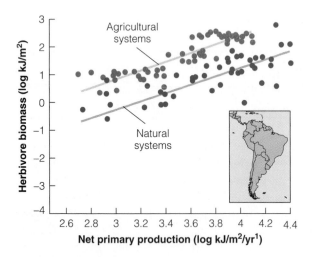

FIGURE 26.15
Relationship between net aboveground primary productivity and herbivore biomass for 51 natural ecosystems and 67 agricultural ecosystems for southern South America. (From Oesterheld et al. 1992.)

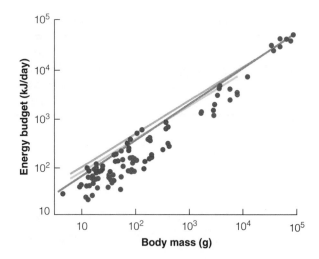

FIGURE 26.16
Field energy budgets of mammals (blue) and birds (purple) compared with three regression estimates of the maximum possible energy budgets for homeotherms. Some animals appear to be operating at maximum intensity, with no safety margin. (From Weiner 1992.)

cattle. The eland has a higher reproductive rate than domestic cattle and has excellent carcass characteristics for marketing. Domesticated herds of eland have been established in Kenya, South Africa, Zimbabwe, Zambia, and the former Soviet Union. A series of studies on meat production and growth efficiency of the eland (Skinner 1972) shows that it is less efficient than cattle at converting plants into meat. If the eland is to be used effectively in game ranching, it should be used in arid regions (King and Heath 1975). In general, wild herbivores are not as efficient as cattle for meat production (Walker 1976).

Game ranching is thus not an efficient way to provide cheap meat to low-income native peoples. From a conservation viewpoint, game ranching may have adverse effects. Geist (1988) points out that the creation of a market for game meat provides an outlet for illegal trade in these species. From an economic viewpoint, game ranching in fenced areas can provide luxury products (meat) and services (tourism) to foreigners at a substantial economic advantage, but private game farms typically do not promote conservation of the entire community. Large predators are not welcome on game farms, and grassland species are often preferred over forest species. The net result is that private game farms conserve only a few species in the ecosystem (MacNab 1991).

Sustainable Energy Budgets

The energy budgets of herbivores and carnivores combine to determine the energy budget of the entire community. Taxonomic approaches to energy transfers have focused on the constraints that determine the maximum energy budgets of animals (Weiner 1992). Food limitations could operate in two quite different ways for animals. The absolute amount of high-quality food may be limited, and the limitations of food could operate on both the time animals have to search and their searching strategy. Alternatively, food may be limiting because of physiological limits on the rate of food conversion into usable energy (Karasov 1986).

The upper limit to sustainable energy budgets is set by the capacity of animals' digestive tracts to assimilate nutrients and energy from food. In vertebrates, the gut consists of several compartments where food can be stored, digested, and absorbed. In ruminants, the fore stomach is a fermentation chamber in which a microbial flora breaks down the cellulose in plant cell walls.

The estimation of maximal rates of energy expenditure in animals is difficult. Drent and Daan (1980) estimated that for birds a maximal energy expenditure was approximately four times the basal metabolic rate;

Kirkwood (1983) estimated maximal energy expenditures of about five times basal metabolic rate; and Weiner (1992) estimated maximal sustained energy budgets of seven times basal metabolism. Figure 26.16 illustrates some field energy budgets for birds and mammals measured by the doubly labeled water technique, in relation to estimated regressions for the maximum sustainable energy budget. Lactating mammals and cold-exposed birds have provided the type of data used to estimate these lines (Weiner 1992,

Kirkwood 1983). It seems likely that maximum energy budgets vary seasonally and perhaps with age, so Figure 26.16 should be viewed as a first approximation only.

The integration of studies on energy use by animals and the physiological constraints that operate on energy acquisition will help community ecologists to understand the mechanisms behind the efficiency of secondary productivity in different ecosystems (Karasov 1986).

Summary

The organic matter produced by green plants is used by a food web of herbivores and carnivores. A great deal of the matter and energy that animals eat is lost in feces and urine, or is used for maintenance metabolism. In warm-blooded vertebrates, often 98% of energy intake is used for maintenance. Invertebrates and fish use less energy for maintenance, typically 60–90% of the energy taken in.

Data on the efficiency of secondary production of individual species is relatively easy to obtain, but the aggregation of this information into trophic levels is problematic. Many species of animals do not feed on only one trophic level. Typical "herbivores" may obtain part of their energy from other animals, and typical "carnivores" may feed on plants, herbivores, and other carnivores. The trophic level concept cannot be mapped directly onto species. Omnivory becomes more common at higher trophic levels.

Considerable energy is lost at each step of the food chain, and thus for a given biomass of green plants, only a much smaller biomass of animals can be supported. Many of the animal species that humans consider important constitute a small component of the energy flow in communities. Herbivores consume a higher fraction of the primary production in aquatic ecosystems than they do in forest or grassland ecosystems. Much of the energy flow in terrestrial systems goes directly from plants to the decomposer food chain.

Secondary production may be limited by a variety of interacting factors. Water and nitrogen limit secondary production in grasslands, and a large fraction of secondary production occurs under ground. Levels of secondary production in agricultural grazing ecosystems are up to ten times higher than those achieved in natural grazing ecosystems. Until we understand the factors that limit primary and secondary production, we cannot predict the effects of environmental changes on a community. To date we have such an understanding for only a few communities.

Key Concepts

1. Energy fixed by green plants flows to either herbivores or detritus, or is lost in respiration.

2. In forest ecosystems, most primary production goes directly into detritus, and only 3–4% goes into the herbivore food web. Aquatic ecosystems are more productive, and typically 20% or more of primary production is consumed by herbivores.

3. Homeotherms use more than 98% of their ingested energy to maintain body temperature, and are much less efficient energy users than are poikilotherms such as insects.

4. Secondary production is limited by primary production and the second law of thermodynamics, which states that no energy transfer is completely efficient.

5. About 5–20% of the energy passes from one trophic level to the next; the remainder is lost to respiration or goes to detritus.

6. Most of the animals humans consider important are a trivial part of an energetics of the ecosystem. Plants and detritus are the main players in all ecosystems.

7. Individual animals have a maximum sustainable energy budget about 5–7 times the basal metabolic rate. This constraint sets an upper limit on energy use.

Selected References

Burns, T. P. 1989. Lindeman's contradiction and the trophic structure of ecosystems. *Ecology* 70:1355–1362.

Cohen, J. E. 1994. Marine and continental food webs: Three paradoxes? *Philosophical Transactions of the Royal Society of London, Series B*, 343:57–69.

Cyr, H. and M. L. Pace. 1993. Magnitude and patterns of herbivory in aquatic and terrestrial ecosystems. *Nature* 361:148–150.

Cousins, S. 1987. The decline of the tropic level concept. *Trends in Ecology and Evolution* 2: 312–316.

French, N. R. 1979. *Perspectives in Grassland Ecology.* Springer-Verlag, New York. 204 pp.

Hairston, N. G., Jr. and N. G. Hairston, Sr. 1993. Cause-effect relationships in energy flow, trophic structure, and interspecific interactions. *American Naturalist* 142:379–411.

MacNab, J. 1991. Does game cropping serve conservation? A reexamination of the African data. *Canadian Journal of Zoology* 69:2283–2290.

McNaughton, S. J., M. Oesterheld, D. A. Frank, and K. J. Williams. 1989. Ecosystem-level patterns of primary productivity and herbivory in terrestrial habitats. *Nature* 341:142–144.

Nagy, K. A. 1987. Field metabolic rate and food requirement scaling in mammals and birds. *Ecological Monographs* 57:111–128.

Pauly, D. and V. Christensen. 1995. Primary production required to sustain global fisheries. *Nature* 374:255–257.

Power, M. E. 1992. Top-down and bottom-up forces in food webs: Do plants have primacy? *Ecology* 73:733–746.

Strayer, D. 1988. On the limits to secondary production. *Limnology and Oceanography* 33:1217–1220.

Weiner, J. 1992. Physiological limits to sustainable energy budgets in birds and mammals: Ecological implications. *Trends in Ecology and Evolution* 7:384–388.

Questions and Problems

26.1 In assessing the ways in which ecologists have studied ecosystems, O'Neill et al. (1986, pg. 67) state:

> It is now well recognized that the trophic-level concept is most useful as a heuristic device and tends to obscure, rather than illuminate, organizational principles of ecosystems.

Read O'Neill et al. (1986) and discuss this claim.

26.2 Does the concept of a maximum sustainable energy budget as illustrated in Figure 26.16 (p. 556) imply an evolutionary assumption that all animals will be selected to operate at maximum energy budgets? Discuss the evolution of assimilation efficiency for a particular species of your choice.

26.3 In discussing the reality of trophic levels, Murdoch (1966a, p. 219) states:

> Unlike populations, trophic levels are ill-defined and have no distinguishable lateral limits; in addition, tens of thousands of insect species, for example, live in more than one trophic level either simultaneously or at different stages of their life histories. Thus trophic levels exist only as abstractions, and unlike populations they have no empirically measurable properties or parameters.

Discuss.

26.4 Compile a list of the efficiency of some of our common physical machines, such as automobiles, electric lights, electric heaters, and bicycles.

26.5 How would it be possible to have an inverted Eltonian pyramid of numbers in which, for example, the standing crop of large animals is larger than the standing crop of smaller animals? In what types of communities could this occur? Do Eltonian pyramids apply to both animals and plants? Del Giorgio et al. (1999) discuss these issues.

26.6 How does the answer to the question *What limits secondary production?* differ from the answer to the question of whether trophic structure is controlled top-down or bottom-up in communities (see p. 496)?

26.7 Why should agricultural grazing systems be more efficient than natural grazing systems at utilizing primary production? Read MacNab (1991) and Oesterheld et al. (1992), and then discuss the problems of measuring efficiency in both kinds of ecosystems.

26.8 The basis for estimating secondary production is the estimation of population size, biomass, and growth, and the accuracy of any estimate of production depends on the accuracy of these three measurements. Read Morgan (1980) and then discuss the relative difficulty of measuring these three variables in freshwater ecosystems for zooplankton, benthic invertebrates, and fish.

26.9 How would you expect trophic level biomass to change as the primary productivity of the community increases? Review the hypotheses of community organization discussed in Chapter 24, and discuss the assumptions underlying your predictions. Compare your expectations with those of Power (1992).

26.10 Lodge et al. (1998) found that in freshwater ecosystems nonvascular plant biomass was reduced nearly 60% by herbivores, whereas vascular plant biomass was reduced only 30% on average. Discuss two reasons why this might occur.

26.11 Could herbivores remove a high fraction of the net primary production in an ecosystem without depressing the standing crop of plants? How might this happen?

26.12 Population density (no. individuals per m^2) of all organisms in all ecosystems falls with increasing body size, so that larger animals are less common. But for species of equal body size, aquatic organisms are 10–20 times more abundant in lakes than terrestrial organisms on land. Suggest two reasons why this might be. Cyr et al. (1997) discuss this issue.

Overview Question
Debate the following resolution: Resolved, that ecological efficiencies between trophic levels cannot in principle be measured, and that ecological efficiencies can be measured only for populations.

CHAPTER 27

Ecosystem Metabolism III:
Nutrient Cycles

L IVING ORGANISMS ARE COMPOSED of chemical elements, and one way to describe an ecosystem is to follow the transfer of chemical elements between the living and the nonliving worlds. Interest in the nutrient content of plants and animals has been an important focus in agriculture for over 100 years. Nutrients often set some limitation on the primary or secondary productivity of a population or a community, and nutrient additions as fertilizer have become increasingly common in agriculture and forestry. In this chapter, we consider how nutrients cycle and recycle in natural systems and in the process link together the living and the dead material in the ecosystem.

Nutrient Pools and Exchanges

Nutrients can be used as an organizing focus in ecosystem studies. We can view the biological community as a complex processor in which individuals move nutrients from one site to another within the ecosystem. These biological exchanges of nutrients interact with physical exchanges, and for this reason nutrient cycles are also called *biogeochemical cycles*. Chemical elements that cycle through living organisms are called *bioelements*. Figure 27.1 illustrates the general pattern of bioelement or nutrient cycles on a global scale. Nutrient cycles are closed on a global scale but are open on a local scale. The individual

atoms that make up the cycle are indestructible and can be recycled in plants and animals. Ecologists are interested in understanding and measuring global nutrient cycles because human activities are altering these cycles, with possible effects on global climate. An analysis of nutrient cycling thus ends with an assessment of human impacts on nutrient cycles and their consequences for animals and plants.

Global nutrient cycles represent the summation of local events occurring in different biotic communities, and to make progress in understanding global nutrient cycles we must begin at the level of the local community. A simple example of a nutrient cycle in a lake is depicted in Figure 27.2. All nutrients reside in *compartments*, which represent a defined space in nature. Compartments can be defined very broadly or very specifically. Figure 27.2 includes all of the plants in the ecosystem as one compartment, but we could recognize each species of plant as a separate compartment, or even the leaves and the stem of a single plant as separate compartments. A compartment contains a certain quantity, or *pool*, of nutrients in the standing crop. In the simple lake ecosystem shown in Figure 27.2, the phosphorus dissolved in the water is one pool, and the phosphorus contained in the bodies of herbivores is another pool.

Compartments exchange nutrients, and thus we must measure the uptake and outflow of nutrients for each compartment. The rate of movement of nutrients between two compartments is called the *flux rate* and is measured as the quantity of nutrient passing from one pool to another per unit of time. The flux

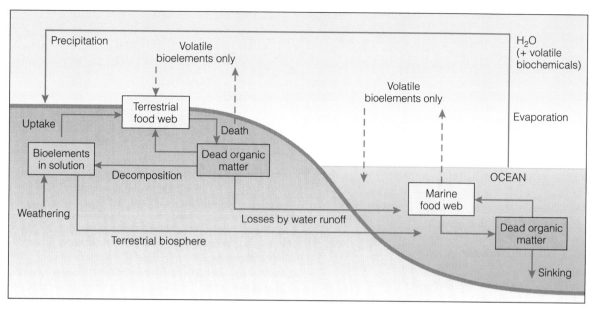

FIGURE 27.1
General schematic of nutrient cycling on a global scale. Movement of nonvolatile elements, such as phosphorus, is largely one way, toward ocean sediments. (From DeAngelis 1992.)

rates and pool sizes together define the nutrient cycle within any particular ecosystem. Ecosystems are not isolated from one another, and nutrients come into an ecosystem through meteorological, geological, or biological transport mechanisms and leave an ecosystem via the same routes. Meteorological inputs include dissolved matter in rain and snow, atmospheric gases, and dust blown by the wind; geological inputs include weathering and elements transported by surface and subsurface drainage; and biological inputs include movements of animals between ecosystems.

Nutrient cycles can be studied by introducing radioactive tracers into laboratory or natural ecosystems. Studies on the movement of radioactive phosphorus in small aquariums illustrate this approach (Whittaker 1975). Figure 27.3 shows the changes that followed the introduction of 100 microcuries (μc) of ^{32}P-labeled phosphoric acid in a 200-liter aquarium. There is a very rapid initial uptake of ^{32}P by phytoplankton. One-half of the ^{32}P had been taken up by the phytoplankton within two hours, and within 12 hours an equilibrium of uptake and excretion of ^{32}P occurred between the phytoplankton and the water. Filamentous algae attached to the sides and bottom of the aquarium slowly picked up ^{32}P, and crustaceans grazing on phytoplankton began to accu-

mulate ^{32}P even more slowly. As the experiment progressed, an increasing proportion of the radioactive tracer began to accumulate in the bottom mud and was tied up in a less active or bound form in the sediments. Some ^{32}P does move from the sediment back into the water column, but more moves down into the sediment and accumulates.

The nutrient cycle of phosphorus in aquariums is generally similar to that in natural lakes. Phosphorus and other nutrients tend to accumulate in the sediment of lakes such that continual nutrient inputs are required to maintain high productivity. The pattern of movement of phosphorus shown in Figure 27.3 helps to explain the results of the lake-fertilization experiments described on page 524. These results are also critical for understanding how lakes can recover from the effects of nutrient additions from pollution (see Figure 23.24). Thus, the sediments of lakes become nutrient-rich deposits.

Nutrient cycles may be subdivided into two broad types. The phosphorus cycle just described is an example of a sedimentary or *local* cycle, which operates within an ecosystem. Local cycles involve the less-mobile elements (nonvolatile elements of Figure 27.1) that have no mechanism for long-distance transfer. By contrast, the gaseous cycles of nitrogen, carbon,

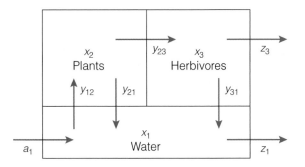

x_1 = amount of P in water
x_2 = amount of P in plants
x_3 = amount of P in herbivores
a_1 = rate of inflow of P in water
z_1 = rate of outflow of P in water
z_3 = rate of outflow of P in herbivores
y_{12} = rate of uptake of P from water by plants
y_{21} = rate of loss of P from plants to water
y_{23} = rate of uptake of P from plants by herbivores
y_{31} = rate of loss of P from herbivores to water

(a)

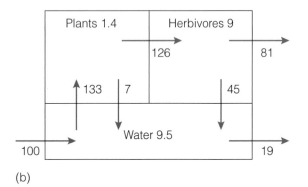

(b)

FIGURE 27.2
Hypothetical nutrient cycle for phosphorus in a simple lake ecosystem composed of three compartments: plants, herbivores, and water. (a) Definition of compartments and flux rates (inflow or outflow). Compartments are standing crops or amounts. (b) Hypothetical distribution (mg) and flux rates (mg/day) of phosphorus after equilibration to a constant input rate of 100 mg/day. (After Smith 1970.)

oxygen, and water (volatile elements) are called *global* cycles because they involve exchanges between the atmosphere and the ecosystem. Global nutrient cycles link together all of the world's living organisms in one giant ecosystem called the *biosphere*, the entire Earth ecosystem.

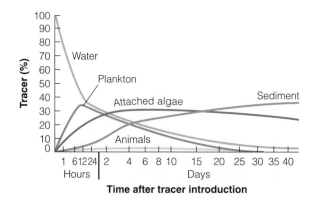

FIGURE 27.3
Movement of radiophosphorus in an aquarium microcosm. Percentage of the tracer present at a given time (after correction for radioactive decay) is on the vertical axis, time after tracer introduction (on a square-root scale) is on the horizontal. Phosphorus accumulates in the sediment in aquaria, as it does in lakes. (After Whittaker 1961.)

Nutrient Cycles in Forests

The harvesting of forest trees removes nutrients from a forest site, and this continued nutrient removal could result in a long-term decline in forest productivity unless nutrients are somehow returned to the system. Because of the economic importance of forest productivity, an increasing amount of work is being directed toward the analysis of nutrient cycles in forests. Figure 27.4 depicts the factors that must be quantified in order to describe the nutrient cycle in a forest. Some examples of nutrient cycles in forest stands will illustrate these concepts.

Nutrient budgets for forest ecosystems attempt to balance the inputs and outputs of nutrients in the system under study. The key to nutrient budgets is the mass balance approach described by the simple equation

$$\text{Inputs} - \text{outputs} = \Delta \text{ storage} \qquad (27.1)$$

The term *change in storage* implies that if the equation is not balanced, then nutrients must either be accumulating somewhere in the ecosystem, such as in leaves or stems, or declining somewhere, such as in the soil. Tracing these inputs and outputs through the ecosystem is the essence of nutrient cycling.

During forest development, nutrients accumulate in leaves and wood. Figure 27.5 illustrates the rapid

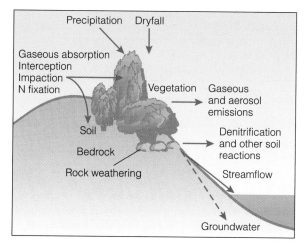

Figure 27.4
A schematic illustration of the pathways of nutrient movement through undisturbed forest ecosystems. To quantify nutrient cycling, all these pathways must be measured. (From Waring and Schlesinger 1985.)

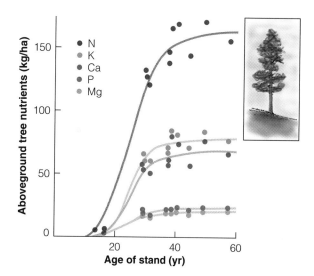

FIGURE 27.5
Accumulation of nutrients during the postfire development of jack pine (Pinus banksiana) *stands in New Brunswick, Canada. More nitrogen accumulates in trees than do other nutrients, and nitrogen is often the limiting factor for forest tree growth. (From MacLean and Wein 1977.)*

TABLE 27.1 **Aboveground accumulation of organic matter and nitrogen in trees for various forest regions.**

Forest region	No. of sites	Organic matter (kg/ha)			Nitrogen (kg/ha)		
		In trees	Total	Above ground (%)	In trees	Total	Above ground (%)
Boreal coniferous	3	51,000	226,000	19	116	3,250	4
Boreal deciduous	1	97,000	491,000	20	221	3,780	6
Temperate coniferous	13	307,000	618,000	54	479	7,300	7
Temperate deciduous	14	152,000	389,000	40	442	5,619	8
Mediterranean	1	269,000	326,000	83	745	1,025	73
Average		208,000	468,000	45	429	5,893	7

Source: From Cole and Rapp (1981).

accumulation of five different nutrients in a stand of jack pine in eastern Canada. As trees increase in size during succession, the soil accumulates nutrients in the surface litter and in the soil organic matter (humus) dispersed through the upper soil horizons. As forest stands age, a systematic change in the uptake of nutrients occurs. Figure 27.6 illustrates changes in nitrogen cycling in a spruce forest in Russia. In this forest the spruce canopy becomes more open after 70 years, and understory vegetation increases in volume and importance in nutrient cycling. Not all forest successions produce the same pattern of changes in nutri-

ent cycling, but in general nutrient cycling varies with forest age. In old growth forests, little accumulation of nutrients occurs, and ecosystem inputs and outputs should be balanced.

The cycling of different nutrients in forest ecosystems is highly variable (Cole and Rapp 1981). Nutrients that are in short supply are recycled far more efficiently than those present in excess of requirements. In most forest sites, nitrogen is a major limiting factor and is present at deficiency levels. Table 27.1 lists the organic matter and nitrogen amounts present in the aboveground component of

B O X   2 7 . 1

ESTIMATING TURNOVER TIME FOR NUTRIENTS

How long does a molecule of phosphorus stay in a lake? This simple question is difficult to answer because ecosystems have many compartments with different turnover times (*residence times*). Estimating turnover time for nutrients in compartments can be achieved most easily using radioactive tracers. It is useful to consider a very simple situation to clarify what ecologists mean by turnover time (DeAngelis 1992).

Consider a single compartment or pool with a constant inflow and outflow (an equilibrial system):

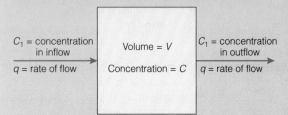

This compartment could represent an entire lake, or an individual organism in a lake. The rate of flow (*q*) is expressed as liters per minute (or some similar rate); concentrations are expressed as grams of nutrient per liter (or some similar measure). Given rapid mixing in the compartment, the outflow concentration will equal that inside the compartment. If the system starts with an inflow concentration of C_1, then it will eventually equilibrate with a concentration in the compartment of C_1.

Given that this is a steady state system, we can ask what is the expected time of residence of a molecule entering the pool? This quantity, the *residence time*, is given by the simple equation

$$T_{res} = \frac{C_1 V}{C_1 q} = \frac{V}{q} \qquad (27.4)$$

where

T_{res} = mean residence time = mean turnover time

V = volume

The mean residence time can be thought of as a measure of the rate of flushing. In particular, if the input of nutrients is stopped such that $C_1 = 0$, the concentration of nutrients in the compartment will decline from that moment according to the equation

$$C_t = C_0 e^{-(q/V)t} \qquad (27.5)$$

where

C_t = concentration of nutrient at time t

C_0 = concentration of nutrient in the compartment at the instant input ceases

t = time after nutrient input ceases

This simple model of exponential decay is useful for obtaining an estimate of recovery time for a compartment that has been subjected to pollution input. If the system is not in equilibrium, turnover time is more difficult to estimate. For lakes, turnover time of nutrients in sediments differs greatly from that of nutrients in the water column. For any ecosystem it is crucial to define compartments carefully.

---

32 forests from five different climatic zones. In the boreal forests of Alaska, only about 20% of the organic matter in the trees is present above ground. Low decomposition rates in these cold Alaskan forests cause most of the nitrogen and organic matter to be tied up in the soil. Coniferous forests have the largest forest floor accumulation of organic matter of all forests, on average about four times the biomass of tropical forests (Waring and Schlesinger 1985).

Nutrient cycles operate more quickly in warmer forests than in colder ones. If we assume as an approximation that forest soil nutrients are in equilibrium for the short time they can be studied (three to ten years), then for each nutrient we can calculate average turnover

time, the time an average atom will remain in the soil before it is recycled into the trees or shrubs (see Box 27.1). Table 27.2 gives the mean turnover times for five elements. All northern forests have very slow turnover of nutrients. On average, boreal conifer forests retain nitrogen 100 times longer than a Mediterranean evergreen oak forest (Cole and Rapp 1981). Deciduous forests turn over nutrients more rapidly than coniferous forests. Coniferous forests use nutrients more efficiently because they retain their needles and do not need to replace all their foliage each year.

Nutrients are lost from forest ecosystems in several ways. Streams transport both dissolved and particulate matter, and measurements of streamwater

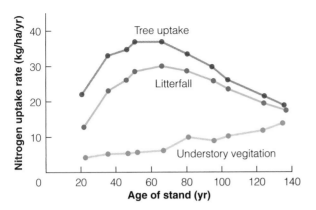

FIGURE 27.6
Uptake and cycling of nitrogen in spruce (Picea abies) *from the ages of 22 years to 138 years. The rate of nitrogen uptake is maximal about 60 years of stand age, when tree growth is maximal. (After Kazimirov and Morozova 1973.)*

TABLE 27.2 Mean turnover time (in years) for five mineral elements in the forest floor for five forest regions.

A steady-state condition is assumed.

Forest region	No. of sites	Mean turnover time (yr)					
		Organic matter	N	K	Ca	Mg	P
Boreal coniferous	3	353	230.0	94.0	149.0	455.0	324.0
Boreal deciduous	1	26	27.1	10.0	13.8	14.2	15.2
Temperate coniferous	13	17	17.9	2.2	5.9	12.9	15.3
Temperate deciduous	14	4	5.5	1.3	3.0	3.4	5.8
Mediterranean	1	3	3.6	0.2	3.8	2.2	0.9
All stands	32	12	34.1	13.0	21.8	61.4	46.0

Source: From Cole and Rapp (1981).

chemistry can provide a good way to monitor overall forest function. Anaerobic soil bacteria produce methane and hydrogen sulfide gases. Plants release hydrocarbons such as terpenes from their leaves, and these compounds may add to atmospheric haze in summer. Both ammonia and hydrogen sulfide can be released from plant leaves (Waring and Schlesinger 1985). During forest fires, nutrients are released in both gases and in particles. Finally, forest harvesting removes nutrients in wood from the ecosystem.

One of the most extensive studies of nutrient cycling in forests was begun by Gene Likens, F. Herbert Bormann, and Noye M. Johnson in 1963 at the Hubbard Brook Experimental Forest in New Hampshire, a nearly mature, second-growth hardwood ecosystem. The area is underlain by rocks that are relatively impermeable to water, so almost all runoff occurs in small streams. The area is subdivided into

Gene E. Likens *(1935–) Director, Institute of Ecosystem Studies, Millbrook, New York*

TABLE 27.3 **Average concentrations of various dissolved substances in precipitation and streamwater for undisturbed watersheds 1–6, Hubbard Brook Experimental Forest, 1963–1969.**

	Precipitation (mg/L)	Streamwater (mg/L)
Calcium	0.21	1.58
Magnesium	0.06	0.39
Potassium	0.09	0.23
Sodium	0.12	0.92
Aluminum	—[a]	0.24
Ammonium	0.22	0.05
Sulfate	3.10	6.40
Nitrate	1.31	1.14
Chloride	0.42	0.64
Bicarbonate	—[a]	1.90
Dissolved silica	—[a]	4.61[b]

[a]Not determined, but very low.
[b]Watershed 4 only.
Source: After Likens and Bormann (1972).

several small watersheds that are distinct yet support similar forest communities, and these watersheds are good experimental units for study and manipulation.

Nutrients enter the Hubbard Brook forest ecosystem in precipitation, and the precipitation input was measured in rain gauges scattered over the study area. Nutrients leave the ecosystem primarily in stream runoff, and this loss was estimated by measuring stream flows. Table 27.3 gives the concentrations of dissolved substances in precipitation and streamwater for the Hubbard Brook watersheds. For most dissolved nutrients, the streamwater leaving the system contains more nutrients than the rainwater entering the system. About 60% of the water that enters as precipitation leaves as stream flow; most of the remaining 40% is transpired by plants or evaporates. The chemical composition of the precipitation and the stream discharges changed very little from year to year.

Annual nutrient budgets for watersheds in the Hubbard Brook system can be estimated from the difference between precipitation input and stream outflow. Eight ions show net losses from the ecosystem: calcium, magnesium, potassium, sodium, aluminum, sulfate, silica, and bicarbonate. Three ions showed an average net gain: nitrate, ammonium, and chloride. If we assume that these nutrient budgets should be in equilibrium in this ecosystem, then the net losses must be made up by chemical decomposition of the bedrock and soil particles.

With this background, Bormann et al. (1974) studied the effect of logging on the nutrient budget of a small watershed at Hubbard Brook. When one 15.6-ha watershed was logged in 1966, the logs and branches were left on the ground; nothing was removed from the area. Great care was taken to prevent disturbance of the soil surface to minimize erosion. For the first three years after logging the area was treated with a herbicide to prevent any regrowth of vegetation. This deforested watershed was then compared with an adjacent, intact watershed.

Runoff in the small streams increased immediately after the logging, and annual runoff in the deforested watershed averaged 32% more than the control watershed in the three years after treatment. Detritus and debris in the stream outflow increased greatly after deforestation, particularly two to three years after logging. Correlated with this increase was a large increase in streamwater concentrations of all major ions in the deforested watershed. Nitrate concentrations in particular increased 40- to 60-fold over the control values (Figure 27.7). For two years the nitrate concentration in the streamwater of the deforested site exceeded the health levels recommended for drinking water. Average streamwater concentrations increased 417% for calcium, 408% for magnesium, 1558% for potassium, and 177% for sodium in the two years after deforestation (Likens et al. 1970).

The net result of deforestation in the Hubbard Brook forest is that the ecosystem is simultaneously irrigated and fertilized, so that for a short time after normal logging occurs, primary production could be stimulated. An array of species has evolved to exploit

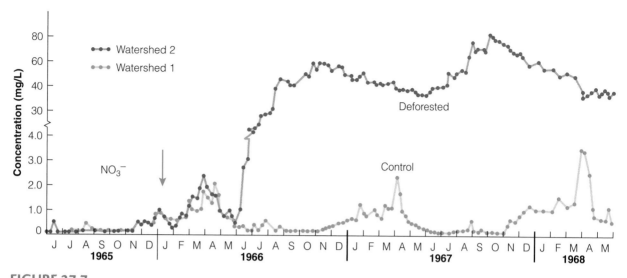

FIGURE 27.7

Streamwater nitrate concentrations in two watersheds in the Hubbard Brook Experimental Forest, New Hampshire. The red arrow marks the completion of the cutting of the trees on watershed 2. Note the change in scale for the nitrate concentration. Nitrate loss was greatly accelerated after logging. (After Likens et al. 1970.)

these transient nutrient-rich situations following a disturbance by fire or logging. These transients help to prevent further nutrient losses and to restore some of the nutrient capital lost by logging or fire.

The Hubbard Brook experiment has been repeated in four other forest types from Costa Rica to British Columbia (Table 27.4). Losses of nutrients in streamwater were high in a tropical rain forest in Costa Rica but were not large at three temperate forest sites. The total nutrient losses at these three sites were mostly due to the harvest of wood. Precipitation input of nutrients must be the main source of nutrient input to these forested systems; another input, rock weathering, operates very slowly. Table 27.4 gives the replacement times required for each nutrient. If harvesting occurs more frequently than nutrient replacement time, the site must undergo a net loss of nutrients unless fertilizers are used.

There is considerable controversy over whether or not harvesting of trees induces a long-term decline in forest productivity (Morris and Miller 1994). Unless the nutrients removed in the harvest of trees are replaced by natural sources or by fertilization, productivity may decline. To date, little evidence indicates that this is happening, but most experiments are of short duration, and the key is the long-term response of the ecosystem (Figure 27.8). Even though many management practices in forestry, such as slash removal or burning, can alter nutrient cycling, at present there is

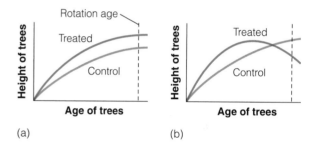

(a) (b)

FIGURE 27.8

Two potential patterns of forest tree growth in relation to management treatments such as fertilization or burning. (a) Long-term improvement in forest productivity. This is the model assumed by foresters. (b) Short-term improvement with long-term detrimental effect. Short-term studies may confuse (a) with (b). (Modified after Morris and Miller 1994.)

no evidence of decline in long-term forest productivity in temperate forests utilized for tree harvest.

Research on nutrient cycling in forests has shown the need for guidelines specifying sound management procedures in forestry. For example, bark is relatively rich in nutrients, and hence timber operations ought to be designed to strip the bark from the trees in the field, not at some distant processing plant. The conservation of nutrients in forest ecosystems can be done intelligently only when we understand how nutrient cycles operate in these systems.

TABLE 27.4 Comparative losses of nutrients following timber harvest in different ecosystems.

Locale/process	N	P	K	Ca	Reference
British Columbia					
(Western hemlock–Douglas fir)					
Losses (kg/ha)					
Harvested products	234	34	168	260	
Streamflow (2 yr)[a]	10	0	8	0	Feller and Kimmins
Total	244	34	172	260	(1984)
Precipitation input (kg/ha/yr)	4	0	1	7	
Replacement time (yr)[b]	61		176	37	
Minnesota (aspen)					
Losses (kg/ha)					
Harvested products	454	43	355	1034	
Streamflow (5 yr)[a]	0	0	0	62	Silkworth and
Total	454	43	355	1096	Grigal (1982)
Precipitation input (kg/ha/yr)	7	3	10	5	
Replacement time (yr)[b]	65	14	36	219	
Florida (slash pine)					
Losses (kg/ha)					
Harvested products	179	15	73	128	
Streamflow[a]	4		1	0	Pritchett and
Total	183	15	74	128	Comerford (1983);
Precipitation input (kg/ha/yr)	5	0	3	5	Riekirk (1983)
Replacement time (yr)[b]	37	75	25	26	
Costa Rica (tropical rain forest)					
Losses (kg/ha)					
Harvested products[c]	111	4		96	
Streamflow (11 mo.)[a]	329	11		392	Ewel et al. (1981);
Total	440	15		488	Lewis (1981)
Precipitation input (kg/ha/yr)	7	2	4	9	
Replacement time (yr)	63	8		54	

[a]Streamflow losses in excess of losses on control watersheds during the interval indicated.
[b]Replacement of losses assuming atmospheric inputs only.
[c]Firewood harvest only.

Efficiency of Nutrient Use

Large areas of the Northern Hemisphere have been glaciated, and the soils, derived from till in which the bedrock has been pulverized, are very fertile, with a high availability of nutrients. Areas of volcanic activity can also have rich soils. But in much of the world, soils are very old, highly weathered, and basically infertile. For example, the continents derived from Gondwanaland—Australia, South America, and India—have large areas covered by very old, poor soils. The vegetation supported on these soils has adapted remarkably well to efficient nutrient use by recycling within the plant and by leaf fall and reabsorption.

Australian soils are typical of very old, highly weathered soils and contain almost no phosphorus, and eucalyptus trees native to them are adapted to grow on soils of low phosphorus content (Attiwill 1981). Table 27.5 (on pages 570–571) gives the biomass and nutrient content of temperate forest ecosystems growing on good soils and from Australian sites on poor soils. There is no suggestion from Table 27.5 that eucalyptus trees growing on poor soils contain fewer nutrients than other tree species from other sites, with the single exception of phosphorus. Euca-

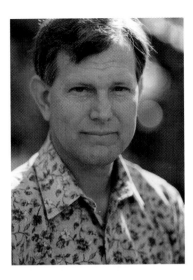

Peter M. Vitousek *(1949–) Professor of Ecosystem Ecology, Stanford University*

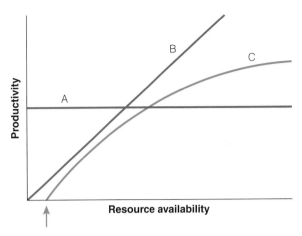

FIGURE 27.9
*Three models of nutrient use efficiency (*NUE*). Model A: Productivity is not related to the amount of nutrient available. This model implies higher nutrient use efficiency at lower resource levels. Model B: Productivity increases linearly with more nutrients. This model assumes a constant nutrient use efficiency at all resource levels. Model C: Productivity increases to an upper limit as resources increase, and there is a minimum level of nutrient needed before there is any production (red arrow). This model implies that nutrient use efficiency is higher at lower resource levels. (After Pastor and Bridgham 1999.)*

lyptus trees have only 20–50% the amount of phosphorus in their tissues compared to trees from Northern Hemisphere forests.

One might expect plants growing in nutrient-poor soils to contain fewer nutrients than plants in fertile soils. In fact, the opposite is true (Chapin 1980). Plants from infertile habitats consistently have higher nutrient concentrations than plants from fertile habitats when grown under the same controlled conditions. Plants from nutrient-poor habitats may achieve this nutrient-rich status by being more efficient than plants from nutrient-rich habitats. Peter Vitousek in 1982 proposed a measure of nutrient use efficiency that captures the different abilities of plants to take up a critical nutrient such as nitrogen and use it for the production of biomass. Vitousek's original formulation of nutrient use efficiency (*NUV*) was generalized by Berendse and Aerts (1987) as

$$NUE = (A)\left(\frac{1}{L}\right) \qquad (27.2)$$

where   *NUE* = nutrient use efficiency

   A = nutrient productivity (dry matter production per unit nutrient in the plant)

   L = nutrient requirement per unit of plant biomass

If we express the nutrient requirement on a relative basis (for example, as the amount of nitrogen uptake

needed to maintain each unit of nitrogen in a plant for a given time period), and if we assume a steady state in the plant community, we can estimate the mean residence time for the nutrient as

$$MRT = \frac{1}{L_n} \qquad (27.3)$$

where   *MRT* = mean residence time for a unit of nutrient in plant tissue

   L_n = relative nutrient requirement for maintenance

Figure 27.9 illustrates the concept of nutrient use efficiency. An abundance of data is available from forests to investigate nutrient use efficiency. We can use litterfall in a forest as a measure of aboveground primary production, and the amount of nutrients in litterfall as a measure of nutrient uptake.

Vitousek (1982, 1984) showed that nutrient use efficiency increases as nutrients become scarcer in tropical forests. Figure 27.10, which plots nitrogen availability against productivity for 90 tropical forest stands, and shows the law of diminishing returns—as

TABLE 27.5 Aboveground Living Biomass and its Nutrient Content of Temperate Forest and Australian Eucalyptus Forest Ecosystems.

Forest	Location	Age (yr)	Biomass (t/ha)	Nutrient Content (kg/ha)					
				N	P	K	Na	Mg	Ca
Coniferous									
1. *Abies amabilis— Tsuga mertensiana*	Washington	175	469	372	67	980	—	160	1046
2. *Abies balsamea— Picea rubens*	Quebec	various	132	387	52	159	—	36	413
3. *Abies mayriana*	Japan	?	129	527	63	278	—	415	515
4. *Cryptomeria japonica*	Japan	?	114	386	36	244	—	118	314
5. *Larix leprolepis*	Japan	?	100	230	28	212	—	40	107
6. *Picea abies*	England	47	140	331	37	161	5	39	212
7. *Picea abies*	Sweden	55	308	770	87	437	38	69	459
8. *Picea glauca*	Minnesota	40	151	383	58	230	—	41	720
9. *Picea mariana*	Quebec	65	107	167	42	84	—	27	276
10. *Pinus banksiana*	Ontario	30	81	171	15	85	—	19	114
11. *Pinus banksiana*	Minnesota	40	150	276	27	106	—	40	226
12. *Pinus maricata*	California	88 (mean)	409	510	418	321	83	140	589
13. *Pinus radiata*	New Zealand	35	305	319	40	324	—	—	187
14. *Pinus resinosa*	Minnesota	40	204	373	46	205	—	62	335
15. *Pinus sylvestris*	Southern	28	21	82	9	40	—	—	50
	Finland	45	79	194	19	98	—	—	115
		47	45	136	14	58	—	—	63
16. *Pinus sylvestris*	England	47	164	364	34	314	8	43	204
17. *Pseudotsuga menziesii*	Washington	36	173	326	67	227	—	—	342
18. *Pseudotsuga menziesii*	Oregon	450	540	371	49	265	—	—	664
19. *Sequoia sempervirens*	California (slopes)	?	1155	1474	616	941	102	232	1068
20. *Sequoia sempervirens*	California (flats)	?	3190	3846	1695	2412	271	569	2574

(continues)

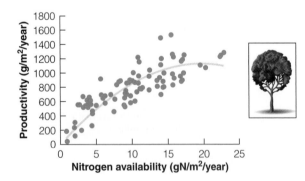

FIGURE 27.10

Nutrient use efficiency in tropical forests. The relationship between the amount of nitrogen available (indexed by nitrogen in the litter) and the productivity of the site (measured by the amount of litterfall) follows a curve of diminishing returns. Model C in Figure 27.9 is supported by these data, which show that tropical forests growing on low-nutrient sites use nitrogen more efficiently. (Data from Vitousek 1984.)

nitrogen becomes abundantly available, productivity reaches a plateau at which some other resource becomes limiting. Similar relationships occur for phosphorus in tropical forests. Forest productivity is limited either by nitrogen or by phosphorus in a variety of tropical and temperate forests, and the pattern of nutrient use efficiency is similar (Bridgham et al. 1995, Waring and Schlesinger 1985).

One consequence of this finding is that forest productivity may be high on soils with low nutrient levels. A classic example is the tropical rain forest of the Amazon Basin, which represents one type of nutrient cycling pattern (Jordan and Herrera 1981). The *oligotrophic* pattern occurs on nutrient-poor soils, such as those in the Amazon Basin, and the *eutrophic* pattern occurs on nutrient-rich soils. In the temperate zone, where most forest research has been done, forests are usually of the eutrophic type on rich soils. Oligotrophic ecosystems on poor soils constitute a much greater proportion of the tropics. But excep-

TABLE 27.5 *(continued)*

		Age (yr)	Biomass (t/ha)	Nutrient Content (kg/ha)					
Forest	Location			N	P	K	Na	Mg	Ca
Hardwood									
1. *Alnus rubra*	Washington	34	185	516	40	224	—	114	334
2. *Betula verrucosa*	England	22	63	264	33	76	3	36	327
3. *Betula verrucosa— Betula pubescens*	Southern Finland	40	91	232	24	136	—	—	185
4. *Betula platyphylla*	Japan	?	114	265	20	84	—	95	401
5. Evergreen broadleaf forest	Japan	?	114	329	52	249	—	94	259
6. *Fagus sylvatica*	England	39	134	285	38	187	4	42	151
7. *Fagus sylvatica*	Sweden	a. 90	314	800	53	458	24	118	980
		b. 90	324	1050	84	452	32	105	602
		c. 100	226	640	65	318	17	85	478
8. *Populus tremuloides*	Minnesota	40	170	383	48	297	—	62	881
9. *Quercus alba*	Missouri	35–92	100	204	20	115	—	35	601
10. *Quercus alba— Quercus rubra— Acer saccharum— Carya glabra*	Illinois	150	190	478	29	310	—	105	1603
11. *Quercus robur*	England	47	130	393	35	246	5	45	257
12. *Quercus robur— Carpinus betulus— Fagus sylvatica*	Belgium	30–75	121	406	32	245	—	81	868
13. *Quercus robur— Fraxinus excelsior*	Belgium	115–160	328	947	63	493	—	126	1338
14. *Quercus stellata— Quercus marilandica*	Oklahoma	various	195	902	75	1093	—	230	3895
Eucalyptus									
1. *E. regnans*	Australia	38	654	399	38	1389	138	192	849
2. *E. regnans*	Australia	27	831	—	17	—	—	—	—
3. *E. oblique—E. dives*	Australia	38	373	426	17	111	103	71	264
4. *E. oblique*	Australia	51	316	—	31	256	—	204	336
5. *E. sieberi*	Australia	27	929	—	14	—	—	—	—
6. Mixed dry sclerophyll	Australia	?	176	395	—	—	—	—	—
7. *E. signata—E. umbra*	Australia	?	104	456	18	192	169	77	344
8. *E. diversicolor*	Australia	37	263	473	27	296	82	211	1133
9. *E. diversicolor—E. calophylla*	Australia	?	305	449	31	424	125	344	1266

Source: After Feller (1980)

tions occur, and not all tropical forests are oligotrophic; nor are all temperate forests eutrophic. Some striking differences exist between these forest types. Oligotrophic systems have a large biomass in the humus layer of the soil, and this layer of fine roots and humus is critical for nutrient cycling and nutrient conservation in these systems.

Productivity and nutrient cycling do not differ greatly in oligotrophic and eutrophic forests, as long as these ecosystems are not disturbed (Jordan and Herrera 1981). But when the forest is cleared for agriculture, the nutrient-poor systems quickly lose their productive potential, whereas the nutrient-rich systems do not. Once the humus and root layer on top of the mineral soil is disturbed in oligotrophic systems, the mechanism of efficient nutrient recycling is lost, and nutrients are leached out of the system. Oligotrophic ecosystems cannot be used for crop production unless critical nutrients are supplied in fertilizers (Sanchez et al. 1982).

FIGURE 27.11

Distribution of acid precipitation in North America and Europe in 1980, and detailed measurements for the United States in 1997. Areas designated 10×, 20×, and 30× received 10, 20, and 30 times more acid in precipitation than expected if the pH were 5.6. (Data from the U.S. Environmental Protection Agency, National Atmospheric Deposition Program, 1997; and from Likens et al. 1981.)

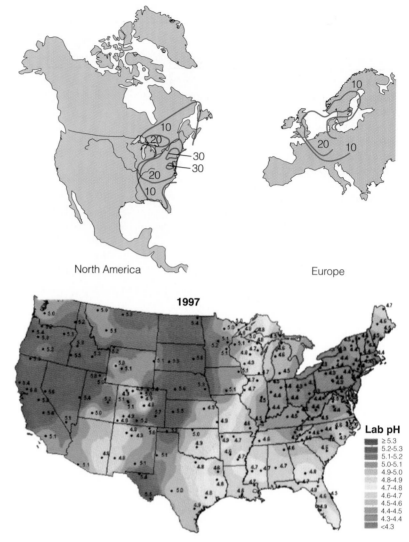

Lab pH

≥ 5.3
5.2-5.3
5.1-5.2
5.0-5.1
4.9-5.0
4.8-4.9
4.7-4.8
4.6-4.7
4.5-4.6
4.4-4.5
4.3-4.4
< 4.3

Acid Rain: The Sulfur Cycle

One human activity—combustion of fossil fuels—has altered the sulfur cycle more than any of the other nutrient cycles. Even though human-produced emissions of carbon dioxide and nitrogen are only about 5–10% of the level of natural emissions, for sulfur humans produce about 160% of the level of natural emissions (Likens et al. 1981). One clear manifestation of this alteration of the sulfur cycle is the widespread problem of acid rain in Europe and North America. Acid precipitation is defined as rain or snow that has a pH lower than 5.6. Low pH values are caused by strong acids (sulfuric acid, nitric acid) that originate as products of the combustion of fossil fuels.

Acid rain emerged as a major environmental problem in the 1960s, when widespread damage to forests and lakes in Europe and eastern North America became apparent. It was one of the first widespread environmental problems recognized because oxides of sulphur and nitrogen can be carried hundreds of kilometers and then deposited in rain and snow. Lakes in eastern Canada were dying because of air pollution produced in the midwestern United States. Lakes in southern Norway were losing fish because of acid rain from England. By 1980, annual pH values of precipitation over large areas of western Europe and eastern North America averaged between 4.0 and 4.5 (Figure 27.11), and individual storms produced acid rain of pH 2 to 3.

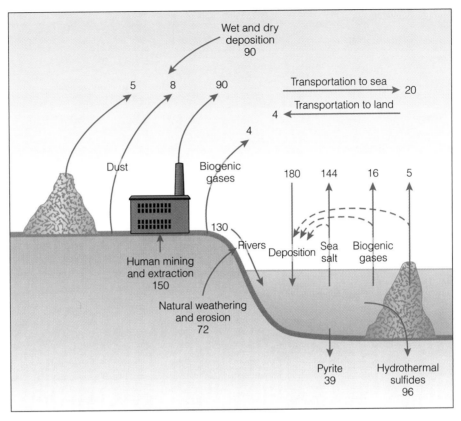

FIGURE 27.12
The global sulphur cycle. The burning of fossil fuels is the major component of atmospheric input of sulphur. All values are 10^{12} g S/yr. (From Brimblecombe et al. 1989.)

Sulfur released into the atmosphere is quickly oxidized to sulfate (SO_4) and is redeposited rapidly on land or in the oceans (Schlesinger 1997). Figure 27.12 illustrates the sources and sinks of the global sulfur cycle. Short-term events like volcanic eruptions contribute to the global sulfur cycle and make it difficult to estimate the equilibrium state of the atmosphere. Human-caused emissions are the largest component of additional sulfur to the atmosphere. Ore smelters and electrical generating plants have increased emissions during the past 100 years. To offset local pollution problems, smelters and generating plants have built taller stacks, which reduce pollution at ground level. Tall stacks (over 300 m), now the standard, have exported the pollution problem downwind. Ice cores from Greenland show large increases in SO_4 deposition from the atmosphere in the past 50 years (Mayewski et al. 1986).

The destructive effect of sulfate pollution on vegetation has been known for over a century. Queenstown, Tasmania, offers a striking example (Figure 27.13). From 1896 to 1922 sulphur-rich smoke poured from the copper smelter of the Mount Lyell Mining and Railway Company, which created Queenstown. A typical ton of ore from the mine consisted of 48% sulphur, 40% iron, and less than 3% copper. Emissions from the stacks contained more than half of the sulphur, and the fumes from the smelter could be toxic enough to cause plant death in one day. The prevailing winds carried the pollution to the east and southeast. Once the vegetation was killed, soil erosion from heavy rains (aided by fire) removed all the topsoil, and now, almost 80 years after the smelter closed, the landscape resembles the moon (Figure 27.13). Acid rain is not a new phenomenon.

FIGURE 27.13

Effects of acid rain from a copper smelter on native vegetation at Queenstown, Tasmania. The copper smelter operated from 1896 to 1922. These photos were taken in October 1992. After almost 80 years there is little sign of any vegetation recovery. (Photos courtesy of A. J. Kenney.)

ESSAY 27.1

ACID RAIN AND THE SUDBURY EXPERIENCE

The recovery of the ecosystems in the vicinity of the nickel and copper smelter at Sudbury, Ontario, is an encouraging sign that ecosystems damaged by air pollution can restore themselves. From 1900 to 1970 the Sudbury smelters spewed over 100 million tons of sulphur dioxide into the atmosphere, as well as thousands of tons of toxic trace metals such as lead and cadmium. At its peak, Sudbury alone accounted for 4% of total global emissions, an amount equal to the current emissions of the entire United Kingdom (Schindler 1997). Within 30 km of the smelter, vegetation was destroyed and thousands of lakes were acidified.

One way to combat acidification of lakes is to add lime (calcium carbonate) to raise their pH. The Swedes have been particularly active in liming lakes affected by acid rain. But because no one would pay for the enormous cost to lime the lakes around Sudbury, they constitute a natural experiment. Since 1980 the Sudbury smelters have emitted less than 10% of their original air pollution, and sources of sulphur dioxide in eastern Canada has been reduced by more than half. During the past 20 years the ecosystem around Sudbury has been recovering via natural processes. Trees and shrubs have recolonized, and the pH in most lakes has risen. Lake trout have colonized many lakes, but not all of them.

The idea that acidified lakes could recover only in geological time has been shown to be false, and there is room for some optimism. But acid rain continues to fall, even if in reduced amounts, and until we have further limited sulfur dioxide and nitrous oxides emissions, lakes will not be able to recover fully. The longer we wait to implement strict air pollution controls, the more damage will accumulate. Acid rain leaches cations such as calcium out of soils and thus reduces the natural ability of the soil to absorb acidity.

The important messages that Sudbury provides are that ecosystems can recover from disturbances, although it may take longer than a few years, and that we should not assume that we can achieve a technological fix for ecological damage. Adding lime to acidified lakes may solve some problems, but it creates another whole set in its wake (Steinberg and Wright 1994).

A net transport of SO_4 occurs from the land to the oceans. The ocean is also a large source of aerosols that contain SO_4. Dimethylsulfide [$(CH_3)_2S$] is the major gas emitted by marine phytoplankton, and it is quickly oxidized to SO_4 and then redeposited in the ocean. Sulfate is abundant in ocean waters (12×10^{20} g of elemental S), and the mean residence time for a sulfur molecule in the sea is over 3 million years (Schlesinger 1997).

The United States and most developing countries have reduced sulfur dioxide emissions during the past 30 years (Figure 27.14). In the United States, sulphur dioxide emissions were reduced 25% between 1990 and 1996. Reduced emissions should reduce surface deposition of acid rain, but the effects of acid rain on the environment do not disappear immediately as sulphur dioxide emissions decline (Likens et al. 1996). A key question remains: *Will forest and aquatic ecosystems recover from the effects of acid rain, and if so, at what rate?*

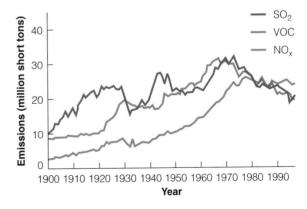

FIGURE 27.14

Emissions of sulphur dioxide (blue), nitrous oxides (green), and volatile organic compounds (such as methane) in the United States from 1900 to 1996. Sulphur dioxide emissions have been falling since the mid-1970s after the Clean Air Act was passed in 1970. (Data from U.S. Environmental Protection Agency, 2000.)

At Hubbard Brook the effect of acid rain has been to leach calcium from the soil to such an extent that available calcium, not nitrogen, now appears to limit forest growth. Streamwater chemistry at Hubbard Brook is slowly recovering from acid rain, and Likens et al. (1996) predict that at least another 10–20 years will be needed for streams to recover, even if sulfur dioxide emissions continue to decrease.

Freshwater ecosystems are particularly sensitive to acid rain. In areas underlain by granite and granitoid rocks, which are highly resistant to weathering, acid rain is not neutralized in the soil, so lakes and streams become acidified. Lakes in these bedrock areas typically contain soft water of low buffering capacity. Thus bedrock can be an initial guide to identifying sensitive areas. The Precambrian Fennoscandian Shield in Scandinavia, the Canadian Shield, all of New England, the Rocky Mountains, and other areas are thus potential trouble spots.

The clearest effects of acid precipitation have been on fish populations in Scandinavia and eastern Canada. Fish populations were reduced or eliminated in many thousands of lakes in southern Norway and Sweden once the pH in these waters fell below pH 5 (Likens et al. 1979). In Canada, lakes containing lake trout have been the principle focus of research on the effect of acid rain. Lake trout disappear in lakes once the pH falls below 5.4, and the cause is reproductive failure because newly hatched trout die (Gunn and Mills 1998). Lake trout, which are a keystone predator in many Canadian lakes, disappear slowly in lakes of low pH. Adult trout do not seem to be affected by low pH, nor is there any food shortage at low pH. The effect is on mortality of the small juveniles, and this causes a slow decline in the trout population over 10–20 years. Once lake trout are gone, acid-tolerant fish such as yellow perch and cisco become more abundant, and the food web shifts dramatically (Figure 27.15).

The effect of acid precipitation on terrestrial ecosystems is more complex and difficult to unravel. A good example of this complexity is found in forest declines in Europe (Schulze 1989). Forest declines have been particularly severe in central Europe, where needle yellowing and needle loss in Norway spruce (*Picea abies*) have brought public attention to the problem. Large-scale damage to spruce trees first became apparent in the late 1970s in Bavaria, Germany, and by the early 1980s about 25% of all European forests were classified as moderately or severely damaged from unknown causes. A variety of interacting factors associated with air pollution have now been shown to cause forest declines (Figure 27.16). There is no one general explanation for all forest declines in Europe or elsewhere; a combination of pollution, climatic events, and insect attacks can lead to forest decline (Freer-Smith 1998).

The forest region of Bavaria in southern Germany is exposed to high concentrations of SO_2, and both ozone and nitrous oxide levels are often high enough to cause direct damage to leaves. But these gaseous pollutants do not affect needles directly. Photosynthetic rates of Norway spruce needles are equal in Germany and in unpolluted New Zealand (Schulze 1989). The main effects of air pollution are on the forest soils. Soil acidification results from nitrate and sulfate deposition, and acid soil water reduces the amount of calcium, magnesium, and potassium that roots can absorb. Interactions among ions in the soil are particularly complex. For example, root uptake of magnesium is suppressed in the presence of aluminum or ammonium ions. Spruce seedlings growing in acid soils develop magnesium deficiencies, especially when ammonium is present. Spruce trees are thus stimulated to increased growth by nitrogen fertilization from the polluted air, but the acidification of the soil that accompanies the added nitrates and sulfates reduces the availability of magnesium and calcium, and the resultant deficiencies of these nutrients cause needle yellowing and loss. Plant diseases seem to have played only a secondary role in forest declines, and once a tree is weakened by nutrient imbalances, fungal diseases or insect attacks may increase.

The human-caused changes in the sulfur cycle thus have the potential to change nutrient cycling in natural ecosystems in a great variety of ways we cannot yet understand, much less predict. We cannot continue this aerial bombardment of ecosystems in the naive belief that nutrient cycles have infinite resilience to human inputs. Recent efforts to curtail sulfate emissions from fossil fuels have reduced the emissions of SO_4, and we must continue to press for further reductions. Once acidic precipitation is reduced, both forest and lake ecosystems can begin to recover from the damage inflicted on them.

The Nitrogen Cycle

The availability of nitrogen is often limiting for both plants and animals, and net primary production is often limited both on land and in the oceans by the

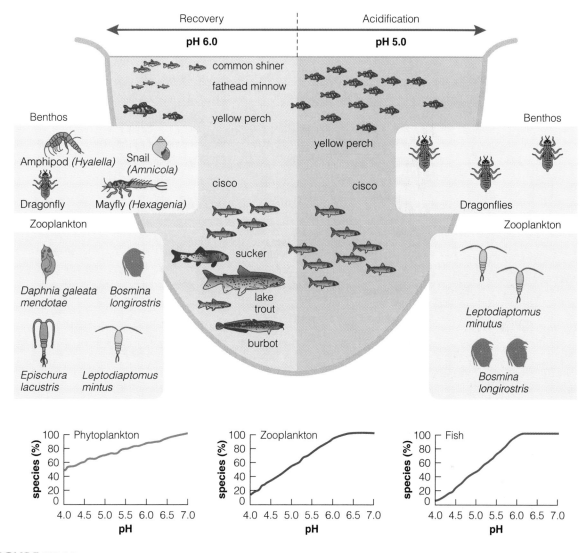

FIGURE 27.15
Schematic illustration of the effects of acidification on eastern Canadian lakes dominated by lake trout. Once the input of acid rain is curtailed, these lakes recover slowly, but full reversibility of the degradation has not yet been seen. (From Gull and Mills 1998.)

amount of nitrogen available. Nitrogen gas is abundant in air (it is 78% nitrogen), but few organisms can use N_2 directly. A small number of bacteria and algae can use nitrogen from the air directly and fix it as nitrate or ammonia. Many of these organisms work symbiotically in the root nodules of legumes to fix nitrogen, and this is a major source of natural nitrogen fixation. Human additions to the global nitrogen cycle have become substantial, particularly with the production and use of nitrogen fertilizers for agriculture.

Figure 27.17 shows the global nitrogen cycle. Human activities add about the same amount of

nitrogen to the biosphere each year as do natural processes, but this human addition is not spread evenly over the globe. The effects of human additions of nitrogen have shown up particularly in changes in the composition of the atmosphere. Nitrogen-based trace gases—nitrous oxide, nitric oxide, and ammonia—have major ecosystem effects. Nitrous oxide is chemically unreactive and persistent in the atmosphere; it traps heat and thus acts as a greenhouse gas that changes climate. Nitrous oxide is increasing in the atmosphere at 0.25% per year (Vitousek et al. 1997). Nitric oxide, by contrast, is

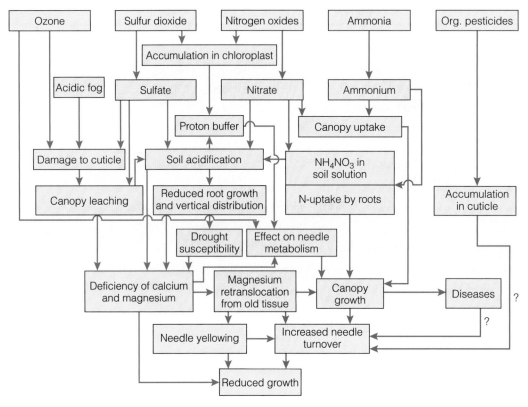

FIGURE 27.16
Processes that contribute to forest decline symptoms in central Europe. Boxes indicate the air pollutants and their effects on soils and plants. (After Schulze 1989.)

highly reactive and contributes significantly to acid rain and smog. Nitric oxide can be converted to nitric acid in the atmosphere, and in western United States acid rain is based more on nitric acid than on sulfuric acid. In the presence of sunlight, nitric oxide and oxygen react with hydrocarbons from auto exhaust to form ozone, the most dangerous component of smog in cities and industrial areas. Nitric oxide is produced by burning fossil fuels and wood. Ammonia neutralizes acids and thus acts to reduce acid rain. Most ammonia is released from organic fertilizers and domestic animal wastes.

The result of human activities on the nitrogen cycle has been an increased deposition of nitrogen on land and in the oceans. Because nitrogen additions are typically coupled with phosphorus additions, the result is eutrophication of freshwater lakes and rivers and coastal marine areas (Carpenter et al. 1998). Phosphorus additions to fresh water typically increase primary production, whereas nitrogen addition to

estuaries increases primary production in marine environments. The adverse effects of eutrophication on aquatic systems are listed in Table 27.6. The North Atlantic Ocean receives nitrogen from many rivers that transport excess nitrogen into the ocean. Nitrogen input to the North Atlantic has increased between twofold and 20-fold since 1750, and inputs from northern Europe are the highest (Figure 27.18). Nitrates in rivers are rising throughout the Northern Hemisphere in proportion to the human population along the rivers. In the Mississippi River, nitrates have more than doubled since 1965. Groundwater concentrations of nitrate are also increasing in agricultural areas, and in some areas are approaching the maximum safe level of nitrate in drinking water (10 mg per liter).

By contrast, the addition of nitrogen to terrestrial ecosystems can have positive effects. Nitrogen deposition on land can relieve the nitrogen limitation of primary production that is common in many terrestrial

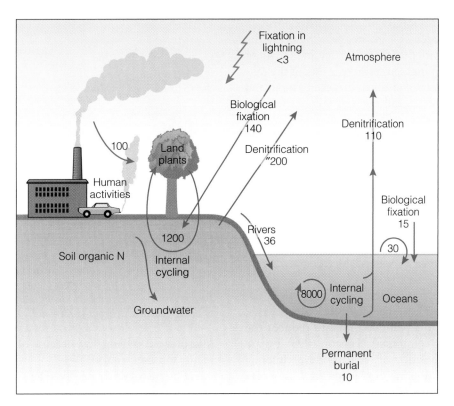

FIGURE 27.17
The global nitrogen cycle. Humans have had a strong effect on the nitrogen cycle. All fluxes are in units of 10^{15} g N/year. (From Schlesinger 1997.)

TABLE 27.6 Adverse effects of nutrient additions of nitrogen and phosphorus on freshwater and coastal marine ecosystems.

Increased biomass of phytoplankton
Shifts in phytoplankton communities to bloom-
  forming species that may be toxic
Increase in blooms of gelatinous zooplankton in
  marine ecosystems
Increased biomass of benthic algae
Changes in macrophytic species composition and
  biomass
Death of coral reefs
Decreases in water transparency
Taste, odor, and water treatment problems for
  domestic water supplies
Oxygen depletion
Increased frequency of fish kills
Loss of desirable fish species
Reductions in harvestable fish and shellfish
Decrease in aesthetic value of bodies of water

Source: From Carpenter et al. (1998).

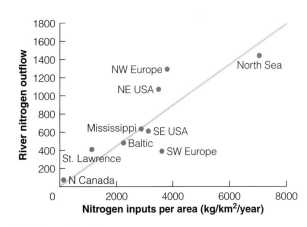

FIGURE 27.18
Nitrogen contribution of rivers draining into the North Atlantic Ocean. Nitrogen inputs are mostly from human activities, particularly fertilizer use. For northern Europe these nitrogen inputs may be 20 times higher than they were in pre-industrial times. River outflow is measured in kg N /km²/year. (From Vitousek et al. 1999a.)

ecosystems. Swedish forests, all of which are nitrogen limited, have averaged 30% greater growth rates in the 1990s compared with the 1950s (Binkley and Högberg 1997). The important concept here is that of the *critical load*—the amount of nitrogen that can be input and absorbed by the plants without damaging ecosystem integrity. When the vegetation can no longer respond to further additions of nitrogen (see Figure 27.9), the ecosystem reaches a state of nitrogen saturation, and all new nitrogen moves into groundwater or stream-flow or back into the atmosphere. Nitrate is highly water soluble in soils, and excess nitrate carries away with it positively charged ions of calcium, magnesium, and potassium. Excess nitrate can thus result in calcium, magnesium, or potassium limiting plant growth, and this is why most commercial garden fertilizers contain more than just nitrogen.

Increasing nitrogen in terrestrial ecosystems can have undesirable effects on biodiversity. In most cases, adding nitrogen to a plant community reduces the biodiversity of the community (Huston 1994). Figure 27.19 illustrates the effect of 12 years of experimental nitrogen fertilization of grasslands in Minnesota (Wedin and Tilman 1996). Species that are nitrogen responsive, often grasses, can take over plant communities enriched in nitrogen. The Netherlands has the highest rates of nitrogen deposition in the world, largely due to intensive livestock operations, a consequence of which has been a conversion of species-rich heathland to species-poor grasslands and forest. The mix of plant and animal species adapted to sandy,

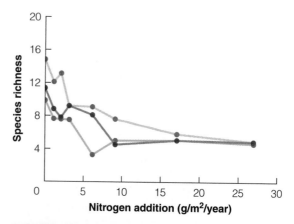

FIGURE 27.19
Vegetation responses to 12 years of nitrogen fertilization in Minnesota grasslands. Three fields were used, and six replicates were used for each level of nitrogen addition. Biodiversity declines dramatically as more nitrogen is added to these grasslands. (From Wedin and Tilman 1996.)

infertile soils is being lost because of nitrogen enrichment (Vitousek et al. 1997).

The nitrogen cycle, like the sulphur cycle, has been heavily affected by human activities during the past 50 years. It is urgent that national and international efforts be directed to reversing these changes and moderating the adverse effects on ecosystems. The most obvious direct effect on humans resulting from these changes in nutrient cycling are now manifest in global climate change, which we discuss in the next (and final) chapter.

Summary

Nutrients cycle and recycle in ecosystems, and tracing nutrient cycles is another way of studying fundamental ecosystem processes. Human activities are changing cycles on a global basis, with consequences for biodiversity, ecosystem function, and climate change. Nutrients reside in *compartments* and are transferred between compartments by physical or biological processes. Compartments can be defined in any operational way to include one or more species or physical spaces in the ecosystem. Nutrient cycles may be *local* or *global*. Global cycles, such as the nitrogen cycle, include a gaseous phase that is transported in the atmosphere. Local cycles include less mobile elements such as phosphorus.

Nutrient cycles in forests have been studied because of nutrient losses associated with logging. If the inputs of nutrients do not equal the outflow for any ecosystem, it will deteriorate over the long run. Logging can result in high nutrient losses even if soil erosion is absent. An undisturbed forest site recycles nutrients efficiently. Nutrient use efficiency is important to ecosystem functioning because plants in poorer soils are more efficient in their nutrient use, such that a greater mass of plant tissue is produced per unit of nutrient.

The sulfur cycle is an example of a global nutrient cycle that is strongly affected by human activities. The burning of fossil fuels adds a large amount of SO_2 to the atmosphere and results in acid rain. Acid rain in combination with other airborne pollutants has caused forest declines in Europe and has eliminated fish populations from many lakes in eastern Canada and Scandinavia. Regulations in North America and Europe have caused SO_2 emissions to decline during the past 30 years, and ecosystems can recover once the inputs are stopped.

The nitrogen cycle is critical because primary production in many terrestrial ecosystems and in coastal waters is limited by nitrogen. Nitrogen emissions by human activity have doubled the input of nitrogen into the air and waters of the globe. Smog in cities and more acid rain are two effects of this added nitrogen. Nitrogen and phosphorus leach from agricultural fertilizer and are a major cause of algal pollution in lakes and rivers. Unless we can curb these emissions of critical nutrients, global ecosystems will continue to be degraded.

Key Concepts

1. Nutrients cycle and recycle in ecosystems. Humans are drastically modifying these nutrient cycles.

2. Global nutrient cycles such as the nitrogen cycle include a gaseous phase. Local nutrient cycles contain nonvolatile elements such as phosphorus.

3. To achieve sustainable harvesting of ecosystems such as forests, nutrient input must equal output.

4. Nutrient use efficiency is higher in low-nutrient systems, and plants have evolved mechanisms to recycle limiting nutrients efficiently.

5. Acid rain is one consequence of human alteration of the sulphur and nitrogen cycles. Reducing sulphur emissions is necessary for ecosystem recovery, but the time scale of recovery is not yet clear.

6. Human additions of nitrogen to the nitrogen cycle have enriched water and land areas and is reducing nitrogen limitation of plant growth in terrestrial ecosystems.

Selected References

Binkley, D. and P. Högberg. 1997. Does atmospheric deposition of nitrogen threaten Swedish forests? *Forest Ecology and Management* 92:119–152.

Carpenter, S. R., N. F. Caraco, D. L. Correll, R. W. Howarth, A. N. Sharpley, and V. H. Smith. 1998. Nonpoint pollution of surface waters with phosphorus and nitrogen. *Ecological Applications* 8:559–568.

DeAngelis, D. L. 1992. *Dynamics of Nutrient Cycling and Food Webs.* Chapman & Hall, New York.

Hooper, D. U. and P. M. Vitousek. 1998. Effects of plant composition and diversity on nutrient cycling. *Ecological Monographs* 68:121–149.

Likens, G. E., C. T. Driscoll, and D. C. Buso. 1996. Long–term effects of acid rain: Response and recovery of a forest ecosystem. *Science* 272:244–245.

McNaughton, S. J., F. F. Banyikwa, and M. M. McNaughton. 1997. Promotion of the cycling of diet–enhancing nutrients by African grazers. *Science* 278:1798–1800.

Newman, E. I. 1997. Phosphorus balance of contrasting farming systems, past and present. Can food production be sustainable? *Journal of Applied Ecology* 34:1334–1347.

Pastor, J. and S. D. Bridgham. 1999. Nutrient efficiency along nutrient availability gradients. *Oecologia* 118:50–58.

Schindler, D. W. 1997. Liming to restore acidified lakes and streams: A typical approach to restoring damaged ecosystems? *Restoration Ecology* 5:1–6.

Schlesinger, W. H. 1997. *Biogeochemistry: An Analysis of Global Change.* 2nd edition. Academic Press, San Diego. 588 pp.

Taylor, G. E. J., D. W. Johnson, and C. P. Andersen. 1994. Air pollution and forest ecosystems: A regional to global perspective. *Ecological Applications* 4:662–689.

Vitousek, P. M., J. Aber, R. W. Howarth, G. E. Likens, P. A. Matson, D. W. Schindler, W. H. Schlesinger, and D. G. Tilman. 1997. Human alteration of the global nitrogen cycle: Causes and consequences. *Ecological Applications* 7:737-750.

Questions and Problems

27.1 In slash-and-burn agriculture, which is common in many tropical countries, forests are cut and burned, and crops are planted in the cutover areas. Yields are usually good in the first year but decrease quickly thereafter. Why should this be? Compare your ideas with those of Tiessen et al. (1994), and evaluate the sustainability of slash-and-burn agriculture.

27.2 Increased nitrogen deposition from the atmosphere may lead to nitrogen saturation of forests. Design a long-term experiment that would address the question of how much nitrogen loading a forest could sustain. Compare your design with that used by the NITREX project in Europe (Gundersen et al. 1998).

27.3 Discuss the relative merits of making a compartment model of a nutrient cycle very coarse (with only a few compartments) versus making it very fine (with many compartments).

27.4 Shallow lakes may have two alternate stable states depending on nutrient influx, one dominated by phytoplankton in turbid water and another dominated by macrophytes in clear water. Discuss how you might determine the critical nutrient loading that would trigger a transition between these states. Would you expect the transition from phytoplankton to macrophytes to occur at the same nutrient loading as the opposite transition? Janse (1997) discusses this problem.

27.5 Schultz (1969, p. 92) states:

> The idea of one cause–one effect is left over from the nineteenth century when physics dominated science. The whole notion of causality is under question in the ecosystem framework. Does it make sense to say that high primary production causes a rich organic soil and a rich organic soil causes high production? This kind of reasoning leads up a blind alley.

Discuss.

27.6 Vitousek (1982) concluded that nitrogen use efficiency was higher in low-nitrogen forests. Chapin (1980, p. 246) concluded that plants from infertile habitats had a lower use efficiency of limiting nutrients than rapidly growing plants. Discuss why these two authors have reached apparently opposite conclusions.

27.7 Large grazing mammals should accelerate the rate of nutrient cycling in ecosystems. McNaughton et al. (1997) showed that foraging by Serengeti ungulates increases nitrogen availability in grasslands. In contrast, Pastor et al. (1993) showed that moose browsing on trees on Isle Royale in Lake Superior decreased primary production. Discuss what nutrient cycling processes might cause these opposite effects of large grazing mammals.

27.8 Soils in Australia contain very low amounts of phosphorus, from 10–50% the amount in North American soils (Keith 1997). Would you predict that eucalypts growing in Australian soils would be phorphorus limited? What adaptations might plants evolve to achieve high nutrient use efficiency when growing on soils of low nutrient content?

27.9 A key question in restoration ecology is how long it will take for an ecosystem to recover from some disturbance caused by humans. Discuss how we might find out what the time frame is for ecological recovery from acid rain.

Overview Question

One suggested ecological response to help restore acidified lakes is to add lime. Explain the mechanisms behind this recommendation, and discuss whether any possible negative effects might result from this amelioration program. How could you determine if this restoration program was effective?

CHAPTER 28

Ecosystem Health:
Human Impacts

Humans, including ecologists, have a peculiar fascination with attempting to correct one ecological mistake with another, rather than removing the source of the problem.
—*D.W. Schindler (1997, p. 4)*

HUMAN IMPACTS ON THE PLANET'S ECOSYSTEMS are significant and growing. During your lifetime these impacts will lead to the most significant problems facing the globe, and our present preoccupation with the stock market and political chicanery will be viewed as the latest example of Nero fiddling while Rome is burning.

Throughout this book we have touched on many of these problems, from overfishing to pest control and the conservation of endangered species. This chapter brings these problems into central focus and asks both *What are the problems?* and *What can we do about them?* The role of ecologists in these matters is much like that of a medical doctor. As scientists our job is to diagnose problems and suggest cures. But just as a patient can ignore a doctor's advice to stop smoking, for example, the public can ignore all the advice and recommendations of ecologists to ameliorate problems. Media discussions over the reality of climate change are a current example. Translating science into policy is an art, and ecological scientists are

not yet very good at it. Our first objective must be to get the science right. The central mandate of applied ecology is to assess these human impacts scientifically and to suggest ways of ameliorating them.

In this chapter we first examine the human population problem, the root cause of all the adverse human impacts on ecosystems. Then we consider two aspects of ecosystem health: the carbon cycle and climate change and species invasions.

Human Population Growth

If growth is good, as some businesspeople seem to believe, then the human beings on the planet ought to be nearing perfection. Global human population has been growing throughout most of recorded history, and probably for more than 2 million years before that. In 2000 an estimated 6.1 billion people inhabited the Earth, and the human population is growing at 1.33% per year, or 214,000 people per day (see Box 28.1). Figure 28.1 shows the growth of the human population over the past 2000 years. The increase appears exponential, but is even more rapid than exponential (Cohen 1995). Because no population can go on increasing without limit, Figure 28.1 immediately

FIGURE 28.1

Human population growth in the past 2000 years. (a) From A.D. 1 to the present. The time scale is logarithmic (to condense the x-axis), the population scale linear. The increase approximately resembles exponential growth. (b) From A.D. 1000 to A.D. 2000. The time scale is linear, and the population scale is logarithmic. Exponential growth would appear a straight line in this graph, so humans have been increasing at a rate that is even greater than exponential. (Data from Cohen 1995, Appendix 2.)

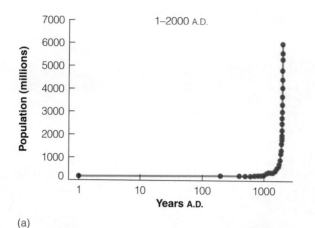

(a)

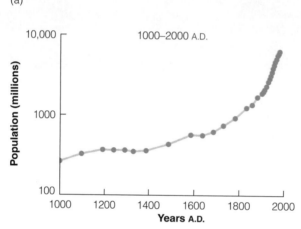

(b)

raises two questions: (1) What is happening now? and (2) Can we estimate the long-term carrying capacity of the Earth for humans?

Current Patterns of Population Growth

The human population can exist in one of two stable configurations that lead to population stability:

$$\underset{\text{growth}}{\text{Zero population}} = \underset{\text{rates}}{\text{high birth}} - \underset{\text{rates}}{\text{high death}} \quad (28.1)$$

or

$$\underset{\text{growth}}{\text{Zero population}} = \underset{\text{rates}}{\text{low birth}} - \underset{\text{rates}}{\text{low death}}$$

The movement between these two states has been called the *demographic transition*. Figure 28.2 illustrates the demographic transition for Sweden and for Mexico. The demographic transition is a descriptive

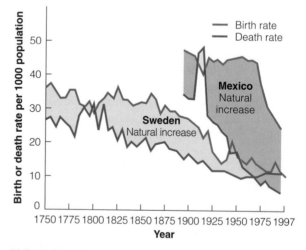

FIGURE 28.2

Demographic transition in Sweden and Mexico, 1750 to 1997. When births exceed deaths, the population grows (natural increase, shaded zones). The transition from high birth rate and high death rate to low birth rate and low death rate took 150 years in Sweden but has been compressed in Mexico to half that time. (From Population Reference Bureau 2000.)

B O X   2 8 . 1

HOW LARGE IS A BILLION ANYWAY?

We are the supreme counters in the universe, and we learn in school how to use decimals and exponents and how to manipulate them correctly. But at some point our numerical system loses meaning, and such is the case of a billion. There are now about 6 billion people on earth, and in 1998 the debt of the U.S. government increased $120 billion to reach a total of $5600 billion. How can we get a grasp on how big a billion is? Try these two simple games:

1. **Question:** How long would it take you to spend $1 billion if you were able to spend $10,000 a day, every day of the year, on anything you wished?
   Answer: To spend $1 billion you would need 274 years with no days off.

2. **Question:** How far would you walk if you took a billion steps?
   Answer: If we assume each of your steps is 61 cm (about 2 feet), you would walk 610 million kilometers (379,000,000 miles). This would mean you could walk to the sun and back twice, or if you are more conservative you could walk to Mars and back when it is farthest from the Earth. Remember to take some water with you.

The point is that while we can talk about and manipulate large numbers like a billion, they exceed our ability to comprehend how large they really are. So when we note that the human population increased 1 billion from 1987 to 1999, the size of this increase is nearly impossible to relate to our normal existence.

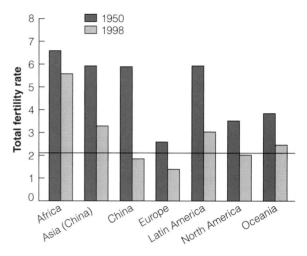

FIGURE 28.3
Decline in fertility rates for seven world regions from 1950 to 1998. Total fertility rate is the average number of children a woman will have at the prevailing birth rates. The horizontal line marks the approximate level of replacement fertility (zero population growth). Fertility rates have fallen in all countries but are still high in Africa and Asia (outside of China). (Data from Population Reference Bureau 2000.)

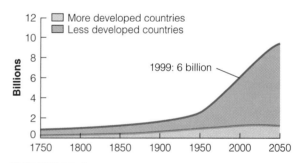

FIGURE 28.4
The contribution to world population growth of the more developed countries (orange) and the less developed countries (blue) since 1750. The major population growth we are now experiencing and will experience for the next 50 years will come from the developing countries of Southeast Asia, Africa, and Latin America. (Data from Population Reference Bureau 2000.)

theory rather than a law of human population growth (Gelbard et al. 1999). After 1950, mortality rates declined rapidly in all countries, but birth rates have declined in a more variable manner. Fertility decline has been most dramatic in China. In 1970 the average Chinese woman could expect to have 5.9 children; by 1999 the expected family size was 1.85 children. In India, fertility rates have fallen more slowly and irregularly. In much of Africa the transition to lower fertility is just beginning (Figure 28.3).

One consequence of variable fertility rates on the growth of the world's population is that the current rapid rise in human population is composed of two quite different elements (Figure 28.4). In the developed nations, populations are near to equilibrium, with net reproductive rates near replacement (total fertility rate = 2.1 or R_0 = 1.0). In many developed countries such as Canada and the United Kingdom, net reproductive rates are in fact below 1, and these populations

will decline in the long term if there is no immigration and if the net reproductive rate does not change. Most developed countries, however, are still increasing in population in the absence of immigration because the age structure is not in equilibrium such that births exceed deaths (Table 28.1) . This increase is due to population momentum and will stop in about 30 years.

TABLE 28.1 Human populations of the different continents and some selected countries as of July 1999.

Population projections are based on United Nations estimates of projected future trends in reproduction and mortality. The rate of increase is the natural rate from births and deaths and excludes immigration and emigration. Doubling times assume that the rate of increase does not change in the future.

Country	Population in July 1999 (millions)	Rate of increase (%/yr)	Doubling time (years)	Projected population 2010	Projected population 2025
World	5982	1.4	49	6883	8054
More developed	1181	0.1	583	1216	1241
Less developed	4801	1.7	40	5667	6813
Africa	771	2.5	28	979	1290
North America	303	0.6	119	333	374
Canada	31	0.4	162	35	39
United States	273	0.6	116	298	335
Central America	135	2.3	30	164	200
South America	339	1.7	41	395	463
Asia	3637	1.5	46	4206	4923
Saudi Arabia	21	3.0	23	29	40
Yemen	16	2.9	24	24	40
Bangladesh	126	1.8	38	150	177
India	987	1.9	37	1167	1414
Pakistan	147	2.8	25	181	225
China	1254	1.0	73	1394	1561
Japan	127	0.2	318	128	121
Europe	728	–0.1	—	731	718
Norway	5	0.3	224	5	5
Sweden	9	–0.1	—	9	9
France	59	0.3	210	62	64
Germany	82	–0.1	—	82	80
Poland	39	0.1	1155	40	41
Russia	147	–0.5	—	145	138
Italy	58	0.0	—	58	55
Spain	39	0.0	1980	40	39

Source: Population Reference Bureau, 1999 World Population Data Sheet.

About 80% of the world's people now live in the less developed countries, and most of the population growth is occurring in these nations (see Figure 28.4).

The projected human population of the globe depends on assumptions about future changes in fertility and mortality. The United Nations projects a population in 2050 that might range from 7.3 billion to 10.7 billion people (Gelbard et al. 1999). No matter what projection is used, without some catastrophe at least 1.3 billion people will be added to the population in the next 25 years because of population momentum. The questions that arise from these projections are, Is the world already overpopulated? If not, will it be overpopulated in 2050? How many humans can the biosphere support?

Carrying Capacity of the Earth

What is the carrying capacity of the Earth for humans? This question has been asked for more than 300 years by scientists interested in demography (Cohen 1995). As a first step, we can ask what range of estimates has so far been produced for how many people the Earth can support. Figure 28.5 shows the ranges of the estimates for the carrying capacity of the Earth, beginning with the first estimate by Leeuwenhoek in 1679. Two aspects of this graph are striking. First, the estimates vary widely, from less than 1 billion to over 1000 billion; second, the estimates do not converge over time but seem to increase in variability, although the median of the estimates seems to stay

E S S A Y   2 8 . 1

THE DEMOGRAPHIC TRANSITION: AN EVOLUTIONARY DILEMMA

The demographic transition involves two puzzling aspects of human fertility. First, a large decrease occurs in the number of children that parents produce, despite the fact that resources are increasing. This change was shown very well in Europe during the nineteenth century. Second, rich families reduce their fertility rate earlier than does the rest of the population, and often they have fewer than the average number of children. Thus a negative correlation exists between wealth and fertility (Mulder 1998). If we assume that humans should follow the principles of natural selection that apply to other animals, these observations create a puzzle. When birds and other animals have more resources, they have more offspring. Why should parents with access to plentiful resources choose to have low fertility rates?

Three hypotheses have been suggested to explain this puzzle of low fertility:

1. *Lowered rates are optimal because of the competitive environment in which offspring are raised.* This idea is the classic view of evolutionary anthropologists, who suggest that high levels of parental investment are critical to a child's success and are costly to the parents. Thus, maximizing fitness is achieved by fewer offspring with higher parental investment, and there is a tradeoff between offspring quality and offspring number.

2. *Lowered rates are a consequence of Darwinian selection, but on nongenetic mechanisms of inheritance.* Cultural selection through imitation drives the demographic transition in fertility because people see that successful people have fewer children. This hypothesis is attractive because it postulates that *ideas*, rather than economic resources, can drive fertility rates, but it suffers some serious flaws. It begs the question of why rich and successful people should reduce their fertility in the first place.

3. *Low fertility is maladaptive, a byproduct of rapid environmental change.* This hypothesis suggests that lowered fertility is indeed an inappropriate evolutionary response not favored by natural selection. The availability of contraceptives is often cited as one explanation of why this maladaptive situation has prevailed in human societies, but it may be difficult to apply this idea to events in the nineteenth century. This explanation is intriguing but remains vague and untestable until the precise environmental changes can be identified.

There is currently much interest in applying the principles of evolutionary ecology to human behavior. However, a shortage of data in human societies at the individual level makes it difficult to find out, for example, whether individuals who have fewer offspring in fact have more grandchildren. At the present time the best guess is that the demographic transition can be explained by the first model, in which people can have more descendents in the long term by pursuing wealth at the cost of immediate reproductive success—so long as wealth is inherited (Penn 1999).

---

around 10–15 billion. Why should these estimates of carrying capacity be so variable?

Carrying capacity is difficult to estimate, and the writers that produced the estimates in Figure 28.5 used quite different methods to get their answers. Some authors such as Raymond Pearl (see p. 163), used curves like the logistic equation to predict the future maximum of the human population. Others general-

ized from existing "maximum" population density and multiplied this by the area of land that could be inhabited. More scientific estimates were made by focusing on a single assumed population constraint such as food, but this promising approach is limited by the assumptions it must make about the amount of available land, average crop yields, the diet to be utilized (vegetarian or meat-eating), and the number of joules

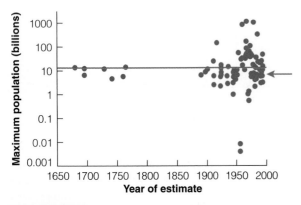

FIGURE 28.5
Estimates of the carrying capacity of the Earth for humans as suggested by various writers from 1679 to the present. Both minimum and maximum estimates are plotted. The median of the estimates is approximated by the blue line around 10–12 billion, and the arrow indicates the human population in 2000. (Data from Cohen 1995, Appendix 3.)

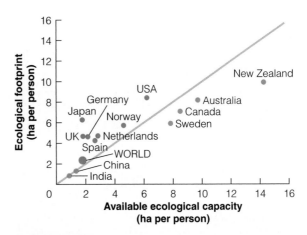

FIGURE 28.6
Ecological footprint in relation to available ecological capacity of several countries and the world as a whole. The ecological footprint expresses in hectares of land per person the current demand of global resources made by each country. The available ecological capacity measures in land area per person the resource base of each country. Countries in red above the diagonal are in an ecological deficit in 1997, countries in green below the diagonal still had resource surpluses. (Data from Wackernagel et al. 1999.)

or calories to be provided to each person each day. The equation for carrying capacity used is

$$\text{Carrying capacity} = \frac{(\text{ha land})\left(\dfrac{\text{yield}}{\text{per ha}}\right)\left(\dfrac{\text{kJ per}}{\text{crop unit}}\right)}{\text{no. of kJ needed per person per year}} \quad (28.2)$$

This equation is a definition, and if we could quantify the terms in it we could identify the carrying capacity of the earth. But because there are many crops grown on different soils, variable yields, losses to pests, and differences in assumed standards of living, it has been impossible to get anyone to agree on how these numbers should be constrained.

A more promising approach to estimating the carrying capacity of the Earth is to recognize that we have multiple constraints because we need food, fuel, wood, and other amenities like clothing and transportation. One approach is to express everything in the amount of land needed to support each activity (for example, wood production) and then to sum these requirements. But this approach also has limits because it is difficult to express energy requirements directly as land areas, and water requirements may be more of a constraint than land areas.

A recent advance in using multiple constraints to estimate carrying capacity is summarized in the concept of an *ecological footprint* (Wackernagel and Rees 1996). For each nation we can calculate the aggregate land and

water area in various ecosystem categories that is appropriated by that nation to produce all the resources it consumes and to absorb all the waste it generates (see Box 28.2). Six types of ecologically productive areas are distinguished in calculating the ecological footprint: arable land, pasture, forest, ocean, built-up land, and fossil energy land. Fossil energy land is calculated on the basis of the land needed to absorb the CO_2 produced by burning fossil fuels. All measures are thus converted to land area per person. If we add up all the biologically productive land on the planet, we find there is about 2 ha of land per person alive in 1997. If we wish to reserve land for parks and conservation, we must reduce this to 1.7 ha per person of land available for human use. This is the benchmark for comparing the ecological footprints of nations. Because different countries differ in their agricultural capacity and resource base, the available ecological capacity for each country must be adjusted for its productivity in each of the six types of productive ecosystems. We end up with a simple comparison of ecological footprints and available ecological capacity. Figure 28.6 illustrates these results for 14 countries. Two things are evident from this figure: The world in general was already in ecological deficit in 1997; and countries vary greatly in their individual foot-

BOX 28.2

HOW TO CALCULATE AN ECOLOGICAL FOOTPRINT

Ecological footprints are estimates of how much area is needed to support a given population. The easiest calculations of footprints are for nations because most data are compiled as national statistics. The first step in calculating an ecological footprint is to compile the total consumption of each commodity used by the population. Here we examine this approach for a single commodity grain consumption in Canada. A more complete example is given in Wackernagel et al. (1999) for all commodities for Italy.

1. Correct consumption data for trade imports and exports:

$$\text{Consumption}_{wheat} = \text{production}_{wheat} + \text{imports}_{wheat} - \text{exports}_{wheat}$$

For Canada in 1993, domestic production of wheat was 51,416,000 tons. Imports of wheat totaled 499,100 tons, and exports were 26,428,400 tons. Thus

$$\text{Consumption}_{wheat} = \text{production}_{wheat} + \text{imports}_{wheat} - \text{exports}_{wheat}$$

$$= 51,416,000 + 499,100 - 26,428,400$$

$$= 25,486,700 \text{ tons}$$

2. Convert the consumption data into the land area required to produce the item:

$$a_{wheat} = \frac{c_{wheat}}{y_{wheat}}$$

where a_{wheat} =   total ecological footprint of wheat in hectares of land

c_{wheat} =   total consumption of wheat in kg

y_{wheat} =   yield of wheat in kg per hectare

For Canada in 1993, the average yield of wheat was 2744 kg/ha. Thus

$$a_{wheat} = \frac{c_{wheat}}{y_{wheat}}$$

$$= \frac{25,486,700}{2.744}$$

$$= 9,288,200 \text{ ha of arable land}$$

3. Obtain the per capita ecological footprint by dividing the total ecological footprint by the human population (N):

$$f_{wheat} = \frac{a_{wheat}}{N}$$

$$= \frac{9,288,200}{28,817,000}$$

$$= 0.32 \text{ ha}$$

This is the per capita ecological footprint attributable to grain consumption for Canada in 1993. To obtain the total ecological footprint, this process would be repeated for each commodity and energy source that is consumed or used.

Most commodities that humans use can be converted into land area in a simple way. The exception is energy use, and to convert these into land area Wackernagel and Rees (1996) first converted all energy use into equivalent amounts of CO_2 production, and then calculated how much land would be needed to absorb this CO_2 in vegetation.

These calculations illustrate the ecological *demand* a country places on the biosphere. The last step is to compile the amount of productive land that a country has within its borders, the ecological *supply*. These values in land area are more readily compiled, and they are adjusted for the productivity of the country's ecosystems. Figure 28.6 illustrates some of the resulting ecological footprint values.

print size and in their available capacity. The utility of these estimates of ecological footprints is that they can direct us toward sustainability in our use of the world's resources by targeting the specific areas in which particular countries overutilize resources.

The analysis of human impacts via ecological footprints suggests that the world is already at or slightly above its carrying capacity. Two other calculations can

be made of human impacts, one for water and one for primary production. Fresh water is an important resource for humans because it has no substitute and is difficult to transport more than a few hundred kilometers. The amount of water on Earth and the volumes moving around are so large that we must discuss the water cycle in units of km^3 (= 10^{12} liters). If all the rainfall on land were evenly distributed, each weather station

FIGURE 28.7
The global water cycle. Pools of water are in units of km³ (black), and flows of water are in km³ per year (red). About half of the pool of soil water is available to plants. (After Schlesinger 1997.)

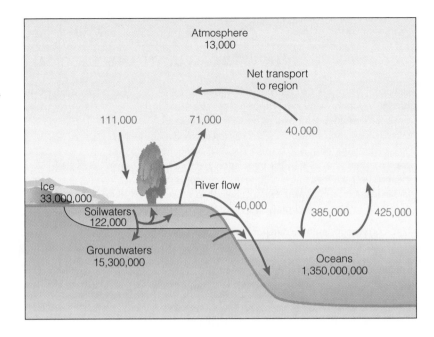

would record about 70 cm of rain per year (Schlesinger 1997). Fresh water constitutes only about 2.5% of the total volume of water on Earth, and two-thirds of it is locked up in glaciers and ice.

Figure 28.7 shows the global water cycle. There is a net transport of water vapor from the oceans to the land that contributes about one-third of the rainfall on land areas. The mean residence time of a water molecule in the ocean is about 3100 years, a tribute to the large volumes of seawater in the ocean. The volume of groundwater is poorly known. It is fossil water because it cannot be used by plants, and it has a mean residence time of over 1000 years (Schlesinger 1997). We depend on fresh water flowing through the hydrological cycle through precipitation for our needs.

Precipitation on land takes one of two routes: evapotranspiration and runoff. The precipitation that is used for vegetative growth of forests, crops, and pastures, and that evaporates back into the atmosphere is called *evapotranspiration*. The remaining precipitation goes to *runoff*, which is the source of water to sustain cities, irrigated crops, industry, and aquatic ecosystems. Postel et al. (1996) estimated that worldwide humans now use 26% of all evapotranspiration on land and 54% of freshwater runoff.

Can we appropriate more of the terrestrial water supply to support an increased population? Postel et al. (1996) argue that we cannot because most land suitable for rain-fed agriculture is already in production. We

could build more dams on rivers to capture more runoff, but the maximum this would permit is estimated at 64% usage of runoff. Given an expected population growth of approximately 33% in the next 25 years (see Table 28.1) something will have to change. If we do not wish to destroy completely all of the Earth's freshwater aquatic ecosystems, we will have to increase our efficiency of water use and reduce the amount of water used for irrigation. Desalination of seawater is an expensive option because it is energy intensive, and it is unlikely to provide a solution for the less developed countries. The message is similar to that we obtained from ecological footprints: Humans are close to the carrying capacity of the earth.

The Carbon Cycle

The most important global issue of our time is climate change, and to begin to understand climate change we must understand the effects humans are having on the global carbon cycle. Plants and animals are primarily composed of carbon, and the global carbon cycle is a reflection of primary and secondary production. The fixation of carbon by plants via photosynthesis over geological time accounts for the oxygen in the Earth's atmosphere. Humans have affected the global carbon cycle nearly as much as they have the sulfur cycle, and intense public interest now focuses on the resulting greenhouse effect and climate change.

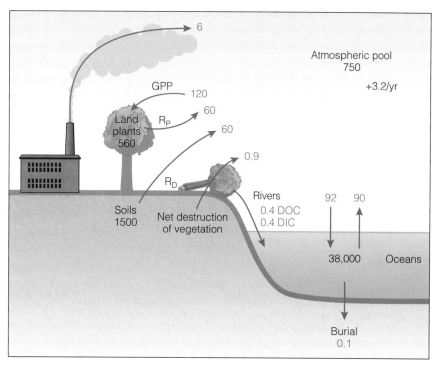

FIGURE 28.8

The present-day global carbon cycle. All pools are expressed in units of 10^{15} g C (black), and all annual fluxes in units of 10^{15} g C/yr (blue). GPP = gross primary production, R_p = respiration from production, R_D = respiration from destruction of vegetation, DOC = dissolved organic carbon, and DIC = dissolved inorganic carbon. (From Schlesinger 1997.)

Figure 28.8 shows the global carbon cycle. The carbon cycle is mostly the carbon dioxide cycle. The largest pool of carbon is dissolved inorganic carbon in the ocean, which contains about 56 times as much carbon as the atmosphere (Schlesinger 1997). The atmospheric pool of carbon is slightly larger than the total carbon bound up in vegetation. The largest fluxes of the global carbon cycle are between the atmosphere and land vegetation, and between the atmosphere and the oceans. These two fluxes are approximately equal, and the mean residence time of a molecule of carbon in the atmosphere is about five years.

The amount of CO_2 in the atmosphere has not been constant. From bubbles of gas trapped in ice cores in Antarctica we know that atmospheric CO_2 was about 270–280 ppm from A.D. 900 to A.D. 1750 (Barnola et al. 1995). The Vostok ice core, which was collected by the Soviet Antarctic Expedition and spans 160,000 years, showed that during the last ice age (20,000 to 50,000 years ago) CO_2 levels were 180–200 ppm, much lower than our current levels

(Barnola et al. 1987). Since 1750 and the start of the Industrial Revolution, CO_2 levels in the atmosphere have risen rapidly and continuously.

Superimposed on these long-term trends is a seasonal trend in CO_2 concentration. The atmospheric oscillations of CO_2 are the result of the seasonal uptake of CO_2 by plants via photosynthesis and seasonal differences in fossil fuel use and CO_2 exchange with the oceans. The majority of terrestrial vegetation occurs in environments with seasonal growth cycles, and plants fix CO_2 so that atmospheric CO_2 levels decline in summer (Figure 28.9). The atmospheric oscillations of CO_2 are superimposed on a long-term increase of 0.4% (1.5 ppm) per year. Some of this increase comes from the burning of fossil fuels. If all the CO_2 released from fossil fuels accumulated in the atmosphere, CO_2 would be increasing about 0.7% per year. But only about 56% of the CO_2 released from fossil fuel is accumulating in the atmosphere (Keeling et al. 1995). What happens to the remainder?

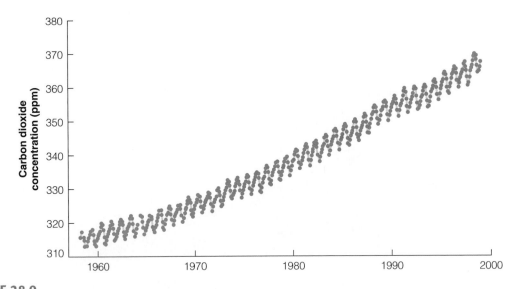

FIGURE 28.9

The concentration of atmospheric CO_2 at Mauna Loa Observatory in Hawaii expressed as a mole fraction in parts per million of dry air. The annual oscillation reflects the seasonal cycles of photosynthesis and respiration by land biota in the Northern Hemisphere; the overall increase is largely due to the burning of fossil fuels. (From Keeling and Whorf 1999 and Scripps Institute of Oceanography.)

We cannot yet balance the global carbon budget to answer this question, and this is one of the most vexing problems in ecosystem ecology today. The sizes of major sources and sinks for carbon are not well understood, and we do not understand the processes that control carbon fluxes (Dewar 1992). Oceanographers believe that about 33% of the CO_2 from fossil fuels enters the ocean each year (Quay et al. 1992). CO_2 is exchanged only at the ocean's surface, and much of the carbon in the oceans is in the deeper waters. Exchange in the oceans between surface waters and deep waters occurs only very slowly. Turnover of carbon for the entire ocean occurs about every 350 years (Schlesinger 1997).

Another source of atmospheric carbon dioxide is the destruction of terrestrial vegetation, often by clearing and burning for agriculture, especially in the tropics (Detwiler and Hall 1988, Schlesinger 1997). Considerable disagreement exists among scientists as to whether terrestrial vegetation is a source of CO_2 or a sink for CO_2. Shifting cultivation is a dominant form of land use in tropical countries, and about 75% of all land-use changes fall under this heading. Shifting cultivation is less destructive of forest because after one to three years the farmers move on and abandon the fields to secondary succession. Such temporary clearing of land contributes less CO_2 to the atmosphere than does permanent conversion of forest land to pastures. In 1990 the best estimate was that tropical areas were a net source of 1.6×10^{15} g carbon because of deforestation (Dixon et al. 1994). At the same time the regrowth of forests in the temperate zones captured about 0.7×10^{15} g of carbon. But any net release of carbon from terrestrial ecosystems complicates the balancing of the carbon budget for the Earth.

The data now available provide the following global budget for carbon (Schlesinger 1997; units = 10^{15} g of carbon per year):

$$\text{Net emissions} = \text{Net changes in the carbon cycle}$$

Fossil fuel	+	destruction of terrestrial plants	=	atmospheric increase	+	oceanic uptake	+	unknown sinks
6.0	+	0.9	=	3.2	+	2.0	+	1.7

We cannot account for almost 25% of global carbon flux, and unknown sinks must be absorbing approximately 1 billion tons of carbon each year! In trying to balance the global carbon budget, it is important to remember that we are focusing on the annual movements of carbon, not on the amount stored in the various reservoirs. The ocean contains the largest pool of carbon, but most of this carbon turns over very slowly. Desert soil carbonates contain more carbon than all

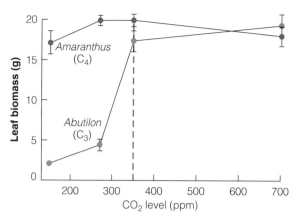

FIGURE 28.10
The response of annual C_3 and C_4 plants to four carbon dioxide levels in a greenhouse. The CO_2 levels were chosen to simulate ambient level during the last ice age (150 ppm), ambient level before the Industrial Revolution (270 ppm), current level (350 ppm, blue line), and a predicted future level (700 ppm). Whereas, the C_4 species Amaranthus retroflexus *showed no response to changes in CO_2 levels, the C_3 species* Abutilon theophrasti *showed a strong CO_2 response, increasing 22% in biomass from current to future doubled CO_2 levels. (Data from Dippery et al. 1995.)*

Fakhri A. Bazzaz *Mallinckrodt Professor of Biology, Harvard University*

terrestrial plants, but virtually no exchange of carbon occurs between the atmosphere and desert soils.

Recent work has focused on terrestrial vegetation as a sink for CO_2 (Idso 1999, Norby et al. 1999). If biomass is increasing in terrestrial vegetation because of a "fertilization" of vegetation by CO_2, and if this increase is rapid enough, we might have located the "unknown sink" of the global carbon cycle. What is the evidence that rising CO_2 levels stimulate plant growth?

Individual Plant Responses to CO_2

The effects of rising CO_2 and temperature have been analyzed in great detail for individual species of plants because this work can be done relatively easily in greenhouses or in open-top chambers in the field. F. A. Bazzaz and his students at Harvard were among the first ecologists to begin to study the effects of increased CO_2 on plant communities. When other resources are available in adequate amounts, addi-

tional CO_2 can increase growth of C_3 plants over a wide range of CO_2 levels. By contrast, C_4 plants do not exhibit increased photosynthetic rates at higher CO_2 levels (Bazzaz 1990). Figure 28.10 illustrates these differences for two species of annual plants. Growth enhancement in C_3 plants is not always a simple phenomenon. Some plants acclimate to high CO_2 levels and show a decline in photosynthetic rate with time. Other resources for growth—water, light, and nutrients—must be present in excess to observe the effect of elevated CO_2 levels. Increased growth with increased CO_2 is also found in coniferous and broadleaved trees, and doubling CO_2 levels increased tree growth an average of 40% (Eamus and Jarvis 1989).

In most plants stomatal conductance declines as CO_2 levels rise. Because stomata are closed more often at high CO_2 levels, water loss through transpiration is reduced, so the overall result is improved water use efficiency (Bazzaz 1990). C_4 crop plants may show improved growth under elevated CO_2 levels because of increased water use efficiency, even when photosynthetic rates are unaffected by CO_2 levels (Rogers et al. 1983).

Many greenhouse studies of plant growth responses to CO_2 enrichment have been of short duration, and dramatic initial growth enhancements may not be sustained. Long-term studies have changed the

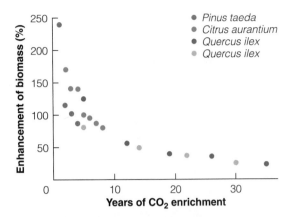

FIGURE 28.11
Percentage enhancement of tree biomass produced by CO_2 enrichment to 650 ppm (~300 ppm above current levels) for three tree species. Short-term enhancement is often over 150% but is not sustained in the long run. Effective enhancement persists after 35 years. The species are loblolly pine from North Carolina, orange trees from Arizona, and evergreen oaks from Italy. (From Idso 1999.)

picture somewhat (Bazzaz et al. 1994, Idso 1999). Figure 28.11 shows the growth enhancement resulting from CO_2 enrichment for three species of trees that grew up to 35 years in an environment enriched to approximately 650 ppm of CO_2. All three species fit a declining exponential curve, but growth enhancement is still 25% even after 30 years.

The effect of CO_2 on plant reproduction has been studied in many herbaceous species. In general, elevated CO_2 levels increase reproductive output. Flower number, fruit number and seed production all increase in herbs exposed to high CO_2 (Ward and Strain 1999). The increase in reproductive output could be due in part to a simple increase in size of individual plants, rather than an increased allocation of resources to reproductive output.

The key to understanding CO_2 effects in long-lived plants like trees is to determine how to extrapolate plant responses measured in the greenhouse on young plants to responses in older, larger plants in natural communities (Bazzaz et al. 1996). Yellow birch trees grown at high density show less enhancement from high CO_2 levels compared with trees grown in individual pots with no root competition (Wayne and Bazzaz 1995). The effect of competition for resources is even more complex in communities of many plant species.

Plant Community Responses to CO_2

To predict the global effects of CO_2 increases, ecologists must take these laboratory studies out into natural communities and expand short-term experiments into longer time frames. This has been done for only a few ecosystems, and in reviewing these findings we can begin to appreciate the task at hand for the coming decade.

Arctic ecosystems may be the most sensitive ecosystems to climate change. Plants in high latitudes may be more severely affected by rising global temperatures than those in lower latitudes (Schneider 1989). Arctic plant communities have several characteristics that make them susceptible to global warming. The active soil layer is shallow because of underlying permafrost, so that most root systems occur in the top 10–15 cm of the soil. Permafrost layers in the soil also contain large amounts of frozen organic matter that is not available to decomposers. Rising global CO_2 levels will lead to increased temperatures and increased evaporation on arctic tundra.

Billings et al. (1983) postulated that rising global temperatures could change arctic tundra communities from a sink for CO_2 to a source of CO_2. To test this idea they took cores of arctic tundra that were 8 cm in diameter into a greenhouse and measured net exchange of CO_2 from these cores at two levels of CO_2 and two levels of water table. Figure 28.12 shows the results of these experiments. Under present conditions in which the water table is at the surface, CO_2 enrichment has a minor effect on the carbon cycle in tundra vegetation, indicating that in this ecosystem CO_2 is not an important limiting factor. By contrast, water table levels strongly affect the carbon cycle of tundra. As the water table falls in the tundra, decomposition of organic matter increases, and this is further exacerbated if temperature rises as well (Billings et al. 1983, 1984). The net result of this work is that a doubling of CO_2 and the associated climatic warming will convert the wet tundra ecosystem of northern Alaska from a CO_2 sink to a CO_2 source. Plant growth in tundra communities is limited more by nitrogen availability, and this limitation prevents rising CO_2 levels from enhancing the growth of individual plants.

The soils of the boreal forest contain one of the largest pools of CO_2 in the terrestrial biosphere—about

---

[1]All tons used in this chapter are metric tons = 1000 kilograms = 2205 pounds.

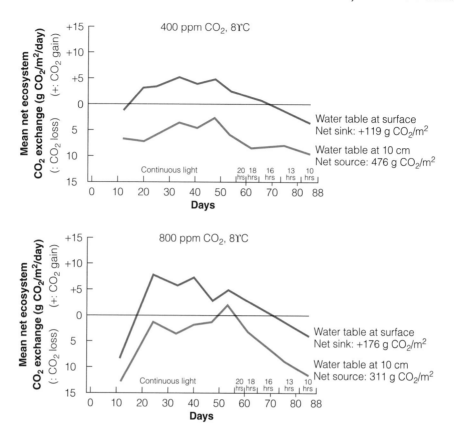

FIGURE 28.12
Net gain or loss of carbon dioxide from cores of arctic tundra from Barrow, Alaska, grown in a greenhouse through a simulated arctic summer at two levels of CO_2 enrichment. The water table is presently at the surface in this tundra community (blue line), but is predicted to drop with global warming (red line). If this happens, the tundra will no longer absorb CO_2 from the atmosphere but will become a source of further additions to increasing atmospheric CO_2. (From Billings et al. 1983.)

150 tons of carbon per ha or a total of 200–500 gigatons (Gt) of carbon (1 Gt = 10^9 tons[1]). If this carbon were released into the atmosphere, it alone could increase atmospheric CO_2 by 50% (Goulden et al. 1998). The stability of this carbon pool, like that of the arctic tundra, critically depends on the depth to which the frozen soils thaw during the growing season. At the present time, black spruce stands in the boreal forest are losing carbon to the atmosphere, probably as a result of soil warming. The increased plant growth associated with CO_2 enrichment in the boreal forest is not enough to compensate for losses from the soil pool.

The ecosystem consequences of CO_2 enrichment for forest communities are difficult to predict because of the time scale of current studies. Long-term plots are needed to measure changes in carbon storage.

Phillips et al. (1998) measured the gain in carbon in tropical forests over the past 25–40 years from basal area measurements on 600,000 individual trees scattered in 478 plots across the tropics. Figure 28.13 shows the biomass change in Amazonian forests from 1975 to 1996. On average, trees in these 97 plots increased in biomass by 1 ton per ha per year. Biomass in trees is accumulating particularly in the neotropical forests of Central and South America. This biomass increase is equivalent to fixing 0.62 tons of carbon per ha per year, and this carbon sink may account for about 40% of the "missing" carbon dioxide in the global carbon cycle.

Few studies with time frames of 5–10 years have yet been done to measure the community responses to CO_2 enrichment. Philip Grime and his research

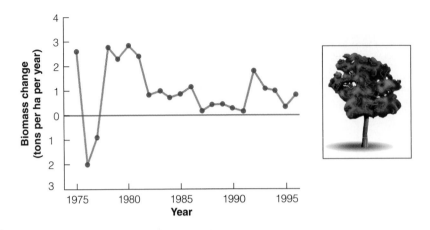

FIGURE 28.13

The annual aboveground biomass change in 97 Amazonian forest plots from 1975 to 1996. Points below the blue line indicate biomass loss. All trees on these plots were measured each year. Biomass increased in almost every year, and thus carbon was being taken up from the atmosphere. Growth in tropical forests could be a large part of the "missing" carbon sink. (From Phillips et al. 1998.)

group working at the University of Sheffield in England have questioned the simple extrapolation of short-term greenhouse experiments to long-term

J. Philip Grime *(1935–) Professor of Plant Ecology, University of Sheffield*

ecosystem effects. Much of the initial research on the effects of CO_2 enrichment has been conducted on crop plants growing on rich soils. Grime has argued that a different model of CO_2 effects is needed for

natural vegetation growing on infertile soils in which resources such as water, nitrogen, and phosphorus are often limiting (Diaz et al. 1994, Grime 1997). The effect of increasing CO_2 levels could be minimal if other resources determine plant community composition. A few experiments in natural ecosystems already suggest that little or no response to CO_2 enrichment occurs. Oechel et al. (1994) showed that enriching a tundra ecosystem with CO_2 had no net effect on carbon flux after three years of enrichment. Increasing temperature had a larger effect. The simple idea that more CO_2 means increased plant production is not correct in natural ecosystems.

Much scientific effort is now being expanded to define the limits of the fluxes of the global carbon cycle, and to define more clearly and more locally the sources and sinks for CO_2. All of this research is important because of the implications of the carbon cycle for climate change, a subject to which we now turn.

Climate Change

Global warming has become one of the major environmental issues of our time. Global warming is a function of the greenhouse effect, one of the most well-established theories of atmospheric chemistry (Schneider 1989). Figure 28.14 illustrates the greenhouse effect, which arises because the Earth's atmosphere traps heat near the surface of the planet. Water vapor, CO_2, and

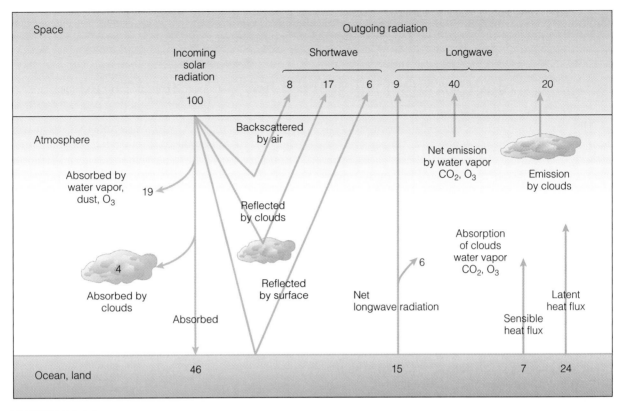

FIGURE 28.14
The greenhouse effect of CO_2 and other trace gases. The sun's radiation is dominated by short wavelengths, which are reflected or absorbed at or near the Earth's surface. The absorbed radiation is re-radiated at longer wavelengths that can be absorbed by atmospheric gases, including CO_2. Higher concentrations of these gases in the atmosphere reduce the net emission of longwave radiation into space, warming the earth. (From MacCracken 1985.)

other trace gases absorb the longer, infrared wavelengths emitted from the Earth, so an increase in the concentration of greenhouse gases tends to warm the Earth. None of this is controversial, and greenhouse principles apply equally well on Earth as they do on Venus, with its dense CO_2 atmosphere and very high temperatures, and on Mars, which has a thin CO_2 atmosphere and very cold temperatures. What is controversial is to predict from the greenhouse effect exactly how much the Earth's temperature will rise for a specified increase in greenhouse gases such as CO_2.

One way to project climate changes is to look back in time. Ice cores have provided one way of doing this. Air is trapped by snow as it is transformed into glacial ice, and by taking ice cores one can sample the atmosphere from an earlier time. The most spectacular example of this method is a 2083-m long ice core collected by the Soviet Antarctic Expedition at Vostok, Antarctica (Barnola et al. 1987). This ice core spans 160,000 years. Temperature at the time of ice formation can be determined by the ratio of oxygen-18 to oxygen-16 in the ice (Lorius et al. 1985). The resulting time series of changes in the Vostok ice core are shown in Figure 28.15. Even though a close correlation exists between carbon dioxide in the air and global temperatures over the past 160,000 years, there are three difficulties in extrapolating these correlations into the future. First, because our present CO_2 levels of 360 ppm exceed the levels found in nature during the past 160,000 years, we do not know if we can extrapolate past relationships into the future. Second, humans are changing atmospheric CO_2 levels very rapidly from year to year, whereas the historical changes in

ESSAY 28.2

EL NIÑO AND THE SOUTHERN OSCILLATION

Climate change on the time scale of months to years affects ecosystems and humans most directly. The most famous of these short-term changes is El Niño, first recognized by South American fishermen as an incursion of warm water off the coast of Peru. Once weather data began to be assembled, meteorologists discovered that El Niño events were correlated with the difference in atmospheric pressure between Tahiti and Darwin, Australia. The Southern Oscillation is the name given to this seesaw of change of atmospheric pressure between these two stations. Once it was realized that these oceanographic and atmospheric processes were coupled, the joint name El Niño–Southern Oscillation (ENSO) was coined.

The Southern Oscillation Index (SOI) is highly negatively correlated with sea surface temperatures in the eastern Pacific:

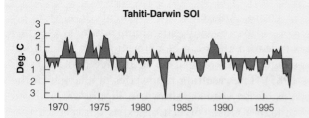

SOI is a relative index, measured by the difference between atmospheric pressure at Tahiti and atmos-

pheric pressure at Darwin, scaled to a long-term average of zero. When the SOI is negative, ocean temperatures are warmer than usual in the eastern Pacific and colder than usual in the western Pacific and Indian Oceans.

The effects of ENSO events are dramatic. When an El Niño event occurs, weather changes are triggered on a global scale, as the following maps show:

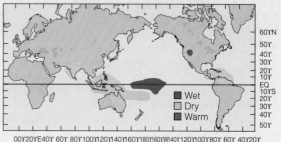

Map of ENSO effects in northern summer

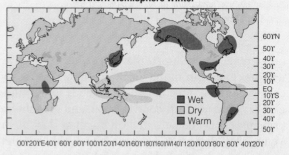

Map of ENSO effects in northern winter

(continued)

CO_2 were very slow. We do not know the rates at which an equilibrium can be established between CO_2 sources and sinks. Third, Figure 28.15 shows only a correlation between CO_2 and temperature; we do not yet know which is cause and which is effect

(Schneider 1989). Greenhouse gases other than CO_2 can contribute to global warming as well. The most important of these gases are methane, nitrous oxides, ozone, and chlorofluorocarbons (CFCs), and together these trace greenhouse gases may be as

Mild winters in the northeastern United States and western Canada are one correlate of El Niño, and severe droughts in Australia, India, Indonesia, Brazil, and Central America are typical of El Niño years.

The ENSO cycle has an average period of four years but varies from two to seven years in length. There is much variation in the strength of the Southern Oscillation for reasons that are not yet clear.

The effects of El Niño on global ecosystems are varied and significant. Warm surface temperatures lead to coral bleaching and the destruction of coral reefs. The pelagic upwelling ecosystem off Peru col-

lapses because the warm water displaces the nutrient-rich cold water. Fisheries such as the Peruvian anchovy collapse, and seabirds, which also depend on these fish, suffer high mortality in El Niño events. Pacific salmon production in the North Pacific is linked to similar changes in oceanographic events (Mantua et al. 1997). It is likely that many ecological changes are driven by large-scale weather changes caused by these oceanographic shifts, and the linkages between large-scale weather changes and ecosystem dynamics are a critical focus of current research.

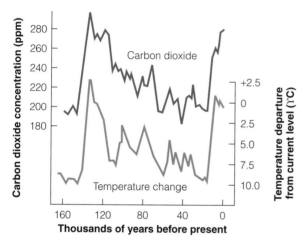

FIGURE 28.15
Long-term variations in global temperature (red) and atmospheric carbon dioxide concentration (blue) determined from the Vostok ice core, Antarctica. Carbon dioxide can be measured from air trapped in the ice as it is formed. A high correlation exists between atmospheric CO_2 levels and global temperature. (From Barnola et al. 1987.)

important or more important than CO_2 in the greenhouse effect of the twenty-first century.

How will changes in greenhouse gas levels and climatic warming affect the Earth's biota? There is an urgent need to answer this question, but it is not the only question ecologists face, and it is important to place this question within the perspective of all the human impacts on the Earth's ecosystems. Figure 28.16 illustrates the interconnecting human impacts on the Earth, many of which we have discussed in bits and pieces throughout this book. In the remainder of this chapter we need to deal with three major human

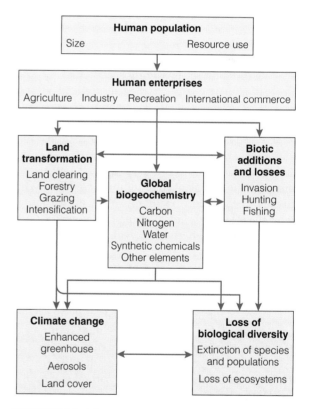

FIGURE 28.16
Human impacts on the Earth's ecosystems. Climate change is only one of five major impacts that interconnect to affect ecosystem integrity and health. (From Vitousek et al. 1996b.)

impacts: climate change, land transformation, and biotic additions and losses. Of these three, climate change gets the most publicity, but the other two changes are more insidious and may be more important (Vitousek et al. 1997b).

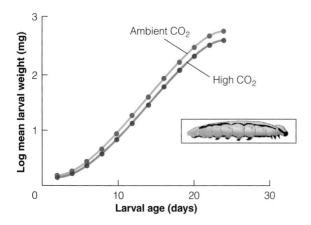

FIGURE 28.17
Growth of larvae of the buckeye butterfly Junonia coenia *on* Plantago lanceolata *grown at ambient and elevated CO$_2$ levels. Plants grown under high CO$_2$ conditions contain less nitrogen in their leaves, which reduces larval growth.(From Fajer et al. 1989.)*

The difficulty of providing specific data on the effects of climate change on agricultural and natural ecosystems is very great (Walker et al. 1999). The sanguine and simple view—that climate change will lead to increased plant growth and more luxuriant natural ecosystems, with no adverse consequences—is certainly not true. We need to determine the effects of climate change on plant communities and then on the animals that depend on them. What do we know about the effects of climate change on animals?

Even though there is no evidence that animals will be affected *directly* by changing CO$_2$ levels in the twenty-first century, there is concern that herbivorous animal populations and communities may be affected indirectly by changes in their food plants. Nitrogen content of plant leaves tends to fall with increased atmospheric CO$_2$ levels, and this may result in a reduction of larval insect growth in nitrogen-limited insect species (Figure 28.17). Insect herbivores are often limited by plant nitrogen in natural communities (White 1974), and reduced nitrogen may reduce insect numbers on CO$_2$-enriched vegetation. To compensate for low nitrogen levels, insect herbivores feeding on plants grown under high CO$_2$ levels may increase their feeding rate by 20–80% (Fajer 1989). Plant damage could increase even if insect numbers fall as CO$_2$ levels rise in the future.

Climatic warming could also disrupt the timing of hatching in insects (Dewar and Watt 1992). Insects that feed on newly emerging foliage are very sensitive to the age of the foliage. Maximum overlap between bud burst and larval emergence results in good survival and growth of the insects. In Scotland the emergence of the winter moth is strongly affected by spring temperatures, and consequently these larvae will emerge earlier if climatic warming occurs. By contrast, the vegetative buds of Sitka spruce, their food plant, are not greatly affected by spring temperatures and will open only slightly earlier when the climate warms. The net result will be a mismatch between the moth and its food plant that will reduce moth survival and growth (Dewar and Watt 1992). These short-term effects could be alleviated by natural selection in the longer term, but the disruption of life cycles in insect herbivores could be a major effect of global warming.

The concentrations of carbon-based secondary compounds (such as phenolics and terpenes) in plants also tend to increase under CO$_2$ enrichment (Peñuelas and Estiarte 1998). Both increased CO$_2$ and decreased nitrogen tend to increase the secondary compounds that deter feeding by herbivores. These compounds may also change the rate of plant decomposition through their effects on fungi and microbes (Ball 1997). As yet the effect of CO$_2$-enriched plants on herbivorous mammals or birds has not been studied, but effects through secondary chemicals might be expected.

Changes in Land Use

Humans convert forests to pastures, and agricultural fields into suburbs, with relatively little thought about the ecological consequences of these land-use changes. But these changes in land use, hectare by hectare, are among the most serious impacts on global ecosystems (Vitousek 1994), and they are difficult to quantify. Unfortunately, the availability of satellite data fails to make global measures of land use changes easy to measure, for two reasons. First, interpreting satellite data is not a simple matter, and for global measurements that demand high resolution, satellite images that are checked with field data collected on the ground are very expensive. Second, different definitions for classifying land cause confusion in measuring land-use changes. For example, two studies in India estimated deforestation rates from 1981 to 1990 with widely different results. One study by FAO (Food and Agricultural Organization of the United Nations) estimated deforestation to occur at 0.6% per year, whereas a simultaneous study by NRSA (National Remote Sensing Agency, India) estimated a deforestation rate of 0.04% per year (Menon and Bawa 1998). Having better satellites will not necessarily solve the second problem of defining ecosystem types that can be seen using satellite imagery. Tropical deforestation is estimated at 8% per decade (Watson 1999), but this is an educated guess rather than an ecological measurement.

Changes in land use have a dual effect: in habitat lost for plant and animal communities, and habitat fragmentation with its associated problems (see p. 361). The conversion of land from forest to agriculture leads to a large increase in CO_2 emissions. Houghton et al. (1999) estimated that until 1960, carbon emissions from land-use changes in the United States were greater than those from fossil fuel combustion. During the past 40 years, forest cover has been increasing in the United States as a result of tree replanting, farms being abandoned, and fire protection. During the 1980s, Houghton et al. (1999) estimated that this increase in forest cover was a net sink for carbon amounting to about 10–30% of the emissions from fossil fuel burning. Changes in land management have long-term consequences for climate change as well as direct effects on biodiversity.

Biotic Invasions and Species Ranges

Introduced species constitute another major impact on the biosphere. Human activity has moved many species across continents and oceans that they could not otherwise cross, with the resulting effects we have already discussed in Chapter 17. The introduction of the rabbit to Australia (see p. 272) and the zebra mussel to the United States (see p. 41) are two of many examples. Table 28.2 gives some examples of the magnitude of these biotic invasions. The economic impact of invasions is one strong reason for attempting to reduce introductions by enacting quarantine laws (Ruesink et al. 1995, Vitousek et al. 1996).

One of the most difficult effects of climate change to alleviate is the resulting changes in species distributions. If species ranges are currently controlled by climate, a shift in climate implies a shift in geographic range. Because shifts in range have occurred many

times in the history of the Earth, at first glance it might appear that this problem will take care of itself. Two factors argue against complacency. First, the speed of climate change is now many times greater than it has ever been in the past, raising the critical question: How fast can species move? Second, human changes in land use have disrupted many possible corridors of movement for both plants and animals, such that dispersal may no longer be possible.

The geographical distribution of many trees is probably limited by climate, and trees are thus a good group to use for studying this question. Figure 28.18 illustrates some projected magnitudes of range shifts that may occur in the next 100 years (Davis 1989). Two different climate change models were used to develop these predictions, which are coarse approximations. The range changes are very large, and there is no past precedent for such a rapid shift. At the end of the last ice age, North American trees dispersed north at rates varying from 10 km to 45 km per 100 years. Beech trees moved at an average rate of 20 km per 100 years (Woods and Davis 1989). But if the predictions indicated in Figure 28.18 are correct, beech trees will move north about 40× faster than they did at the end of the last ice age. This could occur only with the intervention of humans. One of the major conservation problems for the twenty-first century for North America and Europe could be the reestablishment of plant and animal communities at more northerly sites as climatic warming proceeds.

Summer temperatures determine the geographic distribution of C_3 and C_4 grasses in Australia (Hattersley 1983). Figure 28.19 shows the current isoclines for C_4 grasses in Australia and the predicted changes in these isoclines in 2030 under the current model of climatic warming for Australia. The dominance of C_4 grasses will move farther south as the climate warms. Major shifts in grassland species composition will occur in southern Australia in a very short time period, and the ecological consequences of this shift are not easily predicted (Henderson et al. 1993).

If plant communities are disrupted by climatic warming during the next 100 years, the animal communities on which they depend will be disrupted as well. At present few data are available from which to judge these potential effects, and the estimation and monitoring of climatic effects on plants and animals alike are part of the research agenda for all ecologists in the coming years.

TABLE 28.2 Biotic invasions of vascular plants, freshwater fish, and birds.

Island habitats are particularly vulnerable to invasions, but continental areas have also been strongly affected.

Taxa	Locale	Number of native species	Number of nonnative species	Percentage of nonnative species
Vascular plants	Germany	1,718	429	20.0
	Finland	1,006	221	18.0
	France	4,200	438	9.4
	California	4,844	1,025	17.5
	Canada	9,028	2,840	23.9
	Greenland	427	86	16.8
	Australia	20,000	2,000	10.0
	Tanzania	1,940	19	1.0
	South Africa	20,263	824	3.9
	Bermuda	165	303	64.7
	Hawaii	956	861	47.4
	Fiji	1,628	1,000	38.1
	New Zealand	1,790	1,570	46.7
Freshwater fish	California	76	42	35.6
	Canada	177	9	4.8
	Australia	145	22	13.2
	South Africa	107	20	15.7
	Brazil	517	76	12.8
	Hawaii	6	19	76.0
	New Zealand	27	30	52.6
Birds	Europe	514	27	5.0
	South Africa	900	14	1.5
	Brazil	1,635	2	0.1
	Hawaii	57	38	40.0
	New Zealand	155	36	18.8

Source: Data from Vitousek et al. (1996).

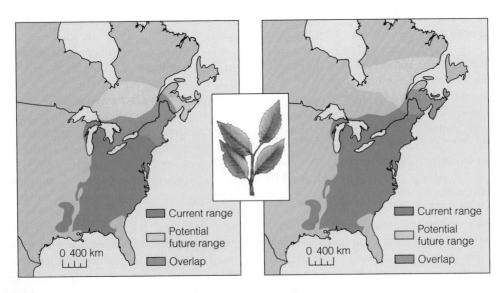

FIGURE 28.18
Current geographic range for American beech (Fagus americana) *in eastern North America, and the potential range in 100 years according to two climatic warming models: (a) a milder scenario, and (b) a more severe scenario. (From Roberts 1989.)*

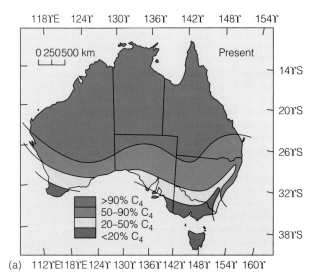

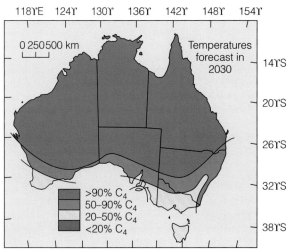

FIGURE 28.19
Map of Australia showing 90%, 50%, and 20% isoclines for C_4 native grass species (a) at the present time and (b) at temperatures forecast for the year 2030. The change in position of the isoclines is based on continental temperatures forecast for 2030 by CSIRO's Climate Impact Group. The geographic distribution of C_4 grasses is limited by summer temperature (see Chapter 7, p. 101). (From Henderson et al. 1993.)

Ecosystem Services

In our technological society we have lost touch with the numerous benefits ecosystems provide to human society. The term *ecosystem services* refers to all the processes through which natural ecosystems and the species they contain help sustain human life on this planet. The following is a list of a few ecosystem services we take for granted:

- Purification of air and water
- Mitigation of droughts and floods
- Generation and preservation of soils and soil fertility
- Detoxification and decomposition of wastes
- Pollination of crops and natural vegetation
- Dispersal of seeds
- Nutrient cycling
- Control of many agricultural pests by natural enemies
- Maintenance of biodiversity
- Protection of coastal shores from erosion
- Protection from ultraviolet rays
- Partial stabilization of climate
- Moderation of weather extremes
- Provision of aesthetic beauty

These ecosystem services are greatly undervalued by human society, as they have no dollar figure attached to them. But human life would cease to exist without these ecosystem services, and in this sense they are of immense value to us (see Box 28.3). Can we quantify this value?

When Robert Costanza and colleagues (1997) attempted to put a dollar figure on ecosystem services, they came up with an estimate of US$33 trillion per year, nearly twice as much as the gross national product of all the countries of the globe ($18 trillion). Many of the valuation techniques that are used in economic analyses of ecosystem services are based on the "willingness to pay" idea. If an individual owns a commercial forest, and ecosystem services provide a $50 increment to timber productivity, that individual should be willing to pay up to $50 for these services. The problem is that willingness to pay may not be a good measure if individuals are ill informed about the actual ecosystem services provided.

There are many uncertainties about this approach to quantifying the value of ecosystem services. The largest service contribution is from nutrient cycling, which makes up about half of the value of ecosystem services. The key point is that the value of ecosystem services is large, and if these services were actually paid for in our economic system, the global market

BOX 28.3

HOW IS BIODIVERSITY RELATED TO ECOSYSTEM FUNCTION?

One of the reasons for preserving species is that biodiversity is related to ecosystem function, so that ecosystem services like water purification and pollination depend on a very broad array of species. This argument for the preservation of biodiversity would be strengthened if native species, rather than introduced species, were the key contributors to ecosystem function. This argument basically centers on the role of rare species in ecosystems. We can envisage two extreme relationships between biodiversity and ecosystem function:

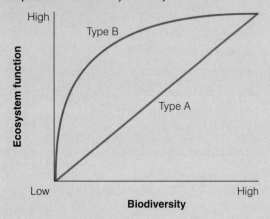

In type A communities, every single species contributes to ecosystem services, even the rare species. By contrast, in type B communities virtually all of the ecosystem services can be provided by relatively few species, and many species are redundant. In the extreme view, if these redundant species went extinct, no loss to ecosystem services would occur.

Many measures of ecosystem function can be made, such as biomass, productivity, stability, nutrient retention, or CO_2 uptake. Schwartz et al. (2000) reviewed the evidence to date from both laboratory and field studies and concluded that few data supported Type A relationships. Only three of 20 experimental tests fit a type A response curve. At present more evidence favors type B relationships.

There is no reason to suspect that all ecosystem functions will fit a single model, and much more work is needed. In particular, ecologists need to focus on ecosystem functions such as water purification, which are of practical significance to human society. The most likely ecosystem function that may fit a type A model is ecosystem variability. Highly variable systems are often unstable, and biodiversity may play an important role in reducing system variability.

The bottom line is that conservation ecologists should be wary of using the relationship between ecosystem function and biodiversity as an argument for preserving biodiversity. For many ecosystem functions, most species may be unnecessary, but this should not be allowed to be an excuse for permitting further extinctions of the Earth's biota.

---

system would be completely different. Many projects like large dams or irrigation projects would no longer be economical because their true cost would exceed social benefits.

Freshwater ecosystem services may be one of the simpler services to quantify, for humans need drinking water, they value lakes and rivers free of serious pollution (Wilson and Carpenter 1999). Specific indicators of water quality such as water clarity or the frequency of algal blooms are simple and are readily understood by the public as important indicators of ecosystem health. Most of the attempts to quantify freshwater ecosystem services have been site specific and cannot readily be extrapolated to larger areas like a state or country. For example, lakefront property values can be shown to decline

substantially as a lake becomes more polluted and algal blooms become frequent. The challenge is to expand these very local economic evaluations to larger scales so that the value of ecosystem services can enter environmental policy debates.

One large-scale experiment illustrates all too well how little we understand and value ecosystem services. The Biosphere 2 in Oracle, Arizona, was an attempt to construct a closed experimental ecosystem covering 1.27 ha. A forest with soil and a miniature ocean were constructed inside the airtight Biosphere, at a cost of over $200 million. From October 1991, eight people began a two-year stint living in isolation inside the Biosphere 2 ecosystem. Unlimited energy and technology were available from the outside to support the project. But the system failed, and the

E S S A Y   2 8 . 3

ON CORALS AND CLIMATE CHANGE

Because the tropical oceans leave a climate record in corals, one of the main techniques used for reconstructing climatic changes over the past several centuries involves coral reefs. Most reef corals live at depths of less than 20 m and grow continuously at rates of 6–20 mm per year. Many coral species produce annual density bands similar to tree rings that can be seen in X-rays or under ultraviolet light, as this example shows:

Many coral records have absolute annual chronologies extending back at least to the fifteenth century. Fossil corals going back several thousand years can be dated with carbon-14 (Gagan et al. 1998). Given that we can detect these annual rings, what can we infer from them?

The skeletons of reef-building corals carry a diverse suite of isotopic and chemical indicators that track water temperature and salinity. For example, the ratio of strontium to calcium traces sea surface temperatures at a time scale of three weeks (data courtesy of M. Gagan, Australian National University):

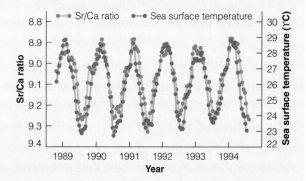

The ecological significance of these methods is that they permit us to reconstruct El Niño–Southern Oscillation events back in time before meteorological measurements were available, enabling us to assess their range of variability in the past. In collaboration with data from tree rings, ice cores, peat bogs, and other biophysical archives, we can begin to construct a picture of how climate has varied globally and locally, and to measure how the Earth's biota has responded to climatic fluctuations (Dunbar and Cole 1999).

experiment had to be stopped after 15 months (Cohen and Tilman 1996). Atmospheric oxygen dropped to 14%, and CO_2 fluctuated irregularly. Most of the vertebrate species in Biosphere 2 went extinct, and all of the pollinators died out. Population explosions of pests such as cockroaches showed clearly that the ecosystem services were impaired even before the experiment was halted. The conclusion is clear: No one knows yet how to engineer a system that will provide humans with all the life-support services that natural ecosystems produce for free. We have no alternative but to maintain the health of ecosystems on Earth, and that is the ecological challenge for the twenty-first century.

ESSAY 28.4

ECONOMICS OF ECOSYSTEM SERVICES

The environment's services are valuable, everyone agrees. The water we drink and the air we breathe are available to us only because of ecosystem services that we take for granted. How can we enroll market forces in the conservation of ecosystems? How can we get corporations and governments to invest in natural capital? Two example show how we might proceed to do this.

New York City's water supply comes from a watershed in the Catskill Mountains, and until recently water was purified by the natural processes of root systems and soil microorganisms, by filtration in the soils of this watershed, and by sedimentation in its streams. But continued additions of sewage to Catskill streams, in addition to fertilizer and pesticide use in local agriculture, had by the early 1990s degraded the Catskill water supplies to standards below those set for drinking water by the Environmental Protection Agency. In 1996 the city had two choices: (1) build and operate a water filtration plant to purify the water at a construction cost of over $6 billion to $8 billion and running costs of $300 million per year, or (2) restore the integrity of the Catskill ecosystem. The city chose to restore ecosystem integrity by buying land in and around the catchment area so that its use could be restricted, and by subsidizing the construction of better sewage treatment plants. By investing $1–1.5 billion in natural capital, New York City has saved a $6–8 billion investment in physical capital (Chichilnisky and Heal 1998.)

Over 90% of plants are pollinated by animals, including about 70% of major crop plants, and agricultural pollination is another clear example of an ecosystem service that we take for granted. The economic value of pollinators can be estimated very easily by looking at crop yields both in the absence of pollinators and in the presence of these animals (Daily 1997). Honeybees have been declining in the United States, partly because of introduced diseases, and in the past they provided much of the pollination of our crops. A 20% decline in the number of honeybee colonies in the United States occurred between 1990 and 1994. The key question is whether or not wild species of pollinators can take up all the slack in pollination when honey bee abundance declines. If they cannot do so, the losses to consumers would range from $1.6 billion to $5.7 billion per year. The important issue becomes how to manage the landscape to maximize the abundance of nonhoneybee pollinators. Pollination is one of the least understood processes in ecosystem functioning, and because of its importance to human food supplies it will be in the spotlight as an ecosystem service with a large yet unappreciated economic value.

Summary

Human impacts on the Earth's ecosystems are rooted in the continuing increase in world population. Six billion people now inhabit the Earth, and the population is growing by 214,000 people every day. The less developed countries of the globe contribute most of this increase, but much of the adverse effects of humans arises in the developed world, with its high demands for energy and materials.

The carrying capacity of the Earth is difficult to estimate. One approach is to estimate the *ecological footprint* of a nation in terms of the amount of land it utilizes to produce the commodities it consumes. By this measure the Earth is already at carrying capacity, and many countries have exceeded their available ecological resources and are in deficit. By no measure can the Earth sustain its current population if all live at the high consumption lifestyle of the developed world.

Human impacts are strongly registered in the carbon cycle. Rising levels of atmospheric CO_2 have occurred for 200 years due to fossil fuel burning and the destruction of native vegetation. The global carbon budget does not balance; a large sink now missing from the overall equation may be located in growing trees in temperate and tropical forests.

Climate change is one crucial result of the altered carbon cycle. Greenhouse gases such as CO_2, methane, and nitrous oxide trap heat at the Earth's surface and increase global temperatures. How increasing greenhouse gases will affect the Earth's biota is a focus of intensive current research. Individual plants typically grow more when CO_2 levels are elevated in a greenhouse, but in natural communities other factors such as nitrogen or water may limit productivity. Some increased plant growth will probably accompany climatic warming, but the ecosystem consequences for plants are far from clear. Arctic ecosystems are particularly susceptible to the effects of climatic warming. Animal communities will be affected indirectly through their food plants.

If geographical distributions are limited by climate, then climatic warming in the twenty-first century will have dramatic effects on the distribution of native animals and plants. Ecosystem restoration may be the major global conservation problem of the next 100 years.

Ecosystems provide services in air and water purification, pollination, and nutrient cycling that are greatly undervalued by society because they are not traded in the marketplace. We do not yet know how to construct intact ecosystems that can support human life so we had best take care of the Earth and safeguard the ecosystem services it provides us for free.

Key Concepts

1. Six billion people now inhabit the Earth, and population growth is the root cause of all our environmental problems.

2. The carrying capacity of the Earth is difficult to estimate, but the best guess is that we are already at or slightly above a sustainable level of world population.

3. The carbon cycle has been greatly altered by CO_2 emissions from fossil fuels and land clearing. Increasing CO_2 and other greenhouse gases will warm the Earth to an unknown extent in the coming centuries.

4. Global warming and atmospheric CO_2 enrichment may increase plant growth, but other limiting factors and community dynamics make predictions of the future difficult.

5. Ecosystem services such as pollination are provided by the Earth's ecosystems for free, and are not valued in our current economic systems.

6. We cannot at present construct an artificial ecosystem that can sustain human life indefinitely, no matter how much money we invest in it.

Selected References

Bazzaz, F. A. 1996. *Plants in Changing Environments: Linking Physiological, Population, and Community Ecology*. Cambridge University Press, Cambridge, England.

Chapin, F. S., III and G. R. Shaver. 1996. Physiological and growth responses of arctic plants to a field experiment simulating climatic change. *Ecology* 77:822–840.

Daily, G. C., S. Alexander, P. Ehrlich, L. Goulder, J. Lubchenco, P. A. Matson, H. A. Mooney, S. Postel, S. H. Schneider, D. Tilman, and G. M. Woodwell. 1997. Ecosystem services: benefits supplied to human societies by natural ecosystems. *Issues in Ecology* 2:1–16.

Grime, J. P. 1997. Climate change and vegetation. Pages 582-594 in M. J. Crawley, ed. *Plant Ecology*. Blackwell Science, Oxford.

Idso, S. B. 1999. The long-term response of trees to atmospheric CO_2 enrichment. *Global Change Biology* 5:493–495.

Laurance, W. F., S. G. Laurance, L. V. Ferreira, J. M. Rankin-De Merona, C. Gascon, and T. E. Lovejoy. 1997. Biomass collapse in Amazonian forest fragments. *Science* 278:1117–1118.

Newman, E. I. 1997. Phosphorus balance of contrasting farming systems, past and present. Can food production be sustainable? *Journal of Applied Ecology* 34:1334–1347.

Norby, R. J., S. D. Wullschleger, C. A. Gunderson, D. W. Johnson, and R. Ceulemans. 1999. Tree responses to rising CO_2 in field experiments: implications for the future forest. *Plant, Cell and Environment* 22:683–714.

Postel, S. L., G. C. Daily, and P. R. Ehrlich. 1996. Human appropriation of renewable fresh water. *Science* 271:785–788.

Vitousek, P. M., H. A. Mooney, J. Lubchenco, and J. M. Melillo. 1997. Human domination of Earth's ecosystems. *Science* 277:494–499.

Wackernagel, M., L. Onisto, P. Bello, A. Callejas Linares, I. S. Lopez Falfan, J. Mendez Garcia, A. I. Suarez Guerrero, and M. G. Suarez Guerrero. 1999. National natural capital accounting with the ecological footprint concept. *Ecological Economics* 29:375–390.

Walker, B. H., W. Steffen, J. Canadell, and J. Ingram, eds. 1999. *The Terrestrial Biosphere and Global Change: Implications for Natural and Managed Ecosystems*. Cambridge University Press, Cambridge, England. 439 pp.

Wilson, M. A., and S. R. Carpenter. 1999. Economic valuation of freshwater ecosystem services in the United States: 1971–1997. *Ecological Applications* 9: 772–783.

Questions and Problems

28.1 Discuss the implications of adopting a "top-down" versus a "bottom-up" view of community organization for evaluating the effects of climatic warming on aquatic and terrestrial ecosystems.

28.2 Peatlands in boreal and subarctic regions store a large pool of as yet undecomposed organic matter. Review the distribution of peatlands and their potential role in climatic change scenarios. Gorham (1991) provides an overview.

28.3 Discuss the implications of the response of C_3 and C_4 plants to changing CO_2 levels (see Figure 28.10) with respect to changes in community composition from the last ice age into the future. Ehleringer et al. (1997) review this question.

28.4 Pastor and Post (1988, p. 55) state that "the carbon and nitrogen cycles are strongly and reciprocally linked." Review the nitrogen cycle (see Chapter 27) and discuss these linkages.

28.5 The concept of *ecosystem health* has been criticized as a concept that applies well to individuals but poorly to a whole ecosystem. Discuss the application of the idea of *health* to populations, communities, and ecosystems. Rapport et al. (1998) discuss this question.

28.6 Discuss what approaches a country might take to eliminate the introduction of nonnative animals and plants.

28.7 Norby et al. (1992) grew yellow poplar (*Liriodendron tulipifera*) trees for three years at three levels of CO_2 enrichment. Photosynthetic rate nearly doubled at high CO_2 levels, but no difference in aboveground biomass of trees grown at normal or elevated CO_2 levels was found after three years. How is this possible?

28.8 In discussing models that will allow us to predict the effect of climatic change on primary production, Ågren et al. (1991, p. 134) state:

> We believe that in spite of gaps in the knowledge used to build the models used to predict the effects of climate change on forests and grasslands, the models are

sufficiently well substantiated and a number of dominant processes well enough understood for the predictions made with the models to be taken seriously.

In contrast, Long and Hutchin (1991, p. 139) conclude that: "there is insufficient information to predict accurately the response of primary production to climate change." Read these papers and discuss why there are such divergent views on this important question.

28.9 In agricultural landscapes, farmers have the choice of managing roadside verges by mowing, burning, grazing, or doing nothing to them. Discuss the implications of these four treatments for the global carbon cycle.

28.10 Discuss ways in which you might extrapolate results of climate change experiments on long-lived trees from the greenhouse to the forest stand. Compare the value of CO_2 growth enhancement from short-term studies in the greenhouse to the few long-term data available in Idso (1999).

28.11 Phosphorus is a major nutrient needed by crops, but it does not occur in the atmosphere and is a sparse element in most rocks. Discuss how you could determine if modern agriculture is sustainable with respect to phosphorus. How could you determine if pre-industrial agriculture was sustainable? Newman (1997) evaluates these questions.

28.12 Discuss the economic and ecological meanings of the word "value." Is there a difference? What activities might an economist include in a valuation of the use of freshwater? Does economics value the nonuse of a resource like lake water? Wilson and Carpenter (1999) discuss these issues.

Overview Question

One suggested ecological response to help arrest increasing atmospheric carbon dioxide levels is to plant trees. Explain the mechanisms behind this recommendation, and discuss how you could calculate how many trees would need to be planted to achieve this policy goal.

EPILOGUE

We have progressed from the simple problem of distribution to the complex problem of how organisms are integrated in biological communities. We have omitted a great deal in our survey. The applied problems of pesticide contamination and pollution have been only briefly discussed; they are dealt with in a wealth of books specifically oriented toward human activities and the environment.

This book has attempted to sketch the framework of ecology as both a pure and an applied science, and its purpose is to develop a particular view of the world, an "ecological consciousness." Two possible reasons for studying ecology are (1) to increase our understanding of the world in which we live and (2) to provide a basis for practical action on ecological problems.

Ecologists are now being called upon for judgments concerning the environmental impacts of a variety of proposals for pipelines, harbors, refineries, ski areas, and housing developments. The specter of global climate change hangs over everyone on Earth. One major task of applied ecology in the years ahead is to make ecological predictions and environmental assessments more a science than an art, and to achieve this goal we need sound ecological theory and solid local data. Human effects on ecosystems can be viewed as large-scale perturbation experiments, and ecologists who have adopted the experimental approach can use these activities to gain critical insights into the behavior of perturbed ecosystems.

We already understand how to ameliorate many of our environmental problems. These problems persist for political and economic reasons because of a lack of clear ecological guidance. Ecologists rarely have much to say in policy decisions, and there is a danger that as scientists we may be relegated to the role of technicians who monitor the demise of the world's ecosystems.

During the past 50 years, ecology has matured very rapidly as a science. Where once finding a single good illustration of some ecological principle required a massive search, now a full complement of examples is available to choose from. We have progressed rapidly from a descriptive science with vague generalizations to an experimental science with precise, quantitative, and testable hypotheses about ecological processes.

At present, theoretical ecology is particularly active and is making additional inroads into dealing with the complex systems ecologists must study. The pages of the *American Naturalist* and *Oikos* testify to the arguments that arise between different theoreticians, and between theoreticians and field ecologists. This is a good sign of vitality: Where there is argument, there is life! Most of the ideas within ecology are still hypotheses with an unknown range of applicability. These hypotheses do have consequences, and the job of ecologists must be to translate these hypotheses into practical suggestions. Some ideas will turn out to be completely wrong; others will withstand practical tests. Through all this, controversies will flare, contradictions will be published, and tempers will explode. The study of ecology is a very human activity and in some respects is a microcosm of the world.

APPENDIX I
A Primer on Population Genetics

Population genetics is the algebraic description of evolution, of how allelic frequencies change over time. Here we consider a very abbreviated basic statement of this approach for diploid organisms. See Tamarin (1999, Chapters 18–20) for more details.

We begin with individuals distinguished by a single locus with two possible alleles (A, B). Every individual is thus one of three genotypes (AA, AB, BB). The most important theorem in population genetics is called the Hardy-Weinberg law. In 1908 a British mathematician, G. H. Hardy, and a German physician, G. Weinberg, independently discovered that an equilibrium will arise and be maintained in both allelic and genotypic frequencies in any diploid population that is large in size, undergoes random mating, has no mutation or migration of individuals, and is subject to no selection. We can illustrate this law simply with an example. Consider a hypothetical population with the following composition:

Genotype	Genotypic frequency (%)	No. individuals
AA	25	1000
AB	0	0
BB	75	3000

We define:

$$\text{Allelic frequency of } A = \frac{\text{Total number of } A \text{ alleles in population}}{\text{Total number of alleles in population}}$$

$$= \frac{2(1000)}{2(1000) + 2(3000)} = \frac{2000}{8000}$$

$$= 0.25$$

$$\text{Allelic frequency of } B = \frac{\text{Total number of } B \text{ alleles in population}}{\text{Total number of alleles in population}}$$

$$= \frac{2(3000)}{2(1000) + 2(3000)} = \frac{6000}{8000}$$

$$= 0.75$$

Under the given assumptions, the Hardy-Weinberg law states that (1) allelic frequencies will not change over time, and (2) genotypic frequencies will come into equilibrium within one generation and will be as follows:

Genotypic frequency of AA = (allelic frequency of A)^2
Genotypic frequency of BB = (allelic frequency of B)^2
Genotypic frequency of AB = 2(allelic frequency of A)
(allelic frequency of B)

For our hypothetical example:

Genotypic frequency of AA = $(0.25)^2$ = 0.0625
Genotypic frequency of BB = $(0.75)^2$ = 0.5625
Genotypic frequency of AB = 2(0.25)(0.75) = 0.3750)

These frequencies sum to 1.00. If we apply the definition again for a second generation, a third generation, and so on, we find that none of these frequencies change. The Hardy-Weinberg law produces a globally stable equilibrium because no matter what the starting genotypic frequencies are, we always come to the same result (for one set of allelic frequencies). We can demonstrate this by doing these same calculations with genotype frequencies of AA = 0%, AB = 50%, and BB = 50%.

The important conclusion is that gene frequencies, or *genetic variability*, in a population will be maintained over time without any change *unless outside forces are applied*. Evolution results from these outside forces. A brief comment on these outside forces follows.

Mutations Genetic mutations are relatively rare events and do not typically cause shifts from Hardy-Weinberg equilibrium. Mutations are important as the source of genetic variation on which natural selection acts.

Migration Migration, or movements of individuals between populations, can be a means of adding or subtracting alleles in a population. Migration can be critical in preventing or aiding adaptation in local populations.

Population size In small populations, chance may be a critical element. For example, if two individuals in the previous example colonize an island, by chance both could be *BB* individuals. This is called random genetic drift and must be considered when populations are small in size.

Random mating Individuals may mate on the basis of similarity (or dissimilarity) so that assortative mating would result. For example, AA individuals may prefer to mate with AA individuals. If the choice of mates involves relatives, then either inbreeding or outbreeding may also be involved. In either case, mating is not random, and the Hardy-Weinberg equilibrium is disturbed.

Natural selection Selection produces adaptation by altering allelic frequencies by eliminating individuals that are less fit. We can introduce the idea of natural selection into our simple model population by defining the notion of *fitness*: the relative reproductive success of a given genotype. The following simple example shows how fitness can be calculated:

Genotype	AA	AB	BB
No. in generation 1	3000	6000	3000
No. in generation 2	2500	6000	3500

$$\text{Relative reproductive success of } AA = \frac{\text{No. in generation 2}}{\text{No. in generation 1}} = \frac{2500}{3000} = 0.83$$

$$\text{Relative reproductive success of } AB = \frac{6000}{6000} = 1.0$$

$$\text{Relative reproductive success of } BB = \frac{3500}{3000} = 1.17$$

By convention, the genotype with the highest relative reproductive success has fitness equal to 1.0, and we thus define *relative fitness* of the three genotypes as follows:

$$\text{Relative fitness of } BB = \frac{\text{Relative reproductive success of } BB}{\text{Highest relative reproductive success observed}}$$

$$= \frac{1.17}{1.17} = 1.0$$

$$\text{Relative fitness of } AB = \frac{1.00}{1.17} = 0.86$$

$$\text{Relative fitness of } AA = \frac{0.83}{1.17} = 0.71$$

We can now define the *selection coefficient* for each genotype:

$$\text{Selection coefficient for genotype } x = 1.0 - \text{relative fitess of genotype } x$$

The selection coefficient against the best genotype is always zero. Note that fitness (equivalent to relative fitness, relative Darwinian fitness, and adaptive value) is always *relative*, and evolution always deals with how fit one genotype is relative to other genotypes. If there is only one genotype in the population, one cannot define fitness.

Data obtained from a multiple census should be cast in the form of a method B table. This table is constructed by asking for each individual caught: (1) Was it marked or unmarked when caught? (2) If marked, when was it last captured? The method B table summarizes the answers to these questions:

		Time of capture				
		1	**2**	**3**	**4**	**5**	**6**
Time of last capture	1		10	3	5	2	2 ←Z_3
	2			34	18	8	4
	3				33	13	8 ←R_3
	4					30	20
	5						43
Total marked		0	10	37	56	53	77
Total unmarked		54	136	132	153	167	132
Total caught		54	146	169	209	220	209
Total released		54	143	164	202	214	207

These data were reported in Jolly (1965), and we will use his formulas to estimate the size of the marked population. We use the following symbols:

M_i = Marked population size at time i

m_i = Marked animals actually caught at time i

S_i = Total animals released at time i

Z_i = Number of individuals marked before time i, not caught in the ith sample but caught in a sample after time i

R_i = Number of the S_i individuals released at time i that are caught in a later sample

Some examples from the table will illustrate these definitions with respect to the method B table:

$$m_2 = 10$$

$$m_5 = 53$$

$$S_3 = 164$$

$$S_4 = 202$$

$$Z_2 = 3 + 5 + 2 + 2 = 12$$

$$Z_4 = 2 + 2 + 8 + 4 + 13 + 8 = 37$$

$$R_1 = 10 + 3 + 5 + 2 + 2 = 22$$

$$R_3 = 33 + 13 + 8 = 54$$

The formula to estimate the size of the marked population is

$$M_i = \frac{S_i Z_i}{R_i} + m_i$$

For example,

$$M_3 = \frac{(164)(39)}{54} + 37 = 155.4$$

$$M_4 = \frac{(202)(37)}{50} + 56 = 205.5$$

From the discussion in Chapter 9, we know that

$$\text{Total population size} = \frac{\text{Marked population size}}{\begin{array}{c}\text{Proportion of}\\\text{animals marked}\end{array}}$$

Consequently, we can calculate estimates for

$$\text{Total population at time } 3 = \frac{155.4}{37/169} = 709.8 \text{ animals}$$

$$\text{Total population at time } 4 = \frac{205.5}{56/209} = 767.0 \text{ animals}$$

Note that this is only one possible way of estimating the size of the marked population, and that it may not be the best technique for all circumstances (Seber 1982).

The concept of *rates* is critical for quantitative work in ecology, and students may find a brief review useful, especially for the discussions in Chapter 10.

A *rate* is a numerical proportion between two sets of things. For example, the number of students failing an examination might be 27 of 350, a failure rate of 7.7%. In ecological usage, a rate is usually expressed with a standard time base. Thus if eight out of 12 seedlings die within one year, the mortality rate is 66.7% *per year*. If a population grows from 100 to 150 within one month, the rate of population increase will be 50% *per month*.

We usually think in terms of *finite rates*, which are simple expressions of observed values. Some ecological examples are

$$\frac{\text{Annual}}{\text{survival}} = \frac{\text{No. alive at end of year}}{\text{No. alive at start of year}}$$

$$\frac{\text{Annual rate}}{\text{of population}} = \frac{\text{Population size at end of year}}{\text{Population size at start of year}}$$

Rates can also be expressed as *instantaneous rates*, in which the time base becomes very short rather than a year or a month. The general relationship between finite rates and instantaneous rates is

$$\text{Finite rate} = e^{\text{instantaneous rate}}$$

$$\text{Instantaneous rate} = \log_e \text{ finite rate}$$

where $e = 2.71828....$

The idea of an instantaneous rate can be explained most simply by the use of compound interest. Suppose we have a population of 100 organisms increas-

ing at a *finite* rate of 10% per year. The population size at the end of year 1 will be

$$100\left(1 + \frac{1}{10}\right) = 110$$

At the end of year 2, it will be

$$110\left(1 + \frac{1}{10}\right) = 121$$

At the end of year 3, it will be

$$121\left(1 + \frac{1}{10}\right) = 133.1$$

In general, for an interest rate of $1/m$ carried out n times,

$$y_n = y_0\left(1 + \frac{1}{m}\right)^n$$

where y_n = Amount at end of the nth operation

y_0 = Amount at start

We can repeat these calculations for a finite interest rate of 5% per half year. Everyone who has invested money in a savings account knows that 5% interest per half year is a *better* interest rate than 10% per year. We can see this quite simply. At six months our population size will be

$$100\left(1 + \frac{1}{20}\right) = 105$$

At one year, it will be

$$105\left(1 + \frac{1}{20}\right) = 110.25$$

and similarly at two years, 121.55, and at three years, 134.01. Compare these values with those obtained earlier for 10% annual interest.

Biological systems often operate on a time schedule of hours and days, so we may be more realistic in using rates that are instantaneous, that divide a year into very many short time periods. Let us repeat the first calculation with an *instantaneous* rate of increase of 10% per year. If we divide the year into 1000 short time periods, each time period having a rate of increase of 0.10/1000, or 0.0001, for the first 1000th of the year we have:

$$100(1 + 0.0001) = 100.01$$

For the second 1000th of the year,

$$100.01(1 + 0.0001) = 100.020001$$

If we repeat this for all 1000 time intervals, we end with 110.5 organisms at the end of one year.

Instantaneous rates and finite rates are nearly complementary when rates are very small. The following table and figure shows how they diverge as the rates become large and illustrates the change in size of a hypothetical population that starts at 100 organisms and increases or decreases at the specified rate for one time period:

Change (%)	Finite rate	Instantaneous rate	Hypothetical population at end of one time period
75	.25	1.386	25
50	.50	0.693	50
25	.75	0.287	75
10	.90	0.105	90
5	.95	0.051	95
0	1.00	0.000	100
5	1.05	0.049	105
10	1.10	0.095	110
25	1.25	0.223	125
50	1.50	0.405	150
75	1.75	0.560	175
100	2.00	0.693	200
200	3.00	1.099	300
400	5.00	1.609	500
900	10.00	2.303	1000

	Decreases	No change	Increases
Finite rates	0 to 1.00	1.00	1.00 to $+\infty$
Instantaneous rates	$-\infty$ to 0.00	0.00	0.00 to $+\infty$

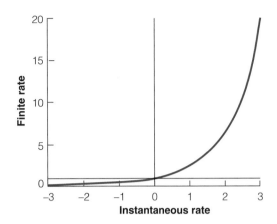

This illustrates one difference between finite rates (always positive or zero) and instantaneous rates (range from $-\infty$ to $+\infty$).

Mortality rates can be expressed as finite rates or as instantaneous rates. If the number of deaths in a short interval of time is proportional to the total population size at that time, then the rate of drop in numbers can be described by the geometric equation

$$\frac{dN}{dt} = iN$$

where N = population size

i = instantaneous mortality rate

t = time

In integral form, we have

$$\frac{N_t}{N_0} = e^{it}$$

where N_0 = starting population size

N_t = population size at time t*

Taking logs, if t = 1 time unit, we obtain:

$$\log_e\left(\frac{N_t}{N_0}\right) = i$$

Note that instantaneous rates are determined for a specific time base (per year, per month, etc.), even though the rate applies to a very short time interval. The examples on page 616 illustrate this.

Since N_t/N_0 is the finite survival rate by definition, we have obtained

$\log_e$(finite survival rate) = instantaneous mortality rate

We thus obtain the following relationships for expressing mortality rates:

Finite survival rate = 1.0 – finite mortality rate
$\log_e$(finite survival rate) = instantaneous mortality rate
Finite survival rate = $e^{\text{instantaneous mortality rate}}$
Finite mortality rate = 1.0 - $e^{\text{instantaneous mortality rate}}$

Why do we need to use instantaneous rates? The principal reason is that instantaneous rates are easier to deal with mathematically. A simple example will illustrate this property. Suppose we have data on an insect population and know that the mortality rate is 50% in the egg stage and 90% in the larval stages. How can we combine these mortalities? If they are expressed as finite mortality rates, we cannot add them because a 50% loss followed by a 90% loss is obviously not 140% mortality, but only 95% mortality. If, however, the mortality is expressed as instantaneous rates, we can add them directly:

	Instantaneous mortality rate
Egg stage (50%)	– 0.693
Larval stages (90%)	–2.303
Combined loss	–2.996

We can convert back to a finite mortality rate by the formula given earlier:

$$\begin{aligned} \text{Finite mortality rate} &= 1.0 - e^{\text{instantaneous mortality rate}} \\ &= 1.0 - e^{-2.996} \\ &= 0.950 \end{aligned}$$

and the combined mortality is seen to be 95 percent.

Four examples will illustrate some of these ideas, and students are referred to Ricker (1975, Chap. 1) for further discussion.

Example 1

A population increases from 73 to 97 within one year. This can be expressed as:

a. Finite rate of population growth = 97/73 = 1.329 per head per year (or, the population grew 32.9% in one year).

b. Instantaneous rate of population growth = $\log_e$(97/73) = 0.284 per head per year.

Example 2

A population decreases from 67 to 48 within one month. This can be expressed as:

a. Finite rate of population growth = 48/67 = 0.716 per head per month (or, the population decreased 28.4% over the month).

b. Instantaneous rate of population growth = $\log_e$(48/67) = -0.333 per head per month.

Example 3

A cohort of trees decreases in number from 24 to 19 within one year. This can be expressed as:

a. Annual survival rate (finite) = 19/24 = 0.792.

b. Annual mortality rate (finite) = 1.0 - annual survival rate = 0.208.

c. Instantaneous mortality rate = $\log_e$(19/24) = -0.234 per year.

Example 4

A cohort of fish decreases in number from 350,000 to 79,000 within one year. This can be expressed as:

a. Annual survival rate (finite) = 79,000/350,000 = 0.2257.

b. Annual mortality rate (finite) = 1.0 - 0.2257 = 0.7743.

c. Instantaneous mortality rate = $\log_e$ 0.2257 = -1.488 per year.

A P P E N D I X   I V
Species Diversity Measures of Heterogenity

There are several different measures of species diversity that are sensitive to both the number of species in the sample and the relative abundances of the species (Krebs 1999). Here we discuss only two of the most commonly used measures of heterogeneity.

The Shannon-Wiener function approaches the measure of species diversity through information theory. We ask the question: How difficult would it be to predict correctly the species of the next individual collected? This is the same problem faced by communication engineers interested in predicting correctly the name of the next letter in a message. This uncertainty can be measured by the Shannon-Wiener function[1]:

$$H = -\sum_{i=1}^{s}(p_i)(\log_2 p_i)$$

where $H =$ information content of sample (bits/individual) = index of species diversity

$S =$ number of species

$p_i =$ proportion of total sample belonging to the i th species

Information content is a measure of the amount of uncertainty, so the larger the value of H, the greater the uncertainty. A message such as *bbbbbbb* has no uncertainty in it, and $H = 0$. For our example on page 000 of two species of 99 and 1 individuals,

$$H = -\left[(p_1)(\log_2 p_1) + (p_2)(\log_2 p_2)\right]$$

$$= -\left[(0.99)(\log_2 0.99) + (0.01)(\log_2 0.01)\right]$$

$$= 0.81 \text{ bit/individual}$$

For a sample of two species with 50 individuals in each,

$$H = -\left[(0.50)(\log_2 0.50) + (0.50)(\log_2 0.50)\right]$$

$$= 1.00 \text{ bit/individual}$$

This agrees with our intuitive feeling that the second sample is more diverse than the first sample.

Two components of diversity are combined in the Shannon-Wiener function: (1) number of species and (2) equitability or evenness of allotment of individuals among the species (Krebs 1999). A greater number of species increases species diversity, and a more even or equitable distribution among species will also increase species diversity measured by the Shannon-Wiener function. Equitability can be measured in several ways. The simplest approach is to ask, What would be the species diversity of this sample if all S species were equal in abundance? In this case,

$$H_{max} = -S\left(\frac{1}{S}\log_2\left[\frac{1}{S}\right]\right) = \log_2 S$$

where $H_{max} =$ species diversity under conditions of maximal equitability

$S =$ number of species in the community

Thus, for example, in a community with two species only,

$$H_{max} = \log_2 2 = 1 \text{ bit/individual}$$

as we observed earlier. Equitability can now be defined as the ratio:

$$E = \frac{H}{H_{max}}$$

where $E =$ equitibility (range 0 - 1)

$H =$ observed species diversity

$H_{max} =$ maximun species diversity $= \log_2 S$

---

[1]This function was derived independently by Shannon and Wiener. It is sometimes mislabeled the Shannon-Weaver function.

Table IV-1 presents a sample calculation illustrating the use of these formulas.

Table IV-1

Sample calculations of species diversity and equitability through the use of the Shannon-Wiener function.
Proportional abundance

Tree species	(p_i)	$(p_i)(log2\ p_i)$a
Hemlock	0.521	0.490
Beech	0.324	0.527
Yellow birch	0.046	0.204
Sugar maple	0.036	0.173
Black birch	0.026	0.137
Red maple	0.025	0.133
Black cherry	0.009	0.061
White ash	0.006	0.044
Basswood	0.00:	0.032
Yellow poplar	0.002	0.018
Magnolia	0.001	0.010
Total	1.000	H = 1.829

$H_{max} = log_2 S = log_2 11 = 3.459$

Notes: Based on the composition of large trees (over 21.5 m tall) in a virgin forest in northwestern Pennsylvania. Note that there is no special theoretical reason to use log_2 instead of log_e or log_{10}. The log_2 usage gives us information units in "bits" (binary digits) and is preferred by information theorists. See Pielou (1969, p. 229). Source: Hough (1936).

Other measures of species diversity can be derived from probability theory. Simpson (1949) suggested the following question: What is the probability that two specimens picked at random in a community of infinite size will be the same species? If a person went into the boreal forest in northern Canada and picked two trees at random, there is a fairly high probability that they would be the same species. If a person went into the tropical rain forest,

by contrast, two trees picked at random would have a low probability of being the same species. We can use this approach to determine an index of diversity:

$$\text{Simpson's index of diversity} = \text{probability of picking two organisms at random that are different species}$$

$$= 1 - (\text{probability of picking two organisms that are the same species})$$

If a particular species i is represented in the community by p_i (proportion of individuals), the probability of picking two of these at random is the joint probability $[(p_i)(p_i)]$ or p_i^2. If we sum these probabilities for all the i species in the community, we get Simpson's diversity (D):

$$D = 1 - \sum_{i=1}^{s}(p_i)^2$$

where D = Simpson's index of diversity

p_i = proportion of individuals of species i in the community

For example, for our two-species community with 99 and 1 individuals,

$$D = 1 - \left[(0.99)^2 + (0.01)^2\right] = 0.02$$

Simpson's index gives relatively little weight to rare species and more weight to common species. It ranges in value from 0 (low diversity) to a maximum of $(1 - 1/S)$, where S is the number of species.

A more detailed discussion of these and other measures of species diversity is given in Magurran (1988) and Krebs (1999).

GLOSSARY

ECOLOGICAL JARGON is often carried to extremes, perhaps in an attempt to confuse amateurs. This book tries to avoid most of this jargon, and the glossary contains words used in the text that might be unfamiliar to students. Lincoln et al. (1998) provide an excellent, complete dictionary of ecological terms.

abiotic factors characterized by the absence of life; include temperature, humidity, pH, and other physical and chemical influences.

accidental species species that occur with a lower degree of fidelity in a community type; are poor species for use in community definition; see *characteristic species*.

aggregation coming together of organisms into a group, as in locusts.

allele one of a pair of characters that are alternative to each other in inheritance, being governed by genes situated at the same locus in homologous chromosomes.

allelopathy influence of plants, exclusive of microorganisms, upon each other, caused by products of metabolism; "antibiotic" interaction between plants.

association major unit in community ecology, characterized by essential uniformity of species composition.

autecology study of the individual in relation to environmental conditions.

autotroph organism that obtains energy from the sun and materials from inorganic sources; contrast with *heterotroph*. Most plants are autotrophs.

biogeography branch of biology that deals with the geographic distribution of plants and animals.

biological control use of organisms or viruses to control parasites, weeds, or other pests.

biosphere the whole-earth ecosystem, also called the *ecosphere*.

biota species of all the plants and animals occurring within a certain area or region.

biotic factors environmental influences caused by plants or animals; opposite of abiotic factors.

bryophytes plants in the phylum Bryophyta comprising mosses, liverworts, and hornworts.

canonical distribution particular configuration of the log-normal distribution of species abundances.

carnivore flesh eater; organism that eats other animals; contrast with *herbivore*.

catastrophic agents term used by Howard and Fiske (1911) to describe agents of destruction in which the percentage of destruction is not related to population density; synonymous with density-independent factors.

characteristic species species that are rigidly limited to certain communities and thus can be used to identify a particular type of community.

climax kind of community capable of perpetuation under the prevailing climatic and edaphic conditions.

closed population in population estimation, a population that is not changing in size during the interval of study, having no natality, mortality, immigration, or emigration.

community group of populations of plants and animals in a given place; an ecological unit used in a broad sense to include groups of various sizes and degrees of integration.

compensation point depth in a body of water at which the light intensity is such that the amount of oxygen produced by a plant's photosynthesis equals the oxygen it absorbs in respiration; the point at which respiration equals photosynthesis such that net production is zero.

competition occurs when a number of organisms of the same or of different species utilize common resources that are in short supply (*exploitation*); if the resources are not in short supply, competition occurs when the organisms seeking that resource harm one another in the process (*interference*).

confidence limits a statistical estimate of the possible range within which one might expect the true parameter value to lie, with a specified probability.

connectance used to describe food web complexity; the fraction of potential interactions in a food web that actually exist.

contingency table frequency distribution of an n-way statistical classification.

continuum index measure of the position of a community on a gradient defined by the species composition.

deme interbreeding group in a population; also known as *local population*.

density number of individuals in relation to the space or volume in which they occur.

deterministic model mathematical model in which all the relationships are fixed and the concept of probability does not enter; a given input produces one exact prediction as output; opposite of *stochastic model*.

diapause period of suspended growth or development and reduced metabolism in the life cycle of many insects, in which the organism is more resistant to unfavorable environmental conditions than during other periods.

dilution rate general term to describe the rate of additions to a population from birth and immigration.

dominance condition in communities or in vegetational strata in which one or more species, by means of their number, coverage, or size, have considerable influence upon or control of the conditions of existence of associated species.

dynamic pool model type of optimum-yield model in which the yield is predicted from the components of growth, mortality, recruitment, and fishing intensity; contrast with *logistic-type model*.

dynamics in population ecology, the study of the reasons for changes in population size; contrast with *statics*.

ecological longevity average length of life of individuals of population under stated conditions.

ecosystem biotic community and its abiotic environment; the whole Earth can be considered as one large ecosystem.

ecotone transition zone between two diverse communities (e.g., the tundra–boreal forest ecotone).

ecotype subspecies or race that is especially adapted to a particular set of environmental conditions.

edaphic pertaining to the soil.

ellfinwood (*Krummholz*) scrubby, stunted growth form of trees, often forming a characteristic zone at the upper limit of tree growth in mountains.

environment all the biotic and abiotic factors that actually affect an individual organism at any point in its life cycle.

epidemiology branch of medicine dealing with epidemic diseases.

epipelic algae algae living in or on the sediments of a body of water.

equitability evenness of distribution of species abundance patterns; maximum equitability occurs when all species are represented by the same number of individuals.

estivation condition in which an organism may pass an unfavorable season and in which its normal activities are greatly curtailed or temporarily suspended.

euryplastic showing a broad developmental response to different environmental conditions.

evapotranspiration sum total of water lost from the land by evaporation and plant transpiration.

experiment a test of a hypothesis, either observational or manipulative. The experimental method is the scientific method.

facultative agents term used by Howard and Fiske (1911) to describe agents of destruction that increase their percentage of destruction as population density rises; synonymous with *density-dependent factors*.

fecundity the potential reproductive capacity of an organism, measured by the number of gametes produced.

fertility an ecological concept of the actual number of viable offspring produced by an organism, equivalent to realized fecundity.

fidelity degree of regularity or "faithfulness" with which a species occurs in certain plant communities, expressed on a five-part scale: (5) exclusive, (4) selective, (3) preferential, (2) companion, indifferent, (1) accidental, strangers.

food chain metaphor for the dependence for food of organisms upon others in a series, beginning with plants and ending with the largest carnivores.

genecology study of population genetics in relation to the habitat conditions; the study of species and other taxa by the combined methods and concepts of ecology and genetics.

genet a unit derived by asexual reproduction from a single original zygote.

genotype entire genetic constitution of an organism; contrast with phenotype.

global stability ability to withstand perturbations of a large magnitude and not be affected; compare with *neighborhood stability*.

gradocoen totality of all factors that impinge on a population, including biotic agents and abiotic factors.

greenhouse gases gases present in the Earth's atmosphere that reflect infrared radiation back to Earth, thus warming it. The most important ones affected by humans are carbon dioxide, methane, nitrous oxide, and chlorofluorocarbons. Water vapor also acts as a greenhouse gas.

gross production production before respiration losses are subtracted; photosynthetic production for plants and metabolizable production for animals.

growth form morphological categories of plants, such as trees, shrubs, and vines.

herbivore organism that eats plants; contrast with *carnivore*.

heterotroph organism that obtains energy and materials by eating other organisms; contrast with *autotroph*.

homeostasis maintenance of constancy or a high degree of uniformity in an organism's functions or interactions of individuals in a population or community under changing conditions; results from the capabilities of organisms to make adjustments.

homeothermic pertaining to warm-blooded animals that regulate their body temperature; contrast with *poikilothermic*.

host organism that furnishes food, shelter, or other benefits to another organism of a different species.

hydrophyte plant that grows wholly or partly immersed in water; compare with *xerophyte* and *mesophyte*.

hypothesis universal proposition that suggests an explanation for some observed ecological situation.

importance value sum of relative density, relative dominance, and relative frequency for a species in the community on a scale from 0 to 300; the larger the importance value, the more dominant a species is in the particular community.

index of similarity ratio of the number of species found in common in two communities to the total number of species that are present in both.

Indifferent species species occurring in many different communities; are poor species for community classification.

interspecific competition competition between members of different species.

intrinsic capacity for increase (r) measure of the rate of increase of a population under controlled conditions, with fixed birth and death rates; also called *innate capacity for increase*.

isotherm line drawn on a map or chart connecting points with the same temperature at a particular time or over a certain period.

keystone species species in a community whose removal has strong effects on community diversity and composition. Keystone species are often top predators and may be relatively rare.

life table tabulation presenting complete data on the mortality schedule of a population.

littoral shallow-water zone of lakes or the sea, with light penetration to the bottom; often occupied by rooted aquatic plants.

local population see *deme*.

logistic equation model of population growth described by a symmetrical S-shaped curve with an upper asymptote.

logistic-type model type of optimum-yield model in which the yield is predicted from an overall descriptive function of population growth without a separate analysis of the components of mortality, recruitment, and growth; contrast with *dynamic pool model*.

log-normal distribution frequency distribution of species abundances in which the *x* axis is expressed on a logarithmic scale; *x* axis is (log) number of individuals represented in sample, *y* axis is number of species.

loss rate general term to describe the rate of removal of organisms from a population by death and emigration.

mesic moderately moist.

mesophyte plant that grows in environmental conditions that include moderate moisture conditions.

metapopulation a set of local populations linked together through dispersal.

monogamy mating of an animal with only one member of the opposite sex.

morphology study of the form, structure, and development of organisms.

multivoltine refers to an organism that has several generations during a single season; contrast with *univoltine*.

mutualism interaction between two species in which both benefit from the association and cannot live separately.

neighborhood stability ability to withstand perturbations of small magnitude and not be affected; compare with *global stability*.

net production production after respiration losses are subtracted.

niche role or "profession" of an organism in the environment; its activities and relationships in the community.

obligate predator or parasite that is restricted to eating a single species of prey.

oligochaetes any of a class or order (Oligochaeta) of hermaphroditic terrestrial or aquatic annelids lacking a specialized head; includes earthworms.

open population in population estimation, a population that has natality, mortality, immigration, or emigration during the interval of study.

optimum yield amount of material that can be removed from a population to maximize biomass (or numbers, or profit, or any other type of "optimum") on a sustained basis.

ordination process by which plant or animal communities are ordered along a gradient.

parasite organism that benefits while feeding upon, securing shelter from, or otherwise injuring another organism (the host); insect parasitoids are usually fatal to their host and behave more like vertebrate predators.

parthenogenesis development of the egg of an organism into an embryo without fertilization.

phenology study of the periodic (seasonal) phenomena of animal and plant life and their relations to the weather and climate (e.g., the time of flowering in plants).

phenotype expression of the characteristics of an organism as determined by the interaction of its genic constitution and the environment; contrast with genotype.

photoperiodism response of plants and animals to the relative duration of light and darkness (e.g., a chrysanthemum blooming under short days and long nights).

photosynthesis synthesis of carbohydrates from carbon dioxide and water by chlorophyll using light as energy and with oxygen as a byproduct.

physiological longevity maximum life span of individuals in a population under specified conditions; the organisms die of senescence.

phytoplankton plant portion of the plankton; the plant community in marine and freshwater environments that floats free in the water and contains many species of algae and diatoms.

poikilothermic of or pertaining to cold-blooded animals, organisms that have no rapidly operating heat-regulatory mechanism; contrast with *homeothermic*.

polyandry mating of a single female animal with several males.

polygyny mating of one male animal with several females.

population group of individuals of a single species.

primary production production by green plants.

production amount of energy (or material) formed by an individual, population, or community in a specific time period; includes growth and reproduction only; see *primary production, secondary production, gross production, net production*.

productivity a general concept denoting all processes involved in production studies—consumption, rejection, respiration and so on; some authors use it as a synonym for production, but this should be avoided because production is a narrower term.

promiscuous not restricted to one sexual partner.

proximate factors in evolutionary terms, the mechanisms responsible for an adaptation with reference to its physiological and behavioral operation; the mechanics of how an adaptation operates; opposite of *ultimate factors*.

ramet a modular unit of a clone that may live independently if separated from the parent organism.

recruitment increment to a natural population, usually from young animals or plants entering the adult population.

resilience magnitude of disturbance that can be absorbed before an ecosystem changes its structure; one aspect of ecosystem stability.

respiration complex series of chemical reactions in all organisms by which energy is made available for use; carbon dioxide, water, and energy are the end products.

saprophyte plant that obtains food from dead or decaying organic matter.

secondary production production by herbivores, carnivores, or detritus feeders; contrast with *primary production*.

self-regulation process of population regulation in which population increase is prevented by a deterioration in the quality of individuals that make up the population; population regulation by adjustments in behavior and physiology within the population rather than by external forces such as predators.

senescence process of aging.

seral referring to a series of stages that follow one another in an ecological succession.

serotinous cones cones of some pine trees that remain on the trees for several years without opening and require a fire to open and release the seeds.

sessile attached to an object or fixed in place (e.g., barnacles).

sigmoid curve S-shaped curve; in ecology, often a plot of time (x-axis) against population size (y-axis); an example is the logistic curve.

stability absence of fluctuations in populations; ability to withstand perturbations without large changes in composition.

standard error a statistical estimate of the precision of an estimate such as the mean.

statics in population ecology, the study of the reasons of equilibrium conditions or average values; contrast with *dynamics*.

stenoplastic having little or no modificational plasticity (steno- means narrow); opposite of *euryplastic*.

steppe extensive area of natural, dry grassland; usually used in reference to grasslands in southwestern Asia and southeastern Europe; equivalent to prairie in North American usage.

sterol any of a group of solid, mostly unsaturated polycyclic alcohols, such as cholesterol or ergosterol, derived from plants and animals.

stochastic model mathematical model based on probabilities; the prediction of the model is not a single fixed number but a range of possible numbers; opposite of *deterministic model*.

sublittoral lower division in the sea from a depth of 40 to 60 meters to about 200 meters; below the littoral zone.

succession replacement of one kind of community by another kind; the progressive changes in vegetation and animal life that may culminate in the climax community.

symbiosis in a broad sense, the living together of two or more organisms of different species; in a narrow sense, synonymous with *mutualism*.

synecology study of groups of organisms in relation to their environment; includes population, community, and ecosystem ecology.

taiga the northern boreal forest zone, a broad band of coniferous forest south of the arctic tundra.

thermoregulation maintenance or regulation of temperature, specifically the maintenance of a particular temperature of the living body.

total fertility rate number of children a woman could expect to produce in her lifetime if the birth rate were held constant at current conditions.

trace element chemical element used by organisms in minute quantities and essential to their physiology.

trophic level functional classification of organisms in a community according to feeding relationships; the first trophic level includes green plants; the second trophic level includes herbivores, and so on.

tundra treeless area in arctic and alpine regions, varying from a bare area to various types of vegetation consisting of grasses, sedges, forbs, dwarf shrubs, lichens, and mosses.

ultimate factors in evolutionary terms, the survival value of a given adaptation; the evolutionary reason for the adaptation; opposite of *proximate factors*.

univoltine refers to an organism that has only one generation per year.

vector organism organism (often an insect) that transmits a pathogenic virus, bacterium, protozoan, or fungus from one organism to another.

wilting point measure of soil water; the water remaining in the soil (expressed as percentage of dry weight of the soil) when the plants are in a state of permanent wilting from water shortage.

xeric deficient in available moisture for the support of life (e.g., desert environments).

xerophyte plant that can grow in dry places (e.g., cactus).

zooplankton animal portion of the plankton; the animal community in marine and freshwater environments that floats free in the water, independent of the shore and the bottom, moving passively with the currents.

BIBLIOGRAPHY

Aarssen, L. W. 1997. On the progress of ecology. *Oikos* 80:177–178.

Abele, L. G. and K. Walters. 1979. Marine benthic diversity: A critique and alternative explanation. *Journal of Biogeography* 6:115–126.

Abrams, P. 1975. Limiting similarity and the form of the competition coefficient. *Theoretical Population Biology* 8:356–375.

———. 1997. Anomalous predictions of ratio-dependent models of predation. *Oikos* 80:163–171.

Abrams, P. A. 1986. Adaptive responses of predators to prey and prey to predators: The failure of the arms-race analogy. *Evolution* 40:1229–1247.

———. 1987. On classifying interactions between populations. *Oecologia* 73:272–281.

———. 1996. Limits to the similarity of competitors under hierarchical lottery competition. *American Naturalist* 148:211–219.

———. 1998. High competition with low similarity and low competition with high similarity: Exploitative and apparent competition in consumer-resource systems. *American Naturalist* 152:114–128.

Abrams, P. A. and C. J. Walters. 1996. Invulnerable prey and the paradox of enrichment. *Ecology* 77:1125–1133.

Adams, J. M., H. Faure, L. Faure–Denard, J. M. McGlade, and F. I. Woodward. 1990. Increases in terrestrial carbon storage from the last glacial maximum to the present. *Nature* 348:711–714.

Agnew, A. D. Q. 1961. The ecology of *Juncus effusus* L. in North Wales. *Journal of Ecology* 49:83–102.

Ågren, G. I., R. E. McMurtrie, W. J. Parton, J. Pastor, and H. H. Shugart. 1991. State-of-the-art of models of production-decomposition linkages in conifer and grassland ecosystems. *Ecological Applications* 1:118–138.

Aguirre, A. A. and E. E. Starkey. 1994. Wildlife disease in U.S. National Parks: Historical and coevolutionary perspectives. *Conservation Biology* 8:654–661.

Akcakaya, H. R., R. Arditi, and L. R. Ginzburg. 1995. Ratio-dependent predation: An abstraction that works. *Ecology* 76:995–1004.

Allen, K. R. 1980. *Conservation and Management of Whales.* University of Washington Press, Seattle.

Allendorf, F. W. and R. F. Leary. 1986. Heterozygosity and fitness in natural populations of birds and mammals. Pages 57–76 in M. E. Soule, ed., *Conservation Biology: The Science of Scarcity and Diversity.* Sinauer Associates, Sunderland, Mass.

Anagnostakis, S. L. and B. Hillman. 1992. Evolution of the chestnut tree and its blight. *Arnoldia* 52:3–10.

Anderbrant, O. and F. Schlyter. 1987. Ecology of the Dutch elm disease vectors *Scolytus laevis* and *S. scolytus* (Coleoptera: Scolytidae) in southern Sweden. *Journal of Applied Ecology* 24:539–550.

Anderson, D. M. 1994. Red tides. *Scientific American* 271:52–58.

Anderson, D. R., Burnham, K. P., White, G. C., and Otis, D. L. 1983. Density estimation of small-mammal populations using a trapping web and distance sampling methods. *Ecology* 64:674–680.

Anderson, G. R. V., A. H. Ehrlich, P. R. Ehrlich, J. D. Roughgarden, B. C. Russell, and F. H. Talbot. 1981a. The community structure of coral reef fishes. *American Naturalist* 117:476–495.

Anderson, R. M. 1991. Populations and infectious diseases: Ecology or epidemiology? *Journal of Animal Ecology* 60:1–50.

Anderson, R. M., H. C. Jackson, R. M. May, and A. M. Smith. 1981b. Population dynamics of fox rabies in Europe. *Nature* 289:765–771.

Anderson, R. M. and R. M. May. 1979. Population biology of infectious diseases: Part I. *Nature* 280:361–367.

———. 1978. Regulation and stability of host-parasite population interactions. *Journal of Animal Ecology* 47:219–247.

———. 1982. Coevolution of hosts and parasites. *Parasitology* 85:411–426.

———. 1991. *Infectious Diseases of Humans: Dynamics and Control.* Oxford University Press, Oxford.

Anderson, S. 1985. The theory of range-size (RS) distributions. *American Museum Novitates* 2833:1–20.

Anderson, W. B. and G. A. Polis. 1998. Marine subsidies of island communities in the Gulf of California: Evidence from stable carbon and nitrogen isotopes. *Oikos* 81:75–80.

Andrén, H. and P. Angelstam. 1988. Elevated predation rates as an edge effect in habitat islands: Experimental evidence. *Ecology* 69:544–547.

Andrewartha, H. G. 1961. *Introduction to the Study of Animal Populations.* University of Chicago Press, Chicago.

Andrewartha, H. G. and L. C. Birch. 1954. *The Distribution and Abundance of Animals.* University of Chicago Press, Chicago.

Angel, M. V. 1993. Biodiversity of the pelagic ocean. *Conservation Biology* 7:760–772.

Antonovics, J., A. D. Bradshaw, and R. G. Turner. 1971. Heavy metal tolerance in plants. *Advances in Ecological Research* 7:1–85.

Aplet, G. H., R. D. Laven, and F. W. Smith. 1988. Patterns of community dynamics in Colorado Engelmann spruce–subalpine fir forests. *Ecology* 69:312–319.

Applegate, V. C. 1950. Natural history of the sea lamprey, *Petromyzon marinus*, in Michigan. U.S. Fish and Wildlife Service, Washington, D.C.

Arditi, R., L. R. Ginzburg, and H. R. Akcakaya. 1991. Variation in plankton densities among lakes: A case for ratio-dependent predation models. *American Naturalist* 138:1287–1296.

Arditi, R., N. Perrin, and H. Saiah. 1991c. Functional responses and heterogeneities: An experimental test with cladocerans. *Oikos* 60:69–75.

Arita, H. T., J. G. Robinson, and K. H. Redford. 1990. Rarity in neotropical forest mamals and its ecological correlates. *Conservation Biology* 4:181–192.

Armbrust, E. J. and G. G. Gyrisco. 1982. Forage crops insect pest management. Pages 443–463 in R. L. Metcalf and W. Luckman, eds. *Introduction to Insect Pest Management.* Wiley, New York.

Armstrong, R. A. and R. McGehee. 1980. Competitive exclusion. *American Naturalist* 115:151–170.

Arrigo, K. R., D. L. Worthen, M. P. Lizotte, P. Dixon, and G. Diekmann. 1997. Primary production in Antarctic sea ice. *Science* 276:394–397.

Arthur, W. 1982. The evolutionary consequences of interspecific competition. *Advances in Ecological Research* 12:127–187.

Ashmole, N. P. 1968. Body size, prey size, and ecological segregation in five sympatric tropical terns (Aves: Laridae). *Systematic Zoology* 17:292–304.

Astrom, M., P. Lundberg, and S. Lundberg. 1996. Population dynamics with sequential density–dependencies. *Oikos* 75:174–181.

Attiwill, P. 1981. Energy, nutrient flow, and biomass. Pages 131–144. *Proceedings of the Australian Forest Nutrition Workshop.* CSIRO, Melbourne.

Ault, T. R. and C. R. Johnson. 1998. Spatially and temporally predictable fish communities on coral reefs. *Ecological Monographs* 68:25–50.

Austin, M. P. 1985. Continuum concept, ordination methods, and niche theory. *Annual Review of Ecology and Systematics* 16:39–61.

———.1990. Community theory and competition in vegetation. Pages 215–238 in J. B. Grace and D. Tilman, eds. *Perspectives on Plant Competition.* Academic Press, San Diego.

———. 1999. A silent clash of paradigms: Some inconsistencies in community ecology. *Oikos* 86:170–178.

Austin, M. P., A. O. Nicholls, and C. R. Margules. 1990. Measurement of the realized qualitative niche: Environmental niches of five *Eucalyptus* species. *Ecological Monographs* 60:161–177.

Austin, M. P. and T. M. Smith. 1989. A new model for the continuum concept. *Vegetatio* 83:35–47.

Bach, C. E. 1991. Direct and indirect interactions between ants (*Pheidole megacephala*), scales (*Coccus viridis*) and plants (*Pluchea indica*). *Oecologia* 87:233–239.

Bacon, P. J. 1985. A systems analysis of wildlife rabies epizootics. Pages 109–130 in P. J. Bacon, ed., *Population Dynamics of Rabies in Wildlife.* Academic Press, London.

Bagenal, T. B. 1973. Fish fecundity and its relations with stock and recruitment. *Rapport du Conseil International pour l'Exploration de la Mer* 164:186–198.

Bailey, N. T. J. 1975. *The Mathematical Theory of Infectious Diseases and its Application.* Griffin, London.

Baker, M. C., D. B. Thompson, G. L. Sherman, M. A. Cunningham, and D. F. Tomback. 1982. Allozyme frequencies in a linear series of song dialect populations. *Evolution* 36:1020–1029.

Ball, A. S. 1997. Microbial decomposition at elevated CO^2 levels: Effect of litter quality. *Global Change Biology* 3:379–386.

Baltz, D. M. 1991. Introduced fishes in marine systems and inland seas. *Biological Conservation* 56:151–177.

Barbehenn, R. V. and E. A. Bernays. 1992. Relative nutritional quality of C_3 and C_4 grasses for a graminivorous lepidopteran (*Paratrytone melane*, Hesperiidae). *Oecologia* 92:97–103.

Barclay-Estrup, P. and C. H. Gimingham. 1969. The description and interpretation of cyclical processes in a heath community. I. Vegetational change in relation to the Calluna cycle. *Journal of Ecology* 57:737–758.

Barlow, N. D. 1995. Critical evaluation of wildlife disease models. Pages 230–259 in B. T. Grenfell and A. P. Dobson, eds. *Ecology of Infectious Diseases in Natural Populations.* Cambridge University Press, Cambridge, England.

Barnola, H. M., D. Raynaud, Y. S. Korotkevich, and C. Lorius. 1987. Vostok ice core provides 160,000-year record of atmospheric carbon dioxide. *Nature* 329:408–414.

Barnola, J. M., M. Anklin, J. Porcheron, D. Raynaud, J. Schwander, and B. Stauffer. 1995. CO_2 evolution during the last millennium as recorded by Antarctic and Greenland ice. *Tellus* 47B:264–272.

Baroudy, E. and J. M. Elliott. 1994. The critical thermal limits for juvenile Arctic charr (*Salvelinus alpinus*). *Journal of Fish Biology* 45:1041–1053.

Bart, J. and E. D. Forsman. 1992. Dependence of northern spotted owls *Strix occidentalis caurina* on old-growth forests in the western USA. *Biological Conservation* 62:95–100.

Bartholomew, G. A. 1958. The role of physiology in the distribution of terrestrial vertebrates. Pages 81–95 in C. L. Hubbs, ed., *Zoogeography*. American Association for the Advancement of Science Publication, Washington, D.C.

Barton, A. M. 1993. Factors controlling plant distributions: Drought, competition, and fire in montane pines in Arizona. *Ecological Monographs* 63:367–398.

Bazzaz, F. A. 1990. The response of natural ecosystems to the rising global carbon dioxide levels. *Annual Review of Ecology and Systematics* 21:167–196.

———. 1991. Habitat selection in plants. *American Naturalist* 137(suppl.):S116–S130.

———. 1996. *Plants in Changing Environments: Linking Physiological, Population, and Community Ecology*. Cambridge University Press, Cambridge, England.

Bazzaz, F. A., S. L. Bassow, G. M. Berntson, and S. C. Thomas. 1996. Elevated CO_2 and terrestrial vegetation: Implications for and beyond the global carbon budget. Pages 43–76 in B. Walker and W. Steffen, eds., *Global Change and Terrestrial Ecosystems*. Cambridge University Press, Cambridge, England.

Bazzaz, F. A., S. L. Miao, and P. M. Wayne. 1994. CO_2-induced enhancements of of co-occurring tree species decline at different rates. *Oecologia* 96:478–482.

Beddington, J. R., C. A. Free, and J. H. Lawton. 1976. Concepts of stability and resilience in predator-prey models. *Journal of Animal Ecology* 45:791–816.

Begon, M. 1979. *Investigating Animal Abundance*. Edward Arnold, London.

Behrenfeld, M. J., A. J. Bale, Z. S. Kolber, J. Aiken, and P. G. Falkowski. 1996. Confirmation of iron limitation of phytoplankton photosynthesis in the equatorial Pacific Ocean. *Nature* 383:508–511.

Beier, P. and R. F. Noss. 1998. Do habitat corridors provide connectivity? *Conservation Biology* 12:1241–1252.

Bell, G. 1980. The costs of reproduction and their consequences. *American Naturalist* 116:45–76.

———.1987. Review of "Evolution through group selection" by V. C. Wynne-Edwards. *Heredity* 59:145–147.

Bell, R. H. V. 1971. A grazing ecosystem in the Serengeti. *Scientific American* 225:86–93.

Belsky, A. J. 1986. Does herbivory benefit plants? A review

———. 1987. The effects of grazing: Confounding of

ecosystem, community, and organism scales. *American Naturalist* 129:777–783.

Benbrook, C. M. 1996. *Pest Management at the Crossroads*. Consumers Union, Yonkers, New York.

Benkman, C. W. 1993. Adaptation to single resources and the evolution of crossbill (*Loxia*) diversity. *Ecological Monographs* 63:305–325.

Benkman, C. W. 1996. Are the ratios of bill crossing morphs in crossbills the result of frequency-dependent selection? *Evolutionary Ecology* 10:119–126.

———. 1999. The selection mosaic and diversifying coevolution between crossbills and lodgepole pine. *American Naturalist* 153:S75–S91.

Bennett, W. A. 1990. Scale of investigation and the detection of competition: An example from the house sparrow and house finch introductions in North America. *American Naturalist* 135:725–747.

Bennett, W. A., R. J. Currie, P. F. Wagner, and T. L. Beitinger. 1997. Cold tolerance and potential overwintering of the red-bellied piranha *Pygocentrus nattereri* in the United States. *Transactions of the American Fisheries Society* 126:841–849.

Bentley, M. D. and J. F. Day. 1989. Chemical ecology and behavioral aspects of mosquito oviposition. *Annual Review of Entomology* 34:401–421.

Berendse, F. and R. Aerts. 1987. Nitrogen-use-efficiency: A biologically meaningful definition? *Functional Ecology* 1:293–296.

Bergelson, J. 1996. Competition between two weeds. *American Scientist* 84:579–584.

Bergerud, A. T. 1980. A review of the population dynamics of caribou and wild reindeer in North America. Pages 556–581 in *2nd International Reindeer/Caribou Symposium*, Roros, Norway.

Bergerud, A. T. and J. P. Elliot. 1998. Wolf predation in a multiple-ungulate system in northern British Columbia. *Canadian Journal of Zoology* 76:1551–1569.

Berry, R. J. 1991. Charles Elton FRS the book that most. *The Biologist* 38:127–129.

Berryman, A. A. 1992. The origins and evolution of predator-prey theory. *Ecology* 73:1530–1535.

Beurden, E. V. 1981. Bioclimatic limits to the spread of *Bufo marinus* in Australia: A baseline. *Proceedings of the Ecological Society of Australia* 1:143–149.

Beverton, R. J. H. 1962. Long-term dynamics of certain North Sea fish populations. Pages 242–259 in E. D. LeCren and M. W. Holdgate, eds., *The Exploitation of Natural Animal Populations*. Blackwell, Oxford, England.

Beverton, R. J. H. and S. H. Holt. 1957. *On the Dynamics of Exploited Fish Populations*. H.M. Stationery Office, London.

Biggins, D. E., J. L. Godbey, L. R. Hanebury, B. Luce, P. E. Marinari, R. Marc, and A. Vargas. 1998. The effect of rearing methods on survival of reintroduced black-footed ferrets. *Journal of Wildlife Management* 62: 643–653.

Billings, W. D. 1938. The structure and development of oldfield shortleaf pine stands and certain associated physical properties of the soil. *Ecological Mongraphs* 8: 437–499.

Billings, W. D., J. O. Luken, D. A. Mortensen, and K. M. Peterson. 1983. Increasing atmospheric carbon dioxide: Possible effects on arctic tundra. *Oecologia* 58: 286–289.

Billings, W. D., K. M. Peterson, J. O. Luken, and D. A. Mortensen. 1984. Interaction of increasing atmospheric carbon dioxide and soil nitrogen on the carbon balance of tundra microcosms. *Oecologia* 65:26–29.

Binkley, C. S. and R. S. Miller. 1983. Population characteristics of the whooping crane, *Grus americana*. *Canadian Journal of Zoology* 61:2768–2776.

Binkley, C. S. and R. S. Miller. 1988. Recovery of the whooping crane *Grus americana*. *Biological Conservation* 45:11–20.

Binkley, D. and P. Högberg. 1997. Does atmospheric deposition of nitrogen threaten Swedish forests? *Forest Ecology and Management* 92:119–152.

Birch, L. C. 1948. The intrinsic rate of natural increase of an insect population. *Journal of Animal Ecology* 17:15–26.

———. 1953a. Experimental background to the study of the distribution and abundance of insects. I. The influence of temperature, moisture, and food on the innate capacity for increase of three grain beetles. *Ecology* 34: 698–711.

———. 1953b. Experimental background to the study of the distribution and abundance of insects. III. The relations between innate capacity for increase and survival of different species of beetles living together on the same food. *Evolution* 7:136–144.

———. 1957. The meanings of competition. *American Naturalist* 91:5–18.

Bjorkman, O. 1975. Inaugural address. Pages 1-16 in R. Marcell, ed. *Environmental and Biological Control of Photosynthesis*. Dr. W. Junk Publishers, The Hague.

Bjorkman, O. and J. Berry. 1973. High efficiency photosynthesis. *Scientific American*. 229:80–93.

Black, C. C. 1971. Ecological implications of dividing plants into groups with distinct photosynthetic production capacities. *Advances in Ecological Research* 7:87–114.

Blackburn, T. M. and K. J. Gaston. 1996. Spatial patterns in the species richness of birds in the New World. *Ecography* 19:369–376.

———. 1997. A critical assessment of the form of the interspecific relationship between abundance and body size in animals. *Journal of Animal Ecology* 66:233–249.

Blaustein, A. R. and D. B. Wake. 1990. Declining amphibian populations: A global phenomenon? *Trends in Ecology & Evolution* 5:203–204.

Boag, P. T. and P. R. Grant. 1981. Intense natural selection in a population of Darwin's finches (Geospizinae) in the Galapagos. *Science* 214:82–84.

Bodenheimer, F. S. 1928. Welche faktoren regulieren die Individuenzahl einer Insektenart in der Natur? *Biologisches Zentralblatt* 48:714–739.

Boerema, L. K. and J. A. Gulland. 1973. Stock assessment of the peruvian anchovy *(Engraulis ringens)* and management of the fishery. *Journal of the Fisheries Research Board of Canada* 30:2226–2235.

Bormann, F. H., G. E. Likens, T. G. Siccama, R. S. Peirce, and J. S. Eaton. 1974. The export of nutrients and recovery of stable conditions following deforestation at Hubbard Brook. *Ecological Monographs* 44:255–277.

Bosch, M. and D. Sol. 1998. Habitat selection and breeding success in yellow-legged gulls *Larus cachinnans*. *Ibis* 140: 415–421.

Botkin, D. B. 1981. Causality and succesion. Pages 36–55 in D. C. West, H. H. Shugart, and D. B. Botkin, eds. *Forest Succession*. Springer-Verlag, New York.

Botzler, R. G. 1991. Epizootiology of avian cholera in wildfowl. *Journal of Wildlife Diseases* 27:367–395.

Boucher, D. H., S. James, and K. H. Keeler. 1982. The ecology of mutualism. *Annual Review of Ecology and Systematics* 13:315–347.

Boutin, S. 1992. Predation and moose population dynamics: A critique. *Journal of Wildlife Management* 56:116–127.

———. 1995. Testing predator-prey theory by studying fluctuating populations of small mammals. *Wildlife Research* 22:89–100.

Boutin, S., C. J. Krebs, R. Boonstra, M. R. T. Dale, S. J. Hannon, K. Martin, A. R. E. Sinclair, J. N. M. Smith, R. Turkington, M. Blower, A. Byrom, F. I. Doyle, C. Doyle, D. Hik, L. Hofer, A. Hubbs, T. Karels, D. L. Murray, V. Nams, M. O'Donoghue, C. Rohner, and S. Schweiger. 1995. Population changes of the vertebrate community during a snowshoe hare cycle in Canada's boreal forest. *Oikos* 74:69–80.

Bouzat, J. L., H. H. Cheng, H. A. Lewin, R. L. Westermeier, J. D. Brawn, and K. N. Paige. 1998a. Genetic evaluation of a demographic bottleneck in the greater prairie chicken. *Conservation Biology* 12:836–843.

Bouzat, J. L., H. A. Lewin, and K. N. Paige. 1998b. The ghost of genetic diversity past: Historical DNA analysis of the greater prairie chicken. *American Naturalist* 152:1–6.

Boveng, P. L., L. M. Hiruki, M. K. Schwartz, and J. L. Bengtson. 1998. Population growth of Antarctic fur seals: Limitation by a top predator, the leopard seal? *Ecology* 79:2863–2877.

Bowen, L. and D. Van Vuren. 1997. Insular endemic plants lack of defenses against herbivores. *Conservation Biology* 11:1249–1254.

Bowers, R. G., M. Begon, and D. E. Hodgkinson. 1993. Host-pathogen population cycles in forest insects? Lessons from simple models reconsidered. *Oikos* 67: 529–538.

Boyce, M. S. 1992. Population viability analysis. *Annual Review of Ecology and Systematics* 23:481–506.

Boyer, J. S. 1982. Plant productivity and environment. *Science* 218:443–448.

Bradshaw, A. D. and K. Hardwick. 1989. Evolution and stress—genotypic and phenotypic components. *Biological Journal of the Linnean Society* 37:137–155.

Brasier, C. M. 1988. Rapid changes in genetic structure of epidemic populations of *Ophiostoma ulmi*. *Nature* 332:538–541.

Brawn, J. D. and S. K. Robinson. 1996. Source-sink population dynamics may complicate the interpretation of long-term census data. *Ecology* 77:3–12.

Breitburg, D. L., T. Loher, C. A. Pacey, and A. Gerstein. 1997. Varying effects of low dissolved oxygen on trophic interactions in an estuarine food web. *Ecological Monographs* 67:489–507.

Brett, M. T. and C. R. Goldman. 1996. A meta-analysis of the freshwater trophic cascade. *Proceedings of the National Academy of Sciences USA* 93:7723–7726.

Brewer, L. G. 1995. Ecology of survival and recovery from blight in American chestnut trees (*Castanea dentata* (Marsh.) Borkh.) in Michigan. *Bulletin of the Torrey Botanical Club* 122:40–57.

Brimblecombe, P., C. Hammer, H. Rodhe, A. Ryaboshapko, and C. F. Boutron. 1989. Human influence on the sulphur cycle. Pages 77–121 in P. Brimblecombe and A. Y. Lein, eds. *Evolution of the Global Biogeochemical sulphur Cycle*. Wiley, New York

Bridgham, S. D., J. Pastor, C. A. McClaugherty, and C. J. Richardson. 1995. Nutrient-use efficiency: A litterfall index, a model, and a test along a nutrient-availability gradient in North Carolina peatlands. *American Naturalist* 145:1–21.

Brock, T. D. and M. T. Madigan. 1988. *Biology of Microorganisms*. Prentice Hall, Englewood Cliffs, New Jersey.

Brokaw, N. V. 1985. Treefalls, regrowth, and community structure in tropical forests. Pages 53–69 in S. T. A. Pickett and P. S. White, eds., *The Ecology of Natural Disturbance and Patch Dynamics*. Academic Press, New York.

Brokaw, N. V. L. 1987. Gap-phase regeneration of three pioneer tree species in a tropical forest. *Journal of Ecology* 75:9–19.

Brooks, J. L. and S. I. Dodson. 1965. Predation, body size, and composition of plankton. *Science* 150:28–35.

Brower, L. P. 1988. Avian predation on the monarch butterfly and its implications for mimicry theory. *American Naturalist* 131 Supp.:S4–S6.

Brown, E. S. 1951. The relation between migration rate and types of habitat in aquatic insects, with special reference to certain species of Corixidae. 121:539–545.

Brown, J. H. 1971. Mammals on mountaintops: Non-equilibrium insular biogeography. *American Naturalist* 105:467–478.

———. 1978. The theory of insular biogeography and the distribution of boreal birds and mammals. *Great Basin Naturalist Memoirs* 2:209–227.

———. 1981. Two decades of homage to Santa Rosalia: Toward a general theory of diversity. *American Zoologist* 21:877–888.

———. 1984. On the relationship between abundance and distribution of species. *American Naturalist* 124:255–279.

———. 1986. Two decades of interaction between the MacArthur-Wilson model and the complexities of mammalian distributions. *Biological Journal of the Linnean Society* 28:231–251.

———. 1995. *Macroecology*. University of Chicago Press, Chicago, Illinois.

Brown, J. H. and D. W. Davidson. 1979. An experimental study of competition between seed-eating desert rodents and ants. *American Zoologist* 19:1129–1145.

Brown, J. H. and E. J. Heske. 1990. Control of a desert-grassland transition by a keystone rodent guild. *Science* 250:1705–1707.

Brown, J. H. and M. V. Lomolino. 1998. *Biogeography*. Sinauer Associates, Sunderland, Mass.

Brown, J. H. and W. McDonald. 1995. Livestock grazing and conservation on southwestern rangelands. *Conservation Biology* 9:1644–1647.

Buckland, S. T., D. R. Anderson, K. P. Burnham, and J. L. Laake. 1993. Distance Sampling: Estimating Abundance of Biological Populations. Chapman & Hall, London.

Buckle, A. P. and R. H. Smith, eds. 1994. *Rodent Pests and Their Control*. CAB International, Wallingford, England.

Buckley, R. C. 1987. Interactions involving plants, homoptera, and ants. *Annual Review of Ecology and Systematics* 18:111–135.

Buffon, G. L. L., Compte de. 1756. *Natural History, General and Particular*. W. Creech, Edinburgh.

Bull, C. M. 1991. Ecology of parapatric distributions. *Annual Review of Ecology and Systematics* 22:19–36.

Burbidge, A. A. and N. L. McKenzie. 1989. Patterns in the modern decline of Western Australia's vertebrate fauna: Causes and conservation implications. *Biological Conservation* 50:143–198.

Burnham, C. R. 1988. The restoration of the American Chestnut. *American Scientist* 76:478–487.

Burns, T. P. 1989. Lindeman's contradiction and the trophic structure of ecosystems. *Ecology* 70:1355–1362.

Burrows, C. J. 1990. *Processes of Vegetation Change*. Unwin Hyman, London.

Busenberg, S. N. and K. L. Cooke. 1993. *Vertically Transmitted Diseases, Models and Dynamics*. Springer-Verlag, Berlin.

Buss, L. W. 1980. Competitive intransitivity and size-frequency distributions of interacting populations. *Proceedings of the National Academy of Science (USA)* 77:5355–5359.

Buss, L. W. and J. B. C. Jackson. 1979. Competitive networks: Nontransitive competitive relationships in cryptic coral reef environments. *American Naturalist* 113:223–234.

Cabana, G. and J. B. Rasmussen. 1994. Modelling food chain structure and contaminant bioaccumulation using stable nitrogen isotopes. *Nature* 372:255–257.

Caddy, J. F. 1999. Fisheries management in the twenty-first century: Will new paradigms apply? *Reviews in Fish Biology and Fisheries* 9:1–43.

Cain, S. A. 1944. *Foundations of Plant Geography*. Harper & Row, New York.

Calderini, D. F. and G. A. Slafer. 1998. Changes in yield and yield stability in wheat during the 20th century. *Field Crops Research* 57:335–347.

Caley, M. J. and D. Schluter. 1997. The relationship between local and regional diversity. *Ecology* 78:70–80.

Caltagirone, L. E. and R. L. Doutt. 1989. The history of the vedalia beetle importation to California and its impact on the development of biological control. *Annual Review of Entomology* 34:1–16.

Cannon, J. R. 1996. Whooping crane recovery: a case study in public and private cooperation in the conservation of endangered species. *Conservation Biology* 10:813:821.

Cao, Q. V. 1994. A tree survival equation and diameter growth model for loblolly pine based on the self-thinning rule. *Journal of Applied Ecology* 31:693–698.

Carey, A. B., S. P. Horton, and B. L. Biswell. 1992a. Northern spotted owls: Influence of prey-base and landscape character. *Ecological Monographs* 62: 223–250.

Carey, C. 1996. Female reproductive energetics. Pages 324–374 in C. Carey, ed., *Avian Energetics and Nutritional Ecology*. Chapman and Hall, London.

Carey, J. R. 1993. *Applied Demography for Biologists With Special Emphasis on Insects*. Oxford University Press, New York.

———. 1995. Insect demography. Pages 289-303 in W. A. Nierenberg, ed., *Encyclopedia of Environmental Biology*. Academic Press, San Diego.

Carey, J. R., P. Liedo, H. G. Muller, J. L. Wang, and J. W. Vaupel. 1998. A simple graphical technique for displaying individual fertility data and cohort survival: Case study of 1,000 Mediterranean Fruit Fly females — technical report. *Functional Ecology* 12:359–363.

Carey, J. R., P. Liedo, D. Orozco, M. Tatar, and J. W. Vaupel. 1995. A male-female longevity paradox in medfly cohorts. *Journal of Animal Ecology* 64:107–116.

Carey, J. R., P. Liedo, D. Orozco, and J. W. Vaupel. 1992b. Slowing of mortality rate at older ages in large medfly cohorts. *Science* 258:457–461.

Cargill, S. M. and R. L. Jefferies. 1984. Nutrient limitation of primary production in a sub-arctic salt marsh. *Journal of Applied Ecology* 21:657–668.

Carlquist, S. 1974. *Island Biology*. Columbia University Press, New York.

Carpenter, S. R. 1996. Microcosm experiments have limited relevance for community and ecosystem ecology. *Ecology* 77:677–680.

Carpenter, S. R., N. F. Caraco, D. L. Correll, R. W. Howarth, A. N. Sharpley, and V. H. Smith. 1998. Nonpoint pollution of surface waters with phosphorus and nitrogen. *Ecological Applications* 8:559–568.

Carpenter, S. R., J. F. Kitchell, and J. R. Hodgson. 1985. Cascading trophic interactions and lake productivity. *BioScience* 35:634–639.

Carpenter, S. R., J. F. Kitchell, J. R. Hodgson, P. A. Cochran, J. J. Elser, M. M. Elser, D. M. Lodge, D. Kretchmer, X. He, and C. N. Von Ende. 1987. Regulation of lake primary productivity by food web structure. *Ecology* 68:1863–1876.

Carson, R. 1962. *Silent Spring*. Houghton Mifflin, Boston.

Carter, R. N. and S. D. Prince. 1988. Distribution limits from a demographic viewpoint. Pages 165–184 in A. J. Davy, M. J. Hutchings, and A. R. Watkinson, eds., *Plant Population Ecology*. Blackwell, Oxford, England.

Case, T. J. 1996. Global patterns in the establishment and distribution of exotic birds. *Biological Conservation* 78: 69–96.

Caswell, H. 1978. Predator-mediated coexistence: A non-equilibrium model. *American Naturalist* 112:127–154.

———. 1989. *Matrix Population Models*. Sinauer Associates, Sunderland, Mass.

Caughley, G. 1970. Eruption of ungulate populations, with emphasis on Himalayan thar in New Zealand. *Ecology* 51: 53–72.

———. 1976a. Plant-herbivore systems. Pages 94–113 in R. M. May, ed. *Theoretical Ecology*. Saunders, Philadelphia.

———. 1976b. Wildlife management and the dynamics of ungulate populations. *Applied Biology* 1:183–246.

———. 1977. *Analysis of Vertebrate Populations*. Wiley, London.

———. 1994. Directions in conservation biology. *Journal of Animal Ecology* 63:215–244.

Caughley, G., H. Dublin, and I. Parker. 1990. Projected decline of the African elehant. *Biological Conservation* 54: 157–164.

Caughley, G., D. Grice, R. Barker, and B. Brown. 1988. The edge of the range. *Journal of Animal Ecology* 57: 771–785.

Caughley, G., G. C. Grigg, J. Caughley, and G. J. E. Hill. 1980. Does dingo predation control the densities of kangaroos and emus? *Australian Wildlife Research* 7:1–12.

Caughley, G. and A. Gunn. 1993. Dynamics of large herbivores in deserts: Kangaroos and caribou. *Oikos* 67:47–55.

———. 1996. *Conservation Biology in Theory and Practice*. Blackwell Science, Oxford.

Caughley, G. and J. H. Lawton. 1981. Plant-herbivore systems. Pages 132–166 in R. M. May, ed., *Theoretical Ecology*. Blackwell, Oxford, England.

Caughley, G., R. Pech, and D. Grice. 1992. Effect of fertility control on a population's productivity. *Wildlife Research* 19:623–627.

Caughley, G., N. Shepherd, and J. Short. 1987. *Kangaroos: Their Ecology and Management in the Sheep Ranglands of Australia*. Cambridge University Press, Cambridge, England.

Caughley, G. and A. R. E. Sinclair. 1994. *Wildlife Ecology and Management*. Blackwell Scientific Publications, Boston.

Cavers, P. B. and J. L. Harper. 1967. Studies in the dynamics of plant populations. I. The fate of seed and transplants introduced into various habitats. *Journal of Ecology* 55:59–71.

Chambers, L., G. Singleton, and G. Hood. 1997. Immunocontraception as a potential control method of wild rodent populations. *Belgian Journal of Zoology* 127: 145–156.

Chapin, F. S., III. 1980. The mineral nutrition of wild plants. *Annual Review of Ecology and Systematics* 11: 233–260.

Chapin, F. S., III, and G. R. Shaver. 1996. Physiological and growth responses of arctic plants to a field experiment simulating climatic change. *Ecology* 77:822–840.

Chapin, F. S., III, L. R. Walker, C. L. Fastie, and L. C. Sharman. 1994. Mechanisms of primary succession following deglaciation at Glacier Bay, Alaska. *Ecological Monographs* 64:149–176.

Chapman, D. G. 1981. Evaluation of marine mammal population models. Pages 277–296 in C. W. Fowler and T. D. Smith, eds., *Dynamics of Large Mammal Populations*. Wiley, New York.

Charnov, E. L. and W. M. Schaffer. 1973. Life-history consequences of natural selection: Cole's result revisited. *American Naturalist* 107:791–793.

Chesson, P. L. 1986. Environmental variation and the coexistence of species. Pages 240–256 in J. Diamond and T. J. Case, eds., *Community Ecology*. Harper & Row, New York.

Chesson, P. L. and T. J. Case. 1986. Overview: Nonequilibrium community theories: Chance, variability, history, and coexistence. Pages 229–239 in J. Diamond and T. J. Case, eds., *Community Ecology*. Harper & Row, New York.

Chesson, P. L. and R. R. Warner. 1981. Environmental variability promotes coexistence in lottery competitive systems. *American Naturalist* 117:923–943.

Chichilnisky, G. and G. Heal. 1998. Economic returns from the biosphere. *Nature* 391:629–630.

Chitty, D. 1955. Allgemeine Gedankengänge über die Dicteschwankungen bei der Erdmaus (*Microtus agrestis*). *Zeitschrift fur Saugetierkunde* 20:55–60.

———. 1960. Population processes in the vole and their relevance to general theory. *Canadian Journal of Zoology* 38:99–113.

Christensen, N. L. and R. K. Peet. 1981. Secondary forest succession on the North Carolina Piedmont. Pages 230–245 in D. C. West, H. H. Shugart, and D. B. Botkin, eds., *Forest Succession: Concepts and Application*. Springer-Verlag, New York.

Chu, K. H., P. F. Tam, C. H. Fung, and Q. C. Chen. 1997. A biological survey of ballast water in container ships entering Hong Kong. *Hydrobiologia* 352:201–206.

Civeyrel, L. and D. Simberloff. 1996. A tale of two snails: Is the cure worse than the disease? *Biodiversity and Conservation* 5:1231–1252.

Clark, C. W. 1990. *Mathematical Bioeconomics: The Optimal Management of Renewable Resources*. Wiley, New York.

———. 1996. Marine reserves and the precautionary management of fisheries. *Ecological Applications* 6:369–370.

Clark, C. W. and M. L. Rosenzweig. 1994. Extinction and colonization processes: Parameter estimates from sporadic surveys. *American Naturalist* 143:583–596.

Clark, J. S. 1998. Why trees migrate so fast: Confronting theory with dispersal biology and the paleorecord. *American Naturalist* 152:204–224.

Clark, J. S., C. Fastie, G. Hurtt, S. T. Jackson, C. Johnson, G. A. King, M. Lewis, J. Lynch, S. Pacala, C. Prentice, E. W. Schupp, T. I. Webb, and P. Wyckoff. 1998. Reid's paradox of rapid plant migration. *BioScience* 48:13–24.

Clark, R. G. and D. Shutler. 1999. Avian habitat selection: Pattern from process in nest-site use by ducks. *Ecology* 80: 272–287.

Clarke, A. 1990. Temperature and evolution: Southern Ocean cooling and the antarctic marine fauna. Pages 9–22 in K. R. Kerry and G. Hempel, eds., *Antarctic Ecosystems*. Springer-Verlag, Berlin.

Clausen, C. P. 1951. The time factor in biological control. *Journal of Economic Entomology* 44:1–9.

———. 1956. *Biological Control of Insect Pests in the Continental United States*. U.S. Department of Agriculture, Washington, D.C.

Clausen, J., D. D. Keck, and W. M. Hiesey. 1948. Experimental Studies on the nature of species. III. Environmental Responses of Climatic Races of *Achillea*. Carnegie Institute, Washington, D.C.

Clausen, T. P., P. B. Reichardt, J. P. Bryant, and R. A. Werner. 1991. Long-term and short-term induction in quaking aspen: Related phenomena? Pages 71–83 in D. W. Tallamy and M. J. Raupp, eds., *Phytochemical Induction by Herbivores*. Wiley, New York.

Clements, F. E. 1916. *Plant Succession: An Analysis of the Development of Vegetation*. Carnegie Institute, Washington, D.C.

———. 1936. Nature and structure of the climax. *Journal of Ecology* 24:252–284.

———. 1949. *Dynamics of Vegetation*. Macmillan (Hafner Press), New York.

Clutton-Brock, T. H., S. D. Albon, and F. E. Guinness. 1989. Fitness costs of gestation and lactation in wild mammals. *Nature* 337:260–262.

Clutton-Brock, T. H., F. E. Guiness, and S. D. Albon. 1982. *Red Deer: Behavior and Ecology of Two Sexes*. University of Chicago Press, Chicago.

Coale, K. H., K. S. Johnson, S. E. Fitzwater, R. M. Gordon, S. Tanner, F. P. Chavez, L. Ferioli, C. Sakamoto, P. Rogers, F. Millero, P. Steinberg, P. Nightingale, D. Cooper, W. P. Cochlan, M. R. Landry, J. Constantinou, G. Rollwagen, A. Trasvina, and R. Kudela. 1996. A massive phytoplankton bloom induced by an ecosystem-scale iron fertilization experiment in the equatorial Pacific Ocean. *Nature* 383:495–501.

Coble, D. W., R. E. Brueswitz, T. W. Fratt, and J. W. Scheirer. 1990. Lake trout, sea lamprey, and overfishing in the upper Great Lakes: A review and reanalysis. *Transactions of the American Fishery Society* 119:985–995.

Cody, M. L. 1985. An introduction to habitat selection in birds. Pages 3–56 in M. L. Cody, ed., *Habitat Selection in Birds*. Academic Press, New York.

Coe, M. J., D. H. Cumming, and J. Phillipson. 1976. Biomass and production of large African herbivores in relation to rainfall and primary production. *Oecologia* 22:341–354.

Cohen, J. E. 1978. *Food Webs and Niche Space*. Princeton University Press, Princeton.

———. 1994. Marine and continental food webs: Three paradoxes? *Philosophical Transactions of the Royal Society of London Series B* 343:57–69.

———. 1995. *How Many People Can the Earth Support?* W.W. Norton, New York.

———. 1997. Population, economics, environment and culture: An introduction to human carrying capacity. *Journal of Applied Ecology* 34:1325–1333.

Cohen, J. E. and D. Tilman. 1996. Biosphere 2 and biodiversity: The lessons so far. *Science* 274:1150–1151.

Cole, D. C., J. Eyles, and B. L. Gibson. 1998. Indicators of human health in ecosystems: What do we measure? *Science of the Total Environment* 224:201–213.

Cole, D. W. and M. Rapp. 1981. Elemental cycling in forest ecosystems. Pages 341–409 in D. E. Reichle, ed. *Dynamic Properties of Forest Ecosystems*. Cambridge University Press, Cambridge, England.

Cole, L. C. 1954. The population consequences of life history phenomena. *Quarterly Review of Biology* 29:103–137.

———. 1958. Sketches of general and comparative demography. *Cold Spring Harbor Symposia on Quantitative Biology* 22:1–15.

Coley, P. D., J. P. Bryant, and F. S. Chapin III. 1985. Resource availability and plant antiherbivore defense. *Science* 230:895–899.

Colinvaux, P. 1973. *Introduction to Ecology*. Wiley, New York.

Collins, S. L., S. M. Glenn, and D. J. Gibson. 1995. Experimental analysis of intermediate disturbance and initial floristic composition: Decoupling cause and effect. *Ecology* 76:486–492.

Collins, S. L. and L. L. Wallace. 1990. *Fire in North American Tallgrass Prairies*. University of Oklahoma Press, Norman.

Condit, R. 1995. Research in large, long-term tropical forest plots. *Trends in Ecology and Evolution* 10:18–22.

Connell, J. H. 1961a. The effects of competition, predation by *Thais lapillus*, and other factors on natural populations of the barnacle, *Balanus balanoides*. *Ecological Monographs* 31:61–104.

———. 1961b. The influence of interspecific competition and other factors on the distribution of the barnacle *Chthamalus stellatus*. *Ecology* 42:710–732.

———. 1970. A predator-prey system in the marine intertidal region. I. *Balanus glandula* and several predatory species of *Thais*. *Ecological Monographs* 40:49–78.

———. 1971. On the role of natural enemies in preventing competitive exclusion in some marine animals and in rain forest trees. Pages 298–312 in P. J. den Boer and G. R. Gradwell, eds., *Dynamics of Numbers in Populations*. Centre for Agricultural Publishing and Documentation, Wageningen, Netherlands.

———. 1978. Diversity in tropical rain forests and coral reefs. *Science* 199:1302–1310.

———. 1987. Change and persistence in some marine communities. Pages 339–352 in A. J. Gray, M. J. Crawley, and P. J. Edwards, eds., *Colonization, Succession and Stability*. Blackwell, Oxford, England.

———. 1990. Apparent versus "real" competition in plants. Pages 9–26 in J. B. Grace and D. Tilman, eds., *Perspectives on Plant Competition*. Academic Press, San Diego.

Connell, J. H., T. P. Hughes, and C. C. Wallace. 1997. A 30-year study of coral abundance, recruitment, and disturbance at several scales in space and time. *Ecological Monographs* 67:461–488.

Connell, J. H. and M. D. Lowman. 1989. Low-diversity tropical rain forests: Some possible mechanisms for their existence. *American Naturalist* 134:88–119.

Connell, J. H. and E. Orias. 1964. The ecological regulation of species diversity. *American Naturalist* 98:399–414.

Connell, J. H. and R. O. Slatyer. 1977. Mechanisms of succession in natural communities and their role in community stability and organization. *American Naturalist* 111:1119–1144.

Connell, J. H. and W. P. Sousa. 1983. On the evidence needed to judge ecological stability or persistence. *American Naturalist* 121:789–824.

Cooch, E. G. and F. Cooke. 1991. Demographic changes in a Snow Goose population: Biological and management implications. Pages 168–189 in C. M. Perrins, J.D. Lebreton, and G. J. M. Hirons, eds., *Bird Population Studies: Relevance to Conservation and Management*. Oxford University Press, Oxford.

Cook, R. E. 1969. Variation in species density of North American birds. *Systematic Zoology* 18:63–84.

Cooke, F. and C. S. Findlay. 1982. Polygenic variation and stabilizing selection in a wild population of lesser snow geese (*Anser caerulescens caerulescens*). *American Naturalist* 120:543–547.

Cooke, F., P. D. Taylor, C. M. Francis, and R. F. Rockwell. 1990. Directional selection and clutch size in birds. *American Naturalist* 136:261–267.

Cooper, W. S. 1939. A fourth expedition to Glacier Bay, Alaska. *Ecology* 20:130–155.

Corbit, M., P. L. Marks, and S. Gardescu. 1999. Hedgerows as habitat corridors for forest herbs in central New York, USA. *Journal of Ecology* 87:220–232.

Costanza, R., R. Darge, R. de Groot, S. Farber, M. Grasso, B. Hannon, K. Limburg, S. Naeem, R. V. O'Neill, J. Paruelo, R. G. Raskin, P. Sutton, and M. van den Belt. 1997. The value of the world's ecosystem services and natural capital. *Nature* 387:253–260.

Cote, I. M. and W. J. Sutherland. 1997. The effectivness of removing predators to protect bird population. *Conservation Biology* 11:395–405.

Coupland, R. T. 1979. *Grassland Ecosystems of the World: Analysis of Grassland and Their Uses.* Cambridge University Press, Cambridge, England.

Courtenay, S. C., T. P. Quinn, H. M. C. Dupuis, C. Groot, and P. A. Larkin. 1997. Factors affecting the recognition of population-specific odours by juvenile coho salmon. *Journal of Fish Biology* 50:1042–1060.

Cousins, S. 1985. Ecologists build pyramids again. *New Scientist* 106:50–54.

———. 1987. The decline of the trophic level concept. *Trends in Ecology and Evolution* 2:312–316.

Cowles, H. C. 1899. The ecological relations of the vegetation on the sand dunes of Lake Michigan. *Botanical Gazette* 27:95–117, 167–202, 281–308, 361–391.

———. 1901. The physiographic ecology of Chicago and vicinity. *Botanical Gazette* 31:73–108, 145–181.

Cramer, W., D. W. Kicklighter, A. Bondeau, B. I. Moore, G. Churkina, B. Nemry, A. Ruimy, A. L. Schloss, and others. 1999. Comparing global models of terrestrial net primary productivity (NPP): Overview and key results. *Global Change Biology* 5:1–15.

Crawley, M. J. 1986. The structure of plant communities. Pages 1–50 in M. J. Crawley, ed. *Plant Ecology.* Blackwell, Oxford, England.

———. 1990. The population dynamics of plants. *Philosophical Transactions of the Royal Society of London, Series B* 330:125–140.

———., ed. 1997. *Plant Ecology.* Blackwell Science, Oxford.

Crocker, R. L., and J. Major. 1955. Soil development in relation to vegetation and surface age at Glacier Bay, Alaska. *Journal of Ecology* 43:427–448.

Crombie, A. C. 1945. On competition between different species of graminivorous insects. *Proceedings of the Royal Society of London* 132:362-395.

Crooks, K. R., M. A. Sanjayan, and D. F. Doak. 1998. New insights on cheetah conservation through demographic modeling. *Conservation Biology* 12:889–895.

Crouse, D. T., B. Crowder, and H. Caswell. 1987. A stage-based population model for loggerhead sea turtles and implications for conservation. *Ecology* 68:1412–1423.

Crow, J. F. and M. Kimura. 1970. *An Introduction to Population Genetics Theory.* Harper & Row, New York.

Crowcroft, P. 1991. *Elton's Ecologists: A History of the Bureau of Animal Population.* University of Chicago Press, Chicago.

Crowder, L. B., D. T. Crouse, S. S. Heppell, and T. H. Martin. 1994. Predicting the impact of turtle excluder devices on loggerhead sea turtle populations. *Ecological Applications* 4:437–445.

Culver, D. C. 1981. On using Horn's Markov succession model. *American Naturalist* 117:572–574.

Cunnington, D. C. and R. J. Brooks. 1996. Bet-hedging theory and eigenelasticity: A comparison of the life histories of loggerhead sea turtles (*Caretta caretta*) and snapping turtles (*Chelydra serpentina*). *Canadian Journal of Zoology* 74:291–296.

Curran, P. J. 1985. *Principles of Remote Sensing.* Longman, London.

Currie, D. J. 1991. Energy and large-scale patterns of animal- and plant-species richness. *American Naturalist* 137:27–49.

Currie, D. J. and V. Paquin. 1987. Large-scale biogeographical patterns of species richness of trees. *Nature* 329:326–327.

Curtis, J. T. 1959. *The Vegetation of Wisconsin.* University of Wisconsin Press, Madison.

Cushing, D. H. and J. G. K. Harris. 1973. Stock and recruitment and the problem of density dependence. Pages 142–155. *Rapport du Conseil International pour l'Exploration de la Mer* 164:141-155.

Cutler, A. 1991. Nested faunas and extinction in fragmented habitats. *Conservation Biology* 5:496–505.

Cyr, H. and M. L. Pace. 1993. Magnitude and patterns of herbivory in aquatic and terrestrial ecosystems. *Nature* 361:148–150.

Cyr, H., R. H. Peters, and J. A. Downing. 1997. Population density and community size structure: Comparison of aquatic and terrestrial systems. *Oikos* 80:139–149.

Daily, G. C., ed. 1997. *Nature's Services: Societal Dependence on Natural Ecosystems.* Island Press, Washington, D.C.

Daily, G. C., S. Alexander, P. Ehrlich, L. Goulder, J. Lubchenco, P. A. Matson, H. A. Mooney, S. Postel, S. H. Schneider, D. Tilman, and G. M. Woodwell. 1997. Ecosystem services: Benefits supplied to human societies by natural ecosystems. *Issues in Ecology* 2:1–16.

Danchin, E. and R. H. Wagner. 1997. The evolution of coloniality: The emergence of new perspectives. *Trends in Ecology and Evolution* 12:342–347.

Danell, K. and K. Huss-Danell. 1985. Feeding by insects and hares on birches earlier affected by moose browsing. *Oikos* 44:75–81.

Danell, K., P. Lundberg, and P. Niemela. 1996. Species richness in mammalian herbivores: Patterns in the boreal zone. *Ecography* 19:404–409.

Darlington, P. J., Jr. 1965. *Biogeography of the Southern End of the World.* Harvard University Press, Cambridge, Mass.

Daubenmire, R. 1966. Vegetation: Identification of typal communities. *Science* 151:291–298.

Daubenmire, R. F. 1954. Alpine timberlines in the Americas and their interpretation. *Butler University Botanical Studies* 2:119–136.

Davis, E. F. 1928. The toxic principle of *Juglans nigra* as identified with synthetic juglone, and its toxic effects on tomato and alfalfa plants. *American Journal of Botany* 15:620.

Davis, M. B. 1986. Climatic instability, time lags, and community disequilibrium. Pages 269–284 in J. Diamond and T. J. Case, eds., *Community Ecology*. Harper & Row, New York.

———. 1989. Lags in vegetation response to greenhouse warming. *Climatic Change* 15:75–82.

Dawkins, R. 1982. *The Extended Phenotype*. Oxford University Press, Oxford, England.

Dawkins, R. and J. R. Krebs. 1979. Arms races between and within species. *Proceedings of the Royal Society of London* 205:489-511.ß

Dayton, P. K. 1971. Competition, disturbance, and community organization: The provision and subsequent utilization of space in a rocky intertidal community. *Ecological Monographs* 41:351–389.

Dayton, P. K., W. A. Newman, and J. Oliver. 1982. The vertical zonation of the deep-sea antarctic acorn barnacle, *Bathylasma corolliforme* (Hoek.): Experimental transplants from the shelf into shallow water. *Journal of Biogeography* 9:95–109.

DeAngelis, D. L. 1992. *Dynamics of Nutrient Cycling and Food Webs*. Chapman & Hall, New York.

DeAngelis, D. L. and J. C. Waterhouse. 1987. Equilibrium and nonequilibrium concepts in ecological models. *Ecological Monographs* 57:1–21.

Death, R. G. and M. J. Winterbourn. 1995. Diversity patterns in stream benthic invertebrate communities: The influence of habitat stability. *Ecology* 76:1446–1460.

DeBach, P., ed. 1964. *Biological Control of Insect Pests and Weeds*. Chapman and Hall, London.

———. 1974. *Biological Control by Natural Enemies*. Cambridge University Press, London.

DeJong, T. J. 1995. Why fast-growing plants do not bother about defence. *Oikos* 74:545–548.

del Giorgio, P. A., J. J. Cole, N. F. Caraco, and R. H. Peters. 1999. Linking plankton biomass and metabolism to net gas fluxes in northern temperate lakes. *Ecology* 80:1422–1431.

del Moral, R. 1999. Plant succession on pumice at Mount St. Helens, Washington. *American Midland Naturalist* 141:101–114.

del Moral, R., J. H. Titus, and A. M. Cook. 1995. Early primary succession on Mount St. Helens, Washington, USA. *Journal of Vegetation Science* 6:107–120.

del Moral, R. and D. M. Wood. 1988. The high elevation flora of Mount St. Helens, Washington. *Madrono* 35:309–319.

del Moral, R. and D. M. Wood. 1993. Early primary succession on the volcano Mount St. Helens. *Journal of Vegetation Science* 4:223–234.

Delcourt, P. A. and H. R. Delcourt. 1987. *Long-Term Forest Dynamics of the Temperate Zone: A Case Study of Late-Quaternary Forest in Eastern North America*. Springer-Verlag, New York.

DeMott, W. R. and W. C. Kerfoot. 1982. Competition among cladocerans: Nature of the interaction between *Bosmina* and *Daphnia*. *Ecology* 63:1949–1966.

DeMott, W. R. and F. Moxter. 1991. Foraging on cyanobacteria by copepods: Responses to chemical defenses and resource abundance. *Ecology* 72:1820–1834.

Dennis, B. 1989. Allee effects: Population growth, critical density, and the chance of extinction. *Natural Resource Modeling* 3:481–538.

Desowitz, R. S. 1991. *The Malaria Capers: More Tales of Parasites and People, Research and Reality*. Norton, New York.

Detwiler, R. P. and C. A. S. Hall. 1988. Tropical forests and the global carbon cycle. *Science* 239:42–47.

Dewar, R. C. 1992. Inverse modeling and the global carbon cycle. *Trends in Ecology and Evolution* 7:105–107.

Dewar, R. C. and A. D. Watt. 1992. Predicted changes in synchrony of larval emergence and budburst under climatic warming. *Oecologia* 89:557–559.

Dhondt, A. A. 1988. Carrying capacity: A confusing concept. *Acta Oecologica* 9:337–346.

Dhondt, A. A., F. Adriaensen, E. Matthysen, and B. Kempenaers. 1990. Nonadaptive clutch sizes in tits. *Nature* 348:723–725.

Diamond, J. 1986. Overview: Laboratory experiments, field experiments, and natural experiments. Pages 3–22 in J. Diamond and T. J. Case, eds., *Community Ecology*. Harper & Row, New York.

———. 1989. Overview of recent extinctions. Pages 37–41 in D. Western and M. Pearl, eds., *Conservation for the Twenty-first Century*. Oxford University Press, New York.

Diamond, J. M. 1975. Assembly of species communities. Pages 342–444 in M. L. Cody and J. M. Diamond, eds., *Ecology and Evolution of Communities*. Belknap Press, Cambridge, Mass.

———. 1997. *Guns, Germs, and Steel: The Fates of Human Societies*. Norton, New York.

Diamond, J. M., K. D. Bishop, and S. Van Balen. 1987. Bird survival in an isolated Javan woodland: Island or mirror? *Conservation Biology* 1:132–142.

Diaz, S., J. P. Grime, J. Harris, and E. McPherson. 1993. Evidence of a feedback mechanism limiting plant response to elevated carbon dioxide. *Nature* 364: ß616–617.

Dickman, C. R. 1989. Patterns in the structure and diversity of marsupial carnivore communities. Pages 241–251 in D. W. Morris, Z. Abramsky, B. J. Fox, and M. R. Willig, eds., *Patterns in the Structure of Mammalian Communities*. Texas Tech University Press, Lubbock.

Diller, J. D. and R. B. Clapper. 1969. Asiatic and hybrid chestnut trees in the eastern United States. *Journal of Forestry* 67:328–331.

Dillon, P. J. and F. H. Rigler. 1975. A simple method for predicting the capacity of a lake for development based on lake trophic status. *Journal of the Fisheries Research Board of Canada* 32:1519–1531.

Dingle, H., ed. 1978. *Evolution of Insect Migration and Diapause*. Springer-Verlag, New York.

Dippery, J. K., D. T. Tissue, R. B. Thomas, and B. R. Strain. 1995. Effects of low and elevated CO^2 on C^3 and C^4 annuals. I. Growth and biomass allocation. *Oecologia* 101:13–20.

Dirzo, R. and M. C. Garcia. 1992. Rates of deforestation in Los Tuxtlas, a neotropical area in southeast Mexico. *Conservation Biology* 6:84–90.

Dittman, A. H. and T. P. Quinn. 1996. Homing in Pacific salmon: Mechanisms and ecological basis. *Journal of Experimental Biology* 199:83–91.

Dixon, R. K., S. Brown, R. A. Houghton, A. M. Solomon, M. C. Trexler, and J. Wisniewski. 1994. Carbon pools and flux in global forest ecosystems. *Science* 263: 185–190.

Doak, D. F., D. Bigger, E. K. Harding, M. A. Marvier, R. E. O'Malley, and D. Thomson. 1998. The statistical inevitability of stability-diversity relationships in community ecology. *American Naturalist* 151:264–276.

Dobson, A. 1995. The ecology and epidemiology of rinderpest virus in Serengeti and Ngorongoro conservation area. Pages 485–505 in A. R. E. Sinclair and P. Arcese, eds., *Serengeti II: Dynamics, Management, and Conservation of an Ecosystem*. University of Chicago Press, Chicago.

Dobson, A. and M. Meagher. 1996. The population dynamics of brucellosis in Yellowstone National Park. *Ecology* 77:1026–1036.

Dobson, A. P., A. D. Bradshaw, and A. J. M. Baker. 1997a. Hopes for the future: Restoration ecology and conservation biology. *Science* 277:515–522.

Dobson, A. P. and P. J. Hudson. 1995. Microparasites: Observed patterns in wild animal populations. Pages 52–89 in B. T. Grenfell and A. P. Dobson, eds., *Ecology of Infectious Diseases in Natural Populations*. Cambridge University Press, Cambridge, England.

Dobson, A. P., J. P. Rodriguez, W. M. Roberts, and D. S. Wilcove. 1997b. Geographic distribution of endangered species in the United States. *Science* 275:550–553.

Dobson, F. S. 1982. Competition for mates and predominant juvenile male dispersal in mammals. *Animal Behaviour* 30:1183–1192.

Dobzhansky, T. 1950. Evolution in the tropics. *American Scientist* 38:209–221.

Dodd, A. P. 1940. The biological campaign against prickly-pear. *Commonwealth Prickly Pear Board, Brisbane* .

———. 1959. The biological control of prickly pear in Australia. Pages 565–577 in A. Keast, R. L. Crocker, and C. S. Christian, eds., *Biogeography and Ecology in Australia*. Dr. W. Junk, Den Haag.

Dodd, J. L. and W. K. Lauenroth. 1979. Analysis of the response of a grassland ecosystem to stress. Pages

43–58 in N. R. French, ed., *Perspectives in Grassland Ecology*. Springer-Verlag, New York.

Doering, P. H., C. A. Oviatt, B. L. Nowicki, E. G. Klos, and L. W. Reed. 1995. Phosphorus and nitrogen limitation of primary production in a simulated estuarine gradient. *Marine Ecology Progress Series* 124:271–287.

Doherty, P. and T. Fowler. 1994. An empirical test of recruitment limitation in a coral reef fish. *Science* 263:935–939.

Downes, S. and R. Shine. 1998. Heat, safety or solitude? Using habitat selection experiments to identify a lizard's priorities. *Animal Behaviour* 55:1387–1396.

Downing, J. A., C. W. Osenberg, and O. Sarnelle. 1999. Meta-analysis of marine nutrient-enrichment experiments: variation in the magnitude of nutrient limitation. *Ecology* 80:1157–1167.

Drent, R. H. and S. Daan. 1980. The prudent parent: Energetic adjustments in avian breeding. *Ardea* 68:225–252.

Drury, W. H. and I. C. T. Nisbet. 1973. Succession. *Journal of the Arnold Arboretum of Harvard University* 54:331–368.

Dublin, L. I., and A. Lotka. 1925. On the true rate of natural increase as exemplified by the population of the United States, 1920. *Journal of the American Statistical Association* 20:305–339.

Dublin, H. T., A. R. E. Sinclair, and J. McGlade. 1990. Elephants and fire as causes of multiple stable states in the Serengeti-Mara woodlands. *Journal of Animal Ecology* 59:1147–1164.

Dunbar, R. B. and J. E. Cole, eds., 1999. *Annual Records of Tropical Systems (ARTS): Recommendations for Research*. International Geosphere Biosphere Program, Past Global Changes.

Dunham, J. B. and G. L. Vinyard. 1997. Relationships between body mass, population density, and the self-thinning rule in stream-living salmonids. *Canadian Journal of Fisheries and Aquatic Sciences* 54:1025–1030.

Durant, S. M. 1998. Competition refuges and coexistence: An example from Serengeti carnivores. *Journal of Animal Ecology* 67:370–386.

Eamus, D. and P. G. Jarvis. 1989. The direct effects of increase in the global atmospheric CO_2 concentration on natural and commercial temperate trees and forests. *Advances in Ecological Research* 19:1–55.

Ebbinge, B. 1992. Population limitation in arctic-breeding geese. Ph.D. thesis, Department of Zoology. University of Groningen, Groningen. 200 pp.

Eberhard, T. 1988. Introduced birds and mammals and their ecological effects. *Swedish Wildlife Research* 13:1–107.

Eberhardt, L. E., L. L. Eberhardt, B. L. Tiller, and L. L. Cadwell. 1996. Growth of an isolated elk population. *Journal of Wildlife Management* 60:369–373.

Eberhardt, L. L. 1988. Using age structure data from changing populations. *Journal of Applied Ecology* 25: 373–378.

Ebert, D. 1998. Experimental evolution of parasites. *Science* 282:1432–1435.

————. 1999. The evolution and expression of virulence. Pages 161–172 in S. C. Stearns, ed., *Evolution of Health and Disease*. Oxford University Press, Oxford.

Eckert, R., D. J. Randall, and G. Augustine. 1988. *Animal Physiology: Mechanisms and Adaptations*. W. H. Freeman, New York.

Edmondson, W. T. 1969. Cultural eutrophication with special reference to Lake Washington. *Mitteilungen Internationale Vereinigung Limnologie* 17:19–32.

Edmondson, W. T. 1991. *The Uses of Ecology*. University of Washington Press, Seattle.

Egerton, F. N. III. 1968a. Ancient sources for animal demography. *Isis* 59:175–189.

————. 1968b. Leeuwenhoek as a founder of animal demography. *Journal of the History of Biology* 1:1–22.

————. 1968c. Studies of animal populations from Lamarck to Darwin. *Journal of the History of Biology* 1:225–259.

————. 1969. Richard Bradley's understanding of biological productivity: A study of eighteenth-century ecological ideas. *Journal of the History of Biology* 2:391–410.

————. 1973. Changing concepts of the balance of nature. *Quarterly Review of Biology* 48:322–350.

Egler, F. E. 1954. Vegetation science concepts. I. Initial floristic composition, a factor in old-field vegetation development. *Vegetatio* 14:412–417.

Ehler, F. E. and R. W. Hall. 1982. Evidence for competitive exclusion of introduced natural enemies in biological control. *Environmental Entomology* 11:1–4.

Ehler, L. E. 1998. Invasion biology and biological control. *Biological Control* 13:127–133.

Ehleringer, J. R., T. E. Cerling, and B. R. Helliker. 1997. C_4 photosynthesis, atmospheric CO^2 and climate. *Oecologia* 112:285–299.

Ehrlich, P. R. and A. Ehrlich. 1981. *Extinction*. Random House, New York.

Ehrlich, P. R. and P. H. Raven. 1964. Butterflies and plants: A study in coevolution. *Evolution* 18:586–608.

Ehrlich, P. R. and E. O. Wilson. 1991. Biodiversity studies: Science and policy. *Science* 253:758–762.

Elson, P. F. 1962. Predator-prey relationships between fish-eating birds and Atlantic salmon. *Bulletin of the Fisheries Research Board of Canada* 133:1–87.

Elton, C. 1927. *Animal Ecology*. Sidgwick and Jackson, London.

Elton, C. and M. Nicholson. 1942. The ten-year cycle in numbers of the lynx in Canada. *Journal of Animal Ecology* 11:215–244.

Elton, C. S. 1958. *The Ecology of Invasions by Animals and Plants*. Methuen, London.

————. 1966. *The Pattern of Animal Communities*. Methuen, London.

Emden, H. F. V., V. F. Eastop, R. D. Hughes, and M. J. Way. 1969. The ecology of *Myzus persicae. Annual Review of Entomology* 14:197–270.

Endler, J. A. 1986. *Natural Selection in the Wild*. Princeton University Press, Princeton, New Jersey.

Enright, J. T. 1976. Climate and population regulation: The biogeographer's dilemma. *Oecologia* 24:295–310.

Epstein, H. E., W. K. Lauenroth, I. C. Burke, and D. P. Coffin. 1997. Productivity patterns of C^3 and C^4 functional types in the U.S. Great Plains. *Ecology* 78:722–731.

Errington, P. L. 1963. *Muskrat Populations*. Iowa State University Press, Ames.

Estes, J. A. and D. O. Duggins. 1995. Sea otters and kelp forests in Alaska: Generality and variation in a community ecological paradigm. *Ecological Monographs* 65:75–100.

Estes, J. A., M. T. Tinker, T. M. Williams, and D. F. Doak. 1998. Killer whale predation on sea otters linking oceanic and nearshore ecosystems. *Science* 282:473–476.

Evans, F. C. 1956. Ecosystem as the basic unit in ecology. *Science* 123:1127–1128.

Ewald, P. W. 1995. The evolution of virulence: A unifying link between parasitology and ecology. *Journal of Parasitology* 81:659–669.

Ewel, J., C. Berish, B. Brown, N. Price, and J. Raich. 1981. Slash and burn impacts on Costa Rican wet forest site. *Ecology* 62:816–829.

Faegri, K. and L. v. d. Pijl. 1971. *The Principles of Pollination Ecology*. Pergamon Press, New York.

Fagan, W. F. and L. E. Hurd. 1994. Hatch density variation of a generalist arthropod predator: Population consequences and community impact. *Ecology* 75:2022–2032.

Fairbairn, D. J. 1986. Does alary dimorphism imply dispersal dimorphism in the waterstrider, *Gerris remigis? Ecological Entomology* 11:481–492.

Fajer, E. D. 1989. The effects of enriched CO_2 atmospheres on plant-insect herbivore interactions: Growth responses of larvae of the specialist butterfly, *Junonia coenia* (Lepidoptera: Nymphalidae). *Oecologia* 81:514–520.

Fajer, E. D., M. D. Bowers, and F. A. Bazzaz. 1989. The effects of enriched carbon dioxide atmospheres on plant-insect herbivore interactions. *Science* 243:1198–1200.

Falkowski, P., R. T. Barber, and V. Smetacek. 1998. Biogeochemical controls and feedbacks on ocean primary production. *Science* 281:200–206.

Falkowski, P. G. 1997. Evolution of the nitrogen cycle and its influence on the biological sequestration of CO_2 in the ocean. *Nature* 387:272–275.

Fauth, J. E. and W. J. Resetarits, Jr. 1991. Interactions between the salamander *Siren intermedia* and the keystone predator *Notophthalmus viridescens. Ecology* 72:827–838.

Feeny, P. P. 1970. Seasonal changes in oak leaf tannins and nutrients as a cause of spring feeding by winter moth caterpillars. *Ecology* 51:565–581.

————. 1976. Plant apparency and chemical defence. *Recent Advances in Phytochemistry* 10:1–40.

————.. 1992. The evolution of chemical ecology: Contributions from the study of herbivorous insects. Pages 1–44 in G. A. Rosenthal and M. Berenbaum, eds., *Her-*

bivores: *Their Interactions with Secondary Plant Metabolites. Vol. II. Evolutionary and Ecological Processes.* Academic Press, San Diego.

Feinsinger, P. 1976. Organization of a tropical guild of nectivorous birds. *Ecological Monographs* 46:257–291.

Feller, M. C. 1980. biomass adn nutrient distribution in two eucalypt forest ecosystems. *Australian Journal of Ecology* 5: 309-333.

Feller, M. ., and J. P. Kimmins. 1984. Effects of clearcutting and slash burning on streamwater chemistry and watershed nutrient budgets in southwestern British Columbia. *Water Resources Research* 20: 29-40.

Fenner, F. and P. J. Chapple. 1965. Evolutionary changes in myxoma virus in England. *Journal of Hygiene* 63:175–185.

Fenner, F. and K. Myers. 1978. Chapter 28. Myxoma virus and myxomatosis in retrospect: The first quarter century of a new disease. Pages 539–570 in E. Kurstak and K. Maramorosch, ed., *Viruses and Environment*, Academic Press, New York.

Fenner, F. and F. N. Ratcliffe. 1965. *Myxomatosis.* Cambridge University Press, Cambridge, England.

Ferretti, M. 1997. Forest health assessment and monitoring: Issues for consideration. *Environmental Monitoring and Assessment* 48:45–72.

Field, C. B., M. J. Behrenfeld, J. T. Randerson, and P. Falkowski. 1998. Primary production of the biosphere: Integrating terrestrial and oceanic components. *Science* 281:237–240.

Finegan, B. 1996. Pattern and process in neotropical secondary rain forests: The first 100 years of succession. *Trends in Ecology and Evolution* 11:119–124.

Fischer, A. G. 1960. Latitudinal variations in organic diversity. *Evolution* 14:64–81.

Fisher, R. A., A. S. Corbet, and C. B. Williams. 1943. The relation between the number of species and the number of individuals in a random sample of an animal population. *Journal of Animal Ecology* 12:42–58.

Fisher, R. N. and H. B. Shaffer. 1996. The decline of amphibians in California's Great Central Valley. *Conservation Biology* 10:1387–1397.

Fleischner, T. L. 1994. Ecological costs of livestock grazing in western North America. *Conservation Biology* 8:629–644.

Fleishman, E., G. T. Austin, and A. D. Weiss. 1998. An empirical test of Rapoport's rule: Elevational gradients in montane butterfly communities. *Ecology* 79:2482–2493.

Fletcher, W. J. 1987. Interactions among subatidal Australian sea urchins, gastropods and algae: Effects of experimental removals. *Ecological Monographs* 57: 89–109.

Flint, M. L. and R. van den Bosch. 1981. A history of pest control. Pages 51–81 in *Introduction to Integrated Pest Management.* Plenum Press, New York.

Flux, J. E. C. 1993. Relative effect of cats, myxomatosis, traditional control, or competitors in removing rabbits from islands. *New Zealand Journal of Zoology* 20:13–18.

Fogleman, J. C., W. B. Heed, and H. W. Kircher. 1982. *Drosophila mettleri* and senita cactus alkaloids: Fitness measurements and their ecological significance. *Comparative Biochemistry and Physiology* 71:413–417.

Ford, E. B. 1931. *Mendelism and Evolution.* Metheun, London.

Forman, R. T. T. 1964. Growth under controlled conditions to explain the hierarchical distributions of a moss, *Tetraphis pellucida. Ecological Monographs* 34:1–25.

Fowler, K. and L. Partridge. 1989. A cost of mating in female fruit flies. *Nature* 338:760–761.

Fox, J. F. 1977. Alternation and coexistence of tree species. *American Naturalist* 111:69–89.

Fraenkel, G. and D. L. Gunn. 1940. *The Orientation of Animals: Kineses, Taxes, and Compass Reactions.* Oxford University Press, New York.

Frank, D. A. and S. J. McNaughton. 1991. Stability increases with diversity in plant communities: Empirical evidence from the 1988 Yellowstone drought. *Oikos* 62:360–362.

Frankham, R. 1998. Inbreeding and extinction: Island populations. *Conservation biology* 12:665–675.

Franklin, I. R. 1980. Evolutionary change in small populations. Pages 135–149 in M. Soule and B. A. Wilcox, eds., *Conservation Biology: An Evolutionary-Ecological Perspective.* Sinauer Associates, Sunderland, Mass.

Franklin, J. F., J. A. MacMahon, F. J. Swanson, and J. R. Sedell. 1985. Ecosystem responses to the eruption of Mount St. Helens. *National Geographic Society Research* 1:198–216.

Franzreb, K. E. 1997. Success of intensive management of a critically imperiled population of red-cockaded woodpeckers in South Carolina. *Journal of Field Ornithology* 68:458–470.

Fraser, L. H. and P. Keddy. 1997. The role of experimental microcosms in ecological research. *Trends in Ecology and Evolution* 12:478–481.

Fraser, R. H. and D. J. Currie. 1996. The species richness-energy hypothesis in a system where historical factors are thought to prevail: Coral reefs. *American Naturalist* 148:138–159.

Fredga, K., A. Gropp, H. Winking, and F. Frank. 1977. A hypothesis explaining the exceptional sex ratio in the wood lemming (*Myopus schisticolor*). *Hereditas* 85:101–104.

Freer-Smith, P. H. 1998. Do pollutant-related forest declines threaten the sustainability of forests? *Ambio* 27:123–131.

French, N. R. 1979. *Perspectives in Grassland Ecology.* Springer-Verlag, New York.

French, N. R., R. K. Steinhorst, and D. M. Swift. 1979. Grassland biomass trophic pyramids. Pages 59–87 in N. R. French, ed., *Perspectives in Grassland Ecology.* Springer-Verlag, New York.

Fretwell, S. D. 1972. *Populations in a Seasonal Environment.* Princeton University Press, Princeton, New Jersey.

———. 1977. The regulation of plant communities by the food chains exploiting them. *Perspectives in Biology and Medicine* 20:169–185.

Futuyma, D. J. 1998. *Evolutionary Biology*. Sinauer, Sunderland, Mass.

Gagan, M. K., L. K. Ayliffe, D. Hopley, J. A. Cali, G. E. Mortimer, J. Chappell, M. T. McCulloch, and M. J. Head. 1998. Temperature and surface-ocean water balance of the Mid-Holocene tropical western Pacific. *Science* 279:1014–1018.

Gaines, S. D. and J. Lubchenco. 1982. A unified approach to marine plant-herbivore interactions. II. Biogeography. *Annual Review of Ecology and Systematics* 13:111–138.

Ganter, B., F. Cooke, and P. Mineau. 1996. Long-term vegetation changes in a snow goose nesting habitat. *Canadian Journal of Zoology* 74:965–969.

Gasaway, W. C., R. D. Boertje, D. V. Grangaard, D. G. Kelleyhouse, R. O. Stephenson, and D. G. Larsen. 1992. The role of predation in limiting moose at low densities in Alaska and Yukon and implications for conservation. *Wildlife Monographs* 120:1–59.

Gaston, K. J. 1988. Patterns in the local and regional dynamics of moth populations. *Oikos* 53:49–59.

———. 1990. Patterns in the geographical ranges of species. *Biological Review* 65:105–129.

———. 1991. How large is a species' geographic range? *Oikos* 61:434–438.

Gaston, K. J., T. M. Blackburn, and J. H. Lawton. 1997. Interspecific abundance-range size relationships: An appraisal of mechanisms. *Journal of Animal Ecology* 66:579–601.

Gaston, K. J., T. M. Blackburn, and J. L. Spicer. 1998a. Rapoport's rule: Time for an epitaph. *Trends in Ecology and Evolution* 13:70–74.

Gaston, K. J., and J. L. Curnutt. 1998. The dynamics of abundance—range size relationships. Oikos 81:38–44.

Gaston, K. J. and J. H. Lawton. 1987. A test of statistical techniques for detecting density dependence in sequential censuses of animal populations. *Oecologia* 74:404–410.

Gaston, K. J., R. M. Quinn, T. M. Blackburn, and B. C. Eversham. 1998b. Species-range size distributions in Britain. *Ecography* 21:361–370.

Gause, G. F. 1932. Experimental studies on the struggle for existence. I. Mixed population of two species of yeast. *Journal of Experimental Biology* 9:389–402.

———. 1934. *The Struggle for Existence*. Macmillan (Hafner Press), New York.

Geist, V. 1978. On weapons, combat and ecology. Pages 1–30 in L. Krames, P. Pliner, and T. Alloway, eds., *Aggression, Dominance and Individual Spacing*. Plenum, New York.

———. 1988. How markets in wildlife meat and parts, and the sale of hunting privileges, jeopardize wildlife conservation. *Conservation Biology* 2:15–26.

Gelbard, A., C. Haub, and M. M. Kent. 1999. World population beyond six billion. *Population Bulletin* 54:1–44.

Giese, R. L., R. M. Peart, and R. T. Huber. 1975. Pest management. *Science* 187:1045–1052.

Gilbert, G. S., S. P. Hubbell, and R. B. Foster. 1994. Density and distance-to-adult effects of a canker disease of trees in a moist tropical forest. *Oecologia* 98:100–108.

Gill, D. E. 1974. Intrinsic rate of increase, saturation density, and competitive ability. II. The evolution of competitive ability. *American Naturalist* 108:103–116.

Giller, P. S. and J. H. R. Gee. 1987. The analysis of community organization: The influence of equilibrium, scale and terminology. Pages 519–542 in J. H. R. Gee and P. S. Giller, eds., *Organization of Communities Past and Present*. Blackwell, Oxford, England.

Gilpin, M. E. and M. E. Soule. 1986. Minimum viable populations: Processes of species extinction. Pages 19–34 in M. E. Soule, ed. *Conservation Biology*. Sinauer Associates, Sunderland, Mass.

Givnish, T. J. 1999. On the causes of gradients in tropical tree diversity. *Journal of Ecology* 87:193–210.

Gleason, H. A., and A. Cronquist. 1964. *The Natural Geography of Plants*. Columbia University Press, New York.

Gliwicz, Z. M. 1990. Food thresholds and body size in cladocerans. *Nature* 343:638–640.

Godfray, H. C. J., L. Partridge, and P. H. Harvey. 1991. Clutch size. *Annual Review of Ecology and Systematics* 22:409–429.

Goldman, C. R. 1968. Aquatic primary production. *American Zoologist* 8:31–42.

———. 1988. Primary productivity, nutrients, and transparency during the early onset of eutrophication in ultra-oligotrophic Lake Tahoe, California-Nevada. *Limnology and Oceanography* 33:1321–1333.

Golley, F. B. 1961. Energy values of ecological materials. *Ecology* 42:581–584.

Gomendio, M., T. H. Clutton-Brock, S. D. Albon, F. E. Guinness, and M. J. Simpson. 1990. Mammalian sex ratios and variation in costs of rearing sons and daughters. *Nature* 343:261–263.

Good, N. F. 1968. A study of natural replacement of chestnut in six stands in the highlands of New Jersey. *Bulletin of the Torrey Botanical Club* 95:240–253.

Good, R. 1964. *The Geography of the Flowering Plants*. Longman's, London.

Gopal, B. and U. Goel. 1993. Competition and allelopathy in aquatic plant communities. *Botanical Review* 59:155–210.

Gordon, H. S. 1954. The economic theory of a common property resource: The fishery. *Journal of Political Economics* 62:124–142.

Gorham, E. 1991. Northern peatlands: Role in the carbon cycle and probable responses to climatic warming. *Ecological Applications* 1:182–195.

Gotelli, N. J. 1998. *A Primer of Ecology*. Sinauer Associates, Sunderland, Mass.

Gould, S. J. 1981. Palaeontology plus ecology as palaebiology. Pages 295–317 in R. M. May, ed. *Theoretical Ecology*. Blackwell, Oxford, England.

Goulden, C. E. and L. L. Hornig. 1980. Population oscillations and energy reserves in planktonic cladocera and their consequence to competition. *Proceedings of the National Academy of Science (USA)* 77:1716–1720.

Goulden, M. L., S. C. Wofsy, J. W. Harden, S. E. Trumbore, P. M. Crill, S. T. Gower, T. Fries, B. C. Daube, S.-M. Fan, D. J. Sutton, A. Bazzaz, and J. W. Munger. 1998. Sensitivity of boreal forest carbon balance to soil thaw. *Science* 279:214–217.

Goward, S. N., C. J. Tucker, and D. G. Dye. 1985. North American vegetation patterns observed with the NOAA-7 advanced very high resolution radiometer. *Vegetatio* 64:3–14.

Gowda, J. H. 1996. Spines of *Acacia tortilis*: What do they defend and how? *Oikos* 77:279–284.

Gower, S. T., R. E. McMurtrie, and D. Murty. 1996. Aboveground net primary production decline with stand age: Potential causes. *Trends in Ecology & Evolution* 11:378–382.

Grace, J. B. 1995. In search of the Holy Grail: Explanations for the coexistence of plant species. *Trends in Ecology & Evolution* 10:263–264.

Graetz, R. D., R. Fisher, and M. Wilson. 1992. *Looking Back: The Changing Face of the Australian Continent 1972–1992*. CSIRO, Canberra, Australia.

Graham, M. 1935. Modern theory of exploiting a fishery, and application to North Sea trawling. *Journal du Conseil Permanente International pour l'Exploration de la Mer* 10:264–274.

Graham, M. 1939. The sigmoid curve and the overfishing problem. *Rapport du Conseil International pour l'Exploration de la Mer* 110:15–20

Grant, B. S., A. D. Cook, C. A. Clarke, and D. F. Owen. 1998. Geographic and temporal variation in the incidence of melanism in peppered moth populations in America and Britain. *Journal of Heredity* 89:465–471.

Grant, B. S., D. F. Owen, and C. A. Clarke. 1996. Parallel rise and fall of melanic peppered moths in America and Britain. *Journal of Heredity* 87:351–357.

Grant, P. R. 1986. *Ecology and Evolution of Darwin's Finches*. Princeton University Press, Princeton, New Jersey.

———. 1999. *Ecology and Evolution of Darwin's Finches*. Princeton University Press, Princeton, New Jersey.

Graunt, J. 1662. *Natural and political observations mentioned in a following index, and made upon the bills of mortality*. Roycroft, London.

Green, M. J. B. and J. Paine. 1997. State of the world's protected areas at the end of the twentieth century. *IUCN World Commission on Protected Areas Symposium* Albany, Australia:1–28.

Green, R. E. 1997. The influence of numbers released on the outcome of attempts to introduce exotic bird species to New Zealand. *Journal of Animal Ecology* 66:25–35.

Greene, H. W., and R. W. McDiarmid. 1981. Coral snake mimicry: does it occur? *Science* 213:1207–1212.

Greenwood, P. J. 1980. Mating systems, philopatry and dispersal in birds and mammals. *Animal Behaviour* 28:1140–1162.

Greenwood, R. J. 1986. Influence of striped skunk removal on upland duck nest success in North Dakota. *Wildlife Society Bulletin* 14:6–11.

Gregory, P. T. 1997. What good is R_0 in demographic analysis? (And what is a generation anyway?). *Wildlife Society Bulletin* 25:709–713.

Grenfell, B. T. and A. P. Dobson, eds., 1995. *Ecology of Infectious Diseases in Natural Populations*. Cambridge University Press, Cambridge, England.

Grime, J. P. 1973. Competitive exclusion in herbaceous vegetation. *Nature* 242:344–347.

———. 1979. *Plant Strategies and Vegetation Processes*. Wiley, New York.

———. 1993. Ecology sans frontieres. *Oikos* 68:385–392.

———. 1997a. Biodiversity and ecosystem function: The debate deepens. *Science* 277:1260–1261.

———. 1997b. Climate change and vegetation. Pages 582–594 in M. J. Crawley, ed. *Plant Ecology*. Blackwell Science, Oxford.

Grosholz, E. D. 1996. Contrasting rates of spread for introduced species in terrestrial and marine systems. *Ecology* 77:1680–1686.

Grubb, P. J. 1987. Global trends in species-richness in terrestrial vegetation: A view from the Northern Hemisphere. Pages 99–118 in J. H. R. Gee and P. S. Giller, eds., *Organization of Communities Past and Present*. Blackwell, Oxford, England.

Guilford, T. 1988. The evolution of conspicuous coloration. *American Naturalist* 131 (Suppl.):S7–S21.

Gulland, F. M. D. 1995. The impacts of infectious diseases on wild animal populations—a review. Pages 20–51 in B. T. Grenfell and A. P. Dobson, eds., *Ecology of Infectious Diseases in Natural Populations*. Cambridge University Press, Cambridge, England.

Gulland, J. A. 1955. Estimation of growth and mortality in commercial fish populations. *U.K. Ministry of Agriculture and Fisheries, Fisheries Investigations, Series 2* 18:1–46.

———. 1988. *Fish Population Dynamics: The Implications for Management*. Wiley, New York.

Gundersen, P., B. A. Emmett, O. J. Kjonaas, C. J. Koopmans, and A. Tietema. 1998. Impact of nitrogen deposition on nitrogen cycling in forests: A synthesis of NITREX data. *Forest Ecology and Management* 101:37–55.

Gunn, J. M. and K. H. Mills. 1998. The potential for restoration of acid-damaged lake trout lakes. *Restoration Ecology* 6:390–397.

Gurevitch, J., L. L. Morrow, A. Wallace, and J. S. Walsh. 1992. A meta-analysis of competition in field experiments. *American Naturalist* 140:539–572.

Gutierrez, L. 1998. Habitat selection by recruits establishes local patterns of adult distribution in two species of damselfishes: *Stegastes dorsopunicans* and *S. planifrons*. *Oecologia* 115:268–277.

Gwynne, M. D. and R. H. Bell. 1968. Selection of vegetation components by grazing ungulates in the Serengeti National Park. *Nature* 220:390–393.

Haber, G. C. 1996. Biological, conservation, and ethical implications of exploiting and controlling wolves. *Conservation Biology* 10:1068–1081.

Haddad, N. M. 1999. Corridor and distance effects on interpatch movements: A landscape experiment with butterflies. *Ecological Applications* 9:612–622.

Haddad, N. M. and K. A. Baum. 1999. An experimental test of corridor effects on butterfly densities. *Ecological Applications* 9:623–633.

Hagen, J. B. 1992. *An Entangled Bank: The Origins of Ecosystem Ecology*. Rutgers University Press, New Brunswick, New Jersey.

Hager, H. A. and K. D. McCoy. 1998. The implications of accepting untested hypotheses: A review of the effects of purple loosestrife (*Lythrum, salicaria*) in North America. *Biodiversity and Conservation* 7:1069–1079.

Hairston, N. G. 1964. Studies on the organization of animal communities. *Journal of Animal Ecology* 33 (Suppl): 227–239.

———. 1980. The experimental test of an analysis of field distributions: Competition in terrestrial salamanders. *Ecology* 61:817–826.

———. 1991. The literature glut: Causes and consequences (reflections of a dinosaur). *Bulletin of the Ecological Society of America* 72:171–174.

Hairston, N. G., F. E. Smith, and L. B. Slobodkin. 1960. Community structure, population control, and competition. *American Naturalist* 94:421–425.

Hairston, N. G. J. and N. G. S. Hairston. 1993. Cause-effect relationships in energy flow, trophic structure, and interspecific interactions. *American Naturalist* 142:379–411.

Hall, R. W. and L. E. Ehler. 1979. Rate of establishment of natural enemies in classical biological control. *Bull Ent. Soc. America* 25:280–282.

Hall, R. W., L. E. Ehler, and B. Bisabri-Ershadi. 1980. Rate of success in classical biological control of arthropods. *Bulletin of the Entomological Society of America* 26:111–114.

Hall, S. J. 1998. Closed areas for fisheries management — the case consolidates. *Trends in Ecology and Evolution* 13:297–298.

Hall, S. J. and D. G. Raffaelli. 1993. Food webs: Theory and reality. *Advances in Ecological Research* 24:187–239.

———. 1997. Food-web patterns: What do we really know? Pages 395–417 in A. C. Gange and V. K. Brown, eds., *Multitrophic Interactions in Terrestrial Systems*. Blackwell Science, London.

Hallegraeff, G. M. 1993. A review of harmful algal blooms and their apparent global increase. *Phycologia* 32:79–99.

Hamilton, N. R., M. C. Sackville, and G. Lemaire. 1995. In defence of the -3/2 boundary rule: A re-evaluation of self-thinning concepts and status. *Annals of Botany* 76:569–577.

Hanski, I. 1982. Dynamics of regional distribution: The core and satellite species hypothesis. *Oikos* 38:210–221.

Hanski, I., J. Koukli, and A. Halkka. 1993. Three explanations of the positive relationship between distribution and abundance of species. Pages 108–116 in R. Ricklefs and D. Schulter, eds., *Species Diversity in Ecological Communities: Historical and Geographical Perspectives*. University of Chicago Press, Chicago.

Hanski, I., M. Kuussaari, and M. Nieminen. 1994. Metapopulation structure and migration in the butterfly *Melitaea cinxia*. *Ecology* 75:747–762.

Hanski, I., A. Moilanen, T. Pakkala, and M. Kuussaari. 1996. The quantitative incidence function model and persistence of an endangered butterfly metapopulation. *Conservation Biology* 10:578–590.

Hanski, I., T. Pakkala, M. Kuussaari, and G. Lei. 1995. Metapopulation persistence of an endangered butterfly in a fragmented landscape. *Oikos* 72:21–28.

Hardin, G. 1960. The competitive exclusion principle. *Science* 162:1292–1297.

———. 1968. The tragedy of the commons. *Science* 162:1243–1248.

Harper, J. L. 1977. *Population Biology of Plants*. Academic Press, New York.

———. 1981. The meanings of rarity. Pages 189–203 in H. Synge, ed. *The Biological Apsects of Rare Plant Conservation*. Wiley, London.

———. 1988. An apophasis of plant population biology. Pages 435–452 in A. J. Davy, M. J. Hutchings, and A. R. Watkinson, eds., *Plant Population Ecology*. Blackwell, Oxford.

Harper, J. L., R. B. Rosen, and J. White. 1986. The growth and form of modular organisms. *Philosophical Transactions of the Royal Society of London, Series B* 313:1–250.

Harris, V. T. 1952. An experimental study of habitat selection by prairie and forest races of the deer mouse, *Peromyscus maniculatus*. *Contributions of the Laboratory of Vertebrate Biology, University of Michigan* 56:1–53.

Harvell, C. D. 1986. The ecology and evolution of inducible defenses in a marine bryozoan: Cues, costs, and consequences. *American Naturalist* 128:810–823.

———. 1998. Genetic variation and polymorphism in the inducible spines of a marine Bryozoan. *Evolution* 52:80–86.

Harwood, J. 1989. Lessons from the seal epidemic. *New Scientist* 121:38–42.

Hassell, M. P. and R. M. May. 1974. Aggregation in predators and insect parasites and its effects on stability. *Journal of Animal Ecology* 43:567–594.

Hattersley, P. W. 1983. The distribution of C_3 and C_4 grasses in Australia in relation to climate. *Oecologia* 57:113–128.

Havens, K. 1992. Scale and structure in natural food webs. *Science* 257:1107–1109.

Hawkins, S. J. and R. G. Hartnoll. 1983. Grazing of intertidal algae by marine invertebrates. *Oceanography and Marine Biology Annual Review* 21:195–282.

Hay, M. E. 1986. Functional geometry of seaweeds: Ecological consequences of thallus layering and shape in contrasting light environments. Pages 635–666 in T. J.

Givnish, ed., *On the Economy of Plant Form and Function*. Cambridge University Press, Cambridge, England.

————. 1996. Marine chemical ecology: What's known and what's next? *Journal of Experimental Marine Biology and Ecology* 200:103–134.

Hayssen, V. and R. C. Lacy. 1985. Basal metabolic rates in mammals: Taxonomic differences in the allometry of BMR and body mass. *Comparative Biochemical Physiology* 81:741–754.

Hayward, T. L. 1991. Primary production in the North Pacific Central Gyre: a controversy with important implications. *Trends in Ecology and Evolution* 6:28–284.

He, F. and P. Legendre. 1996. On species–area relations. *American Naturalist* 148:719–737.

Hecnar, S. J. and R. T. M'Closkey. 1997. Changes in the composition of a ranid frog community following bullfrog extinction. *American Midland Naturalist* 137:145–150.

Hector, A., B. Schmid, C. Beierkuhnlein, M. C. Caldeira, M. Diemer, P. G. Dimitrakopoulos, J. A. Finn, H. Freitas, P. S. Giller, J. Good, R. Harris, P. Hogberg, K. Huss-Danell, J. Joshi, A. Jumpponen, C. Korner, P. W. Leadley, M. Loreau, A. Minns, C. P. H. Mulder, G. O'Donovan, S. J. Otway, J. S. Pereira, A. Prinz, D. J. Read, M. Scherer-Lorenzen, E.-D. Schulze, A.-S. D. Siamantziouras, E. M. Spehn, A. C. Terry, A. Y. Troumbis, F. I. Woodward, S. Yachi, and J. H. P. Lawton. 1999. Plant diversity and productivity experiments in European grasslands. *Science* 286:1123–1127.

Heed, W. B. 1978. Ecology and genetics of Sonoran desert *Drosophila*. Pages 109–126 in P. F. Brussard, ed., *Ecological Genetics: The Interface*. Springer-Verlag, New York.

Heed, W. B. and H. W. Kircher. 1965. Unique sterol in the ecology and nutrition of *Drosophila pachea*. *Science* 149:758–761.

Heesterbeek, J. A. P. and M. G. Roberts. 1995. Mathematical models for microparasites of wildlife. Pages 90–122 in B. T. Grenfell and A. P. Dobson, eds., *Ecology of Infectious Diseases in Natural Populations*. Cambridge University Press, Cambridge, England.

Heide-Jorgensen, M.-P. and T. Harkonen. 1992. Epizootiology of the seal disease in the eastern North Sea. *Journal of Applied Ecology* 29:99–107.

Henderson, S. P., S. Hattersley, S. v. Caemmer, and B. Osmond. 1993. Are C_4 pathway plants threatened by global climatic change? in E.-D. Schulze and M. M. Caldwell, eds., *Ecophysiology of Photosynthesis*. Springer-Verlag, Berlin.

Hengeveld, R. 1989. *Dynamics of Biological Invasions*. Chapman and Hall, London.

Henry, H. and L. Aarssen. 1997. On the relationship between shade tolerance and shade avoidance strategies in woodland plants. *Oikos* 80:575–582.

Herrera, C. M. 1982. Defense of ripe fruit from pests: Its significance in relation to plant-disperser interactions. *American Naturalist* 120:218–241.

Hidore, J. J. and J. E. Oliver. 1993. *Climatology: An Atmospheric Science*. Macmillan, New York.

Hietala, T., P. Hiekkala, H. Rosenqvist, S. Laakso, L. Tahvanainen, and T. Repo. 1998. Fatty acid and alkane changes in willow during frost-hardening. *Phytochemistry* 47:1501–1507.

Hik, D. S., R. L. Jefferies, and A. R. E. Sinclair. 1992. Foraging by geese, isostatic uplift and asymmetry in the development of salt-marsh plant communities. *Journal of Ecology* 80:395–406.

Hilborn, R. and C. J. Walters. 1992. *Quantitative Fisheries Stock Assessment: Choice, Dynamics, and Uncertainty*. Chapman and Hall, New York.

Hilder, V. A. and D. Boulter. 1999. Genetic engineering of crop plants for insect resistance — a critical review. *Crop Protection* 18:177–191.

Hixon, M. A. 1998. Population dynamics of coral-reef fishes: Controversial concepts and hypotheses. *Australian Journal of Ecology* 23:192–201.

Hjort, J. 1914. Fluctuations in the great fisheries of northern Europe, viewed in the light of biological research. *Rapport du Conseil International pour l'Exploration de la Mer* 20:1–228.

Hoar, W. S. 1983. *General and Comparative Physiology*. Prentice Hall, Englewood Cliffs, New Jersey.

Hobson, K. A., J. F. Piatt, and J. Pitocchelli. 1994. Using stable isotopes to determine seabird trophic relationships. *Journal of Animal Ecology* 63:786–798.

Hocker, H. W. J. 1956. Certain aspects of climate as related to the distribution of loblolly pine. *Ecology* 37:824–834.

Hocking, B. 1953. The intrinsic range and speed of flight of insects. *Transactions of the Royal Entomological Society of London* 104:223–345.

Holdaway, R. N. 1989. New Zealand's pre-human avifauna and its vulnerability. *New Zealand Journal of Ecology* 12:11–25.

Holey, M. E., R. E. Rybicki, G. W. Eck, E. H. J. Brown, J. E. Marsden, D. S. Lavis, M. L. Toneys, T. N. Trudeau, and R. M. Horrall. 1995. Progress toward lake trout restoration in Lake Michigan. *Journal of Great Lakes Research* 21:128–151.

Holland, H. D. 1995. Atmospheric oxygen and the biosphere. Pages 127–136 in C. G. Jones and J. H. Lawton, eds., *Linking Species and Ecosystems*. Chapman and Hall, New York.

Holling, C. S. 1959. The components of predation as revealed by a study of small mammal predation of the European pine sawfly. *Canadian Entomologist* 91:293–320.

Holt, R. D. 1977. Predation, apparent competition and the structure of prey communities. *Theoretical Population Biology* 28:181–208.

————. 1985. Population dynamics in two-patch environments: Some anomalous consequences of an optimal habitat distribution. *Theoretical Population Biology* 28:181–208.

Holway, D. A. 1999. Competitive mechanisms underlying the displacement of native ants by the invasive Argentine ant. *Ecology* 80:238–251.

Holzapfel, C. and B. E. Mahall. 1999. Bidirectional facilitation and interference between shrubs and annuals in the Mojave Desert. *Ecology* 80:1747–1761.

Hone, J., R. Pech, and P. Yip. 1992. Estimation of the dynamics and rate of transmission of classical swine fever (hog cholera) in wild pigs. *Epidemiology and Infection* 108:377–386.

Hooper, D. U. and P. M. Vitousek. 1998. Effects of plant composition and diversity on nutrient cycling. *Ecological Monographs* 68:121–149.

Horn, H. S. 1971. *Adaptive Geometry of Trees*. Princeton University Press, Princeton, New Jersey.

———. 1975a. Forest succession. *Scientific American* 232: 90–98.

———. 1975b. Markovian properties of forest succession. Pages 196–211 in M. L. Cody and J. M. Diamond, eds., *Ecology and Evolution of Communities*. Harvard University Press, Cambridge, Mass.

———. 1981. Succession. Pages 253–271 in R. M. May, ed., *Theoretical Ecology*. Blackwell, Oxford, England.

Horne, A. J. and C. R. Goldman. 1994. *Limnology*. McGraw-Hill, New York.

Horppila, J., H. Peltonen, T. Malinen, E. Luokkanen, and T. Kairesalo. 1998. Top-down or bottom-up effects by fish: Issues of concern in biomanipulation of lakes. *Restoration Ecology* 6:20–28.

Howard, L. O. and W. F. Fiske. 1911. The importation into the United States of the parasites of the gypsy-moth and the brown-tail moth. *U.S. Department of Agriculture, Bureau of Entomology Bulletin 91* .

Howard, W. E. 1988. Rodent pest management: The principles. In I. Prakash, ed., *Rodent Pest Management*. CRC Press, Boca Raton, Fl.

Howarth, F. G. 1991. Environmental impacts of classical biological control. *Annual Review of Entomology* 36:485–509.

Hubbell, S. P. and R. B. Foster. 1986. Biology, chance, and history and the structure of tropical rain forest tree communities. Pages 314–329 in J. Diamond and T. J. Case, eds., *Community Ecology*. Harper & Row, New York.

Hudson, P. J., D. Newborn, and A. P. Dobson. 1992. Regulation and stability of a free-living host-parasite system: *Trichostrongylus tenuis* in red grouse. I. Monitoring and parasite reduction experiments. *Journal of Animal Ecology* 61:477–486.

Huffaker, C. B. 1958. Experimental studies on predation: Dispersion factors and predator-prey oscillations. *Hilgardia* 27:343–383.

Huffaker, C. B., K. P. Shea, and S. G. Herman. 1963. Experimental studies on predation: complex dispersion and levels of food in an acarine predator-prey interaction. *Hilgardia* 34:305–330.

Hulme, P. E. 1996. Natural regeneration of yew (*Taxus baccata* L.): Microsite, seed or herbivore limitation? *Journal of Ecology* 84:853–861.

Humphreys, W. F. 1979. Production and respiration in animal populations. *Journal of Animal Ecology* 48:427–453.

———. 1984. Production efficiency in small mammal populations. *Oecologia* 62:85–90.

Huppert, A. and L. Stone. 1998. Chaos in the Pacific's coral reef bleaching cycle. *American Naturalist* 152:447–459.

Huse, G. 1998. Sex-specific life history strategies in capelin (*Mallotus villosus*)? *Canadian Journal of Fisheries and Aquatic Sciences* 55:631–638.

Huston, M. 1979. A general hypothesis of species diversity. *American Naturalist* 113:81–101.

Huston, M. and T. Smith. 1987. Plant succession: Life history and competition. *American Naturalist* 130:168–198.

Huston, M. A. 1994. *Biological Diversity: The Coexistence of Species on Changing Landscapes*. Cambridge University Press, Cambridge, England.

Hutchings, J. A. 1994. Age- and size-specific cost of reproduction within populations of brook trout, *Salvelinus fontinalis*. *Oikos* 40:12–20.

———. 1996. Spatial and temporal variation in the density of northern cod and a review of hypotheses for the stock's collapes. *Canadian Journal of Fisheries and Aquatic Sciences* 53:943–962.

Hutchings, M. J. 1983. Ecology's law in search of a theory. *New Scientist* 98:765–767.

Hutchins, H. E. and R. M. Lanner. 1982. The central role of Clark's nutcracker in the dispersal and establishment of whitebark pine. *Oecologia* 55:192–201.

Hutchinson, G. E. 1959. Homage to Santa Rosalia, or why are there so many kinds of animals? *American Naturalist* 93:145–159.

———. 1961. The paradox of the plankton. *American Naturalist* 95:137–145.

Hutto, R. L. 1985. Habitat selection by nonbreeding migratory land birds. Chapter 16 in M. L. Cody, ed., *Habitat Selection in Birds*. Academic Press, Orlando, Florida.

Idso, S. B. 1999. The long-term response of trees to atmospheric CO_2 enrichment. *Global Change Biology* 5:493–495.

Inderjit, and K. M. M. Dakshini. 1994. Algal allelopathy. *Botanical Review* 60:182–196.

Ishibashi, Y., T. Saitoh, and M. Kawata. 1998. Social organization of the vole *Clethrionomys rufocanus* and its demographic and genetic consequences: A review. *Researches on Population Ecology* 40:39–50.

Jackson, J. B., L. W. Buss, and R. E. Cooke, eds., 1985. *Population Biology and Evolution of Clonal Organisms*. Yale University Press, New Haven, Ct.

Jackson, J. B. C. 1981. Interspecific competition and species' distributions: The ghosts of theories and data past. *American Zoologist* 21:889–901.

Jaenike, J. and R. D. Holt. 1991. Genetic variation for habitat preference: Evidence and explanations. *American Naturalist* 137: S67–S90.

Janes, S. W. 1985. Habitat selection in raptorial birds. Pages 159–188 in M. L. Cody, ed., *Habitat Selection in Birds*. Academic Press, Orlando.

Janse, J. H. 1997. A model of nutrient dynamics in shallow lakes in relation to multiple stable states. *Hydrobiolgia* 342/343:1–8.

Janzen, D. H. 1966. Coevolution of mutualism between ants and acacias in Central America. *Evolution* 20:249–275.

———. 1970. Herbivores and the number of tree species in tropical forests. American Naturalist 104:501–528.

———. 1986. Chihuahuan desert nopaleras: Defaunated big mammal vegetation. *Annual Review of Ecology and Systematics* 17:595–636.

Jedrzejewski, W. and B. Jedrzejewska. 1996. Tawny owl (*Strix aluco*) predation in a pristine deciduous forest (Bialowieza National Park, Poland). *Journal of Animal Ecology* 65:105–120.

Jefferies, R. L. 1988. Vegetational mosaics, plant-animal interactions and resources for plant growth. Pages 341–369 in L. D. Gottlieb and S. K. Jain, eds., *Plant Evolutionary Biology*. Chapman and Hall, London.

Jeffries, M. J., and J. H. Lawton. 1984. Enemy free space and the structure of ecological communities. *Biological Journal of the Linnean Society* 23:269–286.

Jenkins, B., R. L. Kitching, and S. L. Pimm. 1992. Productivity, disturbance and food web structure at a local spatial scale in experimental container habitats. *Oikos* 65: 249–255.

Jermy, T. 1987. The role of experience in the host selection of phytophagous insects. Pages 143–157 in R. F. Chapman, E. A. Bernays, and J. G. J. Stoffolano, eds., *Perspectives in Chemoreception and Behavior*. New York, Springer-Verlag.

Johnson, C. N. 1998. Rarity in the tropics: Latitudinal gradients in distribution and abundance in Australian mammals. *Journal of Animal Ecology* 67:689–698.

Johnson, K. H., K. A. Vogt, H. J. Clark, O. J. Schmitz, and D. J. Vogt. 1996. Biodiversity and the productivity and stability of ecosystems. *Trends in Ecology and Evolution* 11:372–377.

Johnson, M. P., L. G. Mason, and P. H. Raven. 1968. Ecological parameters and plant species diversity. *American Naturalist* 102:297–306.

Jolliffe, P. A. 1997. Are mixed populations of plant species more productive than pure stands? *Oikos* 80:595–602.

Jones, G. P. 1990. The importance of recruitment to the dynamics of a coral reef fish population. *Ecology* 71:1691–1698.

Jordan, C. F. and R. Herrera. 1981. Tropical rain forests: Are nutrients really critical? *American Naturalist* 117:167–180.

Jordano, D., J. Rodriguez, C. D. Thomas, and J. Fernandez Haeger. 1992. The distribution and density of a lycaenid butterfly in relation to *Lasius* ants. *Oecologia* 91:439–446.

Kalisz, S. 1991. Experimental demonstration of seed bank age structure in the winter annual *Collinsia verna*. *Ecology* 72:575–585.

Karasov, W. H. 1986. Energetics, physiology and vertebrate ecology. *Trends in Ecology and Evolution* 1:101–104.

Karban, R. E. and I. T. Baldwin. 1997. *Induced Responses to Herbivory*. University of Chicago Press, Chicago.

Kareiva, P. 1990. Population dynamics in spatially complex environments: Theory and data. *Philosophical Transactions of the Royal Society of London* 330:175–190.

Kashiwagi, A., T. Kanaya, T. Yomo, and I. Urabe. 1998. How small can the difference among competitors be for coexistence to occur? *Researches on Population Ecology* 40:223–226.

Kaufman, D. M. and M. R. Willig. 1998. Latitudinal patterns of mammalian species richness in the New World: The effects of sampling method and faunal group. *Journal of Biogeography* 25:795–805.

Kazimirov, N. I., and R. N. Morozova. 1973. *Biological Cycling of Matter in Spruce Forests of Karelia*. Nauka Publishing House, Leningrad.

Keeling, C. D., T. P. Whorf, M. Wahlen, and J. van der Plicht. 1995. Interannual extremes in the rate of rise of atmospheric carbon dioxide since 1980. *Nature* 375:666–670.

Keeling, C. D., and T. P. Whorf. 1999. Atmospheric CO_2 Concentrations—Mauna Loa Observatory, Hawaii, 1958–1998 (revised July 1999). *Carbon Dioxide Information Analysis Center* U. S. Department of Energy, Oak Ridge, Tennessee: NDP-001.

Keever, C. 1950. Causes of succession on old fields of the Piedmont, North Carolina. *Ecological Monographs* 20: 229–250.

———. 1953. Present composition of some stands of the former oak-chestnut forest in the southern Blue Ridge Mountains. *Ecology* 34:44–54.

Keiner, C. 1998. W.K. Brooks and the oyster question: Science, politics, and resource management in Maryland, 1880–1930. *Journal of the History of Biology* 31:383–424.

Keith, H. 1997. Nutrient cycling in eucalypt ecosystems. Pages 197–226 in J. E. Williams and J. C. Z. Woinarski, eds., *Eucalypt Ecology: Individuals to Ecosystems*. Cambridge University Press, Cambridge, England.

Keith, L. B. 1990. Dynamics of snowshoe hare populations. *Current Mammalogy* 4:119–195.

Kenward, R. E. 1978. Hawks and doves: Factors affecting success and selection in goshawk attacks on woodpigeons. *Journal of Animal Ecology* 47:449–460.

Kerbes, R. H., P. M. Kotanen, and R. L. Jefferies. 1990. Destruction of wetland habitats by lesser snow geese: A keystone species on the west coast of Hudson Bay. *Journal of Applied Ecology* 27:242–258.

Kerfoot, W. C., ed. 1987. *Predation: Direct and Indirect Impacts on Aquatic Communities*. New England University Press, Hanover, N.H.

Kershaw, K. A. and J. H. H. Looney. 1985. *Quantitative and Dynamic Plant Ecology*. Edward Arnold, London.

Ketterson, E. D. and V. J. Nolan. 1982. The role of migration and winter mortality in the life history of a temperate-zone migrant, the dark-eyed junco, as determined from demographic analyses of winter populations. *Auk* 99:243–259.

Keyfitz, N. 1971. On the momentum of population growth. *Demography* 8:71–80.

King, D. A. 1991. The adaptive significance of tree height. *American Naturalist* 135:809–828.

King, J. M. and B. R. Heath. 1975. Game domestication for animal production in Africa. *World Animal Review* 16:23–30.

Kingsland, S. E. 1995. *Modeling Nature: Episodes in the History of Population Ecology.* University of Chicago Press, Chicago.

Kinnear, J. E., M. L. Onus, and N. R. Sumner. 1998. Fox control and rock-wallaby population dynamics. II. An update. *Wildlife Research* 25:81–88.

Kira, T. 1975. Primary production of forests. Pages 5–40 in J. P. Cooper, ed., *Photosynthesis and Productivity in Different Environments.* Cambridge University Press, London.

Kirkpatrick, J. F., J. W. Turner, Jr., I. K. M. Liu, R. Fayrer-Hosken, and A. T. Rutberg. 1997. Case studies in wildlife immunocontraception: Wild and feral equids and white-tailed deer. *Reproduction, Fertility and Development* 9:105–110.

Kirkwood, J. K. 1983. A limit to metabolisable energy intake in mammals and birds. *Comparative Biochemical Physiology* 75:1–3.

Kitching, J. A. and F. J. Ebling. 1967. Ecological studies at Lough Ine. *Advances in Ecological Research* 4:197–291.

Klein, D. R. 1968. The introduction, increase, and crash of reindeer on St. Matthew Island. *Journal of Wildlife Management* 32:350–367.

Klinkhamer, P. G. L., T. Kubo, and Y. Iwasa. 1997. Herbivores and the evolution of the semelparous perennial life-history of plants. *Journal of Evolutionary Biology* 10:529–550.

Klomp, H. 1970. The determination of clutch size in birds: A review. *Ardea* 58:1–124.

Klun, J. A. 1974. Biochemical basis of resistance of plants to pathogens and insects: Insect hormone mimics and selected examples of other biologically active chemicals derived from plants. Pages 463–484 in F. G. Maxwell and F. A. Harris, eds., *Proceedings of the Summer Institute on Biological Control of Plant Insects and Diseases.* University Press of Mississippi, Jackson.

Knoll, A. H. 1986. Patterns of change in plant communities through geological time in J. Diamond and T. J. Case, eds., *Community Ecology.* Harper & Row, New York.

Knox, E. A. 1970. Antarctic marine ecosystems. Pages 69–96 in M. W. Holdgate, ed., *Antarctic Ecology.* Academic Press, London.

Kokkinn, M. J. and A. R. Davis. 1986. Secondary production: Shooting a halcyon for its feathers. Pages 251–261 in P. DeDeckker and W. D. Williams, eds., *Limnology in Australia.* Dr. W. Junk, Dordrecht, Netherlands.

Kotler, B. P. and J. S. Brown. 1988. Environmental heterogeneity and the coexistence of desert rodents. *Annual Review of Ecology and Systematics* 19:281–307.

Kozhov, M. 1963. Lake Baikal and its life. *Monographiae Biologicate* 11:1–344.

Kozlovsky, D. G. 1968. A critical evaluation of the trophic level concept. I. Ecological efficiencies. *Ecology* 49:48–60.

Kozlowski, T. T., P. J. Kramer, and S. G. Pallardy. 1991. *The Physiological Ecology of Woody Plants.* Academic Press, San Diego.

Krebs, C. J. 1999. *Ecological Methodology.* Addison Wesley Longman, Menlo Park, Calif.

Krebs, C. J., S. Boutin, R. Boonstra, A. R. E. Sinclair, J. N. M. Smith, M. R. T. Dale, K. Martin, and R. Turkington. 1995a. Impact of food and predation on the snowshoe hare cycle. *Science* 269:1112–1115.

Krebs, C. J., S. Boutin, R. Boonstra, eds., 2001 *Vertebrate Community Dynamics in the Kluane Boreal Forest.* Oxford University Press, New York.

Krebs, J. R. and N. B. Davies. 1993. *An Introduction to Behavioural Ecology.* Blackwell Scientific Publications, Oxford.

Krebs, J. W., J. S. Smith, C. E. Rupprecht, and J. E. Childs. 1998. Rabies surveillance in the United States during 1997. *Journal of the American Veterinary Medical Association* 213:1713–1728.

Krebs, J. W., M. L. Wilson, and J. E. Childs. 1995b. Rabies-epidemiology, prevention, and future research. *Journal of Mammalogy* 76:681–694.

Krueger, C. C., M. L. Jones, and W. W. Taylor. 1995. Restoration of lake trout in the Great Lakes: Challenges and strategies for future management. *Journal of Great Lakes Research* 21:547–558.

Kruse, G. H. and A. V. Tyler. 1989. Exploratory simulation of English sole recruitment mechanisms. *Transactions of the American Fisheries Society* 118:101–118.

Kruuk, H. 1972. *The Spotted Hyaena: A Study of Predation and Social Behavior.* University of Chicago Press, Chicago.

Kuussaari, M., I. Saccheri, M. Camara, and I. Hanski. 1998. Allee effect and population dynamics in the Glanville fritillary butterfly. *Oikos* 82:384–392.

Lack, D. 1947. The significance of clutch size. *Ibis* 89:302–352.

———. 1954. *The Natural Regulation of Animal Numbers.* Clarendon Press Oxford .

Lago, P. K. and D. F. Stanford. 1989. Phytophagous insects associated with cultivated marijuana *Cannabis sativa* in northern Mississippi. *Journal of Entomological Science* 24:437–445.

LaHaye, W. S. and R. J. Gutierrez. 1999. Nest sites and nesting habitat of the northern spotted owl in northwestern California. *Condor* 101:324–330.

Lambert, B. and M. Peferoen. 1992. Insecticidal promise of *Bacillus thuringiensis*: Facts and mysteries about a successful biopesticide. *Bioscience* 42:112–121.

Lande, R. 1988. Demographic models of the northern spotted owl (*Strix occidentalis caurina*). *Oecologia* 75:601–607.

———. 1993. Risks of population extinction from demographic and environmental stochasticity and random catastrophes. American Naturalist 142:911–927.

Lande, R., B. E. Saether, and S. Engen. 1997. Threshold harvesting for sustainability of fluctuating resources. *Ecology* 78:1341–1350.

Landres, P. B., J. Verner, and J. W. Thomas. 1988. Ecological uses of vertebrate indicator species: A critique. *Conservation Biology* 2:316–328.

Langford, A. N. and M. F. Buell. 1969. Integration, identity and stability in the plant association. *Advances in Ecological Research* 6:83–135.

Larkin, P. A. 1977. An epitaph for the concept of maximum sustained yield. *Transactions of the American Fisheries Society* 106:1–11.

Larkin, P. A., B. Scott, and A. W. Trites. 1990. The red king crab fishery of the southeastern Bering Sea. *Fisheries Management Foundation*, pp. 1–78.

Larsen, D. G., D. A. Gauthier, and R. L. Markel. 1989. Causes and rate of moose mortality in the southwest Yukon. *Journal of Wildlife Management* 53:548–557.

Larsson, S. 1989. Stressful times for the plant stress-insect performance hypothesis. *Oikos* 56:277–283.

Lauck, T., C. W. Clark, M. Mangel, and G. R. Munro. 1998. Implementing the precautionary principle in fisheries management through marine reserves. *Ecological Applications* 8:S72–S78.

Lauenroth, W. K. 1979. Grassland primary production: North American grasslands in perspective. Pages 3–24 in N. R. French, ed., *Perspectives in Grassland Ecology*. Springer-Verlag, New York.

Laurance, W. F., L. V. Ferreira, J. M. Rankin-de Merona, and S. G. Laurance. 1998. Rain forest fragmentation and the dynamics of Amazonian tree communities. *Ecology* 79:2032–2040.

Laurance, W. F., S. G. Laurance, L. V. Ferreira, J. M. Rankin-de Merona, C. Gascon, and T. E. Lovejoy. 1997. Biomass collapse in Amazonian forest fragments. *Science* 278:1117–1118.

Laurance, W. F., K. R. McDonald, and R. Speare. 1996. Epidemic disease and the catastrophic decline of Australian rain forest frogs. *Conservation Biology* 10:406–413.

Law, R. 1979. The cost of reproduction in annual meadow grass. *American Naturalist* 113:3–16.

Lawlor, T. 1998. Biogeography of Great Basin mammals: Paradigm lost? *Journal of Mammology* 79:1111–1130.

Lawlor, T. E. 1986. Comparative biogeography of mammals on islands. *Biological Journal of the Linnean Society* 28:99–125.

Lawrence, D. B. 1958. Glaciers and vegetation in southeastern Alaska. *American Scientist* 46:89–122.

Lawrence, D. B., R. E. Schoenike, A. Quispel, and G. Bond. 1967. The role of *Dryas drummondii* in vegetation development following ice recession at Glacier Bay, Alaska, with special reference to its nitrogen fixation by root nodules. *Journal of Ecology* 55:793–813.

Laws, R. M. 1970. Elephants as agents of habitat and landscape change in East Africa. *Oikos* 21:1–15.

Lawton, J. H. 1984. Non-competitive populations, non-convergent communities, and vacant niches: The herbivores of bracken. Pages 67–100 in D. R. J. Strong, D. Simberloff, L. G. Abele, and A. B. Thistle, eds., *Ecological Communities*. Princeton University Press, Princeton, N.J.

———. 1987. Are there assembly rules for successional communities? Pages 225–244 in A. J. Gray, M. J. Crawley, and P. J. Edwards, eds., *Colonization, Succession and Stability*. Blackwell, Oxford, England.

———. 1990. Red grouse populations and moorland management. *British Ecological Society Ecological Issues* No. 2:1–36.

Leach, M. K. and T. J. Givnish. 1996. Ecological determinants of species loss in remnant prairies. *Science* 273:1555–1558.

Leader-Williams, N. 1988. *Reindeer on South Georgia: The Ecology of an Introduced Population*. Cambridge University Press, Cambridge, England.

Lenton, T. M. 1998. Gaia and natural selection. *Nature* 394:439–447.

Lertzman, K. P. 1992. Patterns of gap-phase replacement in a subalpine, old-growth forest. *Ecology* 73:657–669.

Leslie, P. H. 1966. The intrinsic rate of increase and the overlap of successive generations in a population of guillemots (*Uria aalge* Pont.). *Journal of Animal Ecology* 35:291–301.

Leslie, P. H., and R. M. Ranson. 1940. The mortality, fertility, and rate of natural increase of the vole (*Microtus agrestis*) as observed in the laboratory. *Journal of Animal Ecology* 9:27–52.

Letcher, B. H., J. A. Priddy, J. R. Walters, and L. B. Crowder. 1998. An individual-based, spatially-explicit simulation model of the population dynamics of the endangered red-cockaded woodpecker, Picoides borealis. Biological Conservation 86:1–14.

Lever, C. 1985. *Naturalized Mammals of the World*. Longman, London.

———. 1996. *Naturalized Fishes of the World*. Academic Press, San Diego.

Levin, D. A. 1976. Alkaloid-bearing plants: An ecogeographic perspective. *American Naturalist* 110:261–284.

Levin, S. and D. Pimentel. 1981. Selection of intermediate rates of increase in parasite-host systems. *American Naturalist* 117:308–315.

Levin, S. A. 1970. Community equilibria and stability, and an extension of the competitive exclusion principle. *American Naturalist* 104:413–423.

———. 1974. Dispersion and population interactions. *American Naturalist* 108:207–228.

Lewis, J. R. 1972. *The Ecology of Rocky Shores*. English Universities Press, London.

Lewis, V. R., L. D. Merrill, T. Atkinson, H. and J. S. Wasbauer. 1992. Imported fire ants: Potential risk to California. *California Agriculture* 46:29–31.

Lewis, W. M. 1981. Precipitation chemistry and nutrient loading by precipitation in a tropical watershed. *Water Resources Research* 17:169–181.

Lewontin, R. C. 1965. Selection of colonizing ability. Pages 77–94 in H. G. Baker and G. L. Stebbins, eds., *The Genetics of Colonizing Species*. Academic Press, New York.

———. 1969. The meaning of stability. *Brookhaven Symposia in Biology* 22:13–24.

Liebhold, A. M., J. A. Halverson, and G. A. Elmes. 1992. Gypsy moth invasion in North America: A quantitative analysis. *Journal of Biogeography* 19:513–520.

Liebold, M. A. 1995. The niche concept revisited: Mechanistic models and community context. *Ecology* 76:1371–1382.

Lifson, N. and R. McClintock. 1966. Theory of use and the turnover rates of body water for measuring energy and material balance. *Journal of Theoretical Biology* 12:46–74.

Likens, G. E., ed. 1972. *Nutrients and Eutrophication: The Limiting Nutrient Controversy*. American Society of Limnology and Oceanography, Lawrence, Ks.

Likens, G. E., and F. H. Bormann. 1972. Nutrient cycling in ecosystems. Pages 25–67 in J. A. Wiens, ed., *Ecosystem Structure and Function*. Oregon State University Press, Corvallis.

Likens, G. E., F. H. Bormann, and N. M. Johnson. 1981. Interactions between major biogeochemical cycles in terrestrial ecosystems. Pages 93–123 in G. E. Likens, ed., *Some Perspectives of the Major Biogeochemical Cycles*. Wiley, New York.

Likens, G. E., F. H. Bormann, N. M. Johnson, D. W. Fisher, and R. S. Pierce. 1970. Effects of forest cutting and herbicide treatment on nutrient budgets in the Hubbard Brook watershed-ecosystem. *Ecological Monographs* 40:23–47.

Likens, G. E., C. T. Driscoll, and D. C. Buso. 1996. Long-term effects of acid rain: Response and recovery of a forest ecosystem. *Science* 272:244–245.

Likens, G. E., R. F. Wright, J. N. Galloway, and T. J. Butler. 1979. Acid rain. *Scientific American* 241:43–51.

Lima, S. L. 1998. Nonlethal effects in the ecology of predator-prey interactions. *Bioscience* 48:25–34.

Lincoln, R., G. Boxshall, and P. Clark. 1998. *A Dictionary of Ecology, Evolution and Systematics*. Cambridge University Press, Cambridge, England.

Locke, A. 1996. Applications of the Menge-Sutherland Model to acid-stressed lake communities. *Ecological Applications* 6:797–805.

Lodge, D. M., G. Cronin, E. van Donk, and A. J. Froelich. 1998. Impact of herbivory on plant standing crop: Comparisons among biomes, between vascular and nonvascular plants, and among freshwater herbivore taxa. Pages 149–174 in E. Jeppesen, M. Sondergaard, M. Sondergaard, and K. Christoffersen, eds., *The Structuring Role of Submerged Macrophytes in Lakes*. Springer-Verlag, New York.

Loehle, C. 1987. Hypothesis testing in ecology: Psychological aspects and the importance of theory maturation. *Quarterly Review of Biology* 62:397–407.

———. 1988. Problems with the triangular model for representing plant strategies. *Ecology* 69:284–286.

Long, J. L. 1981. *Introduced Birds of the World*. Davis and Charles, London.

Long, S. P. and P. R. Hutchin. 1991. Primary production in grasslands and coniferous forests with clmate change: An overview. *Ecological Applications* 1:139–156.

Longhurst, A., S. Sathyendranath, T. Platt, and C. Caverhill. 1995. An estimate of global primary production in the ocean from satellite radiometer data. *Journal of Plankton Research* 17:1245–1271.

Lorius, C., J. Jouzel, C. Ritz, L. Merlivat, N. I. Barkov, Y. S. Korotkevich, and V. M. Kotlyakov. 1985. A 150,000-year climatic record from antarctic ice. *Nature* 316:591–596.

Lotka, A. J. 1907. Studies on the mode of growth of material aggregates. *American Journal of Science* 24:199–216.

———. 1913. A natural population norm. *Journal of the Washington Academy of Science* 3:241–248 and 289–293.

———. 1922. The stability of the normal age distribution. *Proceedings of the National Academy of Sciences (USA)* 8:339–345.

———. 1925. *Elements of Physical Biology*. Dover Publications, New York (reprint).

Lovejoy, T. E., J. Bierregaard, R.O. , A. B. Rylands, J. R. Malcolm, C. E. Quintela, L. H. Harper, J. Brown, K.S., A. H. Powell, G. V. N. Powell, H. O. R. Schubart, and M. B. Hays. 1986. Edge and other effects of isolation on Amazon forest fragments. Pages 257–285 in M. E. Soulè, ed., *Conservation Biology*. Sinauer Associates, Sunderland, Mass.

Lovelock, J. 1988. *The Ages of Gaia: A Biography of Our Living Earth*. Norton, New York.

Lubchenco, J. 1978. Plant species diversity in a marine intertidal community: Importance of herbivore food preference and algal competitive abilities. *American Naturalist* 112:23–39.

———. 1986. Relative importance of competition and predation: Early colonization by seaweeds in New England. Pages 537–555 in J. Diamond and T.J. Case (eds.), *Community Ecology*, Harper & Row, New York.

Lubina, J. A., and S. A. Levin. 1988. The spread of a reinvading species: range expansion in the California sea otter. *American Naturalist* 131:526–543.

Ludwig, D., R. Hilborn, and C. Walters. 1993. Uncertainty, resource exploitation, and conservation: Lessons from history. *Science* 260:17.

Ludwig, D. and C. J. Walters. 1985. Are age-structured models appropriate for catch-effort data? *Canadian Journal of Fisheries and Aquatic Sciences* 42:1066–1072.

Lupton, F. G. H. 1977. The plant breeders' contribution to the origin and solution of pest and disease problems. Pages 71–81 in J. M. Cherrett and G. R. Sagar, eds., *Origins of Pest, Parasite, Disease and Weed Problems*. Blackwell, Oxford, England.

Lutz, H. H. 1930. The vegetation of Heart's Content, a virgin forest in northwestern Pennsylvania. *Ecology* 11:1–29.

Macan, T. T. 1974. *Freshwater Ecology*. Wiley, New York.

MacArthur, R. and E. O. Wilson. 1967. *The Theory of Island Biogeography*. Princeton University Press, Princeton, New Jersey.

MacArthur, R. H. 1958. Population ecology of some warblers of northeastern coniferous forests. *Ecology* 39: 599–619.

MacArthur, R. H. 1965. Patterns of species diversity. *Biological Reviews* 40:510–533.

———. 1969. Patterns of communities in the tropics. *Biological Journal of the Linnean Society* 1:19–30.

———. 1972. *Geographical Ecology*. Harper & Row, New York.

MacCracken, M. C. 1985. Carbon dioxide and climate change: Background and overview. Pages 1–23 in M. C. MacCracken and F. M. Luther, eds., *Projecting the Climatic Effects of Increasing Carbon Dioxide*. U.S. Department of Energy, Washington, D.C.

Macdonald, D. W. and D. R. Voigt. 1985. The biological basis of rabies models. Pages 71–108 in P. J. Bacon, ed., *Population Dynamics of Rabies in Wildlife*. Academic Press, London.

MacIsaac, H. 1996. Potential abiotic and biotic impacts of zebra mussels on the inland waters of North America. *American Zoologist* 36:287–299.

MacLean, D.A., and R.W. Wein. 1977. Nutrient accumulation for post fire jack pine and hardwood successional patterns in New Brunswick. *Canadian Journal of Forest Research* 7:562-578.

Macnab, J. 1991. Does game cropping serve conservation? A reexamination of the African data. *Canadian Journal of Zoology* 69:2283–2290.

Macnair, M. R. 1981. Tolerance of higher plants to toxic materials. Pages 177–207. *Genetic Consequences of Man Made Change*. Academic Press, New York.

———. 1987. Heavy metal tolerance in plants: A model evolutionary system. *Trends in Ecology and Evolution* 2:254–259.

Magurran, A. E. 1988. *Ecological Diversity and its Measurement*. Princeton University Press, Princeton, New Jersey.

Majerus, M. E. N. 1998. *Melanism: Evolution in Action*. Oxford University Press, Oxford.

Major, J. 1963. A climatic index to vascular plant activity. *Ecology* 44:485-498.

Malthus, T. R. 1798. *An Essay on the Principle of Population*. Macmillan, New York.

Manne, L. L., S. L. Pimm, J. M. Diamond, and T. M. Reed. 1998. The form of the curves: a direct evaluation of MacArthur & Wilson's classic theory. *Journal of Animal Ecology* 67:784-794.

Mansson, B. A. and J. M. McGlade. 1993. Ecology, thermodynamics and H.T. Odum's conjectures. *Oecologia* 93:582–596.

Mantua, N. J., S. R. Hare, Y. Zhang, J. M. Wallace, and R. C. Francis. 1997. A Pacific interdecadal climate oscillation with impacts on salmon production. *Bulletin of the American Meteorological Society* 78:1069–1079.

Marks, P. L. 1983. On the origin of the field plants of the northeastern United States. *American Naturalist* 122: 210–228.

Marrs, R. H., S. W. Johnson, and M. G. LeDuc. 1998. Control of bracken and restoration of heathland. VIII. The regeneration of the heathland community after 18 years of continued bracken control or 6 years of control followed by recovery. *Journal of Applied Ecology* 35:857–870.

Martin, T. E. 1995. Avian life history evolution in relation to nest sites, nest predation and food. *Ecological Monographs* 65:101–127.

Martinez, N. D. 1991. Artifacts or attributes? Effects of resolution on the Little Rock Lake food web. *Ecological Monographs* 61:367–392.

Masaki, T., Y. Kominami, and T. Nakashizuka. 1994. Spatial and seasonal patterns of seed dissemination of *Cornus controversa* in a temperate forest. *Ecology* 75: 1903–1910.

Massey, A. B. 1925. Antagonism of the walnuts (*Juglans nigra* L. and *J. cinerea* L.) in certain plant associations. *Phytopathology* 15:773–784.

Matthiopoulos, J., R. Moss, and X. Lambin. 1998. Models of red grouse cycles. A family affair? *Oikos* 82:574–590.

Maxwell, F. G. and P. R. Jennings. 1980. *Breeding Plants Resistant to Insects*. Wiley, New York.

May, R. M. 1973. *Stability and Complexity in Model Ecosystems*. Princeton University Press, Princeton, New Jersey.

———. 1974a. Biological populations with nonoverlapping generations: Stable points, stable cycles, and chaos. *Science* 186:645–647.

———. 1974b. On the theory of niche overlap. *Theoretical Population Biology* 5:297–332.

———. 1975. Patterns of species abundance and diversity. Pages 81–120 in M. L. Cody and J. M. Diamond, eds., *Ecology and Evolution of Communities*. Belknap Press, Harvard University, Cambridge, Mass.

———. 1976. Models for two interacting populations. Pages 49–70 in R. M. May, ed., *Theoretical Ecology*. Saunders, Philadelphia.

———. 1978. Host-parasitoid systems in patchy environments: A phenomenological model. *Journal of Animal Ecology* 47:833–844.

———. 1981. Models for single populations. Pages 5–29 in R. M. May, ed., *Theoretical Ecology*. Blackwell, Oxford.

May, R. M. and M. P. Hassell. 1981. The dynamics of multiparasitoid-host interactions. *American Naturalist* 117: 234–261.

Mayewski, P. A., W. B. Lyons, M. J. Spencer, M. Twickler, W. Dansgaard, B. Koci, C. I. Davidson, and R. E. Honrath. 1986. Sulfate and nitrate concentrations from a south Greenland ice core. *Science* 232:975–977.

Maynard Smith, J. 1968. *Mathematical Ideas in Biology*. Cambridge University Press, New York.

Mayr, E. 1982. *The Growth of Biological Thought*. Belknap Press, Harvard University, Cambridge, Mass.

McCann, K., A. Hastings, and G. R. Huxel. 1998. Weak trophic interactions and the balance of nature. *Nature* 395:794–798.

McCarthy, M. A., D. B. Lindenmayer, and M. Dreschler. 1997. Extinction debts and risks faced by abundant species. *Conservation Biology* 11:221–226.

McCarthy, S. A. 1996. Effects of temperature and salinity on survival of toxigenic *Vibrio cholerae* O1 in seawater. *Microbial Ecology* 31:167–175.

McCauley, E. and W. W. Murdoch. 1990. Predator-prey dynamics in environments rich and poor in nutrients. *Nature* 343:455–457.

McClanahan, T. R. 1998. Predation and the distribution and abundance of tropical sea urchin populations. *Journal of Experimental Marine Biology and Ecology* 221:231–255.

McCook, L. J. 1994. Understanding ecological community succession: Causal models and theories, a review. *Vegetatio* 110:115–147.

McCown, R. L. and W. A. Williams. 1968. Competition for nutrients and light between the annual grassland species *Bromus mollis* and *Erodium botrys*. *Ecology* 49: 981–990.

McFalls, J. A. J. 1998. Population: A lively introduction. *Population Bulletin* 53:1–48.

McGowan, D. P. J. and D. L. Otis. 1998. Population demographics of two local South Carolina mourning dove populations. *Journal of Wildlife Management* 62:1443–1451.

McGowan, J. A. and P. W. Walker. 1979. Structure in the copepod community of the North Pacific Central Gyre. *Ecological Monographs* 49:195–226.

———. 1985. Dominance and diversity maintenance in an oceanic ecosystem. *Ecological Monographs* 55:103–118.

McIntosh, R. P. 1985. *The Background of Ecology: Concept and Theory*. Cambridge University Press, Cambridge.

———. 1998. The myth of community as organism. *Perspectives in Biology and Medicine* 41:426–438.

McKey, D. 1979. The distribution of secondary compounds within plants. Pages 55–133 in G. A. Rosenthal and D. H. Janzen, eds., *Herbivores: Their Interaction with Secondary Plant Metabolites*. Academic Press, New York.

McKilligan, N. G. 1987. Causes of nesting losses in the cattle egret *Ardeola ibis* in eastern Australia with special reference to the pathogenicity of the tick *Argas (Persicargas) robertsi* to nestlings. *Australian Journal of Ecology* 12:9–16.

———.. 1996. Field experiments on the effect of ticks on breeding success and health of cattle egrets. *Australian Journal of Ecology* 21:442–449.

McLeod, S. R. 1997. Is the concept of carrying capacity useful in variable environments? *Oikos* 79:529–542.

McNab, B. K. 1994. Energy conservation and the evolution of flightlessness in birds. *American Naturalist* 144:628–642.

McNaughton, S. J. 1976. Serengeti migratory wildebeest: Facilitation of energy flow by grazing. *Science* 191:92–94.

McNaughton, S. J., F. F. Banyikwa, and M. M. McNaughton. 1997. Promotion of the cycling of diet-enhancing nutrients by African grazers. *Science* 278:1798–1800.

McNaughton, S. J., M. Oesterheid, D. A. Frank, and K. J. Williams. 1989. Ecosystem-level patterns of primary productivity and herbivory in terrestrial habitats. *Nature* 341:142–144.

McPeek, M. A. 1998. The consequences of changing the top predator in a food web: A comparative experimental approach. *Ecological Monographs* 68:1–23.

McQueen, D. G., J. R. Post, and E. L. Mills. 1986. Trophic relationships in freshwater pelagic ecosystems. *Canadian Journal of Fisheries and Aquatic Sciences* 43:1571–1581.

McQueen, D. J. 1998. Freshwater food web biomanipulation: A powerful tool for water quality improvement, but maintenance is required. *Lakes and Reservoirs: Research and Management* 3:83–94.

McQueen, D. J., M. R. S. Johannes, J. R. Post, T. J. Stewart, and D. R. S. Lean. 1989. Bottom-up and top-down impacts on freshwater pelagic community structure. *Ecological Monographs* 59:289–309.

Meagher, M. and M. E. Meyer. 1994. On the origin of brucellosis in bison of Yellowstone National Park: A review. *Conservation Biology* 8:645–653.

Menge, B. A. 1995. Joint "bottom-up" and "top-down" regulation of rocky intertidal algal beds in South Africa. *Trends in Ecology and Evolution* 10:431–432.

Menge, B. A. and J. P. Sutherland. 1987. Community regulation: Variation disturbance, competition, and predation in relation to environmental stress and recruitment. *American Naturalist* 130:730–757.

Menges, E. S. 1990. Population viability analysis for an endangered plant. *Conservation Biology* 4:52–62.

Menon, S. and K. S. Bawa. 1998. Deforestation in the tropics: Reconciling disparities in estimates for India. *Ambio* 27:576–577.

Menzel, D. W. and J. H. Ryther. 1961. Nutrients limiting the production of phytoplankton in the Sargasso Sea, with special reference to iron. *Deep Sea Research* 7:276–281.

Mertz, D. B. 1970. Notes on methods used in life-history studies. Pages 4–17 in J. H. Connell, D. B. Mertz, and W. W. Murdoch, eds., *Readings in Ecology and Ecological Genetics*. Harper and Row, New York.

Messier, F. 1994. Ungulate population models with predation: a case study with the North American moose. *Ecology* 75:478-488.

Metcalf, R. L. 1986. Coevolutionary adaptations of rootworm beetles (Coleoptera: Chrysomelidae) to cucurbitacins. *Journal of Chemical Ecology* 12:1109–1124.

Miles, J. 1985. The pedogenic effects of different species and vegetation types and the implications of succession. *Journal of Soil Science* 36:571-584.

Miles, J. 1987. Vegetation succession: Past and present perceptions. Pp. 1–29 in A. J. Gray, M. J. Crawley, and P. J. Edwards, eds., *Colonization, Succession and Stability*. Blackwell, Oxford, England.

Mills, J. N. 1999. The role of rodents in emerging human disease: Examples from the hantaviruses and arenaviruses. Pages 134–160 in G. R. Singleton, L. Hinds, H. Leirs, and Z. Zhang, eds., *Ecologically-based Management of Rodent Pests*. Australian Centre for International Agricultural Research, Canberra.

Mills, L. S., M. E. Soule, and D. F. Doak. 1993. The keystone-species concept in ecology and conservation. *BioScience* 43:219-224.

Milner-Gulland, E. J. and J. R. Beddington. 1993. The exploitation of elephants for the ivory trade: An historical perspective. *Proceedings of the Royal Society of London, Series B* 252:29–37.

Moen, J. 1989. Diffuse competition—a diffuse concept. *Oikos* 54:260–263.

Monaghan, P. and R. G. Nager. 1997. Why don't birds lay more eggs? *Trends in Ecology and Evolution* 12:270–273.

Monson, R. K. 1989. On the evolutionary pathways resulting in C_4 photosynthesis and crassulacean acid metabolism (CAM). *Advances in Ecological Research* 19:57–110.

Montevecchi, W. A. and D. A. Kirk. 1996. Great auk: *Pinguinus impennis*. *Birds of North America* 260:1–20.

Mook, L. J. 1963, Birds and the spruce budworm. Pages 268-291 in R. F. Morris, ed., *The Dynamics of Epidemic Spruce Budworm Populations*. Memoirs of the Entomological Society of Canada.

Moore, P. D. and S. B. Chapman, eds., 1986. *Methods in Plant Ecology*. Blackwell Scientific Publications, Oxford.

Moran, R. J. and W. L. Palmer. 1963. Ruffed grouse introductions and population trends on Michigan islands. *Journal of Wildlife Management* 27:606–614.

Morey, H. F. 1936. A comparison of two virgin forests in northwestern Pennsylvania. *Ecology* 17:43–55.

Morgan, N. C. 1980. Secondary production. Pages 247–340 in E. D. LeCren and R. H. Lowe-McDonnell, eds., *The Functioning of Freshwater Ecosystems*. Cambridge University Press, Cambridge, England.

Moritz, M. A. 1997. Analysing extreme disturbance events: Fire in Los Padres National Forest. *Ecological Applications* 7:1252–1262.

Morris, C. D., V. L. Larson, and L. P. Lounibos. 1991. Measuring mosquito dispersal for control programs. *Journal of the American Mosquito Control Association* 7:608–615.

Morris, L. A. and R. E. Miller. 1994. Evidence for long-term productivity change as provided by field trials. Pages 41–80 in W. J. Dyck, D. W. Cole, and N. B. Comerford, eds., *Impacts of Forest Harvesting on Long-term Site Productivity*. Chapman and Hall, London.

Morris, R. F. 1957. The interpretation of mortality data in studies on population dynamics. *Canadian Entomologist* 89:49–69.

———.1959. Single-factor analysis in population dynamics. *Ecology* 40:580–588.

———. 1963. The dynamics of epidemic spruce budworm populations. *Memoirs of the Entomological Society of Canada* 31:1–332.

Morrow, P. A. and L. R. Fox. 1980. Effects of variation in *Eucalyptus* essential oil yield on insect growth and grazing damage. *Oecologia* 45:209–219.

Morse, D. H. 1980. *Behavioral Mechanisms in Ecology*. Harvard University Press, Cambridge, Mass.

Moss, R., A. Watson, and J. Ollason. 1982. *Animal Population Dynamics*. Chapman–Hall, London.

Moss, R., A. Watson, I. B. Trenholm, and R. Parr. 1993. Caecal threadworms *Trichostrongylus tenuis* in red grouse *Lagopus lagopus scoticus*: Effects of weather and host density upon estimated worm burdens. *Parasitology* 107:199–209.

Moulton, M. P. and S. L. Pimm. 1986. Species introductions to Hawaii. Pages 231–249 in H. A. Mooney and J. A. Drake, eds., *Ecology of Biological Invasions of North America and Hawaii*. Springer-Verlag, New York.

Moutia, L. A. and R. Mamet. 1946. A review of twenty-five years of economic entomology in the islands of Mauritius. *Bulletin of Entomological Research* 36:439–472.

Muirhead-Thomson, R. C. 1951. *Mosquito Behaviour in Relation to Malaria Transmission and Control in the Tropics*. Edward Arnold, London.

Mulder, M. B. 1998. The demographic transition: Are we any closer to an evolutionary explanation? *Trends in Ecology and Evolution* 13:266–270.

Muller, C. H. 1970. Phytotoxins as plant habitat variables. *Recent Advances in Phytochemistry* 3:105–121.

Muller, L. I., R. J. Warren, and D. L. Evans. 1997. Theory and practice of immunocontraception in wild mammals. *Wildlife Society Bulletin* 25:504–514.

Muller-Dombois, D., and H. Ellenberg. 1974. *Aims and Methods of Vegetation Ecology*. John Wiley and Sons, New York.

Murdoch, W. W. 1966. "Community structure, population control, and competition": A critique. *American Naturalist* 100:219–226.

———. 1971. The developmental response of predators to changes in prey density. *Ecology* 52:132–137.

———. 1975. Diversity, complexity, stability, and pest control. *Journal of Applied Ecology* 12:795–807.

———. 1994. Population regulation in theory and practice. *Ecology* 75:271–287.

Murdoch, W. W. and J. Bence. 1987. General predators and unstable prey populations. Pages 17–30 in W. C. Kerfoot and A. Sih, eds., *Predation: Direct and Indirect Impacts on Aquatic Communities*. University Press of New England, Hanover, New Hampshire.

Murdoch, W. W. and C. J. Briggs. 1996. Theory for biological control: Recent developments. *Ecology* 77:2001–2013.

Murdoch, W. W., J. Chesson, and P. L. Chesson. 1985. Biological control in theory and practice. *The American Naturalist* 125:344–366.

Murdoch, W. W., R. F. Luck, S. L. Swarbrick, S. Walde, D. S. Yu, and J. D. Reeve. 1995. Regulation of an insect population under biological control. *Ecology* 76:206–217.

Murdoch, W. W., R. M. Nisbet, S. P. Blythe, W. S. C. Gurney, and J. D. Reeve. 1987. An invulnerable age class and stability in delay-differential parasitoid-host models. *American Naturalist* 129:263–282.

Murdoch, W. W. and A. Oaten. 1975. Predation and population stability. *Advances in Ecological Research* 9:1–131.

Murdoch, W. W. and A. Sih. 1978. Age-dependent interference in a predatory insect. *Journal of Animal Ecology* 47:581–592.

Murdoch, W. W. and A. Stewart-Oaten. 1989. Aggregation by parasitoids and predators: Effects on equilibrium and stability. *American Naturalist* 134:288–310.

Murphy, E. C. and E. Haukioja. 1986. Clutch size in nidiculous birds. *Current Ornithology* 4:141–180.

Murray, B. G. 1979. *Populations Dynamics: Alternative Models*. Academic Press, New York.

Myers, J. H. 1993. Population outbreaks in forest Lepidoptera. *American Scientist* 81:240–250.

Myers, J. H. and D. Bazely. 1991. Thorns, spines, prickles, and hairs: Are they stimulated by herbivory and do they deter herbivores? Pages 325–344 in D. W. Tallamy and M. J. Raupp, eds., *Phytochemical Induction by Herbivores*. Wiley, New York.

Myers, J. H., J. Monro, and N. Murray. 1981. Egg clumping, host plant selection and population regulation in *Cactoblastis cactorum* (Lepidoptera). *Oecologia* 51:7–13.

Myers, K., I. D. Marshall, and F. Fenner. 1954. Studies in epidemiology of infectious myxomatosis of rabbits. III. Observations on two succeeding epizootics in Australian wild rabbits on Riverine Plain of south-eastern Australia. *Journal of Hygiene* 52:337-360.

Myers, R. A. 1998. When do environment-recruitment correlations work? *Reviews in Fish Biololgy and Fisheries* 8:285–305.

Myers, R. A., J. A. Hutchings, and N. J. Barrowman. 1996. Hypotheses for the decline of cod in the North Atlantic. *Marine Ecology Progress Series* 138:293–308.

———. 1997. Why do fish stocks collapse? The example of cod in Atlantic Canada. 7:91–106.

Mysak, L. A. 1986. El Niño, interannual variability and fisheries in the Northeast Pacific Ocean. *Canadian Journal of Fisheries and Aquatic Sciences* 43:464–497.

Nagy, K. A. 1987. Field metabolic rate and food requirement scaling in mammals and birds. *Ecological Monographs* 57:111–128.

———. 1989. Field bioenergetics: Accuracy of models and methods. *Physiological Zoology* 62:237–252.

National Center for Health Statistics. 1998. Health, United States, 1998, with Socioeconomic Status and Health Chartbook. U.S. Department of Health and Human Services, Hyattsville, Maryland.

Nedelman, J., J. A. Thompson, and R. J. Taylor. 1987. The statistical demography of whooping cranes. *Ecology* 68:1401–1411.

Needham, J. 1986. *Science and Civilisation in China, Vol. 6: Part 1. Biology and Biological Technology: Botany*. Cambridge University Press, Cambridge, England.

Newman, E. I. 1997. Phosphorus balance of contrasting farming systems, past and present. Can food production be sustainable? *Journal of Applied Ecology* 34:1334–1347.

Newmark, W. D. 1985. Legal and biotic boundaries of Western North American National Parks: A problem of congruence. *Biological Conservation* 33:197–208.

———. 1995. Extinction of mammal populations in Western North American national parks. *Conservation Biology* 9:512–526.

Newton, I. 1972. *Finches*. Collins, London.

Nichol, S. and W. K. de la Mare. 1993. Ecosystem management and the Antarctic krill. *American Scientist* 81:36–47.

Nicholls, A. O. and C. R. Margules. 1991. The design of studies to demonstrate the biological importance of corridors. Pages 49–61 in D. A. Saunders and R. J. Hobbs, eds., *Nature Conservation 2: The Role of Corridors*. Surrey Beatty and Sons, Chipping Norton, Australia.

———. 1993. An upgraded reserve selection algorithm. *Biological Conservation* 64:165–169.

Nichols, J. D. 1991. Science, population ecology, and the management of the American Black Duck. *Journal of Wildlife Management* 55:790–799.

Nichols. J. D., M. J. Conroy, D. R. Anderson, and K. P. Burnham. 1984. Compensatory mortality in waterfowl populations: a review of the evidence and implications for research and management. *Transactions of the North American Wildlife and Natural Resources Conference* 49:535-554.

Nicholson, A. J. 1933. The balance of animal populations. *Journal of Animal Ecology* 2:132–178.

Nicholson, A. J. and V. A. Bailey. 1935. The balance of animal populations. Part 1. *Proceedings of the Zoological Society of London* 3:551–598.

Nickelson, T. E., and J. A. Lichatowich. 1983. The influence of the marine environment on the interannual variation in coho salmon abundance: an overview. Pages 24-36 in W. Pearcy, ed., *The Influence of Ocean Conditions on the Production of Salmonids in the North Pacific*. Oregon State University Sea Grant Publication, Corvallis.

Nielsen, E. S. and E. A. Jensen. 1957. Primary oceanic production. *Galathea Report* 1:49–136.

Niemi, G. J., J. M. Hanowski, A. R. Lima, T. Nicholls, and N. Weiland. 1997. A critical analysis on the use of indicator species in management. *Journal of Wildlife Management* 61:1240–1251.

Nicklas, K. J., B. H. Tiffney, and A. H. Knoll. 1980. Apparent changes in the diversity of fossil plants. *Evolutionary Biology* 12:1-89.

Noble, I. R. 1981. Predicting successional change. Pages 278–300 in H. A. Mooney, ed., *Fire Regimes and Ecosystem Properties*. U.S. Department of Agriculture, Forest Service, General Technical Report.

Noble, J. C. 1991. On ratites and their interactions with plants. *Revista Chilena de Historia Natural* 64:85–118.

Norby, R. J., C. A. Gunderson, S. D. Wullschleger, E. G. O'Neill, and M. K. McCracken. 1992. Productivity and compensatory responses of yellow-poplar trees in elevated CO_2. *Nature* 357:322–324.

Norby, R. J., S. D. Wullschleger, C. A. Gunderson, D. W. Johnson, and R. Ceulemans. 1999. Tree responses to rising CO_2 in field experiments: Implications for the future forest. *Plant, Cell and Environment* 22:683–714.

Noss, R. F. 1987. Corridors in real landscapes: a reply to Simberloff and Cox. *Conservation Biology* 1:159-164.

Noss, R. F. 1990. Indicators for monitoring biodiversity: A hierarchical approach. *Conservation Biology* 4:355–364.

Noy-Meir, I. 1975. Stability of grazing systems: An application of predator-prey graphs. *Journal of Ecology* 63:459–481.

O'Donoghue, M. 1994. Early survival of juvenile snowshoe hares. *Ecology* 75:1582–1592.

O'Donoghue, M., E. Hofer, and F. I. Doyle. 1995. Predator versus predator. *Natural History* 104:6–9.

Odum, E. P. 1963. *Ecology*. Holt, Rinehart and Winston, New York.

Odum, H. T. 1983. *Systems Ecology: An Introduction*. Wiley, New York.

Oechel, W. C., S. Cowles, N. Grulke, S. J. Hastings, B. Lawrence, T. Prudhomme, G. Riechers, B. Strain, D. Tissue, and G. Vourlitis. 1994. Transient nature of CO_2 fertilization in arctic tundra. *Nature* 371:500–503.

Oesterheld, M., O. E. Sala, and S. J. McNaughton. 1992. Effect of animal husbandry on herbivore-carrying capacity at a regional scale. *Nature* 356:234–236.

Oksanen, L., S. D. Fretwell, J. Arruda, and P. Niemelä. 1981. Exploitation ecosystems in gradients of primary productivity. *American Naturalist* 118:240–261.

Oksanen, L. and T. Oksanen. 1992. Long-term microtine dynamics in north Fennoscandian tundra: The vole cycle and the lemming chaos. *Ecography* 15:226–236.

Olson, J. S. 1958. Rates of succession and soil changes on southern Lake Michigan sand dunes. *Botanical Gazette* 119:125–170.

O'Neill, R. V., D. L. DeAngelis, J. B. Waide, and T. F. H. Allen. 1986. *A Hierarchical Concept of Ecosystems*. Princeton University Press, Princeton, New Jersey.

Orians, G. H. and R. T. Paine. 1983. Convergent evolution at the community level. Pages 431–458 in D. J. Futuyma and M. Slatkin, eds., *Coevolution*. Sinauer Associates, Sunderland, Mass.

Osmond, C. H. and J. Monro. 1981. *Prickly pear*. Academic Press, New York.

Osterhaus, A. D. M. E. and E. J. Vedder. 1988. Identification of virus causing recent seal deaths. *Nature* 335:20.

Ostfeld, R. S. 1997. The ecology of Lyme-disease risk. *American Scientist* 85:338–346.

Overland, L. 1966. The role of allelopathic substances in the "smother crop" barley. *American Journal of Botany* 53:423–432.

Ovington, J. D. 1978. *Australia's Endangered Species: Mammals, Birds, and Reptiles*. Cassell Australia, Melbourne.

Owen, D. F. and R. G. Wiegert. 1981. Mutualism between grasses and grazers: An evolutionary hypothesis. *Oikos* 36:376–378.

Pac, H. I., and K. Frey. 1991. Some population characteristics of the Northern Yellowstone bison herd during the winter of 1988-89. *Montana Department of Fish, Wildlife and Parks*, Bozeman, Montana.

Pacala, S. W. and J. Roughgarden. 1985. Population experiments with the *Anolis* lizards of St. Maarten and St. Eustatius. *Ecology* 66:129–141.

Pace, M. L., S. E. G. Findlay, and D. Fischer. 1998. Effects of an invasive bivalve on the zooplankton community of the Hudson River. *Freshwater Biology* 39:103–116.

Pagel, M. D., R. M. May, and A. R. Collie. 1991. Ecological aspects of the geographical distribution and diversity of mammalian species. *American Naturalist* 137:791-815.

Paine, R. T. 1966. Food web complexity and species diversity. *American Naturalist* 100:65–75.

———. A note on trophic complexity and community stability. *American Naturalist* 104:91–93.

———. 1974. Intertidal community structure: Experimental studies on the relationship between a dominant competitor and its principal predator. *Oecologia* 15:93–120.

———. 1983. On paleoecology: An attempt to impose order on chaos (review). *Paleobiology* 9:86–90.

———. 1984. Ecological determinism in the competition for space. *Ecology* 65:1339–1348.

———. 1992. Food-web analysis through field measurement of per capita interaction strength. *Nature* 355:73–75.

Painter, E. L. and A. J. Belsky. 1993. Application of herbivore optimization theory to rangelands of the western United States. *Ecological Applications* 3:2–9.

Palmer, W. L. 1962. Ruffed grouse flight capability over water. *Journal of Wildlife Management* 26:338–339.

Parer, I., D. Conolly, and W. R. Sobey. 1985. Myxomatosis: The effects of annual introductions of an immunizing strain and a highly virulent strain of myxoma virus into rabbit populations at Urana, N.S.W. *Australian Wildlife Research* 12:407–423.

Park, T., P. H. Leslie, and D. B. Mertz. 1964. Genetic strains and competition in populations of *Tribolium*. *Physiological Zoology* 37:97–162.

Parker, G. G., S. M. Hills, and L. A. Kuehnel. 1993. Decline of understory American chestnut (*Castanea dentata*) in a southern Appalachian forest. *Canadian Journal of Forest Research* 23:259–265.

Pascual, M. A. and M. D. Adkison. 1994. The decline of the Steller sea lion in the northeast Pacific: Demography, harvest, or environment? *Ecological Applications* 4:393–403.

Pastor, J. and S. D. Bridgham. 1999. Nutrient efficiency along nutrient availability gradients. *Oecologia* 118:50–58.

Pastor, J., B. Dewey, R. J. Naiman, P. F. McInnes, and Y. Cohen. 1993. Moose browsing and soil fertility in the boreal forests of Isle Royale National Park. *Ecology* 74:467–480.

Pastor, J. and W. M. Post. 1988. Response of northern forests to CO_2-induced climate change. *Nature* 334:55–58.

Pastoret, P. P. and B. Brochier. 1999. Epidemiology and control of fox rabies in Europe. *Vaccine* 17:1750–1754.

Patten, B. C. 1993. Toward a more holistic ecology, and science: The contribution of H.T. Odum. *Oecologia* 93:597–602.

Patterson, B. D. 1987. The principle of nested subsets and its implications for biological conservation. *Conservation Biology* 1:323–334.

Patterson, R. S., D. E. Weidhaas, H. R. Ford, and C. S. Lofgren. 1970. Suppression and elimination of an island population of *Culex pipiens quinquefasciatus* with sterile males. *Science* 168:1368–1370.

Pauly, D. and V. Christensen. 1995. Primary production required to sustain global fisheries. *Nature* 374:255–257.

Pauly, D., V. Christensen, J. Dalsgaard, R. Froese, and F. Torres Jr. 1998. Fishing down marine food webs. *Science* 279:860–863.

Pearcy, R. W. 1983. The light environment and growth of C3 and C4 tree species in the understory of a Hawaiian forest. *Oecologia* 58:19–25.

Pearcy, R. W. and J. Ehleringer. 1984. Comparative ecophysiology of C3 and C4 plants. *Plant Cell and Environment* 7:1–13.

Pearl, R. 1922. *The Biology of Death*. Lippincott, Philadelphia.

———. 1927. The growth of populations. *Quarterly Review of Biology* 2:532–548.

———. 1928. *The Rate of Living*. Knopf, New York.

Pearl, R. and L. J. Reed. 1920. On the rate of growth of the population of the United States since 1790 and its mathematical representation. *Proceedings of the National Academy of Science (USA)* 6:275–288.

Pearl, R., L. J. Reed, and J. F. Kish. 1940. The logistic curve and the census count of 1940. *Science* 92:486–488.

Pearson, D. L. 1977. A pantropical comparison of bird community structure on six lowland forest sites. *Condor* 79:232–244.

Pech, R., G. M. Hood, J. Mcllroy, and G. Saunders. 1997. Can foxes be controlled by reducing their fertility? *Reproduction, Fertility and Development* 9:41-50.

Pech, R. P., A. R. E. Sinclair, and A. E. Newsome. 1995. Predation models for primary and secondary prey species. *Wildlife Research* 22:55–64.

Peltonen, A. and I. Hanski. 1991. Patterns of island occupancy explained by colonization and extinction rates in shrews. *Ecology* 72:1698–1708.

Penn, D. 1999. Explaining the human demographic transition. *Trends in Ecology and Evolution* 14:32–33.

Peñuelas, J. and M. Estiarte. 1998. Can elevated CO_2 affect secondary metabolism and ecosystem function? *Trends in Ecology and Evolution* 13:20–24.

Peters, R. H. 1977. Unpredictable problems with trophodynamics. *Environmental Biology of Fishes* 2:97–101.

———. 1983. *The Ecological Implications of Body Size*. Cambridge University Press, New York.

Peterson, R. O. 1992. Ecological studies of wolves on Isle Royale Annual Report 1991–1992. *Isle Royale Natural History Association* 16.

Peterson, R. O., N. J. Thomas, J. M. Thurber, J. A. Vucetich, and T. Waite. 1998. Population limitation and the wolves of Isle Royale. *Journal of Mammology* 79:828–841.

Petraitis, P. S. and S. R. Dudgeon. 1999. Experimental evidence for the origin of alternative communities on rocky intertidal shores. *Oikos* 84:239–245.

Petrides, G. A. and W. G. Swank. 1966. Estimating the productivity and energy relations of an African elephant population. *Proceedings of the Ninth International Grassland Congress, Sao Paulo, Brazil* :831–842.

Petrusewicz, K. and A. MacFadyen. 1970. *Productivity of Terrestrial Animals: Principles and Methods*. Blackwell, Oxford, England.

Pettifor, R. A. 1993. Brood-manipulation experiments. I. The number of offspring surviving per nest in blue tits (*Parus caeruleus*). *Journal of Animal Ecology* 62:131–144.

Phillips, J. 1934–1935. Succession, development, the climax, and the complex organism: An analysis of concepts. *Journal of Ecology* 22:554–571; 23:210–246, 488–508.

Phillips, O. L., Y. Malhi, N. Higuchi, W. F. Laurance, P. V. Nunez, R. M. Vasquez, S. G. Laurance, L. V. Ferreira, M. Stern, S. Brown, and J. Grace. 1998. Changes in the carbon balance of tropical forests: Evidence from long-term plots. *Science* 282:439–441.

Pianka, E. R. 1994. *Evolutionary Ecology*. Harper Collins, New York.

Pianka, E. R. and W. S. Parker. 1975. Age-specific reproductive tactics. *American Naturalist* 109:453–464.

Pickering, S. 1917. The effect of one plant on another. *Annals of Botany* 31:181–187.

Pickett, S. T. A. and P. S. White, eds., 1985. *The Ecology of Natural Disturbance and Patch Dynamics*. Academic Press, Orlando, Fl.

Pielou, E. C. 1969. *An Introduction to Mathematical Ecology*. Wiley, New York.

Pielou, E. C. 1979. *Biogeography*. Wiley, New York.

———. 1984. *The Interpretation of Ecological Data*. Wiley, New York.

———. 1991. *After the Ice Age: The Return of Life to Glaciated North America*. University of Chicago Press, Chicago.

Pieper, R. D. 1988. Rangeland vegetation productivity and biomass. Pages 449–467 in P. T. Tueller, ed., *Vegetation Science Applications for Rangeland Analysis and Management*. Kluwer Academic Publishers, Dordrecht, Netherlands.

Pierson, E. A. and R. M. Turner. 1998. An 85-year study of saguaro (*Carnegiea gigantea*) demography. *Ecology* 79:2676–2693.

Pimentel, D. 1961. Species diversity and insect population outbreaks. *Annals of the Entomological Society of America* 54:76–86.

———. 1963. Introducing parasites and predators to control native pests. *Canadian Entomologist* 95:785–792.

———. ed. 1991. *Handbook of Pest Management in Agriculture.* CRC Press, Boca Raton, Fl.

———. 1997. Pest management in agriculture. Pages 1–11 in D. Pimentel, ed., *Techniques for Reducing Pesticide Use: Economic and Environmental Benefits.* John Wiley, Chichester, England.

Pimentel, D., H. Acquay, M. Biltonen, P. Rice, M. Silva, J. Nelson, V. Lipner, S. Giordano, A. Horowitz, and M. D'Amore. 1992. Assessment of environmental and economic impacts of pesticide use. Pages 47–83 in *The Pesticide Question: Environment, Economics, and Ethics.* Chapman and Hall, New York.

Pimental, D., L. McLaughlin, A. Zepp, B. Lakitan, T. Kraus, P. Kleinman, F. Vancini, W. J. Roach, E. Graap, W. S. Keeton, and G. Selig. 1991. Environmental and economic effects of reducing pesticide use. *Bioscience* 41:402-409.

Pimm, S. L. 1982. Food Webs. Chapman and Hall, London.

Pimm, S. L. 1984. The complexity and stability of ecosystems. *Nature* 307:321–326.

Pimm, S. L. 1991. *The Balance of Nature?* University of Chicago Press, Chicago.

Pimm, S. L., J. H. Lawton, and J. E. Cohen. 1991. Food web patterns and their consequences. *Nature* 350:669–674.

Pitcher, T. J., P. Hart, and D. Pauly, eds., 1998. *Reinventing Fisheries Management.* Kluwer Academic Publishers, Boston.

Platt, J. R. 1964. Strong inference. *Science* 146:347–353.

Platt, T. and S. Sathyendranath. 1988. Oceanic primary production: Estimation by remote sensing at local and regional scales. *Science* 241:1613–1620.

Platt, T., S. Sathyendranath, O. Ulloa, W. G. Harrison, N. Hoepffner, and J. Goes. 1992. Nutrient control of phytoplankton photosynthesis in the western North Atlantic. *Nature* 356:229–231.

Polis, G. A. 1991. Complex trophic interactions in deserts: An empirical critique of food-web theory. *American Naturalist* 138:123–155.

———. 1999. Why are parts of the world green? Multiple factors control productivity and the distribution of biomass. *Oikos* 86:3–15.

Polis, G. A. and D. R. Strong. 1996. Food web complexity and community dynamics. *American Naturalist* 147:813–846.

Polis, G. A. and K. O. Winemiller, eds., 1996. *Food Webs: Integration of Patterns and Dynamics.* Chapman and Hall, New York.

Pollock, K. H., J. D. Nichols, C. Brownie, and J. E. Hines. 1990. Statistical inference for capture-recapture experiments. *Wildlife Monographs* 107:1–97.

Ponder, F. J. 1987. Allelopathic interference of black walnut trees with nitrogen-fixing plants in mixed plantings. Pages 195–204 in G. R. Walker, ed., *Allelochemicals: Role in Agriculture and Forestry.* American Chemical Society, Washington, D.C.

Popper, K. R. 1963. *Conjectures and Refutations: The Growth of Scientific Knowledge.* Routledge and Kegan Paul, London.

Porter, S. D., D. F. Williams, R. S. Patterson, and H. G. Fowler. 1997. Intercontinental differences in the abundance of *Solenopsis* fire ants (Hymenoptera: Formicidae): Escape from natural enemies? *Environmental Entomology* 26:373–384.

Porter, W. F. and H. B. Underwood. 1999. Of elephants and blind men: Deer management in the U.S. National Parks. *Ecological Applications* 9:3–9.

Postel, S. L., G. C. Daily, and P. R. Ehrlich. 1996. Human appropriation of renewable fresh water. *Science* 271:785–788.

Pough, F. H. 1988. Mimicry of vertebrates: Are the rules different? *American Naturalist* 131 Supp.:S67–S102.

Poulson, T. L. 1999. Autogenic, allogenic, and individualistic mechanisms of dune succession at Miller, Indiana. *Natural Areas Journal* 19:172–176.

Poulson, T. L. and C. McClung. 1999. Anthropogenic effects on early dune succession at Miller, Indiana. *Natural Areas Journal* 19:177–179.

Poulison, T. L., and Platt. 1996. Replacement patterns of beech and sugar maple in Warren Woods, Michigan. *Ecology* 77:1234-1253.

Power, M. E. 1990. Effects of fish in river food webs. *Science* 250:811–814.

———. 1992a. Habitat heterogeneity and the functional significance of fish in river food webs. *Ecology* 73:1675–1688.

———. 1992b. Top-down and bottom-up forces in food webs: Do plants have primacy? *Ecology* 73:733–746.

Power, M. E., T. D., J. A. Estes, B. A. Menge, W. J. Bond, L. S. Mills, G. Daily, J. C. Castilla, J. Lubchenco, and R. T. Paine. 1996. Challenges in the quest for keystones. *BioScience* 46:609–620.

Pöysä, H., J. Elmberg, K. Sjöberg, and P. Nummi. 1998. Habitat selection rules in breeding mallards (*Anas platyrhynchos*): A test of two competing hypotheses. *Oecologia* 114:283–287.

Pratt, D. M. 1943. Analysis of population development in *Daphnia* at different temperatures. *Biological Bulletin* 85:116–140.

Preston, F. W. 1948. The commonness and rarity of species. *Ecology* 29:254–283.

———. 1960. Time and space and the variation of species. *Ecology* 41:611-627.

———. 1962. The canonical distribution of commonness and rarity. *Ecology* 43:185–215, 410–432.

Price, M. V. 1978. The role of microhabitat in structuring desert rodent communities. *Ecology* 59:910–921.

Price, P. W. 1991. The plant vigor hypothesis and herbivore attack. *Oikos* 62:244–251.

Price, T., M. Kirkpatrick, and S. J. Arnold. 1988. Directional selection and the evolution of breeding date in birds. *Science* 240:798–799.

Primack, R. B. and H. Kang. 1989. Measuring fitness and natural selection in wild plant populations. *Annual Review of Ecology and Systematics* 20:367–396.

Prins, H. H. T. and F. J. Weyerhaeuser. 1987. Epidemics in populations of wild ruminants: Anthrax and impala, rinderpest and buffalo in Lake Manyara National Park, Tanzania. *Oikos* 49:28–38.

Pritchett, W. L., and N. B. Comerford. 1983. Nutrition and fertilization of slash pine. Pages 69-90 in E. L. Stone, ed., *The Managed Slash Pine Ecosystem*. University of Florida, Gainesville.

Pulliainen, E. 1972. Summer nutrition of crossbills (*Loxia pytyopsittacus*, *L. curvirostra* and *L. leucoptera*) in northeastern Lapland in 1971. *Annales Zoologici Fennici* 9:28–31.

Pulliam, H. R. 1988. Sources, sinks, and population regulation. *American Naturalist* 132:652–661.

Pulliam, H. R. and B. J. Danielson. 1991. Sources, sinks, and habitat selection: A landscape perspective on population dynamics. *American Naturalist* 137:S50–S66.

Qi, M. Q. and R. E. Redmann. 1993. Seed germination and seedling survival of C_3 and C_4 grasses under water stress. *Journal of Arid Environments* 24:277–285.

Quay, P. D., B. Tilbrook, and C. S. Wong. 1992. Oceanic uptake of fossil fuel CO_2: Carbon-13 evidence. *Science* 256:74–79.

Quinn, J. F. and A. E. Dunham. 1983. On hypothesis testing in ecology and evolution. *American Naturalist* 122:602–617.

Quinn, R. M., K. J. Gaston, and D. B. Roy. 1998. Coincidence in the distributions of butterflies and their foodplants. *Ecography* 21:279–288.

Rabinowitz, D. 1981. Seven forms of rarity. Pages 205–217 in H. Synge, ed., *The Biological Aspects of Rare Plant Conservation*. John Wiley and Sons, London.

Rapport, D. J., R. Costanza, and A. J. McMichael. 1998. Assessing ecosystem health. *Trends in Ecology and Evolution* 13:397–402.

Rasmussen, D. I. 1941. Biotic communities of Kaibab Plateau, Arizona. *Ecological Monographs* 3:229–275.

Recher, H. F. 1969. Bird species diversity and habitat diversity in Australia and North America. *American Naturalist* 103:75–80.

Reddingius, J. and P. J. den Boer. 1970. Simulation experiments illustrating stabilization of animal numbers by spreading of risk. *Oecologia* 5:240–284.

Reed, J. M. 1999. The role of behavior in recent avian extinctions and endangerments. *Conservation Biology* 13:232–241.

Reeve, J. D. and W. W. Murdoch. 1985. Aggregation by parasitoids in the successful control of the California red scale: A test of theory. *Journal of Animal Ecology* 54:797–816.

———. 1986. Biological control by the parasitoid *Aphytis melinus*, and population stability of the California red scale. *Journal of Animal Ecology* 55:1069–1082.

Reice, S. R. 1994. Nonequilibrium determinants of biological community structure. *American Scientist* 82:424–435.

Reichle, D. E., R. V. O'Neill, and W. F. Harris. 1975. Principles of energy and material exchange in ecosystems. Pages 27–43 in W. H. van Dobben and R. H. Lowe-

McConnell, eds., *Unifying Concepts in Ecology*. Dr. W. Junk Publishers, The Hague.

Reid, C. 1899. *The Origin of the British Flora*. Dulau, London.

Reid, W. V. 1998. Biodiversity hotpots. *Trends in Ecology and Evolution* 13:275–280.

Rejmankova, E., D. R. Roberts, S. Manguin, K. O. Pope, J. Komarek, and R. A. Post. 1996. *Anopheles albimanus* (Diptera: Culicidae) and Cyanobacteria: An example of larval habitat selection. *Environmental Entomology* 25:1058–1067.

Reznick, D. 1985. Costs of reproduction: An evaluation of the empirical evidence. *Oikos* 44:257–267.

Rhoades, D. F. and R. G. Cates. 1976. Towards a general theory of plant antiherbivore chemistry. *Recent Advances in Phytochemistry* 19:168–213.

Rice, E. L. 1984. *Allelopathy*. Academic Press, New York.

Rich, S. M., D. A. Caporale, S. R. Telford, III, T. D. Kocher, D. L. Hartl, and A. Spielman. 1995. Distribution of the *Ixodes ricinus*-like ticks of eastern North America. *Proceedings of the National Academy of Science USA* 92:6284–6288.

Richards, O. W. 1928. Potentially unlimited multiplication of yeast with constant environment and the limiting of growth by changing environment. *Journal of General Physiology* 11:525–538.

———. 1939. An American textbook (review of A. S. Pearse, *Animal Ecology*). *Journal of Animal Ecology* 8:387–388.

Richards, P. W. 1952. *The Tropical Rain Forest*. Cambridge University Press, Cambridge, England.

———. 1969. Speciation in the tropical rain forest and the concept of the niche. *Biological Journal of the Linnean Society* 1:149–153.

Richardson, D. M. and W. J. Bond. 1991. Determinants of plant distribution: Evidence from pine invasions. *American Naturalist* 137:639–668.

Richardson, R. H., J. J. Ellison, and W. W. Averhoff. 1982. Autocidal control of screwworms in North America. *Science* 215:361–370.

Ricker, W. E. 1975. Computation and interpretation of biological statistics of fish populations. *Fisheries Research Board of Canada Bulletin* 191:1–382.

———. 1982. Size and age of British Columbia sockeye salmon (*Oncorhynchus nerka*) in relation to environmental factors and the fishery. *Canadian Technical Reports in Fisheries and Aquatic Sciences* 1115:1–117.

Ricklefs, R. E. and D. Schluter, eds., 1993. *Species Diversity in Ecological Communities: Historical and Geographical Perspectives*. University of Chicago Press, Chicago.

Riekirk, H. 1983. Impacts of silviculture on flatwoods runoff, water quality and nutrient budgets. *Water Resources Bulletin* 19:73-79.

Rijnsdorp, A. D. and R. S. Millner. 1996. Trends in population dynamics and exploitation of North Sea plaice (*Pleuronectes platessa* L.) since the late 1800s. *ICES Journal of Marine Science* 53:1170–1184.

Riney, T. 1964. The impact of introductions of large herbivores on the tropical environment. *International Union for the Conservation of Nature Publications, New Series* 4:261-273.

Risch, T. S., F. S. Dobson, and J. O. Murie. 1995. Is mean litter size the most productive? A test in Columbian ground squirrels. *Ecology* 76:1643–1654.

Ritchie, J. C. 1992. Aspects of the floristic history of Canada. *Acta Botanica Fennica* 144:81–91.

Robbins, C. S., J. R. Sauer, R. S. Greenberg, and S. Droege. 1989. Population declines in North American birds that migrate to the Neotropics. *Proceedings of the National Academy of Sciences (USA)* 86: 7658–7662.

Roberts, C. M. 1997. Ecological advice for the global fisheries crisis. *Trends in Ecology & Evolution* 12:35–38.

Roberts, L. 1989. How fast can trees migrate? *Science* 243:735–737.

Robinson, W. D. 1999. Long-term changes in the avifauna of Barro Colorado Island, Panama, a tropical forest isolate. *Conservation Biology* 13:85–97.

Rodda, G. H., T. H. Fritts, and D. Chiszar. 1997. The disappearance of Guam's wildlife. *BioScience* 47:565–574.

Rodriguez, J. P., D. L. Pearson, and R. R. Barrera. 1998. A test for the adequacy of bioindicator taxa: Are tiger beetles (Coleoptera: Cicindelidae) appropriate indicators for monitoring the degradation of tropical forests in Venezuela? *Biological Conservation* 83:69–76.

Roff, D. A. 1986. The evolution of wing dimorphism in insects. *Evolution* 40:1009–1020.

———. 1992. *The Evolution of Life Histories: Theory and Analysis.* Chapman and Hall, New York.

Rogers, H. H., J. F. Thomas, and G. E. Bingham. 1983. Response of agronomic and forest species to elevated atmospheric carbon dioxide. *Science* 220:428–429.

Rohde, K. 1992. Latitudinal gradients in species diversity: The search for the primary cause. *Oikos* 65:514–527.

———. 1998. Latitudinal gradients in species diversity. Area matters, but how much? *Oikos* 82:184–190.

Rohner, C. and D. W. Ward. 1997. Chemical and mechanical defense against herbivory in two sympatric species of desert Acacia. *Journal of Vegetation Science* 8:717–726.

Rolstad, J. 1991. Consequences of forest fragmentation for the dynamics of bird populations: Conceptual issues and the evidence. *Biological Journal of the Linnean Society* 42:149–163.

Room, P. M. 1981. Biogeography, apparency, and exploration for biological control agents in exotic ranges of weeds. *Proceedings of the 5th International Symposium on the Biological Control of Weeds* Brisbane, 1980:113–124.

———. 1990. Ecology of a simple plant-herbivore system: Biological control of Salvinia. *Trends in Ecology & Evolution* 5:74–78.

Root, R. B. 1967. The niche exploitation pattern of the blue-gray gnatcatcher. *Ecological Monographs* 37:317–350.

———. 1973. Organization of a plant-arthropod association in simple and diverse habitats: The fauna of collards (*Brassica oleracea*). *Ecological Monographs* 43:95–124.

Root, T. 1988. Energy constraints on avian distributions and abundances. *Ecology* 69:330–339.

Rosenberg, A. A., J. R. Beddington, and M. Basson. 1986. Growth and longevity of krill during the first decade of pelagic whaling. *Nature* 324:152–154.

Rosenzweig, M. L. 1968. Net primary productivity of terrestrial communities: Prediction from climatological data. *American Naturalist* 102:67–74.

———. 1969. Why the prey curve has a hump. *American Naturalist* 103:81–87.

———. 1985. Some theoretical aspects of habitat selection. Pages 517–540 in M. L. Cody, ed., *Habitat Selection in Birds.* Academic Press, New York.

———. 1995. *Species Diversity in Space and Time.* Cambridge University Press, Cambridge.

Rosenzweig, M. L. and R. H. MacArthur. 1963. Graphical representation and stability conditions of predator-prey interactions. *American Naturalist* 97:209–223.

Ross, J. and A. M. Tittensor. 1986. The influence of myxomatosis in regulating rabbit numbers. *Mammal Review* 16:163–168.

Ross, R. 1911. *The Prevention of Malaria.* Waterloo, London.

Ross, R. M. and L. B. Quetin. 1986. How productive are Antarctic krill? *BioScience* 36:264–269.

Rothschild, B. J. 1986. *Dynamics of Marine Fish Populations.* Harvard University Press, Cambridge, Mass.

Roughgarden, J. 1974. Niche width: Biogeographic patterns among *Anolis* lizard populations. *American Naturalist* 108:429–442.

———. 1986. A comparison of food-limited and space-limited animal competition communities. Pages 492–516 in J. Diamond and T. J. Case, eds., *Community Ecology.* Harper & Row, New York.

Roughgarden, J. and J. Diamond. 1986. Overview: The role of species interactions in community ecology. Pages 333–343 in J. Diamond and T. J. Case, eds., *Community Ecology.* Harper & Row, New York.

Roughgarden, J., S. D. Gaines, and S. W. Pacala. 1987. Supply side ecology: The role of physical transport processes. Pages 491–518 in J. H. R. Gee and P. S. Giller, eds., *Organization of Communities: Past and Present.* Blackwell, Oxford, England.

Roush, R. T. 1998. Two-toxin strategies for management of insecticidal transgenic crops: Can pyramiding succeed where pesticide mixtures have not? *Philosophical Transactions of the Royal Society of London, Series B* 353:1777–1786.

Rowe, J. S. 1961. The level-of-integration concept and ecology. *Ecology* 42:420–427.

Royama, T. 1970. Factors governing the hunting behaviour and selection of food by the great tit (*Parus major* L.). *Journal of Animal Ecology* 39:619–668.

———. 1984. Population dynamics of the spruce budworm Choristoneura fumiferana. Ecological Monographs 54: 429–462.

———. 1996. A fundamental problem in key factor analysis. *Ecology* 77:87–93.

Ruesink, J. L., I. M. Parker, M. J. Groom, and P. M. Kareiva. 1995. Reducing the risks of nonindigenous species introductions. *BioScience* 45:465–477.

Ruiz, G. M., J. T. Carlton, E. D. Grosholz, and A. H. Hines. 1997. Global invastions of marine and estuarine habitats by non-indigenous species: Mechanisms, extent, and consequences. *American Zoologist* 37:621–632.

Rundel, P. W. 1980. The ecological distribution of C_3 and C_4 grasses in the Hawaiian Islands. *Oecologia* 45:354–359.

Runyon, J., R. H. Waring, S. N. Goward, and J. M. Welles. 1994. Environmental limits on net primary production and light-use efficiency across the Oregon transect. *Ecological Applications* 4:226–237.

Russell, E. S. 1931. Some theoretical considerations on the "overfishing" problem. *Journal du Conseil Permanente International pour l'Exploration de la Mer* 6:3–27.

Russell, G. J., J. M. Diamond, S. L. Pimm, and T. M. Reed. 1995. A century of turnover: Community dynamics at three timescales. *Journal of Animal Ecology* 64:628–641.

Russell, P. F. and T. R. Rao. 1942. On the relation of mechanical obstruction and shade to ovipositing of *Anopheles culicifacies*. *Journal of Experimental Zoology* 91:303–329.

Ryel, L. A. 1981. Population change in the Kirtland's Warbler. *Jack-Pine Warbler* 59:76–91.

Ryther, J. H. 1963. Geographic variation in productivity. Pages 347–380 in M. N. Hill, ed., *The Sea*. Wiley-Interscience, New York.

Ryther, J. H. and W. M. Dunstan. 1971. Nitrogen, phosphorus, and eutrophication in the coastal marine environment. *Science* 171:1008–1013.

Saether, B.-E., T. H. Ringsby, O. Bakke, and E. J. Solberg. 1999. Spatial and temporal variation in demography of a house sparrow metapopulation. *Journal of Animal Ecology* 68:628–637.

Sala, O. E., W. J. Parton, L. A. Joyce, and W. K. Lauenroth. 1988. Primary production of the central grassland region of the United States. *Ecology* 69:40–45.

Sale, P. F. 1977. Maintenance of high diversity in coral reef fish communities. *American Naturalist* 111:337–359.

———. 1982. Stock-recruit relationships and regional coexistence in a lottery competitive system: A simulation study. *American Naturalist* 120:139–159.

———. ed. 1991. *The Ecology of Fishes on Coral Reefs*. Academic Press, San Diego.

Salisbury, E. J. 1961. *Weeds and Aliens*. Collins, London.

Salt, G. and F. S. J. Hollick. 1944. Studies of wireworm populations. I. A census of wireworms in pasture. *Annals of Applied Biology* 31:52–64.

Sanchez, P. A., D. E. Bandy, J. H. Villachica, and J. J. Nicholaides. 1982. Amazon basin soils: Management for continuous crop production. *Science* 216:821–827.

Sanders, H. L. 1968. Marine benthic diversity: A comparative study. *American Naturalist* 102:243–282.

Sand-Jensen, K. and D. Krause-Jensen. 1997. Broad-scale comparison of photosynthesis in terrestrial and aquatic plant communities. *Oikos* 80:203–208.

Sang, J. H. 1950. Population growth in *Drosophila* cultures. *Biological Reviews* 25:188–219.

Sathyendranath, S., T. Platt, E. P. W. Horne, W. G. Harrison, O. Ulloa, R. Outerbridge, and N. Hoepffner. 1991. Estimation of new production in the ocean by compound remote sensing. *Nature* 353:129–133.

Sauer, J. R., J. E. Hines, G. Gough, I. Thomas, and B. G. Peterjohn. 1997. *The North American Breeding Bird Survey Results and Analysis*. Patuxent Wildlife Research Center, Laurel, Maryland.

Savory, A. 1988. *Holistic Resource Management*. Island Press, Washington, D.C.

Schaefer, W. B. 1968. Methods of estimating effects of fishing on fish populations. *Transactions of the American Fisheries Society* 97:231–241.

Schall, J. J. 1983. Lizard malaria: Cost to vertebrate host's reproductive success. *Parasitology* 87:1–6.

Scheffer, V. B. 1951. The rise and fall of a reindeer herd. *Scientific Monthly* 73:356–362.

Schindler, D. W. 1977. Evolution of phosphorus limitation in lakes. *Science* 239:149–157.

———. 1988. Effects of acid rain on freshwater ecosystems. *Science* 239:149–157.

———. 1990. Experimental perturbations of whole lakes as tests of hypotheses concerning ecosystem structure and function. *Oikos* 57:25–41.

———. 1997. Liming to restore acidified lakes and streams: A typical approach to restoring damaged ecosystems? *Restoration Ecology* 5:1–6.

———. 1998. Replication versus realism: The need for ecosystem-scale experiments. *Ecosystems* 1:323–334.

Schindler, D. W. and E. J. Fee. 1974. Experimental Lakes area: Whole-lake experiments in eutrophication. *Journal of the Fisheries Research Board of Canada* 31:937–953.

Schlesinger, W. H. 1997. *Biogeochemistry: An Analysis of Global Change*. Academic Press, San Diego.

Schneider, S. H. 1989. The greenhouse effect: Science and policy. *Science* 243:771–781.

Schneiderhan, F. J. 1927. The black walnut (*Juglans nigra* L.) as a cause of the death of apple trees. *Phytopathology* 17:529–540.

Schoener, T. W. 1986a. Overview: Kinds of ecological communities—ecology becomes pluralistic. Pages 467–479 in J. Diamond and T. J. Case, eds., *Community Ecology*. Harper & Row, New York.

———. 1986b. Resource partitioning. Pages 91–126 in J. Kikkawa and D. J. Anderson, eds., *Community Ecology: Pattern and Process*. Blackwell, Melbourne.

———. 1987a. Axes of controversy in community ecology. Pages 8–31 in W. J. Matthews and D. C. Heins, eds., *Community and Evolutionary Ecology of North American Stream Fishes*. University of Oklahoma Press, Norman.

———. 1987b. The geographical distribution of rarity. *Oecologia* 74:161–173.

Schoener, T. W. and C. A. Toft. 1983. Spider populations: Extraordinarily high densities on islands without top predators. *Science* 219:1353–1355.

Schoenly, K. and J. E. Cohen. 1991. Temporal variation in food web structure: 16 empirical cases. *Ecological Monographs* 61:267–298.

Schoeps, K., P. Syrett, and R. M. Emberson. 1996. Summer diapause in *Chrysolina hyperici* and *C. quadrigemina* (Coleoptera: Chrysomelidae) in relation to biological control of St. John's wort, *Hypericum perforatum* (Clusiaceae). *Bulletin of Entomological Research* 86:591–597.

Schoonhoven, L. M. 1968. Chemosensory bases of host plant selection. *Annual Review of Entomology* 13:115–136.

Schultz, A. M. 1969. A study of an ecosystem: The arctic tundra. Pages 77–93 in G. Van Dyne, ed., *The Ecosytem Concept in Natural Resource Management*. Academic Press, New York.

Schulze, E. D. 1989. Air pollution and forest decline in a spruce (*Picea abies*) forest. *Science* 244:776–783.

Schwartz, M. W., C. A. Brigham, J. D. Hoeksema, K. G. Lyons, M. H. Mills, and P. J. van Mantgem. 2000. Linking biodiversity to ecosystem function: Implications for conservation ecology. *Oecologia* 122:297–305.

Scott, J. A., N. R. French, and J. W. Leetham. 1979. Patterns of consumption in grasslands. Pages 89–105 in N. R. French, ed., *Perspectives in Grassland Ecology*. Springer-Verlag, New York.

Seber, G. A. F. 1982. *The Estimation of Animal Abundance*. Charles Griffin and Company, London.

Seehausen, O., F. Witte, E. F. Katunzi, J. Smits, and N. Bouton. 1997. Patterns of the remnant cichlid fauna in southern Lake Victoria. *Conservation Biology* 11:890–904.

Seigler, D. S. 1996. Chemistry and mechanisms of allelopathic interactions. *Agronomy Journal* 88:876–885.

Selander, R. K. 1965. On mating systems and sexual selection. *American Naturalist* 99:129–141.

Service, M. W. 1997. Mosquito (Diptera: Culicidae) dispersal—the long and short of it. *Journal of Medical Entomology* 34:579–588.

Shaffer, M. L. 1981. Minimum population sizes for species conservation. *Bioscience* 31:131–134.

Sharkey, M. J. 1970. The carrying capacity of natural and improved land in different climatic zones. *Mammalia* 34:564–572.

Sharov, A. A., A. M. Liebhold, and E. A. Roberts. 1998. Optimizing the use of barrier zones to slow the spread of gypsy moth (Lepidoptera: Lymantriidae) in North America. *Journal of Economic Entomology* 91:165–174.

Shaver, G. R. and F. S. Chapin, III. 1995. Long-term responses to factorial, NPK fertilizer treatment by Alaskan wet and moist tundra sedge species. *Ecography* 18:259–275.

Sheail, J. 1987. *Seventy-five Years in Ecology: The British Ecological Society*. Blackwell Scientific Publications, Oxford.

Shepherd, J. G., J. G. Pope, and R. D. Cousens. 1984. Variations in fish stocks and hypotheses concerning their links with climate. *Rapp. P-v Reun. Cons. int. Explor. Mer* 185:255–267.

Short, J. 1998. The extinction of rat-kangaroos (Marsupialia: Potoroidae) in New South Wales, Australia. *Biological Conservation* 86:365–377.

Short, J. and A. Smith. 1994. Mammal decline and recovery in Australia. *Journal of Mammalogy* 75:288–297.

Shugart, H. H. 1984. *A Theory of Forest Dynamics: The ecological Implications of Forest Succession Models*. Springer-Verlag, New York.

Sih, A. 1991. Reflections on the power of a grand paradigm. *Bulletin of the Ecological Society of America* 72:174–178.

Sih, A., P. Crowley, M. McPeek, J. Petranka, and K. Strohmeier. 1985. Predation, competition, and prey communities: A review of field experiments. *Annual Review of Ecology and Systematics* 16:269–311.

Sih, A., G. Englund, and D. Wooster. 1998. Emergent impacts of multiple predators on prey. *Trends in Ecology and Evolution* 13:350–355.

Silkworth, D. R., and D. F. Grigal. 1982. Determining and evaluating nutrient losses following whole-tree harvesting of aspen. *Soil Science Society of America Journal* 46:626-631.

Silliman, R. P. and J. S. Gutsell. 1958. Experimental exploitation of fish populations. *Fisheries Bulletin (U.S.)* 58:215–252.

Silva, M., J. H. Brown, and J. A. Downing. 1997. Differences in population density and energy use between birds and mammals: A macroecological perspective. *Journal of Animal Ecology* 66:327–340.

Silvertown, J. W. 1980. The dynamics of a grassland ecosystem: Botanical equilibrium in the Park Grass Experiment. *Journal of Applied Ecology* 17:491–504.

Simberloff, D. 1998. Flagships, umbrellas, and keystones: Is single-species management passe in the landscape era? *Biological Conservation* 83:247–257.

Simberloff, D. and J. Cox. 1987. Consequences and costs of conservation corridors. *Conservation Biology* 1:63–71.

Simberloff, D. and T. Dayan. 1991. The guild concept and the structure of ecological communities. *Annual Review of Ecology and Systematics* 22:115–143.

Simberloff, D., D. Schmitz, and T. Brown. 1997. *Strangers In Paradise: Impact and Management of Nonindigenous Species in Florida*. Island Press, Washington, D.C.

Simberloff, D. and P. Stiling. 1996. How risky is biological control? *Ecology* 77:1965–1974.

Simberloff, D. S. and W. Boecklen. 1981. Santa Rosalia reconsidered: Size ratios and competition. *Evolution* 35:1206–1228.

Simpson, G. G. 1964. Species density of North American recent mammals. *Systematic Zoology* 13:57–73.

———. 1969. The first three billion years of community evolution. *Brookhaven Symposium in Biology* 22:162–177.

Sinclair, A. R. E. 1975. The resource limitation of trophic levels in tropical grassland communities. *Journal of Animal Ecology* 44:497–520.

———. 1977. *The African Buffalo: A Study of Resource Limitation of Populations*. University of Chicago Press, Chicago.

———. 1989. Population regulation in animals. Pages 197–241 in J. M. Cherrett, ed., *Ecological Concepts*. Blackwell Scientific Oxford.

———. 1997. Fertility control of mammal pests and the conservation of endangered marsupials. *Reproduction, Fertility and Development* 9:1–16.

———. 1998. Natural regulation of ecosystems in protected areas as ecological baselines. *Wildlife Society Bulletin* 26:399–409.

Sinclair, A. R. E. and P. Arcese, eds., 1995. *Serengeti II: Dynamics, Management, and Conservation of an Ecosystem.* University of Chicago Press, Chicago.

Sinclair, A. R. E. and M. Norton-Griffiths. 1982. Does competition or facilitation regulate migrant ungulate populations in the Serengeti? A test of hypotheses. *Oecologia* 53:364–369.

Sinclair, A. R. E. and R. P. Pech. 1996. Density dependence, stochasticity, compensation and predator regulation. *Oikos* 75:164–173.

Sinclair, A. R. E., R. P. Pech, C. R. Dickman, D. Hik, P. Mahon, and A. E. Newsome. 1998. Predicting effects of predation on conservation of endangered prey. *Conservation Biology* 12:564–575.

Sinclair, A. R. E. and J. N. M. Smith. 1984. Do plant secondary compounds determine feeding preferences of snowshoe hares? *Oecologia* 61:403–410.

Singer, F. J. 1998. Thunder on the Yellowstone revisited: An assessment of management of native ungulates by natural regulation. Wildlife Society Bulletin 26:375–390.

Singleton, G. R., L. Hinds, H. Leirs, and Z. Zhang. 1999. *Ecologically-based Management of Rodent Pests.* Australian Centre for International Agricultural Research, Canberra, Australia.

Skaggs, R. W. and W. J. Boecklen. 1996. Extinctions of montane mammals reconsidered: Putting a global-warming scenario on ice. *Biodiversity and Conservation* 5:759–778.

Skellam, J. G. 1951. Random dispersal in theoretical populations. *Biometrika* 38:196–218.

Skellam, J. G. 1955. The mathematical approach to population dynamics. Pages 31-46 in J. B . Cragg and N. W. Pirie, eds. *The Numbers of Man and Animals.* Oliver and Boyd, Edinburgh.

Skelly, D. K. 1996. Pond drying, predators, and the distribution of *Pseudacris* tadpoles. *Copeia* 1996:599–605.

Skinner, J. 1972. Eland vs. beef: Eland cannot compete except in hot semiarid regions. *African Wildlife* 26:4–9.

Skutch, A. F. 1967. Adaptive limitation of the reproductive rate of birds. *Ibis* 109:579–599.

Slatkin, M. 1987. Gene flow and the geographic structure of natural populations. *Science* 236:787–792.

Slobodkin, L. B. 1961. *Growth and Regulation of Animal Populations.* Holt, Rinehart, and Winston, New York.

Smith, D. W. E. 1989. Is greater female longevity a general finding among animals? *Biological Reviews* 64:1–12.

Smith, F. E. 1970. Analysis of ecosystems. Pages 7-18 in D. Reichle, ed. *Analysis of Temperate Forest Ecosystems.* Springer-Verlag, Berlin.

Smith, H. S. 1935. The role of biotic factors in the determination of population densities. *Journal of Economic Entomology* 28:873–898.

Smith, J. N. M. 1988. Determinants of lifetime reproductive success in the song sparrow. Pages 154-172 in T. H. Clutton-Brock, ed., *Reproductive Success.* University of Chicago Press, Chicago.

Smith, S. M. 1980. Responses of naive temperate birds to warning coloration. *American Midland Naturalist* 103:346–352.

Smith, T. B. 1990. Natural selection on bill characters in the two bill morphs of the African finch *Pyrenestes ostrinus.* *Evolution* 44:832–842.

Smith, T. J., III. 1987. Seed predation in relation to tree dominance and distribution in mangrove forests. *Ecology* 68:266–273.

Smith, V. H. 1982. The nitrogen and phosphorous dependence of algal biomass in lakes: An empirical and theoretical analysis. *Limnology and Oceanography* 27:1101–1112.

Smith, V. H. 1983. Low nitrogen to phosphorus ratios favor dominance by blue-green algae in lake phytoplankton. *Science* 221:669-671.

Snaydon, R. W. 1991. The productivity of C^3 and C^4 plants: A reassessment. *Functional Ecology* 5:321–330.

Snyder, G. K. and W. W. Weathers. 1975. Temperature adaptations in amphibians. *American Naturalist* 109:93–101.

Sokal, R. R. and F. J. Rohlf. 1995. *Biometry.* W.H. Freeman, New York.

Soulé, M. and L. S. Mills. 1992. Conservation genetics and conservation biology: A troubled marriage. Pages 55–69 in O. T. Sandlund, K. Hindar, and A. H. D. Brown, eds., *Conservation of Biodiversity for Sustainable Development.* Scandinavian University Press, Oslo, Norway.

Soulé, M. E., ed., 1987. *Viable Populations for Conservation.* Cambridge University Press, Cambridge, England.

Sousa, W. P. 1985. Disturbance and patch dynamics on rocky intertidal shores. Pages 101–124 in S. T. A. Pickett and P. S. White, eds., *The Ecology of Natural Disturbance and Patch Dynamics.* Academic Press, Orlando.

Southwood, T. R. E. 1962. Migration of terrestrial arthropods in relation to habitat. *Biological Reviews* 37:171–214.

———. 1978. *Ecological Methods.* Methuen, London.

Sovada, M. A., A. B. Sargeant, and J. W. Grier. 1995. Differential effects of coyotes and red foxes on duck nest success. *Journal of Wildlife Management* 59:1–9.

Spence, D. H. N. 1982. The zonation of plants in freshwater lakes. *Advances in Ecological Research* 12:37–125.

Spiller, D. A. and T. W. Schoener. 1988. An experimental study of the effect of lizards on web-spider communities. *Ecological Monographs* 58:57–77.

———. 1990. Lizards reduce food consumption by spiders: Mechanisms and consequences. *Oecologia* 83:150–161.

———. 1994. Effects of top and intermediate predators in a terrestrial food web. *Ecology* 75:182–196.

———. 1995. Long-term variation in the effect of lizards on spider density is linked to rainfall. *Oecologia* 103:133–139.

Srivastava, D. S. 1999. Using local-regional richness plots to test for species saturation: Pitfalls and potential. *Journal of Animal Ecology* 68:1–16.

Stamps, J. A. 1991. The effect of conspecifics on habitat selection in territorial species. *Behavioral Ecology and Sociobiology* 28:29–36.

Stearns, S. C. 1989. Trade-offs in life history evolution. *Functional Ecology* 3:259–268.

———. 1992. *The Evolution of Life Histories*. Oxford University Press, Oxford.

Steedman, R. J. and H. A. Regier. 1987. Ecosystem science for the Great Lakes: Perspectives on degradative and rehabilitative transformations. *Canadian Journal of Fisheries and Aquatic Sciences* 44 (Suppl. 2):95–103.

Steinberg, C. E. W. and R. F. Wright, eds., 1994. *Acidification of Aquatic Ecosystems: Implications for the Future*. John Wiley and Sons, Chichester, England.

Stenseth, N. C. 1981. How to control pest species: Application of models from the theory of island biogeography in formulating pest control strategies. *Journal of Applied Ecology* 18:773–794.

Stephens, D. W. and J. R. Krebs. 1986. *Foraging Theory*. Princeton University Press, Princeton, New Jersey.

Stephenson, N. L. 1990. Climatic control of vegetation distribution: The role of water balance. *American Naturalist* 135:649–670.

Sterner, R. W., J. Clasen, W. Lampert, and T. Weisse. 1998. Carbon:phosphorus stoichiometry and food chain production. *Ecology Letters* 1:146–150.

Sterner, R. W. and D. O. Hessen. 1994. Algal nutrient limitation and the nutrition of aquatic herbivores. *Annual Review of Ecology and Systematics* 25:1–29.

Stevens, G. C. 1989. The latitudinal gradient in geographical range: How so many species coexist in the tropics. *American Naturalist* 133:240–256.

Stevens, G. C. and J. F. Fox. 1991. The causes of treeline. *Annual Review of Ecology and Systematics* 22:177–191.

Stockner, J. G. and W. W. Benson. 1967. The succession of diatom assemblages in the recent sediments of Lake Washington. *Limnology and Oceanography* 12:513–532.

Stohlgren, T. J., L. D. Schell, and B. Vanden Heuvel. 1999. How grazing and soil quality affect native and exotic plant diversity in Rocky Mountain grasslands. *Ecological Applications* 9:45–64.

Strauss, S. Y. 1997. Floral characters link herbivores, pollinators and plant fitness. *Ecology* 78:1640–1645.

Strayer, D. 1988. On the limits to secondary production. *Limnology and Oceanography* 33:1217–1220.

Stromayer, K. A. K. and R. J. Warren. 1997. Are overabundant deer herds in the eastern United States creating alternate stable states in forest plant communities? *Wildlife Society Bulletin* 25:227–234.

Strong, D. R. 1984. Density-vague ecology and liberal population regulation in insects. Pages 313–327 in P. W. Price, C. N. Slobodchikoff and W. S. Gaud, ed., *A New Ecology*. Wiley, New York .

———. 1986. Density-vague population change. *Trends in Ecology and Evolution* 1:39–42.

———. 1992. Are trophic cascades all wet? Differentiation and donor-control in speciose ecosystems. *Ecology* 73:747–754.

Strong, D. R., J. H. Lawton, and T. R. E. Southwood. 1984. *Insects on Plants*. Harvard University Press, Cambridge, Mass.

Strong, D. R. J., E. D. McCoy, and J. R. Rey. 1977. Time and the number of herbivore species: The pests of sugarcane. *Ecology* 58:167–175.

Sugihara, G. 1980. Minimal community structure: An explanation of species abundance patterns. *American Naturalist* 116:770–787.

Sultan, S. E. 1995. Phenotypic plasticity and plant adaptation. *Acta Botanica Neerlandica* 44:363–383.

Sutherland, J. P. 1990. Perturbations, resistance, and alternative views of the existence of multiple stable points in nature. *American Naturalist* 136:270–275.

Sutherland, W. J., ed., 1996a. *Ecological Census Techniques: A Handbook*. Cambridge University Press, Cambridge.

———. 1996b. *From Individual Behaviour to Population Ecology*. Oxford University Press, New York.

Sutherst, R. W., R. B. Floyd, and G. F. Maywald. 1996. The potential geographical distribution of the cane toad, *Bufo marinus* L. in Australia. *Conservation Biology* 10:294–299.

Swan, L. A., and C. S. Papp. 1972. *The Common Insects of North America*. Harper and Row, New York.

Swetnam, T. W. and A. M. Lynch. 1993. Multicentury, regional-scale patterns of western spruce budworm outbreaks. *Ecological Monographs* 63:399–424.

Swinton, J. and R. M. Anderson. 1995. Model frameworks for plant-pathogen interactions. Pages 280–294 in B. T. Grenfell and A. P. Dobson, eds., *Ecology of Infectious Diseases in Natural Populations*. Cambridge University Press, Cambridge, England.

Swinton, J. and C. A. Gilligan. 1996. Dutch elm disease and the future of the elm in the U.K.: A quantitative analysis. *Philosophical Transactions of the Royal Society of London, Series B* 351:605–615.

Swinton, J., J. Harwood, B. T. Grenfell, and C. A. Gilligan. 1998. Persistence thresholds for phocine distemper virus infection in harbour seal *Phoco vitulina* metapopulations. *Journal of Animal Ecology* 67:54–68.

Sydeman, W. J., J. H. R. Huber, S. D. Emslie, C. A. Ribic, and N. Nur. 1991. Age-specific weaning success of northern elephant seals in relation to previous breeding experience. *Ecology* 72:2204–2217.

Tabashnik, B. E., N. L. Cushing, N. Finson, and M. W. Johnson. 1990. Field development of resistance to *Bacillus thuringiensis* in diamondback moth (Lepidoptera: Plutellidae). *Journal of Economic Entomology* 83:1671–1676.

Tabashnik, B. E., Y.-B. Liu, T. Malvar, D. G. Heckel, L. Masson, and J. Ferre. 1998. Insect resistance to *Bacillus thuringiensis*: Uniform or diverse? *Phiolsophical Transactions of the Royal Society of London, Series B* 353:1751–1756.

Takasu, F. 1998. Modeling the arms race in avian brood parasitism. *Evolutionary Ecology* 12:969–987.

Talbot, F. H., B. C. Russell, and G. R. V. Anderson. 1978. Coral reef fish communities: Unstable, high-diversity systems. *Ecological Monographs* 48:425–440.

Tamarin, R. H. 1999. *Principles of Genetics*. McGraw Hill College Division, New York.

Tanner, E. V. J., P. M. Vitousek, and E. Cuevas. 1998. Experimental investigation of nutrient limitation of forest growth on wet tropical mountains. *Ecology* 79:10–22.

Tanner, J. E., T. P. Hughes, and J. H. Connell. 1996. The role of history in community dynamics: A modelling approach. *Ecology* 77:108–117.

Tansley, A. G. 1939. *The British Islands and Their Vegetation*. Cambridge University Press, Cambridge, England.

———. 1947. Obituary notice: Frederic Edward Clements 1874–1945. *Journal of Ecology* 34:194–196.

Taylor, G. E. J., D. W. Johnson, and C. P. Andersen. 1994. Air pollution and forest ecosystems: A regional to global perspective. *Ecological Applications* 4:662–689.

Taylor, O. R. J. 1985. African bees: Potential impact in the United States. *Bulletin of the Entomological Society of America* 31:15–28.

Taylor, R. J. 1984. *Predation*. Chapman and Hall, New York.

Teeri, J. A. and L. G. Stowe. 1976. Climatic patterns and the distribution of C_4 grasses in North America. *Oecologia* 23:1–12.

Temple, S. A. 1987. Do predators always capture substandard individuals disproportionately from prey populations? *Ecology* 68:669–674.

Templeton, A. R. 1986. Coadaptation and outbreeding depression. Pages 105–116 in M. E. Soulé, ed., *Conservation Biology*. Sinauer Associates, Sunderland, Mass.

Ter Braak, C. J. F. and I. C. Prentice. 1988. *A Theory of Gradient Analysis*. Academic Press, London.

Terborgh, J., R. B. Foster, and P. V. Nunez. 1996. Tropical tree communities: A test of the nonequilibrium hypothesis. *Ecology* 77:561–567.

Terrenato, L., M. F. Gravina, A. S. Martini, and L. Ulizzi. 1981. Natural selection associated with birth weight. III. Changes over the last twenty years. *Annals of Human Genetics* 45:267–278.

Thacker, R. W., M. A. Becerro, W. A. Lumbang, and V. J. Paul. 1998. Allelopathic interactions between sponges on a tropical reef. *Ecology* 79:1740–1750.

Thomas, A. S. 1960. Changes in vegetation since the advent of myxomatosis. *Journal of Ecology* 48:287–306.

———. Further changes in vegetation since the advent of myxomatosis. *Journal of Ecology* 51:151–183.

Thomas, M. B. and W. A. J. 1998. Biocontrol—risky but necessary. *Trends in Ecology and Evolution* 13:325–329.

Thompson, D. Q. 1991. History of purple loosestrife (*Lythrum salicaria* L.) biological control efforts. *Natural Areas Journal* 11:148–150.

Thompson, J. N. 1994. *The Coevolutionary Process*. University of Chicago Press, Chicago.

Thompson, S. D. 1982. Structure and species composition of desert heteromyid rodent species assemblages: Effects of a simple habitat manipulation. *Ecology* 63:1313–1321.

Thompson, S. K. 1992. *Sampling*. John Wiley and Sons, New York.

Thornthwaite, C. W. 1948. An approach toward a rational classification of climate. *Goegraphical Review* 38:55–94.

Thornton, I. W. B., T. R. New, R. A. Zann, and P. A. Rawlinson. 1990. Colonization of the Krakatau Islands by animals: A perspective from the 1980s. *Philosophical Transactions of the Royal Society of London, Series B* 328:131–165.

Tiessen, H., E. Cuevas, and P. Chacon. 1994. The role of soil organic matter in sustaining soil fertility. *Nature* 371:783–785.

Tilghman, N. G. 1989. Impacts of white-tailed deer on forest regeneration in northwestern Pennsylvania. *Journal of Wildlife Management* 53:524–532.

Tilman, D. 1977. Resource competition between planktonic algae: An experimental and theoretical approach. *Ecology* 58:338–348.

———. 1982. *Resource Competition and Community Structure*. Princeton University Press, Princeton, New Jersey.

———. 1985. The resource-ratio hypothesis of plant succession. *American Naturalist* 125:827–852.

———. 1986. Resources, competition and the dynamics of plant communities. Pages 51–75 in M. J. Crawley, ed., *Plant Ecology*. Blackwell, Oxford.

———. 1987. The importance of the mechanisms of interspecific competition. *American Naturalist* 129:769–774.

———. 1989. Ecological experimentation; strengths and conceptual problems. Pages 136–157 in G. E. Likens, ed., *Long term Studies in Ecology*. Springer-Verlag, New York.

———. 1990. Mechanisms of plant competition for nutrients: The elements of a predictive theory of competition. Pages 117–141 in J. B. Grace and D. Tilman, eds., *Perspectives on Plant Competition*. Academic Press, San Diego.

———. 1996. Biodiversity: Population versus ecosystem stability. *Ecology* 77:350–363.

Tilman, D., S. S. Kilham, and P. Kilham. 1982. Phytoplankton community ecology: The role of limiting nutrients. *Annual Reviews of Ecological Systematics* 13:349–372.

Tilman, D., J. Knops, D. Wedin, P. Reich, M. Ritchie, and E. Siemann. 1997a. The influence of functional diversity and composition on ecosystem processes. *Science* 277:1300–1302.

Tilman, D., C. L. Lehman, and C. E. Bristow. 1998. Diversity-stability relationships: Statistcal inevitability or ecological consequence? *American Naturalist* 151:277–282.

Tilman, D., C. L. Lehman, and K. T. Thomson. 1997b. Plant diversity and ecosystem productivity: Theoretical considerations. *Proceedings of the National Academy of Sciences of USA* 94:1857–1861.

Tilman, D., R. M. May, C. L. Lehman, and M. A. Nowak. 1994. Habitat destruction and the extinction debt. *Nature* 371:65–66.

Tollrian, R. and C. D. Harvell, eds., 1999. *The Ecology and Evolution of Inducible Defenses*. Princeton University Press, Princeton, N.J.

Tomback, D. F. 1982. Dispersal of whitebark pine seeds by Clark's nutcracker: A mutualism hypothesis. *Journal of Animal Ecology* 51:451–467.

Tranquillini, W. 1979. *Physiological Ecology of the Alpine Timberline*. Springer-Verlag, Berlin.

Tregenza, T. 1995. Building on the ideal free distribution. *Advances in Ecological Research* 26:253–307.

Trout, R. C., J. Ross, A. M. Tittensor, and A. P. Fox. 1992. The effect on a British wild rabbit population (*Oryctolagus cuniculus*) of manipulating myxomatosis. *Journal of Applied Ecology* 29:679–686.

Turchin, P. 1999. Population regulation: A synthetic view. *Oikos* 84:160–163.

Turesson, G. 1922. The genotypical response of the plant species to the habitat. *Hereditas* 3:329–346.

———. 1925. The plant species in relation to habitat and climate. *Hereditas* 6:147–236.

———. 1930. The selective effect of climate upon the plant species. *Hereditas* 14:99–152.

Turner, J. R. G., J. J. Lennon, and J. A. Lawrenson. 1988. British bird species distributions and the energy theory. *Nature* 335:539–541.

Twigg, L. E. and C. K. Williams. 1999. Fertility control of overabundant species: Can it work for feral rabbits? *Ecology Letters* 2:281–285.

Ueno, O. and T. Takeda. 1992. Photosynthetic pathways, ecological characteristics, and the geographical distribution of the Cyperaceae in Japan. *Oecologia* 89:195–203.

Underwood, A. J. and E. J. Denley. 1984. Paradigms, explanations, and generalizations in models for the structure of intertidal communities on rocky shores. Pages 153–180 in D. R. J. Strong, D. Simberloff, L. G. Abele, and A. B. Thistle, eds., *Ecological Communities*. Princeton University Press, Princeton, N.J.

Underwood, A. J. and P. G. Fairweather. 1989. Supply-side ecology and benthic marine assemblages. *Trends in Ecology and Evolution* 4:16–20.

Utida, S. 1957. Cyclic fluctuations of population density intrinsic to the host-parasite system. *Ecology* 38:442-449.

Uvarov, B. P. 1931. Insects and climate. *Transactions of the Entomological Society of London* 79:1–247.

van Breemen, N. 1995. How *Sphagnum* bogs down other plants. *Trends in Ecology & Evolution* 10:270–275.

van den Bosch, R., P. S. Messenger, and A. P. Gutierrez. 1982. *An Introduction to Biological Control*. Plenum, New York.

van der Maarel, E. 1996. Pattern and process in the plant community: Fifty years after A.S. Watt. *Journal of Vegetation Science* 7:19–28.

Van Der Werf, H. M. G., E. W. J. M. Mathijssen, and A. J. Haverkort. 1996. The potential of hemp (*Cannabis sativa* L.) for sustainable fibre production: A crop physiological approach. *Annals of Applied Biology* 129:109–123.

Van Emden, H. F. 1991. The role of host plant resistance in insect pest mis-management. *Bulletin of Entomological Research* 81:123–126.

Van Lenteren, J. C. and W. A. Overholt. 1994. Ecology and integrated pest management. *Insect Science and its Application* 15:557–582.

Van Riper, C., III, S. G. Van Riper, M. L. Goff, and M. Laird. 1986. The epizootiology and ecological significance of malaria in Hawaiian land birds. *Ecological Monographs* 56:327–344.

Van Valen, L. 1973. A new evolutionary law. *Evolutionary Theory* 1:1–30.

Vanderwerf, E. 1992. Lack's clutch size hypothesis: An examination of the evidence using meta-analysis. *Ecology* 73:1699–1705.

Varley, G. C. and G. R. Gradwell. 1960. Key factors in population studies. *Journal of Animal Ecology* 29:399–401.

Varley, G. C., G. R. Gradwell, and M. P. Hassell. 1973. *Insect Population Ecology: An Analytical Approach*. Blackwell Scientific Publications, Oxford.

Vaupel, J. W., J. R. Carey, K. Christensen, T. E. Johnson, A. I. Yashin, N. V. Holm, I. A. Iachine, V. Kannisto, A. A. Khazaeli, P. Liedo, V. D. Longo, Y. Zeng, K. G. Manton, and J. W. Curtsinger. 1998. Biodemographic trajectories of longevity. *Science* 280:855–859.

Vekemans, X. and C. Lefebvre. 1997. On the evolution of heavy-metal tolerant populations of *Armeria maritima*: Evidence from allozyme variation and reproductive barriers. *Journal of Evolutionary Biology* 10:175–191.

Verboom, B. and R. Van Apeldoorn. 1990. Effects of habitat fragmentation on the red squirrel, *Sciurus vulgaris* L. *Landscape Ecology* 4:171–176.

Verwijst, T. 1989. Self-thinning in even-aged natural stands of *Betula pubescens*. *Oikos* 56:264-268.

Vezina, A. and R. Peterman. 1985. Tests of the role of nuclear polyhedrosis virus in the population dynamics of its host, Douglas-fir tussock moth, *Orgyia pseudotsugata* (Lepidoptera: Lymantriidae). *Oecologia* 67:260–266.

Vinson, S. B. 1997. Invasion of the red imported fire ant (Hymenoptera: Formicidae): Spread, biology, and impact. *American Entomologist* 43:23–39.

Vitousek, P. M. 1982. Nutrient cycling and the nutrient use efficiency. *American Naturalist* 119:553–572.

———. 1984. Litterfall, nutrient cycling, and nutrient limitation in tropical forests. *Ecology* 65:285–298.

———. 1994. Beyond global warming: Ecology and global change. *Ecology* 75:1861–1876.

Vitousek, P. M., J. Aber, R. W. Howarth, G. E. Likens, P. A. Matson, D. W. Schindler, W. H. Schlesinger, and D. G. Tilman. 1997a. Human alteration of the global nitrogen cycle: Causes and consequences. *Ecological Applications* 7:737–750.

Vitousek, P. M., C. M. D'Antonio, L. L. Loope, and R. Westbrooks. 1996. Biological invasions as global environmental change. *American Scientist* 84:468–478.

Vitousek, P. M., H. A. Mooney, J. Lubchence, and J. M. Melillo. 1997b. Human domination of earth's ecosystems. *Science* 277:494–499.

Vollenweider, R. A. 1974. *A Manual on Methods of Measuring Primary Production in Aquatic Environments*. Blackwell, Oxford, England.

Volterra, V. 1926. Fluctuations in the abundance of a species considered mathematically. *Nature* 118:558–560.

Waage, J. 1990. Ecological theory and the selection of biological control agents. Pages 135–157 in M. Mackauer, L. Ehler, and J. Roland, eds., *Critical Issues in Biological Control*, Intercept Ltd, Andover, U.K.

Wackernagel, M., L. Onisto, P. Bello, A. Callejas Linares, I. S. Lopez Falfan, J. Mendez Garcia, A. I. Suarez Guerrero, and M. G. Suarez Guerrero. 1999. National natural capital accounting with the ecological footprint concept. *Ecological Economics* 29:375–390.

Wackernagel, M. and W. E. Rees. 1996. *Our Ecological Footprint: Reducing Human Impact on the Earth*. New Society Publishers, Gabriola Island, B.C.

Wake, D. B. 1998. Action on amphibians. *Trends in Ecology and Evolution* 13:379–380.

Walk, J. W. and R. E. Warner. 1999. Effects of habitat area on the occurrence of grassland birds in Illinois. *American Midland Naturalist* 141:339–344.

Walker, B. H. 1976. An assessment of the ecological basis of game ranching in southern African savannas. *Proc Grassld Soc Sth Afr* 11:125–130.

———. 1992. Biodiversity and ecological redundancy. *Conservation Biology* 6:18–23.

Walker, B. H., W. Steffen, J. Canadell, and J. Ingram, eds., 1999. *The Terrestrial Biosphere and Global Change: Implications for Natural and Managed Ecosystems*. Cambridge University Press, Cambridge, England.

Walker, L. R. and F. S. Chapin, III. 1987. Interactions among processes controlling successional change. *Oikos* 50:131–135.

Walkinshaw, L. H. 1983. Kirtland's Warbler: The natural history of an endangered species. *Cranbrook Institute of Science, Bloomfield Hills, Mich. Bulletin* 58.

Wallace, A. R. 1878. *Tropical Nature and Other Essays*. Macmillan, New York.

Walters, C. 1986. *Adaptive Management of Renewable Resources*. Macmillan, New York.

Walters, C. and J. Korman. 1999. Linking recruitment to trophic factors: Revisiting the Beverton-Holt recruitment model from a life history and multispecies perspective. *Reviews in Fish Biology and Fisheries* 9:187–202.

Walters, C. and J. J. Maguire. 1996. Lessons for stock assessment from the northern cod collapse. *Reviews in Fish Biology and Fisheries* 6:125–137.

Walters, C. J., C. E. Robinson, and T. G. Northcote. 1990. Comparative population dynamics of *Daphnia rosea* and *Holopedium gibberum* in four oligotrophic lakes. *Canadian Journal of Fisheries and Aquatic Sciences* 47:401–409.

Walters, J. R. 1991. Application of ecological principles to the management of endangered species: The case of the red-cockaded woodpecker. *Annual Review of Ecology and Systematics* 22:505–523.

Ward, J. and G. P. Adams. 1998. Body condition and adjustments to reproductive effort in female moose (*Alces alces*). *Journal of Mammology* 79:1345–1354.

Ward, J. K. and B. R. Strain. 1999. Elevated CO^2 studies: Past, present and future. *Tree Physiology* 19:211–220.

Ward, J. P. J., R. J. Gutierrez, and B. R. Noon. 1998. Habitat selection by northern spotted owls: The consequences of prey selection and distribution. *Condor* 100:79–92.

Wardle, P. 1965. A comparison of alpine timberlines in New Zealand and North America. *New Zealand Journal of Botany* 3:113–135.

Wardle, P. and M. C. Coleman. 1992. Evidence for rising upper limits of four native New Zealand forest trees. *New Zealand Journal of Botany* 30:303–314.

Waring, R. H. and W. H. Schlesinger. 1985. *Forest Ecosystems: Concepts and Management*. Academic Press, Orlando.

Warming, J. E. B. 1896. *Lehrbuch der ökologischen Pflanzengeographie*, Gebruder Bornträger, Berlin.

———. 1909. *Oecology of Plants*. Oxford University Press, Oxford.

Warner, R. E. 1968. The role of introduced diseases in the extinction of the endemic Hawaiian avifauna. *Condor* 70:101–120.

Warren, M. S. 1994a. The UK status and suspected metapopulation structure of a threatened European butterfly, the marsh fritillary *Eurodryas aurinia*. *Biological Conservation* 67:239–249.

Warren, P. H. 1994b. Making connections in food webs. *Trends in Ecology and Evolution* 9:136–140.

Waters, T. F. 1957. The effects of lime application to acid bog lakes in northern Michigan. *Transactions of the American Fisheries Society* 86:329–344.

Watkinson, A. R. 1997. Plant population dynamics. Pages 359–400 in M. J. Crawley, ed., *Plant Ecology*. Blackwell Science, Oxford.

Watson, A. and R. Moss. 1970. Dominance, spacing behavior and aggression in relation to population limitation in vertebrates. Pages 167–220 in A. Watson, ed., *Animal Populations in Relation to Their Food Resources*. Blackwell, Oxford.

Watson, N. H. F. 1974. Zooplankton of the St. Lawrence Great Lakes—species composition, distribution, and abundance. *Journal of the Fisheries Research Board of Canada* 31:783–794.

Watson, R. 1999. Common themes for ecologists in global issues. *Journal of Applied Ecology* 36:1–10.

Watson, R. T., M. C. Zinyowere, and R. H. Moss, eds., 1998. *The Regional Impacts of Climate Change: An Assessment of Vulnerability*. Cambridge University Press, Cambridge, England.

Watt, A. S. 1940. Contributions to the ecology of bracken (*Pteridium aquilinum*). I. The rhizome. *New Phytologist* 39:401–422.

———. 1947a. Contributions to the ecology of bracken (*Pteridium aquilinum*). IV. The structure of the community. *New Phytologist* 47:97–121.

———. 1947b. Pattern and process in the plant community. *Journal of Ecology* 35:1–22.

Watt, A. S. 1955. Bracken versus heather, a study in plant sociology. *Journal of Ecology* 43: 490-506.

Wayne, P. M. and F. A. Bazzaz. 1995. Seedling density modifies the growth responses of yellow birch maternal families to elevated carbon dioxide. *Global Change Biology* 1:315–324.

Wecker, S. C. 1963. The role of early experience in habitat selection by the prairie deer mouse, *Peromyscus maniculatus bairdi*. *Ecological Monographs* 33:307–325.

Wedin, D. A. and D. Tilman. 1996. Influence of nitrogen loading and species composition of the carbon balance of grasslands. *Science* 274:1720–1723.

Weeks, J. R. 1996. *Population: An Introduction to Concepts and Issues*. Wadsworth, San Francisco.

Weidenhamer, J. D. 1996. Distinguishing resource competition and chemical interference: Overcoming the methodological impasse. *Agronomy Journal* 88:866–875.

Weiner, J. 1992. Physiological limits to sustainable energy budgets in birds and mammals: Ecological implications. *Trends in Ecology and Evolution* 7:384–388.

Weller, D. E. 1987. A re-evaluation of the –3/2 power rule of plant self-thinning. *Ecological Monographs* 57:23–43.

———. 1991. The self-thinning rule: Dead or unsupported?—a reply to Lonsdale. *Ecology* 72:747–750.

Westemeier, R. L., J. D. Brawn, S. A. Simpson, T. L. Esker, R. W. Jansen, J. W. Walk, E. L. Kershner, J. L. Bouzat, and K. N. Paige. 1998. Tracking the long-term decline and recovery of an isolated population. *Science* 282:1695–1697.

Western, D. and M. C. Pearl, eds., 1989. *Conservation for the Twenty-first Century*. Oxford University Press, New York.

Westoby, M. 1984. The self-thinning rule. *Advances in Ecological Research* 14:167–225.

Weston, L. A. 1996. Utilization of allelopathy for weed management in agroecosystems. *Agronomy Journal* 88:860–866.

Wetzel, R. G. 1983. *Limnology*. Saunders, Philadelphia.

Wetzel, R. G. and H. L. Allen. 1971. Functions and interactions of dissolved organic matter and the littoral zone in lake metabolism and eutrophication. in Z. Kajak and A. Hillbricht-Ilkowska, eds., *Productivity Problems of Freshwaters*. Proceedings of IBP-UNESCO Symposium, Poland, May 1970, Warsaw.

White, G. C., D. R. Anderson, K. P. Burnham, and D. L. Otis. 1982. Capture-recapture and removal methods for sampling closed populations. *Los Alamos National Laboratory* LA-8787-NERP, 235 pp.

White, G. C. and R. A. Garrott. 1990. *Analysis of Wildlife Radio-tracking Data*. Academic Press, New York.

White, G. G. 1981. Current status of prickly pear control by *Cactoblastis cactorum* in Queensland. *Proceedings of the 5th International Symposium on the Biological Control of Weeds* Brisbane, 1980:609-616.

White, H. C. 1939. Bird control to increase the Margaree River salmon. *Fisheries Research Board of Canada Bulletin* 58:1–30.

White, T. C. R. 1974. A hypothesis to explain outbreaks of looper caterpillars, with special reference to populations of *Selidosema suavis* in a plantation of *Pinus radiata* in New Zealand. *Oecologia* 16:279–310.

———. 1993. *The Inadequate Environment: Nitrogen and the Abundance of Animals*. Springer-Verlag, New York.

Whitehead, D. R. 1973. Late-Wisconsin vegetational changes in unglaciated eastern North America. *Quaternary Research* 3:621–631.

———. 1981. Late-Pleistocene vegetional changes in northeastern North Carolina. *Ecological Monographs* 51:451–471.

Whitmore, T. C. 1998. *An Introduction to Tropical Rain Forests*. Oxford University Press, Oxford.

Whittaker, R. H. 1953. A consideration of climax theory: The climax as a population and pattern. *Ecological Monographs* 23:41–78.

———. 1956. Vegetation of the Great Smoky Mountains. *Ecological Monographs* 26:1–80.

———. 1960. Vegetation of the Siskiyou Mountains, Oregon and California. *Ecological Monographs* 30:279–338.

———. 1962. Classification of natural communities. *Botanical Review* 28:1–239.

———. 1961. Experiments with radiophosphorus tracer in aquarium microcosms. *Ecological Monographs* 31: 157-158.

———. 1967. Gradient analysis of vegetation. *Biological Reviews* 42:207–264.

———. 1972. Evolution and measurement of species diversity. *Taxon* 21:213–251.

———. 1975. *Communities and Ecosystems*. Macmillan, New York.

Whittaker, R. H., and P. P. Feeny. 1971. Allelochemics: chemical interactions between species. *Science* 171:757-770.

Whittaker, R. H., S. A. Levin, and R. B. Root. 1973. Niche, habitat, and ecotope. American *Naturalist* 107:321-338.

Whittaker, R. J., M. B. Bush, and K. Richards. 1989. Plant recolonization and vegetation succession on the Krakatau Islands, Indonesia. *Ecological Monographs* 59:59–123.

Wiens, J. A. 1977. On competition and variable environments. *American Scientist* 65:590–597.

———. 1984a. On understanding a non-equilibrium world: Myth and reality in community patterns and processes. Pages 439–457 in D. R. Strong Jr., D. Simberloff, L. G. Abele, and A. B. Thistle, eds., *Ecological Communities*. Princeton University Press, Princeton, N.J.

———. 1984b. Resource systems, populations, and communities. Pages 397–436 in P. W. Price, C. N. Slobodchikoff and W. S. Gaud, ed., *A New Ecology*. Wiley, New York.

———. 1985. Habitat selection in variable environments: Shrub-steppe birds. Pages 227–251 in M. L. Cody, ed., *Habitat Selection in Birds*. Academic Press, New York.

———. 1989. *The Ecology of Bird Communities*. Cambridge University Press, Cambridge.

Wiens, J. A., J. F. Addicott, T. J. Case, and J. Diamond. 1986. Overview: The importance of spatial and temporal scale in ecological investigations. Pages 145–153 in J. Diamond and T. J. Case, eds., *Community Ecology*. Harper & Row, New York.

Wiens, J. A., R. L. Schooley, and R. D. J. Weeks. 1997. Patchy landscapes and animal movements: Do beetles percolate? *Oikos* 78:257–264.

Wilcove, D. S. and S. K. Robinson. 1990. The impact of forest fragmentation on bird communities in eastern North America. Pages 319–331 in A. Keast, ed., *Biogeography and Ecology of Forest Bird Communities*. SPB Academic Publishing bv, The Hague, Netherlands.

Wilkinson, D. M. 1999. Is Gaia really conventional ecology? *Oikos* 84:533–536.

Williams, C. B. 1964. *Patterns in the Balance of Nature*. Academic Press, London.

Williams, G. D. 1966. *Adaptation and Natural Selection*. Princeton University Press, Princeton, N.J.

Williams, P. J. le B. 1998. The balance of plankton respiration and photosynthesis in the open oceans. *Nature* 394:55–57.

Williamson, G. B. and E. M. Black. 1981. High temperature of forest fires under pines as a selective advantage over oaks. *Nature* 293:643–644.

Williamson, M. 1989. The MacArthur and Wilson theory today: True but trivial. *Journal of Biogeography* 16:3–4.

Williamson, M. and A. Fitter. 1996. The varying success of invaders. *Ecology* 77:1661–1666.

Willig, M. R. and S. K. Lyons. 1998. An analytical model of latitudinal gradients of species richness with an empirical test for marsupials and bats in the New World. *Oikos* 81:93–98.

Wills, C. 1996. *Yellow Fever, Black Goddess: The Coevolution of People and Plagues*. Addison Wesley, Reading, Mass.

Wills, C., R. Condit, R. B. Foster, and S. P. Hubbell. 1997. Strong density- and diversity-related effects help to maintain tree species diversity in a neotropical forest. *Proceedings of the National Academy of Sciences of the USA* 94:1252–1257.

Wilson, E. O. 1994. *Naturalist*. Island Press, Washington, D.C.

Wilson, J. B., H. Gitay, S. H. Roxburgh, W. M. King, and R. S. Tangney. 1992. Egler's concept of "Initial floristic composition" in succession — ecologists citing it don't agree what it means. *Oikos* 64:591–593.

Wilson, J. W. I. 1974. Analytical zoogeography of North American mammals. *Evolution* 28:124–140.

Wilson, M. A. and S. R. Carpenter. 1999. Economic valuation of freshwater ecosystem services in the United States:1971–1997. *Ecological Applications* 9:772–783.

Winkler, D. W. and J. R. Walters. 1983. The determination of clutch size in precocial birds. *Current Ornithology* 1:33–68.

Winston, M. R. 1995. Co-occurrence of morphologically similar species of stream fishes. *American Naturalist* 145:527–545.

Winters, G. H. and J. P. Wheeler. 1985. Interaction between stock area, stock abundance, and catchability coefficient. *Canadian Journal of Fisheries and Aquatic Sciences* 42:989–998.

Wold, N. and R. J. Marquis. 1997. Induced defense in white oak: Effects on herbivores and consequences for the plant. *Ecology* 78:1356–1369.

Wolda, H. and B. Dennis. 1993. Density dependence tests, are they? *Oecologia* 95:581–591.

Wolff, J.O. 1997. Population regulation in mammals: an evolutionary perspective. *Journal of Animal Ecology* 66:1-13.

Wood, D. M. and R. del Moral. 1987. Mechanisms of early primary succession in subaline habitats on Mount St. Helens. *Ecology* 68:780–790.

———. 1988. Colonizing plants on the pumice plains, Mount St. Helens, Washington. *American Journal of Botany* 75:1228–1237.

Woodroffe, R. and J. R. Ginsberg. 1998. Edge effects and the extinction of populations inside protected areas. *Science* 280:2126–2128.

Woods, K. D. and M. B. Davis. 1989. Paleoecology of range limits: Beech in the upper peninsula of Michigan. *Ecology* 70:681–696.

Wootton, J. T. 1998. Effects of disturbance on species diversity: A multitrophic perspective. *American Naturalist* 152:803–825.

Wootton, J. T., M. S. Parker, and M. E. Power. 1996. Effects of disturbance on river food webs. *Science* 273:1558–1561.

Wright, H. E., Jr. 1968. The roles of pine and spruce in the forest history of Minnesota and adjacent areas. *Ecology* 49:937–955.

Wu, L., A. D. Bradshaw, and D. A. Thurman. 1975. The potential for evolution of heavy metal tolerance in plants. III. The rapid evolution of copper tolerance in *Agrostis stolonifera*. *Heredity* 34:165–187.

Wynne-Edwards, V. C. 1962. *Animal Dispersion in Relation to Social Behavior*. Oliver and Boyd, Edinburgh.

———. 1986. *Evolution Through Group Selection*. Blackwell, Oxford, England.

Yoda, K., T. Kira, H. Ogawa, and K. Hozumi. 1963. Self-thinning in overcrowded pure stands under cultivated and natural conditions. *Journal of the Institute of Polytechnics, Osaka City University, Series D* 14:107-129.

Yodzis, P. 1986. Competition, mortality, and community structure. Pages 480–491 in J. Diamond and T. J. Case, eds., *Community Ecology*. Harper & Row, New York.

Young, A., T. Boyle, and T. Brown. 1996. The population genetic consequence of habitat fragmentation for plants. *Trends in Ecology & Evolution* 11:413–418.

Young, B. E. 1996. An experimental analysis of small clutch size in tropical house wrens. *Ecology* 77:472–488.

Young, T. P. 1990. Evolution of semelparity in Mount Kenya lobelias. *Evolutionary Ecology* 4:157–171.

Zanette, L., P. Doyle, and S. M. Tremont. 2000. Food shortage in small fragments: Evidence from an area-sensitive passerine. *Ecology* 81:1654–1666.

Zar, J. H. 1996. *Biostatistical Analysis*. Prentice-Hall, London.

CREDITS

ART CREDITS

Chapter 1 Reprinted with permission. Figure 1.2 G. Caughly, N. Shepherd, and J. Short. *Kangaroos: Their Ecology and Management in the Sheep Rangelands of Australia*, p. 12. Copyright © 1987 Cambridge University Press, Cambridge. Reprinted with permission.

Chapter 2 Figure 2.2 Robet H. Tamarin. *Principles of Genetics*, 4th ed., p. 548. Copyright © 1993 Wm. C. Brown Communications Inc., Dubuque, Iowa. All rights reserved Reprinted with permission. Figure 2.3 P. R. Grant. *Ecology and Evolution of Darwin's Finches*, p. 213. Copyright © 1986 Princeton University Press, Princeton, New Jersey. Reprinted with permission. Figure 2.5 F. Cooke and C. S. Findlay. Polygenic variation and stabilizing selection in a wild population of lesser snow geese (*Anser Caerulescens.*) In *American Naturalist*, Vol. 120, p. 544. Copyright © 1982 the University of Chicago Press, Chicago. Reprinted with permission. Figure 2.8 B. E. Young. An experimental analysis of small clutch size in tropical house wrens. *Ecology*, Vol. 77, p. 479. Copyright © 1996 *Ecology*. Reprinted with permission.

Chapter 3 Figure 3.3 E. Baroudy and J. M. Elliott. The critical thermal limits for juvenile Arctic charr (*Salvelinus alpinus*). *Journal of Fish Biology*, Vol. 45, p. 1049. Copyright © 1994 The Fisheries Society of the British Isles. Reprinted with permission. Figure 3.5 M. R. MacNair. Tolerance of higher plants to toxic material. In *Genetic Consequences of Man-Made Change*, ed. J. A. Bishop and L. M. Cook, p 199. Copyright © 1981 Academic Press, London. Reprinted with permission.

Chapter 4 Figure 4.4 J. A. Lubena and S. A. Levin. The spread of a reinvading species: Expansion in the California sea otter. *American Naturalist*, Vol. 131, Fig. 1, p. 529, Fig. 2, p. 535. Copyright © 1988 The University of Chicago Press, Chicago. Reprinted with permission. Figure 4.7 E. C. Pielou. *Biogeography*, p. 247. Copyright © 1979 John Wiley and Sons, Inc. Reprinted with permission. Figure 4.8

L. A. Swan and C. S. Papp. *The Common Insects of America*, p. 586. Copyright © 1972 by L. A. Swan and C. S. Papp. Reprinted with permission of HarperCollins Publishers Inc.

Chapter 5 Figure 5.2 J. A. Wiens. Habitat selection in variable environments: Shrub-steppe birds. *Habitat Selection in Birds*, ed. M. L. Cody, pp. 227-251, Fig. 5. Copyright © 1985 Academic Press, Orlando. Reprinted with permission. Figure 5.3 M. C. Baker, D. B. Thompson, G. L. Sherman, M. A. Cunningham, and D. F. Timback. Allozyne frequencies in a linear series of song dialect populations. *Evolution*, Vol. 36, pp. 1020-1029. Copyright © 1982 Society of the Study of Evolution. Reprinted with permission. Figure 5.5 S. D. Thompson. Structure and species composition of desert hetermyid rodent species assemblages: Effects of a simple habitat manipulation, Fig. 1, p. 1314; Fig. 2, p. 1316. Copyright © 1982 by Ecological Society of America.. Reprinted with permission.

Chapter 6 Figures 6.1, 6.2 J. A. Kitching and F. J. Ebling. Ecological studies at Lough Ine. *Advances in Ecological Research*, Vol. 4, pp. 259-260. Copyright © 1967 Academic Press, Inc. Reprinted with permission. Figure 6.3 J. E. Kinnear, M. L. Onus, and N. R. Sumner. Fox control and rock-wallaby population dynamics. II. An update. *Wildlife Research* 25, Fig. 1, page 83. Copyright © 1998 Wildlife Research. Reprinted with permission. Figure 6.6 C. Van Riper III, S. G. Van Riper, M. L. Goff, and M. Laird. The epizootiology and ecological significance of malaria in Hawaiian land birds. *Ecological Monographs*, Vol. 56, Fig. 2, p. 331; Fig. 4, .p. 333; Fig. 11, p. 338. Copyright © 1986 Ecological Society of America. Reprinted with permission. Figure 6.8 R. W. Thacker, M. A. Becerro, W. A. Lumbang, and V. J. Paul. Allelopathic interactions between sponges on a tropical reef. *Ecology* 79, Fig. 4, page 1745 Copyright © 1998 by Ecological Society of America. Reprinted with permission. Figure 6.9 J. M. Diamond. Assembly of species communities, Fig. 22, p. 390. Reprinted with permission of the publishers from *Ecology and Evolution of Communities*, ed. M. L. Cody and J. M. Diamond, Cambridge, MA: The Belknap Press of Harvard University Press. Copyright © 1975 by the President and Fellows of Harvard College. Figure 6.10 I. New-

Figures 13.15 and 13.17 F. Messier Ungulate population models with predation: A case study with the North American moose. *Ecology*, Vol. 75, pp. 482, 483. Copyright © 1994 by Ecological Society of America. Reprinted with permission. Figure 13.18 W.W. Murdoch and A. Sih. Age-dependent interference in a predatory insect. *Journal of Animal Ecology*, Vol. 47, p. 586. Copyright © 1978 Blackwell Scientific Publications Ltd. Reprinted with permission. Figure 13.19 L. J. Mook. Birds and the spruce budworm. In R. F. Morris, ed. *The Dynamics of Epidemic Spruce Budworm Populations*, p. 270. Copyright © 1963 Memoirs of the Entomological Society of Canada. Reprinted with permission. Figure 13.20 W. Jedrzejewski and B. Jedrzejewska. Tawny owl (*Strix aluco*) predation in a pristine deciduous forest (Bialowieza National Park, Poland). *Journal of Animal Ecology*, Vol. 65, p. 109. Copyright © Blackwell Scientific Publicions Ltd. Reprinted with permission. Figure 13.22 D.A. Spiller and T. W. Schoener. An experimental study of the effect of lizards on web-spider communities. *Ecological Monographs*, Vol. 58, p. 63. Copyright © 1988 Ecological Society of America. Reprinted with permission. Figure 13.24 R. E. Kenward. Hawks and doves: factors affecting success and selection in goshawk attacks on woodpigeons. *Journal of Animal Ecology*, Vol. 47, p. 455. Copyright © 1978 Blackwell Scientific Publications Ltd. Reprinted with permission.

Chapter 14   Figure 14.1 R. H. Whittaker and P. P. Feeny. Allelochemics: Chemical interactions between species. *Science*, Vol. 17, p. 760. Copyright © 1971 by the AAAS. Reprinted with permission. Figure 14.2 P. D. Coley, J. P. Bryant, and F. S. Chapin, III. Resource availability and plant antiherbivore defense. *Science*, Vol. 230, Fig. 1, p. 896. Copyright © 1985 by the AAAS. Reprinted with permission. Figures 14.3 and 14.4 P. P. Feeny, Seasonal changes in oak leaf tannins and nutrients as a cause of spring feeding by winter moth caterpillars. *Ecology*, Vol. 51, pp. 565, 571. Copyright © 1970 Ecological Society of America. Reprinted with permission. Figure 14.6 C. D. Harvell. The ecology and evolution of inducible defenses in a marine bryozona: Cues, costs, and consequences. *American Naturalist*, Vol. 128, p. 812. Copyright © 1986 the University of Chicago Press. Reprinted by permission; Photo of bryozoan spines from *Ecology*, Vol. 73, p. 1568. Copyright © C. Drew Harvell. Reprinted by permission. Figure 14.7 J. H. Myers and D. Bazely. Thorns, spines, prickles, and hairs: Are they stimulated by herbivory and do they deter herbivores? *Phytochemical Induction by Herbivores*, ed. T. W. Tallamy and M. J. Raupp, p. 332. Copyright © 1991 John Wiley and Sons, Inc. Reprinted with permission. Figure 14.9 R. H. V. Bell. A grazing ecosystem in the Serengeti. *Scientific American*, Vol. 255, pp. 92-93. Copyright © 1971 by Scientific American, Inc. All rights reserved. Reprinted with permission from W. H. Freeman and Company. Figure 14.10 M. D. Gwynne and R. H. V. Bell. Selection of vegetation components by grazing ungulates in the Serengeti National Park. *Nature*, Vol.

220, p. 391. Copyright © 1968 Macmillan Magazines Ltd., London, England. Reprinted with permission. Figure 14.11 S. J. McNaughton. Serengeti migratory wildebeest: facilitation of energy flow by grazing. *Science*, Vol. 191, p. 93. Copyright © 1976 by the AAAS. Reprinted with permission. Figure 14.16 N. Leader-Williams. *Reindeer on South Georgia: The Ecology of an Introduced Population*, p. 76. Copyright © 1988. Cambridge University Press, Cambridge, England. Reprinted with permission. Figure 14.22 C. E. Bach. Direct and indirect interactions between ants (*Pheidole megacephala*), scales (*Coccus viridis*) and plants (*Pluchea indica*). *Oecologia*, Vol. 87, p. 235. Copyright © 1991 Springer-Verlag, Heidelberg. Reprinted with permission.

Chapter 15   Figure 15.1 R. M. Anderson and R. M. May. Population biology of infectious diseases: Part I. *Nature*, Vol. 280, p. 364. Copyright © 1979 Macmillan Magazines Ltd., London, England. Reprinted with permission. Figure 15.9 A. Dobson and M. Meagher. The population dynamics of brucellosis in Yellowstone National Park. *Ecology*, Vol. 77, p. 1029. Copyright © 1996 Ecological Society of America. Reprinted with permission. Figure 15.14 D. W. Macdonald and D. R. Voigt. The biological basis of rabies models. In P. J. Bacon, ed. *Population Dynamics of Rabies in Wildlife*, p. 93. Copyright © 1985 Academic Press, London. Reprinted with permission. Figure 15.19 F. Fenner and K. Myers. Myxoma virus and myxomatosis in retrospect: The first quarter century of a new disease. *Viruses and Environment*, ed. E. Kurstak and K. Maramorosch, p. 555. Copyright © 1978 Academic Press, Inc. Reprinted with permission of the publisher and the authors.

Chapter 16   Figure 16.5 D. L DeAngelis and J. C. Waterhouse. Equilibrium and nonequilibrium concepts in ecological models. *Ecological Monographs*, Vol. 57, p. 4. Copyright © 1987 Ecological Society of America. Reprinted with permission. Figure 16.13 T. R. E. Southwood. *Ecological Methods*, p. 381. Copyright © 1978 Methuen, London. Reprinted with permission. Figure 16.16 B. E. Saether, T. H. Ringsby, O. Bakke, and E. J. Solberg. Spatial and temporal variation in demography of a house sparrow metapopulation. *Journal of Animal Ecology*, Vol. 68, p. 632. Copyright © 1999 Blackwell Scientific Publications Ltd. Reprinted with permission. Figure 16.17 J. O. Wolff. Population regulation in mammals: an evolutionary perspective. *Jounal of Animal Ecology*, Vol. 66, p. 8. Copyright © 1997 Blackwell Scientific Publications Ltd. Reprinted with permission.

Chapter 17   Figure 17.3 L. K. Boerema and J. A. Gulland. Stock assessment of the Peruvian anchovy (*Engraulis ringens*) and management of the fishery. *Journal of the Fisheries Research Board of Canada*,.Vol. 30. Copyright 1973 (c) *Fisheries Research Board of Canada*. Reprinted with permission. Figure 17.10 T. E. Nickelson and J. A. Lichatowich. The influence of the marine environment on the interannual variation in coho salmon abundance: An overview. In W. Pearcy, ed. *The Influence of Ocean Conditions on the Production of Salmonids in the North Pacific*, p. 32. Copyright © 1983 Oregon State University Sea Grant Publication, Corvallis. Reprinted with permission. Figure 17.11 G. H. Kruse and A. V. Tyler. Exploratory simulation of English sole recruitment mechanisms. *Transactions of the American Fisheries Society*, Vol. 118, p. 104. Copyright © 1989 *Transactions of the American Fisheries Society*. Reprinted with permission. Figure 17.21 K. R. Allen. *Conservation and management of whales*, p. 13. Copyright © 1980 University of Washington Press, Seattle. Reprinted with permission. Figure 17.22 D. G. Chapman Evaluation of marine mammal population models. *Dynamics of Large Mammal Populations*, C. W. Fowler and T. D. Smith, eds., p. 289. Copyright © 1981 Wiley, New York. Reprinted with permission.

Chapter 18 Figure 18.2 P. DeBach *Biological Control by Natural Enemies*, p. 4. Copyright © 1974. Cambridge University Press, London. Reprinted with permission. Figure 18.3 R. van den Bosch, P. S. Messenger, and A. P. Gutierrez. *An Introduction to Biological Control*, p. 2. Copyright © 1982 Plenum, New York. Reprinted with permission. Figure 18.4 G. G. White. Current status of prickly pear control by *Cactoblastis cactorum* in Queensland. *Proceedings of the 5th International Symposium on the Biological Control of Weeds* Brisbane, *1980*, p. 610. Copyright © 1981 CSIRO. Reprinted with permission. Figures 18.7 and 18.8 J. D. Reeve and W. W. Murdoch. Biological control by the parasitoid Aphytis melinus, and population stability of the California red scale. *Journal of Animal Ecology*, Vol. 55. p. 1077. Copyright © 1985 Blackwell Scientific Publications Ltd. Reprinted with permission. Figure 18.9 W. W. Murdoch, R. F. Luck, S. L. Swarbrick, S. Walde, D. S. Yu, and J. D. Reeve. Regulation of an insect population under biological control. *Ecology*, Vol. 76, p. 211. Copyright © 1995 Ecological Society of America. Reprinted with permission. Figure 18.13 G. R. Caughley, Pech, and D. Grice. Effect of fertility control on a population's productivity. *Wildlife Research*, Vol. 19, p. 626. Copyright © 1992 Wildlife Research. Reprinted with permission. Fig. 18.18 F. E. Ehler and R. W. Hall. Evidence for competitive exclusion of introduced natural enemies in biological control. *Environmental Entomology*, Vol. 11, p. 2. Copyright © 1982 Entomological Society of America. Reprinted with permission.

Chapter 19 Figure 19.2 R. L. Westemeier, J. D. Brawn, S. A. Simpson, T. L. Esker, R. W. Jansen, J. W. Walk, E. L. Kershner, J. L. Bouzat, and K. N. Paige. Tracking the long-term decline and recovery of an isolated population. *Science*, Vol. 282, p. 1696. Copyright © 1998 by the AAAS. Reprinted with permission. Figure 19.4 B. H. Letcher, J. A. Priddy, J. R. Walters, and L. B. Crowder. An individual-based, spatially-explicit simulation model of the population dynamics of the endangered red-cockaded woodpecker,

Picoides borealis. *Biological Conservation*, Vol. 86, p. 4. Copyright © 1998 *Elsevier Applied Science Publishers Ltd.*, *Barking England*. Reprinted with permission. Figure 19.5 J. T. Curtis. *The Vegetation of Wisconsin*, p. 20. Copyright © 1959 University of Wisconsin Press, Madison. Reprinted with permission. Figure 19.6 R. Dirzo and M. C. Garcia. Rates of deforestation in Los Tuxtlas, a Neotropical area in southeast Mexico. *Conservation Biology*, Vol. 6, p. 88. Copyright © 1992. Blackwell Scientific Publications Ltd., Cambridge MA. Reprinted with permission of the Society for Conservation Biology and Blackwell Scientific Publications, Inc. Figure 19.7 A. Peltonen and I. Hanski. Patterns of island occupancy explained by colonization and extinction rates in shrews. *Ecology*, Vol. 72. Copyright © 1991 Ecological Society of America. Reprinted with permission. Figure 19.8 B. Verboom and R. Van Apeldoorn. Effects of habitat fragmentation on the red squirrel, *Sciurus vulgaris* L. *Landscape Ecology*, Vol. 4, p. 174. Copyright © 1990 Academic Publishing bv, the Netherlands. Reprinted with permission. Figure 19.9 H. Andren and P. Angelstam. Elevated predation rates as an edge effect in habitat islands: experimental evidence. *Ecology*, Vol. 69, p. 546. Copyright © 1988 Ecological Society of America. Reprinted with permission. Figure 19.10 N. M. Haddad. Corridor and distance effects on interpatch movements: A landscape experiment with butterflies. *Ecological Applications*, Vol. 9, p. 617. Copyright © 1999 *Ecological Applications*. Reprinted with permission. Figure 19.11 M. Corbit, P. L. Marks, and S. Gardescu. Hedgerows as habitat corridors for forest herbs in central New York, USA. *Journal of Ecology*, Vol. 87, p. 228. Copyright © 1999 Blackwell Scientific Publications Ltd. Reprinted with permission. Figure 19.12 A. O. Nicholls and C. R. Margules. The design of studies to demonstrate the biological importance of corridors. In D. A. Saunders and R. J. Hobbs, eds. *Nature Conservation 2: The Role of Corridors*, p. 53. Copyright © 1991 Surrey Beatty and Sons, Chipping Norton, Australia. Figure 19.15 A. A. Burbidge and N. L. McKenzie. Patterns in the modern decline of Western Australia's vertebrate fauna: causes and conservation implications. *Biological Conservation*, Vol. 50, p. 158. Copyright © 1989 *Elsevier Applied Science Publishers Ltd.*, *Barking England*. Reprinted with permission. Figure 19.16 W. D. Newmark Legal and biotic boundaries of Western North American National Parks: A problem of congruence. *Biological Conservation*, Vol. 33, p. 199. Copyright © 1985 *Elsevier Applied Science Publishers Ltd.*, *Barking England*. Reprinted with permission. Figure 19.18 and 19.19 E. S. Menges. Population viability analysis for an endangered plant. *Conservation Biology*, Vol. 4, pp. 53, 58. Copyright © 1990 Blackwell Scientific Publications Ltd., Cambridge MA. Reprinted with permission of the Society for Conservation Biology and Blackwell Scientific Publications, Inc. Figure 19.20 A. B. Carey, S. P. Horton, and B. L. Biswell. Northern spotted owls: influence of prey-base and landscape character. *Ecological Monographs*, Vol. 62, p. 229. Copyright © 1992 Ecological Society of America.

J. Gibson. 1995. Experimental analysis of intermediate disturbance and initial floristic composition: decoupling cause and effect. *Ecology*, Vol. 76, p. 488. Copyright © 1995 Ecological Society of America. Reprinted with permission.

Chapter 23 Figure 23.2 J. A. Wiens. Resource systems, populations, and communities. In P. W. Price, C.N. Slobodchikoff, and W.S. Gaud, ed. *A New Ecology*, p. 402. Copyright © 1984 Wiley, New York. Reprinted with permission. Figure 23.5 D. L. Breitburg, T. Loher, C. A. Pacey, and A. Gerstein. Varying effects of low dissolved oxygen on trophic interactions in an estuarine food web. *Ecological Monographs*, Vol. 67, p. 490. Copyright © 1997 Ecological Society of America. Reprinted with permission. Figure 23.8 S. J. Hall and D. G. Raffaelli. Food-web patterns: What do we really know? In A. C. Gange and V. K. Brown, eds. *Multitrophic Interactions in Terrestrial Systems*, p. 402. Copyright © 1997 Blackwell Science, London. Reprinted with permission. Figure 23.9 B. Jenkins, R. L. Kitching, and S. L. Pimm. Productivity, disturbance and food web structure at a local spatial scale in experimental container habitats. *Oikos*, Vol. 65, p. 252. Copyright © 1992 *Oikos*, Sweden. Reprinted with permission. Figure 23.11 R. B. Root. Organization of a plant-arthropod association in simple and diverse habitats: the fauna of collards (*Brassica oleracea*). *Ecological Monographs*, Vol. 43, p. 98. Copyright © 1973 Ecological Society of America. Reprinted with permission. Figure 23.12 J. A. Estes, M. T. Tinker, T. M. Williams, and D. F. Doak. Killer whale predation on sea otters linking oceanic and nearshore ecosystems. *Science*, Vol. 282, p. 474. Copyright © 1998 by the AAAS. Reprinted with permission. Figure 23.15 L. W. Buss and J. B. C. Jackson. Competitive networks: Nontransitive competitive relationships in cryptic coral reef environments. *American Naturalist*, Vol. 113, p. 232. Copyright © 1979 the University of Chicago Press, Chicago. Reprinted with permission. Figure 23.16 T. J. Smith III. Seed predation in relation to tree dominance and distribution in mangrove forests. *Ecology*, Vol. 68. Copyright © 1987 Ecological Society of America. Reprinted with permission. Figure 23.17 J. L. Brooks and S. I. Dodson. Predation, body size, and composition of plankton. *Science*, Vol. 150, p. 33. Copyright © 1965 by the AAAS. Reprinted with permission. Figure 23.18 Z. M. Gliwicz. Food thresholds and body size in cladocerans. *Nature*, Vol. 343, p. 640. Copyright © 1990 Macmillan Magazines Ltd., London, England. Reprinted with permission. Figure 23.19 D. Tilman. Biodiversity: population versus ecosystem stability. *Ecology*, Vol. 77, p. 359. Copyright 1996 © Ecological Society of America. Reprinted with permission. Figure 23.21 A. P. Dobson, A. D. Bradshaw, and A. J. M. Baker. Hopes for the future: restoration ecology and conservation biology. *Science*, Vol. 277, p. 515. Copyright © 1997 by the AAAS. Reprinted with permission. Figures 23.23 and 23.24 W. T. Edmondson. Cultural eutrophication with special reference to Lake Washington. *Mitteilungen Internationale Vereinigung Limnologie*, Vol. 17, pp. 24, 27. Copyright © 1969 *Mitteilungen Internationale Vereinigung Limnologie*. Reprinted with permission. Graph for Essay 23.1 G. Cabana and J. B. Rasmussen. Modelling food chain structure and contaminant bioaccumulation using stable nitrogen isotopes. *Nature*, Vol. 372, p. 255. Copyright © 1994 Macmillan Magazines Ltd., London, England. Reprinted with permission. Graph for Essay 23.2 T. J. Pitcher, P. Hart, and D. Pauly, eds. *Reinventing Fisheries Management*, p. 314. Copyright © 1998 Kluwer Academic Publishers, Boston. Reprinted with permission.

Chapter 24 Figure 24.1 J. H. Connell, T. P. Hughes, and C. C. Wallace. A 30-year study of coral abundance, recruitment, and disturbance at several scales in space and time. *Ecological Monographs*, Vol. 67, Fig. 2, p. 472. Copyright © 1997 Ecological Society of America. Reprinted with permission. Figure 24.3 G. P. Jones. The importance of recruitment to the dynamics of a coral reef fish population. *Ecology*, Vol. 71, p. 1693. Copyright © 1990 Ecological Society of America. Reprinted with permission. Figure 24.5 J. Lubchenco. Relative importance of competition and predation: Early colonization by seaweeds in New England. In J. Diamond and T. J. Case, eds. *Community Ecology*, p. 549. Copyright © 1986. Harper & Row, New York. Reprinted with permission. Figure 24.6 R. T. Paine. Ecological determinism in the competition for space. *Ecology*, Vol. 65, p. 1345. Copyright © 1984 Ecological Society of America. Reprinted with permission. Figure 24.7 D. L. DeAngelis and J. C. Waterhouse. Equilibrium and nonequilibrium concepts in ecological models. *Ecological Monographs*, Vol. 57, p. 8. Copyright © 1987 Ecological Society of America. Reprinted with permission. Figure 24.8 B. A. Menge and J. P. Sutherland. Community regulation: variation disturbance, competition, and predation in relation to environmental stress and recruitment. *American Naturalist*, Vol. 130, p. 739. Copyright © 1987 the University of Chicago Press, Chicago. Reprinted with permission. Figure 24.10 D. J. McQueen, M. R. S. Johannes, J. R. Post, T. J. Stewart, and D. R. S. Lean. Bottom-up and top-down impacts on freshwater pelagic community structure. *Ecological Monographs*, Vol. 59, p. 301. Copyright © 1989 Ecological Society of America. Reprinted with permission. Figure 24.12 M. E. Power. Effects of fish in river food webs. *Science*, Vol. 250, p. 1560. Copyright © 1990 by the AAAS. Reprinted with permission. Figure 24.14 F. W. Preston. The canonical distribution of commonness and rarity. *Ecology*, Vol. 43, p. 196. Copyright © 1962 Ecological Society of America. Reprinted with permission. Figures 24.15, 24.17, and 24.18 R. MacArthur and E. O. Wilson. *The Theory of Island Biogeography*, pp. 8, 21, 22. Copyright © 1962 Princeton University Press, Princeton, New Jersey. Reprinted with permission. Figures 24.20 and 24.21 T. E. Lawlor, T. E. Comparative biogeography of mammals on islands. *Biological Journal of the Linnean Society*, Vol. 28, pp. 100, 106. Copyright 1986 Linnean Society. Reprinted with permission.

Chapter 25   Figure 25.2 D. L. DeAngelis. 1992. *Dynamics of Nutrient Cycling and Food Webs*, p. 25. Copyright © 1992 Chapman & Hall, New York. Reprinted with permission. Figure 25.3 C. B. Field, M. J. Behrenfeld, J. T. Randerson, and P. Falkowski. Primary production of the biosphere: integrating terrestrial and oceanic components. *Science*, Vol. 281, p. 238. Copyright © 1998 by the AAAS. Reprinted with permission. Figures 25.6 and 25.7 T. L. Hayward. Primary production in the North Pacific Central Gyre: A controversy with important implications. *Trends in Ecology and Evolution*, Vol. 6, p. 283. Copyright © 1991 *Trends in Ecology and Evolution*. Reprinted with permission. Figure 25.8 J. H. Ryther and W. M. Dunstan. Nitrogen, phosphorus, and eutrophication in the coastal marine environment. *Science*, Vol. 171, p. 1011. Copyright © 1971 by the AAAS. Reprinted with permission. Figure 25.9 J. A. Downing, C. W. Osenberg, and O. Sarnelle. Meta-analysis of marine nutrient-enrichment experiments: variation in the magnitude of nutrient limitation. *Ecology*, Vol. 80, p 1163. Copyright © 1999 Ecological Society of America. Reprinted with permission. Figure 25.10 S. Sathyendranath, T. Platt, E. P. W. Horne, W. G. Harrison, O. Ulloa, R. Outerbridge, and N. Hoepffner. Estimation of new production in the ocean by compound remote sensing. *Nature*, Vol. 353, p. 129. Copyright © 1991 Macmillan Magazines Ltd., London, England. Reprinted with permission. Figure 25.11 D. W. Schindler and E. J. Fee. Experimental Lakes area: Whole-lake experiments in eutrophication. *Journal of the Fisheries Research Board of Canada*, Vol. 31. Copyright © 1974 Fisheries Research Board of Canada. Reprinted with permission. Figure 25.13 D. W. Schindler. Evolution of phosphorus limitation in lakes. *Science*, Vol. 195, p. 261. Copyright © 1977 by the AAAS. Reprinted with permission. Figure 25.14 V. H. Smith. Low nitrogen to phosphorus ratios favor dominance by blue-green algae in lake phytoplankton. *Science*, Vol. 221, p. 670. Copyright © 1983 by the AAAS. Reprinted with permission. Figures 25.17 and 25.18 T. Kira. Primary production of forests in J. P. Cooper, ed. *Photosynthesis and Productivity in Different Environments*, pp. 7, 15. Copyright © 1975 Cambridge University Press, London. Reprinted with permission. Figure 25.19 H. E. Epstein, W. K. Lauenroth, I. C. Burke, and D. P. Coffin. Productivity patterns of C_3 and C_4 functional types in the U.S. Great Plains. *Ecology*, Vol. 78, p. 728. Copyright © 1997 Ecological Society of America. Reprinted with permission. Figure 25.20 S. M. Cargill and R. L. Jefferies. Nutrient limitation of primary production in a sub-arctic salt marsh. *Journal of Applied Ecology*, Vol. 21, p. 663. Copyright © 1984 *Journal of Applied Ecology*. Reprinted with permission. Figure 25.22 D. Tilman, J. Knops, D. Wedin, P. Reich, M. Ritchie, and E. Siemann. The influence of functional diversity and composition on ecosystem processes. *Science*, Vol. 277, Fig. 1, p. 1300. Copyright © 1997 by the AAAS. Reprinted with permission.

Chapter 26   Figure 26.1 K. A. Nagy. Field metabolic rate and food requirement scaling in mammals and birds. *Ecological Monographs*, Vol. 57, p. 120. Copyright © 1987 Ecological Society of America. Reprinted with permission. Figure 26.4 D. Pauly and V. Christensen. Primary production required to sustain global fisheries. *Nature*, Vol. 374, p. 257. Copyright © 1995 Macmillan Magazines Ltd., London, England. Reprinted with permission. Figure 26.5 N. G. J. Hairston and N. G. S. Hairston. Cause-effect relationships in energy flow, trophic structure, and interspecific interactions. *American Naturalist*, Vol. 142, p. 389. Copyright © 1993 the University of Chicago Press, Chicago. Reprinted with permission. Figure 26.6 H. Cyr and M. L. Pace. Magnitude and patterns of herbivory in aquatic and terrestrial ecosystems. *Nature*, Vol. 361, Fig. 1, p. 148. Copyright © 1993 Macmillan Magazines Ltd., London, England. Figure 26.7 S. Cousins. Ecologists build pyramids again. *New Scientist*, Vol. 106, Fig. 1, p. 51. Copyright © 1985 *New Scientist*. Reprinted with permission. Figure 26.8 S. J. McNaughton, M. Oesterheid, D. A. Frank, and K. J. Williams. Ecosystem-level patterns of primary productivity and herbivory in terrestrial habitats. *Nature*, Vol. 341, p 143. Copyright © 1989 Macmillan Magazines Ltd., London, England. Figure 26.9 N. R. French. *Perspectives in Grassland Ecology*, p. 13. Copyright © 1979 Springer-Verlag, New York. Reprinted with permission. Figure 26.10 W. K. Lauenroth. Grassland primary production: North American grasslands in perspective. *Perspectives in Grassland Ecology*, N. R. French, ed., p. 18. Copyright © 1979 Springer-Verlag, New York. Reprinted with permission. Figure 26.11 N. R. French, R. K. Steinhorst, and D. M. Swift. Grassland biomass trophic pyramids. *Perspectives in Grassland Ecology*, N. R. French, ed. p. 67. Copyright © 1979 Springer-Verlag, New York. Reprinted with permission. Figure 26.12 J. A. Scott, N. R. French, and J. W. Leetham. Patterns of consumption in grasslands. *Perspectives in Grassland Ecology*, N. R. French, ed., p. 18. Copyright © 1979 Springer-Verlag, New York. Reprinted with permission. Figure 26.13 J. L. Dodd and W. K. Lauenroth. Analysis of the response of a grassland ecosystem to stress. . *Perspectives in Grassland Ecology*, N. R. French, ed., p. 48. Copyright © 1979 Springer-Verlag, New York. Reprinted with permission. Figure 26.14 J. MacNab, J. Does game cropping serve conservation? A reexamination of the African data. *Canadian Journal of Zoology*, Vol. 2284. Copyright © 1991 National Research Council of Canada. Reprinted with permission. Figure 26.15 M. Oesterheld, O. E. Sala, and S. J. McNaughton. Effect of animal husbandry on herbivore-carrying capacity at a regional scale. *Nature*, Vol. 356, p. 235. Copyright © 1992 Macmillan Magazines Ltd., London, England. Figure 26.16 J. Weiner. Physiological limits to sustainable energy budgets in birds and mammals: Ecological implications. *Trends in Ecology and Evolution*, Vol. 7, p. 385.

Chapter 27 Figure 27.1 D. L. DeAngelis *Dynamics of Nutrient Cycling and Food Webs*, p. 5. Copyright © 1992 Chapman & Hall, New York. Reprinted with permission. Figure 27.2 F. E. Smith. Analysis of ecosystems. *Analysis of Temperate Forest Ecosystems*, ed. D. Reichle, p. 12. Copyright © 1970 Springer-Verlag, Berlin. Reprinted by permission. Figure 28.3 R. H. Whittaker. Experiments with radiophosphorus tracer in aquarium microcosms. *Ecological Monographs*, Vol. 31, p. 164. Copyright © 1961 Ecological Society of America. Reprinted with permission. Figure 27.4 R. H. Waring and W H. Schlesinger. *Forest Ecosystems: Concepts and Management*, p. 123. Copyright © 1985 Academic Press, Orlando. Reprinted by permission. Figure 27.5 D. A. MacLean and R. W. Wein. Nutrient accumulation for post fire jack pine and hardwood successional patterns in New Brunswick. *Canadian Journal of Forest Research*, Vol. 7, p. 569. Copyright 1977 *Canadian Journal of Forest Research*. Reprinted with permission. Figure 27. 7 G. E. Likens, F. H. Bormann, N. M. Johnson, D. W. Fisher, and R. S. Pierce. Effects of forest cutting and herbicide treatment on nutrient budgets in the Hubbard Brook watershed-ecosystem. *Ecological Monographs*, Vol. 40, p. 33. Copyright © 1970 Ecological Society of America. Reprinted with permission from Duke University Press. Figure 27.9 J. Pastor and S. D. Bridgham. Nutrient efficiency along nutrient availability gradients. *Oecologia*, Vol. 118, p. 53. Copyright © 1999 Copyright © 1983 Springer-Verlag, Berlin. Reprinted by permission. Figure 27.11 G. E. Likens, F. H. Bormann, and N. M. Johnson. Interactions between major biogeochemical cycles in terrestrial ecosystems. *Some Perspectives of the Major Biogeochemical Cycles*, ed. G. E. Likens, p. 103, Copyright © 1981 John Wiley and Sons, New York. Reprinted by permission. Figure 27.12 W. H. Schlesinger. *Biogeochemistry: An Analysis of Global Change*, p. 96. Copyright © 1997 Academic Press, San Diego. Reprinted with permission. Fig. 27.15 J. M. Gunn and K. H. Mills. The potential for restoration of acid-damaged lake trout lakes. *Restoration Ecology*, Vol. 6, p. 391. Copyright © 1998 *Restoration Ecology*. Reprinted with permission. Figure 27.16 E. D. Schulze. Air pollution and forest decline in a spruce *(Picea abies)* forest. *Science*, Vol. 24, p. 777 Copyright © 1989 by the AAAS. Reprinted by permission. Figure 27.17 W. H, Schlesinger. *Biogeochemistry: An Analysis of Global Change*, p. 309. Copyright © 1991 Academic Press, San Diego. Reprinted by permission. Figure 27.18 P. M. Vitousek, J. Aber, R. W. Howarth, G. E. Likens, P. A. Matson, D. W. Schindler, W. H. Schlesinger, and D. G. Tilman. Human alteration of the global nitrogen cycle: causes and consequences. *Ecological Applications*, Vol. 7, p. 743. Copyright © 1997 *Ecological Applications*. Reprinted with permission. Figure 27.19 D. A. Wedin and D. Tilman. Influence of nitrogen loading and species composition of the carbon balance of grasslands. *Science*, Vol. 274, Fig. 1, p. 1721. Copyright © 1996 by the AAAS. Reprinted with permission.

Chapter 28 Figures 28.7 and 28.8 W. Schlesinger. *Biogeochemistry: An Analysis of Global Change* pp. 133, 197. Copyright © 1997 Academic Press, San Diego. Reprinted with permission. Figure 28.11 S. B. Idso. The long-term response of trees to atmospheric CO_2 enrichment. *Global Change Biology*, Vol. 5, Fig. 1, p. 493. Copyright © 1999 *Global Change Biology*. Reprinted with permission. Figure 28.12 W. D. Billings, J. O. Luken, D. A. Mortensen, and K. M. Peterson. Increasing atmospheric carbon dioxide: possible effects on arctic tundra. *Oecologia*, Vol. 58, p. 288. Copyright © 1983 Springer-Verlag, Heidelberg. Reprinted with permission. Figure 28.13 O. L. Phillips, Y. Malhi, N. Higuchi, W. F. Laurance, P. V. Nunez, R. M. Vasquez, S. G. Laurance, L. V. Ferreira, M. Stern, S. Brown, and J. Grace. Changes in the carbon balance of tropical forests: evidence from long-term plots. *Science*, Vol. 282, Fig. 1, p. 439. Copyright © 1998 by the AAAS. Reprinted with permission. Figure 28.14 M. C. MacCracken. Carbon dioxide and climate change: background and overview. *Projecting the Climatic Effects of Increasing Carbon Dioxide*, M. C. MacCracken and F. M. Luther, eds., p. 410. Copyright © 1985 U.S. Department of Energy, Washington, D.C. Figure 28.15 H. M. Barnola, D. Raynaud, Y. S. Korotkevich, and C. Lorius. Vostok ice core provides 160,000-year record of atmospheric carbon dioxide. *Nature*, Vol. 329, p. 494. Copyright © 1987 Macmillan Magazines Ltd., London, England. Reprinted with permission. Figure 28.16 P. M. Vitousek, J. Aber, R. W. Howarth, G. E. Likens, P. A. Matson, D. W. Schindler, W. H. Schlesinger, and D. G. Tilman. Human alteration of the global nitrogen cycle: causes and consequences. *Ecological Applications*, Vol. 7, p. 494. Copyright © 1997 *Ecological Applications*. Reprinted with permission. Figure 28.17 E. D. Fajer, M. D. Bowers, and F. A. Bazzaz. The effects of enriched carbon dioxide atmospheres on plant-insect herbivore interactions. *Science*, Vol. 243, p. 1199. Copyright © 1989 by the AAAS. Reprinted with permission. Figure 28.18 L. Roberts. How fast can trees migrate? *Science*, Vol. 243, p. 736. Copyright © 1989 by the AAAS. Reprinted with permission. Figure 28.19 S. P. Henderson, S. Hattersley, S. von Caemmer, and B. Osmond. Are C_4 pathway plants threatened by global climatic change? *Ecophysiology of Photosynthesis*, E. D. Schulze and M. M. Caldwell, eds., p. 269. Copyright © 1993 Springer-Verlag, Berlin. Reprinted with permission. Coral cross section and temperature plot for Essay 28.3 R. B. Dunbar and J. E. Cole, eds. *Annual Records of Tropical Systems (ARTS): Recommendations for Research*. Copyright © 1999 International Geosphere Biosphere Program, Past Global Changes.

TABLE CREDITS

Table 7.1 C. C. Black. Ecological implications of dividing plants into groups with distinct photosynthetic production capacities. *Advances in Ecological Research,* Vol. 7, p. 100. Copyright © 1971 Academic Press Ltd., London. Reprinted with permission. Table 10.1 J. N. M. Smith. Determinants of lifetime reproductive success in the song sparrow. *Reproductive Success,* T. H. Clutton-Brock, ed., p. 159. Copyright © 1988 University of Chicago Press, Chicago. Table 12.1 J. A. Wiens. *The Ecology of Bird Communities: Processes and Variation,* Vol. 2, p. 17. Copyright © 1989 Cambridge University Press, Cambridge. Table 14.1 P. D. Coley, J. P. Bryant, and F. S. Chapin, III. Resource availability and plant antiherbivore defense. *Science,* Vol. 230, p. 897. Copyright © 1985 by the AAAS. Reprinted with permission. Table 19.2 J. L. Bouzat, H. A. Lewin, and K. N. Paige. The ghost of genetic diversity past: historical DNA analysis of the greater prairie chicken. *American Naturalist,* Vol. 152. Copyright © 1998 the University of Chicago Press, Chicago. Reprinted with permission. Table 19.3 L. Rolstad, J. Consequences of forest fragmentation for the dynamics of bird populations: Conceptual issues and the evidence. *Biological Journal of the Linnean Society,* Vol. 42, p. 151. Copyright © 1991 The Linnean Society of London. . Reprinted by permission of Academic Press Ltd. Table 19.4 R. F. Noss. Corridors in real landscapes: a reply to Simberloff and Cox. *Conservation Biology,* Vol. 1, p. 160. Copyright © 1987 Blackwell Scientific Publications, Cambridge. Reprinted by permission of the Society for Conservation Biology and Blackwell Scientific Publications, Inc. Table 21.1 M. Huston and T. Smith. Plant succession: life history and competition. *American Naturalist,* Vol. 130, Tbl. 2, p. 170. Copyright © 1987 the University of Chicago Press, Chicago. Reprinted with permission. Table 24.3 A. Sih, P. Crowley, M. McPeek, J. Petranka, and K. Strohmeier. Predation, competition, and prey communities: A review of field experiments. *Annual Review of Ecology and Systematics,* Vol. 16, p. 289. Copyright © 1985 by Annual Reviews, Inc. Reproduced with permission, from the *Annual Review of Ecology and Systematics.* Table 25.1 C. B. Field, M. J. Behrenfeld, J. T. Randerson, and P. Falkowski. Primary production of the biosphere: integrating terrestrial and oceanic components. *Science,* Vol. 281, p. 239. Copyright © 1998 by the AAAS. Reprinted with permission. Table 26.3 D. Pauly and V. Christensen. Primary production required to sustain global fisheries. *Nature,* Vol. 374, p. 256. Copyright © 1995 Macmillan Magazines Ltd., London, England. Reprinted with permission. Table 26.4 J. A. Scott, N. R. French, and J. W. Leetham. Patterns of consumption in grasslands. *Perspectives in Grassland Ecology,* N. R. French, ed., p. 101. Copyright © 1979 Springer-Verlag, New York. Reprinted with permission. Tables 27.1 and 27.2 D. W. Cole and M. Rapp. Elemental cycling in forest ecosystems. *Dynamic Properties of Forest Ecosystems*, D. E. Reichle, ed., pp. 354, 357. Copyright © 1981 Cambridge University Press, Cambridge, England. Table 27.6 S. R. Carpenter, N. F. Caraco, D. L. Correll, R. W. Howarth, A. N. Sharpley, and V. H. Smith. Nonpoint pollution of surface waters with phosphorus and nitrogen. *Ecological Applications,* Vol. 8, p. 561. Copyright © 1998 *Ecological Applications.* Reprinted with permission.

PHOTO CREDITS

Unless otherwise indicated, all photos provided by the author.

PART OPENER PHOTO CREDITS

Part One: Francois Gohier / Photo Researchers, Inc. Part Two: Bruce Heinemann / Photodisc. Part Three: Geostock / Photodisc. Part Four: Amanda Clement / Photodisc.

COVER PHOTO CREDITS

Front cover from left to right: Tom McHugh / Photo Researchers Inc., Ander Gallant / Image Bank, Francois Gohier / Photo Researchers Inc., Ryan McVay / Photodisc, and Geostock / Photodisc.

Back cover from left to right: Amanda Clement / Photodisc, B & C Alexander / Photo Researchers Inc., Bruce Heinemann / Photodisc, Jeremy Woodhouse / Photodisc, and Joseph Van O's / Image Bank.

SPECIES INDEX

Page numbers followed by *f* and *t* indicate figures and tables, respectively.

A

Acacia spp.
    mutualism with ants, 239–241
    spines/thorns on, 242
Acacia collinsii, mutualism with ants, 240*f*
Acacia cornigera, mutualism with ants, 240*f*
Acacia tortilis, spines on, 242
Accipiter gentilis, predation on flocks, 231*f*
Achatina fulica, 351*f*
    control of, 351–352
Achillea spp., ecotypic races of, 96–97
Achillea borealis, ecotypic races of, 97
Achillea lanulosa, ecotypic races of, 96*f*, 97
African buffalo, mortality rate for, 137, 138*f*
African elephant, 473*f*
    as keystone species, 472–473
    overkill of, 361–362, 361*f*
    productivity of, 540–542, 541*t*
African finch, natural selection in, 21
African honey bee, distribution of, 32, 33*f*
Agriotes spp., population estimation for, 121–122
Agropyron boreale, as modular organism, 117*f*
Agrostis tenuis, heavy-metal tolerance of, 37–38, 37*f*
Alder, 416*f*
    in glacial moraine succession, 413–416, 415*f*
Alewife, and zooplankton community, 475
Alfalfa, pest control on, 347–349
Alfalfa weevil
    control of, 348–349
    life cycle of, 348–349, 348*f*
Alnus spp., in glacial moraine succession, 414–417, 416*f*
Alosa pseudoharengus, and zooplankton community, 475
Amazilia saucerottei, in montane forest guild, 470
Ambrosia dumosa, 196*f*
    competition/facilitation with, 194*f*, 195–196, 195*f*
American beech

codominance in, 428–429, 429*f*
    distribution of, 396*f*
        climate change and, 102–103, 103*f*, 601, 602*f*
American chestnut
    chestnut blight on, 44, 45*f*, 475
    distribution of, 44, 45*f*
    as dominant species, 475
American grape, genetic manipulation of, 342
Ammophila breviligulata, in sand-dune succession, 418
Anas discors, nest-site habitat selection by, 65, 66*f*
Andropogon scoparius, in sand-dune succession, 418
Andropogon virginicus, 421*f*
    in abandoned farmland succession, 421
    life history of, 421
Anolis spp.
    niche breadth of, species diversity and, 448, 448*f*
    and spider populations, 227
Anolis cybotes, niche breadth of, species diversity and, 448, 448*f*
Anolis marmoratus, niche breadth of, species diversity and, 448, 448*f*
Anopheles albimanus, habitat selection by, 58
Anopheles culicifacies, habitat selection by, 58
Anopheles gambiae, habitat selection by, 58
Anopheles maculatus, habitat selection by, 58
Anser caerulescens caerulescens
    natural selection in, 21, 22*f*
    in trophic cascades, 497–499, 500*f*
Ant
    in biological control, 5
    homopterous insects and, 254–255
    introduction of, 197–198
    latitude and biodiversity in, 438
    mutualism with plants, 239–241, 254–255
Antarctic beech, distribution of, 52, 52*f*, 53
Anthus pratensis
    habitat selection by, 60
Anthus trivialis, 61*f*
    habitat selection by, 60
Aonidiella aurantii

biological control of, 339–342, 340*f*–342*f*
    pesticide use on, 332, 332*f*
Aphid
    dispersal of, 55
    habitat recognition in, 63
    plant stress and, 251
Aphytis melinus, biological control with, 339–342, 340*f*–342*f*
Apis mellifera ligustica, interspecific competition with, 32
Apis mellifera scutellata, distribution of, 32, 33*f*
Apple tree, as allelopathic agent, 78, 79*f*
Arctic charr, temperature tolerance of, 35–36, 35*f*
Arctostaphylos spp., in heathland dynamics, 426
Arenaria laricifolia, as modular organism, 117*f*
Argas robertsi, and host reproduction, 264–265, 265*f*
Arisaema triphyllum, distribution of, 370*f*
Aspen, as modular organism, 117
    Aspen tortrix moth, population irruptions of, 250–251
Aster
    in abandoned farmland succession, 420
    life history of, 421
Aster ledophyllus, after volcanic eruption, 406
Aster pilosus, in abandoned farmland succession, 420
Atlantic cod, harvesting of, 321–323
Atlantic salmon, predation on, 216
Atriplex rosea
    leaf structure of, 99*f*
    photosynthetic production in, 99*f*
Atriplex triangularis
    leaf structure of, 99*f*
    photosynthetic production in, 99*f*
Azuki bean weevil, predation on, 215, 215*f*

B

Bacillus anthracis, anthrax caused by, 279
Bacillus thuringiensis (Bt), for insect-resistant crops, 343

Backswimmer, response to prey density, 223, 224*f*

Balanus balanoides
distribution of, 94, 105
interspecific competition with, 137

Baleen whale, harvesting of, 324, 324*f*

Barley, as smother crop, 78–79

Barnacle
distribution of, 94, 95*f*, 105
survivorship curves of, 137*f*

Barnacle goose, population growth in, 165

Bat, rabies in, 269, 270, 270*f*

Bay-breasted warbler
feeding positions of, 193*f*, 194
responses to prey abundance, 224, 225*f*

Bearberry, in heathland dynamics, 426

Beech, timberline for, 94

Bettongia lesueur, predation and distribution of, 74, 75*f*

Betula glandulosa, snowshoe hare preference for, 248

Betula pubescens, self-thinning line for, 298, 299*f*

Bison, brucellosis in, 268–269, 268*f*

Biston betularia, natural selection in, 17–18, 18*f*

Bitternut hickory, range of, 396*f*

Black-footed ferret, decline of, 373

Black noddy tern, resource partitioning in, 198*f*

Black oak
range of, 396*f*
in sand-dune succession, 419

Black-throated green warbler, feeding positions of, 193, 193*f*

Black walnut, as allelopathic agent, 78

Blackburnian warbler, feeding positions of, 193, 193*f*

Blowfly, control of, 343–344

Blue-green algae, in eutrophication, 525–526, 527*f*

Blue-grey gnatcatcher, realized feeding niche of, 191*f*

Blue-grey noddy tern, resource partitioning in, 198*f*

Blue tit, clutch size in, 24, 24*f*

Blue whale, harvesting of, 324

Blue wildebeest, grazing by, 243*f*

Blue-winged teal, nest-site habitat selection by, 66, 66*f*

Bluegill sunfish, catch curve for, 129, 129*f*

Bonas umbellus, dispersal of, 48–49

Bracken fern, patch dynamics with, 427, 428*f*

Brain coral, 117*f*

Branta bernicla bernicla, population growth in, 165, 165*f*

Brassica oleracea var. *acephala*, in food web, 469, 470*f*

Brent goose, population growth in, 165, 165*f*

Brevicoryne brassicae, in collardstands, 469, 470*f*

Broken-striped newt, as keystone predator, 473

Bromus mollis, competition with, 198–199

Broomsedge, 421*f*
in abandoned farmland succession, 420
life history of, 420

Brown-headed cowbird, nest parasitism by, 26

Brown noddy tern, resource partitioning in, 198*f*

Brown-tail moth, control of, with parasites, 282

Brucella abortus, brucellosis caused by, 267–268

Bryozoan, marine, spines in, 241–242, 241*f*

Bt. *See Bacillus thuringiensis*

Buckeye butterfly, carbon dioxide enrichment and, 600*f*

Bufo americanus, predation on, 473

Burroweed, 196*f*
competition/facilitation with, 194*f*, 195–196, 195*f*

Burrowing rat kangaroo, distribution of, predation and, 74, 75*f*

Buteo, habitat selection by, 60

Buteo jamaicensis
habitat selection by, 60
prey quality in predation by, 229, 229*t*

Butterfly
food sources and distribution of, 76*f*, 77
in fragmented habitats, 369, 369*f*
metapopulations of, 289–290
temporary habitats of, 375

Butterfly fish, feeding specializations of, 488, 488*t*

C

Cabbage aphid, in collardstands, 469, 470*f*

Cacospongia spp., effects of allelopathy on, 80, 80*f*

Cactoblastis cactorum, biological control with, 334–335

Calamovilfa longifolia, in sand-dune succession, 418–419

Calandra spp., population growth in, 163–164

Calandra oryzae
capacity for increase in, 142–144, 143*f*, 144*f*
population growth in, 164
competition and, 188*f*, 189, 189*f*

Calanoid copepod, diversity gradient for, 440, 440*f*

California red scale
biological control of, 339–342, 340*f*–342*f*
pesticide use on, 332, 332*f*

California sea otter, dispersal of, 45–46, 46*f*

Callosobruchus chinensis, predation on, 215, 215*f*

Calluna spp., patch dynamics with, 426–427

Calluna vulgaris, 426*f*

Canada lynx, density cycles in, 226–227, 227*f*

Canis familiaris dingo, predation by, 218–219, 219*f*

Cape may warbler, feeding positions of, 193*f*, 194

Capelin, 153*f*
reproduction in, 152–153

Caretta caretta, population dynamics for, 174*t*, 175–176, 175*f*

Caribou, abundance of, predation and, 220

Carpodacus mexicanus, competition with, 205

Carrion fly, reproductive rate of, 3

Carya cordiformis, range of, 396*f*

Carya ovata, range of, 396*f*

Castanea dentata
chestnut blight on, 44, 45*f*, 475
distribution of, 44, 45*f*
as dominant species, 475

Castanea mollissima, resistance to chestnut blight in, 44

Cat, feral
control of, 374
petrel predation by, 374

Cattle, brucellosis in, 268

Cattle egret, tick infestations on, 264–265, 265*f*

Centrostephanus rodgersii, and kelp populations, 73

Ceratitis capitata, mortality rates for, 138, 138*f*

Chaetodipus (Perognathus) fallax, habitat selection by, 61, 64*f*

Chamaecyparis lawsoniana, gradient analysis of, 393

Cheetah, interspecific competition with, 84

Chenopodium album, self-thinning line for, 298*f*

Chestnut blight
dispersal of, 44
resistance to, 44

Chickweed, control of, with smother crop, 79

Chinese chestnut, resistance to chestnut blight in, 44

Chionochloa spp., grazing on, 249

Chionoecetes opilio, harvesting of, 320

Chipping sparrow, habitat selection by, 61

Chlorostilbon canivetti, in montane forest guild, 470

Cholera bacterium, marine dispersal of, 47

Choristoneura fumiferana
abundance of, predation and, 224, 225*f*
population irruptions of, 250, 251*f*

Chrysaora quinquecirrha, in food chain, 464, 465*f*

Chrysolina quadrigemina, host plant and distribution of, 76–77

Chthamalus stellatus

distribution of, 94
   survivorship curve of, 137f
Cichlid fish, extinction of, 372
Cladonia spp., in heathland dynamics, 426
Clam, zebra mussels and, 42
Clark's nutcracker, seed dispersal by, 254
Coccus viridis, mutualism with ants, 255, 255f
Coho salmon, environmental variables in
   production of, 315, 315f
Colibri thalassinus, in montane forest
   guild, 470
Collard, in food web, 469, 470f
Columba palumbus, value of flocking for, 231f
Common crossbill, 82f
   diet of, 82
      and population density, 252,
         253, 253f
   emigration by, 252–253, 252f
Common dandelion, dispersal of, 54
Common grackle, range size for, 112f, 113
Common mussel, distribution of, 71, 72f
   predation and, 71–72, 73f
Common oak, defenses against herbivores,
   238–239, 239f, 240t
Coral
   bleaching of, 489, 489f
   climate change and, 605
   coverage by, disturbance and,
      486–488, 487f
   recruitment rate of, 487, 488f
Coral snake, warning coloration of, 230f, 231
Coralline algae, competition of, 492, 493f
Corixa falleni, dispersal of, 54
Corixa nigrolineata, dispersal of, 54
Corixidae, dispersal of, 54
Cormorant, competition with, 192
Cornus controversa, seed dispersal by, 48f
Cotton, selective breeding of, 342
Cottonwood, in sand-dune succession, 419
Cottony-cushion scale, 333f
   biological control of, 333–334
Cowbird, nest parasitism by, 26
Coyote, and fox populations, 218
Crane fly, 54f
   flightless, 53, 54f
Cricket frog, interaction with keystone
   species, 473
Crossbill, competition among, and diet
   shifts, 81–82
Cryphonectria parasitica
   dispersal of, 44
   resistance to, 44
Cryptochaetum iceryae, biological control
   with, 333
Cuckoo, nest parasitism by, 26
Culex pipiens quinquefasciatus, sterilization
   of, 343–344, 344t
Culex quinquefasciatus, avian malaria
   transmission by, 77–78

Cyrtobagous salviniae, biological control
   with, 337
Cyrtobagous singularis, biological control
   with, 337
Cytisus scoparius, cottony-cushion scale
   on, 334

D

Dactylopius indicus, biological control
   with, 334
Damselfish, recruitment levels for,
   population size and, 490–491, 490f, 491f
Daphnia spp., population growth in, 167,
   167f, 169, 170f
Daphnia magna, population growth in, 170f
Daphnia rosea, population density in, 167f
Deer mouse, habitat selection by, 61–62, 65t
Dendroica spp., competition among, feeding
   positions and, 193–194, 193f
Dendroica castanea
   feeding positions of, 193f, 194
   responses to prey abundance, 224, 225f
Diabrotica undecimpunctata, habitat selection
   by, 58
Diamondback moth, resistance to Bt, 343
Didinium nasutum, predation by,
   212–213, 213f
Dingo, predation by, 218–219, 219f
Dipodomys spp., as keystone guild, 473
Dipodomys merriami, habitat selection by,
   61–62, 63f
Dog, rabies in, 270
Dogwood, seed dispersal by, 48f
Dragonfly, habitat selection by, 58
Dreissena polymorphis, dispersal of, 41–42, 42f
Dromaius novaehollandiae, predation and
   abundance of, 219
Drosophila melanogaster, population growth
   in, 163, 164f
Drosophila pachea
   distribution of, 76
   food source of, 76
Drosophila serrata, capacity for increase in,
   145, 146f
Dryas integrifolia, 415f
Duck, predation on, 217–218, 218f
Dwarf birch, snowshoe hare preference
   for, 248
Dysidea spp., allelopathic effects of, 80

E

Eastern hare wallaby, 373f
Eastern meadowlark
   distribution and abundance of, 7–8, 8f
   range size for, 112f, 113
Eastern phoebe, distribution of,
   temperature and, 90, 91f
Eastern yellow robin, habitat fragmentation
   and, 367
Eland, production efficiency of, 555–556

Elephant, and multiple stable states, 509
Elminius modestus, distribution of, 105
Empoasca fabae, control of, 343
Emu, abundance of, predation and, 219
Engelmann spruce, age structure of,
   48, 150f
English sole, recruitment in, 315–316, 316f
English walnut, lack of toxic chemicals
   from, 78
English yew, distribution of, 84
Engraulis ringens, yields in harvest of,
   308–309, 308f, 309f
Enhydra lutris, dispersal of, 45–46, 46f
Eotetranychus sexmaculatus, predation on,
   214, 214f, 215f
Epilobium angustifolium, after volcanic
   eruption, 406
Erigeron canadensis, 421f
   in abandoned farmland succession,
   419–421
Erodium botrys, competition with, 198–199
Escherichia coli, competition in, 196
Eucalyptus
   herbivory on, 257
   nutrients in, 568–569
Euglandina rosea, introduction of, 352
Euphausia superba, 324f
   harvesting of, 324–325, 325f
Euptoieta claudia, in fragmented habitats,
   369, 369f
European corn borer, habitat selection by, 58
European cuckoo, nest parasitism by, 26
European grape, genetic manipulation of,
   for pest control, 342
European rabbit
   myxomatosis in, 272–275, 274f,
      75f, 277
   and grassland succession, 424
   sterilization of, 344–346, 346f, 347f
European robin, life expectation for, 128

F

Fagus grandifolia
   codominance in, 428, 429f
   distribution of, 396f
      climate change and, 102–103, 103f,
         601, 602f
Fairy tern, resource partitioning in, 198f
Felis concolor, corridors for, 368–369
Feral cat
   control of, 374
   petrel predation by, 374
Ferruginous hawk, habitat selection by, 60
Fescue grass, as modular organism, 117f
Festuca brachyphylla, as modular
   organism, 117f
Festuca ovina, in grassland cyclic changes, 427
Fin whale, harvesting of, 324, 325f
Finch
   character displacement in, 201–202, 202f

mass emigration by, 252–253
natural selection in, 19, 21
noninteractive grazing by, 251–253
population irruptions of, 252–253
Floating fern, biological control of, 336–337, 338f
Florida panther, corridors for, 369
Fox, rabies in, 269, 270f
Fruit fly
    fertility and survival of, 145, 146f
    mortality rates for, 138, 138f
Fruit pigeon, checkerboard distribution of, 80, 81f
Furbish's lousewort
    conservation of, 377–379
    geographic range of, 377, 377f
    population growth for, 377–378, 378f
    probabilities of survival for, 378, 380t

G

Gadus morhua, harvesting of, 321–323
Gal·pagos ground finch
    character displacement in, 201, 202f
    natural selection in, 19, 20f
Geospiza fortis
    character displacement in, 201, 202f
    natural selection in, 19, 20f
Geospiza fuliginosa, character displacement in, 202f
Geospiza magnirostris, character displacement in, 202f
Glanville fritillary butterfly, metapopulations of, extinction of, 290–292, 291f
Glaucomys sabrinus, owl predation on, 379
Glaucomys volans, competition with, 362
Goose
    natural selection in, 21, 22f
    population growth in, 165
Gorgonid coral, as modular organism, 118f
Goshawk, predation on flocks, 231f
Grain beetle, competition and population growth in, 188f, 189, 189f
Grape phylloxera, control of, 342
Grasshopper, grass consumption by, 246
Great auk, 362f
    extinction of, 361
Great-tailed grackle, sex ratio for, 130
Great-winged petrel, predation on, 374
Green scale, mutualism with ants, 255, 255f
Grey seal, phocine distemper virus in, 267
Grizzly bear, legal *vs.* biotic boundaries for, 375, 376f
Grus americana, population growth in, 166f, 167
Guppy, in laboratory studies on harvesting, 317, 317f, 318f
Gypsy moth, 44f
    control of, with parasites, 282
    dispersal of, 42–44, 43f

H

Harbor seal, phocine distemper virus in, 265–267, 266f
Harp seal
    cod predation by, 323
    phocine distemper virus in, 267
Harpagornis moorei, extinction of, 373
Heather, patch dynamics with, 426–427
Hemlock, shade tolerance of, 97–98
Herpestes auropunctatus, 351f
    introduction of, 351
Heterospilus prosopidis, predation by, 215, 215f
Himalayan thar, population irruptions of, 249
Honey bee, decline of, 606
Hordeum vulgare, as smother crop, 78–79
Horseweed, 421f
    in abandoned farmland succession, 419–421
House finch, competition with, 205
House mouse, trophic level of, 544
House sparrow
    competition with, 205
    population growth rates of, 300, 300f
House wren, clutch size in, 25, 25f
Hummingbird, montane forest guild of, 470
Hyla crucifer, interaction with keystone species, 473
Hypera postica
    control of, 347–349
    life cycle of, 347–349, 348f
Hypericum perforatum, control of, 76

I

Icerya purchasi, 333f
    biological control of, 333–334
Indian fleabane, mutualism with ants, 255, 255f
Indian mongoose, 351f
    introduction of, 351
Italian honey bee, interspecific competition with, 32

J

Jack pine
    postfire nutrient cycle in, 562–563, 563f
    in sand-dune succession, 419
Juglans californica, lack of toxic chemicals from, 78
Juglans hindsii, lack of toxic chemicals from, 78
Juglans nigra, as allelopathic agent, 78
Juglans regia, lack of toxic chemicals from, 78
Juncus effusus, community composition with, 389, 390f
Junonia coenia, carbon dioxide enrichment and, 600f

K

Kangaroo, distribution of abundance and, 8, 9f
    predation and, 74
Kangaroo rat
    habitat selection by, 61–62, 64f
    as keystone guild, 473
Killer whale, and sea otter decline, 472
King crab
    harvesting of, 319–321
    life cycle of, 320, 321f
Klamath weed, control of, 76
Krill, 324f
    harvesting of, 324–325, 325f

L

Ladybird beetle
    biological control with, 333–334
    melanic frequency of, 29
Lagorchestes leporides, 373f
Lake trout
    acid rain and, 576, 577f
    predation on, 220–222, 221f
Larus cachinnans, habitat selection by, 67–68, 68f
Laurel tree, mortality rates for, biodiversity gradients and, 450, 450f
Leaf beetle, habitat selection by, 58
Lebistes reticulatus, in laboratory studies on harvesting, 317, 317f, 318f
Lepidoptera spp.
    feeding on common oak by, 238, 238f, 240t
    relative abundance of, 435f
Lepomis macrochirus, catch curve for, 129, 129f
Leporipoxvirus spp., myxomatosis caused by, 272–275
Lepus americanus
    in boreal forest food chain, 462, 464f
    density cycles in, 226–227
    selective feeding by, 248, 248f
Lesser snow goose
    natural selection in, 21, 22f
    in trophic cascades, 497–499, 500f
Lichen, in heathland dynamics, 426
Linepithema humile, introduction of, 198, 199f
Little bluestem, in sand-dune succession, 418
Littorina littorea, and algae diversity, 453, 453f
Lizard
    effect on spider populations, 227
    lizard malaria in, and reproduction, 264, 264f
Lobelia deckenii deniensis, reproduction in, 152
Lobelia telekii, 152f
    reproduction in, 151–152

Loblolly pine
   climatic limits for, 92f, 93
   distribution of, 92–93, 92f
Locust
   biological control of, 5
   plagues of, 3
Loggerhead sea turtle, population
   dynamics for, 174t, 175–176, 175f
Lophocereus schottii, and fruit fly
   distribution, 76
Loxia curvirostra, diet of, 82
Lupine, after volcanic eruption, 406
Lupinus lepidus, after volcanic eruption, 406
Lymantria dispar, 44f
   control of, with parasites, 282
   dispersal of, 42–44, 43f
Lynx canadensis, density cycles in,
   226–227, 227f
Lyssavirus spp., rabies caused by, 269

M

Macropus sulinginosus, abundance of, 110f
Mallard, habitat selection by, 68
Mallotus villosus, 153f
   reproduction in, 152–153
Marram grass, in sand-dune succession, 418
Meadow pipit
   habitat selection by, 60
Melitaea cinxia, metapopulations of,
   extinction of, 290–292, 291f
Membranipora membranacea, inducible
   spines on, 241, 241f
Mephitis mephitis
   predation by, 217, 218f
   rabies in, 269
Metepeira datona, population density of,
   predation and, 228f
Miconia argentea, in gaps, 428–429
Microtus agrestis, age distribution in, 148f
Micrurus spp., warning coloration of, 230f
Micrurus diastema, warning coloration
   of, 230f
Micrurus elegans, warning coloration of, 230f
Mink, muskrat predation by, 218
Moose
   abundance of, predation and, 219–220
   plant preferences of, 251
Morone saxatilis, in food web, 465f
Mosquito
   distribution of, 50
   habitat selection by, 58
   malaria transmission by, 5
      to birds, 77–78
Moth, distribution and abundance
   of, 111, 111f
Mountain birch, self-thinning line
   for, 298, 299f
Mouse, plagues of, 3
Mouse typhoid bacterium, evolution of,
   276f, 277

Mus musculus, trophic level of, 544
Muskrat, predation on, 218
Mynah bird, as biological control, 5
Myopus schisticolor, sex ratio for, 130
Myrtle warbler, feeding positions
   of, 193f, 194
Mytilus spp., predation by, and
   biodiversity, 449
Mytilus californianus, predation by, and
   biodiversity, 471
Mytilus edulis, distribution of, 71, 72f
   predation and, 71–72, 73f
Myzus persicae, habitat recognition in, 63–65

N

Nile perch, introduction of, 372
North Sea haddock, recruitment in, 313
Northern cod
   age structure of, 323, 323f
   collapse of, 323
   harvesting of, 321–323, 321f, 322f
Northern flying squirrel, owl predation
   on, 379
Northern red oak, range of, 396f
Northern spotted owl
   conservation of, 379–380
   diet of, 379–380
   home range of, 379–380, 379f
   importance of habitat for, 381, 381f
Norway spruce
   acid rain on, 576
   nitrogen cycle in, 563, 565f
   primary production by, 532
Nothofagus spp.
   distribution of, 52, 52f, 53
   timberline for, 94
Notonecta hoffmanni, response to prey
   density, 223, 224f
Notophthalmus viridescens, as keystone
   predator, 473
Nucifraga columbiana, seed dispersal
   by, 254

O

Oak, in abandoned farmland succession,
   421–423, 422f
Ocotea whitei, mortality rates for,
   biodiversity gradients and, 450, 450f
Odocoileus virginianus
   immunocontraception for, 345, 345f
   and multiple stable states, 509
Ondatra zibethicus, predation on, 218
Opuntia spp.
   biological control of, 335–336, 336f
   distribution in Australia, 334f, 335
   photosynthesis in, 101
Opuntia stricta
   biological control of, 335, 336f
   spines on, 242, 242f
Opuntia vulgaris, biological control of, 334

Organ-pipe gorgonian coral, as modular
   organism, 118f
Oryctolagus cuniculus
   myxomatosis in, 272–275, 274f,
      275f, 277
   and grassland succession, 424
   sterilization of, 346–347, 346f, 347f

P

Pacific madrone, gradient analysis of, 393
Pacific salmon
   distribution of, 32
   habitat recognition by, 65
   harvesting of, 318
Paralithodes camtschatica
   harvesting of, 319–321, 320f
   life cycle of, 320, 321f
Paramecium aurelia
   coexistence in, 190
   population growth in, 163, 163f
Paramecium bursaria, coexistence in, 190
Paramecium caudatum
   population growth in, 163, 163f
   predation on, 212–213, 213f
Paraphrys vetulus, recruitment
   in, 315–316, 316f
Parrot crossbill, 82f
   diet of, 82
Parus caeruleus, clutch size in, 24, 24f
Passer domesticus
   competition with, 205
   population growth rates of, 300, 300f
Pasteurella multocida, avian cholera caused
   by, 279
Pedicularis furbishiae
   conservation of, 377–379
   geographic range of, 377, 377f
   population growth for, 377–378, 378f
   probabilities of survival for, 378, 380t
Peppered moth, natural selection
   in, 17–18, 18f
Periwinkle snail, and algae diversity, 453, 453f
Perognathus fallax, habitat selection
   by, 61, 64f
Peromyscus maniculatus, habitat selection
   by, 61
Peromyscus maniculatus bairdi, habitat
   selection by, 61–62, 65t
Peruvian anchovy, yields in harvest of,
   308–309, 308f, 309f
Petrel, recovery of, 374
Petrogale lateralis, red fox predation
   and, 74, 74f
Petromyzon marinus, introduction of,
   220–222, 221f
Phalacrocorax aristotelis, 192f
   competition with, 192
Phalacrocorax carbo, competition with, 192
Pheidole megacephala, mutualism with,
   255, 255f

Philodice bryantae, in montane forest guild, 470

Phoca groenlandica
cod predation by, 323
phocine distemper virus in, 267

Phytophthora infestans, control of, 343

Picea abies
acid rain on, 576
nitrogen cycle in, 563, 565f
primary production by, 532

Picea sitchensis, 417f
in glacial moraine succession, 415–417, 416f

Pilocercus elaphoides, warning coloration of, 230f

Pinguinis impennis, 362f
extinction of, 361

Pinus spp., gradient analysis of, 393, 393f

Pinus albicaulis, seed dispersal in, 254

Pinus banksiana
postfire nutrient cycle in, 562–563, 563f
in sand-dune succession, 419

Pinus echinata, in abandoned farmland succession, 421–422, 422f

Pinus pungens, gradient analysis of, 393, 393f

Pinus rigida, gradient analysis of, 393, 393f

Pinus taeda
climatic limits for, 92f, 93
distribution of, 92–93, 92f

Pinus virginiana, gradient analysis of, 393, 393f

Pisaster spp.
and biodiversity, 448–449
food web of, 449f

Pisaster ochraceous, as keystone species, 471

Pitch pine, gradient analysis of, 393, 393f

Plaice
stock and recruitment relationship in, 313–314, 314f
yields in harvest of, 310–312, 310f, 311t

Plantago lanceolata, carbon dioxide enrichment and, 600f

Plantago maritima, genetic variation within, 94–96

Plasmodium mexicanum, effect on reproduction, 264, 264f

Plethodon glutinosus, distribution of, competition and, 80–81

Plethodon jordani, distribution of, competition and, 80–81

Pleuronectes platessa
stock and recruitment relationship in, 313–314, 314f
yields in harvest of, 310–312, 310t, 311f

Pliocercus spp., warning coloration of, 230f

Pluchea indica, mutualism with ants, 255, 255f

Pocket mouse, habitat selection by, 63f, 64

Polioptila caerulea, realized feeding niche of, 191f

Pomacentrus amboinensis, recruitment levels for, 490, 490f

Pomacentrus moluccensis, recruitment levels for, 490–491, 491f

Populus deltoides, in sand-dune succession, 419

Populus tremuloides, as modular organism, 117

Porcellio scaber, habitat selection by, 57–58

Port Orford cedar, gradient analysis of, 393

Potato leafhopper, control of, 343

Prairie chicken, decline of, 357–358, 358f, 358t

Prairie dog, decline of, 373

Prickly pear
biological control of, 335–336, 336f
distribution in Australia, 335f, 335
photosynthesis in, 101

Prunus pumila, in sand-dune succession, 419

Pseudomyrmex spp., mutualism with acacias, 239–241

Pteridium aquilinum, patch dynamics with, 427, 428f

Pterodroma macroptera, predation on, 374

Ptilinopus pivoli, checkerboard distribution of, 80, 81f

Ptilinopus solomonensis, checkerboard distribution of, 80, 81f

Q

Quercus alba, range of, 396f

Quercus borealis, range of, 396f

Quercus robur, defenses against herbivores, 238–239, 239f, 240t

Quercus veluntina
range of, 396f
in sand-dune succession, 419

Quiscalus mexicanus, sex ratio for, 130

R

Raccoon, rabies in, 269, 270, 270f

Rana utricularia, predation on, 473

Red-cockaded woodpecker
habitat destruction and, 362
life cycle of, 362, 363f

Red crossbill, diet of, 82

Red deer, reproductive costs for, 153, 153f

Red fox
duck predation by, 218
extinction caused by, 372
rabies in, 269, 270–272, 271f, 273f
trophic level of, 544

Red kangaroo
abundance of, 8, 9f
predation and, 218–219, 219f
distribution of, 8, 9f

Red king crab
harvesting of, 319–321, 320f
life cycle of, 320, 321f

Red locust, biological control of, 5

Red squirrel
in boreal forest food chain, 462, 464f
distribution of, 364, 366f

Red-tailed hawk
habitat selection by, 60
prey quality in predation by, 229, 229t

Red-tide dinoflagellates, marine dispersal of, 47

Reindeer, population growth in, 165–167, 165f
irruptions in, 249, 249f

Rhizopertha dominica
capacity for increase in, 143–144, 144f
competition and population growth in, 188f, 189, 189f

Rock wallaby, red fox predation and, 74, 74f

Rodolia cardinalis, 334f
biological control with, 333–334

Ruffed grouse, dispersal of, 48–49

Rumex crispus var. *littoreus*, distribution of, 50

S

Saccharomyces spp., competition and population growth in, 187–189, 87f, 188f

Saccharomyces cervisiae, competition and population growth in, 187–189

Salamander, distribution of, competition and, 80–81

Salix spp.
in sand-dune succession, 418–419
temperature tolerance of, 90

Salix glauca, in snowshoe hare diet, 248

Salmo fario, population estimation for, 122–123

Salmo salar, predation on, 216

Salmon, distribution of, 32

Salmonella typhimurium, evolution of, 276f, 277

Salvelinus alpinus, temperature tolerance of, 35–36, 35f

Salvinia auriculata, predators on, 337

Salvinia molesta, biological control of, 336–337, 338f

Sand cherry, in sand-dune succession, 419

Sand reed grass, in sand-dune succession, 418–419

Sandwort, as modular organism, 117f

Sargassum spp., adaptations to light by, 98–99, 98f

Sayornis phoebe, distribution of, temperature and, 90, 91f

Schizosaccharomyces kephir, competition and population growth in, 187–188, 187f, 188f

Sciurus vulgaris
in boreal forest food chain, 462, 464f
distribution of, 364, 366f

Scleroporus occidentalis, lizard malaria in, and reproduction, 264, 264f

Scotch broom, cottony-cushion scale on, 334

Screwworm, control of, 343–344

Sea lamprey, introduction of, 220–222, 221f

Sea nettle, in food chain, 464, 465f

Sea otter
decline of, 471–472
as keystone predator, 471–472, 472f

Sea urchin, 73*f*
   distribution of, and algae populations, 72–73
Sei whale, harvesting of, 324
Senita cactus, and fruit fly distribution, 76
Shag, 192*f*
   competition with, 192
Shagbark hickory, range of, 396*f*
Shortleaf pine, in abandoned farmland succession, 421–422, 422*f*
Shrew, incidence functions for, 364, 365*f*
Sitka spruce, 417*f*
   in glacial moraine succession, 415–417, 414*f*
Skunk, rabies in, 269, 270*f*
Smilacina racemosa, distribution of, 370*f*
Snake, warning coloration in, 230*f*, 231
Snow crab, harvesting of, 320
Snow tussocks, grazing on, 249
Snowshoe hare
   in boreal forest food chain, 462, 464*f*
   density cycles in, 226–227
   selective feeding by, 248, 248*f*
Sockeye salmon, stock and recruitment relationship in, 314*f*
Song sparrow, cohort life table for, 134*t*
Sooty tern, resource partitioning in, 198*f*
Sorex araneus, incidence functions for, 364, 365*f*
Sorex caecutiens, incidence functions for, 364, 365*f*
Southern corn rootworm, habitat selection by, 58
Southern flying squirrel, competition with, 362
Soybean, pest control on, 343
Sperm whale, harvesting of, 324
Sphagnum spp.
   in glacial moraine succession, 415
   prevalence of, 417
Spider, population density of, predation and, 227
Spizella passerina, habitat selection by, 61
Spruce, nitrogen cycle in, 563, 565*f*
Spruce budworm
   abundance of, predation and, 224, 225*f*
   population irruptions of, 250, 251*f*
Stellaria media, control of, with smother crop, 79
Striped bass, in food web, 465*f*
Striped skunk
   predation by, 217, 218*f*
   rabies in, 269
Strix aluco, population density for, 226, 227*f*
Strix occidentalis caurina
   conservation of, 379–380
   diet of, 379–380
   home range of, 379–380, 379*f*
Sturnella magna
   distribution and abundance of, 7–8, 8*f*
   range size for, 112*f*, 113
Subalpine fir, age structure of, 148, 150*f*

Sugar maple, dominance of, 428, 429*f*, 474
Swainson's hawk, habitat selection by, 60
Sylvilagus brasiliensis, myxomatosis in, 272
Syncerus caffer, mortality rate for, 137, 138*f*

T
Table-mountain pine, gradient analysis of, 393, 393*f*
Taraxacum officinale, dispersal of, 54
Taurotragus oryx, production efficiency of, 555–556
Tawny owl, population density for, 226, 227*f*
Taxus baccata, distribution of, 84
Tern, resource partitioning by, 197, 198*f*
Tetraphis spp., distribution of, hierarchy of scales for, 107*f*
Thomson's gazelle
   grazing by, 242–246, 244*f*
   population trends for, 245–246, 246*f*
Tiger beetle, as indicator species, 399*f*, 400
Tipula subaptera, 54*f*
Tipula trivittata, 54*f*
Tree pipit, 61*f*
   habitat selection by, 60
Trema micrantha, in gaps, 428–429
Tribolium spp., population growth in, 163–164, 164*f*
Tribolium castaneum, population growth in, 164*f*
Trichostrongylus tenuis, 279
Troglodytes aedon, clutch size in, 25, 25*f*
Trout, population estimation for, 122–123
Tympanuchus cupido, decline of, 357–358, 357*f*, 358*t*
Typhlodromus occidentalis, predation by, 214, 214*f*, 215*f*

U
Ursus arctos, legal *vs*. biotic boundaries for, 375, 376*f*

V
Vedalia, 334*f*
   biological control with, 333–334
Vibrio cholerae, marine dispersal of, 47
Virginia pine, gradient analysis of, 393, 393*f*
Vitis labrusca, genetic manipulation of, 342
Vitis vinifera, genetic manipulation of, for pest control, 342
Vulpes fulva
   duck predation by, 218
   extinction caused by, 372
   rabies in, 269, 270–272, 271*f*, 273*f*
   trophic level of, 544

W
Walnut tree, secondary plant substances of, 235
Warbler, competition among, feeding positions and, 193–194, 193*f*

Water boatman, dispersal of, 54
Western fence lizard, lizard malaria in, and reproduction, 264, 264*f*
Western gray kangaroo, abundance of, 110*f*
Whale, harvesting of, 324–325
Wheatgrass, as modular organism, 117*f*
White-crowned sparrow
   habitat selection by, 61, 62*f*
   song dialect populations of, 61–62, 61*f*
White-fronted goose, population growth in, 165
White oak, range of, 396*f*
White pine, in sand-dune succession, 419
White tailed deer
   immunocontraception for, 345, 345*f*
   and multiple stable states, 509
White-winged crossbill, 82*f*
   diet of, 81–82
Whitebark pine, seed dispersal in, 254
Whooping crane, population growth in, 166*f*, 167
Wildebeest
   grazing by, 242–246, 243*f*–245*f*, 247
   and multiple stable states, 509
   population boundaries for, 130
   population trends for, 245–246, 246*f*
Willow
   in sand-dune succession, 418–419
   temperature tolerance of, 90
Winter moth
   density dependence in, 295, 296*f*
   feeding on oak by, 238–239
   key factor analysis of, 295*f*, 295
   life table for, 294–295, 294*t*
Wireworm, population estimation for, 121–122
Wolf
   predation by, 219–220
   response to prey density, 222–223, 222*f*, 223*f*
Wood lemming, sex ratio for, 130
Wood pigeon, value of flocking for, 231*f*
Woodrat, owl predation on, 380

Y
Yarrow, ecotypic races of, 96–97
Yeast, competition and population growth in, 187–189, 187*f*
Yellow-legged gull, habitat selection by, 67–68, 68*f*

Z
Zebra
   grazing by, 242–246, 244*f*, 245*f*
   population trends for, 245–246, 246*f*
Zebra mussel, dispersal of, 41–42, 42*f*
Zonotrichia leucophrys nuttali
   habitat selection by, 61, 62*f*
   song dialect populations of, 61, 62*f*

SUBJECT INDEX

Page numbers followed by *f*, *n*, and *t*
  indicate figures, footnotes, and tables,
  respectively.

A

α-selection, 200
Abandoned farmland, succession on,
  419–422, 422*f*, 423*f*
Abortion, contagious, 268–269
Absolute density, measurement of, 120–125
Abundance
  in biodiversity measurement, 435–438
  body size and, 119–120, 120*f*, 549
  in definition of ecology, 2
  distribution and, 7–9, 111–113
  intrinsic capacity for increase in, 144
  of keystone species, 473
  population regulation and, 300
  of prey, 216–222
    predator response to, 222–224, 222*f*
  range size and, 111–113
  during succession, 410–411, 412*t*
  variation in, 280, 280*f*, 282, 282*f*, 300
Acid-bog lakes, bound nutrients in, 525
Acid rain, 572–576
  distribution of, 572, 572*f*
  nitric acid and, 578
  recovery from, 575
  sulfur cycle and, 572–576
Actual evapotranspiration
  in climate classification, 88
  definition of, 88, 528
  and primary production, 528
Adaptation(s), 18–25
  of bird clutch size, 22–25
  imperfection of, 97–98
  limitations on, 21–22
  natural selection in, 18–21
Adaptive value (fitness), 19, 611
Additive mortality, *vs.* compensatory
  mortality, 292–293, 293*f*
Additivity, in population regulation, 283
Africa, multiple stable states in, 508–509, 508*f*
Age(s)
  distribution of, 146–149
    stable, 140–141, 141*f*, 142*f*, 147–148
    stationary, 147–148

  in life tables, 133–138
  and primary production, 532
  variation within populations, 130–131
Aggregative response, of predators to prey
  density, 222
Agriculture
  as applied ecology, 3
  genetic manipulation in, 342–343
  habitat reduction for, 363, 364*f*
  loss to pests in, 331, 331*f*
Alaska, glacial moraine succession in,
  413–416
Algae
  predation on
    and distribution, 72–73
    and diversity, 453, 453*f*
Alkaloids, incidence in plants, 257
Allee effect, 292
Allelopathy
  and distribution, 78–80
  for weed control, 78–79
Alpha-selection, 200
α (within-habitat) diversity, 445–446, 447
Alpine ecosystems, timberlines in, 93–94
Altitude, of timberlines, 93, 93*f*
Altruism, in kin selection, 27
Amphibians, species-area curve
  for, 503, 503*f*
Anderson, Roy, 259, 259*f*
Andrewartha, H. G., 106, 106*f*, 285–286
Animal ecology, *vs.* plant ecology, 11
Animal Ecology (Elton), 2
Antarctic marine food chain, 462*f*, 463
Anthrax, 279
Aposematic (warning) coloration, 229–231
Apparency theory of plant defense,
  236–237
Apparent competition, 203
Aquatic communities. *See also specific types*
  dominant species in, 475
  food web in, 462, 463*f*, 477
  keystone species in, 471–473
  primary production in
    limitations on, 518–526
    measurement of, 516
  restoration of, 480–482
  trophic efficiencies for, 545–546, 545*f*

Arctic communities, climate change in,
  594, 595*f*
Aristotle, 3, 268
Arms race analogy for coevolution, 26
  for host-parasite systems, 276
  for predator-prey systems, 228–229
Artifacts, as index of relative density, 127
Assimilation rate, measurement of, 538
Australia
  mammal extinctions in, 372, 373*f*
  marsupial biodiversity in, 440, 441*f*
Autotrophs, 513
Avian cholera, 279

B

Baikal, Lake, biodiversity in, 444
Bailey, V., 284
Balance of nature
  vs. evolutionary theory, 4
  in understanding of ecology, 3, 282
Ballast water, marine dispersal with, 47
Barriers
  to distribution, 53
  to jump dispersal, 48–49, 52
Barro Colorado Island, cyclic changes in,
  428–429, 430*f*, 430*t*
Basal metabolic rate, body size and, 538
Basal species, 465*t*, 468
Batesian mimicry, 231
Bazzaz, F. A., 593, 593*f*
Behavioral ecology, 9–10, 11
Between-habitat (β) diversity, 445
Beverton, Ray, 309
Bible, plagues in, 3
Big-bang reproduction, 149
  evolution of, 149–153, 151*f*
Biocontrol. *See* Biological pest control
Biodiversity, 434–457
  and community stability, 478–480
  ecosystem function and, 604
  gradients in, 438–440
    climate and, 450–452
    competition and, 447–448, 447*f*, 450
    disturbance and, 452–453
    historical factors in, 444–445, 444*f*
    model of, 443
    predation and, 448–450

productivity and, 452
   spatial heterogeneity and, 445–447
history of study of, 436
local, 454–456, 455f
measurement of, 434–438, 617–618
nitrogen deposition and, 580
preservation of, values in, 12
and primary production, 533–534, 533f
regional, 454–456, 455f
Bioelement cycles. *See* Nutrient cycles
Biogeochemical cycles. *See* Nutrient cycles
Biogeography, 41
   island, theory of, 503–508
Biological pest control
   agents of, properties of, 337
   definition of, 332–333, 333f
   density-dependence in, 339, 342
   early attempts at, 4–5
   examples of, 333–337
   with genetics, 342–344
   with immunocontraception,
      344–347, 345f
   risks of, 351–352
   success rates for, 349–351, 350t, 351f
   theory of, 337–342, 340t
Biomanipulation, lake restoration
   with, 499
Biomass
   grazing and, 246–248
   as measure of metabolism, 513
   vs. yield, 513, 514f
Biosphere, 10
Biosphere 2 project, 604–605
Biotic school of population regulation,
   284, 285
Birch, Charles, 142, 144, 144f
Birch, L. C., 106, 285–286
Bird(s)
   abundance of, body size and,
      119–120, 120f
   biodiversity in
      geographic pattern of, 439f
      local and regional, 455, 456f
   clutch size in
      adaptation of, 22–25
      cost-benefit model for, 23, 24f, 25
   flightlessness in, 53
   habitat selection by
      behavioral mechanisms of, 58–62
      decision making process of, 59f
      evolution of, 66
      models for, 68
      for nesting sites, 66, 66f
   introduced species of
      number of, 602t
      success rates for, 49, 49t
   production efficiency of, 545, 545t
   range size for, 108f
   species-area curve for, 503, 504f
Birth rate. *See* Natality

Birth weight, in humans, stabilizing
   selection for, 21, 21f
Blast zone, volcanic, 403
Bodenheimer, F. S., 284
Bog(s), 417
Bogor Botanical Garden, 364–365
Boreal forests
   carbon dioxide in, 594–595
   food chain in, 462, 464f
Bormann, F. Herbert, 565
Bottom-up model of community
   organization, 496–497
Boundaries
   of communities, 392–394
   of ranges, 111, 130
Bradley, Richard, 5
Braun-Blanquet, J., 388
Breeding, selective, for pest control,
   342–343
British Isles, intertidal zonation in, 94, 95f
Brown, Jim, 112
Brown's model of distribution-abundance
   correlation, 112
Brucellosis, in ungulates, 268–269

C

C-S-R model of plant strategies, 200f, 201
C_3 photosynthesis pathway, 99
C_3 plants
   carbon dioxide levels and, 593, 593f
   carbon fixation in, 100–101
   characteristics of, 101t
   distribution of, 101, 601
   leaf anatomy of, 99–100, 99f
   photosynthetic rate in, 99, 99f
   prevalence of, 99, 101, 102
   productivity of, 530
C_4 photosynthesis pathway, 99
C_4 plants
   carbon dioxide levels and, 593, 593f
   carbon fixation in, 100–101
   characteristics of, 101t
   distribution of, 100f, 101
      climate change and, 601, 604f
   leaf anatomy of, 99–100, 99f
   photosynthetic rate in, 99
   prevalence of, 101, 102
   productivity of, 530
Cabot, John, 321
Caffeine, 235
Calcium, acid rain and, 576
California, climatic diagram for, 89f
Calvin cycle. *See* C_3 photosynthesis
   pathway
CAM. *See* Crassulacean acid metabolism
Canada, acid rain in, 575
Cannibalism, 206, 465t
Capture-recapture method, for population
   estimation, 122–125, 126, 612–613
Carbon

cycle of
   global, 591, 591f
   human influence on, 590–596
   isotopes of, in food chain analysis, 466
   in *Sphagnum* bogs, 417
Carbon dioxide
   animal responses to, 600
   atmospheric levels of, 591–592, 592f,
      593
   in climate change, 596–600, 597f
   deforestation and, 592, 601
   and global temperatures, 598, 599f
   in metabolism estimates, 515
   plant responses to, 593–596
Carbon fixation, in C_3 *vs.* C_4 plants,
   100–101
Carnivory, 206. *See also* Predation
Carrying capacity
   definition of, 554
   ecological *vs.* economic, 554
   for human population, 586–590, 588f
Carson, Rachel, 331–332, 332f
Cascade model of community organization,
   trophic, 497–500
Catastrophic agents
   and population loss, 356
   in population regulation, 283, 285
Catch curves, in mortality estimation, 129
Catch per unit fishing effort, as index of
   relative density, 127
Catnip, 235
Caughley, Graeme, 249, 250f
Census, multiple, for population
   estimation, 124
Central Gyre, North Pacific, primary
   productivity in, 518, 520f
Certainty, in science, 14
Chains of extinction, 373
Character displacement, competition and,
   201–202, 202f
Checkerboard distribution, 80, 81f
Chemical defenses, in plants, 235–237
Chesapeake Bay
   food web in, 463, 464, 465f
   marine invasions in, 47
Chihuahuan Desert, keystone species in, 473
Chitty, Dennis, 126, 286–287
Chitty, Helen, 126, 286
Chlorophyll, in C_3 *vs.* C_4 plants, 99
Cholera, avian, 279
Cinnamon, 235
Classical competition theory, of community
   organization, 460–461
Clements, Frederick, 408f
   on communities, 387–388, 389
   monoclimax hypothesis of, 407–408, 423
   on vegetation development, 5
Climate(s), 86–88
   and biodiversity gradients, 450–452
   classification of, 88

diagrams of, 88, 89*f*
and population densities, 285
in population regulation, 284
variation within, tolerance for, 108
Climate change
and coral reefs, 489, 605
human impact on, 596–600
and species distribution, 102–103, 601
Climate school of population regulation, 284
Climatic climax, 407–408
Climatology, 86–88
Climax, successional, 423–424
monoclimax hypothesis of, 407–408, 423
Climax-pattern hypothesis of succession, 423–424
Closed communities, 391
Closed populations, density estimation for, 123
Cloves, 235
Clutch size, bird
adaptation of, 22–25
cost-benefit model for, 23, 24*f*, 25
Coarse-grained habitats, 364
Coastlines. *See also* Intertidal zones
distributional boundaries in, 94, 95*f*
Coevolution, 25–26
of crop-pest systems, 351
of plant-herbivore systems, 253–254
of predator-prey systems, 228–231
Coexistence
diet specialization and, 194
in host-parasite systems, 276
in laboratory populations, 189–190
Cohort life tables, 133, 134*t*, 135–137
Colonization
dispersal in, 50–53
of fragmented patches, 364–365
selective, and nested subsets, 371
Coloration, warning, 229–231
Common garden, in ecotype experimentation, 94–97
Commons, tragedy of, 321
Communities, 386–400. *See also under*
Equilibrium; Nonequilibrium models
boundaries of, 392–394
changes in
cyclic, 426–429
directional (*See* Succession)
characteristics of, 392
classification of, 388, 389
definition of, 386
distributional relations in, 395–398
dynamic relations in, 386–392
early studies of, 5
fundamental unit of, 387–388, 389
importance of species in
definition of, 459
measurement of, 461
indicator species in, 398–400
individualistic view of, 387–390
organismic view of, 387

organization of, 459
models of, 495–501
theories of, 459–462
persistence of, 494–495, 494*f*
redundancy model of, 387
stability in, 392, 459–460, 477–480
local *vs.* global, 460, 477
multiple states of, 508–509
Compartment(s)
in food webs, 465*t*
in nutrient cycles, 560
Compartment models of disease, 259
with constant population size, 259–262, 259*f*, 262*f*
for rabies, 272, 272*f*, 273
with variable population size, 262–264, 262*f*
Compensation, in population regulation, 283
Compensation point, in metabolism, 515
Compensatory mortality, *vs.* additive mortality, 292–293, 293*f*
Competition, interspecific, 179–204. *See also specific types*
adaptations to, 194, 196–197
character displacement as, 201–202, 202*f*
in competitive ability, 199–201
and biodiversity gradients, 447–448, 447*f*, 450
vs. coexistence, 189–190
and community organization, 459, 460–462
criteria for occurrence of, 197, 198*t*
definition of, 82, 179
diffuse, 202–203
and distribution, 34–35, 80–83
range size of, 111
with dominant species, 474–476
effects of, 196, 196*f*
indirect, 202–203
and evolution, 196–197
vs. facilitation, 194*f*, 195–196
through interference, 180
in laboratory populations, 185–190
in natural populations, 190–199
outcome of, 81
paradox of, 194–195
pathways of, 202–203, 203*f*
among predators, 206, 207*f*
prevalence of, 192, 196
as regulator of populations, 284
for resources, 180–185
types of, 82, 180
Competition-predation theory of community organization, 462
Competition-spatial patchiness theory of community organization, 462
Competition theory, classical, of community organization, 460–461
Competitive displacement models of communities, slow, 494

Competitive exclusion, 190
in biological control agents, 350
critique of, 191–192, 196
disturbances and, 452–453
by dominant species, 474, 474*f*
in natural populations, 191–192
Connectance, in food webs, 465*t*, 467, 467*f*
Connell, Joseph H., 449*f*
Conservation biology
declining-population paradigm of, 360–361
definition of, 355
problems in, examples of, 377–380
reserves in, 373–375
rise of, 11
small-population paradigm of, 355–359, 355*f*
values in, 12
Conservation Biology (journal), 12
Constant escapement strategy for fisheries, 327
Constant exploitation rate strategy for fisheries, 327
Constant quota strategy for fisheries, 327
Contagious abortion, 268–269
Contest (interference) competition, 180, 203
Continental climates, 86
Continental drift, and distribution, 52, 52*f*
Continuum of community equilibrium, 460, 461*f*
Continuum view of vegetation, 393, 394, 394*f*
Control. *See* Pest control
Copper, tolerance for, in grasses, 37, 37*f*
Coral reef communities
allelopathy in, 80
biodiversity in, 451–452, 452*f*
climate change and, 605
disturbances in, 486–491
organization of fish in, 488–491
Cormack, Richard, 125, 126
Corridors
advantages/disadvantages of, 368–370, 368*t*
evaluation of effectiveness of, 370, 371*f*
Cost(s)
of grazing, 247
of pest damage, 331
of plant defense mechanisms, 236
of reproduction, 23, 25, 153
Cost-benefit model, for bird clutch size, 23, 24*f*, 25
Costanza, Robert, 603
Count(s)
roadside, for relative density estimation, 127–128
total, for absolute density measurement, 121
Cover, as index of relative density, 127
Cowles, H. C., 407*f*
on plant succession, 5, 406–407

on sand-dune succession, 416–417, 419
Crassulacean acid metabolism (CAM), 99
Crassulacean acid metabolism (CAM)
  plants, characteristics of, 101, 101*t*, 336
Critical load, for nitrogen in terrestrial
  ecosystems, 580
Critical period hypothesis, of fish
  sensitivity, 314
Critical threshold, for diseases, 268
Crops. *See also* Agriculture
  genetic manipulation of, 342–343
  loss to pests, 331, 331*f*
Crustacean zooplankton
  distribution in Great Lakes, 397, 398*t*
Cultural pest control, 332–333
Cycles, in food webs, 465*t*
Cyclones, and coral reefs, 486–487, 487*f*

D

Dams, river, food webs with, 501, 501*f*
Darling, Fraser, 554
Darwin, Charles
  on adaptation, 94
  on diversity and productivity, 533
  as ecologist, 17
  on flightless animals, 53
  on natural selection, 17
Darwinian selection. *See* Individual
  selection
Davis, Margaret, 102, 102*f*
DDT
  and cottony-cushion scale, 334
  and reappearance of pests, 332, 334
Deciduous forests, temperate, energy flow
  in, 546, 547*f*
Declining population
  diagnosis of, 360
  paradigm in conservation biology,
    360–361
Deep ecology, *vs.* ecology, 6
Defense mechanisms, plant, 235–242
  costs of, 236
  examples of, 238–242
  with secondary substances, 235–237
  theories on, 236–237
  types of, 237
Deforestation
  and carbon cycle, 592, 601
  and nutrient loss, 466–467, 467*f*, 468*f*
  rates of, 600–601
  runoff after, 466
Demes, 116
Demographic stochasticity, and population
  loss, 356
Demographic transition, 584, 587
Demography
  early, 3–4
  evolution of traits in, 149–154
  individual variation in, 145
  projections *vs.* predictions in, 143
  statistics in, 133–149

age distributions, 146–149
  intrinsic capacity for increase,
    138–145
  life tables, 133–138
  reproductive value, 145–146
Density, population
  absolute, measurement of, 120–125
  equilibrium in, 281–282, 285
  estimation of, 119–120
    problems with, 125, 125*f*
  factors in, 116, 119, 128–130
  and habitat suitability, 67, 67*f*
  net reproductive rate and, 158, 158*f*
  population size and, 125
  relative, indices of, 125–128
  self-thinning rule of, 297–298
Density-dependent rates, 281–282, 285
  in biological control, 339, 342
  types of, 292, 292*f*
Density-independent models of
  communities, 493
Density-independent rates, 281–282, 285
Descriptive ecology, 7
Deterministic extinctions, 360–361
Detritus, trophic level of, 544
Developmental response, of predators to
  prey density, 222, 223, 224*f*
Dialects, bird song, and habitat selection,
  61–62
Diet(s)
  coexistence and, 194
  of herbivores, selective feeding in, 248
  of predators, optimal, 225–226
  shifts in, competition and, 81–82
  stable isotope analysis of, 466
Diffuse community boundaries, 392
Diffuse competition, 202–203
Diffusion, definition of, 47
Dilution rate. *See* Natality
Direct competition, 203
Directional changing environment models
  of communities, 493–494
Directional selection, 19, 20*f*
Discrete generations
  population growth with, 157–160,
    158*f*, 159*f*
  predation in, 207–209, 208*f*
Disease, 258–275
  and biodiversity, 450
  definition of, 179, 258
  and distribution, 77–78
  effects of, 258
    on individuals, 264–267
    on populations, 267–275
  mathematical models of, 258–264
  vs. parasitism, 258
  transmission of, 5
Dispersal(s), 41–55
  in colonization, 50–53
  evolutionary advantages of, 53–55
  examples of, 41–46

vs. inbreeding, 301
  measurement of, 129–130
  modes of, 46–50
  seed
    in corridors, 369
    as mutualism, 253–254
    after volcanic eruption, 406
  study of, 41
Disruptive selection, 20*f*, 21
Distribution(s)
  abundance and, 7–9, 111–113
  analysis of, 32–39
  boundaries of, 111
  in communities, 395–398
  continental drift and, 52, 52*f*
  in definition of ecology, 2
  limitations on
    climate change as, 102–103, 601
    dispersal as, 41–55
    habitat selection as, 57–69
    interspecific relations as, 71–83
    light as, 97–102
    temperature/moisture as, 88–97
  range size of
    and abundance, 111–113
    spatial scale of, 106–108, 107*f*
    variation in, 108–111
  tolerance and, 35–38
  transplantation and, 32–35, 35*f*
The Distribution and Abundance of Animals
  (Andrewartha and Birch), 106, 285
Disturbance(s)
  and biodiversity gradients, 452–453
  in coral reef communities, 486–491
  definition of, 485–486
  exogenous *vs.* endogenous, 486
  measurement of, 486, 486*t*
  in plant competition theory, 200*f*, 201
  recovery after
    spatial scale and, 480, 480*f*
    and stability, 460, 477–480
  in rocky intertidal communities,
    491–492
Disturbance hypothesis of biodiversity,
  intermediate, 453, 453*f*
Divergence, of communities, 419
Dominant species, 474–476
  vs. keystone species, 474, 474*f*
Dominant year-classes, 148
Doubly labeled water, energy expenditure
  estimation with, 538, 540
Downgrade series, in cycles of change, 427
Drought
  and distribution, 92
  resistance to, 90–92
    biodiversity and, 479
  types of, 92
Dynamic pool models, for optimum yields,
  309–316
Dynamic stability hypothesis of food chain
  length, 468

E

Ecological footprint, 588–589, 588f
Ecological genetics, 9
Ecological specialization model of distribution-abundance correlation, 112
Ecological stability. *See also* Stability
  factors in, 288–289, 289f
Ecology
  approaches to, 7
  definitions of, 2
  disciplines related to, 2, 9–10
  history of, 3–7
  integration in, 10–11
  methodology of, 11–15
  problems of, 7–9
  reasons for studying, 609
  scientific method in, 11–15
  theoretical framework of, 10–11
Economic(s)
  carrying capacity, 554
  of ecosystem services, 603–606
  of fisheries, 318f, 319
  of grazing, 247
  in optimum yields, 318–319
Economic rent, maximum, *vs.* maximum sustained yield, 319
Ecosphere, 10
Ecosystem, in study of ecology, 10
Ecosystem services, 603–606
Ecotypes
  climate change and, 103
  examples of, 94–97
  origin of term, 94
Edges, in fragmented habitats, 367
Effective population size, 358, 359
Ehrlich, Anne, 386
Ehrlich, Paul, 360, 361f, 386
El Niño
  and coral reef bleaching, 489
  effects of, on fisheries, 308–309
  and Southern Oscillation, 598–599
El Niño-Southern Oscillation (ENSO), 598–599
  coral reef records of, 605
Elevation, in gradient analysis, 393, 394f
Elton, Charles, 6f, 309f
  on animal ecology, 6
  definition of ecology by, 2
  on diversity and stability, 480
  on food chain length, 468
  on lynx-hare cycle, 227
  on prey-predator abundances, 549
Eltonian pyramid, 549
Emigration
  measurement of, 129–130
  as population parameter, 116, 119
  in source-sink populations, 299–300
Emissions, and sulfur cycle, 573–574, 575f
Endemic phase, 225

Endogenous disturbances, 486
Energetic hypothesis of food chain length, 467–468, 468f
Energy
  allocation of, clutch size and, 23, 25
  and biodiversity, 450–452
  expenditure of, estimation of, 538, 540, 556–557
  flow of, in food webs, 514, 514f
  partitioning of, 537–538
  sustainable budgets for, 556–557, 556f
ENSO. *See* El Niño-Southern Oscillation
Environmental heterogeneity, in predator-prey theory, 214–215
Environmental physiology, 9
Environmental stochasticity, and population loss, 356
Environmental studies, *vs.* ecology, 6
Environmentalism, *vs.* ecology, 6
Epidemic phase, 225
Epidemiology, 5
Epizootic, 265n
Equilibrium. *See also* Nonequilibrium models
  in communities, 391–392
    continuum of, 460, 461f
    disturbances and, 452–453
    stability of, 459–460, 460f, 477–478
  in population density, 281–282, 285
Errington, Paul, 218
Escapement, 326n
Escapement strategy for fisheries, constant, 327
Essay on Population (Malthus), 3–4
Estuaries, primary production in, 526
Eucalyptus forests, nutrient content of, 568–569, 570t–571t
Eutrophic nutrient cycling, 570–571
Eutrophication, 523
  control of, 524–525
  effects of, 578, 579t
  in lakes, 518, 523–525, 527f
Evaporation rate, temperature and, 87
Evapotranspiration
  in climate classification, 88
  and primary production, 528
  in water cycle, 590
Evolution, 17–28. *See also* Coevolution
  adaptation in, 18–25
  vs. balance of nature, 4
  competition and, 196–197
  of demographic traits, 149–154
  in ecological studies, 11
  hypotheses in, 14
  process of, 17–18
  units of selection in, 26–27
Evolutionary ecology, 7, 10, 11
Exclusion, competitive. *See* Competitive exclusion
Exogenous disturbances, 486

Experiment(s), definition of, 12, 13
Experimental analysis of population regulation, 296–297
Experimentation
  design of, 12
  role of, 11
Exploitation. *See also* Harvest(s)
  of common-property resources, 321
  principles of, 317
Exploitation rate strategy for fisheries, constant, 327
Exploitative competition. *See* Resource competition
Extinction(s)
  causes of, 361
    habitat loss, 362–371
    introduced species, 371–372
    overkill, 361–362
    secondary, 372–373
    stochasticity, 356
  chains of, 373
  deterministic, 360–361
  of local populations, after volcanic eruptions, 51
  measurement of, 507
  of metapopulations, 289–290
  nonrandom, 370–371
  selective, 371
  stochastic models of, 172, 356
Extinction coefficient for light, and primary productivity, 518
Extinction vortex, of small-population paradigm, 355–356, 355f
Extrinsic factors in population regulation, 286, 286f

F

Facilitation
  in grazing, 245–246
  among plants, 194f, 195–196
Facilitation model of succession, 408, 409
Facts, definition of, 13
Facultative agents, in population regulation, 282–284, 285
Farmland, abandoned, succession on, 420–423, 422f, 423f
Fast-growing plants, characteristics of, 237t
Fecal pellets, as index of relative density, 127
Fecundity
  definition of, 128
  potential, *vs.* realized fertility, 128
Feeding capacity, as index of relative density, 127
Feeny, Paul, 236
Fenner, Frank, 272, 275f
Fertility
  definition of, 128
  early theories on, 3–4
  of pests, reduction of, 344–347
  realized, *vs.* potential fecundity, 128

schedule of, 139
Field studies
  advantages/disadvantages of, 217
  role of, 11
Fine-grained habitats, 364
Finite rates, 614–616
Fire(s)
  control of, and species loss, 366
  and plant diversity, 453, 454f
Fish
  acid rain and, 576
  in coral reef communities, 488–491
  introduced species of, 602t
  latitude and biodiversity in, 440
  plastic growth rates of, 199
Fisheries
  dynamic pool models for, 309–316
  economic model of, 318f, 319
  and food webs, 477
  king crab, 319–321, 320f
  laboratory studies on, 317–318
  logistic models for, 307–309, 307f
  no-take zones for, 327–328
  northern cod, 321–323, 321f
  optimum yields in, 318–319
  primary production required for,
    546, 547t
  stock in, 305–306
  strategies for, 327–328
  whaling, 324–325
  yield in, 305–306
Fiske, W. F., 282–284
Fitness, definition of, 19, 611
Flagship species, 400
Flightlessness, evolution of, 53
Floristic composition, initial, 407f, 408
Floristic provinces, 395, 395f
Floristics, relay, 407–408, 407f
Floristics model of succession, preemptive
  initial, 408
Fluctuating environment models of
  communities, 493
Flux rate, 560–561
Food
  in energy budgets, 556
  as regulator of populations, 288
Food chains
  analysis of, with stable isotopes, 466
  definition of, 463
  length of, 467–468
Food webs, 463–469
  and biodiversity, 448–449
  complexity of, 467, 478
    stress and, 495–496, 495f
  definitions of terminology in, 465t
  energy flow in, 514, 514f
  overfishing and, 477
  simplification of, 462–463
  stability of, 478
  structure of, 465–469, 465f

trophic levels in, 463–464
Footprint, ecological, 588–589, 588f
Foraging theory, optimal, 225–227
Forbes, Edward, 5
Ford, E. B., 287
Forests. See also specific types
  acid rain in, 576, 578f
  carbon dioxide enrichment in, 594–595
  fragmentation of, 363–364, 364f, 365f
  multiple stable states in, 508–509, 508f
  nutrient cycles in, 531, 562–567, 563f
  production in
    efficiency of, 546, 547f
    primary, 528–530
Fossil pollen
  community history in, 389–391, 391f
  as record of distribution, 102
Foster, Robin, 449
Founder effect, 41
Foyn, Svend, 324
Fragmentation. See Habitat destruction and
  fragmentation
Freshwater communities
  acid rain in, 576
  acidification of, 575, 576
  food webs in, 500–501
  nitrogen cycle in, 578, 579f
  nutrient loss in, 566–567
  phosphorus cycle in, 561, 562f
  primary production in, 518, 519f
    limitations on, 523–526
  restoration of, 480–482, 499
Fretwell-Oksanen model of community
  organization, 499
Frost drought
  and distribution, 92, 93–94
  and timberlines, 93–94
Fruits
  defenses against damage of,
    253–254, 254t
  seed dispersal with, 253–254
Fugitive species, 54
Functional ecology, 7
Functional response, of predators to prey
  density, 222–223, 222f–224f
Fundamental niche
  definition of, 190
  vs. realized niche, 191, 192f
Fundamental unit, of communities,
  387–388, 389, 394, 394f
Fynbos vegetation, spines on, 242

G

Gaia hypothesis, 388
Galápagos Islands, plant biodiversity on,
  502–503, 502f
Game ranching, secondary production in,
  554–556, 555t
Gametic selection, 27
Gaps, colonization of, 428–429, 430f, 430t

Garden, common, in ecotype
  experimentation, 94–97
Gas exchange, in metabolism estimates,
  515, 516
Gause, Georgyi Frantsevich, 187, 189f, 217
Gause's hypothesis of competition. See
  Competitive exclusion
Gene flow, dispersal and, 41
Generalists
  in distribution-abundance
    correlation, 112
  effects on prey populations, 226
Generation(s)
  definition of, 140n
  discrete
    population growth with, 157–160,
      158f, 159f
    predation in, 207–209, 208f
  mean length of, 141
  overlapping
    population growth with, 160–162
    predation in, 209–212, 209f
Generation (cohort) life tables, 133, 134t,
  135–137
Genet(s), 118–119
Genetic(s)
  as adaptation limitation, 22
  ecological, 9
  of habitat selection, 60–61
  in pest control, 342–344
  population, 9
  in population regulation, 300
  variation in, 611
    and population loss, 356
    within species, 94–97, 103
Genetic drift
  definition of, 359
  minimum population for prevention
    of, 358–359
Genetic engineering, for pest resistance, 343
Genetic individuals, 118–119
Genotypes
  phenotypes from, 18–19
  variation in, 94–97, 286–287
Geologic events, and geographic
  distributions, 52–53
Glacial moraines, succession on, 413–416
Glaciation
  and community changes, 390–391
  and geographic distributions, 52–53
  and range size, 109–111
  tree dispersal after, 48, 102
Glacier Bay
  ice recession rate for, 414, 414f
  moraine succession in, 413–416, 416f
  soil pH change at, 415, 415f
Gleason, H. A., 387, 388, 389f
Global nutrient cycles, 560, 561f, 562
Global stability, 460, 477–478
Global warming

in coral reef bleaching, 489
human impact on, 596–600
and species distribution, 102–103
Glossary, 619–622
Gondwana, 52, 52*f*
Gordon, H. Scott, 318, 319
Gossypol, 343
Gradient analysis, 393–394, 393*f*
  models for, 394, 394*f*
Grant, Peter, 19, 19*f*
Grant, Rosemary, 19, 19*f*
Grass
  allelopathic effect of apple trees
    on, 78, 79*f*
  distribution of C$_4$ species, 100*f*, 101
  as modular organism, 116–117, 119
Grasslands
  cyclic changes in, 427, 427*f*
  distribution of, 550*f*
  fragmentation of, 367
  multiple stable states in, 508–509, 508*f*
  primary production in, 530
    biodiversity and, 533–534, 533*f*
    climate and, 550, 550*f*
  secondary production in, limitations on,
    549–554
  stability and biodiversity in,
    478–480, 479*f*
Grazing
  costs/benefits of, 247
  effects on plants of, 246–248
  facilitation in, 245–246
  interactive systems of, 248–251
  models for, 250, 250*f*, 251*f*
  noninteractive systems of, 251–253
  in Serengeti Plains, 242–246, 244*f*
Great Basin, nested subsets in, 370, 371,
  371*f*, 372*f*
Great Lakes, 397*f*
  index of similarity for, 395–398, 398*t*
Great Plains, primary production in,
  530, 531*f*
Great Smokey Mountain National Park,
  climatic diagram for, 89*f*
Green world hypothesis, 551
Greenhouse effect, 596–600, 597*f. See also*
  Climate change
Grime, Philip, 595–596, 596*f*
Grime's theory of plant competition,
  200*f*, 201
Grinnell, Joseph, 190
Gross production
  primary, 515, 516–517
  secondary, 528
Ground cover, as index of relative
  density, 127
Group living, 231
Group selection, 27
  in Gaia hypothesis, 388
Growth

individual, plasticity in, 199
population (*See* Population growth)
Guilds, 469–471

H

Habitat(s)
  definition of, 57, 65
  fine-grained *vs.* coarse-grained, 364
  spatial heterogeneity in, and
    biodiversity, 445–447
  specificity of, with rare species, 356–357
  suitability of, 65–66, 67*f*
Habitat destruction and fragmentation
  extinction caused by, 362–371
  and population dynamics, 364–371, 366*t*
  and source/sink populations, 299
Habitat selection, 57–69
  approaches to study of, 57
  behavioral mechanisms of, 57–65
    in birds, 58–62
    genetics in, 60–61
    hierarchical decision making process
      in, 59, 60*f*
    in insects, 58
    in rodents, 62–65
  changes in habitat and, 60, 68
  evolution of preferences in, 62–65
  models of, 65–68
Haeckel, Ernst, 2
Hairston-Smith-Slobodkin (HSS) model of
  community organization, 496, 496*f*
Handling time, 223
Hanski, Ilkka, 111
Hanski's rule, 111
Hardy, G. H., 610
Hardy-Weinberg law, 610
Harmony, in understanding of ecology, 3
Harvest(s), population, 305–328
  dynamic pool models for, 309–316
  of king crab, 319–321, 320*f*
  laboratory studies on, 316–317
  logistic models for, 307–309, 307*f*
  of northern cod, 321–323, 321*f*
  nutrient removal by
    in forests, 562, 567
    and primary production, 531, 567
  optimum yields in, 318–319
  risk-aversive management strategies for,
    325–328
  of whales, 324–325
Harvest method of measuring primary
  production, 515–516
Hawaii, endemic birds of, introduced
  diseases and, 77–78, 77*f*
Heathlands, cyclic changes in, 426–427, 426*f*
Heavy-metal toxicity, adaptation in
  tolerance for, 36–38
Hedgerows, as plant colonization corridors,
  369, 370*f*
Herbivory, 235–254

and community organization, 459, 462
competitive effects of, 196, 196*f*
defense mechanisms of plants against,
  235–242
definition of, 179, 206
effect on plants of, 236, 246–248
metabolism in, 537
mutualism in, seed dispersal as,
  253–254
population dynamics of, 248–253
  in interactive systems, 248–251
  in noninteractive systems, 251–253
primary production and, 546, 548*f*,
  549, 549*f*
selective feeding in, 248
in Serengeti Plains, 242–246
Heterogeneity
  in biodiversity, 434, 445–447
    measurement of, 438, 617–618
  in predator-prey theory, 214–215
  spatial, 445–447
Heterotrophs, 513
Historia Animalium (Aristotle), 3
Holling, Buzz, 222
Hollow curve, in species abundances, 435
Holt, Sydney, 309
Homopterous insects, ants and, 254–255
Honesty, in science, 12, 14
Horizontal (cohort) life tables, 133, 134*t*,
  135–137
Howard, L. O., 282–284
HSS model. *See* Hairston-Smith-Slobodkin
  model
Hubbard Brook Experimental Forest,
  565–567
Hubbell, Steve, 449
Huffaker, Carl, 214, 219*f*
Human(s)
  age structure in, 148, 149*f*
  birth weight in, stabilizing selection
    for, 21, 21*f*
  and carbon cycle, 590–596
  and climate change, 596–600
  ecosystem services for, 603–605
  fertility rates of
    decline in, 585, 585*f*
    schedule for, 139*t*
    socioeconomic status and, 4, 4*t*, 587
  impacts on ecosystems, 599, 599*f*
  land use by, 600–601
  life expectation for, 128
  life tables for, 133, 136–137, 136*n*
  population growth in, 583–590
    carrying capacity *vs.*, 586–590, 588*f*
    current patterns of, 584–586,
      585*f*, 586*t*
    early theories on, 3–4
    history of, 583, 584*f*
    logistic curve for, 167–168, 168*f*
  as predators, 206

rabies in, 269, 270
reproductive rate of, 140f
survivorship curve/schedule for,
124f, 139t
Human Genome Project, 10
Humboldt, Alexander von, 502
Hummock-and-hollow cycle, 427, 427f
Hypotheses
definition of, 12, 13
vs. ecological truth, 14
uses of, 12–15

I

IBP. See International Biological Program
Ice ages
and community changes, 390–391
and geographic distributions, 52–53
and range size, 109–111
tree dispersal after, 48, 102
Ice cores, climate change records
in, 597, 599f
Ideal despotic distribution model of habitat
selection, 67
Ideal free distribution model of habitat
selection, 66–67, 67f
Immigration
measurement of, 129–130, 507
as population parameter, 116, 119
in predation, 213, 214
in source-sink populations, 299–300
Immunocontraception, 344–347, 345f
Inbreeding
dispersal as counterstrategy to, 301
minimum population for prevention
of, 358–359
Incidence functions, 364, 365f
Indicator species, 398–400
selection of, 399
types of, 399–400
Individual(s)
effects of disease on, 264–267
genetic, 118–119
phenotypic and genotypic changes in,
286–287
regulation by, 286–288
evolution of, 300–302, 301f
variation among, 130–131
and demography, 145
Individual optimization hypothesis, on
clutch size, 25
Individual selection, 26
and kin selection, 27
Individualistic view of communities,
387–390
Inducible defenses
in plants, 237, 242
spines as, 241–242
Infanticide, 301
and territoriality, 301
Inhibition model of succession, 408, 409

Initial floristic composition, 407f, 408
Initial floristics model of succession,
preemptive, 408
Insect(s)
climate change and, 600
control of (See Pest control)
distribution of
dispersal in, 50, 54–55
habitat selection in, 58
flightlessness in, 53, 54f
fugitive species of, 54
irruptions of, 250–251
parasitoid, 206, 284, 337–339
production efficiency of, 545, 545t
weather as regulation on, 284
Instantaneous rates, 614–616
Integrated pest management (IPM),
347–349, 347f
definition of, 332
Integration, levels of, in ecology, 10–11
Interactions, species. See also specific types
classification of, 179–180
complex, 254–255
in food webs, 465t
mechanism vs. effect of, 179
Interactive herbivore systems, 248–251
Interference competition
definition of, 180
as direct competition, 203
Intermediate disturbance hypothesis of
biodiversity, 453, 453f
Intermediate species, in food webs, 465t, 468
International Biological Program (IBP),
549, 550
Intertidal zones
distributional boundaries in, 94, 95f
disturbances in, 491–492
food webs in, and biodiversity, 448–449
keystone species of, 471
organization of, 492
Intransitive competition, 474, 474f
Intrinsic capacity for increase, 138–145
vs. realized population growth rate, 162
repeated production and, 149–151
Introduced species, 601
biological control with, risks of,
351–352
competition from, 197–198
definitions of terms for, 50
extinction caused by, 371–372
as pests, 331
predation by, 220–222
prevalence of, 602t
stages of, 49, 50
success rates for, 49
Inversely density-dependent rates, 281–282
Invertebrates
abundance of, size and, 548f
biodiversity of, disturbances
and, 453, 454f

distribution of, predation and, 71–72
habitat selection by, 57–58
IPM. See Integrated pest management
Iron, as limitation on productivity, 520–522
Irruption, population
in insects, 250–251
in ungulates, 248–251, 248f
Islands
biodiversity on, 434, 503–505
biogeography theory of, 503–508
community organization on, 501–508
equilibrium model for, 504, 504f
flightlessness on, 53
introduced species on, success rates
for, 49
predation on, 227
Isoclines
predator, 210–212, 210f, 211f
prey, 209–210, 209f, 211f, 212
zero growth, 184
Isotopes, stable, in food chain analysis, 466

J

Janzen-Connell model of high tropical-
forest diversity, 449–450, 449f
Johnson, Noye M., 565
Jolly, George, 125, 126, 127f
Journal of Animal Ecology (journal), 284
Juglone, 78, 235
Jump dispersal
barriers to, 48–49, 52
continental drift and, 52
definition of, 47

K

K-selection, 199–201
Kelp, distribution of, predation
and, 72–73
Key factor, definition of, 295
Key factor analysis, 293–296, 297f
Keystone species, 400, 471–474
abundance of, 473
definition of, 471
vs. dominant species, 474, 474f
examples of, 471–473
identification of, 473–474
in intertidal zones, 449
in tropical communities, 449
Kin selection, 27
and warning coloration, 230–231
Koebele, Albert, 333
Krakatau, recolonization of, 50–51, 51f
Krantz anatomy, 99

L

Laboratory studies
advantages and disadvantages of, 217
role of, 11
Lack, David, 23–25, 23f, 190
Lack clutch size, 25

Lakes
acid-bog, bound nutrients in, 525
acidification of, 575, 576
phosphorus cycle in, 561, 562*f*
primary production in, 518, 519*f*
limitations on, 523–526
restoration of, 480–482
with biomanipulation, 499
Land(s)
heat absorption on, 86
precipitation on, 87
public, grazing on, 247
Land use, changes in, 600–601
Landbridge islands, species-area curve for, 505–507
Landscapes, integration in, 10
Latitude(s)
biodiversity gradients by, 438–440
climate and, 450–452
competition and, 447–448, 447*f*, 450
disturbance and, 452–453
historical factors in, 444–445, 444*f*
model of, 443
predation and, 448–450
productivity and, 452
spatial heterogeneity and, 445–447
and climatic variability, 108
and primary production, 518–519
and range size, 108–111
and solar radiation, 86, 86*f*
temperatures by, 86, 87*f*
and timberlines, 93, 93*f*
Law(s)
of -3/2 power, 297–298
definition of, 13
Hardy-Weinberg, 610
of minimum, 34, 40
of tolerance, 35–36, 36*f*
Leaf area duration, primary production and, 530, 531*f*
Lenoble, F., 388
Leslie, George, 126
Leslie, Patrick H., 173, 173*f*
Leslie matrix, 173–175, 173*f*
Liebig, Justus von, 34
Liebig's law of minimum, 34
critique of, 40
Life cycles
evolution of, 149–154
population projection matrices for, 173–176, 173*f*
Life-dinner principle, of evolution in predation, 228
Life table(s), 133–138
data collection for, 135, 137–138
examples of, 133–134, 134*t*
in key factor analysis, 293–295
types of, 135–137
Life-table age distribution, 147–148
Lifson, N., 540
Light

adaptations to, 97–98
competition for, in succession models, 412
and distribution, 97–102
and productivity, in aquatic communities, 518, 518*f*, 519*f*
roles of, 97
Likens, Gene, 565, 565*f*
Lime, for lake acidification, 575
Limitations, population, 280, 280*f*, 283, 288
experimental analysis of, 296–297
Limiting factors in population regulation
definition of, 288
in experimental analysis, 296–297
Lincoln, F. C., 122*n*
Lincoln method, 122*n*
Linkage density, 465*t*
food web size and, 467, 467*f*
Lizard malaria, effect on reproduction, 264, 264*f*
Local biodiversity
definition of, 455
vs. regional biodiversity, 454–456, 455*f*
Local nutrient cycles, 561
Local population(s)
sink, 299
size of, for rare species, 356–357
source, 299
Local population model of distribution-abundance correlation, 112–113
Local stability, 460, 477
Log-normal distribution of relative abundances, 436–437, 437*f*
Logging. *See* Deforestation
Logistic curve for population growth, 161–162
development of, 4, 161
for humans, 167–168, 168*f*
laboratory tests of, 163–164
in Lotka-Volterra equations, 180–181
problems with, 163, 166, 168
Logistic models for optimum yield, 307–309, 307*f*
recruitment curve in, 312*f*, 313
Longevity, types of, 128
Loss rate, estimation of, 124
Lotka, Alfred J., 6, 6*f*, 138, 140
Lotka-Volterra equations, for resource competition, 180–182, 182*f*, 183*f*
Lottery hypothesis of community organization, 488–491
Lough Ine, distribution of invertebrates in, 71–72, 72*f*
Lovelock, James, 388
Lubchenco, Jane, 492, 492*f*
Lying-in-wait strategy of pest predation, 341

M

MacArthur, Robert, 193, 195*f*, 436
MacArthur-Wilson model, 503–508
Macroparasites, 258

Malaria
lizard, reproductive effects of, 264, 264*f*
transmission of, 5
to birds, 77–78
Malthusian parameter. *See* Intrinsic capacity for increase
Mammals
abundance of, body size and, 119–120, 120*f*
distribution of, 51
diversity gradients for, 440, 442*f*
production efficiency of, 545, 545*t*
range sizes for, 108, 109*f*
species-area curves for, 505, 506*f*, 507
Management
of pests (*See* Pest control)
resource, 326–328
Mangrove forests, zonation in, predation and, 474, 475*f*
Marine communities
biodiversity gradients in, 446–447
currents in, 519*f*
food chain in, 462, 463*f*
heat absorption by, 86
keystone species in, 471–472
precipitation over, 87
primary production in, 516, 517*f*, 518–522, 526
temperature variation in, 108–109, 110*f*
surface, 489
Marine dispersal, human factors in, 47
Mark-recapture model. *See* Capture-recapture method
Marsupials, biodiversity of, in Australia, 440, 441*f*
Match/mismatch hypothesis, of fish sensitivity, 314
Mathematical ecology, groundwork for, 6
Matrices, projection, for population growth, 173–176, 173*f*
Maximum economic rent, *vs.* maximum sustained yield, 319
Maximum power principle of ecosystem evolution, 543
Maximum reproduction, theory of, 25
Maximum sustained yield, 305
vs. maximum economic rent, 319
vs. optimum yield, 318
May, Robert, 160, 163*f*, 259
McClintock, R., 540
McCormack, George, 126*f*
Menge-Sutherland (MS) model of community organization, 495–496, 495*f*, 496*f*
Metabolism. *See also* Production
body size and, 538, 539*f*
of communities, 513
of ecosystems, 513–514
Metapopulations
dynamics of, 289, 289*f*
extinction of, 289–290

Methodology, of ecology, 11–15
Mexico, demographic transition in, 584*f*
Michigan, Lake, 418*f*
    sand-dune succession on, 418–419, 420*f*
Microcosm studies, *vs.* field studies, 217
Microparasites, 258
    models of infection with, 259, 259*f*
Migration
    between fragmented habitats, 368–369
    in grazing, 242–243, 244*f*
    in population genetics, 611
Mimicry, of warning coloration, 230*f*, 231
Minimum, law of, 34
    critique of, 40
Minimum viable population (MVP),
    356–359
    estimation of, 358–359
    for rare species, 356–358
Mining wastes, adaptation in tolerance for,
    37, 37*f*
Mobbing, 231
Mobility, of organism, and population
    estimation, 119
Möbius, Karl, 5
Models, definition of, 13
Modular organisms, 116–119
    examples of, 117*f*, 118*f*
    population structure of, 118–119
Modules, 117–118
Moisture
    adaptations to, 94–97
    cycle of, 87
    and distribution, 88–97
    temperature and, 91–94
    tolerance for, 88–91, 90*f*
Monoclimax hypothesis of succession,
    407–408, 423
Monocultures
    patch dynamics and, 427
    pests on, 350–351
Monophagous predators
    distribution of, 76–77
    effects on prey populations, 226
Montane forests, guilds in, 470
Moraines, glacial, succession on, 414–417
Morphine, 235
Mortality
    density-dependent rates of, 281
    disease and, 265–267
    measurement of, 124, 128–129
        life tables for, 133–138
        as per capita rate, 135
    as population parameter, 116, 119
Mosaic community boundaries, 392
MS model. *See* Menge-Sutherland model
Multiple stable states, 508–509
Multiplication rate, for discrete
    generations, 157–158
Murdoch, Bill, 339, 339*f*
Muskeg (*Sphagnum* bog), 415, 417
Mutations, 611

as limitation on adaptation, 22
Mutualism
    between ants and acacias, 239–241
    and community organization, 459, 462
    definition of, 179
    seed dispersal as, 253–254
MVP. *See* Minimum viable population
Myxoma virus
    evolution of, 272–275, 274*t*, 277
    myxomatosis caused by, 272–275
Myxomatosis, in European rabbits,
    272–275, 274*f*, 275*f*, 277

N

Natality
    definition of, 128
    density-dependent rates of, 281
    measurement of, 124, 128
        in fertility schedule, 139
        as per capita rate, 135
    as population parameter, 116, 119
Natural history, in ecology, 2, 3, 7
Natural History (Buffon), 3
Natural History of Animals (Aristotle), 268
Natural pest control, 332
Natural selection
    in population genetics, 611
    process of, 17–18
    role in evolution, 17
    types of, 19–21, 20*f*
    units of, 26–27
Naturalist (Wilson), 436
Nature, balance of
    vs. evolutionary theory, 4
    in understanding of ecology, 3, 282
NDVI. *See* Normalized difference
    vegetation index
Neighborhood (local) stability, 460, 477
Nest(s), selection of sites for, 65, 66*f*
Nest parasitism, and coevolution, 26
Nested subsets, 371, 371*f*
Net production
    primary, 515
        annual global, 516, 517*f*
        variation in, 516, 516*t*
    secondary, 539
Net reproductive rate, 139–140
    of disease agents, 260, 263
    population density and, 158, 158*f*
    with two stable points, 220, 220*f*
New York City, water supply for, 606
New Zealand
    distribution in, 52
    timberlines in, 94
Niche(s)
    breadth and overlap of, and biodiversity,
        447–448, 447*f*
    in competitive exclusion principle, 190
    definitions of, 190
    diagram of, 190, 191*f*
    evolution of, under competition, 197, 448

fundamental *vs.* realized, 190–191, 192*f*
Niche-diversification hypothesis of
    equilibrium, 488–491
Niche-preemption hypothesis of
    biodiversity, 435
Nicholson, A. J., 284, 284*f*
Nicotine, 235
Niño, El. *See* El Niño
NIS (nonindigenous species). *See*
    Introduced species
Nitrate, 580
Nitric oxide, 578
Nitrogen
    competition for, in succession
        models, 412
    cycle of, 576–580
        in forests, 563–564, 563*t*
        global, 577, 579*f*
    isotopes of, in food chain analysis, 466
    and primary production, in oceans,
        519–520, 522
    in soil, during glacial moraine
        succession, 415–416, 416*f*
Nitrous oxide, 577–578
No-take zones for fisheries, 327–328
Nonequilibrium models. *See also*
    Equilibrium
    of communities, 391–392, 493–495
        continuum of, 460, 461*f*
        patches and disturbance in, 485–486
        stability in, 460
        types of, 493–495
    of pest control, 339, 342
    of predation, 339
Nonindigenous species (NIS). *See*
    Introduced species
Noninteractive herbivore systems, 248,
    251–253
Nonnative species. *See* Introduced species
Nonrandom extinction, 370–371
Normalized difference vegetation index
    (NDVI), 528
North America, dispersal of mammals in, 51
North Carolina, abandoned farmland
    succession in, 420–423, 422*f*, 423*f*
North Pacific Central Gyre
    primary production in, 518, 519*f*, 520*f*
Numerical response, of predators to prey
    density, 222–223, 222*f*
Nutrient(s), and primary production
    in lakes, 523–526
    in oceans, 519–522
    in terrestrial ecosystems, 530–531
Nutrient cycles, 560–581
    deforestation and, 466–467, 467*f*, 468*f*
    exchanges in, 560–562
    in forests, 531, 562–567, 563*f*
    global, 560, 561*f*, 562
    local, 561
    pools in, 560–562
    turnover times in

estimation of, 564
in forests, 564, 565*t*
use efficiency in, 568–571, 569*f*
Nutrient ratios, 526
Nutrient use efficiency (NUV), 568–571, 569*f*

O

Objectivity, in science, 12
Obligate associations, in communities, 389
Ocean(s)
biodiversity gradients in, 446–447
currents in, 519*f*
heat absorption by, 86
precipitation over, 87
primary production in, 516, 517*f*, 518–522
surface temperature of, increases in, 489
Oceanic islands, species-area curve for, 505–507
Odum, Eugene, 2, 537, 538*f*, 543
Odum, Howard, 537, 543
Old-field succession, 420–423, 422*f*, 423*f*
Oligotrophic lakes, 518
Oligotrophic nutrient cycling, 570–571
Omnivory, 465*t*
prevalence of, 468–469
Ontario, Lake, food chain in, 466
Open communities, 391–392
Open populations, density estimation for, 123
Optimal foraging theory, 225–227
Optimality models, 23
Optimum yield, 318–319
development of concept of, 5
dynamic pool models for, 309–316
economic factors in, 318–319
logistic models for, 307–309, 307*f*
Ore smelters, and acid rain, 573
Organismic view of communities, 387
The Origin of Species (Darwin)
on capacity for increase, 144
on competition, 205
on rarity, 304
Overcompensation hypothesis of grazing, 246, 247, 247*f*
Overfishing, 305
and food webs, 477
Overkill, extinctions caused by, 361–362
Overlapping generations
population growth with, 160–162
predation in, 209–212, 209*f*
Oviposition site selection, 58
Ozone, 578

P

Paine, Bob, 471, 471*f*
Paradox of plankton, 194–195
Parasitism
for control of insect pests, 282

definition of, 179, 206
vs. disease, 258
and distribution, 77–78
effects of, 258
on individuals, 264
evolution of systems in, 275–277
models of, 276
nest, 26
Parasitoids, insect, 206
biological control with, 337–339
as facultative agents of regulation, 284
Park, Thomas, 164
Park Grass Experiments, 533
Patches
definition of, 485
movements between, 368–369
succession dynamics in, 424–429
Patchiness
in community organization, 462, 485–486
definition of, 485
Patuxent Wildlife Research Center (PWRC), 40
Pearl, Raymond, 134, 136*f*, 166, 587
Pelt records
as index of relative density, 127
lynx-hare cycle in, 227
Penicillin, and local distribution, 78
Peninsulas, biodiversity on, 440
PEP carboxylase, in carbon fixation, 100–101
Peppermint oil, 235
Per capita rates
calculation of, 135
in survivorship curves, 134
Persistence, of communities, and stability, 460
Pest(s)
and biodiversity, 449–450
cost of damage by, 331
definition of, 331
Pest control, 331–352
biological
examples of, 333–337
with genetics, 342–344
with immunocontraception, 344–347, 345*f*
with parasites, 282
risks of, 351–352
success rates for, 349–351, 350*t*, 351*f*
theory of, 337–342, 340*t*
definition of, 331
integrated, 347–349, 347*f*
strategies for, 332
Pesticides
ecological consequences of, 331–332
effectiveness of, 331
resistance to, 332
Petersen, C. G. J., 122

Petersen method, 122–123
pH, of soil
in glacial moraine succession, 415, 415*f*
in sand-dune succession, 419
Phase plane, 184
Phenotypes
changes in, 286–287
natural selection of, 18–21, 20*f*
Phenotypic variation, 96
Phocine distemper virus, effect on seals, 265–267, 266*f*
Phospho-enolpyruvate. *See* PEP
Phosphorus
cycle of, 561, 562*f*
as limitation on productivity
in lakes, 524–525, 524*f*, 525*f*, 527*f*
in oceans, 519–520
Photoperiodism, definition of, 97
Photosynthesis
definition of, 97, 515
and distribution, 102
light variation and, 97
and primary production, 515
rates of, 97
types of, 99–102
Physiological ecology, 35–36
Physiology, environmental, 9
Phytoplankton
competition among, 194–195
nutrient ratios in, 525–526
predation on, 41–42
productivity of, 520, 521, 521*f*
Pielou, Chris, 52, 52*f*
Pit feeders, in collardstands, 469, 470*f*
Plagues, of animals, 3
Plankton, paradox of, 194–195
Plant(s). *See also* Herbivory
biodiversity in
disturbances and, 453, 454*f*
historical pattern of, 444, 445*f*
and productivity, 533–534, 533*f*
carbon dioxide enrichment and, 593–596
communities of, organismic *vs.* individualistic views of, 387
competition among, 194–196
evidence of, 198–199
evolution of, 200*f*, 201
pathways of, 203*f*
in population regulation, 297–298
defense mechanisms of, 235–242
distribution of
dispersal in, 49–50
range size for, 108*f*
introduced species of, number of, 602*t*
as modular organisms, 116–118
population regulation in, 297–298
productivity of, 533–534, 533*f*
quadrat sampling of, 122

secondary substances of, 235–236, 236*f*, 239
seed dispersal by, 253–254, 369, 406
succession of
early description of, 5
early *vs.* late, 409, 409*t*
water and
need for, 86–87
tolerance for, 90–91
Plant apparency theory of defense, 236–237
Plant ecology
vs. animal ecology, 11
early studies of, 5
Plant stress hypothesis of herbivory, 251
Plant vigor hypothesis of herbivory, 251
Plasticity, in growth rates, 199, 298
Policy, ecological research and, 7, 12
Pollen, fossil
community history in, 389–391, 391*f*
as record of distribution, 102
Pollination, value of, 606
Pollution
and coral reef bleaching, 489
sulfate, 573, 574*f*, 576
Polyclimax hypothesis of succession, 423–424
Ponds, temporary, keystone predators in, 473
Pools, nutrient, 560–562
Population(s), 116–131
characteristics of, 116
composition of, 130–131
decline in, 360–361
definition of, 116, 130
density of (*See* Density)
disease and, effects of, 267–275
dynamic relations between, 386–392
growth of (*See* Population growth)
limitations on study of, 130
minimum viable, 356–359
modular, 116–119
parameters of, 116, 119
estimation of, 119, 128–130
regulation of (*See* Population regulation)
size of, 125, 358, 359, 611
as unit of study, 116
unitary, 116, 117
Population ecology
early, 3–4
experimentation in, 11
Population genetics, 9, 610–611
Population growth, 157–176
with discrete generations, 157–160, 158*f*, 159*f*
early theories on, 3–4
field data on, 164–168
limitations on, 280, 280*f*, 283
logistic theory of (*See* Logistic curve)

mathematical theories of, 157–162
natural selection during, lack of, 287
with overlapping generations, 160–162
projection matrices for, 173–176, 173*f*
rate of, in logistic models, 307
selection of model for, 166
stochastic models of, 169–172, 171*f*, 172*f*
time-lag models of, 168, 169, 169*f*
Population regulation, 280–302
approaches to study of, 293–297
controversies over, 290
definitions in, 283, 288
evolutionary implications of, 300–302
habitat loss and, 364–371, 366*t*
historical views of, 282–288
vs. limitation, 283, 288
modern synthesis of, 288–293
in plants, 297–298
self, 286–288, 300–302, 301*f*
simple model of, 281–282, 281*f*
in sink populations, 299–300
in source populations, 299–300
Population size, 125
effective, 358, 359
in population genetics, 611
Possible interactions, in food webs, 465*t*
Potato blight, control of, 343
Potential evapotranspiration
in climate classification, 88
definition of, 88
Potential fecundity
costs of, 153
vs. realized fertility, 128
Potential longevity, 128
Power, Mary, 500, 500*f*
Prairies, habitat fragmentation and species loss in, 366–367
Precautionary principle, 14
Precipitation
acid, 573–576
climate classification by, 88
global distribution of, 87, 88*f*
nutrients in, 566, 566*t*
and species distributions, 90
in water cycle, 590
Predation, 206–232. *See also* Predator; Prey
and abundance of prey, 216–222
predator response to, 222–224, 222*f*
and biodiversity gradients, 448–450
and community organization, 459, 462
definition of, 179
and distribution of prey/predators, 71–77
effects of, 206
evolution in, 228–231
group living as, 231
warning coloration as, 229–231
field studies of, 216–224
habitat fragmentation and, 367–368, 367*f*
laboratory studies of, 212–216
mathematical models of, 207–212

mortality from, additive *vs.* compensatory, 292–293
nonequilibrium model of, 339
optimal foraging theory of, 225–227
types of, 206
Predator(s)
biological control with, 337
competition among, 206, 207*f*
in food webs, 465*t*, 468
introduction of, extinction caused by, 372
keystone, 471–473
optimal diet for, 225–227
and prey abundance, 216–222
and prey distribution, 71–75
responses of, 222–224, 222*f*
restriction of, by prey, 75–77
Predator isocline, 210–212, 210*f*, 211*f*
Predator pit, 220
Predator removal experiments, on community organization, 496, 497*t*
Predictions
vs. demographic projections, 143
of hypotheses, 12, 13
Preemptive initial floristics model of succession, 408
Prey
abundance of, 216–222
and predator distribution, 75–77
profitability of, 225–226
quality of, 229
restriction of, by predators, 71–75
warning coloration in, 229–231
Prey isocline, 209–210, 209*f*, 211*f*, 212
Prey-predator ratio, in food webs, 468, 468*f*
Primary population parameters, 119
Primary production, 515–534
efficiency of, 516–517
global, 516, 517*f*, 529*f*
limitations on, 517–531
in freshwater communities, 523–526
in marine communities, 518–522
in terrestrial communities, 527–531
measurement of, 515–516
plant diversity and, 533–534, 533*f*
variation in, 516, 516*t*
Primary succession, 403
on Mount St. Helens, 403–406, 405*f*, 406*f*
Principles, definition of, 13
Probabilistic model. *See* Stochasticity
Production. *See also* Primary production; Secondary production
and biodiversity gradients, 452
food chain length and, 468
Production ecology, 5
Profitability, of prey type, 225–226
Projection(s), demographic, 143
Projection matrices for population growth, 173–176, 173*f*
Provinces, floristic, 395, 395*f*

Proximal approach to habitat selection, 57
Proximate factors in adaptation, 22–23
Public lands, grazing on, 247
Pure reciprocal model of community organization, 497
PWRC. *See* Patuxent Wildlife Research Center
Pyroclastic flows, 403

Q

Quadrat(s), definition of, 121
Quadrat sampling, for population estimation, 121–122
Questionnaires, as index of relative density, 127
Quota strategy for fisheries, constant, 327

R

r-selection, 199–201
Rabbit hemmorhagic disease, 277
Rabies, 269–272
    models for, 272, 272*f*, 273
    spread of, 269, 270–272, 271*f*
    vaccination for, 270–272
    vectors of, 269, 269*f*, 270*f*
Rain forests, tropical
    cyclic changes in, 428–429
    deforestation of, 364, 365*f*
    nutrient cycling in, 570
Rainfall. *See* Precipitation
Ramensky, L. G., 388
Ramets, 118–119
Ranching, game, secondary production in, 554–556, 555*t*
Random colonization model of succession, 407*f*, 408–409
Random mating, 611
Ranges, geographic. *See also* Distribution
    human impact on, 601
    for rare species, 356–357
    size variation in, 108–111
    spatial scale of, 106–108, 107*f*
Rapoport, Eduardo, 108
Rapoport's rule, 108
    ecological explanations for, 108–111
Rare species
    characteristics of, 356–357
    classification of, 356–357, 357*t*
    minimum viable population for, 356–359
    prevalence of, 355
Ratchet effect in harvesting, 319, 319*f*
Rates, 614. *See also specific types*
    finite, 614–616
    instantaneous, 614–616
Ratio-dependent model of predation, 212
Realized fertility, *vs.* potential fecundity, 128
Realized longevity, 128
Realized niche, 190–191
    vs. fundamental niche, 191, 192*f*

Reciprocal replacement, 428
Recolonization, of fragmented patches, 364–365
Recruitment, in exploited populations, 312–313, 312*f*, 313*f*
    environmental factors in, 314–315
    of fisheries, 305*n*
Recruitment hypothesis of community organization, variable, 488–491
Red Queen Hypothesis, 277
Red tides, 535
Redfield, A. C., 526
Redfield ratio, 526
Redundancy model of communities, 387, 392
Reefs, tropical. *See* Coral reef communities
Regional biodiversity
    definition of, 455
    vs. local biodiversity, 454–456, 455*f*
Regulating factors in population regulation, definition of, 288
Regulation, population. *See* Population regulation
Reid, Charles, 48
Reid's paradox, 48
Relationships
    classification of, 179–180
    in definition of ecology, 2
Relative abundance
    log-normal distribution of, 436–437, 437*f*
    measurement of, 435–438
Relative density, indices of, 125–128
Relative fitness, 611
Relay floristics, 406–408, 407*f*
Remote sensing, of primary production, 522, 523*f*
Repeated reproduction, 149
    evolution of, 149–154, 151*f*
Replacement, reciprocal, 428
Replacement probabilities, in succession models, 409–412
Reproduction
    big-bang *vs.* repeated, evolution of, 149–154, 151*f*
    carbon dioxide enrichment and, 594
    costs of, 23, 25, 153
    disease and, 264–265
    fitness and, 19
    maximum, theory of, 25
    random, 611
    rates of, 139–140
    selection in, group *vs.* individual, 27
    suppression of, 301
    after transplantation, 34–35
Reproductive values, 145–146
    definition of, 145–146
Reptiles, species-area curve for, 503, 503*f*
Reserves
    current size/number of, 375, 376*f*
    design of, 373–375
    goals of, 373–374

    oceanic, 327–328
    selection of, 373–375
Residence times, nutrient, 564
Residual reproductive value, 146
Resource availability hypothesis of plant defense, 237, 238*f*
Resource (exploitative) competition
    definition of, 180
    as indirect competition, 203
    theories on, 180–185
        Lotka-Volterra equations, 180–182, 182*f*, 183*f*
        Tilman's model, 182–185, 185*f*, 186*f*
Resource concentration hypothesis, of herbivory on monocultures, 350–351
Resource management
    principles of, 326
    risk-aversive strategies for, 326–328
Resource-partitioned continuum model of vegetation, 394, 394*f*
Resource partitioning
    competition and, 197
    by grazers, 243, 245*f*
Resource-utilization curves, 196–197, 197*f*
Respiration
    measurement of, 538
    in primary production, 515, 517
    in secondary production, 538
Responses, of predators to prey density, 222–224
Restoration ecology, 480–482
    biomanipulation in, 499
Ribulose-diphosphate. *See* RuDP
Richness, species, 434–435. *See also* Biodiversity
Ricker, Bill, 313, 313*f*
Risk-aversive management strategies, 326–328
Rivers
    food webs in, 500–501, 500*f*, 501*t*
        with dams, 501, 501*f*
    in nitrogen cycle, 578, 579*f*
Rivet model of communities, 386–387, 392
Roadside counts, as index of relative density, 127–128
Rocky intertidal communities. *See* Intertidal zones
Rodents, habitat selection by, 61–62, 63*f*
Rosenzweig, Michael, 503, 503*f*
Rosenzweig-MacArthur model of predatory-prey interaction, 209–212, 209*f*, 211*f*
    for grazing systems, 250
Ross, Robert, 5
Ruderals, 201
RuDP carboxylase, in carbon fixation, 100–101
Runoff
    after deforestation, 466
    in water cycle, 590

S

Salt marsh communities, primary production in, 530–531, 531*f*

Sampling methods, for absolute density measurement, 121–125

Sampling model, of distribution-abundance correlation, 111–112

San Francisco Bay, marine invasions in, 47

Sand dunes, succession on, 418–419, 420*f*

Sap feeders, in collardstands, 469, 470*f*

Sargasso Sea, productivity in, 520–521

Satellite remote sensing, of primary production, 522, 523*f*

Saturation, local, 455–456, 455*f*

Schaefer models. *See* Logistic models

Schindler, David, 524, 527*f*

Scientific method, 11–15
  definitions in, 13
  schematic illustration of, 13*f*

Scotland, successional transitions in, 424, 425*f*

Scramble competition. *See* Resource competition

Search-and-destroy strategy of pest predation, 341–342

Seaweed
  abundance of, predation and, 492, 492*f*
  adaptations to light by, 98–99, 98*f*
  morphology of, 98, 98*f*

Seber, George, 125, 126, 127*f*

Secondary plant substances, 235–237
  evolutionary origin of, 235–236, 236*f*
  in oak, 239

Secondary production, 537–557
  efficiency of, 544–549, 545*t*
  limitations on, 549–556
    in game ranching, 554–556
    in grasslands, 549–554
  measurement of, 537–544, 540*f*
    problems with, 544
  sustainable energy budgets in, 556–557, 556*f*

Secondary succession, 403

Secular dispersal, definition of, 47–48

Seed(s)
  dispersal of
    in corridors, 370
    as mutualism, 253–254
    after volcanic eruption, 406
  size of, and succession after volcanic eruption, 406

Seed shadows, 450

Segregation, ecological, to avoid competition, 197, 199

Selection, natural. *See* Natural selection

Selective breeding, for pest control, 342–343

Selective colonization, and nested subsets, 371

Selective extinction, and nested subsets, 371

Self-regulation, 286–288
  evolution of, 300–302, 301*f*

Self-thinning rule of competition, 297–298

Sequential breakage hypothesis of relative abundance patterns, 437–438, 437*f*

Serengeti Plains
  herbivory in, 242–246
  predation in, 219

Serial passage experiments, on evolution of virulence, 276–277

Seroprevalence, definition of, 267*n*

Sex ratios, variation in, 130

Shade intolerance, 97

Shade tolerance, 97

Shannon-Wiener function, 617–618

Sharp community boundaries, 392–393

Shelford, Victor, 35, 36*f*

Shelford's law of tolerance, 35–36, 36*f*

Ships, marine dispersal by, 47

Shores. *See* Intertidal zones

Shortgrass prairie, energy flow in, 550–554, 552*f*, 553*f*, 553*t*

Sigmoid-curve theory, of population exploitation, 306–307, 306*f*

Silent Spring (Carson), 332

Simberloff, Dan, 368, 370*f*

Similarity, index of, for communities, 395–398

Simpson's index of diversity, 618

Sink populations, 299–300

SIR (susceptible, infected, recovered) model of disease, 260

Size
  body
    and abundance, 119–120, 120*f*, 549
    basal metabolic rate and, 538
    and density, 121*t*, 298
    and food chain organization, 464
    and population estimation technique, 119
    variation of, 130–131
  population, 125, 358, 359, 611

Size-based matrix, 174*f*, 175

Size-efficiency hypothesis of dominance, 475–476, 476*f*

Slow competitive displacement models of communities, 494

Slow-growing plants, characteristics of, 237*t*

Small-population paradigm of conservation biology, 355–359
  extinction vortex of, 355–356, 355*f*
  minimum viable population in, 356–359

Smelters
  and acid rain, 573
  recovery from, 575

Smith, H. S., 284–285

Smith, Maynard, 169, 208

Smother crops, 78–79

Snow, and timberlines, 93

Social movements, *vs.* ecology, 6–7

Socioeconomic status, and fertility rate, 4, 4*t*, 587

SOI. *See* Southern Oscillation Index

Soil(s)
  acidification of, 576
  during succession
    on abandoned farmland, 422
    on glacial moraines, 415, 415*f*
    on sand dunes, 419
  texture of, with C_3/C_4 plants, 102

Soil drought. *See* Drought

Soil sickness, 78

Solar radiation
  latitude and, 86, 86*f*
  and temperatures, 86

Song dialect, and habitat selection, 61

Soulé, Michael, 356, 356*f*

Source populations, 299–300

South America
  dispersal of mammals in, 51
  productivity of wildlife *vs.* livestock in, 556*f*

Southern Oscillation, El NiÒo and, 598–599

Southern Oscillation Index (SOI), 598

Spatial heterogeneity, and biodiversity, 445–447

Spatial patchiness, in community organization, 462

Specialist(s)
  in distribution-abundance correlation, 112
  predator
    distribution of, 76–77
    effects on prey populations, 226

Speciation, 18

Species-area curve, 502–503

Species richness. *See also* Biodiversity
  definition of, 434
  measurement of, 435

Species richness-energy model of biodiversity, 450–451, 451*f*

Spectral reflectance pattern of green plants, 527–528

Sphagnum bogs, 415, 417

Spines, as defense mechanism, 241–242

St. Helens, Mount, 404*f*
  primary succession on, 403–406, 405*f*, 406*f*, 413
  recolonization of, 51

Stability
  community, 477–480
    biodiversity and, 478–480
    with equilibrium, 459–460, 460*f*
    local *vs.* global, 460, 477
    measurement of, 460, 479
    multiple states of, 508–509
  ecological, 288–289, 289*f*

Stability hypothesis of food chain length, dynamic, 468
Stability-time hypothesis of biodiversity, 450
Stabilizing selection, 20*f*–22*f*, 21
Stable age distribution, 140–141, 141*f*, 142*f*, 147–148
Stable isotopes, in food chain analysis, 466
Stage-based matrix, 174*f*, 175
Starvation, as facultative agent of regulation, 284
Static life tables, 135–137, 136*t*
Stationary age distribution, 147–148
Statistical hypotheses, 14
Stem canker disease, biodiversity gradients and, 450
Sterilization, for pest control, 343–344
    effectiveness of, 345–346
Stochasticity
    in models of population growth, 169–172, 171*f*, 172*f*
    in population loss/extinction, 356
Stock
    definition of, 305
    in fisheries, 305–306
Stock production models. *See* Logistic models
Storms, and coral reefs, 486–487, 487*f*
Streams, nutrient loss through, 566–567
Stress
    food web complexity and, 495–496, 495*f*
    in Grime's theory of plant competition, 200*f*, 201
Stress hypothesis, plant, 251
Strip feeders, in collardstands, 469, 470*f*
Structural defenses, in plants, 237
Subclimaxes, 408
Subtidal communities, marine, keystone species in, 471–472
Succession
    on abandoned farmland, 420–423
    climax state of, 423–424
    definition of, 403
    early description of, 5
    on glacial moraines, 414–417
    mathematical model of, 409–412
    patch dynamics in, 426–429
    primary, 403–406, 405*f*, 406*f*
    in restoration ecology, 480
    on sand-dunes, 416–419, 420*f*
    secondary, 403
    theories of, 406–409, 407*f*, 410*f*
Sudbury, Ontario, acid rain in, 575
Suitability, habitat, 67, 67*f*
Sulfur cycle
    and acid rain, 573–576
    global, 573, 573*f*
Superorganismic view of communities, 387, 388
Supply-side ecology, 491
Surplus yield models. *See* Logistic models

Survival
    observation of, for life tables, 137
    rate of, estimation of, 124–125
    after transplantation, 34–35
Survivorship curves, 134–137, 134*f*
    types of, 134–135, 135*f*
Sustainable energy budgets, 556–557, 556*f*
Sustained yield, maximum, 305, 318, 319
Sweden, demographic transition in, 584*f*
Swine fever, 261
Switching, prey, 226

T

Tallgrass prairie
    energy flow in, 550–554, 552*f*, 553*t*
    trophic pyramid for, 550, 551*f*
Tannin, in oak leaves, 239
Tansley, A. G., 387–388, 387*f*, 423
Temperate forests
    energy flow in, 546, 547*f*
    nutrient content of, 568–569, 570*t*–571*t*
Temperature(s)
    adaptations to, 94–97
    annual, average ranges of, 87*f*
    and distribution, 88–97
    and evaporation/transpiration rates, 87
    global, carbon dioxide and, 598, 599*f*
    moisture and, 91–94
    and primary production, 530
    solar radiation and, 86
    tolerance for, 88–90, 90*f*
Temporary ponds, keystone predators in, 473
Tennessee, climatic diagram for, 89*f*
Tens rule of species introduction, 49, 50
Tension zones, 395, 397*f*, 491
Terrestrial communities. *See also specific types*
    acid rain in, 576, 578*f*
    nitrogen cycle in, 578–580
    production in
        efficiency of, 546, 547*f*
        primary, 527–531
    water needs in, 86–87
Territoriality
    as counterstrategy to infanticide, 301
    and population densities, 288
Theories, definition of, 13
The Theory of Island Biogeography (Wilson and MacArthur), 436
Thermodynamics, in ecology, 543
Thorns, as defense mechanism, 242
-3/2 power rule, 297–298
Threshold population size, in harvesting, 326–327
Tillers, 119
Tilman, David, 478, 479*f*
Tilman's model of resource competition, 182–185, 185*f*, 186*f*
Timberlines, 93–94

Time-lag models of population growth, 168, 169, 169*f*
Time series, 184
Tolerance
    adaptations in, 36–38
    for climate variation, 108
    determination of, 35–36
    and distribution, 35–36, 108
    law of, 35–36, 36*f*
    for moisture, 88–91, 90*f*
    range sizes and, 108
    for shade, 97
    for temperature, 88–90, 90*f*
Tolerance model of succession, 407*f*, 408, 409
Top-down model of community organization, 496–499
Top predators, 465*t*, 468
Topography, and biodiversity, 445, 446
Total response, of predators to prey density, 224, 225*f*
Tragedy of commons, 321
Transition matrix, 411*t*
Transition probabilities, in succession models, 411–412
Transitive competition, 474, 474*f*
    in rocky intertidal zones, 492
Transmission coefficient, 261
Transpiration rate, temperature and, 87
Transplantation, species
    dangers of, 38
    distribution analysis with, 32–35, 35*f*
Traps, as index of relative density, 127
Tree(s)
    as allelopathic agents, 78
    biodiversity in
        available energy and, 450–451, 451*f*
        latitude and, 438, 438*f*
        vs. vertebrates, 446, 446*f*, 450–451, 451*f*
    carbon dioxide enrichment and, 594, 594*f*
    dispersal of
        after ice age, 48, 102
        rates for, 48, 48*f*
    primary production by, age and, 532
    range shifts in, 601
Tree line, 93–94
Trophic cascade model of community organization, 497–500
Trophic efficiency, 545–546
Trophic levels, 463–464
    guilds within, 469–471
    isotopic signatures of, 466
Trophic species, 465*t*
Tropical forests
    carbon dioxide enrichment in, 595, 596*f*
    nutrient use efficiency in, 569–570, 569*f*
    rain, 363, 365*f*, 428–429, 570
True Law of Population (Doubleday), 4
Truth, ecological, 14

Tundra, arctic, climate change and, 594, 595*f*
Turesson, Göte, 94–96
Turnover times, nutrient
    estimation of, 564
    in forests, 564, 565*t*

U

Ultimate approach to habitat selection, 57
Ultimate factors in adaptation, 22–23
Ultraviolet (UV) radiation, and coral reef
    bleaching, 489
Umbrella species, 399–400
Uncertainty, in science, 14
Ungulates
    brucellosis in, 268–269
    grazing by, in Serengeti Plains, 242–246
    habitats of, 554*f*
    population irruption of,
        248–251, 248*f*
    secondary production by, 554–556
Unitary organisms, 116, 117
Upgrade series, in cycles of change, 427
Upwelling, and primary production, 522
UV (ultraviolet) radiation, and coral reef
    bleaching, 489
Uvarov, B. P., 284

V

Vaccination, rabies, 270–272
Values, in science, 12
Variable recruitment hypothesis of
    community organization, 488–491
Vegetation. *See also specific types*
    climate and, 87–88
Verhulst, Pierre-François, 4
Vertebrates, biodiversity in
    available energy and, 450–451, 451*f*
    vs. trees, 446, 446*f*, 450–451, 451*f*

Vesijärvi, Lake, biomanipulation of, 499
Vicarious distribution, 53
Vigor hypothesis, plant, 251
Viruses, evolution of, 277
Vitousek, Peter, 569, 570*f*
Vocalization, frequency of, as index of
    relative density, 127
Volcanic eruptions
    local extinctions after, 51
    primary succession after, 403–406
    recolonization after, 50–51
Volterra, Vito, 209, 210*f*

W

Walker, Brian, 387
Wallace, A. W., 434
Wallace, Alfred, 17
Warming, J. E. B., 5, 406–407
Warning coloration, 229–231
    mimicry of, 230*f*, 231
Washington, Lake, restoration of,
    480–482, 480*f*
Water. *See also* Moisture
    cycle of, 87
        global, 590, 590*f*
    doubly labeled, energy expenditure
        estimation with, 538, 540
    role of, for terrestrial species, 86–87
    supply of
        in carrying capacity, 589–590
        economics of, 606
Watt, A. S., 424, 425, 427
Weeds, control of. *See also* Pest control
    with smother crops, 78–79
Weinberg, G., 610
Whittaker, Robert, 393, 393*f*
Wiens, John, 58, 60*f*, 64*f*
Wilson, Edward O., 436, 436*f*

Wind(s)
    on sand dunes, 418
    and timberlines, 93
Winter (frost) drought, 92, 93–94
Within-habitat (α) diversity, 445–446, 447
Woodlands, multiple stable states in,
    508–509, 508*f*
Wright, Stewart, 359
Wynne-Edwards, V. C., 288

X

Xerophytes, adaptations of, 91

Y

Year-classes
    dominant, 148
    failures of, 314–315
Yield(s)
    vs. biomass, 513, 514*f*
    definitions of, 305
    dynamic pool models for, 309–316
    for fisheries, 305–306
    logistic models for, 307–309, 307*f*
    maximum sustained, 305, 318, 319
    optimum, 5, 318–319
    of wildlife *vs.* livestock, 554
Yoda's law, 297–298
Ythan Estuary, food chain length
    in, 467, 467*f*

Z

Zero growth isocline, 184
Zona pellucida glycoproteins (ZPG), in
    immunocontraception, 344–345
Zooplankton
    distribution in Great Lakes, 397, 398*t*
    dominant species of, 475–476, 476*f*
    predation on, 42